D1776554

THE STELLAR POPULATIONS OF GALAXIES

INTERNATIONAL ASTRONOMICAL UNION

UNION ASTRONOMIQUE INTERNATIONALE

THE STELLAR POPULATIONS OF GALAXIES

PROCEEDINGS OF THE 149TH SYMPOSIUM OF THE INTERNATIONAL ASTRONOMICAL UNION, HELD IN ANGRA DOS REIS, BRAZIL, AUGUST 5–9, 1991

EDITED BY

B. BARBUY

Instituto Astrônomico e Geofísico, Universidade de São Paulo, Brazil

and

A. RENZINI

Department of Astronomy, University of Bologna, Italy

KLUWER ACADEMIC PUBLISHERS

DORDRECHT / BOSTON / LONDON

Library of Congress Cataloging-in-Publication Data

International Astronomical Union. Symposium (149th : 1991 : Angra dos Reis, Brazil)
The stellar populations of galaxies : proceedings of the 149th Symposium of the International Astronomical Union, held in Angra dos Reis, Brazil, August 5-9, 1991 / edited by Beatriz Barbuy and Alvio Renzini.
p. cm.
Includes bibliographical references and indexes.
ISBN 0-7923-1698-3 (hb : acid-free paper)
1. Milky Way--Congresses. 2. Galaxies--Congresses. 3. Stars--Populations--Congresses. I. Barbuy, Beatriz, 1950- . II. Renzini, Alvio. III. Title.
QB857.7.I58 1991
523.1'12--dc20 92-2211

ISBN 0-7923-1698-3 (HB)

Published on behalf of
the International Astronomical Union
by
Kluwer Academic Publishers, P.O. Box 17, 3300 AA Dordrecht, The Netherlands.

Kluwer Academic Publishers incorporates
the publishing programmes of
D. Reidel, Martinus Nijhoff, Dr W. Junk and MTP Press.

Sold and distributed in the U.S.A. and Canada
by Kluwer Academic Publishers,
101 Philip Drive, Norwell, MA 02061, U.S.A.

In all other countries, sold and distributed
by Kluwer Academic Publishers Group,
P.O. Box 322, 3300 AH Dordrecht, The Netherlands.

Printed on acid-free paper

Printed in the Netherlands

TABLE OF CONTENTS

PREFACE xv
THE ORGANIZING COMMITTEES xvii
CONFERENCE PHOTOGRAPH xviii
LIST OF PARTICIPANTS xix

I. THE STELLAR POPULATIONS IN THE MILKY WAY

J.E. Hesser
The Halo Populations 1

B.W. Carney
The Ages of Globular Clusters 15

N.B. Suntzeff
Population II in the Milky Way Galaxy and The LMC 23

R.M. Rich
The Stellar Population of the Galactic Bulge 29

E.M. Sadler
Metal-rich Stars in the Galactic Bulge and their Implications for Elliptical Galaxies 41

S. Ortolani
Globular Clusters of the Inner Galactic Bulge 47

T. de Zeeuw
Dynamics of the Galactic Spheroid 51

S.R. Majewski
Structure and Kinematics at the North Galactic Pole 61

K.C. Freeman
The Disk of the Galaxy 65

B. Gustafsson, B. Edvardsson, P. Nissen, D.L. Lambert, J. Tomkin, & J. Andersen
The Chemical and Dynamical Evolution of the Galactic Disk 75

M.W. Feast, P.A. Whitelock, & R. Sharples
The Galactic Bulge and the Thick Disc 77

R. Wielen, C. Dettbarn, B. Fuchs, H. Jahreiss, & G. Radons
Dynamics of Stellar Populations in Galactic Disks 81

P.S. Conti
The Stellar Content of Spiral Arms 93

J.M. Nemec & A.F. Linnell Nemec
Mixture Models for Studying Stellar Populations: Bivariate Relationships and Posterior Mixing Proportions Applicable to the Thick-Disk Problem 103

A. Maeder
Stellar Evolution and Stellar Populations in Galaxies 109

G. Hensler, A. Burkert, J.W. Truran, H. Dünhuber, & C. Theis
The Formation of Disks in Galaxies 119

M. Spite
Trends of Element Abundances in the Stars of Our Galaxy 123

B.E.J. Pagel
Chemical Evolution of Stellar Populations 133

B. Barbuy
The [O/Fe] Ratio in Halo, Disk and Bulge Stars 143

II. THE STELLAR POPULATIONS OF NEARBY RESOLVED GALAXIES

M. Mateo
The Stellar Populations of the Magellanic Clouds 147

D. Hatzidimitriou & L.T. Gardiner
Survey of Stellar Populations in the SMC 157

S. van den Bergh & C.J. Pritchet
Stellar Populations in M31 and M33 161

W.L. Freedman
The Dwarf Elliptical Galaxies of the Local Group & the Stellar Populations and Age of M32 169

J.R. Mould
Intermediate Age Populations 181

G.S. Da Costa
Dwarf Spheroidal Galaxies 191

M. Azzopardi & J. Lequeux
Surveying Carbon Stars in the Dwarf Spheroidal Galaxies 201

M. Tosi, L. Greggio, P. Focardi, & G. Marconi
Stellar Populations in Dwarf Irregular Galaxies of the Local Group 207

N. Bergvall & J. Rönnback
Blue Low Surface-Brightness Galaxies 211

E. Bica
Star and Star Cluster Spectral Libraries 215

R.L. Kurucz
Model Atmospheres for Population Synthesis 225

III. THE STELLAR POPULATIONS OF NON-RESOLVED GALAXIES

R.W. O'Connell
Ultraviolet Observations of Stellar Populations in Globular Clusters and Galaxies 233

J.A. Frogel
The Stellar Content of the Bulges of Spiral Galaxies:
What do We Know and How Do We Know It? 245

S.M. Faber, G. Worthey, & J.J. Gonzales
Absorption-Line Spectra of Elliptical Galaxies and their Relation to Elliptical Formation 255

R. Bender
1. Evidence for Merger Signatures in the Line-Strength Gradients of Elliptical Galaxies.
2. A Universal relation Between Metallicity and Potential Depth for
All Hot Stellar Systems? 267

R. Terlevich
Young Massive Elliptical Galaxies: Where are They? 271

M.G. Edmunds
Metallicity of Unresolved Stellar Populations 277

M.D. Gregg
Stellar Populations and Peculiar Velocities of Elliptical Galaxies 281

I. Pronik, N. Merkulova, L. Metik, & V. Pronik
The Stellar Population in Galaxies Observed on the Astrophysical Station "ASTRON" 287

H.C. Ferguson, A.F. Davidsen, & G.A. Kriss
The Hot Stellar Component in Elliptical Galaxies and Spiral Bulges 291

R.S. Ellis
Galaxies at Intermediate Redshifts 297

G. Bruzual A.
Evolutionary Population Synthesis 311

C. Chiosi, G. Bertelli, & A. Bressan
Models of Population Synthesis 321

A. Renzini
The Age Ladder from Low- to High-Redshift Populations 325

S. Djorgovski & D.J. Thompson
Searches for Primeval Galaxies 337

P. Traat
Some Insights into the Photometric Evolution of Galaxies 349

A. Buzzoni, G. Chincarini, & E. Molinari
Testing the World Model Through High Redshift Galaxies in Clusters 353

B. Rocca-Volmerange
UV to IR Models of Galaxy Evolution and Cosmology 357

J. Silk
Dark Populations 367

G.S. Bisnovatyi-Kogan
The Neutron Star Population in the Galaxy 379

POSTER PAPERS

A. Acker, J. Köppen, B. Stenholm & G. Jasniewicz
Chemical abundances of galactic planetary nebulae 383

M. Albrow & P.L. Cottrell
Line asymmetry and projection factors in Cepheid variables 384

C. Allen, W.J. Schuster & A. Poveda
Orbital characteristics of high-velocity stars in two galactic mass distributions 385

E.J. Alfaro, J. Cabrera-Caño & A.J. Delgado
Three-dimensional picture of the galactic disk as defined by the young stellar component 386

L. I. Arany-Prado & R. de la Reza
Astration and production in chemical evolution 387

J.C.N. de Araujo & R. Opher
Can the first stars formed be pre-galactic? 388

B.P. Artamanov & P. Traat
How many bursts of star formation in M82? 389

C.C. Batalha
Pre main sequence chromospheric activity 390

M.A. Bershady
Morphology and colors of distant galaxies 391

E. Bica, J.J. Clariá, H. Dottori, J.F.C. Santos Jr. & A. Piatti
Integrated UBV photometry of 624 LMC clusters 392

D.J. Bomans, A. Vallenari, W. Seggewiss, T. Richtler, E.K. Grebel & K.S. de Boer
Photometric results for stellar fields inside supergiant shell LMC4 393

A.C. Borges & J.A. de Freitas Pacheco
Synthetic M/L_B ratios for E and S galaxies 394

A. Bressan, G. Bertelli, C. Chiosi, F. Fagotto & E. Nasi
New theoretical isochrones 395

A. Bressan, G. Bertelli & C. Chiosi
The HR diagram of LMC supergiant stars 396

A. Buzzoni, G. Gariboldi & L. Mantegazza
Metallicity distribution of elliptical galaxies through a quantitative calibration of the Magnesium Mg_2 index 397

L. Carigi & J. Franco
The evolution of the galaxy: the ^{16}O gradient and the surface gas density 398

B.W. Carney & P. Seitzer
The outer disk of the Galaxy 399

R.R. de Carvalho & S. Djorgovski
Systematic differences between the field and cluster ellipticals 400

S. Castro, B. Barbuy, S. Ortolani & E. Bica
Metallicity of the star III-17 in NGC 6553 401

M. Catelan & M.L. Quarta
Synthetic horizontal branches for galactic globular clusters 402

M. Catelan
A possible reconcilliation among different RR Lyrae absolute magnitude calibrations and implications for the age-metallicity relation 403

S.A. Cellone, J.C. Forte & D. Geisler
Structure and colors of dwarf elliptical galaxies in the Fornax cluster 404

A. Chokshi
Galaxy formation study using QSO absorbers 405

R. Cid Fernandes, H.A. Dottori, R.B. Gruenwald & S.M. Viegas
Warmers: source of ionization and N enrichment in AGN 406

G. Clementini, C. Cacciari & J.A. Fernley
The M_V(RR)-[Fe/H] relation from the Baade-Wesselink method applied for RR Lyrae stars 407

J. Colin, B. Dauphole-Fouillet & C. Ducourant
Separation of halo and thick disc stars in two catalogs 408

C.J. Corbally & R.F. Garrison
Spectral characteristics of early G-dwarf stars towards the galactic poles 409

M. Crézé, B. Chen, A.C. Robin & O. Bienaymé
Searching significant signatures of stellar population characteristics in multivariate star count samples 410

A. Danks, M. Pérez & B. Altner
Ultraviolet studies of the face-on galaxy NGC 2217 411

M. Della Valle
Nova LMC 1991: a super-bright nova in the Large Magellanic Cloud 412

S. Djorgovski & G. Piotto
Stellar population changes in post-core-collapse globular clusters 413

N. Epchtein, F. Guglielmo & W.B. Burton
A deep near infrared southern sky survey: A new probe for hidden stellar populations in our galaxy 414

M. Erdelyi-Mendes, B. Barbuy & A. Milone
Synthetic Mg_1, Mg_2 and Mgb indices 415

D.A. Forbes
Infrared detection of supernova remnants in the nucleus of NGC 253? 416

P. François
Relative abundances of Thorium and Europium in halo stars 417

J.A. de Freitas Pacheco, R.D.D. Costa
He abundance in red giants 418

J.A. de Freitas Pacheco & R.D.D. Costa
Mass loss from central stars of planetary nebulae with WC spectrum 419

A.C.S. Friaça
Models of emission line regions around central cluster galaxies 420

U. Fritze - v. Alvensleben, H. Krüger & K.J. Fricke
QSO absorption lines and the gas and star content of high redshift galaxies 421

D. Geisler, N. Suntzeff, M. Mateo & J. Graham
Metal abundances of Magellanic Cloud clusters 422

L. Girardi & E. Bica
A comparison between a color evolution model and a new sample of LMC clusters: formation rate 423

P. Goudfrooij & B. van den Hoek
Radial population synthesis and the ionization of gas in elliptical galaxies 424

P.J. Green, B. Margon, S.F. Anderson, P.M. Garnavich, K. Cook & D.J. MacConnell
A CCD search for faint high-latitude carbon stars: dwarfs among the giants 425

M.D. Gregg
A spectrophotometric investigation of dwarf elliptical galaxies 426

R.B. Gruenwald. S.M. Viegas & G. Detthow
QSO absorption-line systems and star formation 427

H.C. Harris
Stellar abundances in the outer galactic disk 428

E.V. Held, T. de Zeeuw, J. Mould & A. Picard
NGC 185 and the extended Faber-Jackson relation 429

M. Hernandez-Pajares & E. Monte
Study of stellar populations using neural network techniques 430

M. Hernandez-Pajares & J. Nuñez
The spherical harmonics as an alternative tool for determining the kinematical parameters of the local Milky Way 431

R. Hill, B.F. Madore & W.L. Freedman
Luminosity functions and mass functions for massive stars: associations in the Large and Small Magellanic Clouds 432

J.E. Horvath & G.A. Foglia
The stellar populations of neutron and strange stars in the Galaxy 433

S.M.G. Hughes & P.R. Wood
Abundances of long-period variables: initial results 434

V.A. Hughes
The time dependent radio sources in Cepheus A 435

P. Jablonka, D. Alloin & E. Bica
Ages and metallicities of M31 star clusters 436

Jablonka & N. Arimoto
Dark-to-luminous mass ratio in spiral galaxies 437

S.A. Josey & N. Arimoto
The colour gradient in M31: evidence for disc formation by biased infall? 438

J.M. Juan-Zornoza & J. Sanz i Subirana
Moments of peculiar velocities in stellar systems with point axial symmetry 439

W. Kollatschny, A. Goerdt & K.J. Fricke
Circumnuclear star formation in Seyfert galaxies 440

J.K. Kotilainen
Stellar populations in Seyfert 1 galaxies 441

H. Krüger, U. Fritze-v. Alvensleben, H.-H. Loose & K.J. Fricke
Stellar populations of BCD galaxies from spectrophotometric evolutionary synthesis 442

J.B. Laird, B.W. Carney & D.W. Latham
The halo metallicity distribution 443

M.G. Lee, W.L. Freedman & B.F. Madore
Blue stars in the dwarf elliptical galaxy NGC 205 444

M.G. Lee, W.L. Freedman & B.F. Madore
Stellar populations in the local group galaxy NGC 185 445

Y.-W. Lee
Evidence for an old galactic bulge from RR Lyrae stars in Baade's window: the inside-out picture of galaxy formation 446

C. Leitherer, L. Drissen & C. Robert
Injection of mass and energy into the ISM by massive stars 447

P. te Lintel Hekkert & H. Dejonghe
The distribution function for a new sample of OH/IR stars 448

X. Luri, J. Torra & F. Figueras
UBVRI distances and metallicities for a sample of late-type Hipparcos stars 449

D. Maccagni, B. Garilli, D. Bottini & G. Vettolani
Color distribution of galaxies in the core of S0400 450

W.J. Maciel & C.C.M. Leite
Planetary nebulae and stellar populations 451

G. Magris C. & G. Bruzual A.
Evolutionary spectral synthesis and the UV upturn in elliptical galaxies 452

A.P. Marston, P.N. Appleton, M. Lysaght & C. Struck-Marcell
Star formation and stellar populations in ring galaxies 453

M. Mateo & D. Hatzidimitriou
The evolution of the red giant clump and the structure of the Small Magellanic Cloud 454

M. Mateo, N. Suntzeff, J. Nemec, D. Terndrup, W. Weller, E. Olszewski, M. Irwin & R. McMahon
The stellar population and internal kinematics of the Sextans dwarf spheroidal Galaxy 455

B.R. McNamara
Color mapping cluster cooling flow galaxies 456

G.R. Meurer & G. Mackie
The morphology and stellar populations of the dwarf amorphous galaxies NGC 216 and NGC 2915 457

A. Milone, B. Barbuy, M. Spite & F. Spite
CNO overabundances in 6 stars of ω Centauri 458

D. Minniti, S.D.M. White, E. Olszewski, J. Hill & M. Irwin
Kinematics of the galactic bulge 459

E. Molinari, D. Pedrana, M. Banzi, A. Buzzoni & G. Chincarini
Evolutionary status of galaxy population in clusters at intermediate redshift ($z \approx 0.2$) 460

A.A. Myullyari, Yu.V. Nechitajlov, S. Ninković & V.V. Orlov
A study of the properties of galactocentric orbits 461

Y.K. Ng, G. Bertelli, A. Bressan & C. Chiosi
A color-magnitude diagram in field #3 of the Palomar-Groningen Survey 462

S. Ninković
On the local velocity dispersions 463

S. Ninković
On the total kinetic energy of our Galaxy with the contribution of the populations 464

R.P. Ortiz & J.R.D. Lépine
A model of the galaxy for star counts in the infrared 465

A. Paquet, R. Bender & W. Seifert
Metallicity gradients in the disks of S0 galaxies 466

J.W. Parker
30 Doradus: the stellar content and IMF 467

M.G. Pastoriza, H. Dottori, E. Terlevich, R. Terlevich & A. Diaz
The RSG and WR content of the circumnuclear HII regions of NGC 3310 468

R.F. Peletier & S.P. Willner
How transparent are spiral galaxies? An analysis from a near-infrared perspective 469

M. Peña, S. Torres-Peimbert & M.T. Ruiz
The halo planetary nebulae M2-29 and BB-1 470

A. Poveda, C. Allen & W. Schuster
Moving clusters among galactic halo stars 471

M.A. Prieto
Population synthesis in starburst galaxies 472

E. Rebeirot, M. Azzopardi & B.E. Westerlund
Carbon stars in the Small Magellanic Cloud: positions, finding charts and spectrophotometry 473

E. Recillas-Cruz, A. Serrano P.-G., L. Carrasco & V. Ortega
Infrared properties of galaxies in clusters: Abell 194, Abell 426 (Perseus) and Abell 2151 (Hercules) 474

A. Renzini & L. Stanghellini
Synthetic P-AGB populations 475

R. de la Reza & L. da Silva
Analysis of very rich K giant stars 476

T. Richtler, E.K. Grebel & W. Seggewiss
Photometry of Galactic globular clusters of the Disk system 477

C. Robert, L. Drissen & C. Leitherer
The Wolf-Rayet and O star content of violent starbursts 478

S.C.F. Rossi, P. Benevides-Soares, B. Barbuy & G. Pineau des Forêts
Grain cooling in collapsing clouds 479

A. Ruelas-Mayorga & P. Teague
Baade's window photometry and spectroscopy 480

S.D. Ryder
Massive star formation and chemical evolution in NGC 1313 481

J.J. Salzer & R. Elston
Star-formation histories of blue compact dwarf galaxies 482

J.F.C. Santos Jr., E. Bica & H. Dottori
CNO excess in 47 Tucanae from the integrated spectrum synthesis 483

A.A. Schmidt, M.V.F. Copetti, D. Alloin & P. Jablonka
Tests and discussion on the solution uniqueness of population synthesis methods 484

R.A. Schommer, E.W. Olszewski & M.A. Aaronson
On main sequence distances and the local distance scale 485

W.J. Schuster, L. Parrao & M.E. Contreras Martínez
Kinematics of halo and high-velocity disk stars 486

W.J. Schuster, C. Allen & A. Poveda
Halo and high-velocity disk stars 487

F. Schweizer & P. Seitzer
Correlations between UBV colors and fine structure in E+S0 galaxies 488

R.E. de Souza, B. Barbuy, S. dos Anjos & M. Erdelyi-Mendes
A method to derive velocity dispersions from composite spectra 489

F. Spite, M. Spite, R. Cayrel & S. Huille
Lithium in population II stars 490

S.A. Stanford & H.A. Bushouse
A near-IR imaging survey of interacting galaxies 491

G. Stasińska, A. Acker, A. Fresneau, J.F. Gameiro, J. Köppen, B. Stenholm & R. Tylenda
The population of planetary nebulae in the galactic bulge 492

E. Telles, R. Terlevich & B.E.J. Pagel
CCD photometry of HII galaxies 493

E. Terlevich, R. Terlevich, A.I. Diaz & M.L.G. Vargas
Observations of Ly_α emission in young galaxies 494

F. Thévenin, G. Jasniewicz & A. Bijaoui
Tools for a new approach of stellar populations 495

C.G. Tinney
The bolometric luminosity function for the lowest mass stars 496

W.D. Vacca & P.S. Conti
Analysis of the optical spectra of Wolf-Rayet galaxies 497

B. van den Hoek & P. Goudfrooij
Evolutionary population synthesis 498

J.M. van der Hulst, E.D. Skillman, G.D. Bothun & T.R. Smith
The HI surface density in low surface brightness galaxies 499

M.B. Vila & M.G. Edmunds
Abundance gradients and physical properties of spiral galaxies 500

S.J. Wagner
Colour-differences among globular cluster systems 501

N.A. Walton, M.J. Barlow, R.E.S. Clegg & D.J. Monk
Stellar evolution in the Magellanic Clouds from studies of planetary nebulae 502

P. A. Whitelock & R. Catchpole
The shape of the bulge from Iras Miras 503

P. A. Whitelock
The periods of miras in the bulge and their latitude dependence 504

A.J. Willis, H. Schild & L.J. Smith
New WHT+FOS optical spectra of WR stars in M33 & M31 505

C.D. Wilson
OB associations in NGC 6822 506

G. Worthey
Chemical abundances in old populations 507

Y. Wu, M. Huang & J. He
Morphological and spatial distributions of high velocity molecular outflows 508

S.E. Zepf
The stellar populations and dynamics of elliptical galaxies in compact groups 509

H. Zinnecker
The pre-main sequence stellar population in galaxies 510

ABSTRACTS

R. Cubarsí
Stellar populations of a local sample from its velocity distribution 511

E. Janot-Pacheco
The group of Be/X-ray sources in the Galaxy 511

P. te Lintel Hekkert & A.A. Zijlstra
The progenitors of planetary nebulae 512

C. de Loore & D. Vanbeveren
A binary evolutionary model for the progenitor of SN 1987A 512

M.-N. Perrin, R. Cayrel, B. Barbuy & R. Buser
Stellar parameters in the Basel field SA 141 513

A.H. Prestwich
Near infrared imaging of a giant red envelope galaxy 513

C.A. Torres, G.R. Quast & R. de la Reza
The space distribution of low mass star forming regions 514

Subject Index 515
Element Index 521

PREFACE

The IAU Symposium No. 149 "THE STELLAR POPULATIONS OF GALAXIES" was held near Angra dos Reis, Brazil, from August 5 to 9, 1991, and was attended by 211 registered participants from 26 countries.

The meeting was sponsored by the following IAU Commissions: Commission 28 (Galaxies), Commission 29 (Stellar Spectra), Commission 33 (Structure and Dynamics of the Galactic System), Commission 35 (Stellar Constitution), and Commission 37 (Star Clusters and Associacions).

Besides the IAU support, additional financial support was provided by the following institutions: Projeto Banco Interamericano de Desenvolvimento/Universidade de São Paulo (BID/USP), Fundação de Amparo à Pesquisa do Estado de São Paulo (FAPESP), Fundação de Amparo à Pesquisa do Estado do Rio Grande do Sul (FAPERGS), Conselho Nacional de Desenvolvimento Científico e Tecnológico (CNPq), Financiadora de Estudos e Projetos (FINEP); Instituto Astronômico e Geofísico da Universidade de São Paulo (IAG-USP), and Observatório Nacional (ON/CNPq). The contribution of these institutions provided the overall support to the Symposium, and thanks to it partial or full support to 98 participants was provided.

It was a broad Symposium, covering the whole subject of stellar populations from the solar neighborhood to the most distant radiogalaxies, and the meeting was correspondingly structured in three phases. The first phase focussed on our own Galaxy, with main reviews devoted to the stellar populations in the Galactic Bulge, Halo, Disk and Spiral Arms, then touching upon their age, composition, and kinematics. A classical approach to stellar populations, and yet one which is still plenty of surprise and discoveries.

The second phase of the meeting was devoted to nearby galaxies that observations are able to resolve into individual stars, such as the Magellanic Clouds, the Dwarf Spheroidals, Andromeda, and the other members of the Local Group. When the Symposium was first planned the intent was to devote ample time to the results from the Hubble Space Telescope, which would have enormously increased the number of *resolved* galaxies. Unfortunately, for the well known reasons little relevant HST data became available.

The third and last phase was devoted to distant, non-resolved galaxies, whose stellar populations can only be studied in integrated light. Here the main reviews were devoted to low, intermediate, and high-redshift galaxies, as well as to the ongoing search for primeval galaxies. Through the whole Symposium theoretical papers innervated the complex wealth of observational evidences being provided, thus addressing topics such as the dynamics of the stellar populations in the Galaxy, stellar evolution, stellar spectra and model atmospheres as main ingredients in the construction of the population synthesis tools which are necessary to study unresolved galaxies, etc. The question of how galaxies formed, evolved, and acquired their presesent morphology has percurred the whole meeting, leading to much discussion and excitement. To make progress in this direction the Symposium was indeed conceived.

An impressive number of poster papers were also presented, too many for allowing an even cursory discussion of them during the formal sessions of the meeting. But spirited discussions took place before the posters themselves, during one week of splendid isolation, next to the uncontaminated seashore of Angra Dos Reis, the ideal *anchorage* for such a populous symposium.

The Editors

Alvio Renzini — Scientific Organizing Committee

Beatriz Barbuy — Local Organizing Committee

SCIENTIFIC ORGANIZING COMMITTEE

G. Bruzual, Universidad de Mérida, Venezuela
J. Clariá, Universidad Nacional de Cordoba, Argentina
K. Freeman, Australian Nacional University, Australia
J. de Freitas Pacheco, Universidade de Saõ Paulo, Brazil
G. Gilmore, University of Cambridge, UK
B. Gustafsson, Astronomiska Observatoriet, Uppsala, Sweden
R. Kudritzki, Universität München, Germany
J. Lequeux, Observatoire de Meudon, France
H. Maehara, Okayama Astrophysical Observatory, Japan
J. Melnick, European Southern Observatory, Chile
J. Mould, Caltech, USA
B. Pagel, Nordita, Denmark
A. Renzini (Chairman), Università di Bologna, Italy
J. Silk, University of California, Berkeley, USA
M. Spite, Observatoire de Meudon, France
S. Torres-Peimbert, UNAM, Mexico
A. Tutukov, Academy of Sciences, USSR

LOCAL ORGANIZING COMMITTEE

B. Barbuy (Chairperson), Universidade de São Paulo
E. Bica, Universidade Federal do Rio Grande do Sul
A. Borges, Universidade de São Paulo
M. Erdelyi-Mendes, Universidade de São Paulo
S. Angelica Gonçalves, Universidade de São Paulo
W. Maciel, Universidade de São Paulo
S. Rossi, Universidade de São Paulo
L. da Silva, Observatório Nacional

LIST OF PARTICIPANTS

Michael D. Albrow: Department of Physics, University of Canterbury, New Zealand
Emilio J. Alfaro: Instituto de Astrofisica de Andalucia, Granada, Spain
Elhanan Almoznino: Wise Observatory, Tel-Aviv University, Tel-Aviv, Israel
Sandra dos Anjos: Instituto Astronômico e Geofísico, Usp, São Paulo, Brazil
José C. de Araujo: Instituto Astronômico e Geofísico, Usp, São Paulo, Brazil
Taft Armandroff: Kitt Peak National Observatory, Tucson, Arizona, USA
Marc Azzopardi: Observatoire de Marseille, Marseille, France
Beatriz Barbuy: Instituto Astronômico e Geofísico, Usp, São Paulo, Brazil
Celso Batalha: Observatório Nacional, Rio de Janeiro, Brazil
Ralf Bender: Landessternwarte Königstuhl, Heidelberg, Germany
Nils Bergval: Astronomiska Observatoriet, Uppsala, Sweden
Matthew Bershady: University of Chicago, Chicago, Illinois, USA
Eduardo Bica: Universidade Federal do Rio Grande do Sul, Porto Alegre, Brazil
Gena Bisnovatyi-Kogan: Space Research Institute, Moscow, USSR
Bruce Bohannan: Kitt Peak National Observatory, Tucson, Arizona, USA
Richard Boyle: Vatican Observatory Group, Steward Observatory, Tucson, Arizona, USA
Alessadro Bressan: Osservatorio Astronomico di Padova, Padova, Italy
Gustavo Bruzual: Centro de Investigaciones de Astronomia, Merida, Venezuela
Alberto Buzzoni: Osservatorio Astronomico di Brera-Milano, Milano, Italy
Carla Cacciari: Osservatorio Astronomico, Bologna, Italy
Hugo Capelato: Instituto de Pesquisas Espaciais, São José dos Campos, Brazil
Leticia Carigi: Centro de Investigaciones de Astronomia, Merida, Venezuela
Bruce W. Carney: Department of Physics, University of North Carolina, Chapel Hill, North Carolina, USA
Luis Carrasco: Instituto de Astronomia, Universitad Autonoma de Mexico, Mexico
Reinaldo de Carvalho: Observatório Nacional, Rio de Janeiro, Brazil
Sandra Castro: Instituto Astronômico e Geofísico, Usp, São Paulo, Brazil
Márcio Catelán: Instituto Astronômico e Geofísico, Usp, São Paulo, Brazil
Jenai O. Calzetta: Instituto Astronômico e Geofísico, Usp, São Paulo, Brazil
Sergio Cellone: Facultad Ciencias Astronomicas y Geofisica, La Plata, Argentina
Arati Chokshi: Ipac/Caltech, Pasadena, California, USA
Cristina Chiappini Leite: Instituto Astronômico e Geofísico, Usp, São Paulo, Brazil
Cesare Chiosi: Osservatorio Astronomico di Padova, Padova, Italy
Juán Clariá: Observatorio Astronomico Nacional, Cordoba, Argentina
Gisella Clementini: Osservatorio Astronomico, Bologna, Italy
Peter Conti: Jila, University of Colorado, Boulder, Colorado, USA
Christopher Corbally: Vatican Observatory Group, Steward Observatory, Tucson, Arizona, USA
Gary S. Da Costa: Anglo Australian Observatory, Epping, Australia
Roberto D. D. Costa: Instituto Astronômico e Geofísico, Usp, São Paulo, Brazil
Jacques Colin: Observatoire de Bordeaux, Floirac, France
Michel Crézé: Observatoire de Strasbourg, Strasbourg, France
Rafael Cubarsí: Universitad Politecnica de Catalunya, Barcelona, Spain
Augusto Damineli Neto: Instituto Astronômico e Geofísico, Usp, São Paulo, Brazil
Anthony C. Danks: SFX/GSFC, Greenbelt, Maryland, USA
Massimo Della Valle: European Southern Observatory, La Silla, Chile
Gustavo Detthow: Instituto Astronômico e Geofísico, Usp, São Paulo, Brazil
Arjun Dey: Berkeley Astronomy Department, University of California, Berkeley, USA
S. George Djorgovski: Caltech, Pasadena, California, USA
Horácio Dottori: Universidade Federal do Rio Grande do Sul, Porto Alegre, Brazil
Michael G. Edmunds: Department of Physics, University of Wales, Cardiff, UK
Richard S. Ellis: Department of Physics, University of Durham, Durham, UK
Richard Elston: Kitt Peak National Observatory, Tucson, Arizona, USA

Sandra M. Faber: Lick Observatory, Santa Cruz, California, USA
Michael W. Feast: South African Astronomical Observatory, Cape, South Africa
Henry C. Ferguson: Institute of Astronomy, University of Cambridge, Cambridge, UK
Patrick François: Observatoire de Paris, Paris, France
Kenneth C. Freeman: Mount Stromlo and Side Spring Observatories, Canberra, Australia
José A. de Freitas Pacheco: Instituto Astronômico e Geofísico, Usp, São Paulo, Brazil
Amâncio C. Friaça: Instituto Astronômico e Geofísico, Usp, São Paulo, Brazil
Uta Fritze-v. Alvensleben: Universitatssternwarte, Göttingen, Germany
Duncan Forbes: Institute of Astronomy, University of Cambridge, Cambridge, UK
Wendy L. Freedman: Carnegie Observatories, Pasadena, California, USA
Jay A. Frogel: Department of Astronomy, Ohio State University, Columbus, Ohio, USA
Bianca Garilli: Istituto di Fisica Cosmica del CNR, Milano, Italy
P. Doug Geisler: Cerro-Tololo Interamerican Observatory, La Serena, Chile
Leo Girardi: Universidade Federal do Rio Grande do Sul, Porto Alegre, Brazil
Ian S. Glass: South African Astronomical Observatory, Cape, South Africa
Sandra A. Gonçalves: Instituto Astronômico e Geofísico, Usp, São Paulo, Brazil
Paul J. Green: Department of Astronomy, University of Washington, Seattle, Washington, USA
Michael Gregg: Mount Stromlo Observatory, Canberra, Australia
Ruth B. Gruenwald: Instituto Astronômico e Geofísico, Usp, São Paulo, Brazil
François Guglielmo: Observatoire de Paris-Meudon, Meudon, France
Ravi K. Gulati: Osservatorio Astronomico di Trieste, Trieste, Italy
Bengt Gustafsson: Astronomiska Observatoriet, Uppsala, Sweden
Hugh C. Harris: U.S. Naval Observatory, Flagstaff, Arizona, USA
Despina Hatzidimitriou: Anglo-Australian Observatory, Epping, Australia
Peter te Lintel Hekkert: Mount Stromlo and Siding Spring Observatories, Canberra, Australia
Enrico Held: Osservatorio Astronomico, Bologna, Italy
Gerhard Hensler: Physikzentrum, Universität Kiel, Kiel, Germany
Manuel Hernandez-Pajares: Universitad Politecnica de Catalunya, Barcelona, Spain
James E. Hesser: Dominion Astrophysical Observatory, Victoria, Canada
David Hollowell: Los Alamos National Laboratory, Los Alamos, New Mexico, USA
Jorge Horvath: Instituto Astronômico e Geofísico, Usp, São Paulo, Brazil
Shaun M.G. Hughes: Anglo-Australian Observatory, Coonabavabran, Australia
Victor A. Hughes: Queen's University, Astronomy Group, Kingston, Ontario, Canada
Thais Idiart: Instituto Astronômico e Geofísico, Usp, São Paulo, Brazil
Pascale Jablonka: Observatoire de Paris-Meudon, Meudon, France
Eduardo Janot-Pacheco: Instituto Astronômico e Geofísico, Usp, São Paulo, Brazil
Gerard Jasniewicz: Observatoire de Strasbourg, Strasbourg, France
Dean Johnson: Department of Physics, University of Wales, Cardiff, UK
Simon A. Josey: Astronomy Center, University of Sussex, Brighton, UK
J.M. Juan-Zornoza: Universitad Politecnica de Catalunya, Barcelona, Spain
Wolfram Kollatschny: Universitätssternwarte, Göttingen, Germany
Jari K. Kotilainen: Institute of Astronomy, University of Cambridge, Cambridge, UK
Harald Krüger: Universitätssternwarte, Göttingen, Germany
Robert Kurúcz: Harvard-Smithsonian Center for Astrophysics, Cambridge, Massachussets, USA
John Laird: Bowling Green State University, Bowling Green, Ohio, USA
Henny J. Lamers: Space Research Laboratory, Utrecht, Netherlands
Myuong G. Lee: Carnegie Observatories, Pasadena, California, USA
Y.-W. Lee: Yale Astronomy Department, New Haven, Connecticut, USA
Claus Leitherer: Space Telescope Science Institute, Baltimore, Maryland, USA
Carol Lonsdale: Ipac/Caltech, Pasadena, California, USA
H. Lorenz: Zentralinstitut für Astrophysik, Potsdam, Germany
Dario Maccagni: Istituto di Fisica Cosmica del CNR, Milano, Italy
Maria A. D. Machado: Instituto Astronômico e Geofísico, Usp, São Paulo, Brazil

Walter J. Maciel: Instituto Astronômico e Geofísico, Usp, São Paulo, Brazil
Barry F. Madore: Ipac/Caltech, Pasadena, California, USA
André Maeder: Observatoire de Génève, Sauverny, Switzerland
Brian McNamara: Kapteyn Laboratorium, Gröningen, Netherlands
Gladis C. Magris: Centro de Investigaciones de Astronomia, Merida, Venezuela
Márcio A. G. Maia: Observatório Nacional, Rio de Janeiro, Brazil
Steve Majewski: Carnegie Institute of Washington, Pasadena, California, USA
Gianni Marconi: Dipartimento di Astronomia, Università degli Studi di Bologna, Bologna, Italy
Anthony P. Marston: Drake University, Des Moines, Iowa, USA
Mario L. Mateo: Carnegie Institute of Washington, Pasadena, California, USA
Jorge Melnick: European Southern Observatory, La Silla, Chile
Márcia Erdelyi Mendes: Instituto Astronômico e Geofísico, Usp, São Paulo, Brazil
T.K. Menon: Department of Geophysics & Astronomy, University of British Columbia, Vancouver, Canada
Leon Mestel: Astronomy Center, University of Sussex, Brighton, UK
Gerhardt Meurer: Anglo-Australian Observatory, Epping, Australia
André Milone: Instituto Astronômico e Geofísico, Usp, São Paulo, Brazil
Dante Minniti: University of Steward Observatory, Tucson, Arizona, USA
Jeremy Mould: Caltech, Pasadena, California, USA
Rundsthen V. de Nader: Observatório Nacional, Rio de Janeiro, Brazil
Julio Navarro: Institute of Astronomy, University of Cambridge, Cambridge, UK
James Nemec: Department of Astronomy, University of British Columbia, Vancouver, Canada
Yuen K. Ng: Leiden Observatory, Leiden, Netherlands
S. Ninković: Astronomical Observatory, Belgrade, Yugoslavia
Robert W. O'Connell: Astronomy Department, University of Virginia, Charlottesville, Virginia, USA
Roberto Ortiz: Instituto Astronômico e Geofísico, Usp, São Paulo, Brazil
Sergio Ortolani: Osservatorio Astronomico, Padova, Italy
Jeremiah P. Ostriker: Princeton University Observatory, Princeton, New Jersey, USA
Bernard Pagel: Nordita, Köbenhavn, Denmark
Alexander Paquet: Landessternwarte Königstuhl, Heidelberg, Germany
Joel W. Parker: University of Colorado, Boulder, Colorado, USA
Reynier F. Peletier: Harvard-Smithsonian Center for Astrophysics, Cambridge, Massachussets, USA
Paulo Pellegrini: Observatório Nacional, Rio de Janeiro, Brazil
Mario R. Perez: IUE Observatory, Greenbelt, Maryland, USA
Ruth Peterson: Steward Observatory, University of Arizona, Tucson, Arizona, USA
Daniel Pfenniger: Observatoire de Génève, Sauverny, Switzerland
Marc Pinsonneault: Department of Astronomy, Yale University, New Haven, Connecticut, USA
Lilia Prado: Observatório Nacional, Rio de Janeiro, Brazil
Andrea Prestwich: Nasa-Marshall Space Flight Center, Huntsville, Alabama, USA
Almudena Prieto: Space Telescope, Garching bei München, Germany
Irmia J. Pronik: Crimean Astrophysical Observatory, Crimea
Philippe Prugniel: Observatoire de Haute-Provence, St.-Michel, France
Germano Quast: Laboratório Nacional de Astrofísica, Itajubá, Minas Geraes, Brazil
Hernán Quintana: Pontificia Universitad Catolica de Chile, Santiago, Chile
Alak Ray: Tata Institute of Fundamental Research, Bombay, India
Elsa Recillas-Cruz: Instituto de Astronomia, Universitad Autonoma de Mexico, Mexico
Alvio Renzini: Dipartimento di Astronomia, Università degli Studi di Bologna, Bologna, Italy
Ramiro de la Reza: Observatório Nacional, Rio de Janeiro, Brazil
R. Michael Rich: Columbia University, New York, USA
Carmelle Robert, Université de Montreal, Quebec, Canada
Brigitte Rocca-Volmerange: Institut d'Astrophysique, Paris, France
Jari Rönnback: Astronomiska Observatoriet, Uppsala, Sweden
Silvia C. F. Rossi: Instituto Astronômico e Geofísico, Usp, São Paulo, Brazil
Alex Ruelas-Mayorga: Instituto de Astronomia, Observatorio Astronomico Nacional, Mexico

Stuart Ryder: Mount Stromlo Observatory, Canberra, Australia
Elaine M. Sadler: Anglo-Australian Observatory, Epping, Australia
John Salzer: Kitt Peak National Observatory, Tucson, Arizona, USA
João F. Santos Jr.: Universidade Federal do Rio Grande do Sul, Porto Alegre, Brazil
Jaume Sanz i Subirana: Universitad Politecnica de Catalunya, Barcelona, Spain
Ricardo Schiavone: Observatório Nacional, Rio de Janeiro, Brazil
Alex Schmidt: Astronomy Center, University of Sussex, Brighton, UK
Henrique Schmitt: Universidade Federal do Rio Grande do Sul, Porto Alegre, Brazil
Robert Schommer: Cerro-Tololo Interamerican Observatory, La Serena, Chile
William J. Schuster: Instituto de Astronomia, Universitad Autonoma de Mexico, Mexico
François Schweizer: Carnegie-DTM, Washington DC, USA
Wilhelm Seggewiss: Astronomisches Institut, Universität Bonn, Bonn, Germany
Alfonso Serrano: Instituto de Astronomia, Universitad Autonoma de Mexico, Mexico
Joseph Silk: Astronomy Department, University of California, Berkeley, California, USA
Licio da Silva: Observatório Nacional, Rio de Janeiro, Brazil
Paulo R. da Silva: Instituto Astronômico e Geofísico, São Paulo, Brazil
Ronaldo de Souza: Instituto Astronômico e Geofísico, São Paulo, Brazil
Denis Spergel: Princeton University, Princeton, New Jersey, USA
Monique Spite: Observatoire de Paris-Meudon, Meudon, France
François Spite: Observatoire de Paris-Meudon, Meudon, France
Adam Stanford: Astronomy Department, University of California, Berkeley, California, USA
Letizia Stanghellini: Osservatorio Astronomico, Bologna, Italy
Grazyna Stasińska: Observatoire de Paris-Meudon, Meudon, France
Nathalie Stout: Observatório Nacional, Rio de Janeiro, Brazil
Nicholas Suntzeff: Cerro-Tololo Interamerican Observatory, La Serena, Chile
Massimo Tarenghi: European Southern Observatory, Garching bei München, Germany
Peter Teague: Laboratory for Experimental Astrophysics, Livermore, California, USA
José Eduardo Telles: Institute of Astronomy, University of Cambridge, Cambridge, UK
Elena Terlevich: Institute of Astronomy, University of Cambridge, Cambridge, UK
Roberto Terlevich: Institute of Astronomy, University of Cambridge, Cambridge, UK
Fréderic Thévenin: Observatoire de la Côte d'Azur, Nice, France
Christopher G. Tinney: California Institute of Technology, Pasadena, California, USA
J. Torra: Departamento d'Astronomia i Meteorologia, Universitad de Barcelona, Barcelona, Spain
Carlos A. Torres: Laboratório Nacional de Astrofísica, Itajubá, Minas Geraes, Brazil
Silvia Torres-Peimbert: Instituto de Astronomia, Universitad Autonoma de Mexico, Mexico
Monica Tosi: Osservatori Astronomico, Bologna, Italy
Peeter Traat: Tartu Observatory, Tartu, Estonia
William Vacca: Jila, University of Colorado, Boulder, Colorado, USA
Dany Vanbeveren: Department of Physics, Vrije Universiteit Brussel, Brussels, Belgium
Sidney van den Bergh: Dominion Astrophysical Observatory, Victoria, Canada
L. Bob van den Hoek: Sterrenkundig Institut "Anton Pannekoek", Amsterdam, Netherlands
Thijs van der Hulst: Kapteyn Astronomical Institut, Gröningen, Netherlands
Rúben A. Vasquez: Observatorio Astronomico, La Plata, Argentina
M. Begoña Vila: Department of Physics & Astronomy, University of Wales, Cardiff, UK
Stefan Wagner: Landessternwarde Heidelberg-Königstuhl, Heidelberg, Germany
Nicholas A. Walton: Department of Physics & Astronomy, University College London, London, UK
Roland Wielen: Astronomisches Rechen-Institut, Heidelberg, Germany
Patricia Whitelock: South African Astronomical Observatory, Cape, South Africa
Allan J. Willis: Department of Physics & Astronomy, University College London, London, UK
Christine D. Wilson: Astronomy Program, University of Maryland, College Park, Maryland, USA
Guy Worthey: Lick Observatory, Santa Cruz, California, USA
Yuefang Wu: Peking University, Beijing, China
P. Tim de Zeeuw: Sterrenwacht Leiden, Leiden, Netherlands
Steve E. Zepf: Space Telescope Science Institute, Baltimore, Maryland, USA
Hans Zinnecker: Institut für Astronomie und Astrophysik, Wurzburg, Germany

The Halo Populations

JAMES E. HESSER
Dominion Astrophysical Observatory, Herzberg Institute of Astrophysics, National Research Council of Canada, Victoria, B.C., V8X 4M6 Canada

Abstract.
Understanding the halo populations of the Milky Way impacts upon a vast landscape of stellar, Galactic and extragalactic astrophysics. Topics likely to play important roles at this meeting are introduced, including aspects of properties of the outer halo, the halo-to-disc transition, globular cluster binary stars and dynamics, chemistry, and age determinations.

1. Overview

"Stellar populations in the Galactic halo" is an exceedingly broad topic affecting nearly all aspects of this Symposium in one way or another. The primary goal of this introduction is to identify themes likely to be amplified and, with fortune, clarified by others. The topics were chosen based on my perception of their ultimate relevance to our understanding of galaxies and galaxy formation. While the emphasis is on work of the last two years or so, it is so extensive that many worthy achievements must be ignored due to space limitations.

To set the scene, it is well known but easy to forget that the Galaxy has undergone dramatic evolution and is, in every sense of the words, a very dynamic entity. Its halo *appears* to be sparsely populated. The known mass associated with visible light accounts for only a few percent of the total estimated mass of the Galaxy. A series of major questions, however, remain unanswered about the mysterious (but see Kurucz 1991) dark matter commongly thought to dominate the mass and dynamics of the Galaxy. In particular, what is its spatial distribution? Does *visible* halo matter trace *dark* halo matter? Is dark matter baryonic, after all?

How big is our halo? We know of individual stars at $R_{gc} \gtrsim 50$ kpc (§2.1), which is comparable to the distances of the Magellanic Clouds, who themselves are contained in the Galactic halo. We know of globular star clusters extending out to some 120 kpc, and we know of dwarf spheroidal galaxies at ~250 kpc. Certainly all of these objects are current constituents of the Galactic halo (but remember the initial caveat about not viewing the Galaxy as a static entity). *If* the bound halo extends to such great distances, is it reasonable for galaxy formation theories to use R_{max} values of 30 kpc?

Throughout the halo, chemical composition, as characterized by the logarithmic metals-to-hydrogen ratio relative to the Sun, [M/H] (where M refers to generic heavy elements, usually of the Fe peak), spans three or more dex, with a wide range at each R_{gc} zone. Does the halo possess an [M/H] gradient? Are there element abundance-ratio differences between halo and disc populations, or between halo cluster and halo field stars? Are there zero-metal stars in the halo?

To what extent are the halo populations we observe today representative of the original halo populations? How similar were the latter to those probed by absorption line systems of distant quasars? Are globular cluster and field halo stars drawn from the same parent population? What roles do dynamical processes play in cores of star

B. Barbuy and A. Renzini (eds.), The Stellar Populations of Galaxies, 1–14.

clusters (and are they perhaps relevant to activity hidden from view in the cores of, say, elliptical galaxies)? What roles did R_{gc} or z-height play in star formation within the halo and its globular clusters? How, where, and at what [M/H] does the halo join onto other Galactic components? Is a 'thick disc' unambiguously present and, if so, is there significant overlap between it and the halo?

Perhaps when more of these questions are answered, we will know what is meant by the title of my talk. In the meantime, my aim is to examine some recent efforts that address aspects of these questions. The interested reader will find considerable recent information in Janes (1991), and will find Gilmore, King and van der Kruit (1990) or Larson (1990) invaluable for placing it in context.

2. The Outer Halo

The very sparseness of the visible halo makes its study particularly challenging. There are simply too few globular star clusters to trace all aspects of early Galactic evolution, especially when, of the total ~150 objects (see, *e.g.*, Webbink 1985), perhaps only ~75% can be identified with the 'true' halo (§3.1). In order to make progress, field halo stars must be identified. In the past decade, increasing emphasis has been placed upon *in situ,* chemical and kinematical surveys of the halo field populations. The power of such surveys to resolve questions of size, shape, parentage and role of the 'second parameter' in the halo populations was driven home to me by Freeman at Patras (see, *e.g.*, Freeman 1983, Ratnatunga and Freeman 1985), but many others were then, and now, pursuing surveys of enormous importance (*e.g.*, Beers, Preston and Shectman 1985, 1991, Bothun, *et al.* 1991, Croswell, *et al.* 1991, Grenon 1989, 1991, Rose and Agostinho 1991, Ryan and Norris 1991a,b, Sandage and Fouts 1987, Schuster and Nissen (1989), Suntzeff, Kinman and Kraft 1991...). The importance of understanding selection biases in such surveys cannot be overstressed. Moreover, 'consumers' should appreciate that the surveys are difficult and time consuming: for instance, only one star in a thousand examined by Beers, *et al.* turns out to have low [M/H].

2.1. Spatial Extent

Individual stars are known at distances of 50 kpc (Ciardullo, Jacoby and Bond 1989), globular clusters at more than 100 kpc and dwarf spheroidals (dSphs) at more than 200 kpc. It seems probable that individual stars will ultimately be identified at distances intermediate between 50 kpc and the outermost globulars or dSphs. Most analyses of velocity ellipsoids and of field star and cluster counts infer the halo to be nearly round in shape for $R_{gc} \gtrsim 20$ kpc. They also find that the stellar volume density varies as $R_{gc}^{-3.5}$. As R_{gc} decreases in the halo, its subsystems become increasingly velocity anisotropic and flattened (see, *e.g.*, Hartwick 1987, Preston, Shectman and Beers 1991, Sommer-Larsen, Christensen and Flynn 1991, and Vedel and Sommer-Larsen 1990).

2.2. Chemical Composition

Deciding whether or not there is a gradient in chemical composition as a function of R_{gc} within the Galactic halo requires field star samples: the globular clusters are again too few. In the seminal Lick RR Lyrae survey, Suntzeff, Kinman and Kraft (1991) find no evidence for such a gradient with respect to R or |z| outside the solar circle, but within it find Δ[M/H]~ -0.06 dex kpc^{-1} (see their figures 4, 5). They find the metallicity distribution function of the field and cluster RR Lyrae stars to be the same over the entire range of R_{gc} sampled. However, they note some differences when comparing the metallicity distribution of the entire globular cluster system to that of field RR Lyraes: for R>R_o, [M/H] and its dispersion are indistinguishable, while for R<R_o the field RR Lyraes are more metal rich than the globular clusters. They also find that the field variables obey a period-shift, metallicity relation similar to that of the cluster variables. Assuming the populations to be the same, they estimate that 2% of the halo mass is locked up in globulars and that the mass of the luminous halo within 4-25 kpc is $\sim 9 \times 10^8 M_\odot$.

A sample of 372 field subdwarfs with [M/H] and v_r determinations has been extensively modelled by Ryan and Norris (1991a,b); their database includes the stars studied by Laird, *et al.* (1988). For stars with [M/H]< -1.4 (presumably a nearly pure halo sample), they find no correlation between kinematics and [M/H] except for an increase in σ_W as [M/H] decreases. The metallicity distribution appears to be in good agreement with a simple model (Hartwick 1976) over 2.5 decades in [M/H]. They argue that Searle's (1977) stochastic model for the formation of the Galaxy does not appear to match the field star data closely, but could account for the globular cluster properties if some ten enrichments per fragment had occurred. They further argue that it is possible, but not established, that field halo stars and globular clusters have different parent populations.

A characteristic long associated with the outer halo is the so-called 'second parameter effect': a discord between [M/H]s inferred from the form of the horizontal and giant branches for individual globular clusters and dSphs. During the past two years, age differences have received significant observational support as constituting a major component of the second parameter phenomenon in star clusters (see, *e.g.*, Bolte 1989, Green and Norris 1990, Lee, Demarque and Zinn 1990, VandenBerg, Bolte and Stetson 1990, Sarajedini and Demarque 1990). These studies suggest that there *may* be a modest age, metallicity relation in the Galaxy, with the oldest objects being more deficient in heavy elements.

Another advance stemming from increased survey activity has been the alleviation of the apparent paucity of extremely metal-deficient G dwarfs in the solar neighborhood described provocatively by Bond (1981). Beers (1991) and Ryan and Norris (1991a,b), among others, suggest that the metallicity distribution of the Galactic halo stars remains rather flat to [M/H]~ -4.5, the lowest levels measured. Such results seem consistent with a simple model of Galactic chemical evolution (Hartwick 1976). For me, recent research has strengthened evidence that globular clusters and field stars in the Galactic halo may be from the same parent populations. However, recall that colors of globular cluster systems around ellipticals suggest that their average composition at a particular radius is lower than that of the underlying halo light (Harris 1991).

2.3. Luminosity Functions

Studies by Richer, Fahlman and their collaborators have apparently pushed I-band luminosity functions to within hundreths of a solar mass of the expected brown dwarf domain (*e.g.*, Richer, *et al.* 1991). Some, and possibly all, of the clusters they have observed exhibit very steep mass functions below $\sim 0.4\ M_\odot$, from which they suggest that perhaps the mysterious halo dark matter is composed of Population II objects. Fall and Rees (1985), Aguilar, Hut and Ostriker (1988), Chernoff and Weinberg (1990) and others suggest that the present day clusters may be but a small fraction of the original population. Such concepts may be consistent with recent observations suggesting that weak clustering and/or dissolution is occurring in the halo. Sommer-Larsen and Christensen (1987) believe they have discovered a distant group of stars in the process of dissolving, while Doindus and Beers (1989) find statistical evidence for field horizontal branch candidate stars lying 5-8 kpc from the Sun to cluster on $r \lesssim 25$ pc scales. Thus, if copious low-mass star formation occurred in metal deficient material (Zinnecker 1987), as Richer, *et al.* 's data suggest, the halo dark matter may be composed of low-mass stars originally formed in clusters (see also Ashman 1990). Richer, *et al.* note that their results are subject to at least one important caveat: the upturn in inferred mass-function slope occurs disturbingly close to the slope change in the theoretical M/L relation. Nonetheless, this is an observational opportunity clearly deserving attention in the era of giant ground-based telescopes (and of second-generation HST cameras).

3. Transition from Halo to Disc

If the thick disc exists as a discrete component of the Galaxy, was it formed by mergers, by collapse, or by upwards scattering? Are the halo and the thick disc distinct and, if so, what values of [M/H], R_{gc}, z, etc. characterize the transition region? What effect does superbubble ejection of disc gas into the halo and thick disc (Heiles 1991, De Geus 1991) have on Galactic evolution? In my mind, such intriguing questions form the backdrop to any discussion of the inner halo.

3.1. Spatial Extent

Empirically, the systemic luminosity function for the Galactic globular clusters, as for the cluster systems of other galaxies, is reasonably well described by a gaussian with $<M_V> \sim -7.4$ and $\sigma \sim 1.3$ (see, *e.g.*, Harris 1991). However, at least one important subsystem, first suspected by Kinman (1959), exists. Zinn (1985), Hesser, Shawl and Meyer (1986) and Armandroff and Zinn (1988) demonstrated that about a quarter of the globulars form a separate, disc-like subsystem. An excellent review by Zinn (1991) summarizes present understanding of this division, and the possibility that a bulge subsystem may also be present. The disc globulars exhibit average features similar to those commonly associated with the stellar thick disc (Table 1).

TABLE I
Globular Cluster and Stellar Populations (Zinn 1991)

	Halo		Thick Disk	
	Stellar	Cluster	Stellar	Cluster
<[Fe/H]>~	-1.7	-1.6	-0.5	-0.5
V_{rot} (km/s)	30 ± 10	43 ± 29	180 ± 10	193 ± 29
σ_{los} (km/s)	110 ± 10	116 ± 11	60 ± 10	59 ± 11

An important constraint on formation and evolution within the halo subsystems may be emerging from the range of slopes of the initial mass functions inferred for upper main sequence stars in globular clusters. The inference of a true initial mass function from observational data requires correction for internal (*e.g.*, Pryor, Smith and McClure 1986) and external (*e.g.*, Stiavelli, *et al.* 1991) dynamical effects, which is not necessarily straightforward. Those corrections notwithstanding, the range of inferred slopes was recently found to correlate better with |z| or R_{gc} (Capaccioli, Ortolani and Piotto 1991), than with [M/H] (McClure, *et al.* 1986, Hesser 1988). Such inferences suggest that gradients within the halo and transition regions may have been set up by a combination of external and internal dynamical processes.

3.2. Chemical Composition

The difficulty of cleanly identifying possible components of the field star populations, even when good kinematic and chemical data are available, is impressive. In particular, the characterization of roughly where the true halo is overtaken by the purported thick disc remains ill defined.

Morrison, Flynn and Freeman (1990) find many stars with thick disc kinematics at [M/H]~ -1.6, i.e., nearly a dex more metal deficient than the mean thick disc population. This may be related in some fashion to Carney, Storm and Jones' (1991) evidence among RR Lyrae stars for a sudden population change at [M/H]~ -1.7. Rees and Cudworth (1991) showed that the globular cluster M28 (NGC 6626) has a disc-like orbit, lies at z$\sim$ 0.3 kpc, yet has [M/H]~ -1.7. Latham (1991) reports that he and his collaborators find stars in their *in situ* surveys of v_rs and [M/H]s for ~1300 objects that exhibit solar metallicities and halo kinematics, and vice versa, leading naturally to the question, 'what are they?'. The extensive Geneva photometric and radial velocity surveys summarized by Grenon (1991) find halo and bulge stars in the solar neighborhood, with the clear implication that models of the Galaxy must be modified to include *radial* interchange of stellar populations.

For field horizontal branch stars, Preston, Shectman and Beers (1991) interpret a small color gradient, in the sense that $<(B-V)_o>$ becomes redder by ~0.025 mag per 2 kpc as one proceeds from 2 to 12 kpc in the halo, as evidence for a decrease in the mean age of such stars as one proceeds outward into the halo. They also suspect an increase in the range of ages as one moves outwards over that interval.

In general the stars in the above mentioned surveys can be reasonably assumed to be relatively old. Fitzsimmons, *et al.* (1991), Conlon, *et al.* (1990, 1991), and Quin *et al.* (1991) all report high latitude B stars, some of which appear to be normal Population I objects at $2 \lesssim z \lesssim 25$ kpc. If they are not nearby, subluminous objects with spectra mimicing normal stars (*e.g.*, Waelkins, *et al.* 1987, Tobin 1987), were they ejected from disc clusters or formed at high Galactic latitude? Evidence for recent star formation in the vicinity of the optical 'jet' in the halo of NGC 5128 (Graham 1975, Osmer 1978) remains for me a sober reminder that galaxy halos may represent more complex mixtures than some of us enjoy contemplating.

The confusing evidence regarding the characterization of the transition region notwithstanding, the data presently available admit the concept of a metal-deficient halo joining onto a more enriched thick disc subsystem (see especially Majewski (1991), of which I was unaware when I spoke). However, the mixture models of Nemec and Nemec (1991 and this volume) illustrate how very difficult, indeed, it is to formulate unambiguous descriptive parameters for Galactic subsystems.

4. Binaries, Dynamics and Exotica

Besides their frequency being almost certainly a fundamental characteristic of a stellar population, binaries are an important source of energy that can retard or prevent dynamical collapse in the cores of globular clusters, and that provide channels to form unusual stars. When integrated over the Galactic lifetime, internal and external dynamical processes (*e.g.*, collisions, mergers, tidal shocks) are clearly relevant to understanding how observations today relate to properties of the original halo population (see Djorgovski 1991 for a clear review). Moreover, the study of present-day 'exotica' in globular clusters may bear strongly upon our ultimate understanding of the field halo populations, particularly if theoretical arguments (§2.3) for a much larger cluster system in the early Galaxy are valid. Finally, such physics seems intrinsically worthy of study, for how can we hope to understand, say, elliptical galaxies, if we cannot realistically model the simplest dynamical systems known, the Galactic globular clusters?

4.1. Field Halo Binaries

In Buenos Aires, Latham (1991) announced that his radial velocity surveys carried out over a decade with Carney and collaborators now yield ~150 halo binary systems for which they have been able to determine orbits. This is quite simply a *tour de force* observational result! They find a similar, ~20% binary frequency among their Population II samples as in their studies of Population I samples. They also have reasons to infer in all of their samples that the secondaries are drawn from normal field populations, which suggests that the predominant formation mechanism is likely to be capture and/or collision, rather than fission.

4.2. Evidence for Binaries in Globular Clusters

The dramatic rise in evidence for a significant binary component in at least some globular star clusters continues unabated. Due to their potential influence on cluster evolution (*e.g.*, Goodman and Hut 1989, McMillian, Hut and Makino 1990), it seems appropriate to update that evidence.

Again in Buenos Aires, Meylan (1991; see also Paresce, *et al.* 1991) reported evidence for 21 blue stragglers within a $26 \times 26''$ frame of the center of 47 Tucane secured with HST's Faint Object Camera. This is, of course, a region of the cluster that remains largely unexplored from the ground, due to image crowding; obvious blue straggler candidates at larger radii are essentially lacking in published color-magnitude diagrams (*e.g.,* Hesser, *et al.* 1987). Meylan and his collaborators argue that these newly discovered blue stragglers, the central X-ray sources, the millisecond pulsars (Manchester, *et al.* 1991), and two stars thought to be in the process of ejection from the core (Meylan, Dubath and Mayor 1991), are all consistent with close encounters with binaries in a dense core.

In NGC 5466, Mateo, *et al.* (1990, and Nemec's contribution elsewhere herein) have identified nine variable blue stragglers, three of which are eclipsing binaries, with the remaining six being SX Phe pulsators. They suggest that all non-eclipsing blue stragglers in this cluster (and others) may have been formed by mergers of single stars. Astro I observations of ω Centauri, by O'Connell and his colleagues (1991) reveal numerous ultraviolet bright stars, whose nature remains to be explained.

Bailyn (1991) thoroughly reviews the arguments that neutron star binaries (low mass X-ray binaries: LMXRBs) are more common in globular clusters than in the field, thus leading to the expectation that cataclysmic variables (CVs) will be numerous in clusters. However, Cederbloom, *et al.* (1991) find the inner $6.6''$ of M15 (thought to be a post-core-collapse cluster) to be bluer in B−V yet stronger in Hα absorption than at larger radii. This seems to imply that the excess blue light does not arise from CVs (see §4.3)

In a burgeoning field of research, available data led Kulkarni, Narayan and Romani (1990) to estimate that there may be as many as $\sim 10^4$ millisecond pulsars in Galactic globulars. Prince, *et al.* (1991) postulate that 2127+11C, the eight hour binary pulsar in M15, is being ejected following a close encounter with another binary, a mechanism that may populate the field with such pulsars. Relativistic winds from millisecond pulsars may be responsible for ejecting cold gas from globulars (Spergel 1991); note that HI gas may have been, at last, detected in the process of ejection from NGC 2808 (Faulkner, *et al.* 1991).

As reviewed elsewhere (Hesser 1988, 1991), some globular cluster color-magnitude diagrams exhibit photometric evidence for a sequence of roughly equal-mass binaries more luminous and cooler than the single-star main sequence. While residual widths in the observed main sequences of M92 and M30 may arise in part from crowding (see, *e.g.,* Stetson and Harris 1988), modelling is consistent with a binary frequency of 9 and 4%, respectively (Romani and Weinberg 1991) in the outer regions where the data were obtained. Murray, Clarke and Pringle (1991) have suggested that if all stars formed simultaneously as single stars, the protostellar discs would increase the interaction cross sections during cluster collapse, thus leading to the establishment of at least a few percent of 'primordial' binaries. The radial velocity surveys of Pryor, *et al.* (1989) appear to be consistent with such percentages. On the other hand, Bolte (1991) reports a 10% main-sequence binary fraction with primary to secondary mass ratio ≤ 1.4 at $3r_c$ in NGC 288. This implies a much larger total binary percentage if all mass ratios are considered.

4.3. Color Gradients in Clusters

One of the most puzzling, and potentially important, observations regarding the structure of globular clusters is the finding by Djorgovski, *et al.* (1991) that clusters with pronounced central cusps (presumably reflecting their condition as post core-collapsed objects) are bluer towards the center, resulting from a demise of red giant branch stars and/or of subgiants, although an increase in the number of blue stragglers could contribute, as well. On the other hand, in NGC 6171, Ferraro, *et al.* (1991) find red giant branch stars to be more centrally concentrated!

4.4. General Remarks

In his review at Buenos Aires, Meylan (1991) noted that the latest determinations of velocity dispersions, masses and M_Vs for old clusters in the Milky Way, the Large Magellanic Cloud and the Fornax dwarf spheroidal now overlap, thus stressing the similarities of clusters found (and probably formed) in quite different environments.

5. Halo Chemistry and Ages

5.1. Chemical Abundance Ratios

Unravelling globular cluster ages is tied to understanding element ratios as a function of [M/H], and to knowing how measurements of the brighter stars relate to the ratios, presumably primordial, exhibited by main sequence stars too faint for study. Alternatively, it is essential to understand if it is meaningful to adopt trends found among more readily measured halo field stars for application to age determinations of globular clusters. The role of [O/Fe] in globular cluster ages has been stressed by VandenBerg (1988) and VandenBerg and Stetson (1991), while another point of view is presented by Straniero and Chieffi (1991) and Chieffi, Straniero and Salaris (1991). Characterizing the selective [O/Fe] enhancement relative to [M/H] in field stars throughout the halo has been a focus of recent research, but [O/Fe] is very difficult to measure in globular cluster stars.

From measurements of the permitted, high-excitation OI triplet at 7771-75Å for 30 metal-deficient G dwarfs, Abia and Rebolo (1989) inferred a strong, nearly monotonic increase in [O/Fe] as [M/H] decreases. Their results differ from those obtained by Barbuy (1988 and this volume) and by Barbuy and Erdelyi-Mendes (1989) using the forbidden [OI] line at 6300Å in K giants. Spiesman and Wallerstein (1991) and Brown, *et al.* (1991) present evidence from both field and cluster stars that also counters the Abia and Rebolo results. Specifically, analysis of ω Cen giants allowed Brown, *et al.* to demonstrate that, over the range $-2.0 \leq$[Fe/H]≤ -1.0, [O/Fe] vs. [M/H] is very similar to that found by Barbuy for field halo giants; similar results were obtained from lower signal-to-noise data by Paltoglou and Norris (1989). Interestingly, Brown, *et al.* remark that the metallicity enrichment processes appear to have been similar in ω Cen ($M \sim 3 \times 10^6 M_\odot$) and the Galactic halo ($\sim 10^9 M_\odot$).

A particularly compelling study was described at Buenos Aires by Bessell (see Bessell, Sutherland and Ruan 1991). They have determined [O/Fe] for a sample of metal-poor G dwarfs in two independent ways: via the near ultraviolet OH lines and via the permitted OI triplet. Their OH results agree well with those of, *e.g.*, Barbuy and her collaborators, while the OI lines agree with Abia and Rebolo. The OH lines are formed in the same layers of the atmospheres as the majority of the metal lines. Consequently, they believe the lower oxygen enhancements deduced from them represent the actual abundances, while they attribute the OI results to real metal-poor stellar atmospheres being hotter in deeper layers than the models. In a potentially related study by Lambert (1991), evidence that forbidden and molecular lines of C and O yield more reliable results than permitted ones leads him (and others) to suspect systematic errors in the model atmospheres for metal-poor stars. Should that be so, *all* abundance determinations using those atmospheres may well be in error and Galactic chronology made even more uncertain.

Bessell and collaborators draw two important conclusions regarding the halo populations. First, that [O/Fe]$\sim$[α/Fe] throughout (Lambert (1991) summarized additional support for this result). Second, that the rise in oxygen enhancement from zero near solar metallicities reaches a plateau, [O/Fe]$\sim$ +0.5-0.6, around [M/H]~ -1.6 instead of at ~ -1.0, as commonly reported.

Another constraint on Galactic formation is inherent in the observation by Sneden and his collaborators (see, *e.g.*, Gilroy, *et al.* 1988 and the review by Cowan, Thielemann and Truran 1991) that, among metal deficient stars, heavy element formation has been predominantly by the r-process, whereas among disc stars it has been by a mixture of r- and s-processes. Interestingly, the dispersion in heavy element abundances is generally found to be less than observational errors for stars with [M/H]≤ -1, except for the rare earth elements, where the dispersion is an order of magnitude greater than the observational errors. Within individual globular clusters the small spread in abundances for elements heavier than the CNO group constitutes a very strong constraint on the formation and subsequent evolution of such clusters (Murray and Lin 1990, Truran, Brown and Burkert 1991).

5.2. Ages

Renzini (1991) and Rood (1990), among others, emphasize that uncertainities in (m$-$M) determinations dominate uncertainties in cluster ages. [M/H] uncertainties are important for the handful of subdwarfs used for comparison with cluster color-magnitude diagrams. All calibrations of M_v^{RR} as a function of [M/H] depend intimately upon theory (Renzini), while model colors remain uncertain (VandenBerg 1991). Significant theoretical progress by Pinsonneault, Kawaler and Demarque (1990) demonstrated that rotation is unlikely to affect present age determinations. However, He diffusion can lower the inferred ages by 5-10% with respect to those estimated using standard models (Deliyannis and Demarque 1991, Proffitt and VandenBerg 1991). Deliyannis and Demarque showed from consideration of Li that neglecting He diffusion has probably not caused significant age errors.

A major conceptual breakthrough occurred in 1989 regarding techniques for *relative* age determinations throughout the Galactic halo (VandenBerg, Bolte and Stetson 1990, Sarajedini and Demarque 1990). By use of the color difference between the turnoff region and the base of the red giant branch, relative ages can be determined accurately within modest ranges of [M/H]. Paltoglou and Bell (1991) have shown that with optimized filter selection the [M/H] range may be extended. Hatzidimitriou (1991) suggests that a color difference between the mean color of the red horizontal and of the red giant branches at the same luminosity may serve the same purpose for clusters with $[M/H] \gtrsim -1.7$.

The key results of the recent *relative* age determinations are: a) Among metal-poor clusters, age differences are $\lesssim 0.5$ Gyrs (VandenBerg, Bolte and Stetson 1990; van Albada, Dickens and Wevers 1981; Heasley and Christian 1991), with the possible exception of Ru106, which Buonanno, *et al.* (1991) find to be younger. (I suspect there is some inconsistency between this result, and that of Preston, Shectman and Beers (1991) regarding the field horizontal branch stars.) b) Among globulars of intermediate [M/H], there are objects exhibiting a spread in ages of ~3 Gyrs (VandenBerg, Bolte and Stetson). The key pair of clusters here is NGC 288 and NGC 362, which have identical properties to within the errors of present measurements. However, Dickens, *et al.* (1991) suggest that the stars of NGC 362 exhibit more extensive mixing than do those of NGC 288, which I find quite worrisome. c) Among the metal-rich clusters, Hatzidimitriou (1991) reports a surprisingly large range of ages (~6 to 16 Gyrs). If true, this would eliminate any age gap between 'halo' (thick disc?) clusters and the oldest open clusters. Furthermore, in the younger clusters one would expect to see carbon stars. However, it would seem worthwhile to examine whether differential reddening and/or field star contamination might be affecting her initial age inferences with this promising technique.

It is rather discouraging that with all the improvement in color-magnitude diagrams and theory, there is still such a large uncertainity in the cosmologically important age of the oldest clusters in the Galaxy. Most experienced practitioners agree that with available models the minimum age for metal-poor globulars is not likely to be less that 14 Gyr. But the real errors are large, systematic and difficult to quantify. Hence, we probably cannot decide at this time if the formation of the halo was by a Searle-Zinn (1979), Tinsley-Larson (1979), Toomre (1977) type of subunit merger; by a modified Eggen, Lynden-Bell and Sandage (1962; see also Sandage 1990) collapse; or, as seems likely to me, by some combination thereof.

6. 'Final' Impressions

'Final' impressions from my study of the recent literature are the 'initial' ones with which I go into this conference (my feelings about one of the following was changed during the meeting, but this text closely reflects what I said):

Extragalactic globular cluster systems suggest there may be important relative differences in formation paths. Couture, Harris and Allwright (1991) find that the relative importance of mergers and dissipative collapse apparently can vary among elliptical galaxies that today look structurally similar.

In spite of impressive new data sets and sophisticated analyses, I cannot tell if the halo-to-disc transition is smooth, or clearly separable into discrete (but extensively overlapping?) components; further *in situ* surveys are essential.

Exploration of the effects of the new Livermore interior, and Kurucz astmospheric, opacities deserves priority in the next few years, and is rumored likely to modify significantly some cherished impressions.

Globular cluster cores are 'gold mines' for study of exotic stars and dynamical astrophysics; they are natural targets for adaptive optics and HST.

One believable, precise E(B−V) determination for a globular cluster is worth ten color magnitude diagrams for it; the larger E(B−V) is, the more weight this statement deserves. Attempts to decide the chronology of early Galactic formation may ultimately hinge on determining reliable reddenings for bulge clusters.

Once upon a time, I thought globular clusters and stellar populations in the Galactic halo were relatively simple...how naive I was! But if they are much more complicated than they once appeared (at least to me), it raises an interesting query at the outset of our meeting: how much fundamental astrophysics is being 'swept under the rug' in our studies of halo and bulge properties of the nearest galaxies, when even nearer, purportedly simpler, globular clusters hold so many surprises?

References

Abia, C., Rebolo, R.: 1989, *ApJ* **347**, 186

Aguilar, L., Hut, P., Ostriker, J.P.: 1988, *ApJ* **335**, 720

Armandroff, T.E., Zinn, R.: 1988, *AJ* **96**, 92

Ashman, K.M.: 1990, *MNRAS* **247**, 662

Bailyn, C.D.: 1991, in Janes, K.A., ed(s)., *ASPCS v. 13, The Formation and Evolution of Star Clusters*, ASP, San Francisco, 307

Barbuy, B.: 1988, *A&A* **191**, 121

Barbuy, B., Erdelyi-Mendes, M.: 1989, *A&A* **214**, 239

Beers, T.C.: 1991, in Terlevich, R., ed(s)., *Elements and the Cosmos,* Cambridge U. Press, in press

Beers, T.C., Preston, G.W., Shectman, S.A.: 1985, *AJ* **90**, 2089

Beers, T.C., Preston, G.W., Shectman, S.A., Kage, J.A.: 1990, *AJ* **100**, 849

Bessell, M.S., Sutherland, R., Ruan, K.: 1991, preprint

Bolte, J.: 1989, *AJ* **97**, 1688

Bolte, M.J.: 1991, *ApJ,* in press

Bond, H.E.: 1981, *ApJ* **248**, 606

Bothun, G., Elias, J.H., MacAlpine, G., Matthews, K., Mould, J.R., Neugebauer, G., Reid, I.N.: 1991, *AJ,* in press

Brown, J.A., Wallerstein, G., Cunha, K., Smith, V.V.: 1991, *ApJ,* in press

Buonanno, R., Buscema, G., Fusi Pecci, F., Richer, H.B., Fahlman, G.G.: 1991, *AJ* **100**, 1811

Capacciolio, M., Ortolani, S., Piotto, G.: 1991, *A&A* **244**, 298

Carney, B.W., Storm, J., Jones, R.V.: 1991, preprint

Cederbloom, S.E., Moss, M.J., Cohn, H.N., Lugger, P.M., Bailyn, C.D., Grindlay, J.E., McClure, R.D.: 1991, in Janes, K.A., ed(s)., *ASPCS v. 13, The Formation and Evolution of Star Clusters*, ASP, San Francisco, 246

Chernoff, D.F., Weinberg, M.D.: 1990, *ApJ* **351**, 121

Chieffi, A., Straniero, O., Salaris, M.: 1991, in Janes, K.A., ed(s)., *ASPCS v. 13, The Formation and Evolution of Star Clusters*, ASP, San Francisco, 219

Ciardullo, R., Jacoby, G.H., Bond, H.E.: 1989, *AJ* **98**, 1648

Conlon, E.S., Dufton, P.L., Keenan, F.P., Leonard, P.J.T.: 1990, *A&A* **236**, 357

Conlon, E.S., Dufton, P.L., Keenan, F.P., McCausland, R.J.H.: 1991, *MNRAS* **248**, 820

Couture, J., Harris, W.E., Allwright, J.W.B.: 1991, *ApJ* **372**, 97

Cowan, J.J., Thielemann, F.-K., Truran, J.W.: 1991, *ARA&A* **29**, 447

Croswell, K., Latham, D.W., Carnery, B.W., Schuster, W. Aguilar, L.: 1991, *AJ* **101**, 2078

De Geus, E.J.: 1991, in Janes, K.A., ed(s)., *ASPCS v. 13, The Formation and Evolution of Star Clusters*, ASP, San Francisco, 40

Deliyannis, C.P., Demarque, P.: 1991, *ApJ* **379**, 216

Dickens, R.J., Croke, B.F.W., Cannon, R.D., Bell, R.A.: 1991, *Nat* **351**, 212

Doinidis, S.P., Beers, T.C.: 1989, *ApJ* **340**, L57

Djorgovski, S.: 1991, in Janes, K.A., ed(s)., *ASPCS v. 13, The Formation and Evolution of Star Clusters*, ASP, San Francisco, 112

Djorgovski, S., Piotto, G., Phinney, E.S., Chernoff, D.F.: 1991, *ApJ* **372**, L41

Eggen, O.J., Lynden-Bell, D., Sandage, A.R.: 1962, *ApJ* **136**, 748

Fall, M., Rees, M.: 1985, *ApJ* **298**, 18

Faulkner, D.J., Scott, T.R., Wood, P.R., Wright, A.E.: 1991, *ApJ* **374**, L45

Ferraro, F.R., Clementini, G., Fusi Pecci, F., Buonanno, R.: 1991, *MNRAS,* in press

Fitzsimmons, A., Keenan, F.P., Conlon, E.S., Dufton, P.L., Williams, P.M.: 1991, *MNRAS* **249**, 336

Freeman, K.C.: 1983, in West, R.M., ed(s)., *IAU: Highlights of Astronomy*, Reidel, Dordrecht, 201

Gilroy, K.K., Sneden, C., Pilachowski, C., Cowan, J.J.: 1988, *ApJ* **327**, 298

Gilmore, G., King, I.R., van der Kruit, P.C.: 1990, *The Milky Way as a Galaxy*, Univ. Science Books, Mill Valley

Goodman, J., Hut, P.: 1989, *Nat* **339**, 40

Graham, J.A.: 1975, *BAAS* **7**, 414

Green, E.M., Norris, J.E.: 1990, *ApJ* **353**, L17

Grenon, M.: 1989, *Ap&SS* **156**, 29

Grenon, M.: 1991, Invited Review, IAU General Assembly, Buenos Aires

Harris, W.E.: 1991, *ARA&A* **29**, 543

Hartwick, F.D.A.: 1976, *ApJ* **209**, 418

Hartwick, F.D.A.: 1987, in Gilmore, G., Carswell, B., ed(s)., *The Galaxy*, Reidel, Dordrecht, 281

Hatzidimitriou, D.: 1991, *MNRAS,* in press

Heasley, J.N., Christian, C.A.: 1991, *AJ* **101**, 967

Heiles, C.: 1991, UCB RAL Preprint 179

Hesser, J.E.: 1988, in Blanco, V.M., Phillips, M.M., ed(s)., *ASPCS v. 1, Opportunities in Southern Hemisphere Astronomy*, ASP, San Francisco, 161

Hesser, J.E.: 1991, in Lambert, D.L., ed(s)., *ASPCS, Frontiers of Stellar Evolution*, ASP, San Francisco, in press

Hesser, J.E., Harris, W.E., VandenBerg, D.A., Allwright, J.W.B., Shott, P., Stetson, P.B.: 1987, *PASP* **99**, 739

Hesser, J.E., Shawl, S.J., Meyer, J.E.: 1986, *PASP* **98**, 403

Janes, K.A., editor: 1991, *ASPCS v. 13, The Formation and Evolution of Star Clusters*, ASP, San Francisco

Kinman, T.D.: 1959, *MNRAS* **119**, 559

Kulkarni, S.R., Narayan, R., Romani, R.W.: 1990, *ApJ* **356**, 174

Kurucz, R.J.: 1991, *Comments on A & A,* in press

Laird, J.B., Rupen, M.P., Carney, B.W., Latham, D.W.: 1988, *AJ* **96**, 1908

Lambert, D.L.: 1991, Invited Review, IAU General Assembly, Buenos Aires

Larson, R.B.: 1990, *PASP* **102**, 709

Latham, D.W.: 1991, Invited Review, IAU General Assembly, Buenos Aires

Lee, Y-W, Demarque, P., Zinn, R.: 1980, *ApJ* **350**, 155

Majewski, S.R.: 1991,*ApJS,* in press

Manchester, R.N., Lyne, A.G., Robinson, C., D'Amico, N., Bailes, M., Lim, J.: 1991, *Nat* **352**, 219

Mateo, M., Harris, H.C., Nemec, J., Olszewski, E.W.: 1990, *AJ* **100**, 469

McClure, R.D., VandenBerg, D.A., Smith, G.H., Fahlman, G.G., Richer, H.B., Hesser, J.E., Harris, W.E., Stetson, P.B., Bell, R.A.: 1986, *ApJ* **307**, L49

McMillian, S., Hut, P., Makino, J.: 1990, *ApJ* **362**, 522

Meylan, G.: 1991, Invited Review, IAU General Assembly, Buenos Aires

Meylan, G., Dubath, P., Mayor, M.: 1991, *ApJ,* in press

Morrison, H.L., Flynn, C., Freeman, K.: 1990, *AJ* **100**, 1191

Murray, S.D., Clarke, C.J., Pringle, J.E.: 1991, *ApJ,* in press

Murray, S.D., Lin, D.N.C.: 1990, *ApJ* **357**, 105

Nemec, J., Nemec, A.F.L.: 1991, *PASP* **103**, 95

O'Connell, R.W.: 1991, Invited Review, IAU General Assembly, Buenos Aires

Osmer, P.S.: 1978, *ApJ* **226**, L79

Paltoglou, G., Bell, R.A.: 1991, preprint

Paltoglou, G., Norris, J.E.: 1989, *ApJ* **336**, 185

Paresce, F., *et al.* : 1991, *Nat* **352**, 297

Pinsonneault, M.H., Kawaler, S.D., Demarque, P.: 1990, *ApJS* **74**, 501

Preston, G.W., Shechtman, S.A., Beers, T.C.: 1991, *ApJ* **375**, 121

Prince, T.A., Anderson, S.B., Kulkarni, S.R., Wolszczan, A.: 1991, *ApJ* **374**, L41

Proffitt, C.R., VandenBerg, D.A.: 1991, *ApJS,* December

Pryor, C., McClure, R.D., Hesser, J.E., Fletcher, J.M.: 1989, in Merritt, D., ed(s)., *Dynamics of Dense Stellar Systems*, Cambridge U. Press, 175

Pryor, C., Smith, G.H., McClure, R.D.: 1986, *AJ* **92**, 1358

Quin, D.A., Brown, P.J.F., Conlon, E.S., Dufton, P.L., Keenan, F.P.: 1991, *ApJ* **375**, 342

Ratnatunga, K., Freeman, K.C.: 1985, *ApJ* **291**, 260

Rees, R.F., Cudworth, K.M.: 1991, *AJ* **102**, 152

Renzini, A.: 1991, in Banday, T., Shanks, T., ed(s)., *Observational Tests of Inflation*, Kluwer, Dordrecht, in press

Richer, H.B., Fahlman, G.G., Buonanno, R., Fusi Pecci, F., Searle, L.,Thompson, I.B.: 1991, *ApJ,* in press

Romani, R.W., Weinberg, M.D.: 1991, *ApJ* **372**, 487

Rood, R.T.: 1990, in Vangioni-Flam, E., Cassé, M., Audouze, J., Tran Thanh Van, J., ed(s)., *Astrophysical Ages and Dating Methods*, Editions Frontières, Gif sur Yvette, 313

Rose, J.A., Agostinho, R.: 1991, *AJ* **101**, 950

Ryan, S.G., Norris, J.E.: 1991a, *AJ* **101**, 1835

Ryan, S.G., Norris, J.E.: 1991b, *AJ* **101**, 1865

Sandage, A.: 1990, *JRASC* **84**, 70

Sandage, A., Fouts, G: 1987, *AJ* **93**, 74

Sarajedini, A., Demarque, P.: 1990, *ApJ* **365**, 219

Schuster, W.J., Nissen, P.E.: 1989, *A&A* **222**, 69

Searle, L.: 1977, in Tinsley, B.M., Larson, R.B., ed(s)., *The Evolution of Galaxies and Stellar Populations*, Yale Univ. Obs., New Haven, 219

Searle, L., Zinn, R.: 1978, *ApJ* **225**, 357

Sommer-Larsen, J., Christensen, P.R.: 1987, *MNRAS* **225**, 499

Sommer-Larsen, J., Christensen, P.R., Flynn, C.: 1991, *MNRAS,* in press

Spergel, D.N.: 1991, *Nature* **352**, 221

Spiesman, W.J., Wallerstein, G.: 1991, preprint

Stetson, P.B., Harris, W.E.: 1988, *AJ* **96**, 909

Stiavelli, M., Piotto, G., Capaccioli, M., Ortolani, S.: 1991, in Janes, K.A., ed(s)., *ASPCS v. 13, The Formation and Evolution of Star Clusters*, ASP, San Francisco, 449

Straniero, O., Chieffi, A.: 1991, *ApJS* **76**, 525

Suntzeff, N.B., Kinman, T.D, Kraft, R.P.: 1991, *ApJ* **367**, 528

Tobin, W.: 1987, *I.A.U. Colloq.* **95**, 503

Tinsley, B.M., Larson, R.B.: 1979, *MNRAS* **186**, 503

Toomre, A.: 1977, in Tinsley, B.M., Larson, R.B., ed(s)., *The Evolution of Galaxies and Stellar Populations*, Yale Univ. Obs., New Haven, 401

Truran, J.W., Brown, J., Burkert, A.: 1991, in Janes, K.A., ed(s)., *ASPCS v. 13, The Formation and Evolution of Star Clusters*, ASP, San Francisco, 78

van Albada, T.S., Dickens, R.J., Wevers, B.H.M.R.: 1981, *MNRAS* **196**, 823

VandenBerg, D.A.: 1988, in Grindlay, J.E., Philip, A.G.D., ed(s)., *IAU Symp. 126, Globular Cluster Systems in Galaxies*, Kluwer, Dordrecht, 107

VandenBerg, D.A.: 1991, in Janes, K.A., ed(s)., *ASPCS v. 13, The Formation and Evolution of Star Clusters*, ASP, San Francisco, 183

VandenBerg, D.A., Bolte, M., Stetson, P.B.: 1990, *AJ* **100**, 445

VandenBerg, D.A., Stetson, P.B.: 1991, *AJ* **102**, 1043

Vedel, H., Sommer-Larsen, J.: 1990, *ApJ* **359**, 104

Waelkens, C., Waters, L.B.F.M., Cassatella, A., Le Bertre, T., Lamers, H.J.C.L.M.: 1987, *A&A* **181**, L5

Webbink, R.F.: 1985, in Goodman, J., Hut, P., ed(s)., *IAU Symp. 113, Dynamics of Star Clusters*, Reidel, Dordrecht, 541

Zinn, R.J.: 1985, *ApJ* **293**, 424

Zinn, R.: 1991, in Janes, K.A., ed(s)., *ASPCS v. 13, The Formation and Evolution of Star Clusters*, ASP, San Francisco, 532

Zinnecker, H.: 1987, in Azzopardi, M., Matteucci, F., ed(s)., *ESO Workshop on Stellar Evolution and Dynamics in the Outer Halo of the Galaxy*, ESO, Garching, 599

DISCUSSIONS

Mateo: You comment that the steep mass function slopes in globular clusters may help explain the dark matter problem with a baryonic solution. However, the M/L's of globular clusters are always among the *lowest* observed in halo environments (*e.g.,* M/L_V's > 10 are common for the outer parts of galaxies and for dSph galaxies). Thus, these steep slopes seem to make the problem worse, implying an even steeper slope for these non-globular cluster populations if you want to use baryons to explain the dark matter.
Hesser: You may well be correct. However, over $\sim$ 15 Gyrs we have reasons to believe that many globulars have been dissolved (*cf.* Fall and Rees 1985, Aguilar, Hut and Ostriker 1988; Chernoff and Weinberg 1990). Moreover, internal dynamical evolution may have modified the original properties of the clusters substantially.
Mould: If the mass function of globular clusters has been modified by dynamical processes, we should be counting halo M dwarfs at 25th magnitude in the field in order to determine the initial mass function of the halo.
Hesser: That would be an excellent thing to do! I assume that in about a year, when the Keck Telescope goes into routine operation, you will be doing it?
Mould: You might assume that!
Ostriker: Note that not only have we lost a signficant fraction of the globular clusters initially in the Galaxy, but also, in the surviving clusters, it is possible that a large fraction of the low mass stars have been lost, especially from the inner parts from which most of our data come.
Hesser: You drive home how much we have to learn about the dynamical history of globulars.
Zinnecker: I am wondering about the effects of stellar binarity in color-magnitude diagrams and luminosity functions. If, as you have quoted, 20% of Population II stars are spectroscopic binaries, the total binary frequency (extrapolating to wider separations) may be as high as 80%. Surely, if almost every Population II star were a binary, we should explore the errors of ignoring binarity!
Hesser: There are many ramifications to the increasing evidence for binaries in Population II, and I wholeheartedly agree we must explore them. Perhaps it is premature to conclude what the total binary frequency is, however. Pryor, McClure, Fletcher and I are finding that some 6% of globular cluster stars are in binaries with orbital periods between 0.2 and 20 years. This frequency appears to be significantly below that found for stars near the Sun, but it lends support to theoretical predictions that binaries will be destroyed in globular clusters by close encounters with other stars.
Pagel: A philosophical point about stellar populations: A population should be defined by a common history (*e.g.,* common effective yield) and may have a large range in kinematics and chemical composition as the halo and the bulge both do. Conversely, it is important to realize that properties of different populations may overlap, so they cannot be placed in a one-dimensional sequence. Cats and dogs may have the same age and metallicity, but they are still cats and dogs!

The Ages of the Globular Clusters

BRUCE W. CARNEY
University of North Carolina

Abstract. The recent results for the relation between M_V and $[Fe/H]$ for RR Lyrae variables and horizontal branch stars are reviewed: $< M_V(RR) > = 0.15[Fe/H] + 1.08$. If $[O/Fe] = 0.3$, the oldest clusters' ages approach 20 Gyrs, and the the most metal-poor clusters are older than the most metal-rich by several Gyrs. If $[O/Fe] \propto -0.4[Fe/H]$, the most metal-poor clusters have ages of about 14 Gyrs and the most metal-rich clusters are not much younger. In both cases, there are age spreads among the intermediate metallicity clusters of up to 6 Gyrs.

Key words: Globular Clusters - RR Lyrae Variables

1. Introduction

Relics of the Galaxy's earliest years, the globular clusters probe not only the date of the very first star formation events, but also the timescale and coherence of the Galaxy's formation. Precise relative and absolute cluster ages have not yet been achieved, but considerable progress has been made. The basic technique is, of course, the comparison of cluster color-magnitude diagrams with theoretical isochrones. A full match requires knowledge, preferably empirical, of bolometric corrections to convert V into m_{bol}, relations between color indices and $T_{\rm eff}$, distances, and chemical compositions, including separate handling of helium, CNO elements, and iron-peak elements. The isochrones also require accurate treatment of convection, which for low-mass stars is confined to the outer envelope, plus all the other details involved in solving the differential equations of stellar structure and nucleosynthesis. It is our opinion that the excellent matches obtained between isochrones and color-magnitude diagrams (e.g., Hesser et al. 1987; Richer and Fahlman 1987) do not provide optimum age estimates because of the reliance upon convection theory. We prefer to estimate ages solely from the turn-off luminosity, which is insensitive to convection conditions for low-mass stars and to outer surface boundary conditions.

One must still deal with distance and composition determinations for clusters. This paper deals primarily with the former problem, but illustrates the importance of the second as well.

2. Cluster Distances

The most direct method of measuring the distances to globular clusters is main sequence fitting to similar-metallicity field stars with accurate trigonometric parallaxes. One of the key reasons for deriving ages is to test the rapidity of the formation of the halo, and to discern 10% effects in ages requires a minimum precision (all other factors being equal) of 10% in distance. Only one field halo dwarf, HD 103095, has a trigonometric parallax with such precision; hence we can estimate distances only to those clusters with metallicities like that of HD 103095. Until improved parallax data are available, either from HIPPARCOS or HST, we will continue to rely instead on the absolute magnitudes of horizontal branch stars, especially the RR Lyrae variables. The basic idea is to estimate $< M_V(RR) >$ as a function of $[Fe/H]$. Measurement of the gap in magnitude between a cluster's observed

B. Barbuy and A. Renzini (eds.), The Stellar Populations of Galaxies, 15–21.

horizontal branch and its main sequence turn-off yields the turn-off luminosity, independent of distance and reddening. Knowledge of the slope of the relation yields relative cluster ages, whereas additional knowledge of the zero point yields absolute ages, assuming the full chemical composition profile is also known.

2.1. The Slope

A wide variety of techniques have been used to estimate M_V vs. $[Fe/H]$ (see Carney et al. 1992; hereafter CSJ). Four merit special attention. Lee et al. (1990) and Lee (1990) have published theoretical models of horizontal branch evolution, and argued that the slope of the relation is 0.17. This shallow slope conflicts with that determined by Sandage (1990 and references therein), who used the relation between the fundamental period, the mass, the luminosity, and the temperature for RR Lyraes derived by van Albada and Baker (1971). Neglecting the minor mass effect, the essence of the method is to compare variables in one cluster with those in a "fiducial" cluster at equal temperatures. Variations in periods at the same temperatures should then reflect variations in intrinsic luminosities. The slope found from this "period shift" technique was found by Sandage to be 0.37, over twice that found from theory. The difference is profound, for at face value the steep slope implies all the globular clusters have the same age (assuming, incorrectly, they all have the same gap between their horizontal branches and their main sequence turn-offs), whereas the shallower slope implies a significant age-metallicity relation, with the metal-rich clusters being several Gyrs younger than the metal-poor clusters. Resolution of this quandry has not been easy. Buonanno et al. (1989; hereafter BCF) utilized model isochrones to determine relative cluster distances by main sequence fitting. The relative distances then yielded relative $< M_V(RR) >$, and the resultant slope was found to be 0.37 ± 0.14. An ensuing debate (King et al. 1988; Buonanno et al. 1990) did not change this result significantly. Finally, several groups (e.g., Jones et al. 1988; Cacciari et al. 1989a,b; Fernley et al. 1990; Liu and Janes 1990; and references within each paper) have been applying variants of the Baade-Wesselink technique to derive distances to individual field RR Lyraes. A recent critical review by Jones et al. (1992), wherein all the systematic effects are taken into account and highly-evolved stars removed from the sample, yielded a slope of 0.16 ± 0.03. How do we resolve this bimodality in the results for the slope of the $< M_V(RR) >$ vs. $[Fe/H]$ relation?

There is an independent test of the Baade-Wesselink results. As part of the analyses, as summarized by Jones et al. (1992), a relation between $< M_K(RR) >$ and log P is obtained, which has a slope of -2.33 ± 0.20. Longmore et al. (1990) obtained the slope of the M_K vs. log P relation in each of 8 globular clusters with a wide range of metallicities. Upon eliminating the two clusters with 5 or fewer variables, leaving six clusters with 20 or more variables, the slope is found to be -2.31 ± 0.06. The field stars, which lie at a wide variety of derived distances, thus yield the same slope as the variables within any one cluster. Hence the slope of the M_K vs. log P, and by inference the slope of the $< M_V(RR) >$ vs. $[Fe/H]$ relation, obtained from the Baade-Wesselink analyses must be considered secure.

The focus must now be upon the main sequence fitting results of BCF and the period shifts analyses done by Sandage (1990). CSJ argue that the main sequence results are affected by a subtle but crucial metallicity effect. The transformation from $T_{\rm eff}$ into $B-V$ necessary for the isochrone-based main sequence fits relies upon the synthetic colors of model atmospheres computed by Kurucz (1979). Since these synthetic colors predict colors for the Sun that are too blue, it has long been thought that the synthetic flux distributions lack opacity, probably due to atomic and molecular line blanketing, in the blue and violet regions. CSJ tested this hypothesis by using the large sample of proper motion stars being studied by Carney, Latham, and Laird. Considering only those stars with $[Fe/H] \leq -0.45$, $M_{\rm V} \geq +5.0$, $E(B-V) \leq 0.05$ and which are neither evolved nor double-lined, CSJ found 355 stars for which they could determine temperatures using infrared (metallicity insensitive) $V-K$ color indices, based ultimately in fact on Kurucz's predictions of the slope of the flux distribution in the Paschen continuum, a region also insensitive to line blanketing. The metallicities were determined from the echelle spectra (cf. Laird et al. 1988). CSJ could therefore estimate the $B-V$ value *predicted* by the isochrones for a star of known metallicity and temperature and compare that the each star's *observed* value. The difference was found to be a function of metallicity, and in a sense consistent with increasingly deficient line blanketing in the models as the metallicities increase. While not a large effect, only 0.034 mag per 1 dex change in $[Fe/H]$, the steep slope of the main sequences resulted in a decrease in BCF's derived $< M_{\rm V}(RR) >$ vs. $[Fe/H]$ slope from 0.37 to 0.12!

CSJ also found the period shift analysis to require revision. The major problem was found to be the proper definition of the temperature. Sandage (1990) had relied upon either the magnitude-averaged or intensity-averaged $B-V$ color index and the conversion to $T_{\rm eff}$ given by Butler et al. (1978). As Jones (1988) has shown, however, the Baade-Wesselink photometric angular diameters derived using any blue color index ($B-V$ or $b-y$) fails to predict the correct temperatures of RR Lyraes during their expansion phases. Thus any mean blue color index will fail to predict the correct temperature, and the failure will be greater for larger amplitude stars (zero amplitude non-pulsating stars' temperatures may, of course, still measured by blue color indices). Even were the temperatures derived from the color indices correct, however, the mean color yields an average of the temperature over the pulsation cycle. This is not what is required for the analysis. Instead, one needs the equilibrium temperature, $T_{\rm eq}$, which is defined by $[L_{eq}/4\pi\sigma R_{eq}^2]^{1/4}$. The Baade-Wesselink results yield L_{eq} and R_{eq}, so new relations between observable parameters and $T_{\rm eq}$ may be obtained and the period shift analysis redone. CSJ found that the $< M_{\rm V}(RR) >$ vs. $[Fe/H]$ slope then became 0.14, based on variables in 8 globular clusters, and 0.16, based on 141 field stars studied by Suntzeff et al. (1991). (They noted analyses based on Lub's 1977 sample are inappropriate since, by selecting stars to fully sample both metallicity and period space, Lub built in a bias toward evolved, brighter stars.) An indication that the new temperature calibrations are correct has been pointed out by Storm et al. (1991) and Carney et al. (1991). Their respective studies of the light curves of RR Lyraes in the globular clusters M5 and M92 showed that a V vs. $B-V$ plot showed the non-variables, fundamental mode, and first overtone mode pulsators overlap, but not in the V vs. $T_{\rm eq}$ plane.

When these four major techniques are now compared, and a few other lower-weight results are added in, CSJ found the final slope of the $< M_V(RR) >$ vs. $[Fe/H]$ relation to be 0.15 ± 0.01.

2.2. The Zero Point

As CSJ discuss, the zero point determination remains somewhat uncertain. It is based on the statistical parallax analyses of Strugnell et al. (1986) and Barnes and Hawley (1986), and the main sequence fit of HD 103095 to the similar-metallicity, only slightly reddened main sequence obtained by Richer and Fahlman (1987). The formal relation is then

$$< M_V(RR) > \; = 0.15(\pm 0.01)[Fe/H] + 1.01(\pm 0.08). \tag{1}$$

3. Cluster Ages

CSJ compiled apparent magnitudes of horizontal branches and main sequence turn-offs, plus $[Fe/H]$ values, for 24 clusters, and used Equation 1 and the empirical bolometric corrections of Carney (1983) to determine turn-off luminosities. Before they could derive ages, however, they had to adopt helium abundances and estimate the effects of the CNO elements. Caputo et al. (1987) and Steigman (1989) have, for different reasons, urged adoption of $Y = 0.23$. However, it is inappropriate to adopt $[Z] = [Fe/H]$, for in halo stars it is well known that the lighter elements do not scale with iron abundance in solar proportions. CSJ adopted two cases. In Case A, they assumed that all the "α-rich" species (such as Ne, Mg, Ca, Si, S) *and* oxygen are enhanced by 0.3 dex relative to iron. Since oxygen is such an important contributor (roughly 50% of the heavy element atoms), the net effect is to increase the $[Z_{eff}]$ value over $[Z]$ by typically 0.2 dex. In Case B, CSJ assumed the Abia and Rebolo (1989) results for $[O/Fe]$ vs. $[Fe/H]$ are correct, with enhancements in oxygen increasing as iron abundance declines. This raises $[Z_{eff}]$ by 0.3 to 0.8 dex over $[Z]$. As noted already by Stranierro and Chieffi (1991), the use of $[Z_{eff}]$ is the appropriate metallicity for use in isochrones computed with scaled solar abundances, $[Z]$.

The results for Cases A and B differ dramatically, showing we can make no claims for accurate relative or absolute ages of globular clusters until the oxygen abundances are better known. In Case A, the most metal poor clusters have ages of roughly 20 Gyrs, whereas the most metal rich clusters are about 15 Gyrs old. In the intermediate metallicity domain, however, some clusters, such as Palomar 12 and Ruprecht 106, are found to be 6 Gyrs younger than comparable metallicity clusters like NGC 288. There may be a crude age-metallicity relation, in other words, but it is clearly not applicable to all clusters. The derived age spreads are in excellent cluster-by-cluster agreement with the results of VandenBerg et al. (1990), who estimated relative ages for clusters of similar metallicity by comparing the color differences between their turn-offs and their giant branches. In Case B, the most metal-poor clusters have ages of about 16 Gyrs, and the most metal-rich clusters are not obviously much younger than that. Again, however, the 6-Gyrs age spread among the intermediate-metallicity clusters is found.

4. Future Work

The tasks ahead of us are clearly defined. We still need to improve our zero points of the $< M_V(RR) >$ vs. $[Fe/H]$ relation, perhaps by using statistical parallaxes of field halo dwarfs and, of course, using trigonometric parallaxes as they become available. We obviously must improve the precision of the $[O/Fe]$ abundance ratios in halo stars, preferably unevolved halo dwarfs and preferably by using means that do not rely on the high excitation (9.15 eV) O I lines used by Abia and Rebolo (1989). We recommend, instead, a program using the ultraviolet electronic or the infrared vibration-rotation transitions of the OH molecule.

Acknowledgements

I thank the National Science Foundation for support through grant AST89-20742 to the University of North Carolina.

References

Abia, C., and Rebolo, R. 1989, ApJ, 347, 186.
Barnes, T. G., and Hawley, S. L. 1986, ApJ, 307, L9.
Buonanno, R., Cacciari, C., Corsi, C. E., and Fusi Pecci, F. 1990, A&A, 230, 315.
Buonanno, R., Corsi, C. E., and Fusi Pecci, F. 1990, A&A, 216, 80. (BCF)
Butler, D., Dickens, R. J., and Epps, E. 1978, ApJ, 225, 148.
Cacciari, C., Clementini, G., and Buser, R. 1989b, A&A, 209, 154.
Cacciari, C., Clementini, G., Prevot, L., and Buser, R. 1989a, A&A, 209, 141.
Caputo, F., Martinez Roger, C., and Paez, E. 1987, A&A, 183, 228.
Carney, B. W. 1983, AJ, 88, 623.
Carney, B. W., Storm, J., & Jones, R. V. 1992, ApJ, in press. (CSJ)
Carney, B. W., Storm, J., Trammel, S. R., and Jones, R. V., 1991, PASP, in press.
Fernley, J. A., Skillen, I., Jameson, R. F., Marang, F., Kilkenny, D., and Longmore, A. J. 1990, MNRAS, 247, 287.
Hesser, J. E., Harris, W. E., VandenBerg, D. A., Allwright, J. W. B., Shott, P., and Stetson, P. B. 1987. PASP, 99, 739.
Jones, R. V. 1988, ApJ, 326, 305.
Jones, R. V., Carney, B. W., and Latham, D. W. 1988, ApJ, 332, 206.
Jones, R. V., Carney, B. W., Storm, J., and Latham, D. W. 1992, ApJ, in press.
King, C. R., Demarque, P., and Green, E. M. 1988, in Calibration of Stellar Ages, ed. A. G. D. Philip, (L. Davis Press: Schenectady), p. 211.
Kurucz, R. L. 1979. ApJS, 40, 1.
Laird, J. B., Carney, B. W., and Latham, D. W. 1988, AJ, 95, 1843.
Lee, Y.-W. 1990, ApJ, 363, 159.
Lee, Y.-W., Demarque, P., and Zinn, R. 1990, ApJ, 350, 155.
Liu, T., and Janes, K. A. 1990, ApJ, 354, 273.
Longmore, A. J., Dixon, R., Skillen, I., Jameson, R. F., and Fernley, J. A. 1990, MNRAS, 247, 684.
Lub, J. 1977, Ph.D. Thesis, University of Leiden.
Richer, H. B., and Fahlman, G. G. 1987, ApJ, 316, 517.
Sandage, A. 1990, ApJ, 350, 631.
Steigman, G., Gallagher, J. S., and Schramm, D. N. 1989, Comm. Astrophys., 14, 97.
Storm, J., Carney, B. W., and Beck, J. A. 1991, PASP, in press.
Straniero, O., and Chieffi, A. 1991. ApJS, 76, 525.
Strugnell, P., Reid, N., and Murray, C. A. 1986, MNRAS, 220, 413.
Suntzeff, N. B., Kinman, T. D., and Kraft, R. P. 1991, ApJ, 367, 528.
van Albada, T. S., and Baker, N. 1971, ApJ, 169, 311.
VandenBerg, D. A., Bolte, M., and Stetson, P. B. 1990, AJ, 100, 445.

HESSER: I would like to make two comments on your excellent review of this very active area. First, the development of the relative age-dating technique via the color difference technique (which has been one of the most exciting ideas in many years for those of us who struggle to understand the age profile of the Galactic halo) is limited to certain metallicity ranges, as you stressed. Very recently Paltoglou and Bell have shown that the $[Fe/H]$ range over which reliable relative ages may be determined can be extended considerably by use of more appropriate filters, a set of which exists on the FOC aboard the HST. Second, some of our most convincing data for a range of ages among intermediate metallicity globulars comes from the NGC 288/NGC 362 comparison. Dickens et al., in a recent Nature paper, have shown from high dispersion spectra of numerous giants that these two clusters have virtually identical abundances. (Note, however, in the impressive recent study by Chieffi and Straniero of cluster color-magnitude diagrams interpreted through their new isochrone grid that they may infer an $[M/H]$ difference of 0.3 to 0.4 dex between them.) I am disturbed by Dickens et al.'s demonstration that the mixing history of NGC 288 and NGC 362 have been quite different, and wonder if this is affecting the relative age determinations in some way not presently appreciated.

CARNEY: As I recall, the Paltoglou and Bell filter choice is helped by minimizing line-blanketing effects. However, metallicity still must play a role since convection determines the red giant branch temperature relative to the turn-off at a fixed age. So even were convection wholly understood, we would be unable, I believe, to encompass the whole metallicity range of globulars. The extension of the technique is certainly welcome, but I maintain that the turn-off luminosity must be the fundamental basis for age estimates. As for mixing, it should not in principle affect ages estimates from turn-off luminosities, unless we are seriously in error about the relative $[O/Fe]$ values adopted for the two clusters. I am actually somewhat more worried about the Briley et al. findings of variations of CN strengths on the main sequence of 47 Tuc. Your excellent color-magnitude diagram of the cluster seems to rule out primordial abundance variations, leaving us with the unpalatable idea of mixing on the main sequence.

PINSONNEAULT: I don't believe that main sequence mixing can explain the observed CN anomalies in metal-poor clusters. Evolutionary models including rotation do not predict substantial main sequence mixing of CNO elements, and also predict less mixing in metal-poor than in metal-rich systems. As far as the accuracy of the interior models is concerned, I'd like to note that solar models constructed with the best available physics (Livermore interior opacities, Kurucz molecular opacities) match helioseismology observations to a high degree of accuracy. This indicates that the basic physics in the models is already very good.

CARNEY: But if it's not mixing, a conclusion I would be happy to agree to, I presume the variations are primordial, yet somehow 47 Tuc's main sequence and turn-off widths are very slender, implying very small overall metallicity variations. Perhaps it has something to do with pre-main sequence or binary star evolution.

RENZINI: Could you give us an estimate of the sizes of the systematic errors affecting the Baade-Wesselink method?

CARNEY: The most serious sources of systematic errors are, in my opinion, the conversion from $V-K$ color index into temperature and the conversion of radial velocity in pulsational velocity, which is a function of limb darkening and instrumental resolution. While I believe these are under reasonable control, it means that absolute distances are less well known than relative distances, when the zero points drop out. That's why we seek alternatives for calibrating the luminosity-metallicity zero points. (In fact, the current Baade-Wesselink results yield the "correct" results in zero point to ± 0.1 mag or better, if you accept the HD 103095 vs. M5 and statistical parallax results.) There is one additional problem, that the relative depths of the formations of the spectral lines (velocities) and continuum (colors/temperatures) are constant through the critical parts of the pulsational cycle. Work by us and by Abi Saha measuring lines that form at a wide variety of depths suggest this is not a problem for RR Lyraes. Cepheids may be a different matter.

KURUCZ: I worry about your implicit assumption that interior models and evolutionary tracks are correct. 1. There are new opacities by Iglesias and Rogers at Livermore that are as much as four times higher than the Los Alamos opacities that were used at Yale. They also can be sensitive to individual elemental abundances. 2. The surface boundary condition in all interior models and envelope calculations is wrong. The diffusion approximation and the opacities are not correct. They have to use a model atmosphere boundary condition. 3. There are systematic errors in VandenBerg's work because he used Bell's predicted colors which Buser has shown have systematic errors because of missing opacity. 4. The opacities for RR Lyraes are wrong in the atmospheres and in the envelopes. There are strong effects of microturbulent velocity with phase in these atmospheres. The new Livermore opacities strongly affect the theoretical calculations of pulsation. I think it would be useful to assume that all the halo clusters are the same age and then see how the interior calculations would have to be changed to accomplish that.

CARNEY: I see no reason to believe *a priori* that all globular clusters have the same age, and I'd hate to tune physics on the basis of astronomical ideology. Responses to your detailed comments: 1. New opacities are obviosuly important, but the effects are mostly in the 10^5-10^6 K domain, so they won't affect interiors solutions in net luminosity much. Generally higher opacities will only mean greater derived ages. 2. The surface boundary conditions do not affect my conclusions at all since we rely only on the turn-off luminosities. 3. I agree. That was the point of our tests of the isochrones to seek metallicity effects on the $B-V$ vs. temperature relation. 4. Yes, that must be true, but the effect occurs near minimum radius, a phase of strong decelerations and accelerations. We do not rely on those phases in Baade-Wesselink analyses for those and other obvious reasons.

POPULATION II IN THE MILKY WAY GALAXY AND THE LMC

NICHOLAS B. SUNTZEFF
Cerro Tololo Inter-American Observatory
Casilla 603, La Serena, CHILE

Abstract.

The RR Lyrae and globular cluster populations are used to study the Pop II in the Galaxy and the LMC. The metallicity gradient in the Galaxy changes abruptly at the position of the solar circle. The statistics of field RR Lyraes imply that the Pop II field population is older towards the Galactic center. The density of RR Lyraes in the solar neighborhood implies a luminous mass density of $6.4 \pm 1.8 \times 10^{-5} M_{\odot}$ pc^{-3} for the Galactic halo. The total luminosities (M_V) for the Galactic and LMC halos are -18.4 and -15.1, and the ratio of globular cluster to luminous Pop II mass is 0.02 in both cases. The agreement of this ratio in two systems with very different tidal fields argues against the formation of the field population as the disruption of many smaller systems.

1. Introduction

Clues to the formation of galaxies lie in the characteristics of the oldest population of stars, traditionally called Population II after Baade. This ancient population remains today as a ghost-like remnant of the proto-galaxy, and through a study of the kinematics and chemistry of its constituent stars, we can gain insight into the initial conditions for galaxy formation.

The studies of the Pop II in the Galaxy rely primarily on two types of samples: *local* samples in the solar neighborhood; and *in situ* samples. Both samples suffer from the extreme low density of Pop II stars. A recent estimate for ratio of the density of young and old disk stars to Pop II stars is 1200 (Gilmore and Wyse 1985). The local sample technique gets around this problem by sampling stars based on proper motion and radial velocity data. This approach was pioneered by Eggen, Sandage, and collaborators (Eggen, *et al.* 1962), and Schmidt (1975), and later extended by Carney and collaborators (Carney and Latham 1987). The main disadvantage of this approach is the kinematical bias of the sample selection. For example, the sample of stars discovered by Morrison, Flynn, and Freeman (1990) which have low abundances but disk-like kinematics are selected against in local samples.

The *in situ* selection is hampered by the difficulty of picking out the Pop II objects from the tremendous number of local field stars. The globular cluster sample does not have this problem, but there are only a limited number of clusters ($\sim$ 120), and there is no guarantee that the stars in globular clusters have the same history as the field population. Indeed, there is evidence that in M87 the cluster and field spheroid populations are different (Strom, *et al.* 1981). Other evidence that the cluster stars are different than the halo field is the radial gradient in CN-band strength seen in 47 Tuc giants (Paltoglou 1990) and the paucity of field giants (at [Fe/H] $\sim$ -1.6) with strong CN bands (Langer, *et al.* 1991).

In order to select stars in the *in situ* samples, some property must be used to weed out the local disk stars. Various attempts to select out the Pop II sample have used metallicity indices based on zero-proper motion stars selected by color

B. Barbuy and A. Renzini (eds.), The Stellar Populations of Galaxies, 23–28.

(Friel 1988), objective dispersion spectra (Ratnatunga and Freeman 1989, Preston, Shectman, and Beers 1991), and variable stars (Kinman 1965). These latter two approaches have their disadvantages also in that stars selected are biased in terms of metallicity.

In this short contribution, I will summarize the nature of the Pop II in the Galaxy and the LMC, using the RR Lyrae giants as probes of the density and metallicity. RR Lyraes are arguably the stellar constituent with the most precisely determined spatial densities of any class of stars. This is because the blue amplitudes for the fundamental pulsators (RRab) are typically 0.8 mag or greater, which is large enough that simple blink techniques can discover them. Whether a star at the red giant branch tip becomes an RR Lyrae is a poorly understood function of primarily age and metallicity, and also a host of other parameters. But by relying on the RR Lyraes in globular clusters, where the underlying stellar population is much better understood, we can convert the RR Lyrae number densities in the field into precise properties of the field stars in general.

2. The Galaxy

The metallicity and density distribution for the RR Lyraes discovered in the Lick Astrographic survey by Kinman are discussed in Suntzeff, Kinman, and Kraft (1991) (SKK). This sample of 171 stars, which includes the RR Lyraes discovered by Saha (1984), has [Fe/H] measured from ΔS , with an accuracy of $\sim$ 0.15 dex. The distribution is plotted as a function of galactocentric distance in Figure 1. One can see that there is both a real scatter of about 0.3 dex in [Fe/H] at any galactocentric distance, and that the abundance gradient changes abruptly at the solar circle. As discussed in SKK, the globular cluster abundance, when weighted by the total number of RR Lyraes in each cluster, shows the *same* average metallicity distribution over all galactocentric distances as the field stars.

We can convert the spatial distribution of field RR Lyraes into total *luminous* mass distribution by calculating the average number of RR Lyraes per unit luminosity in globular clusters and assuming the typical M/L for clusters is 2. In SKK, we calculate the average N_{RR} (the number of RR Lyraes in a cluster scaled to M_V=–7.5) at various radial bins, and use this to integrate the density distribution of field RR Lyraes from 1 to 35 kpc. The total Pop II luminosity, luminous mass outside of 1 kpc, and a summary of the cluster properties are given in Table 1. The local space density of RR Lyraes implies a luminous mass density of $6.4 \pm 1.8 \times 10^{-5} M_\odot$ pc^{-3} for the Pop II in the solar neighborhood.

3. The LMC

In Table 2 we list all the known Pop II clusters in the LMC, along with the metallicities from Olszewski, *et al.* (1991) and Suntzeff, *et al.* (1991). Kinman has estimated that the total number of field RRab stars is $\sim$ 10000. Using the average N_{RR}=12 (175 RR Lyraes in 9 clusters with ΣM_V=–10.4), we derive the total Pop II luminosity, luminous mass, and average Pop II cluster properties, which are listed in Table 1. The average properties of the two systems are quite similar, expect for the

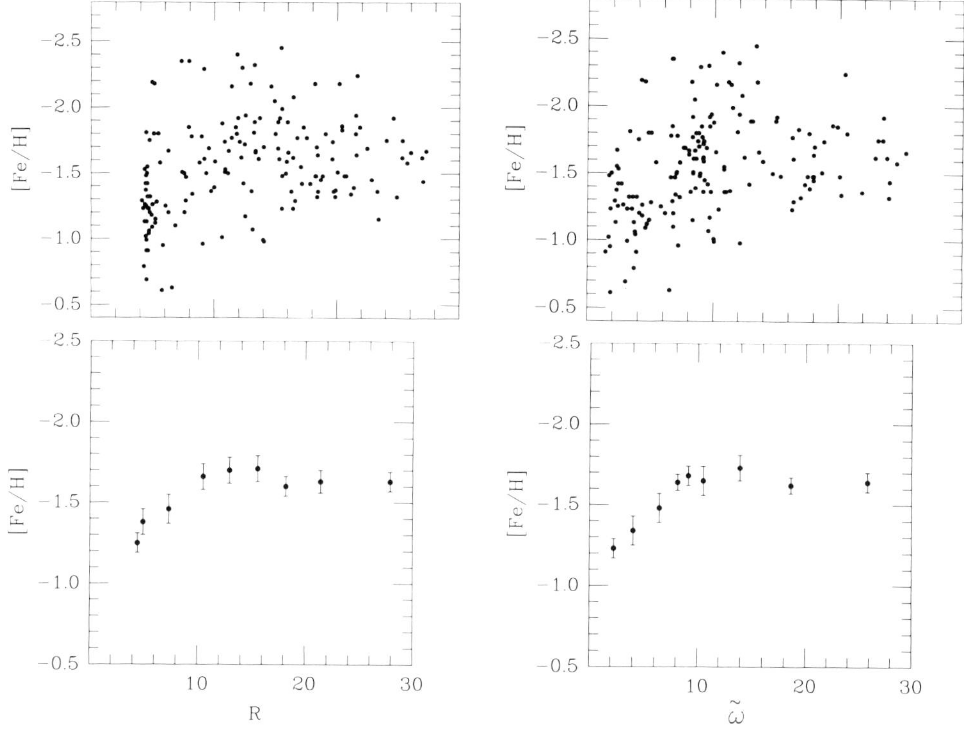

Fig. 1. The metallicity distribution of RR Lyraes in the Galactic halo plotted as a function of galactocentric distance (R) and galactocentric distance projected onto the Galactic plane ($\tilde{\omega}$). The upper panels show the individual values and the lower panels the averaged values in bins of 20. Note the abrupt change in the metallicity gradient at the solar position.

mean metallicities, which appear to reflect the trend seen in other galaxies that the more luminous galaxies have more metal-rich cluster systems.

The ratio of cluster to total Pop II mass is 0.02, which is the same as the Galaxy (SKK). It is unlikely that two galaxies, with such different tidal fields, would have the same ratio if the field stars were remnants of tidally disrupted clusters.

Finally, we note that Schommer, *et al.* (1991) have calculated that the dynamical mass of the LMC out to Reticulum is $\sim 1.3 \times 10^{10} M_{\odot}$. If we subtract off a disk mass (to 6°radius) of $\sim 3 \times 10^{9} M_{\odot}$ estimated from the exponential profile (Bothun and Thompson 1988), using the LMC luminous mass given in Table 1 we find that M(dynamical)/M(luminous) > 50. This indicates that a large amount of dark matter is present in the LMC, a result also reported by Schommer, *et al.* (1991) using the rotation curve fit to the outer clusters.

I would like to thank Dr. Tom Kinman for the communications of his results on the field RR Lyrae population in the LMC prior to publication. I also thank Bob Schommer and Ed Olszewski for a critical reading of this manuscript.

TABLE I

A comparison of the properties of the Population II in the Galaxy and the LMC

Globular Cluster	$< M_v >$	σ	N	$< [Fe/H] >$	σ	N_{RR}	M_v
Galaxy	-7.2	1.4	102	-1.57	0.30	10-18	-13.0
LMC	-7.8	0.8	13	-1.86	0.23	12	-10.9

Population II	Mass ($M_\odot$)	M_v	
Galaxy	4×10^9	-18.4	($1 < R < 35$ kpc)
LMC	1.9×10^8	-15.1	

TABLE II

The LMC population II globular clusters

Cluster	Radius(°)	[Fe/H]	n(RR)	V	E(B–V)	M_v	N_{RR}
Hodge 11	4.7	-2.06	0	11.32	0.03	-7.18	0.0
NGC 1466	8.4	-2.17	38	10.69	0.03	-7.81	28.7
NGC 1754	2.6	-1.54	...	11.20	0.07	-7.4	...
NGC 1786	2.5	-1.87	9	10.49	0.07	-8.13	5.0
NGC 1835	1.4	-1.79	31	9.52	0.10	-9.20	6.5
NGC 1898	0.6	-1.37	...	10.90	0.03	-7.6	...
NGC 1916	0.2	-2.08	...	9.88	0.07	-8.74	...
NGC 2005	0.9	-1.92	...	10.20	0.07	-8.42	...
NGC 2019	1.3	-1.81	0	10.55	0.07	-8.07	0.0
NGC 2210	4.4	-1.97	12	10.36	0.07	-8.26	5.9
NGC 2257	8.4	-1.8	31	11.48	0.04	-7.05	47.0
NGC 1841	14.9	-2.11	22	10.98	0.18	-8.00	13.9
Reticulum	11.4	-1.71	29	12.40	0.03	-6.10	105.7

References

Bothun, G.D., and Thompson, I.B.: 1988, *AJ* **96**, 877

Carney, B.W., and Latham, D.W.: 1987, *AJ* **93**, 116

Eggen, O.J. Lynden-Bell, D., and Sandage, A.R.: 1962, *ApJ.* **136**, 749

Friel, E.D.: 1988, *AJ* **95**, 1727

Gilmore, G., and Wyse, R.F.G.: 1985, *AJ* **90**, 2015

Kinman, T.D.: 1965, *Ap.J.Suppl* **11**, 199

Langer, G.E., Suntzeff, N.B., and Kraft,R.P.: 1991, *PASP* , submitted

Morrsion, H.L., Flynn, C., and Freeman, K.C.: 1990, *AJ* **100**, 1191

Olszewski, E.W., Schommer, R.A., Suntzeff, N.B., and Harris, H.C.: 1991, *AJ* **101**, 515

Paltoglou, G.: 1990, *BAAS* **22**, 1289

Preston, G.W., Shectman, S.A., and Beers, T.C.: 1991, *Ap.J.* **375**, 121

Ratnatunga, K.U., and Freeman, K.C.: 1989, *Ap.J.* **339**, 126

Saha, A.: 1984, *Ap.J.* **283**, 580

Schmidt, M.: 1975, *Ap.J.* **202**, 22

Schommer, R.A., Olszewski, E.W., Suntzeff, N.B., and Harris, H.C.: 1991, *AJ* , accepted

Strom, S.E., *et al.* : 1981, *Ap.J.* **245**, 416

Suntzeff, N.B., Kinman, T.D., and Kraft, R.P.: 1991, *Ap.J.* **367**, 528 SKK

Suntzeff,N.B., *et al.* : 1991, *AJ* , submitted

M. Edmunds: Isn't it dangerous to try and deduce total "halo" masses (or perhaps "spheroid" might be a better term) from globular clusters or RR Lyraes, since they may only trace a metal-poor component? Particularly density distributions as a function of radius may be unreliable, if applied to the whole halo.

B. Carney: In a poster, Laird, Latham, and I note that in sample of stars in retrograde Galactic orbits, which ought to define a "pure" halo sample, there are near-solar metallicity stars, but they are so rare that corrections to halo mass densities derived using RR Lyraes and globular clusters are negligible.

R. Kurucz: There is another parameter that varies with radius: the stellar accretion and loss rate from clusters. It increases inward because the collision rate increases inward. Clusters nearer the center will accrete field stars with higher abundances so there will be an apparent shift in cluster abundance with radius.

N. Suntzeff: Accretion of field stars by globular clusters would imply intra-cluster metallicity variations which are not seen.

J. Nemec: Concerning your conclusion that there is a metallicity gradient in the Galactic halo, how can you be sure that the non-zero slope seen at small $\tilde{\omega}$ in your [Fe/H] – $\tilde{\omega}$ diagram is not due to "contamination" from the thick disk or old disk component?

B. Carney: I share Bernard Pagel's concern about using one parameter to define a population - both chemistry and kinematics, at least, are necessary. You noted, in fact, that both the halo and the thick disk produce RR Lyraes. So how do you know that in the inner Galaxy your mean metallicities of RR Lyraes aren't affected by mixing stars from different populations.

N. Suntzeff: We, in fact, have used a second parameter, the *position* of the star in the Galaxy, to define the population. The RR Lyraes are an *in situ* sample that are not contaminated by the thick disk simply because the search technique finds RR Lyraes in a given solid angle to a given limiting magnitude. The volume searched (a cone) contains a negligible amount of disk and thick disk volume compared to the volume searched in the halo.

Y.-W. Lee: There is now some evidence (see my paper in this volume) that the systematic variation in HB morphology with galactocentric distance continues to vary into the center of the Galaxy. This would imply that the bulge is indeed older than the halo by ~1.5 Gyr. Other possibilities, such as high Y or high core rotation rates in the bulge, can be ruled out from the observed period distributions of RR Lyrae stars.

T. Armandroff: I'd like to call your attention to a poster by Da Costa, Zinn and myself that re-examines the question of a metallicity gradient in the halo globulars. We have determined spectroscopic abundances for stars in the very distant globulars Pal 3, 4, and 14. We find no metallicity gradient in the outer halo (R > $R_\odot$). A weak gradient is present inside the solar circle.

E. Bica: We have recently enlarged the sample of LMC clusters with integrated UBV photometry, in a collaboration with colleagues from Cordoba and Porto Alegre (poster in this meeting). Many new SWB Class VII clusters were detected. In particular Hodge 7 is located in the same locus as Hodge 11 or NGC 2257 in the (U-B),(B-V) diagram. Have you checked Hodge 7 for RR Lyrae stars? It would be a good candidate.

Map of the Milky Way at 1 micron produced by the DIRBE experiment on board the Cosmic Background Explorer satellite (COBE). The nuclear Bulge is clearly visible as a flattened central structure of about one kpc size.

THE STELLAR POPULATION OF THE GALACTIC BULGE

R.M. RICH*
Astronomy Department and Columbia Astrophysics Laboratory
Columbia University
538 West 120th Street
New York, NY 10027

ABSTRACT. The central kiloparsec of the Milky Way contains a distinct stellar population that resembles distant ellipticals and bulges. An abundance range from 1/10 to nearly 10 times the solar metal abundance produces evolved stars ranging from RR Lyraes to late M giants, OH/IR stars, and Miras. I describe the kinematics, structure, chemical evolution and possible age range of the galactic bulge. Infrared imaging of the M31 bulge reveals a population of luminous AGB stars that may have progenitors younger than 15 Gyr. It is important to measure the age of bulge populations relative to the old globular clusters, and to consider a range of formation scenarios.

1. Introduction

It is especially appropriate to discuss the galactic bulge at a meeting on stellar populations, since Baade's (1944) study of the bulge populations in M31 and the Galaxy helped to found the subject. Giants at a luminosity of $M_v = -3$ were discovered in the bulge of M31; shortly thereafter, Baade (1951) also discovered the RR Lyrae population in the galactic bulge, proving the bulge's existence. It was logical to unify the properties of spheroidal populations with the globular clusters: old, and metal poor. The impact of Baade's advance was such that descriptions of the Milky Way bulge as being old and metal poor persist to this day.

There were a few voices crying in the wilderness that, based on the strong Fe and Na lines in its integrated spectrum, the bulge must be more metal rich. Morgan had obtained spectra of the bulge's integrated light in the 1950's (Morgan & Osterbrock, 1969) and found it strong-lined, and comparable to that of the M31 bulge. Baum (1959) also noted that ellipticals were much redder than globular clusters and therefore could not contain the same stellar population.

Bulge fields are crowded and reddened, and the unresolved turnoff population creates a high background light that must be subtracted. Consequently, spectroscopy of large samples of bulge occurred only after large aperture telescopes and linear detectors were operating in the Southern Hemisphere.

Nassau & Blanco (1958) discovered the key stellar tracer of the bulge population, luminous late M giants. In the following landmark grism surveys using the CTIO 4 m (Blanco, 1988 and references therein) these stars were used to trace the structure of the bulge. In fact, the late M giants present in the bulge in such large numbers are not present in the globular clusters. In principle, one could now make a firm link between these stars, and the cool, luminous giants that Johnson's (1966) IR photometry concluded must be an important component of giant elliptical galaxies. In fact, the presence of M giants evolved from stars of high metallicity with RR Lyraes evolved from stars of low metallicity is *prima facie* evidence of the bulge's wide range in abundance. Arp's (1965) photometry, though hampered by severe crowding, also supported a wide abundance range.

The bulge has been most intensively studied along the minor axis. The region surrounding NGC 6522 ($b = -4°$) known as Baade's Window, has been specially targeted, but one can work over a wide region, even as close as $-2.5°$ from the nucleus (Baade, 1963). For $R_0 = 8$kpc, a field 8 degrees from the nucleus has an impact parameter of 1 kpc.

* Alfred P. Sloan Foundation Fellow

B. Barbuy and A. Renzini (eds.), The Stellar Populations of Galaxies, 29–40.

Modern spectroscopy by Whitford & Rich (1983) and Rich (1988) confirms both the wide abundance range and the presence of some very metal rich stars. The initial picture of the old, metal poor bulge is now shown to be incorrect. The bulge may even have an age range as well. Continuing study of the bulge is inspired by the resemblance of its stellar population to that of unresolved bulges and ellipticals (Fig. 1)

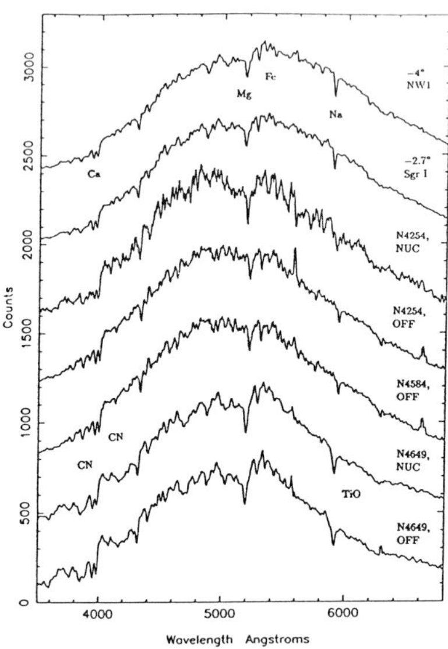

Fig. 1. Integrated spectra of patches of the galactic bulge near Baade's Window ($b = -4°$) and near Sgr I ($b = -2.6°$); these compared with the integrated spectra of other galaxies. Notice that the line strength does not change between the two latitudes in the inner bulge. Hβ is strong in the bulge fields, but CN and the HK break are noticeably weak relative to the nuclei.

Following this section, we consider the age, abundance, kinematics, and structure of the bulge population. Looking beyond these dominant descriptive parameters, §3 considers the bulge's chemical and dynamical evolution as tracing its history of formation. We turn in §4 to the study of other bulge populations, and conclude in §5 by posing problems in the study of the bulge.

Recent reviews discussing the bulge population include that of Frogel (1988) and the proceedings of the first ESO/CTIO conference on Galactic Bulges, held in 1990. In 1992, IAU Symposium 153 will be devoted to the subject of galactic bulges.

2. Description of the Stellar Population

The wide range of abundance in bulge stars gives rise to a rich variety of evolved progeny, including late M giants, OH/IR stars, RR Lyrae stars, red and blue horizontal branch stars (Fig. 2). It is noteworthy that Ortolani (1990 and this meeting) finds a red horizontal branch clump in metal rich globular clusters such as NGC 6553. It is likely that UV-bright stars responsible for the far-UV rising flux in external galaxies are also present. The wide abundance range is the key factor responsible for the bulge's diversity. The bulge may contain an age spread as well, giving rise to the 800 day Miras and OH/IR stars (see §2.3 below). Blanco's surveys of giants in the bulge and Magellanic Clouds find that luminous carbon stars must comprise $< 10^{-3}$ of the bulge AGB stars, by far lower than the clouds or solar neighborhood. The lack of carbon stars is generally ascribed to the high metal abundance of the stars (see Rich, 1988 for other implications).

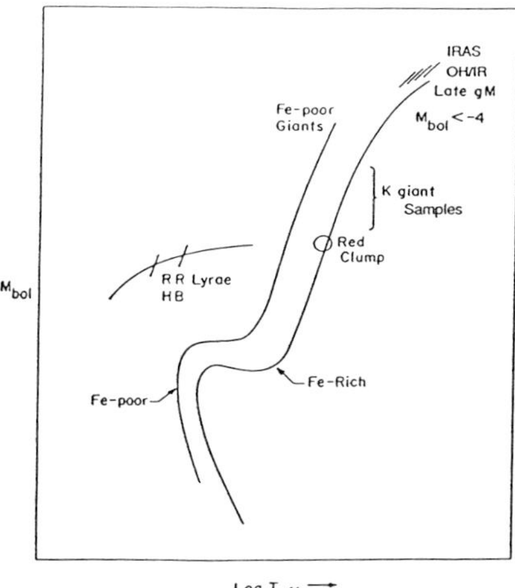

Fig. 2. The wide abundance range in the galactic bulge causes an unusual mix of evolved stars. The progenitors of the late M giants, Miras, OH/IR and IRAS stars are metal rich. On the other hand, the RR Lyraes and HB stars are evolved from the metal poor population (certain to be very old). The red "clump" giants are evolved from the metal rich population.

The disk globular cluster system is also found in the same volume as the bulge. By and large, these clusters are of solar abundance and lower (Armandroff, 1989); direct comparison of bulge field and cluster member spectra show that the bulge field is much more metal rich than the clusters.

2.1. Abundances

Ideally, one wants to measure the abundance distribution of long-lived stars, but this is not possible because the bulge turnoff is too faint for spectroscopy. Instead, we must search among the more luminous evolved stars for a population unbiased with respect to abundance. The K giants offer such a population. Virtually every star must evolve through the K giant phase; M giants have metal rich progenitors while the progenitors of RR Lyraes are metal poor. These other populations give good insight into the stellar evolution process, but are not suitable for the study of chemical evolution.

Modern digital spectroscopy shows that K giants in Baade's Window (500 pc from the nucleus) range in abundance from 1/10 to nearly 7 times the Solar abundance (Figure 3; Rich, 1988), with a mean of twice solar. Geisler & Friel (1990) employ CCD imaging and Washington photometry to confirm this result, and enlarged the original 88 K giant sample to a few hundred. In collaboration with A. McWilliam and R. Luck, I have begun to study bulge giants at high resolution.

One expects the RR Lyrae stars to be generally metal poor. Walker & Terndrup (1991) find that the Baade's window RR Lyrae stars peak in abundance at −1 dex; RR Lyraes of solar metallicity, commonly found in the solar neighborhood, are absent.

Turning to the M giants, it was long expected that the most luminous M giants would be extremely metal rich. Frogel & Whitford (1987) proposed that high metallicity might extend the lifetimes of turnoff stars, permitting massive progenitors to live long enough to be the source of the brightest M giants (Renzini & Greggio, 1990 calculate that a 15 Gyr old population could be as high as $1.1 M_\odot$ at the turnoff).

Surprisingly, it has instead been found that even the most luminous M giants are not extremely metal rich (Sharples *et al.*, 1990; Terndrup *et al.*, 1991), with abundances of 2-3 times solar. It is of great importance to measure even crude abundances for the most metal rich late M giants; V.

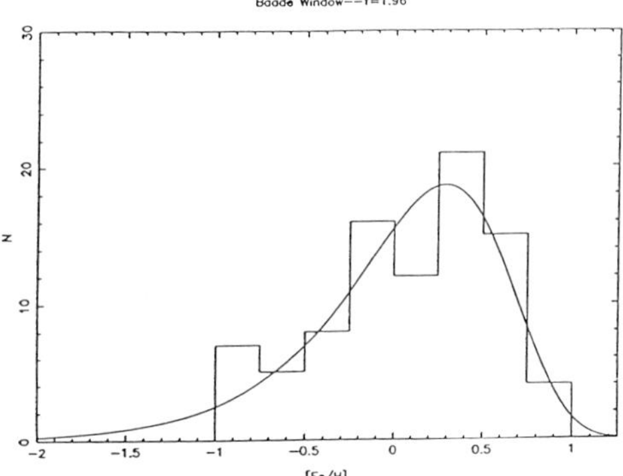

Fig. 3. Abundance distribution for 88 K giants in Baade's Window from Rich (1988). Theoretical curve represents the simple model of chemical evolution with $y = 2.0$ (see text). If the evolution were dominated by infall, there would be fewer metal poor stars. If wind outflow dominated the evolution, the theoretical curve would be the same but shifted to lower abundance. If the wind catastrophically terminated evolution, the metal rich end would be sharply cut off.

Smith is analyzing echelle spectra of a relatively clean continuum point in the M giant spectra near 7400Å.

If the bulge's abundance picture were not already complex enough, the planetary nebulae present a further contradictory picture. Ratag (1991) finds that bulge planetaries have normal (disk) N O, Ne, and Ar abundances, but are enhanced in He. These data appear to conflict with McWilliam & Rich's (1991) finding that the bulge giants are enhanced in alpha-capture elements.

In collaboration with N. Tyson I am investigating the abundance distribution and abundance gradient (first seen by van den Bergh 1971) in the galactic bulge. Figure 4 gives the preliminary result, that the bulge's abundance is constant within 1 kpc, declining quickly beyond that. It is not clear whether there is a gradient or population transition. Blanco (1988) finds a remarkable drop in the number count of M giants. Gradients are seen in the M giants using CO (Frogel *et al.* 1990) and in TiO (Terndrup *et al.* 1990). It will be of great interest to determine if Fe and and α-capture elements (Si,Mg, and Ca) show differing gradients in abundance.

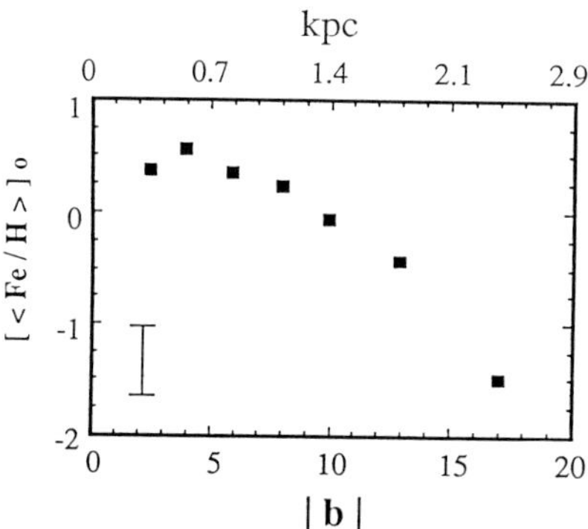

Fig. 4. Abundance gradient in the galactic bulge measured from Washington photometry of the giants (Tyson, 1991) . The zero-point of the abundance scale has been set by requiring Baade's Window to have an abundance of $\langle[\mathrm{Fe/H}]\rangle = +0.3$, in agreement with Rich's (1988) spectroscopy.

The contradictory abundance picture derived from the evolved stars in the bulge is of sufficient concern that it will be worthwhile to measure abundances of the last unbiased stellar sample–the bulge main sequence.

2.2. Kinematics

The $10^{10}M_{\odot}$ of the bulge dominates the inner kpc, supporting the galactic rotation curve there (Bahcall *et al.* 1983). While less massive than the disk, the bulge is 5-10 times as massive as the stellar halo. Dramatic progress can be expected soon in defining the global kinematics of the bulge. It also appears that kinematics depend on metal abundance, though this question remains plagued by small number statistics. If we believe that Rich's (1988) K giant sample is representative of the bulge, then stars more metal rich than -0.5dex comprise more than 80% of the bulge's mass.

The metal rich stellar population of the bulge appears to rotate at about 100 km/sec. Efforts by Feast *et al.* (1990) find rotation of 10 km/sec/deg in late M giants and IRAS sources in the bulge; the OH/IR stars (te Lintel Hekkert 1990) and planetary nebulae (Kinman *et al.*, 1989) follow the general rotation field.

Like the halo, the bulge has a large radial velocity dispersion. Sharples *et al.* 1990 report radial velocities for over 300 of Blanco's M giants (in Baade's Window), finding a non-gaussian velocity distribution, but confirming Mould's (1983) dispersion of 113 km/sec. Rich (1990b) finds K giants in Baade's Window to have $\sigma = 104$km/sec, confirmed in the Sadler *et al.* (1992) 500 K giant sample. A key question is whether the 1/10 solar stars have a higher dispersion than the metal rich stars (as suggested by the small sample of Rich 1990b). Rich *et al.* (1992) finds that 33 bulge RR Lyraes have a dispersion of 127 km/sec, in agreement with the metal poor bulge K giants. Minnitti (these proceedings) also finds this effect, and further suggests that metal poor K giants have less rotation support. The bulge's velocity dispersion declines with increasing galactocentric distance (Tyson & Rich, 1991) and this can be used to show that the bulge has no missing mass (Kent *et al.* , 1991).

Recently, it has become possible to measure proper motions from plates obtained in the 1950s. Spaenhauer *et al.* (1991) find that the w velocity dispersion is 66 km/sec, compared with a radial velocity dispersion of 110 km/sec (as one might expect looking at the *COBE* image). They also find that low abundance stars have a larger w velocity dispersion, though this is only a 2σ result (the 500 K giant sample mentioned above consists of these proper motion stars and should clarify the issue).

Several recent lines of evidence suggest that the bulge may be a bar (see §2.4 below); measurement of bulge minor axis rotation may confirm or refute this idea. The possibility that the bulge may be a bar also bears on the interpretation of proper motion data, which frequently is analyzed under the assumption that velocities in the plane are isotropic.

2.3. Age

The discovery of RR Lyrae stars shows that at least part of the bulge is very old; Lee (these proceedings) suggests that the bulge may be the oldest stellar population in the Galaxy. The situation is complicated by the presence of very luminous evolved stars.

Bolometric measurements of Blanco's latest bulge M giants reveal the extremely luminous termination of the bulge AGB, fully 1.5 mag brighter than the He core flash (Frogel & Whitford, 1982). Wood & Bessell (1983) derive large formal pulsation masses for bulge Miras, suggesting a progenitor population as young as 1-2 Gyr.

Presently, there is a clear conflict between Frogel *et al.*'s assertion that the distribution function of M_{bol} for bulge LPV's is identical to that of the globular cluster LPV's. If we take the Mira P-L relationship of Whitelock (1990) and the bulge IRAS Mira period distribution of Whitelock *et al.* 1990, we find that most bulge LPV's must be brighter than $M_{bol} = -5.0$, given that the majority of bulge LPV's exceed 400 days in period. The longest period Miras (770 days) must be as bright as $M_{bol} = -5.4$ (coincidentally, the brightest M31 bulge giants are also -5.4). Given that extinction and the bulge's spatial thickness potentially render the measurement of absolute magnitude of an individual star impossible, it becomes attractive to compare the period distributions of the bulge and cluster LPV's. In this case, the very long periods (high luminosities) of the bulge stars stand out.

Revision of the bulge distance to 8.1 kpc by Walker & Terndrup (1991) and others requires that the Frogel & Whitford (1987) bulge luminosity function of M giants from the Blanco grism surveys be brightened by 0.31 magnitudes. One is now placed in the uncomfortable position of explaining how old low mass stars can ascend to $M_{bol} = -4.7$, yet also exhibit vigorous mass loss (OH/IR and IRAS phases) that should terminate this evolution at fainter magnitudes. This mass loss is estimated by Whitelock (1990) to be as high as $10^{-4} M_{\odot} yr^{-1}$. Van der Veen (1989) proposes that low mass Miras evolve toward longer periods without increasing in luminosity: low mass stars could have long periods, but the P-L relationship would be invalid.

Direct application of Iben & Renzini's (1983) theory requires a $1.5 M_{\odot}$ progenitor mass for stars at $M_{bol} \approx -5$. Greggio & Renzini offer two explanations. First, the metallicities may be greatly enhanced, allowing massive stars to live 15 Gyr. As discussed in §2.1, the M giant abundances are not high enough to extend the lifetimes, and if $\Delta Y/\Delta Z > 0$, then we are almost forced to conclude that the progenitors must be younger than the globular clusters (Renzini & Greggio, 1990). Blue straggler progeny may be present, but probably cannot account for the rank and file 400 day bulge Mira that attains $M_{bol} = -5$.

We can look to the color-magnitude diagrams of bulge fields and easily conclude that the very young (1-2 Gyr) progenitors of Wood & Bessell (1983) are not present (Rich, 1985). If the modulus is now 14.57, however, the CMD's of Rich (1985) and Terndrup (1988) cannot rule out the presence of a substantial population of intermediate age stars; note that Terndrup's isochrones must be faded by 0.31 mag.

A further smoking gun is the lack of metal rich RR Lyrae stars in the bulge. These stars are found in large numbers in the much more metal poor solar neighborhood. Is the bulge too young to make RR Lyrae stars? One compromise is that the metal rich bulge is slightly yonger than the globular clusters.

We can collect the problems with age into a set of questions.

1. How can the bulge be older than the globular clusters, yet have an AGB that terminates at M_{bol} =-5 and contain Miras in excess of 700 days (along with very luminous OH/IR stars)? How can the AGB attain such high luminosities and have large mass loss, yet also be evolved from a 15 Gyr old population?
2. If the He abundance is normal (or likely enhanced, if $\Delta Y/\Delta Z > 0$), stars of high metallicity may have *decreased* lifetimes, if H is no longer the majority element. If the luminous M giants have only 2-3 times the solar metals, the lifetimes will not be so extended, anyway.
3. The IR colors of the bulge giant branch are similar to those of 47 Tuc, despite the high derived metal abundances for the stars.
4. Metal rich ([Fe/H] > -0.3) RR Lyraes comprise approximately 1/3 of the known RR Lyraes in the solar neighborhood (and are thought to be members of the thick disk). Why are they absent from the bulge?

2.4. Structure

If pictures are worth a thousand words, the recent image of the bulge obtained with the COBE satellite would qualify. Seen in the 1 μm light of K giants which trace the mass, the bulge is revealed to be a compact, peanut-shaped system contained within 1 kpc. Not surprisingly, Blanco (1988) finds the M giants dramatically concentrated to the plane, and the highest abundances are found within 1kpc as well. The flattened nature of the bulge is quantitatively illustrated by Kent's

(1991) measurement of a (350pc=2.5°) scale height at 2 μm . The M giant counts of Blanco and Terndrup (1989) also hint at a flattened, box-shaped structure (as does Harmon & Gilmore's (1989) image constructed from IRAS sources). One feature conclusively evident from the *COBE* map is that bulge is not smoothly connected to the $R^{-3.5}$ spheroid. Whitelock *et al.* (1990) use Mira luminosities from the P-L relation to show that the bulge is too deep in the line of sight to fit the $R^{-3.65}$ power law for M giants found by Blanco & Terndrup (1989). This conflict may be resolved if the bulge is in fact triaxial (or even a bar).

Recently, several lines of evidence independently support the idea that the bulge is in fact a bar, pointing toward $b > 0$ (Blitz & Spergel, 1991); Weinberg, 1991; Whitelock, 1991). At this meeting, Whitelock showed that the bulge Miras appear to be more numerous and closer at positive galactic latitudes; there is evidence that the bulge Miras have a wide spatial distribution as well. While the above authors concur on the issue of triaxiality, they suggest that there may be triaxial components in addition to the bulge (bar) that may be connected with the thick disk, and they are not all in agreement on the size and orientation of these additional components. The problem arises as to what an edge-on bar should look like. It has been suggested (Pfenniger & Norman, 1990) that peanut-shaped bulges might actually be edge-on bars.

A central bar may scatter bulge stars into the solar vicinity, thus providing an explanation for the enigmatic high velocity metal rich stars found in the solar neighborhood by Grenon (1989). Barbuy & Grenon (1990) report these stars to have high O abundances as would be expected for bulge members (Matteuci & Brocato, 1990).

Searches for minor axis rotation and quantitative analysis of the *COBE* data should resolve the question of the bulge's geometry.

Massive molecular clouds of CO reside near the nucleus (Bally *et al.* , 1988). Presumably, these clouds survive the hot gas ejected by Type I SNe, and are probably a source of the star formation known to occur in the nuclear region. They may be a source of secular acceleration, propelling stars into orbits above the plane on timescales of a few Gyr.

3. Chemical Evolution of the Bulge

The early history of the bulge's formation is written in the ages and abundances of its stars. The compact appearance of the bulge suggests that it may satisfy the requirements of the simple model of chemical evolution–no significant infall or outflow, effectively instantaneous reclycling, and have largely formed in a free-fall time. Indeed, Rich (1990a,b) finds that the abundance distribution is fit by the simple model with a yield of twice solar (See Figure 3). Tyson's (1991) survey using Washington Photometry finds that the bulge's mean abundance remains constant within 750 pc; perhaps the evolution of this volume closely satisfied the requirements of the simple model.

We can use the abundance distribution to determine if wind outflow affected the bulge's chemical evolution. In the strict simple model of chemical evolution, the abundance distribution is described by a single parameter, the yield. For the case in which gas is removed from the system by outflow, the abundance distribution can vary in two ways. If the outflow is steady, the yield is reduced and the abundance distribution shifts toward lower abundances. If the wind outflow occurrs suddenly (for example, when the thermal energy of the gas exceeds the binding energy of the galaxy) then the abundance distribution is truncated, and lacks the metal rich end (Searle & Zinn, 1978; Rich, 1990a). Using Washington photometry, N. Tyson and I are surveying 8 fields from 250 to 1500 pc on the minor axis. We hope to determine the abundance distribution function in each field and study its changes as one leaves the bulge. We expect to be able to determine if the gradient is really a transition to the halo, or is due to a wind outflow.

3.1. Chemical Abundance Profiles

If we can measure the ratios of Fe, O, r-process, and s-process elements, we may gain insight into the earliest era of the bulge's formation. Short-lived (10^6yr) massive star SNe produce O, α-capture, and r-process elements, but little Fe. Core deflagration SNe (likely linked to white dwarfs, hence a 10^8 yr lifetime) contribute most of the Fe. If we can measure the abundances of O, Fe, and Eu (an r-process element) in individual bulge giants, we can determine the timescale for enrichment. Barbuy (these proceedings) finds that O is enhanced in metal rich stars in the solar neighborhood. A. McWilliam and I have been measuring abundances in stars covering the full abundance range in the bulge. We have found that Mg, Si, and Eu are enhanced relative to Fe. The presence of strong Eu ($\lambda\lambda$6645Å lines in bulge giants with [Fe/H] = +0.5 indicates that massive star SNe remained important in the enrichment well above the solar abundance. The analysis of 4 bulge giants is given in Table 1; all are enhanced in Eu, Mg, and Si. We have not yet measured the O abundance; spectrum synthesis will be required.

TABLE 1. COMPOSITION OF BULGE GIANTS
A. McWilliam, R.M. Rich (1991)

	BW IV - 3	I - 141	I - 194	IV - 25
Fe/H	−1.13	−0.90	−0.16	+0.60
Ca/Fe	+0.30	+.42	+.13	+.14
Si/Fe	+0.23	+.52	+.47	+.25
Mg/Fe	+0.31	+.84	+.64	+.51
Na/Fe	+0.22	+.40	+.44	+.80
Al/Fe	+0.02	+.40	+.41	+.74
Zr/Fe	...	−.78	+.26	+.03
Y/Fe	+0.09	+.12	−.16	+0.10
Ba/Fe	−0.32	−.95	−.26	−0.74:
La/Fe	+0.41	+.33	+.10	+0.18
Eu/Fe	+0.83	+.74	+.63	+0.36

3.2 Formation of the Bulge (Bar?)

The chemical evolution and dynamical data provide support to two broad classes of formation scenarios. In the collapse and spin-up picture, the bulge forms in a free-fall time (the classic ELS picture). Violent star formation enriches the ISM, and gas dissipation causes the latest (most metal rich) stellar generations to be flattened and rotation supported. In this picture, it is difficult to see how the bar forms.

Alternatively, the bulge could have formed from in a violent starburst occurring in gas already in a massive disk near the nucleus (analagous to some luminous IRAS galaxies observed today). If the disk dominated the potential, it would be unstable to bar formation. Acceleration mechanisms of Pfenniger & Norman (1990) could heat the bar to produce today's observed thickness. It seems difficult to understand how the classic collapse and spin-up could produce a bar. If the bulge formed in a free-fall time (the canonical scenario) how was there also sufficient time to enrich Fe (presumably from Type I SNe) to the highest observed values? Was there enough time for the dissipational processes that would be required if successive metal rich populations were "spun up" relative to the first generation of metal poor stars?

4. Extension to Other Galaxies

To what extent does our bulge resemble ellipticals and distant bulges? Morgan & Osterbrock (1969) noted the strong-lined integrated spectrum of Baade's Window and compared it to that of distant galaxies. Whitford (1978) used an early spectrum scanner at the CTIO 1.5 m to quantitatively show that Mg, Na, and Fe were enhanced in the integrated light of Baade's Window as they are in distant galaxies. Figure 1 shows that the integrated light of the bulge has weaker metals than the giant ellipticals; note particularly the weakness of the blue CN bands (also seen by Rich (1988) in individual bulge stars).

In recent years, it has become possible to extend the study of bulge populations to M31 and M32. The most luminous bulge giants 750pc from the M31 nucleus are late M giants, just as has been found by Blanco in the Milky Way. Perhaps that is not surprising, given the long-known presence of TiO bands in galaxy spectra. Rich & Mould (1991) resolve the M31 bulge in the IR, 500 pc from the nucleus, and measure a giant branch extended to $M_{bol} = -5$ (Figure 5). Davies *et al.* (1991) suggest that the giants brighter than −4.75 are contamination from an inner disk population thought to resemble the LMC Bar West. Even using Kent's (1989) maximum disk model, they cannot account for the large number of bright giants in these fields. Further, their model succeeds only because they shift the original Frogel & Whitford (1987) luminosity function 0.31 magnitude brighter, due to a change in the adopted distance modulus. Their revised bulge luminosity function now extends to $M_{bol} = -4.75$, *a full 1.1 magnitudes above the He core flash luminosity.* Surprisingly, the elliptical companion M32 has a nearly identical population of luminous giants; Freedman (these proceedings) reports on a population of IR luminous giants even brighter than those in the M31 bulge. Freedman and I have begun to take spectra of these unusual stars in M32; we are interested to see if any are luminous carbon stars.

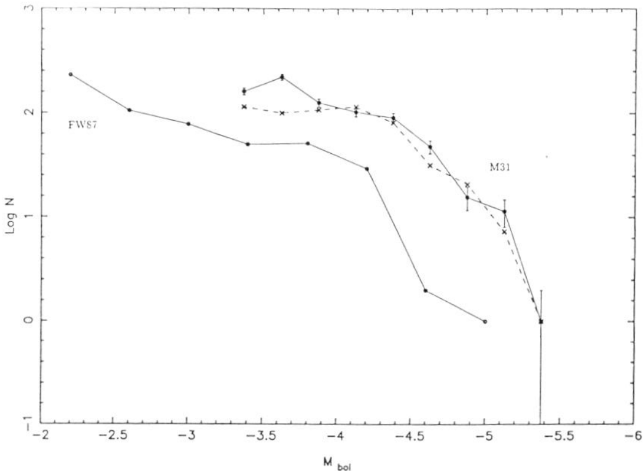

Fig. 5. Luminosity functions for the M31 bulge (Rich & Mould 1991) corrected for completeness and compared with Frogel & Whitford's (1987) luminosity function of Blanco's Baade Window M giants. The dashed line function is the result of applying bolometric corrections dependent on color; the solid line (a larger sample) employs $BC_K = 3.1$ for all stars. A color-magnitude diagram is published in Rich & Mould 1991; a similar J, K CMD for a field in M32 may be found in Freedman's paper in these proceedings.

Of course, disk contamination and blue straggler progeny (Greggio & Renzini, 1990) may be sufficient to explain the extended giant branches. Further, the extended giant branches consist largely of variable stars observed at one epoch, and the luminous tip may be enhanced with giants at maximum light. It is interesting that the observed termination of the M31 AGB at −5.4 is consistent with the bolometric luminosities of the Milky Way bulge's 770 day Miras. If these stars are the relics of intermediate age and young populations, how did it come to be that the bulges of the Galaxy, M31, and M32 come to have the same extended star formation histories? In order to address these questions in depth, Mould, Graham and myself are surveying the M31 bulge in the IR.

It is important to recall that the disk population in photometrically deconvolved disk/spheroid models extends to the very nucleus. If this is indeed an intermediate age stellar population, like the LMC bar, then we must wonder how this population formed deep within the bulge's potential well, and whether some bulges might be closely related to these inner disks, as Kormendy (1991) has suggested.

5. Challenges for Future Bulge Research

The foregoing is an attempt to describe our present state of knowledge in the study of the galactic bulge. The following questions and comments illustrate where we may deepen our understanding.

1. Could the bulge have a complex star formation history, like the Magellanic Clouds or the Carina Dwarf? Specifically, could the metal rich bulge have formed in a burst taking place after the formation of the globular clusters?
2. Luminous IR galaxies are observed with $\approx 10^{10} M_{\odot}$ of molecular gas in their centers. Could such gas experience a starburst and later form a bulge?
3. Is the galactic bulge actually a bar? If so, how does it connect to the nucleus and to the halo?
4. Are there genuine kinematic differences as a function of abundance? Do these represent different populations (bar vs. halo), or the evolutionary history of the galactic bulge?
5. Are secular accelaration mechanisms important in producing the bulge's present structure, especially spatial thickness?
6. Given that the bulge has spatial thickness and a range in abundance, can one hope to derive the age range from turnoff photometry?
7. What is the connection between the bulge and the central molecular gas and activity?

References

Arp, H. 1965, ApJ, 141, 45.
Baade, W. 1944, ApJ, 100, 137.
Baade, W. 1951, Pub.Obs.Univ.Mich., 10, 7.
Baade, W. 1963 in The Evolution of Galaxies and Stellar Populations (Harvard U. Press: Cambridge) p.279.
Bahcall, J.N., Schmidt, M., and Soneira, R.M. 1983, ApJ, 265, 730.
Barbuy, B., and Grenon,M. 1990 in Bulges of Galaxies, B. Jarvis and D. Terndrup, eds. p. 47.
Baum, W.A. 1959, PASP, 71, 106.
Bally, J., Stark, A.A, Wilson, R.W., and Henkel, C. 1988, ApJ, 324, 223.
Blanco, V. 1988, AJ, 95, 1400.
Blanco, V., and Terndrup, D. 1989, AJ, 98, 843 .
Blitz, L., and Spergel, D.N. 1991, Ap.J., 379, 631.
Davies, R., Frogel, J.A., and Terndrup, D.M. 1991, AJ, in press.
Feast, M.W., Glass, I.S., Whitelock, P.A. and Catchpole, R.M. 1989, MNRAS, 241, 375.
Frogel, J.A. 1988, ARAA, 26, 51.
Frogel, J.A., Blanco, V.M., Terndrup, D.M., and Whitford, A.E. 1990, ApJ, 353, 494.
Frogel, J.A., and Whitford, A.E. 1987, ApJ, 1987; ApJ, 320, 199.
Geisler, D. and Friel, E. 1990, in Bulges of Galaxies, B. Jarvis and D. Terndrup, eds. p. 77.
Grenon, M. 1989, Astrophys. Space Science, 156, 29.
Harmon, R.T., and Gilmore, G. 1988, MNRAS, 235, 1025.
Iben,I. and Renzini, A. 1983, ARAA, 21, 271.
Johnson, H.L. 1966, ARAA, 4, 193.
Kent, S.M. 1989, AJ, 97, 1614.
Kent, S.M., Dame, T.M., and Fazio, G. 1991, ApJ, 378, 496.
Kinman, T.D., Feast, M.W., and Lasker, B.M. 1989, AJ, 95, 804.
Kormendy, J. *et al.* 1991, Private communication.
Matteuci, F., and Brocato, E. 1990 ApJ, 365, 539.

McWilliam, A., and Rich, R.M. 1991, in preparation.
Morgan, W., and Osterbrock, D. 1969, AJ, 74, 515.
Mould, J. 1983, ApJ, 266, 255.
Nassau, J.J. and Blanco, V.M. 1958, ApJ, 128, 46.
Ortolani, S., Barbuy, B., and Bica, E. 1990, AA, 236, 362.
Pfenniger,D., and Norman,C. 1990, ApJ, 363, 391.
Ratag, M. 1991, Ph.D. Thesis, Leiden University.
Renzini, A., and Greggio, L. 1990, in Bulges of Galaxies, B. Jarvis and D. Terndrup, eds. p. 47.
Rich, R.M. and Mould, J.R. 1991, AJ, 101, 1286.
Rich, R.M. 1990a, in Bulges of Galaxies, B. Jarvis and D. Terndrup, eds. p. 65.
Rich, R.M. 1990b, ApJ, 362, 604.
Rich, R.M. 1988, AJ, 95, 828.
Rich, R.M. 1985, Mem.S.A.It., 56, 23.
Searle, L., and Zinn, R. 1978, ApJ, 225, 357 .
Sharples, R., Walker, A., and Cropper, M. 1991, MNRAS, 246, 54.
Spaenhauer, A., Jones, B., and Whitford, A.E. 1991, ApJ, in press.
Suntzeff, N. *et al.* 1991, ApJ, 367, 568.
Te Lintel Hekkert, P. 1990, Ph.D. Thesis, Leiden University.
Terndrup, D.M. 1988, AJ, 96, 884.
Terndrup, D.M. *et al.* 1990, 357, 453 .
Terndrup, D.M., Frogel, J.A., and Whitford, A.E. 1991, ApJ, 378, 742.
Tyson, N.D. 1991, Ph.D. Thesis, Columbia University.
Tyson, N.D., and Rich, R.M. 1991, ApJ, 367, 547.
van den Bergh, S. 1971, AJ, 76, 1082.
Walker, A., and Terndrup, D.M. 1991, ApJ, 378, 1991.
Weinberg, M. 1991, preprint.
Whitelock, P.A. in Confrontation Between Stellar Evolution and Pulsation C. Cacciari and G. Clementini, eds. p. 365 .
Whitelock, P.A. *et al.* 1991, MNRAS, 248, 276 .
Whitford, A.E. 1978, ApJ, 226, 777.
Whitford, A.E., and Rich, R.M. 1983, ApJ, 274, 723.
Wood, P.R., and Bessell, M.S. 1983, ApJ, 265, 748.

Question & Answer Section

Freedman: I wanted to comment (and I will discuss this in more detail in my talk tomorrow) that there is a population of luminous stars in M32 that are more luminous than $M_{bol} = -4.2$ and unlike for the bulge of M31, cannot be explained by M31 disk contamination.

Rich: Mould and I proposed disk contamination as one possible explanation for the most luminous M31 bulge stars. Frogel's model depends on use of the Frogel & Whitford (1987) luminosity function *shifted 0.3 mag brighter* due to his new adopted distance modulus. FW87's old luminosity function dropped at -4.2 and now that drop occurs at -4.5. If we can make such a large (0.3) magnitude change in the luminosity of the AGB termination point and not affect conclusions about the age, then we might just as well not bother to continue these observations.

Te Lintel Hekkert: The OH/IR stars in the bulge are not extreme in terms of luminosity and periods (and thus mass). They lie well within the ranges for the Mira variables from Whitelock *et al.*

Rich: Using a new distance modulus of 14.5 kpc, we have good agreement between the luminosities of the longest period Miras (800 days), the AGB tip, and the OH/IR stars. I think we must now ask whether even the 400 day Miras would be extreme progeny for a 15 Gyr old population.

Pagel: A comment on helium: if a substantial amount of He comes from intermediate-mass stars, there could be an analog of the O/Fe effect in the sense of there being less helium in old populations even if metal-rich.

Suntzeff: I believe that there is no obvious problems with the lack of metal rich RR Lyraes in the bulge. Assuming the galactic globular clusters are good templates for the formation of RR Lyraes we (Suntzeff *et al.* 1991) showed that the rate of the probablility of formation of metal-rich to metal-poor RR Lyraes is 1:50. Thus, the lack of metal rich RR Lyraes may not be very surprising unless there is a large old metal rich population.

Rich: More than 80% of the K giants in Baade's Window exceed -0.5 dex in abundance. The bulge is far more metal rich than the solar neighborhood, yet in the solar vicinity, approximately 1 out of 3 RR Lyraes are metal rich. I think we will attain the deepest understanding by stressing the bulge/solar vicinity comparison.

Frogel: 1. It doesn't matter if a luminosity function with C stars is used as a representative intermediate-age population since the fraction of C stars depends on [Fe/H]. Change [Fe/H] and C stars turn into M stars without changing the luminosity function substantially. 2. There may well abe a small fraction of the bulge that is relatively young but the fact that the L.F. of the LPVs in the bulge is identical to that for globular clusters means that based on LPVs, most of the bulge population must have an age similar to that of the metal rich globulars. 3. While we have shown that (J-K) appears to be a good termperature indicator for non-variable M giants, I would hesitate to use this color for LPVs for which blanketing problems are very severe.

Rich: (1) The maximum luminosity attainable by an AGB star depends on how mass loss terminates the AGB. I do not think we understand the final stages of C and M star AGB evolution sufficiently to assert that AGB termination point is not a function of metallicity, as well as age. (2) I think you will be hard pressed to show me examples of 600 day Miras in Globular Clusters.

Faber: How does the luminosity function change as a function of latitude?

Frogel: My IR photometry of Blanco's optically selected M giants finds no variation in the luminosity function with latitude, even beyond 1 kpc. In M31, I find a larger number of stars with $M_{bol} < -4.75$ at greater distances from the nucleus; I attribute these stars to contamination by a disk population similar to the LMC Bar West.

Renzini: Has Whitelock actually determined luminosities for the individual bulge Miras?

Whitelock: The longest period Miras in our bulge sample are around 700 days but there are only a few this long and they can plausibly be explained as binary mergers as you (Renzini) had suggested. The ongest period single stars are around 600 days and will have luminosities about $M_{bol} = -5$ fromt he PL relations. We have no easy way of determining absolute luminosities of individual stars because of the large line of sight depth of the bulge.

Mateo: 1. Exactly what are the pulsation masses Wood & Bessell (1983) derive for the 500-day Miras? 2. Assuming the answer is about 2 $M_{\odot}$and your claim that there is certainly no 1-2 Gyr population in the bulge, then whey are these pulsation masses so large?

Whitelock (responding to Mateo): The effective temperatures of the Miras are really very difficult to measure, but with plausible values the pulsation masses of the longest period objects in the bulge are in the range 1.0-1.5 $M_{\odot}$. It is of course the initial masses that are important and these may be significantly larger than the pulsation masses if the star has experienced much mass loss.

METAL–RICH STARS IN THE GALACTIC BULGE AND THEIR IMPLICATIONS FOR ELLIPTICAL GALAXIES

ELAINE M. SADLER
Anglo–Australian Observatory
P.O. Box 296
Epping, NSW 2121
Australia

ABSTRACT. A new study of more than 400 K giants in the inner Galactic bulge confirms that many have high metallicity ([Fe/H] $\geq$ 0). Such stars are valuable templates in understanding the stellar populations of elliptical galaxies, and may help explain some of the puzzling line–strength anomalies seen in the integrated spectra of giant ellipticals.

1. INTRODUCTION

Understanding the stellar population of elliptical galaxies is a difficult and long–standing problem, since we can study giant ellipticals only in their integrated light rather than focussing on individual components of their population.

Whitford (1978) suggested that giants in the Galactic nuclear bulge might be useful templates in constructing population models of unresolved galaxies, and studies by Rich (1988) and Terndrup (1988) confirm that the bulge has many old, metal–rich stars. Here I describe a new study of K giant stars in the Galactic bulge, discuss ways of calibrating [Fe/H] for above–solar abundances, and compare the line indices measured in bulge giants with those observed in elliptical galaxies, both within their nuclei and further out.

2. NEW OBSERVATIONS OF BULGE GIANTS

Over the past three years, Don Terndrup, Michael Rich and I have observed a large sample of Galactic bulge K giants with fibre–fed spectrographs at the AAT and CTIO. Our aim is to probe the dynamics and abundance distribution in the inner bulge in detail, and to identify metal–rich stars which may be good templates for elliptical galaxies.

Our sample comprises 440 stars in Baade's Window (BW) originally catalogued by Arp (1965) for which proper motions have been measured by Jones et al. (1991, in preparation). We have measured radial velocities for all 440 stars and line–strength indices (Faber et al. 1985) for most, so that in principle we know both the metallicity and the three–dimensional space motion of each star in the sample.

B. Barbuy and A. Renzini (eds.), The Stellar Populations of Galaxies, 41–46.

3. MODEL SPECTRA FOR METAL-RICH STARS

In studying large samples of stars, it is obviously useful to be able to estimate the metal abundance reliably from medium or low-resolution spectra. In studying the integrated light of elliptical galaxies this is essential, since line broadening from the internal velocity dispersion of the galaxy rules out high-resolution spectral studies. The usual approach is an empirical calibration of several line-strength indices via a grid of standard stars whose abundance is well determined from high-resolution studies. This, however, may be difficult to extend to very metal-rich stars, since few of the well-studied stars in the solar neighbourhood have abundances much above solar and the exact abundance of the so-called 'super metal rich' (SMR) stars is still under debate (e.g. Taylor 1991).

I describe here an alternative approach based on a grid of model spectra constructed for metal-rich stars by Roger Bell and Bob Dickens (private communication). It is possible to produce models for any desired combination of T_{eff}, log g and [Fe/H], and to study in a quantitative way how various line strength indices respond to changes in each of these parameters. This in turn can show which indices are most useful in studying the integrated light of ellipticals.

The model spectra discussed here cover the temperature range T_{eff} 4400–5400 K, surface gravity log g 1.5–4.5 and [Fe/H] −0.5 to +1.0. They do not yet include the TiO bands which become important at temperatures below 4000 K (Bell and Gustafsson 1989), and the predicted Balmer lines are too weak, but nevertheless the models give useful insights into the behaviour of the observed indices.

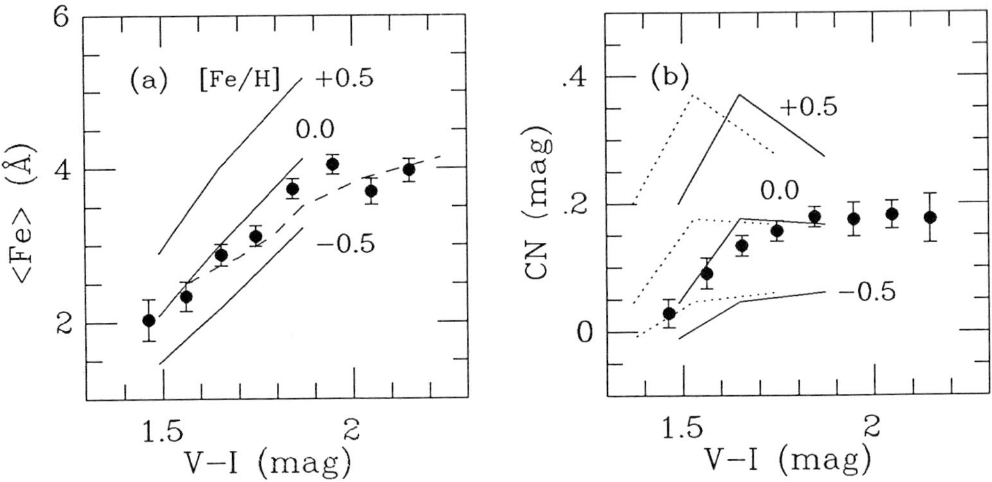

Figure 1: Predicted line strengths for models with [Fe/H] = −0.5, 0.0 and +0.5 (solid lines) and mean data for BW giants (V≥15.5 mag) binned by V−I colour. (a) <Fe> index: the dashed line shows the empirical relation for solar abundance derived by Faber et al. (1985); (b) CN index: the dotted and solid lines correspond to E(V−I)=0.58 and 0.70 mag respectively.

V magnitudes and V−I colours have been measured for our sample by Terndrup (1988), while Bessell (1979) gives an empirical relation between V−I colour and T_{eff} and shows that V−I is not seriously affected by blanketing or gravity effects. Thus a plot of index against V−I is essentially a plot against T_{eff}. For a first look, I have simply binned the sample in 0.1 mag bins of V−I and calculated the mean line strength index in each bin. Figure 1 shows the results for the <Fe> and CN indices. <Fe> is relatively insensitive to changes in surface gravity, while CN shows only small variations with temperature for all but the hottest K stars.

The abundance derived from <Fe> depends sensitively on the assumed reddening. In Figure 1a, an error of 0.1 mag in E(V−I) corresponds to ± 300 K in T_{eff}, and this in turn means a change of 0.3 dex in [Fe/H], or a factor of two in abundance. The extinction in BW is not uniform and taking the same mean E(V−I) over the whole field may not be appropriate but I have adopted E(V−I)=0.70 mag for now, following Walker and Mack (1986), because this seems to fit the 'kink' in the CN distribution in Figure 1b.

Preliminary results from this study suggest that <[Fe/H]>$\sim$ 0 in Baade's Window, in agreement with previous work, and that about 40% of K giants with V $>$ 15.5 have above–solar abundance.

4. STELLAR POPULATIONS IN ELLIPTICAL GALAXIES

Extensive studies of line–strength indices and line–strength gradients in ellipticals have been carried out by Burstein et al. (1984) and Gorgas et al. (1990). Here, I discuss two of the puzzles raised by these studies, the plots of Hβ and <Fe> indices versus Mg_2.

Burstein et al. (1984) pointed out that the Hβ absorption lines in elliptical nuclei are stronger than predicted by extrapolation from globular clusters. However Figure 2a shows that the mean indices for bulge stars overlap with Galactic globular clusters at the low abundance end and with the brightest elliptical galaxies at higher abundances. Since the bulge, like Galactic globulars, lacks an intermediate age population (Terndrup 1988), it is tempting to identify a single sequence of old populations extending from globular clusters to the brightest elliptical nuclei and differing only in metallicity. The radial Hβ gradients within ellipticals are flat or very shallow (Gorgas et al. 1990, Faber et al. 1992), so the outer parts of giant ellipticals also lie on this sequence.

The Hβ index is largely insensitive to both metallicity and surface gravity for $Mg_2 > 0.15$ mag, so the only way to move a population to higher Hβ strength in Figure 2a is to increase the mean temperature, i.e. to add either an intermediate–age population or (perhaps less plausibly) a significant fraction of extremely metal–poor stars.

Figure 2b shows the Mg_2 versus <Fe> plot. Here, bulge K stars extend the sequence described by Galactic globular clusters and low–luminosity ellipticals, which is tightly defined because changes in temperature and abundance have roughly equal and opposite effects. Giant ellipticals lie well off the relation defined by the other objects. Increasing surface gravity will move a star off the sequence, since dwarfs have stronger Mg_2 than giants, but dwarfs would have to contribute almost 50% of the V light in the nuclei of the most metal–rich galaxies to fit the data, and this appears unlikely.

Another possibility is that M giants affect the line–strength indices measured from integrated light in metal–rich ellipticals. Late M giants in BW have very high Mg_2 and low <Fe> because of gross continuum effects from their strong TiO bands. Simulations adding M–giant light (using a template made from bulge M giants) to a pure K giant population

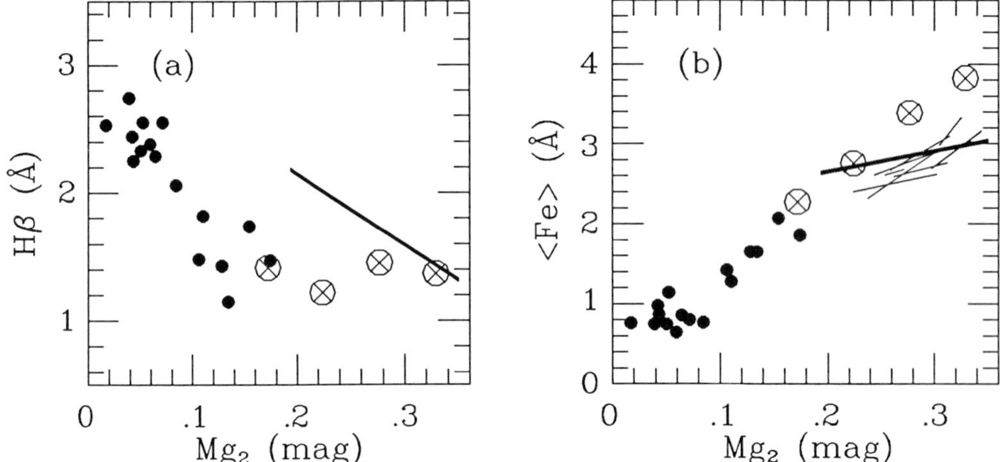

Figure 2: (a) Hβ index versus Mg_2 for bulge K giants from our study (⊗), and for Galactic globulars (•) and elliptical nuclei (solid line) from Burstein et al. (1984). (b) <Fe> versus Mg_2 with the same symbols, thin lines show gradients within individual ellipticals observed by Davies and Sadler (1992).

show that the Mg_2 and <Fe> strengths in nearby ellipticals observed by Davies and Sadler (1992) can be reproduced if M giants contribute roughly 10% of the V light in the nucleus and 5% at the half–light radius R_e. If this is correct, these galaxies probably have [Fe/H] in the range 0.0 to +0.3 in the nucleus and −0.3 to 0.0 at R_e.

In summary, it appears that optical spectra of the brightest ellipticals can be understood in terms of an old population alone. This does not appear to be true of less luminous ellipticals, and I suspect that as we go to lower luminosities we see evidence of progressively more recent star formation.

References

Arp, H., 1965. Astrophys. J., **141**, 43.

Bell, R.A. and Gustafsson, B., 1989. MNRAS, **236**, 653.

Bessell, M.S., 1979. PASP, **91**, 589.

Davies, R.L. and Sadler, E.M., 1992, in preparation.

Burstein, D., Faber, S.M., Gaskell, C.M. and Krumm, N., 1984. Astrophys. J., **287**, 586.

Faber, S.M., Friel, E.D., Burstein, D. and Gaskell, C.M., 1985. Astrophys. J. Suppl. Ser., **57**, 711.

Faber, S.M., Worthey, G. and Gonzalez, J.J., 1992, this meeting.

Gorgas, J., Efstathiou, G. and Aragon Salamanca, A., 1990. MNRAS, **245**, 217.

Rich, R.M., 1988. Astron. J., **95**, 828.

Taylor, B.J., 1991. Astrophys. J. Suppl. Ser., **76**, 715.

Terndrup, D.M., 1988. Astron. J., **96**, 884.

Walker, A.R. and Mack, P., 1986. MNRAS, **220**, 69.

Whitford, A.E., 1978. Astrophys. J., **226**, 777.

Discussion

RICH: I see two possibilities to account for the differences in mean metallicity between our two samples. First, the mean log abundance is lower than the log mean abundance. Second, my use of J−K, combined with the Cayrel standards, was a completely empirical approach. It will be interesting to see what abundances you derive for stars measured at high resolution. Finally, Terndrup's mean abundance was based on the giant branch colour. Bulge giants, however, are known to have colours too blue for their metal abundance.

SADLER: Yes, it is important to remember that $<[Fe/H]> \neq \log <z>$, where z is the abundance. I don't think the published <[Fe/H]> of roughly +0.08 for your sample (Rich 1988) is inconsistent with the value of 'close to zero' estimated from the models, but I have made the check you suggest by comparing abundance standard stars with the models. The agreement is good for hotter stars, but the models may underestimate [Fe/H] for the standards by 0.2–0.3 dex for T_{eff} below about 4800 K. Figure 1a compares a solar abundance model with the Faber et al. (1984) observed mean relation for solar abundance standards.

GUSTAFSSON: We certainly know that the models you have used are underblanketed in the UV and blue. This may cause scale errors in the abundances, possibly also depending on T_{eff}. In using saturated lines these problems are more severe. For more detailed future work on SMR stars, more recent opacity data such as those of Kurucz should be used.

FROGEL: (1) The model Whitford and I made predicted ~10% V light from M giants, in agreement with your estimate. (2) Terndrup, Whitford and I have rederived <[Fe/H]> for bulge M giants using atomic Na and Ca lines at 2.7 μm. We get ~ +0.3 in agreement with Rich and Whitford, but find no spread. As others have commented, one must allow for effects of variable reddening and selection biases between M and K stars.

MOULD: Just a reminder to observers of the abundance distribution in the bulge that the distribution you find will depend on selection effects. If you observe M stars you will tend to find metal–rich stars and if you observe blue stars you will tend to find low metallicity, just because this is the part of the HR diagram where these stars live. To estimate the distribution in the population, you have to correct for these colour or type selection effects.

PELETIER: One of the parameters that could move objects away from the relation that you showed between Mg_2 and <Fe> is the [O/Fe] ratio in galaxies, since Mg_2 is correlated with the O abundance. The [O/Fe] ratio is determined primarily by the timescale of formation. Have you thought about this?

SADLER: No, not in any detail. I made some simple calculations for M stars because we know they must be present in any old metal–rich population, but certainly there could be other effects as well.

PETERSON: The O/Fe ratio affects K giant spectra significantly at the metal–rich end. (1) O correlates with Mg. An Mg excess alters ionization fraction by its contribution of excess electrons, and makes neutral lines appear stronger. (2) O alters CN by greater CO formation. N/Fe also changes CN strength, and N/O is known to vary from HII work. So CN is very sensitive to abundance ratios of CNO elements.

SADLER: Your cautions are certainly valid. The attraction of CN is that it is much less affected by temperature than the other indices.

GREGG: (1) I agree with your assertion that as one goes from high to low luminosity ellipticals the amount of young or intermediate–age population increases, but this may be due in part to the presence of both high and low surface brightness ellipticals in the sample. The low surface brightness objects contain a younger population than the M32–like dwarfs which form a continuum with the more luminous high surface brightness ellipticals (see my contribution for details). (2) On the other hand, the work of Francois Schweizer shows that galaxies with more fine structure probably have larger proportions of young or intermediate age populations and these objects also tend to be the most luminous objects, in conflict with your suggestion that the brightest ellipticals have the oldest populations.

SADLER: The low luminosity galaxies I was discussing all fall into your high surface brightness category, so I don't think this is a problem. Most of the ellipticals which Roger Davies and I studied were in clusters, so this may have some effect.

SCHWEIZER: Whether one finds or does not find intermediate–age populations in ellipticals may depend on how one selects one's sample. In our study of correlations between line-strength indices and fine structure (Schweizer et al. 1990 Ap. J. Lett **364**, L33) we concentrated on field galaxies and avoided rich clusters. We find evidence for enhanced intermediate–age populations (enhanced $H\beta$, weak CN and Mg_2) in luminous ($M_B < -21.4$) ellipticals with much fine structure (e.g. NGC 3610, 5018, 3640, 596). There is some evidence that ellipticals in clusters may have formed earlier and may, therefore, have little or no intermediate–age population, as Elaine found.

GLOBULAR CLUSTERS OF THE INNER GALACTIC BULGE

Sergio Ortolani
Osservatorio Astronomico di Padova
Vicolo dell'Osservatorio 5
35122 Padova Italy

Abstract. Deep CCD photometry for a number of bulge-projected metal rich globular clusters and their nearby field background have been obtained . The V/V-I or I/V-I color-magnitude diagrams of the observed clusters are similar to the background fields and show high metallicity peculiar features. They are all very compact and slightly elongated.

1. Introduction

We recently started a program (Ortolani et al. 1990), to study, photometrically and spectroscopically the obscured, metal rich globular clusters projected in the direction of the inner galactic bulge clusters with the aim to investigate the high metallicity effects on the c-m diagrams morphology and to check the properties of the metal rich old population of the bulge. Here we present only the photometric results while the spectroscopy is discussed in the same volume by Barbuy and Bica.

Only little information is so far available for these clusters because of the difficulties in observing very dense stellar fields. Since the reddening in the direction of the Bulge is quite high, we took advantage of the high red sensitibity of the recent CCD detectors available at ESO La Silla to observe the most reddened clusters in the V and I bands using the color index V-I as temperature indicator instead of the classical B-V which is heavily affected by the B extinction. In some cases the observations have been carried out also in the extreme red band Gunn z.

In our previous study of the BVRI c-m diagrams of the bulge cluster NGC 6553 (Ortolani et al., 1990) we found a number of peculiar features connected with high metallicity effects such as the curvature of the red giant branch with a faint tip going down, in the visual band, almost at the level of the horizontal branch. This feature has been interpreted as due to blanketing effects in the cool metal rich giants. Recent high dispersion spectroscopic observations of a relatively bright giant in NGC 6553 (Barbuy et al., 1991) gave an almost solar metallicity, confirming the hypothesis that we are dealing with high metallicity clusters. The interpretation of the tilted, red horizontal branch, is, on the contrary, not yet clarified.

2. Observations

The observations have been obtained at La Silla European Southern Observatory with the 1.5 m Danish telescope equipped with the RCA CCDs ESO n.8 and n.5 which provide the highest near infrared quantum efficiency. The field of view projected on the sky of the single CCD image is about 3 x 4 arcminutes. The I band observations have been tied in the Cousins system through Landolt standard star observations.

Eight globular clusters and two open clusters have been observed in the direction of the inner bulge at galactic latitudes ranging from -4 to +4 degrees, with some of them within two degrees from the galactic plane, selected on the base of high metallicity indications coming from the literature or from the integrated spectra obtained from one of us (E.B.).

B. Barbuy and A. Renzini (eds.), The Stellar Populations of Galaxies, 47–50.

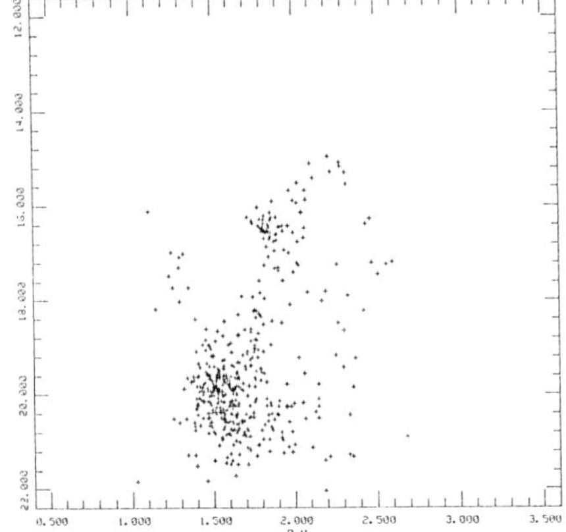

Fig. 1. NGC 6553 c-m deep diagram. Only the best 500 fitted stars have been selected.

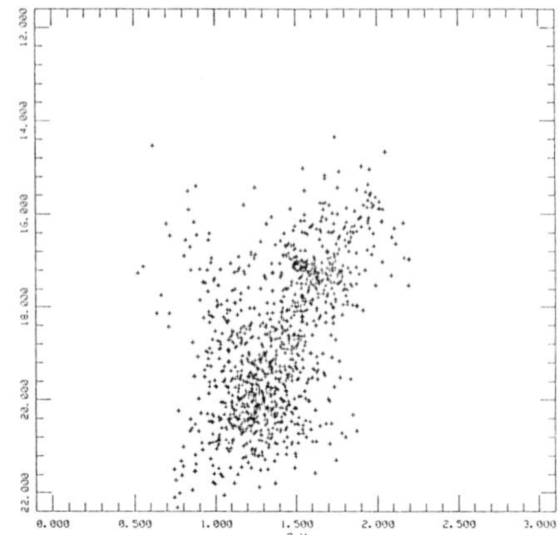

Fig. 2. NGC 6528 c-m diagram. Circular extraction of the inner 90" radius circle.

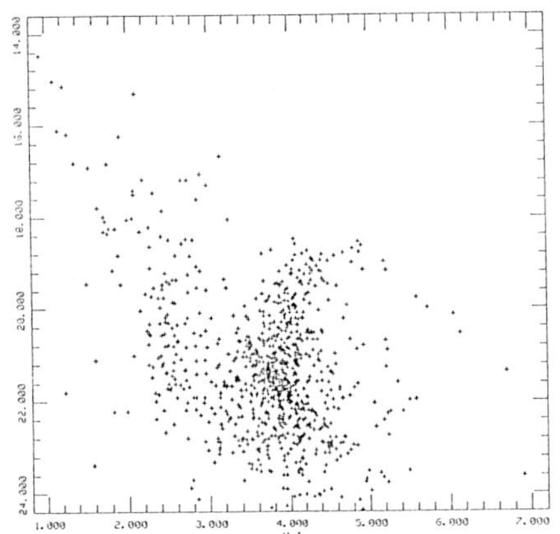

Fig. 3. Terzan 1 V/V-I c-m diagram.

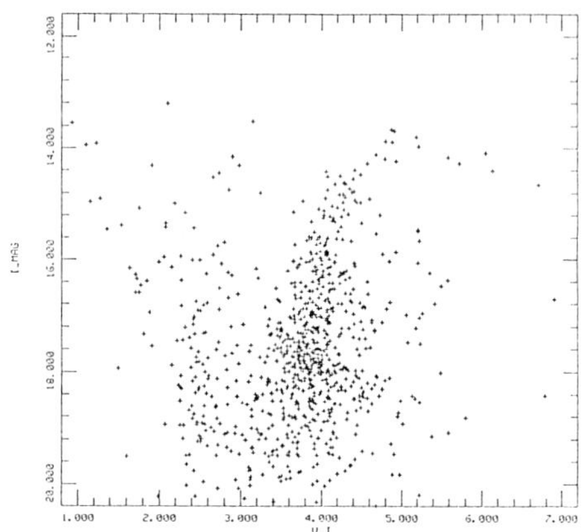

Fig. 4. Terzan 1 I/V-I diagram.

One degree projected on the sky corresponds at the distance of the Galactic center (8 kpc) to about 140 pcs.
Here we will discuss the results from three clusters (NGC 6553, NGC 6528 and Terzan 1) which have been completely reduced.

3. Results

In Fig. 1 a deep, selected c-m diagram of NGC 6553, containing only good quality measurements, is plotted. The main features of a very metal rich population are present: in particular the red giant turnover at about V=15, the B-V color saturation at about 2.4 due to blanketing effects and the red, tilted horizontal branch. The main sequence is also detected with the turnoff at about V=20 and B-V=1.5. The blue bright stars above the turnoff are main sequence stars from the contaminating foreground galactic disk field population. This diagram can be used as observational template for old metal rich populations.
The age can be obtained from the comparison with the c-m diagrams of other metal rich clusters (Ortolani et al., 1990) or from isochrone fitting.
Solar abundance isochrones have been recently calculated by VandenBerg and Laskarides (1987) and, using new overshooting models, by Bressan et al. (1991). While VandenBerg and Laskarides isochrones are calculated only in effective temperatures and bolometric luminosities, Bressan et al. transformed them into the observative plane in the UBVRI colors. The simultaneous fitting of the complete solar metallicity isochrone with t=15 BY and Y=0.28 is excellent from the main sequence up to the turnover of the giants where the transformation equations do not seem to be able to take into account the strong blanketing effects due to the molecular bands. At present there are no isochrones able to reproduce this feature. From the fitting a distance modulus (m-M)v=15 and a reddening E(B-V)=0.8 is obtained, almost coincident with the values found by empirical comparison with other metal rich cluster c-m diagrams. Due to the photometric spread at the turnoff and the uncertainties due to the assumption in the models (in particular the helium content and the CNO abundances) this result does not esclude a slightly younger age as found by Terndrup (1988) in his analysis of bulge field images.
In Fig. 2 the analogous c-m diagram for the cluster NGC 6528 is presented. This cluster is projected on the direction of the Baade Window, it is more concentrated than NGC 6553 and, at the same time, intrinsically fainter. Even if the images are better than those of NGC 6553 the diagram is not very well defined due to the more crowded field. The c-m diagram shows the same features of NGC 6553 and it is very similar to the diagrams obtained by Terndrup (1988) from images of selected fields in the Baade Window. Again the fitting with solar isochrones gives the same results previously obtained from empirical comparisons (Ortolani et al., 1991). For this cluster we have a distance modulus of about (m-M)v=15.9 and a reddening around E(B-V)=0.6. A distance of 7.5 Kpc from the Sun is derived, locating NGC 6528 in the inner bulge. It should be noted that NGC 6553, at 4.1 Kpc from the Sun is not so close to the Galactic Center.
The highly reddened cluster Terzan 1 is projected at only 1 degree from the Galactic plane. It is severely contaminated by field stars and, in particular, three bright, young disk main sequence stars are superimposed on the cluster at a few arcseconds from its center. For this cluster the integrated spectra or photometry in the visual-red bands can be seriously contaminated and the deduced reddening and metallicity should be considered with some care. Infrared measurements, where the flux from blue stars are reduced, are more reliable. The V, I c-m diagrams are presented in Fig. 3 and 4. The strong blanketing effect of the coolest giant stars, showing the turnover in both the diagrams, seems to suggest that we are dealing with a cluster probably more metallic than NGC 6553 and 6528. Another peculiarity is the considerable luminosity difference between the disk main sequence upper limit and the red giant turnover, indicating a strong absorption of the cluster stars. If the concentration of points located at about V=21.4 and V-I=3.8 is assumed as horizontal branch of the cluster, adopting an absolute magnitude of the HB around I=0 (Bressan et al., 1991), and the reddening derived from integrated IR photometry E(B-V)=1.5 (Malkan, 1982), we get a distance d=8.5 kpc, locating Terzan 1 approximately at the distance of the Galactic Center. These values are, however, incompatible with the fit of solar metallici-

ty isochrones, suggesting that the assumed HB position or other parameters (metallicity, helium abundance, reddening) could be wrong.
An alternative possibility is that the main concentration of points is just due to a crowding effect of the field population while the true horizontal branch of the cluster is considerably brighter, corresponding to the sparse clump located at about V=20.0 and V-I=3.8. In this case the isochrone fitting is quite good. The derived distance from the Sun is reduced to 2-3 Kpc and the reddening rises to about E(B-V)=2.3 (Av=7), clearly much higher than the currently adopted Malkan's value. If the higher value is adopted, considering that the reddest giants of Terzan 1 have B-V=4.1, a dereddened saturation color value of the giants of B-V=1.8 is obtained. It is easy to see from Fig. 1 and 2 that this value is almost identical for NGC 6553 and 6528.

4. Conclusions

From the analysis of three clusters projected on the direction of the inner Galactic bulge we found evidence of a number of common c-m diagram peculiar features due to blanketing effects connected with high metallicity (probably around the solar value), in spite of their line of sight distance spread. They are relatively old, heavily reddened and show a peculiar tilted HB. The dereddened color index saturation, for the giants, occurs at B-V=1.7-1.8 while the luminosity of the giants reaches about Mv=-0.1, -0.5.
The turnover of the red giants implies that the brightest stars of the old metal rich population, in the visual band, are not the coolest, as predicted by the current theoretical models. In the Cousins I band the luminosity of the coolest giants as a function of the temperature is almost constant because the reduced blanketing effects in this spectral range are not strong enough to produce the turnover at solar metallicity. For higher values, however, it seems that the turnover can be produced also in the I band (Ortolani et al., 1991).

REFERENCES

Barbuy B., Castro S., Ortolani S., Bica E. 1991. A & A submitted
Bressan A., Bertelli G., Chiosi C. 1991. This volume
Malkan M. 1982, in *Astrophysical Parameters for Globular Clusters*, eds. Hayes D.S., Philip A.G.D., Davis Press, p. 533
Ortolani S., Barbuy B., Bica E. 1990. A & A, 236, 362
Ortolani S., Barbuy B., Bica E. 1991. A & A, in press
Ortolani S., Barbuy B., Bica E. 1991. A & A Letters, in press
VandenBerg D.A., Laskarides P.G. 1987. Ap J Suppl., 64, 103
Terndrup D.M. 1988. A J, 96, 884

DYNAMICS OF THE GALACTIC SPHEROID

TIM DE ZEEUW
Sterrewacht Leiden
The Netherlands

ABSTRACT. The intrinsic shape and the kinematics of the Galactic bulge and the metal–poor stellar halo are discussed. There is strong evidence that the Galactic bulge is triaxial, and is rotating rapidly. The stellar halo is nearly round, rotates slowly, and has an anisotropic velocity distribution.

1. Introduction

The Galactic bulge and the metal–poor stellar halo are the two major constituents of the Galactic spheroid (Freeman 1987). A careful study of their photometric, kinematic and dynamical properties is required to answer questions such as: What kind of spiral galaxy do we live in? Can we think of the bulge as a small elliptical galaxy? What is the shape and extent of the dark halo? How was the Galaxy formed? An extensive and up–to–date discussion of these and related issues can be found in Freeman (1987), and in the Proceedings of the 1989 Saas Fee Advanced Course on *The Milky Way as a Galaxy* (Gilmore, King & van der Kruit 1990). In this review we concentrate on two specific questions that have received considerable attention in the past few years, namely, i) what is the intrinsic shape of each of the two components of the spheroid, and ii) what is their internal velocity structure?

2. The Galactic Bulge

2.1. INTEGRATED LIGHT

The Galactic bulge is heavily obscured at optical wavelengths, and is best studied in the near–infrared. Early measurements showed that the integrated surface brightness distribution between 2 and 2.4 micron is flattened, and has a central cusp (Becklin & Neugebauer 1968; Maihara *et al.* 1978; Matsumoto *et al.* 1982; Allen, Hyland & Jones 1983). A variety of functions have been proposed as fits to the observed surface brightness profile, including power laws, exponentials and de Vaucouleurs' profiles. Sellwood & Sanders (1988) reviewed much of the early work, and showed that the volume emissivity profile in the inner 500 parsec is well–described by a power–law with slope −1.8, while at larger radii the profile steepens considerably, and the slope approaches −3.7. Kent (1992) analyzed the Spacelab 2.4 micron data (Kent, Dame & Fazio 1991) and corrects for foreground contamination by the disk, which may amount to 40% of the observed light. The resulting bulge profile is similar to the

B. Barbuy and A. Renzini (eds.), The Stellar Populations of Galaxies, 51–60.

one proposed by Sellwood & Sanders, and has a characteristic major axis scale–length of about 900 pc. The bulge isophotes have an axis ratio of 0.61, and are slightly box–shaped. Inside one kpc the bulge may be more flattened.

Blitz & Spergel (1991b) have reanalyzed the balloon measurements of Matsumoto *et al.* (1982), and find that the bulge is in fact brighter on one side than on the other. They argue that this is not due to extinction variations, but is caused by the bulge being triaxial rather than oblate, with the near side at positive galactic longitude. Their analysis does not constrain the axis ratio of the bulge in the plane of the disk, nor does it give an accurate value for the angle ϕ between the long axis of the bulge and the line–of–sight to the Galactic center. Analysis of the COBE map of the Galaxy might improve this situation.

2.2. GAS KINEMATICS

It has been known for a long time that the inner rotation curve of the Galaxy displays a prominent hump. If ascribed to circular motion, it requires a very strong central density concentration (Bahcall, Schmidt & Soneira 1982). A more natural explanation is to assume that the central part of the Galaxy is not axisymmetric, but contains a triaxial bulge. The simple closed orbits available to the gas are now elongated, and the gas velocity varies along the orbit. When viewed from the proper direction, i.e., by proper orientation of the bulge, one may therefore observe gas velocities that are substantially higher than the local circular velocity, and hence see a hump in the rotation curve (e.g., Lake & Norman 1983). Gerhard & Vietri (1986) showed that the inner rotation curve can indeed be reproduced in this manner, for example by a prolate bulge with a stationary figure, seen nearly broadside on. Evidence for non–circular motions in the central regions of the Galaxy had been discussed earlier by Liszt & Burton (1980) in terms of a simple kinematic model which is in fact equivalent to motion on elliptic orbits (see Kent 1992).

Binney *et al.* (1991) have constructed the most comprehensive model to date of the motions of both the atomic and the molecular gas in the inner kpc of the Galaxy. These authors assume the gas moves on non–selfintersecting closed orbits, and show that a rapidly rotating inner bar with its corotation radius at 2.4 ± 0.5 kpc, an axis ratio in the plane of the Galaxy of 0.75, and seen nearly end–on at an angle ϕ to the line–of–sight equal to $16° \pm 2°$, provides a natural explanation for the observed kinematics of the gas. Designating the elongated central component as a bar or as a triaxial bulge may be a matter of semantics: the gas motions in the Galactic plane do not constrain the density distribution of the bar outside the plane, and, as Binney *et al.* point out, it is possible that the rapidly rotating bar is identical to the triaxial box–shaped bulge evident in the integrated light (§2.1). The bar and the bulge may also be separate Galactic components: an initially flat bar in a disk galaxy may grow fatter in time, become box–shaped, and either form a bulge, or a metal–rich component in a pre–existing bulge (Combes *et al.* 1990; Pfenniger & Norman 1990). Thus, the bulge and/or the bar may have formed separately from the halo.

Proper treatment of the gas kinematics probably requires use of a two–dimensional hydrodynamical code, because the closed orbit approximation breaks down when the orbits become very elongated, or when the gas switches from one orbit family to another. Early hydrodynamical studies that discuss the effects of a triaxial bulge on the Galactic HI kinematics include van Albada (1985) and Mulder & Liem (1986). The former author requires a bar which is rather similar to the one found by Binney *et al.* (1991), wheareas the latter authors require a bar which has its nearest side at negative rather than at positive longitude. Both studies use a rather crude approximation to the Galactic potential. It clearly is worthwhile to redo this kind of study, but now with a more up–to–date potential.

Much of the classical work on stars in the bulge has been restricted to a small number of special areas that are not heavily obscured by interstellar extinction, and hence gives only limited information on the shape and structure of the entire bulge (e.g., Frogel 1988). The IRAS mission has improved this situation dramatically. Habing *et al.* (1985) showed that a simple criterion based on the observed flux densities at 12 and 25 micron allows one to select evolved stars located near the tip of the AGB from the IRAS point source catalog. The resulting distribution of sources on the sky clearly shows the disk of the Galaxy, and the bulge (see also Habing 1988). The inferred luminosity profile (corrected for the effects of confusion near the Galactic plane) is consistent with that deduced from the integrated light measurements, and the measured axis ratio is about 0.7 (Harmon & Gilmore 1988).

The recent interest in triaxiality has spurred re-analysis of the properties of various populations of bulge stars. Nakada *et al.* (1991) investigated the luminosity function of a subsample of the IRAS AGB stars in the bulge, and found that the stars at positive longitudes are brighter on average than those at negative longitudes. Whitelock & Catchpole (1992) similarly find that the bulge Mira variables at positive longitudes have a mean distance smaller than those at negative longitudes, just as expected. The RR Lyrae stars show marginal evidence for a triaxial distribution also, but with its near end at negative longitude (Wesselink 1987; Le Poole & Habing 1990). It will be interesting to see whether this result holds up when a larger sample is studied, since unlike the Miras and the AGB stars, the RR Lyraes presumably belong to the metal-poor halo. This may itself be triaxial, and indeed elongated in a direction opposite to the bulge (Blitz & Spergel 1991a; §3.1).

Weinberg (1992) has re-analyzed the IRAS point source catalog, considered only objects with galactic latitude $|b| < 3°$, and used selection criteria based on the IRAS colors which differ slightly from those employed by Habing *et al.* (1985). He assumes that all these stars have the same absolute bolometric luminosity, calculates individual bolometric corrections based on the observed IRAS colors and some assumptions about the spectral energy distribution, and furthermore assumes a uniform extinction throughout the Galaxy. This yields distances for all the objects, and results in a map of AGB stars in the Galaxy. The distribution is not axisymmetric, but appears to be lopsided. Weinberg argues that this is due to the apparent luminosity cutoff of his sample, which means that hardly any stars beyond 10 kpc are included. Analysis of a deeper, but incomplete, sample shows a more symmetric bar-like distortion, again with the near side at positive longitudes. Instead of fitting a specific model to the distribution of stars, Weinberg calculates the coefficients of the harmonic expansion that fits the data best. He reconstructs a smooth density distribution from his expansion coefficients, ignoring the asymmetric terms. This yields a symmetric bar, out to about 5 kpc, with the near side at positive longitude, an axis ratio in the plane of the Galaxy of about 0.6, and seen at an angle ϕ of 36°. Varying the prescriptions for the bolometric correction and for the extinction, and allowing for a spread in intrinsic luminosities of the AGB stars, does not change the main result—the existence of a bar—but does influence its properties. Specifically, the viewing angle ϕ may well be somewhat smaller, meaning the bar is seen more end-on, in agreement with the rotating bar of Binney *et al.* (1991). It remains to be seen whether the spatial extent of 5 kpc found by Weinberg can be reconciled with the much smaller size found by Binney *et al.* It is also not clear to what extent confusion of sources in the Galactic plane influences this result.

Weinberg's analysis is restricted to the Galactic plane. It should be extended to three dimensions as soon as larger samples of stars in the bulge are available. The proposed 2MASS (Kleinmann 1991) and DENIS (Epchtein & Burton 1992) sky surveys at 2 micron will see every AGB star in the Galaxy, and are the ideal datasets for doing this problem.

2.4. STELLAR KINEMATICS

Stars at the tip of the AGB are surrounded by an expanding dust shell, and can be readily identified by means of their OH emission at 18 cm. The spherical geometry of the shell causes the line profiles to be double–peaked, so the measured velocity width is twice the expansion velocity of the shell, and the mean velocity is the radial velocity of the embedded star. Because radio measurements are not hampered by Galactic obscuration, these OH/IR stars may well be the best set of tracers to study the kinematics of the entire bulge.

Early radio surveys of the inner bulge (Habing *et al.* 1983; Winnberg *et al.* 1985) found a few dozen OH/IR stars, and showed that the bulge may well rotate rapidly, with rotation velocities of about 100 km/s at less than 100 pc from the center, and with a radial velocity dispersion between 60 and 140 km/s. Lindqvist *et al.* (1991) have recently completed a large VLA survey, and now have 134 OH/IR stars in the central 100 pc. The earlier fast rotation is confirmed, and the dispersion about the mean motion is found to be of the order of 80 km/s (Habing *priv. comm.*). This value for the dispersion is lower than the 110 km/s for the giants in Baade's window (Sharples, Walker & Cropper 1990; Tyson & Rich 1990), but compares well with that derived for Mira variables (Catchpole 1990, Menzies 1990) and planetary nebulae (Kinman, Feast & Lasker 1988), which are in evolutionary stages bracketing the OH/IR stage. Both the Miras and the planetary nebulae show evidence for rotation, but these populations reach rotational velocities of 80 km/s only at about 750 pc from the Galactic Center. This appears to differ from the fast rotation of the OH/IR stars, but may be due to the customary but suspect fitting of linear rotation curves to samples that go out to different distances from the center. It is also possible that the OH/IR sample includes a population of disk objects. The rotation of the Miras and the planetary nebulae is consistent with the figure rotation expected for the bar invoked by Binney *et al.* (1991) to explain the gas kinematics in the inner kpc.

A number of OH 18 cm surveys of IRAS point sources have been carried out (Eder, Lewis & Terzian 1988; te Lintel–Hekkert *et al.* 1989; Sivagnanam *et al.* 1990), resulting in radial velocities for about 1500 OH/IR stars in the Galaxy. This sample is incomplete near the Galactic plane. Te Lintel–Hekkert & Dejonghe (see te Lintel–Hekkert 1990) have modeled the observed kinematics by choosing a potential for the Galaxy, and then asking what intrinsic velocity distribution is consistent with the observed kinematics. So far, they have considered oblate models in which the velocity dispersion is isotropic in the meridional plane. They find evidence for two components in their models, and identify one with the old disk, and the other with a "thick disk plus bulge". Although the sampling of the bulge is incomplete, the models give mean streaming motions (rotation) of about 100 km/s at 1 kpc, and a velocity dispersion of well over 100 km/s.

The method used by te Lintel–Hekkert & Dejonghe is quite flexible. Radial velocities from different samples—notably the Lindqvist *et al.* (1991) data—can be included, provided the selection effects are understood. The effects of an anisotropic velocity distribution can be incorporated easily, and may be tested by use of proper motion surveys such as done in Baade's window (Spaenhauer, Jones & Whitford 1992). Models with a stationary triaxial bulge can be considered also, but inclusion of figure rotation will require a considerable numerical effort. Ideally, one would like to take the mass model of Binney *et al.* (1991) and then investigate whether the radial velocities of different populations are consistent with each other, or whether they indicate truly different Galactic components (see Rich 1990). This should also settle the issue of whether the bulge formed directly by gaseous infall from the halo, or is an inward extension of the (thick) disk (see e.g., Lewis & Freeman 1989; Carney, Latham & Laird 1990). Such a project is feasible, and would be helped considerably by a complete radio survey of the OH/IR stars in the bulge.

2.5. OTHER BULGES

Various lines of evidence suggest that spiral bulges as a class are not oblate. The position angles of the apparent major axis of the bulge and the disk of spiral galaxies often differ from each other. This is a natural consequence of triaxiality, and is caused by a projection effect. A well-known example is the bulge of M31 (Stark 1977). Bertola, Vietri & Zeilinger (1991) studied a sample of 32 bulges, and showed that if disks are round, then bulges as a class are indeed triaxial, and have shapes similar to elliptical galaxies. The derived distribution of shapes may be incorrect, however, as it is now clear that photometrically the disks of spirals are not round, but instead are slightly elongated, with an axis ratio close to 0.9. This issue was reviewed recently by Kuijken & Tremaine (1991; see also Franx & de Zeeuw 1992). Derivation of the intrinsic shapes of bulges will require inclusion of kinematic data, just as was done for ellipticals (Binney 1985; Franx, Illingworth & de Zeeuw 1991).

Stellar absorption line measurements of spiral bulges are consistent with rotationally supported axisymmetric models, when the disk potential is taken into account (Jarvis & Freeman 1985; Kent 1989). However, the data are consistent also with triaxial shapes with substantial internal streaming, and/or figure rotation. Individual bulges also show signs of triaxiality. The regular gas velocity field of NGC 4845 is well-fit by motion on elongated closed orbits in a triaxial bulge, with axis ratios $b/a = 0.74 \pm 0.06$ and $c/a = 0.60 \pm 0.06$ (Bertola, Rubin & Zeilinger 1989; Gerhard, Vietri & Kent 1989). The various recent indications that the Galaxy contains a triaxial bulge are therefore fully in line with what we know about other spiral bulges.

3. The Metal-Poor Halo

3.1. STARCOUNTS

The Galactic stellar halo consists of metal-poor stars, with [Fe/H] $\lesssim -1.0$. Starcounts have shown that the density profile of the stellar halo is proportional to $r^{-3.5}$ out to about 20 kpc, and is possibly steeper beyond this distance (e.g., Saha 1985). The flattening of the halo deduced from starcounts is modest, with an axis ratio c/a between 0.6 and 0.8 (Gilmore, Wyse & Kuijken 1989).

Blitz & Spergel (1991a) suggested that many of the asymmetries in the Galactic HI velocity field can be explained by assuming that the spheroid is triaxial, and has a slowly rotating figure. The inferred axis ratio b/a in the plane of the disk is larger than 0.87, and the near side is at negative longitudes, with the long axis at roughly 45° from the line-of-sight to the Galactic center. This spheroid is therefore distinct from the central triaxial bar which is aligned in the opposite direction (§2). On the other hand, Kuijken (1991) has argued convincingly that a model in which the Galaxy is slightly lopsided, and has a round central bulge, gives at least as good a description of the HI velocities (see Kuijken & Tremaine 1991). It will be interesting to see whether detailed starcounts in a number of well-chosen directions will be able to confirm or rule out either of these suggestions.

3.2. STELLAR KINEMATICS

Ratnatunga & Freeman (1985, 1989) showed that the velocity dispersion of K giants in the direction of the South Galactic Pole remains constant out to large distances. This is a surprising result. In the solar neighbourhood the observed line-of-sight velocity dispersion comes from the motion perpendicular to the Galactic plane. As the distance above the

Galactic plane increases, the radial motion contributes more and more to the observed velocity dispersion. Since there is no reason to assume that the velocity dispersions in the radial direction and the direction perpendicular to the Galactic plane are equal—they differ in the solar neighborhood—one expects the observed velocity dispersion to vary. Not surprisingly, a model with constant anisotropy and a velocity ellipsoid aligned with spherical coordinates is inconsistent with the data. Ratnatunga & Freeman concluded that a constant anisotropy model can give a good fit to the data, but only if the velocity ellipsoid is aligned with cylindrical coordinates. This would imply that the halo is very flat, with $c/a < 0.5$, i.e., much flatter than deduced from starcounts (see also Binney, May & Ostriker 1987).

The value of c/a deduced from kinematic observations has gone up in the past few years, mainly due to corrections to the measurements. Morrison, Flynn & Freeman (1990) now take the local dispersion in the perpendicular direction to be 100 km/s, and argue that the lower value of 75 km/s used earlier was caused by inclusion of thick disk stars. Van der Marel (1991) has used the Jeans equations, considers both cylindrical and spherical alignment of the velocity ellipsoid, and assumes that the velocity anisotropy of the halo does not vary much with position. He finds that the starcounts and the kinematic data are consistent if the dark halo, which provides the gravitational potential, has $c/a > 0.34$. If the stellar and the dark halo have the same shape the constraint is $c/a > 0.53$.

A number of authors have constructed detailed dynamical models for the metal-poor halo based on phase-space distribution functions (White 1985; Sommer-Larsen 1987; Levison & Richstone 1986; Dejonghe & de Zeeuw 1988; Sommer-Larsen & Christensen 1989; Arnold 1990; Vedel & Sommer-Larsen 1990). Both spherical and oblate axisymmetric models with anisotropic velocity distributions have been considered, and they have been compared with kinematic data along a number of directions out of the Galactic plane. The best fitting models have a halo velocity distribution which is roughly isotropic in the central regions, somewhat radially anisotropic between 6 and 10 kpc, and tangentially anisotropic beyond 10 kpc. The evidence for tangential anisotropy at very large radii ($\sim$60 kpc) is less convincing (Arnold 1991; Norris & Hawkins 1991). The halo density distribution in these models has two components, one nearly spherical, with $c/a \gtrsim 0.7$, and the other more flattened, with a scale-height of less than 3 kpc. The distribution of RR Lyrae stars also shows evidence for two such halo components (Hartwick 1987).

Instead of measuring radial velocities for stars at large distances in the halo, one can also do a more comprehensive study of the metal-poor stars in the solar vicinity, by measuring not only their radial velocities but also their proper motions. The resulting space motions can be used to calculate the associated stellar orbits in a plausible Galactic potential. The fraction of its orbit that each star spends in the solar neighbourhood is then known, so that the observed number density of halo stars can be corrected to yield the total density distribution of all halo stars represented by the sample observed near the Sun. This approach was advocated by May & Binney (1986), and is sometimes referred to as doing halo observations from an armchair. It has recently been implemented by Sommer-Larsen & Zhen (1990). These authors use a separable Galactic potential, so that three integrals of motion are known analytically, and computation of the orbits and the correction factors is straightforward. They find that the stars that pass through the solar neighbourhood are representative of about 90% of the known halo population between 8 and 20 kpc, that these indeed have the $r^{-3.5}$ density law derived from direct starcounts, and that they can be divided into two components, one nearly spherical with $c/a \sim 0.85$ and the other substantially flattened, and possibly with a higher mean metallicity. The velocity anisotropy is found to vary with radius. These results agree remarkably well with the earlier studies using more distant samples. It will be interesting to include other local samples in this analysis (e.g., Sandage & Fouts 1987; Norris & Ryan 1989; Nissen & Schuster 1991).

The metal–poor halo rotates rather slowly, with $v_{rot} \leq 40$ km/s (Norris 1986). This is indicated also by the kinematics of globular clusters. Whereas the disk globular clusters (with [Fe/H]> −0.8) form a reasonably fast rotating dynamical subsystem, the halo globular clusters (with [Fe/H]< −0.8) show almost no evidence for rotation (Zinn 1985; Armandroff 1989). This is consistent with a nearly spherical rotationally supported halo, but is equally consistent with a triaxial halo that is pressure supported, and has a nearly stationary figure (see also Long, Ostriker & Aguilar 1991).

If the stellar halo is triaxial, then the lack of figure rotation means that one can construct extensions of the available oblate dynamical models by using a separable triaxial potential for the Galaxy. The tools for doing this are available (e.g., Dejonghe & Laurent 1991). Such a modeling effort must be supported by additional kinematic data. The most recent dynamical models for the stellar halo are constrained by radial velocity measurements in a modest number of directions, and in some of these by data in only one or two distance bins (e.g., Arnold 1990; Vedel & Sommer–Larsen 1990). A major improvement in this situation will come from the inclusion of proper motion measurements for distant objects, such as carried out by Majewski (1991) and others. The triaxial models may also shed light on Majewski's suggestion that the stellar halo appears to be counter–rotating.

It is not evident that the customary procedure of first binning the kinematic data and then comparing the resulting averages with dynamical models is the best approach. There have been persistent suggestions that there is velocity clumping in the halo (e.g., Doinidis & Beers 1989), possibly due to the fact that small stellar groups drizzle in from larger radii, and leave their kinematic signature on the observed halo populations (Freeman 1987). Although some theoretical work has been done on this issue (Quinn & Goodman 1986), there is room for further careful N–body simulations, and also for an investigation of the consequences of velocity clumping on kinematic studies of the halo, and the associated dynamical modeling with smooth distributions.

3.3 OTHER GALAXIES

The properties of the Galactic metal–poor halo are consistent with what we know about the stellar halos of nearby normal galaxies. The metal poor globular clusters in M31 form a slowly rotating nearly round system (Elson & Walterbos 1988). The globular cluster systems of nearby elliptical galaxies are also nearly round and show marginal evidence for rotation (Mould *et al.* 1990). Radial velocity measurements of planetary nebulae in the halos of nearby galaxies promise to improve our knowledge of the velocity distribution in these systems substantially (Ford *et al.* 1989).

The dark halos of spiral galaxies, which dominate their gravitational potential, are likely to be triaxial. This is suggested by numerical simulations of galaxy formation (Frenk *et al.* 1988; White & Ostriker 1990; Dubinski & Carlberg 1991), which invariably result in prolate/triaxial dark halos. Luminous elliptical galaxies are triaxial as a class, but the majority of them is oblate/triaxial (Franx, Illingworth & de Zeeuw 1991). This may indicate limitations in the numerical simulations, but may also be caused, for example, by the effects of dissipation during the formation of the luminous part of a galaxy. By analogy, one expects that the stellar halos of spiral galaxies are triaxial, but that the dark halo simulations may be a poor guide to their shape. It is unlikely that we will be able to establish the triaxiality of nearby stellar halos directly any time soon, although misalignments between the kinematic and photometric axes of, e.g., the planetary nebula systems would be a strong indication. Spiral disks which are embedded in a triaxial halo are slightly elongated. Better constraints on the shapes of the dark halos therefore come from a careful analysis of the photometry and kinematics of the disks (see Franx & de Zeeuw 1992).

4. Conclusions

Observations of the integrated light, starcounts, and measurements of the kinematics of the atomic and molecular gas in the inner region of the Galaxy all indicate that the Galactic bulge is triaxial, with its near side at positive longitude, and its long axis close to the line-of-sight to the Galactic center. The COBE observations, and the starcounts to be done with the proposed 2 micron surveys, will further delineate the shape and orientation of the bulge. The consequences of triaxiality for the stellar kinematics of the bulge have not been explored yet, mainly because at present velocities are available for a relatively modest number of bulge stars. A systematic survey of OH/IR stars in the entire bulge will improve this situation dramatically.

Starcounts indicate that the metal-poor halo is slightly flattened, and has an axis ratio of about 0.8. Despite earlier reports to the contrary, the measurements of radial velocities in a number of directions out of the Galactic plane are consistent with a nearly round stellar halo, in which the velocity anisotropy varies with radius. Further improvements in the dynamical models for the halo are possible, notably inclusion of the effects of triaxiality, but they will require a substantial observational effort, including starcounts, radial velocity measurements, and proper motion studies. It is not clear whether the halo population is dynamically well-mixed. This issue needs urgent attention.

When combined with abundance measurements, kinematic studies will provide significant constraints on the formation history of the various components of the Galactic spheroid.

It is a pleasure to acknowledge enlightening discussions with Richard Arnold, Joris Blommaert, Marijn Franx, Ken Freeman, and Roeland van der Marel.

References

Allen, D.A., Hyland, A.R., & Jones, T.J., 1983. MNRAS, **204**, 1145.
Armandroff, T.E., 1989. AJ, **97**, 375.
Arnold, R., 1990. MNRAS, **244**, 465.
Arnold, R., 1991. *PhD Thesis*, Cambridge.
Bahcall, J.N., Schmidt, M., & Soneira, R., 1982. ApJ, **258**, L23.
Becklin, E.E., & Neugebauer, G., 1968. ApJ, **151**, 145.
Bertola, F., Rubin, V.C., & Zeilinger, W.W., 1989. ApJ, **345**, L29.
Bertola, F., Vietri, M., & Zeilinger, W.W., 1991. ApJ, **374**, L13.
Binney, J.J., 1985. MNRAS, **212**, 767.
Binney, J.J., May, A., & Ostriker, J.P., 1987. MNRAS, **226**, 156.
Binney, J.J., Gerhard, O.E., Stark, A.A., Bally, J., & Uchida, K.I., 1991. MNRAS, **252**, 210.
Blitz, L., & Spergel, D.N., 1991a. ApJ, **370**, 205.
Blitz, L., & Spergel, D.N., 1991b. ApJ, **379**, 631.
Carney, B.W., Latham, D.W., & Laird, J.B., 1990. AJ, **99**, 572.
Catchpole, R.M., 1990. In Proceedings of the ESO/CTIO Workshop on *Bulges of Galaxies*, eds B.J. Jarvis & D.M. Terndrup, p. 111,
Combes, F., Debbash, F., Friedli, D., & Pfenniger, D., 1990. A&A, **233**, 82.
Dejonghe, H., & de Zeeuw, P.T., 1988. ApJ, **329**, 720.
Dejonghe, H., & Laurent, D., 1991. MNRAS, **252**, 606.
Doinidis, S.P., & Beers, T.C., 1989. ApJ, **340**, L57.
Dubinski, J., & Carlberg, R.G., 1991. ApJ, **378**, 496.
Eder, J., Lewis, B.M., & Terzian, Y., 1988. ApJS, **66**, 183.

Elson, R.A.W., & Walterbos, R.A.M., 1988. ApJ, **333**, 594.

Epchtein, N., & Burton, W.B., 1992. This volume.

Ford, H.C., Ciardullo, R., Jacoby, G.H., Hui, X., 1989. In *IAU Symposium 131, Planetary Nebulae*, ed. S. Torres–Peimbert (Dordrecht: Reidel), p. 335.

Franx, M., Illingworth, G.D., & de Zeeuw, P.T., 1991. ApJ, **383**, 112.

Franx, M., & de Zeeuw, P.T., 1992. ApJ, submitted.

Freeman, K.C., 1987. ARAA, **25**, 603.

Frenk, C.S., White, S.D.M., Davis, M., & Efstathiou, G., 1988. ApJ, **327**, 507.

Frogel, J.A., 1988. ARAA, **26**, 51.

Gerhard, O.E., & Vietri, M., 1986. MNRAS, **223**, 377.

Gerhard, O.E., Vietri, M., & Kent, S.M., 1989. ApJ, **345**, L33.

Gilmore, G., Wyse, R.F.G., & Kuijken, K., 1989. ARAA, **27**, 555.

Gilmore, G., King, I.R., & van der Kruit, P.C., 1990. *The Milky Way as a Galaxy*, 19th Saas Fee Advanced Course, eds R. Buser & I.R. King (Mill Valley: Univ. Science Books).

Habing, H.J., 1988. A&A, **200**, 40.

Habing, H.J., Olnon, F.M., Winnberg, A., Matthews, H.E., & Baud, B., 1983. A&A, **128**, 230.

Habing, H.J., Olnon, F.M., Chester, T., Gillett, F., Rowan–Robinson, M., & Neugebauer, G., 1985. A&A, **152**, L1.

Harmon, R., & Gilmore, G., 1988. MNRAS, **235**, 1025.

Hartwick, F.D.A., 1987. In *The Galaxy*, eds G. Gilmore & B. Carswell (Dordrecht: Reidel), p. 281.

Jarvis, B.J., & Freeman, K.C., 1985. ApJ, **295**, 324.

Kent, S.M., 1989. AJ, **97**, 1614.

Kent, S.M., 1992. ApJ, submitted.

Kent, S.M., Dame, T.M., & Fazio, G., 1991. ApJ, **378**, 131.

Kinman, T.D., Feast, M.W., Lasker, B.M., 1988. AJ, **95**, 804.

Kleinman, S.G., 1991. In *Robotic Telescopes in the 1990s*, ASP Conference Series, ed. A.V. Filippenko, in press.

Kuijken, K., 1991. In *Warped Disks and Inclined Rings around Galaxies*, eds S. Casertano, P.D. Sackett, & F. Briggs (Cambridge University Press), p. 159.

Kuijken, K., & Tremaine, S.D., 1991. In *Dynamics of Disk Galaxies*, ed. B. Sundelius (Göteborg, Sweden), p. 71.

Lake, G., & Norman, C.A., 1983. ApJ, **270**, 51.

Le Poole, R.S., & Habing, H.J., 1990. In Proceedings of the ESO/CTIO Workshop on *Bulges of Galaxies*, eds B.J. Jarvis & D.M. Terndrup, p. 33.

Levison, H.F., & Richstone, D.O., 1986. ApJ, **308**, 627.

Lewis, & Freeman, K.C., 1989, AJ, **97**, 139.

Liszt, H.S., & Burton, W.B., 1980. ApJ, **236**, 779.

Long, K., Ostriker, J.P., & Aguilar, L., 1991. Preprint.

Lindqvist, M., Winnberg, A., Habing, H.J., & Matthews, H.E., 1991. Preprint.

Maihara, T., Oda, N., Sugiyama, T., Okuda, H., 1978. PASJ, **30**, 1.

Majewski, S.R., 1991. *PhD Thesis*, University of Chicago.

Matsumoto, T., Hayakawa, S., Koizumi, H., Murakawa, H., 1982. In *The Galactic Centre*, AIP Conf. 83, eds G.R. Riegler & R.D. Blandford (New York: Am. Inst. of Physics), p. 48.

May, A., & Binney, J.J., 1986. MNRAS, **221**, 857.

Menzies, J.W., 1990. In Proceedings of the ESO/CTIO Workshop on *Bulges of Galaxies*, eds B.J. Jarvis & D.M. Terndrup, p. 115.

Morrison, H.L., Flynn, C., & Freeman, K.C., 1990. AJ, **100**, 1191.

Mould, J.R., Oke, J.B., Nemec, J.M., de Zeeuw, P.T., 1990. AJ, **99**, 1823.

Mulder, W.A., & Liem, B.T., 1986. A&A, **157**, 148.

Nakada, Y., Deguchi, S., Hashimoto, O., Izumiura, H., Onaka, T., Sekiguchi, K., & Yamamura, I., 1991. Nature, **353**, 140.
Nissen, P.E., & Schuster, W.J., 1991. A&A, **251**, 457.
Norris, J., 1986. ApJS, **61**, 667.
Norris, J., & Ryan, S.G., 1989. ApJ, **340**, 739.
Norris, J., & Hawkins, M.R.S., 1991. ApJ, **380**, 104.
Pfenniger, D., & Norman, C., 1990. ApJ, **363**, 391.
Quinn, P.J., & Goodman, J., 1986. ApJ, **309**, 427.
Rich, R.M., 1990. ApJ, **362**, 604.
Ratnatunga, K.U., & Freeman, K.C., 1985. ApJ, **291**, 260.
Ratnatunga, K.U., & Freeman, K.C., 1989. ApJ, **339**, 126.
Saha, A., 1985. ApJ, **289**, 310.
Sandage, A., & Fouts, G., 1987. AJ, **93**, 74.
Sellwood, J.A., & Sanders, R.H., 1988. MNRAS, **233**, 611.
Sharples, R., Walker, A., Cropper, M., 1990. MNRAS, **246**, 54.
Sivagnamam, P., Braz, M.A., Le Squeren, A.M., & Tran Minh, F., 1990. A&A, **233**, 112.
Sommer–Larsen, J., 1987. MNRAS, **227**, 1P.
Sommer–Larsen, J., & Christensen, P.R., 1989. MNRAS, **239**, 441.
Sommer–Larsen, J., & Zhen, C., 1990. MNRAS, **242**, 10.
Spaenhauer, A., Jones, B.F., & Whitford, A.E., 1992. AJ, **103**, 297.
Stark, A.A., 1977. ApJ, **213**, 368.
te Lintel–Hekkert, P., 1990. *PhD Thesis*, University of Leiden.
te Lintel–Hekkert, P., Versteege–Hensel, H.A., Habing, H.J., & Wiertz, M., 1989. A&AS, **78**, 399.
Tyson, N.D., & Rich, R.M., 1990. In Proceedings of the ESO/CTIO Workshop on *Bulges of Galaxies*, eds B.J. Jarvis & D.M. Terndrup, p. 119.
van Albada, G.D., 1985. In *IAU Symposium 106, The Milky Way Galaxy*, eds H. van Woerden, R.J. Allen, & W.B. Burton (Dordrecht: Reidel), p. 547.
van der Marel, R.P., 1991. MNRAS, **248**, 515.
Vedel, H., & Sommer–Larsen, J., 1990. ApJ, **359**, 104.
Weinberg, M.D., 1992. ApJ, **384**, 81.
Wesselink, Th., 1987. *PhD Thesis*, Nijmegen.
White, S.D.M., 1985. ApJ, **294**, L99.
White, S.D.M., & Ostriker, J.P., 1990. ApJ, **349**, 22.
Whitelock, P., & Catchpole, R., 1992. This volume.
Winnberg, A., Baud, B., Matthews, H.E., Habing, H.J., & Olnon, F.M., 1985. ApJ, **291**, L45.
Zinn, R., 1985. ApJ, **293**, 424.

Discussion

Ostriker: If the bulge is triaxial then low angular momentum orbits will be box like and at some point pass close to the Galactic Center. This destroys globular clusters in the inner few kiloparsecs with box–like orbits and then the survivors will be, as observed, a relatively rapidly rotating population (i.e., on tube orbits).

de Zeeuw: Agreed.

STRUCTURE AND KINEMATICS AT THE NORTH GALACTIC POLE

S. R. MAJEWSKI
The Observatories of the Carnegie Institution of Washington
813 Santa Barbara Street
Pasadena, CA 91101 U.S.A.

ABSTRACT. Results from a complete survey of proper motions to $B = 22.5$ at the North Galactic Pole are summarized. Evidence from this and other surveys indicates that (1) the thick disk may extend to as much as 5.5 kpc above the Galactic plane at the solar radius, and (2) the thick disk may contain stars with metallicities as low as [Fe/H] = -1.6 or lower. These two properties of the thick disk mean that surveys of halo stars risk serious contamination by thick disk stars unless very conservative selection criteria are used. Applying these conservative selection criteria to existing surveys of halo stars reveals a surprising result - namely, that the halo is in retrograde rotation.

1. The Mayall 4-m Proper Motion Survey

We are conducting a deep astrometric survey (Majewski 1990, 1992) which relies on KPNO 4-m photographic plates to obtain high quality ($\sigma_\mu < 0.10''$ cent^{-1} to $B = 22.0$) proper motions for magnitude limited ($B \leq 22.5$) samples of stars in key Galactic fields. We have now completed work on a 0.3 deg^2 field at the North Galactic Pole (NGP). Multicolor UBV photometry has been used to provide estimates of abundance (through uv-excess) and distance (through photometric parallaxes) for a subsample ($U < 21.5$) of blue ($0.3 \leq B-V \leq 1.1$) dwarf stars extending to distances of tens of kiloparsecs above the Galactic plane. The proper motions have been tied to an extragalactic reference frame of 139 galaxies and QSOs to better than $0.01''$ cent^{-1}. Most surveys of halo dwarfs have been limited to the solar neighborhood, with samples defined by high proper motions or low metallicity. However, it has become clear (Carney *et al.* 1990, Norris & Ryan 1989b, Morrison *et al.* 1990, Majewski 1990) that separation of the various stellar populations on the basis of kinematics and/or abundance is fraught with difficulties because the distributions of these properties for the thin disk, thick disk, and halo have significant overlap. Our deep survey allows study of the kinematical and chemical properties of the Galactic thick disk and halo from a complete sample of field dwarfs *in situ*.

2. Properties of the Thick Disk

An important result from this survey is that a relatively sharp break is seen in the distribution of ultraviolet excesses at a distance of about 5.5 kpc above the Galactic plane. A similar metallicity break is also seen in the K giant data of Ratnatunga & Freeman (1989). Furthermore, the distribution of u and v velocities (stellar motions parallel to the Galactic plane) in our survey also show a break at this same distance. Together, these data suggest a sudden change in the mix of stellar populations from thick disk to halo, and that the thick disk extends to a relatively large distance from the Galactic plane, $z \sim 5.5$ kpc. On the near side of this break, the thick disk stars are observed to have a linear variation in asymmetric

B. Barbuy and A. Renzini (eds.), The Stellar Populations of Galaxies, 61–64.

drift with z, but no metallicity gradient. Although it is beyond the scope of this discussion, we note that this is a strange result since it seems to imply a dissipational origin for the thick disk on the one hand, and a more chaotic origin on the other. A key point for the present discussion is that the thick disk is found to have a *broad* distribution in metallicity, with stars as metal-poor as [Fe/H] ~ -1.6, and possibly even more metal weak. The presence of these metal-weak thick disk stars was first demonstrated by Morrison *et al.* (1990). Recently, Rees & Cudworth (1991) have determined that the metal weak ([Fe/H] < -1.3) globular cluster M28 has thick-disk-like kinematics; if the globular cluster system has anything to do with the field star population, this might be further evidence of a metal-weak tail in the thick disk.

3. Properties of the Halo

In keeping with a number of other recent studies of the halo, our data show no metallicity gradient to distances tens of kiloparsecs above the Galactic plane. The dispersion in metallicity is quite large, and, as found by Carney *et al.* (1990), there exist in our sample halo stars with metallicities as high as [Fe/H] ~ -0.5. The most surprising result of our astrometric survey, however, is that the mean reflex velocity of the halo is measured to be approximately -275 km s^{-1}, with no variation with z-distance. Best estimates of the rotational velocity of the Local Standard of Rest (LSR) yield values near 220 km s^{-1}. The H I velocity work of Gunn *et al.* (1979) gives 220 ± 10 km $s^{-1}$for this value and is generally regarded as definitive. Thus our measured reflex velocity implies that the halo is in *retrograde* rotation about the Galaxy (at least in the direction of the NGP).

4. Is the Halo in Retrograde Rotation?

Until recently, two types of surveys have been used to derive the value of the rotational velocity of the Galactic halo. The first type of survey selects halo stars from the solar neighborhood based on either kinematics or metallicity. Great care must be used in interpreting velocity data from kinematically-selected samples, as has been stressed by Norris & Ryan (1989b). Unless the selection bias is well understood, in general it is preferable to rely on non-kinematically biased samples. However, selection of halo stars on the basis of low metallicity must take into account the presence of the metal-weak thick disk stars. *Unless the selection threshold is extremely conservative, metallicity-selected samples risk severe contamination by the thick disk population.* Morrison *et al.* point out that in the solar neighborhood, thick disk stars with $-1.0 \leq$[Fe/H]≤ -1.6 outnumber halo stars in the same metallicity range by a factor of two. Thus, in order to select a local sample of stars with pure halo kinematics, a metallicity limit of at least [Fe/H]≤ -1.6 must be used.

The second type of halo survey makes use of luminous stars at great distances as tracers of the halo population. A number of such surveys have been conducted using RR Lyrae stars, blue horizontal branch (BHB) stars, carbon stars, and giant branch stars, and all have relied exclusively on radial velocity information. Based on a compilation of data from these surveys, Norris (1986) concluded that the halo is in prograde rotation with a speed of 37 ± 10 km s^{-1}, in contradistinction to the retrograde result we have obtained. However, it should be pointed out that, in general, these tracer surveys have probed to distances only a few kiloparsecs from the sun, less than the $z \sim 5.5$ kpc extent of the thick disk. *Halo tracer studies which probe to within only 5 kpc or so from the Galactic plane may be severely contaminated by evolved thick disk stars.* Since the thick disk has a significant rotational velocity compared to the halo, the presence of even a small number of such stars in a halo star sample could greatly influence any derivation of the rotational velocity.

It is worthwhile to review the various halo surveys while remaining cognizant of these two important sources of contamination by thick disk stars. In Table 1 we summarize the derived halo reflex velocities from halo surveys, concentrating only on those stars in these surveys which lie beyond the $z \sim 5$ kpc spatial extent of the thick disk or, if drawn more locally, more metal poor than the metal-weak thick disk stars. The number of stars in each sample is given as N. The Pier (1984) results are listed with the caveat that the distribution of his stars are not ideal for measuring the rotational velocity of the halo. The value listed for Ratnatunga & Freeman (1989) is based on a naive trigonometric deprojection of the mean radial velocity, 168 ± 25 km s^{-1}, found for the most distant stars in their field at ($l = 272$, $b = +38$) with the assumption of no net motions other than pure cylindrical rotation. The resulting value of -213 km s^{-1}is an *upper* limit, since some of the rotational motion is transverse to the line of sight. The result for Majewski (1992) is based on an average of the three most distant halo bins in his Table 10. The Allen *et al.* (1991) stars are those for which they have determined a maximum z-distance greater than 4 kpc; the reflex velocity listed is obtained by converting their mean angular momenta with $v_{LSR} = 225$ km s^{-1}and $R_o = 8.0$ kpc. For the Norris & Ryan (1989b) data, we list three values for the halo reflex velocity: their result using both proper motions and radial velocities, their result after correction for selection biases (in particular the high proper motion selection bias), and their result using only radial velocity data (but based on the proper motion-selected sample). The Norris & Ryan correction makes the reflex velocity more positive, but we stress that this correction was formulated with the *a priori* assumption that the halo is in *prograde* rotation by 40 km s^{-1}.

Table 1. "Pure" Halo Samples

Sample	Reference	N	v_{reflex} $(km\,s^{-1})$
Blue Horizontal Branch	Pier (1984)	150	(-272 ± 41)
K giants SA127, $< z >= 12.8$ kpc	Ratnatunga & Freeman (1989)	14	$< -213 \pm 32$
NGP dwarfs ($< z >\sim 13$ kpc)	Reid (1990)	~ 200	-240 ± 30
NGP dwarfs	Majewski (1990, 1992)	111	-267 ± 9
[Fe/H]≤ -2.0, $z_{max} > 4$ kpc	Allen *et al.* (1991)	13	-317
[Fe/H]≤ -1.8, uncorrected	Norris & Ryan (1989b)	254	-254 ± 6
[Fe/H]≤ -1.8, corrected	Norris & Ryan (1989b)	254	-236 ± 6
[Fe/H]≤ -1.8, V_{radial} only	Norris & Ryan (1989b)	254	-214 ± 14

If v_{LSR} is taken to be 220 $km\,s^{-1}$, then the results listed in Table 1 are relatively consistent in predicting a halo in retrograde rotation. Only the v_{reflex} values from the Ratnatunga & Freeman K giants and the radial velocity solution of Norris & Ryan are prograde results, and only barely. The former is an *upper* limit, and allows for a retrograde halo. The latter result is based on a sample of stars selected by high proper motion; this bias can select against stars with high *radial* velocities for some halo phase space distributions, and this bias, in turn, will yield a more positively rotating solution. The remaining entries in Table 1 are more extreme by many tens of kilometers per second, and give evidence for a retrograde halo, *if* our assumed value of v_{LSR} is correct (in light of the present discussion, however, it may be timely to "re-open the case" on v_{LSR}).

This rather surprising result may find a natural explanation by way of the satellite accretion models of Quinn & Goodman (1986). Retrograde moving satellites are more stable against tidal effects than prograde satellites, which are found to rapidly sink towards the Galactic center. If the halo field stars have come from disrupted satellites (as suggested earlier by Searle & Zinn 1978), then we might expect a preponderance of retrograde orbiting halo stars. Norris & Ryan (1989a) have found such an asymmetry towards retrograde orbits and have appealed to the Quinn & Goodman picture as an explanation. This satellite accretion picture can also explain the net retrograde motion of the 30 globular clusters with $-1.3 \geq$ [Fe/H]

≥ -1.7 (Rodgers & Paltoglou 1984), the presence of halo moving groups (cf. Doinidas & Beers 1989, and references therein), the high velocity A stars (Rodgers *et al.* 1981), and the existence of halo stars with high metal abundances (Carney *et al.* 1990, Majewski 1990).

References

Allen, C., Schuster, W., J. & Poveda, A. 1991, *Astr. Ap.*, , **244**, 280.
Carney, B. W., Latham, D. W., & Laird, J. B. 1990, *A.J.*, **99**, 572.
Gunn, J., Knapp, J., and Tremaine, S. 1979, *A.J.*, **84**, 1181.
Majewski, S. R. 1990, *Ph.D. Thesis*, University of Chicago.
Majewski, S. R. 1992, *Ap. J. Suppl.*, in press.
Morrison, H. L., Flynn, C., & Freeman, K. C. 1990, *A.J.*, **100**, 1191.
Norris, J. 1986, *Ap. J. Suppl.*, **61**, 667.
Norris, J. & Ryan, S. G. 1989a, *Ap. J.*, **336**, L17.
Norris, J. & Ryan, S. G. 1989b, *Ap. J.*, **340**, 739.
Pier, J. 1984, *Ap. J.*, **281**, 260.
Quinn, P. J. & Goodman, J. 1986, *Ap. J.*, **309**, 472.
Ratnatunga, K. U. & Freeman, K. C. 1985, *Ap. J.*, **291**, 260.
Rees, R. & Cudworth, K. M. 1991, *A.J.*, **102**, 152.
Reid, N. 1990, *M.N.R.A.S.*, **247**, 70.
Rodgers, A. W. & Paltoglou, G. 1984, *Ap. J.*, **283**, L5.
Rodgers, A. W., Harding, P. & Sadler, E. 1981, *Ap. J.*, **244**, 912.
Searle, L. & Zinn, R. 1978, *Ap. J.*, **225**, 357.

Discussion

Nemec: You have chosen to fit the stars with $z > 1$ kpc with a two-component 'halo plus thick disk' model. Based on this model you find for the 'thick disk' component a gradient [in asymmetric drift], which you interpret as due to dissipational collapse. Can you be sure that contamination from the high-z tail of the thin disk component, and contamination from the low-z side of the halo component, are not causing the apparent gradient?

Majewski: For $1 \leq z \leq 5.5$ kpc, the distributions for the metallicity and asymmetric drift appeared double-peaked and a two-Gaussian model seemed most appropriate for fitting. In this distance range, the tail of the halo distribution does overlap the distribution of thick disk points. However, if the halo velocity distribution is Gaussian for $z < 5$ kpc, as it is beyond this, then our technique should be robust against halo contamination. Contamination by thin disk stars could contribute to a gradient, but only within a few thin disk scale heights. Of course, an important question which remains to be answered is whether in fact the thin and thick disks are discrete components at all, or whether, as Norris suggests, the thick disk is simply a continuous and extended tail of the thin disk.

Corbally: How do you select for F and G dwarfs rather than giants? How might giant contamination affect your results?

Majewski: We assume *a priori* that all stars are dwarfs. At the faint magnitudes of the survey we would expect to find practically no giant contamination and subgiant contamination at a level of less than 10%. Note that assuming every star is a dwarf results in the *most conservative* kinematics. Converting proper motions to velocities for subgiant stars, which are at greater distances than dwarfs of the same apparent magnitude and color, results in a halo rotation even more extreme in the retrograde sense!

Corbally: Note that Corbally & Garrison (cf. poster contribution at this conference) find that about 7% of late F and G stars in the inner halo are in fact giants or subgiants rather than dwarfs, and this is in the region of the Hertsprung gap and so a lower bound.

THE DISK OF THE GALAXY

K.C. FREEMAN
Mount Stromlo and Siding Spring Observatories
The Australian National University
Canberra, Australia

ABSTRACT. In this review, I will concentrate on the observational aspects of the dynamical evolution of the disk population. First I will discuss some general properties of disks, such as their structure and kinematics and the thick disks. The next section is on the kinematical properties of the galactic disk near the sun, including some important new results on the age-velocity dispersion relation and the age-metallicity relation. Finally I will discuss some properties of the galactic thick disk. Two important questions, chemical gradients and dark matter in the disk, could not be included in this review.

1. The Structure and Kinematics of Disks

1.1. THE STRUCTURE OF DISKS

For a sample of edge-on galaxies, van der Kruit and Searle (1981,1982) showed that the light distribution has the form $L(R, z) = L_\circ e^{-R/h}\text{sech}^2(z/z_\circ)$. The scale height $z_\circ$ is typically about 700 pc, and is independent of radius. This $L(R, z)$ brightness distribution is truncated at about 4 to 5 radial scalelengths. A surface density distribution of this form represents a locally isothermal sheet (ie the vertical velocity dispersion σ_z is independent of z) with ${\sigma_z}^2 = 2\pi G\rho_\circ {z_\circ}^2$. For $z \gg z_\circ$, $\text{sech}^2(z/z_\circ) \to \exp(-2z/z_\circ)$.

Galactic dust is a problem in determining the form of L(R,z) near the plane of external galaxies. In our Galaxy, star counts towards the galactic poles (Gilmore and Reid 1983; Pritchet 1983) indicate that the z-distribution more nearly exponential, with a scaleheight of about 300 pc. The effects of dust in external galaxies can be greatly reduced by infrared observations, and 2.2μ surface photometry of edge-on galaxies also shows that the vertical surface brightness distribution in disks is nearly exponential (eg Wainscoat *et al.* 1989). Such a vertically exponential disk is not locally isothermal; the velocity dispersion increases by a factor $\sqrt{2}$ from its value at z = 0 to the asymptotic value at large z. New observations of the vertical velocity dispersion for K giants near the SGP (Flynn and Freeman, preprint) show that the old disk *is* indeed isothermal up to at least 450 pc above the plane. There is no inconsistency here; it just means that the colder young disk and the hotter old disk together produce the observed vertical exponential density distribution.

B. Barbuy and A. Renzini (eds.), The Stellar Populations of Galaxies, 65–73.

1.2. THE VELOCITY DISPERSION OF DISKS

The vertical velocity dispersion of the constant scaleheight van der Kruit and Searle disks is given by hydrostatic equilibrium as $\sigma_z(R) \propto \exp(-R/2h)$. This is observed directly in a few face-on spirals (*e.g.* van der Kruit and Freeman 1986). Then, *if* the anisotropy σ_R/σ_z is approximately constant with radius, we would expect that $\sigma_R(R)$ is also $\propto \exp(-R/2h)$. In the Galaxy, σ_R would then rise to about 100 km s^{-1} near the center of the disk.

It is difficult to measure $\sigma_R(R)$ in the disks of external galaxies (see however Bottema 1988). In the Galaxy, $\sigma_R(R)$ can be measured directly, using velocities of individual stars. Lewis and Freeman (1989) measured velocities for about 600 K giants of the old disk, out to about 18 kpc from the galactic center. The variation of σ_R with R follows closely the expected exponential law given above; in the inner parts of the disk, σ_R is about 110 km s^{-1} and the stability parameter Q is roughly constant with radius, in the range 1.5 to 2. The kinematically determined scalelength $h = 4.4 \pm 0.3$ kpc, which is close to the mean scalelength determined directly from the distribution of various kinds of objects (de Vaucouleurs and Pence, 1978; van der Kruit, 1986; Habing 1988). From the above equations, it follows that the anisotropy σ_R/σ_z is indeed approximately constant with radius. This result means that the processes which heat the disk in the radial and vertical directions keep the ratio σ_R/σ_z approximately constant (≈ 2), despite the large variation in density and velocity dispersion over the observed range in radius. The reason for this is not yet fully understood: see Carlberg (1987), Wielen and Fuchs (1988).

To summarize: the kinematics of galactic disks show that σ_R and $\sigma_z \propto \exp(-R/2h)$, as expected. The disk heating processes lead to constant scaleheight with radius, and constant anisotropy σ_R/σ_z.

1.3. THICK DISKS

Following the early work on the thick disk component of disk galaxies (Tsikoudi 1979, 1980; Burstein 1979), van der Kruit and Searle (1981,1982) showed that some edge-on galaxies with bulges have a *second* flattened component, the thick disk, in addition to the usual thin disk. Star counts by Gilmore and Reid (1983), confirmed by Yoshii *et al.* (1987), showed that the Galaxy also has a thick disk, with scale height of about 1000 pc and column density of order 10 percent of that of the thin disk.

The thick disk raises some interesting questions about its origin and its implications for galaxy formation:

- Are thick disks found only in galaxies with significant central bulges ? Some disk galaxies do not have thick disks (eg van der Kruit and Searle, 1981, 1982), so thick disk formation is not an essential part of the formation and evolution of disk galaxies.
- Are thick disks in rotational equilibrium, like thin disks (in which case they could have come from heated thin disks) or are they more slowly rotating (as if they were an intermediate population formed during the collapse) ?
- How old are the stars of the thick disk ? Did the thick disk form very early in the life of the parent galaxy (like the metal-weak halo) or later (like much of the thin disk) ?

Observations of stellar kinematics in the Galaxy are needed to understand the nature of the thick disk, because such measurements cannot be made yet for external galaxies. The next section is on the relationships between the kinematics of nearby stars (including thick disk stars) and their ages and metallicities. Here are two preliminary points about the thick disk near the sun, which will be useful in interpreting the kinematics of the nearby stars:

1. Yoss *et al.* (1987) showed that there is a population of G and K giants at the galactic poles in the metallicity range $-0.5 >$ [Fe/H]> -1 with a vertical velocity dispersion of about 40

km s^{-1} and therefore a scaleheight of about 1000 pc. This identifies the metallicity range in which most of the nearby thick disk stars are found.

2. Many stars in this metallicity range are included in the study of nearby high proper motion stars by Laird *et al.* (1988). These stars belong to a rapidly rotating population, with an asymmetric drift less than 50 km s^{-1}. From their color distribution, these stars appear to be as old as the disk globular clusters.

There will be more discussion of the galactic thick disk in §3.

2. The Nearby Disk

The chemical and kinematical properties of nearby disk stars depend on their age. The older disk stars have lower metallicities in the mean, which reflects the chemical evolution of the Galaxy, although the details are not fully understood. This age-metallicity relation (AMR) is usually presented as a fairly smooth relationship between the mean metallicity and the age (eg Strömgren 1987). The velocity dispersion σ of nearby stars increases with stellar age. This is believed to result from heating of the disk by interactions of disk stars with spiral arms, giant molecular clouds, and possibly massive black holes in the galactic halo. The age - velocity dispersion relation (AVR) is usually presented as a smooth increase of the velocity dispersion with stellar age: see Wielen (1977). However, for solar abundance stars, Strömgren's (1987) Table 1 shows that the velocity dispersion does not change much between ages of about 4 to 13 Gyr, which indicates that the dynamical heating may saturate at some level.

A new kinematically unbiased sample of nearby F stars (Edvardsson *et al.* 1991), with very precise ages, velocities and metallicities, shows that there is significant structure in the age - σ - [Fe/H] relations. See Nissen (1990) for a preliminary report. Here are some of the important results:

- The age - metallicity relation: Figure 1 shows that the AMR is broad. The younger stars (age < 3 Gyr) have higher mean metallicity and the older stars (age > 10 Gyr) have lower mean metallicity, as previously recognized. However, the stars with ages between 3 and 10 Gyr cover the entire [Fe/H] range from +0.2 to -0.6, with no obvious trend of metallicity with age.
- The age - velocity relation show three age zones:
  1. The youngest stars, with ages < 3 Gyr, have a low velocity dispersion, $\sigma_W = 10 \pm 1$ km s^{-1}.
  2. For ages between 3 and 10 Gyr, $\sigma_W = 19 \pm 2$ km s^{-1}, *with no trend of velocity dispersion with age.* These stars have [Fe/H] in the range +0.2 to -0.6 and represent the old disk.
  3. The oldest stars, with ages > 10 Gyr, have $\sigma_W = 42 \pm 3$ km s^{-1}. These old stars have [Fe/H] values mostly below -0.5, so their velocity dispersion and metallicities indicate that they belong to the thick disk; this velocity dispersion corresponds to a scaleheight of about 1 kpc.

The other two velocity components show similar structure. See Freeman (1991) for a more detailed discussion of this age - velocity dispersion relation.

The Edvardsson *et al.* data suggest that:

- thin disk heating saturates after about 3 Gyr, at $\sigma_W \approx 20$ km s^{-1}, because the velocity dispersion does not increase between ages of 3 and 10 Gyr (see also Carlberg *et al.* 1985; Strömgren 1987, Table 1).
- The thick disk is very old, because the characteristic metallicity range and velocity dispersion of the thick disk are seen only among the stars older than 10 Gyr.
- the energy of the thick disk ($\sigma_W \approx 40$ km s^{-1}) does not come from the processes that heat the thin disk, which appear to saturate at about 20 km s^{-1}. Its velocity dispersion must come from some other kind of process.

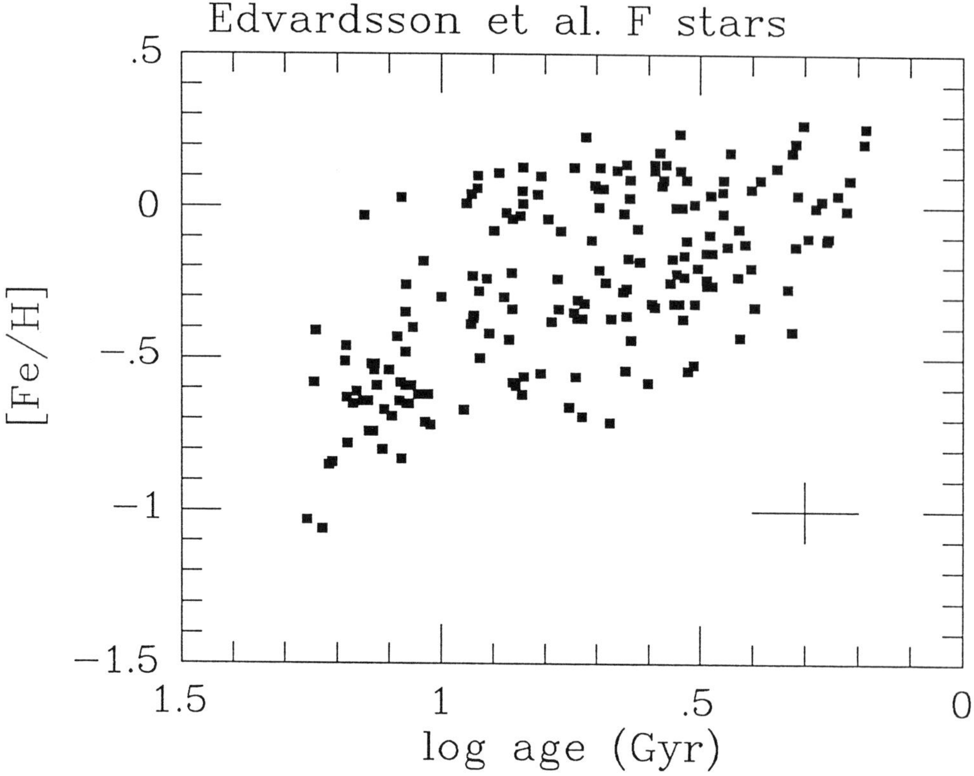

Figure 1:–The metallicity - age distribution for the Edvardsson *et al.* (1991) F stars. The error bars are shown in the lower right corner. Note the large spread in metallicity for stars with ages between about 3 and 10 Gyr.

This new F star sample shows that there is a homogeneous old disk population with stellar ages between about 3 and 10 Gyr, [Fe/H] between +0.2 and -0.6, and $\sigma_W \approx 20$ km s^{-1}. This old disk population shows no obvious internal correlations of velocity dispersion, abundance and age. The interesting point here is that the old disk near the sun is not just some mean point on the AVR and AMR, but is in fact a well defined and homogeneous population in which there is little evidence for dynamical and chemical evolution.

How do we reconcile the well-known correlations between age, velocity dispersion and metallicity with the picture given by the new F star data ?

1. The correlation of velocity dispersion with metallicity comes primarily from the young stars (ages $<$ 3 Gyr) of high [Fe/H] and the old (ages $>$ 10 Gyr) thick disk stars of lower [Fe/H]. There is no trend of velocity dispersion with metallicity for stars in the 3 to 10 Gyr (old disk) interval.

2. The AVR given by the new F star data, which suggest the thin disk - old disk - thick disk structure described above, is rather different from the usual smooth version of the solar neighborhood AVR. It seems likely that the high internal precision of the new F star ages (± 0.1 in log age) from Strömgren photometry reveals this structure. The ages used for the older AVR determinations come mostly from less direct methods (Ca II emission and mean main sequence lifetimes) which may have somewhat larger errors that smooth out the structure in the AVR.
3. The age - metallicity relation shown in Figure 1 is intrinsically quite broad, and shows little trend between ages of 3 and 10 Gyr. The trends usually associated with the AMR are again produced mainly by the presence of the relatively metal rich stars younger than 3 Gyr and the relatively metal poor (thick disk) stars older than 10 Gyr.

What does the breadth of the AMR imply about the chemical evolution of the thin disk ? We see from Figure 1 that the younger stars are metal rich and have a relatively small spread in metallicity. These young stars have a smaller velocity dispersion, so they come from a relatively small region of the Galaxy. The older stars of the thin disk (the 3 to 10 Gyr sample) have a larger velocity dispersion and include objects that were born anywhere in a broad annular region that extends right around the Galaxy; their metallicity distribution is an average over this annulus. It appears that (i) chemical enrichment is rather inhomogeneous, as Nissen *et al.* (1985) suggest, and (ii) the younger F stars in the solar neighborhood come from a region that just happens to have relatively high metallicity. The metallicity of the younger stars is not exceptionally high; the sample includes some much older stars with similarly high metallicities.

In summary, this new sample of F stars suggests that the well known trends of age, metallicity and velocity dispersion among disk stars are produced primarily by stars younger than 3 Gyr and older than 10 Gyr. Stars with ages between 3 and 10 Gyr come from a homogeneous old disk population which shows no such trends.

The F stars in the Edvardsson *et al.* sample were older than about 1 Gyr. Wilson (1990) has recently measured the kinematics and abundances for a sample of about 500 nearby K giants, which includes some younger objects. For solar neighborhood F stars with ages < 1 Gyr, the (U,V) velocity distribution is quite clumpy (Eggen, 1969); it is dominated by two major clumps which represent the Hyades and Sirius moving groups in the solar neighborhood. These groups are believed to come from dissolving giant molecular clouds or spiral arm segments. Unbound aggregates in the galactic disk dissolve along precessing ellipses (the Lindblad dispersion orbits). From the dispersion orbits for the Hyades and Sirius groups, Wilson calculated the expected radial velocities of stars on these dispersion orbits away from the solar neighborhood. He then tested this dispersion orbit picture by observing about 500 disk K giants out to about 1 kpc from the sun, in the predicted direction of the dispersion orbits ($l = 270 \pm 30°$). The ages of the Hyades and Sirius group stars are only a few galactic years so, from the discussion of the Edvardsson *et al.* F stars, we would expect to see evidence for their dispersion orbits only among the more metal rich stars. It turned out that the dispersion orbits are seen very clearly for stars with metallicities [Fe/H]$>$ -0.25, and not for more metal weak stars.

Wilson's sample covers the metallicity range $0.3 >$ [Fe/H]> -1. The distribution of stars with [Fe/H]> -0.25 is dominated by the groups. In the (V component of velocity) - metallicity plane shown in Figure 2, this population, which has an age of a few galactic years, is obviously unmixed. For the more metal weak stars, the populations are similar to those seen in the Edvardsson *et al.* sample. Stars with $-0.3 >$ [Fe/H]> -0.6 belong to the old disk, with velocity dispersion $\sigma_V = 27 \pm 1$ km s^{-1} and asymmetric drift 8 ± 2 km s^{-1}. The K giants with [Fe/H]< -0.6 belong to the thick disk, with $\sigma_V = 38 \pm 4$ km s^{-1} and asymmetric drift 27 ± 5 km s^{-1}.

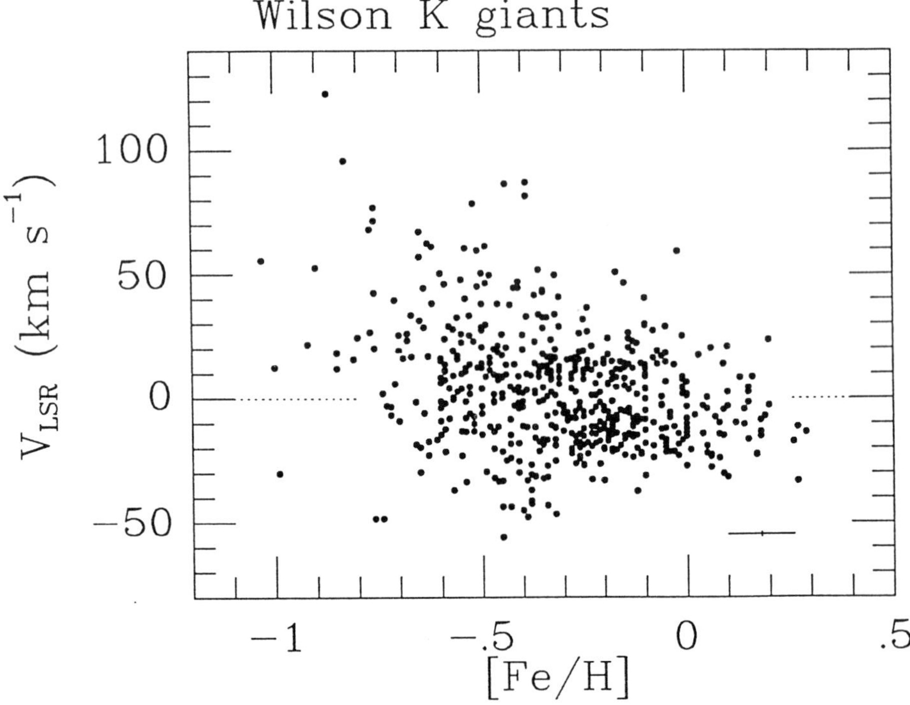

Figure 2:–The metallicity - velocity relation for Wilson's K giants. V_{LSR} is the V-component of the stellar velocity relative to the local standard of rest. (Here V is positive towards $l = 270°$). For [Fe/H] > -0.25, the distribution is dominated by the two velocity clumps which correspond to the young Hyades and Sirius moving groups. These stars are obviously unmixed dynamically. Stars with $-0.3 >$ [Fe/H] > -0.6 belong to the old disk. Stars with [Fe/H] < -0.6 are predominantly thick disk objects, with a larger velocity dispersion and asymmetric drift. Error bars are shown in the lower right-hand corner.

3. The Galactic Thick Disk

In this section, I return to the questions of the age and rotation of thick disks, which can be measured only for the galactic thick disk; even there the situation is not entirely clear. I also discuss the discreteness of the thick disk, and then the recently discovered metal-weak extension of the galactic thick disk, which may have some interesting implications for the early history of the Galaxy.

3.1. THE ROTATION AND AGE OF THE THICK DISK.

The kinematical properties of the galactic thick disk can be measured from observations of stars with metallicities between about −0.6 and −1.0. For these stars, the vertical velocity dispersion is about 40 km s^{-1} and their scaleheight is therefore about 1 kpc. In the solar neighborhood, there are many estimates of the mean rotation of the thick disk: see Sandage and Fouts (1987), Norris (1987), Laird *et al.* (1988), among others. There are also several studies of the kinematics of the thick disk *in situ*: *e.g.* Yoss *et al.* (1987), Friel (1988), Ratnatunga and Freeman (1989), and Armandroff (1989). The solar neighborhood and the *in situ* data give similar results for the velocity dispersion and asymmetric drift of the thick disk. The radial component of its velocity dispersion is about 65 km s^{-1} and the asymmetric drift is about 30 km s^{-1} relative to the LSR. The thick disk is clearly a rapidly rotating population.

Most recent work indicates that the thick disk near the sun is very old, *i.e.* at least as old as the disk globular clusters. See for example the work of Edvardsson *et al.* discussed above and also Nissen (1990) for the nearby F stars, Rose and Agostinho (1991) for more distant F stars, and Gilmore *et al.* (1989) and Carney *et al.* (1989) for age estimates from nearby high proper motion stars. For an alternative view, see Norris and Green (1989).

3.2. THE DISCRETENESS OF THE THICK DISK.

Is the thick disk a separate component of the Galaxy, or just the higher energy, metal weaker tail of the old disk. This problem has been widely and inconclusively discussed (see Freeman 1987; Gilmore *et al.* 1989). It remains important because it has obvious implications about galaxy formation and the origin of the thick disk. For example, did the stars of the thick disk form during galactic collapse, before thin disk star formation had started ? Or did the thick disk form after thin disk formation was already under way ?

In the second category of formation pictures, we can consider some specific possibilities:

1. The stars that we now recognize as members of the thick disk were the first stars to form in the early thin disk, and have therefore suffered the most heating by the same processes that continue to heat the thin disk to the present time. This picture, in which the thick disk is not a discrete component, now seems unlikely from the discussion in §2, which indicates that the heating of the thin disk saturates when σ_W reaches about 20 km s^{-1}. (We recall that $\sigma_W \approx 40$ km s^{-1} for the thick disk).
2. The stars that are now in the thick disk formed early in the life of the disk, and were heated by some short-lived phenomenon, such as a transient bar or strong spiral structure in the early disk, or by an early epoch of satellite accretion (see Freeman 1990 for more discussion of this possibility). In this case, the thick disk would be a discrete component, with some interesting information content about the early Galaxy.

We do not not yet understand how the thick disk fits into the galaxy formation picture. However thick disks are clearly not an essential feature of the formation and subsequent evolution of disk galaxies, because many disk galaxies do not have thick disks.

3.3. THE METAL-WEAK THICK DISK.

Norris *et al.* (1985) showed that a significant fraction of spectroscopically selected metal weak stars near the sun have kinematics like the thick disk, although their metallicities are in the range usually associated with the slowly rotating metal-weak halo of the Galaxy ([Fe/H]< -1). Morrison *et al.* (1990) then investigated the kinematics of spectroscopically selected G and K giants near the sun and also in a more distant field at about the same galactocentric radius. Their most metal weak stars, with [Fe/H]< -1.6, clearly belong to the slowly rotating halo.

Stars in the metallicity range -1.0 to -1.6 and with $|z| > 1$ kpc are also slowly rotating. However, in the same metallicity range, the stars near the galactic plane with $|z| < 1$ kpc include a rapidly rotating disk component with an asymmetric drift of 50 ± 15 km s^{-1} and velocity dispersion $\sigma_W = 40 \pm 13$ km s^{-1}; its density is comparable to that of the metal weak halo. This confirms the result by Norris *et al.* that the thick disk extends in metallicity to [Fe/H] $\approx$ -1.6.

The metal weak extension of the thick disk is not so evident in the RR Lyrae stars, although it is clearly seen among the G and K giants. The implications of this observation are not clear. One possibility is that the metal weak thick disk may be somewhat younger than the rest of the thick disk. For example, the metal weak thick disk stars may be the debris of younger metal weak satellites accreted towards the end of the satellite accretion phase. We note that the entire mass of the metal weak extension of the thick disk is only of order 10^8 $M_\odot$.

Acknowledgements

I am very grateful to B. Edvardsson, B. Gustafsson, D. Lambert, P. Nissen, J. Tomkin and J. Andersen for allowing me to use their new F star data before publication, and I thank B. Gustafsson, P. Nissen and C. Norman for interesting discussions and encouragement.

References

Armandroff, T. 1989. Astron.J., **97**, 375.
Bottema, R. 1988. Astron.Astrophys., **197**, 105.
Burstein, D. 1979. Astrophys.J., **234**, 829.
Carlberg, R. 1987. Astrophys.J., **322**, 59.
Carlberg, R., Dawson, P., Hsu, T., VandenBerg, D. 1985. Astrophys.J., **294**, 674.
Carney, B., Latham, D., Laird, J. 1989. Astron.J., **97**, 423.
de Vaucouleurs, G., Pence, W. 1978. Astron.J., **83**, 1163.
Edvardsson, B., Gustafsson, B., Lambert, D., Nissen, P., Tomkin, J., Andersen, J. 1991. In preparation.
Eggen, O.J. 1969. Astrophys.J., **155**, 701.
Freeman, K.C. 1987. Ann.Rev.Astron.Astrophys., **25**, 603.
Freeman, K.C. 1990. In *Dynamics and Interactions of Galaxies*, ed R. Wielen, (Springer), p. 36.
Freeman, K.C. 1991. In *Dynamics of Disk Galaxies*, ed B. Sundelius, (University of Goteborg), in press.
Friel, E.D. 1988. Astron.J., **95**, 1727.
Gilmore, G., Reid, N. 1983. Mon.Not.R.Astron.Soc., **202**, 1025.
Gilmore, G., Wyse, R., Kuijken, K., 1989. Ann.Rev.Astron.Astrophys., **27**, 555.
Habing, H. 1988. Astron.Astrophys., **200**, 40
Laird, J., Carney, B., Latham, D. 1988. Astron.J., **95**, 1843.
Lewis, J.R., Freeman, K.C. 1989. Astron.J., **97**, 139.
Morrison, H.L., Flynn, C., Freeman, K.C. 1990. Astron.J., **100**, 1191.
Nissen, P. 1990. In *Elements and the Cosmos*, ed R. Terlevich, (Cambridge: Cambridge University Press).
Nissen, P., Edvardsson, B., Gustafsson, B. 1985. In *Production and Distribution of C,N,O Elements*, ed I. Danziger *et al.* , (ESO), p. 131.
Norris, J.E. 1987. Astrophys.J., **314**, L39.
Norris, J.E., Bessell, M.F., Pickles, A. 1985. Astrophys.J.Suppl., **58**, 463.
Norris, J.E., Green, E.M. 1989. Astrophys.J., **337**, 272.

Pritchet, C. 1983. Astron.J., **88**, 1476.
Ratnatunga, K., Freeman, K.C. 1989. Astrophys.J., **339**, 126.
Rose, J., Agostinho, R. 1991. Astron.J., **101**, 950.
Sandage, A., Fouts, G. 1987. Astron.J., **92**, 74.
Strömgren, B. 1987. In *The Galaxy*, ed G. Gilmore and R. Carswell, (Dordrecht: Reidel), p. 229.
Tsikoudi, V. 1979. Astrophys.J., **234**, 842.
Tsikoudi, V. 1980. Astrophys.J.Suppl., **43**, 365.
van der Kruit, P.C. 1986. Astron.Astrophys., **157**, 230.
van der Kruit, P.C., Freeman, K.C. 1986. Astrophys.J., **303**, 556.
van der Kruit, P.C., Searle L. 1981. Astron.Astrophys., **95**, 105, 116.
van der Kruit, P.C., Searle L. 1982. Astron.Astrophys., **110**, 61, 79.
Wainscoat, R., Freeman, K.C.,, Hyland, A.R. 1989. Astrophys.J., **337**, 163.
Wielen, R. 1977. Astron.Astrophys., **60**, 263.
Wielen, R., Fuchs, B. 1988. In *The Outer Galaxy*, ed L. Blitz and F. Lockman, (Springer), p. 100.
Wilson, G.A. 1990. Thesis, Australian National University.
Yoshii, Y., Ishida, K., Stobie, R.S. 1987. Astron.J., **93**, 323.
Yoss, K., Neese, C., Hartkopf, W. 1987. Astron.J., **94**, 1600.

Discussion

Ellis: You commented that thick disks are not universal and therefore that they are not an essential feature of galaxy evolution. If, however, mergers are a common feature of galaxy evolution, then one would expect at least a major subset (say 50 percent) of luminous disk galaxies to have thick disks. Can you rule this out ?
Freeman: No, not at this stage. However, if mergers are a common feature of galaxy evolution, then the present frequency of thick disks would depend on the evolutionary phase at which the merging occurred. For example, if almost all the merging occurred while the fragments were almost entirely gaseous, then stellar thick disks need not be very common now. Conversely, it appears that thick disks typically contribute less than 10 percent of the light of the thin disk in large spirals; this means that there cannot have been much merger activity after the first one or two Gyr of star formation had taken place in the thin disk. Quinn, Hernquist and Fullagar (preprint) show that any such post-thin-disk accretion is limited to a few percent of the mass of the thin disk.
Carney: In a binary system, tidal effects act to turn eccentric orbits into circular ones, with close pairs affected most strongly. Thus the period at which the break between circular and eccentric orbits occurs is a chronometer. In the Hyades, this is 5.7 days, and in M67 it is 10.5 days. The survey of proper motion stars which Dave Latham, John Laird and I have been studying shows the transition at about 20 days in the halo and roughly the same for the thick disk stars. So star formation in both started at about the same time. But who knows where ?
Carney: Do you have an idea of the thick disk's scale length ?
Freeman: No; at this time there are arguments for the thick disk having a longer scale length than the thin disk, and also for a shorter scale length.

THE CHEMICAL AND DYNAMICAL EVOLUTION OF THE GALACTIC DISK

B. GUSTAFSSON and B. EDVARDSSON
Astronomical Observatory
Box 515
S-751 20 Uppsala
Sweden

P. NISSEN
Institute of Physics and Astronomy
University of Aarhus
Dk-8000 Aarhus C
Denmark

D. L. LAMBERT and J. TOMKIN
Department of Astronomy
University of Texas
Austin, TX 78712-1083
Texas

and

J. ANDERSEN
Copenhagen University Observatory
Brorfeldevej 23
DK-4340 Tollose
Denmark

ABSTRACT. For a sample of 189 disk stars, essentially dwarfs of spectral type F, we have determined chemical abundances of O, Na, Al, Mg, Si, Ca, Ti, Fe, Ni, Y, Zr and Ba, as well as ages and space motions. These stars have deep enough convection zones so that the surface layers may be presumed to be well mixed and thus representative of the chemical composition of the pre-stellar cloud. The sample was selected without kinematical bias such that the stars represent different metallicity groups ranging from [Fe/H]=-1.0 to 0.3 and such that ages, varying from 1 to 15 Gyears, can be derived from isochrones. Thus, the chemical composition of the interstellar medium, as a function of age and galactocentric distance (approximated by R(m) = 0.5*{R(apogal)+R(perigal)}) may be mapped. The following basic results are obtained:

1. The alpha-element (O, Mg, Si, Ca, Ti) abundances vary relative to Fe in a well defined way as a function of [Fe/H] with an excess of [alpha/Fe] for the more metal-poor stars.The excess is significantly greater for O than for the other alpha elements, which is not expected from recent SN II model calculations. There is, for the most metal-poor stars, a tendency for the alpha excess to decrease as a function of R(m).This probably indicates a more rapid star formation in the interior parts of the early Galactic disk. The individual scatter around these relations is about 0.05 dex, as expected from the observational errors, indicating an efficient mixing of gas enriched by alpha elements and Fe, respectively, in the disk at all times.

B. Barbuy and A. Renzini (eds.), The Stellar Populations of Galaxies, 75–76.

2. The abundances of the "odd" elements Na and Al, relative to Fe, show a less pronounced variation with [Fe/H], Al yet being more similar to the alpha elements in this respect.This puts further constraints on supernova models.

3. The Fe abundance relative to H shows a general tendency to decrease with increasing age. The scatter around this relation (0.2 dex) is, however, significantly greater than the expected observational differential uncertainty in [Fe/H] and is also hard to explain as a a result of bias in the selection procedure or of orbital diffusion, and even as a combination of these. This suggests that mixing of metal-rich and metal-poor gas in the Galaxy is significantly less efficient than that of SN II (alpha-rich) and SN Ia (Fe-rich) gas. This might be explained as a result of infall of metal-poor gas; however, a very efficient triggering of star-formation by this infall is then necessary if current estimates of the infall rate are typical for the history of the Galactic disk. Alternatively, one has to invoke identical or similar sites for the production of alpha elements and Fe, yet giving relative contributions of these elements that vary as a function of Galactic age. The scatter in [Fe/H] at a given age and R(m) could then be explained as the result of star formation occuring both in well mixed interstellar gas and in gas enriched by local supernovae. Another, more speculative explanation would be that the Galactic disk is the result of the merging of two different galaxy populations.

4. The abundances of Ba and other s-elements relative to Fe show a significant variation with age at a given [Fe/H], indicating a long characteristic time scale for the formation of these elements, yet different from that of the formation of Fe.

5. The three most s-element rich stars were found to have very similar R(m). We suggest an s-element rich gas cloud to be the common origin of these.

6. There is a significant tendency for some additional grouping in the dynamic-chemical space for our sample of field stars. We tentatively suggest a group of less than 10 sample stars to which the Sun belongs to be enriched in Ca and Fe, relative to other elements like Na, Mg, Al and Si, as compared with other stars of solar overall metallicity.

The details of this study will be published in forthcoming papers in Astronomy & Astrophysics.

G. Hensler: Do your results concerning the metal content of the old disk stars with respect to their radial positions allow the conclusion that a radial propagation of metal enrichment must have happened during disk formation?

B. Gustafsson: I think our variation of the alpha-element abundance relative to iron, at a given iron abundance, as a function of galactocentric mean distance, may indicate a radial propagation of the alpha-element enrichment.

THE GALACTIC BULGE AND THE THICK DISC

M W FEAST and P A WHITELOCK
South African Astronomical Observatory

R SHARPLES
University of Durham UK

ABSTRACT. The large reported discrepancy between the scale height in the Galactic Bulge for objects such as RR Lyrae variables on the one hand and M-giants or IRAS sources on the other is primarily due to these scale heights being determined at different mean galactocentric distances. Mira variables apparently provide a Bulge-like population in the solar vicinity and show that the thick disc consists of a number of subsystems. There are M-giants in the extended solar neighbourhood, away from the Galactic plane, which seem similar to M-giants in the Bulge. Photometric and kinematic studies on some of these stars are in progress.

1. Gradients in the Bulge

Amongst questions of relevance to our understanding of the formation and evolution of our Galaxy are: (1) Can the Galactic Bulge population be divided into subpopulations of different metallicities, ages or kinematics? (2) How is the Bulge population related to other galactic populations? (3) Is there a significant 'Bulge type' population in the solar neighbour-hood? (4) If so, is this population to be identified with another galactic component (e.g. old disc, thick disc or halo)?

Evidence has been presented (e.g. Terndrup *et al.* 1990) for a radial metallicity gradient in the Bulge. Whilst such a gradient may well exist it is important to notice that it is quite modest in the region between galactic latitudes ($|b|$) of 4° and 8° where the Bulge is perhaps best seen free from contamination by either the disc or the halo (cf. also Whitelock 1991). It has sometimes been suggested that different types of object have radically different distributions in the Bulge. The evidence for this however generally depends on a comparison of estimates of scale heights which are made at different mean distances from the Galactic Centre for different types of objects. Such comparisons can be misleading unless the distributions are exponential. For instance it seems widely believed that the Bulge defined by M-giants or IRAS sources is much smaller than that defined by the distribution of optical light or the number density of RR Lyrae variables, with scale heights of ~1500pc for the latter and a few hundred parsecs for the former (see for example Harmon and Gilmore 1988). These scale heights however apply to quite different distances from the Galactic Centre as can be seem from figure 1. de Vaucouleurs and Pence (1978) note

B. Barbuy and A. Renzini (eds.), The Stellar Populations of Galaxies, 77–79.

that their estimates of optical surface brightness are uncertain for $|b| < 15°$ so that there is no useful overlap between their measures and those of Blanco (1988) for M-giants. The distribution of RR Lyrae variables is shallow in the outer parts of the Bulge but steepens markedly between $|b| = 8°$ and $4°$ so that the difference in distribution from the M-giants is much less than would be deduced from the scale heights quoted above. The RR Lyrae data is from Oort and Plaut (1975) with completeness corrections from Wesselink (1987). Note that the RR Lyrae point at $|b| = 3°.9$ has not been corrected for completeness and is therefore a lower limit. A similar comparison of the distribution of IRAS sources (Harmon and Gilmore 1988 figure 3) with that of the RR Lyraes shows no significant differences.

The above discussion shows that our current understanding of the Bulge does not allow us to predict which components of it could form a significant population in the extended solar neighbourhood, for that we must turn to the observations.

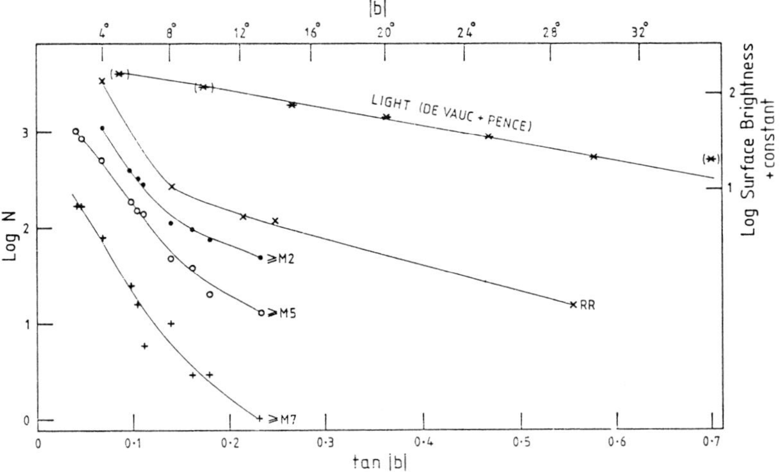

Figure 1: Distribution of surface brightness or surface number density with galactic latitude ($|b|$) for various objects. Note that log N for the RR Lyrae variables is normalized to a different unit area from the M-giants. Uncertain points are enclosed in brackets.

2. The Bulge and the Thick Disk

One of the distinctive features of the Bulge is the presence of Mira variables with a wide range of periods (Lloyd Evans 1976; Whitelock *et al.* 1991). Evidence from general galactic kinematics and from Miras in globular clusters indicates that the period range corresponds to a range in metallicities and initial masses (cf. Feast 1989 and references there). There is at present no evidence for a difference, at a given period, between Miras in the Bulge and in the solar neighbourhood. Miras can therefore be considered as tracing (part of) the local Bulge type population. Determinations of asymmetric drifts, velocity dispersions and (inferred) scale heights as a function of period show that local Miras with periods near 250 days are to be associated with a Gilmore-type thick disc population. Miras of longer period have kinematic properties near those of the old disc whilst the kinematics of shorter period stars are intermediate between the halo and thick disc (Feast 1989). These results suggest a range of disc-like systems between the halo and the old disc rather than single homogeneous thick disc. One may wish to consider these subsystems together as a 'generalized thick disc'.

Our recent work has concentrated on groups of relatively nearby M-giants which may be similar to those in the Bulge.

A characteristic of Bulge M-giants is that they are displaced in a J-H/H-K diagram from local bright M-giants (Frogel and Whitford 1987; Feast *et al.* 1990). This is plausibly a metallicity effect; the Bulge stars having [Fe/H] $\sim +0.3$. However, since the bright local M-giants are more massive and much younger than those in the Bulge the effect could be at least partly due to a difference in surface gravity (cf. Feast *et al.* 1990). A sample of local M-giants of intermediate type (generally M2 and M3) at a mean height of $\sim$ 500pc above the Galactic plane show that there are M-giants in our region of the Galaxy in the same position in the J-H/H-K diagram as Bulge stars (Feast *et al.* 1990 figure 7). This reinforces the conclusions drawn for the Mira results that there exists a 'generalized thick disc ' component which can be regarded as the local Bulge type population.

The Bulge contains a large number of late M-giants (M5 and later). The many late M-giants found by Stephenson (1986) in the extended solar neighbourhood with $|b| > 10^{\circ}$ are candidates for local representatives of this Bulge population (cf. Feast *et al.* 1990). Many of them are in the same region of the J-H/H-K diagram as Bulge stars. Adopting a visual absolute magnitude for these local stars similar to that of the late M-giants in the Bulge, Stephenson estimated a scale height (z) for his stars of $\sim$900pc. There are however some problems with both his system of magnitudes and his method of analysis and a revision is in process. We have also obtained radial velocities of 228 of Stephenson's stars and these data should allow us to compare their kinematics with those of other populations.

Studies of local Miras and M-giants away from the Galactic plane seem a promising way of investigating the more metal-rich components of the local Bulge type populations.

References

Blanco, V.M. (1988) *A. J.,* **95**, 1400-1403.

Feast, M.W. (1989) 'Mira Variables, Stellar Evolution and Galactic Structure', in E. G. Schmidt (ed.), *The Use of Pulsating Stars in Fundamental Problems of Astronomy,* IAU Colloq. 111, Cambridge University Press, pp. 205-213.

Feast, M.W., Whitelock, P.A. & Carter, B.S. (1990) *M.N.R.A.S.,* **247**, 227-236.

Frogel, J.A. & Whitford, A.E. (1987) *Ap. J.,* **320**, 199-237.

Harmon, R. & Gilmore, G. (1988) *M.N.R.A.S.,* **235**. 1025-1047.

Lloyd Evans, T. (1976) *M.N.R.A.S.,* **174**, 169-184.

Oort, J.H. & Plaut, L. (1975) *Astr. Ap.,* **41**, 71-86.

Stephenson, C.B. (1986) *Ap. J.,* **301**, 927-937.

Terndrup, D.M., Frogel. J.A. & Whitford, A.E. (1990) *Ap. J.,* **357**, 453-476.

de Vaucouleurs, G. & Pence, W.D. (1978) *A. J.,* **83**, 1163-1173.

Wesselink, T.J.H. (1987) Thesis Nijmegen.

Whitelock, P.A. (1991) This volume.

Whitelock, P.A., Feast, M.W. & Catchpole, R.M. (1991) *M.N.R.A.S.,* **248**, 276-312.

DYNAMICS OF STELLAR POPULATIONS IN GALACTIC DISKS

ROLAND WIELEN, CHRISTIAN DETTBARN, BURKHARD FUCHS, HARTMUT JAHREISS, GUNNAR RADONS
Astronomisches Rechen-Institut
Moenchhofstrasse 12-14
D-6900 Heidelberg
Germany

ABSTRACT. We discuss observational data and dynamical considerations on the age dependence of the spatial distribution and of the velocities of stars in disk galaxies (thick disk, thin disk, spiral arms). We present new results on the age-velocity relation, on the heating of a galactic disk by massive black holes, and on theoretically predicted luminosity and colour profiles in edge-on galaxies.

1. Introduction

We shall concentrate in this paper on those effects in the spatial distribution and in the kinematics of stars which depend on the ages of the stars. This means that we consider the behaviour of different stellar generations born at various epochs. We do not discuss global phenomena such as galactic bars or warps.

It is an important question for our understanding of the evolution of galaxies which of the following processes are mainly reflected in the directly observed age dependence of the spatial distribution and kinematics of stars:

(a) the varying physical conditions in a galaxy at the epochs of formation of the different stellar generations,

(b) dynamical processes which occur after formation, during the lifetime of a stellar generation, or

(c) merger processes, during which (smaller) galaxies are captured by and dissolved in the (main) galaxy under consideration.

Probable answers to this question are different for various components of galaxies:

Halo/Bulge :	(a) formation or (c) merging
Thick disk :	(a) formation or (c) merging
Thin disk :	(b) mainly dynamics
Spiral arms :	(b) dynamics

B. Barbuy and A. Renzini (eds.), The Stellar Populations of Galaxies, 81–92.

2. Thick Disk

The 'thick disk' is mainly seen in the density distribution of stars, derived in our Galaxy from star counts towards the galactic poles at distances z between 1 kpc and 4 kpc (Gilmore and Reid, 1983). The main question is whether the thick disk is really a distinct galactic component or a continuous transition between the halo and the thin disk. The relation between stellar metallicities and velocities, especially in the direction of galactic rotation, can also shed light on this problem (Gilmore et al., 1989; Carney et al., 1990).

It seems to us essentially unavoidable in any scenario in which the formation of a galaxy starts with the formation of a nearly round halo and proceeds later to the formation of a thin disk, to have a transitional phase in this process which leads to the formation of stars with spatial distributions, kinematics, and metallicities intermediate between the (extreme) halo and the (thin) disk. One would otherwise have to postulate an interruption of star formation between the halo and disk phases.

The thick disk cannot have been 'puffed up' from an old thin disk by normal dynamical processes in such a thin disk. It is already difficult to understand the dynamical formation of the old parts of the thin disk, with a scale height of about 0.5 kpc, from a thin gaseous disk.

The merging of one or more smaller galaxies with our Galaxy is a possible explanation for the formation of a thick disk. In such a scenario, the thick disk could be either the direct relict of the merged galaxy, dissolved and smoothly distributed in our Galaxy, or an old thin disk, formed before the merger event(s), could be 'puffed up' by the gravitational perturbations of the merging galaxy.

At present, no clear decision on the nature of the thick disk is possible. Our personal impression is that the thick disk is more likely the last phase of halo formation. If this conjecture is correct, the name 'thin halo' would be more appropriate than 'thick disk' for this phenomenon.

3. Thin Disk

The main observational facts on the age dependence of the spatial distribution and kinematics of stars in the (thin) disk of galaxies are:

(a) The age-velocity relation (AVR): the dispersion σ of the peculiar space velocities increases strongly with age τ.

(b) The scale height h_z in the z direction, perpendicular to the galactic plane, increases strongly with the age of a stellar generation. This is, of course, a dynamical consequence of (a).

(c) Young stars are mainly concentrated to (i.e. they are mainly born in) spiral arms; older stars are distributed smoothly over the whole disk (due to dynamical migration processes).

3.1. AGE-VELOCITY RELATION

The age-velocity relation for nearby stars has been derived by Wielen (1974, 1977) and Jahreiss and Wielen (1983), based mainly on the second Catalogue of Nearby

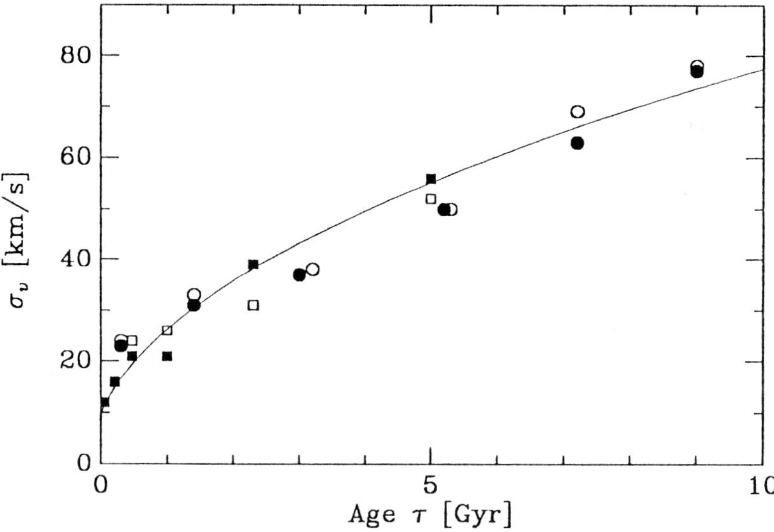

Fig. 1. Age-velocity relation for nearby stars. The total velocity dispersion σ_v is shown as a function of age τ. The data refer to a cylinder perpendicular to the galactic plane at the position of the sun. The filled symbols and the fitting line are taken from Wielen (1977), and are based on the 1969 version of the Catalogue of Nearby Stars. The open symbols are derived from the 1991 data of the Catalogue of Nearby Stars, for the same sample of stars. Circles refer to McCormick K and M dwarfs with measured Ca H and K emission intensities. Squares refer to cepheids or groups of main sequence stars.

Stars (Gliese, 1969). The main observational result of these studies was that the total velocity dispersion σ_v, derived from the three components $\sigma_U, \sigma_V, \sigma_W$ by $\sigma_v^2 = \sigma_U^2 + \sigma_V^2 + \sigma_W^2$, increases with the age τ roughly according to $\sigma_v^2 = \sigma_{v,0}^2 + C\tau$, i. e. $\sigma_v \propto \sqrt{\tau}$ for older stars. The total velocity disperson σ_v increases from $\sigma_{v,0} = 10$ km/s for the youngest stars up to about 80 km/s for the oldest disk stars. The velocity dispersion of the older stars is largely based on the McCormick K and M dwarfs with weak CaII H and K emission intensities. Since these dwarfs have been detected on objective-prism plates, they are free from kinematical selection effects.

Our results on the AVR were in good agreement with a study by Mayor (1974), based on evolved F stars. In recent years, however, a number of investigations (Carlberg et al., 1984; Knude et al., 1987; Stroemgren, 1987; Meusinger et al., 1991) have come to results which strongly deviate from ours. While the young part of the AVR is mostly in good agreement, many of these other studies propose an essentially flat AVR for older stars and smaller total velocity dispersions for the oldest disk stars, sometimes as low as about 40 km/s. The ages and distances of the evolved stars considered in these studies have been derived from photometric data based on the Stroemgren system.

We have confirmed our AVR by using the data from the latest (third) version of the Catalogue of Nearby Stars (Gliese and Jahreiss, in preparation). Figure 1 shows that the AVR based on the new data (1991) is in very good agreement with the AVR based on the old data (1969), although the distances of many stars have been significantly improved during this time. Especially the high velocity dispersions for the older stars are confirmed. Even if our ages for the oldest groups of stars were wrong, the high velocity dispersions of these groups would remain unexplained from the point of view of the other studies. It is also not conceivable that our oldest groups of stars are significantly contaminated by stars from the halo or even the thick disk, since our sample is located near $z = 0$, where the density of the halo and of the thick disk is very much smaller than that of the thin disk. It is also directly visible from the observed velocity distribution of our stars (e.g. Wielen, 1982) that the velocity dispersions are not governed by a few stars with extremely high space velocities.

The discrepancies in the AVRs cannot be explained, as already mentioned, by errors in the age determinations alone. If our AVR is correct, the other studies would have either underestimated the stellar distances (needed for calculating tangential velocities from proper motions, on the basis of the photometric data in the Stroemgren system), or the samples of stars used in these studies would not be representative of disk stars, essentially by avoiding old (and metal-weak) stars. The latter possibility, coupled with a large uncertainty in the ages of the 'older' stars, would explain both a flat AVR and smaller velocity dispersions for older stars found in these studies. Since we have confirmed our AVR using the most recent data on nearby late dwarfs, we shall continue to use this AVR as the observational basis for our dynamical considerations.

3.2. DIFFUSION MECHANISMS

The increase in the velocity dispersion of disk stars with increasing age is very probably due to the gravitational acceleration of stars after their birth by massive objects. Such an interpretation has been described by us in detail in a number of papers and reviews (Wielen, 1977; Fuchs and Wielen, 1987; Wielen and Fuchs 1983, 1985, 1988, 1989, 1990). Three different sources of the diffusion of stellar orbits have been proposed:

(a) Giant Molecular Clouds (GMCs): While GMCs have sufficiently high masses, of the order of 10^5 solar masses, their number at the distance of the sun from the galactic center is too small to explain the locally observed AVR. This has been shown by analytical means by Lacey (1984) and by numerical simulations by Dettbarn (ongoing Ph.D. work). GMCs are, however, highly effective in scattering stars. During such 'deflections', the stellar velocity changes its direction but not its amount. This process is probably important in transferring orbital energy from the motions parallel to the galactic plane into those perpendicular to it, and vice versa.

(b) Spiral arms, wavelets, and other instabilities in the galactic disk: Long-lived spiral density waves are not able to provide the required heating mechanism. Short-lived wavelets may, however, supply sufficient gravitational 'heating' for the galactic disk (e.g. Jenkins and Binney, 1990). GMCs are then required in addition to transfer energy from σ_U and σ_V into σ_W. It is, however, still uncertain whether or not

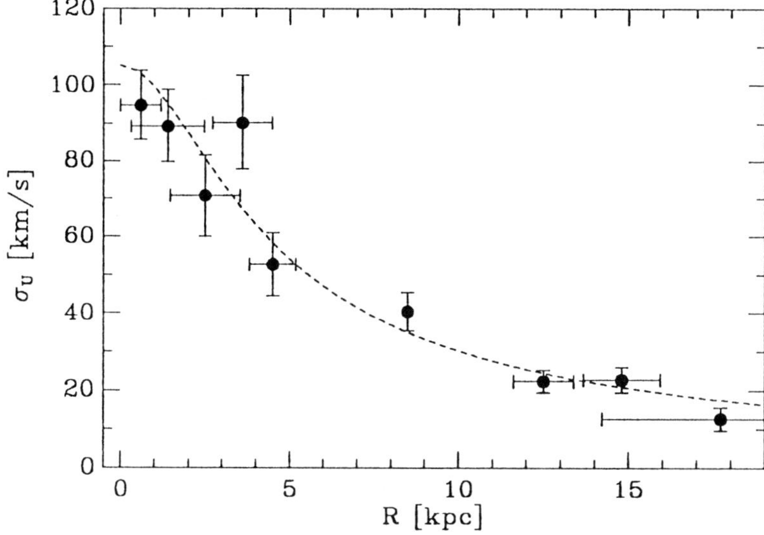

Fig. 2. The variation of the stellar velocity dispersion σ_U as a function of the distance R from the center of our Galaxy. Dots with error bars: Observational data from Lewis and Freeman (1989). Dashed curve: Theoretical prediction for $\sigma_U(R)$, based on the heating of the disk by massive black holes.

the lifetime and strength of a typical wavelet in our galactic disk is adequate to explain the observed AVR. One would need a rather noisy galactic disk ($\beta = 90$ in the terminology of Jenkins and Binney).

(c) Massive black holes (MBHs): Lacey and Ostriker (1985) and Ipser and Semenzato (1985) have proposed that massive black holes are the constituents of the dark coronae of galaxies and are responsible for the gravitational heating of galactic disks. MBHs explain nicely the observed property of the local AVR, namely $\sigma_v(\tau) \propto \sqrt{\tau}$. The individual mass of one typical MBH can be derived from the local AVR and from an estimate of the local mass density of the dark corona, based on the observed (flat) rotation curve. The result is about $3 \cdot 10^6$ solar masses, implying about 10^5 to 10^6 MBHs in the corona of our Galaxy. Heating of the galactic disk by MBHs would also explain the observed behaviour of the velocity dispersion $\sigma_U(R)$, in the radial direction as a function of the distance R from the galactic center, and a nearly constant thickness of the galactic disk, $h_z(R)$. This is shown in Figs. 2 and 3 which are based on formulae given by Wielen and Fuchs (1988), on a softened isothermal dark galactic corona with a density $\rho_{MBH} \propto (a^2 + R^2)^{-2}$ and $a = 3\ kpc$, a self-gravitating disk with a surface density $\mu_d(R) \propto exp(-h_R/R)$, and a flat rotation curve. MBHs would destroy globular clusters very effectively (Wielen 1985, 1987, 1988, 1991; Wielen and Fuchs, 1990). Hence the initial number of globular clusters in our Galaxy and in other galaxies would have been higher by a factor of about 10 or so compared to now, which does not seem impossible.

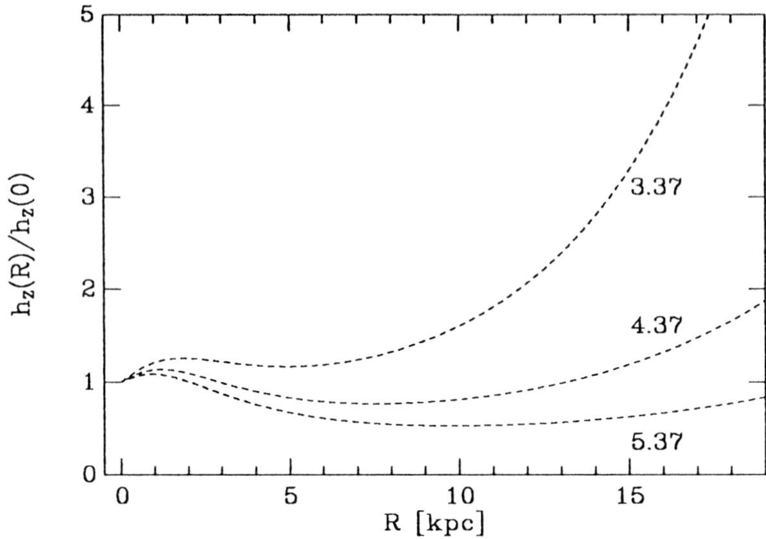

Fig. 3. Theoretical prediction for the scale height $h_z(R)$ in the z direction as a function of the distance R from the center of our Galaxy, based on the heating of the disk by massive black holes. The dashed curves show $h_z(R)/h_z(0)$ for various values of the exponential scale height h_R in the radial direction. The value of $h_R = 4.37$ *kpc*, derived by van der Kruit (1986), leads to a nearly constant value of h_z with respect to R, in fair agreement with observational data.

In summary, MBHs explain many observational results nicely, there are no firm observations which rule out the existence of MBHs, but, of course, MBHs are still quite speculative objects.

3.3. EDGE-ON GALAXIES

The increase in the perpendicular scale height, $h_z(\tau)$, with increasing age τ of a stellar generation is directly observed in the solar neighbourhood only. In other parts of our Galaxy and in external galaxies, no quantitative confirmation of this effect is available at present, although qualitative indications imply its general occurence in galactic disks. The most suitable candidates for studying the age-scale height-relation, $h_z(\tau)$, in external galaxies are those galaxies which are seen edge-on. Unfortunately, it is presently impossible to derive $h_z(\tau)$ from the distribution of individual objects of various ages, e.g. open clusters, in edge-on galaxies. At present, we can only investigate indirect consequences of $h_z(\tau)$, such as colour gradients in the z direction or the luminosity profile in the z direction. In both cases, the data are integrated over all ages τ (and also integrated along the line-of-sight).

The number of observational studies of the luminosity and colour profiles in edge-on galaxies with high photometric accuracy and high resolution in z is still rather limited but hopefully rapidly increasing due to the use of CCDs. A prototype

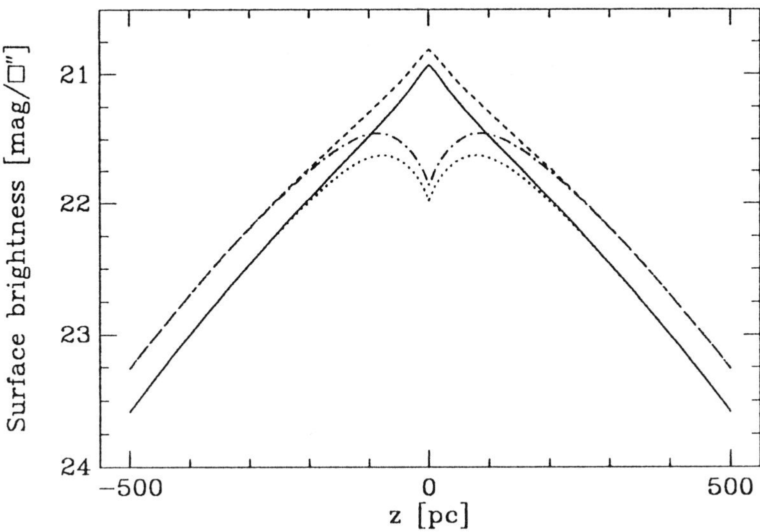

Fig. 4. Theoretical prediction for a luminosity profile in an edge-on galaxy, based on the diffusion theory of stellar orbits. The curves show the run of the surface brightness in the V filter (in magnitudes per unit area, with an arbitrary zero point) as a function of the height z above the galactic plane. Full curve: no dust, constant metallicity; dashed curve: no dust, age-dependent metallicity; dotted curve: with dust, constant metallicity; dash-dotted curve: with dust and age-dependent metallicity.

of such an investigation are the observations carried out by Wainscoat et al. (1989).

We have started to predict the luminosity and colour profiles in edge-on galaxies theoretically on the basis of a phenomenological theory of the diffusion of stellar orbits (G. Radons, ongoing Ph.D. work). In Figs. 4 and 5, we present some results which are based on a constant diffusion coefficient, on a constant star formation rate, and which would apply for an extragalactic observer who looks at our Galaxy edge-on and measures the profiles in the z direction at the solar distance ($R_\odot = 8.5\,kpc$).

Let us first neglect the absorbing effect of dust and the change of stellar metallicities with age. Then the luminosity profile, integrated over all ages, is extremely well-fitted by an exponential law. The colour profile shows the young blue stars at z = 0 and the redder colours of older stars at higher distances z. If we include an age-metallicity relation, the colour profile gets slightly flatter, because the older stars at higher z are not as red as before. The reddening by a dust layer with an exponential scale height of 50 pc has, however, a dramatic effect. The luminosity profile then shows a dip at z = 0, and the colour profile becomes flatter. If we add up all the effects, the colour profile varies only slightly with z, by about $0\overset{m}{.}1$. The exponential decline of the surface brightness is also strongly perturbed by dust absorption. In summary, our theoretical predictions for the luminosity and colour profiles in the z direction in edge-on galaxies seem to indicate that it should be

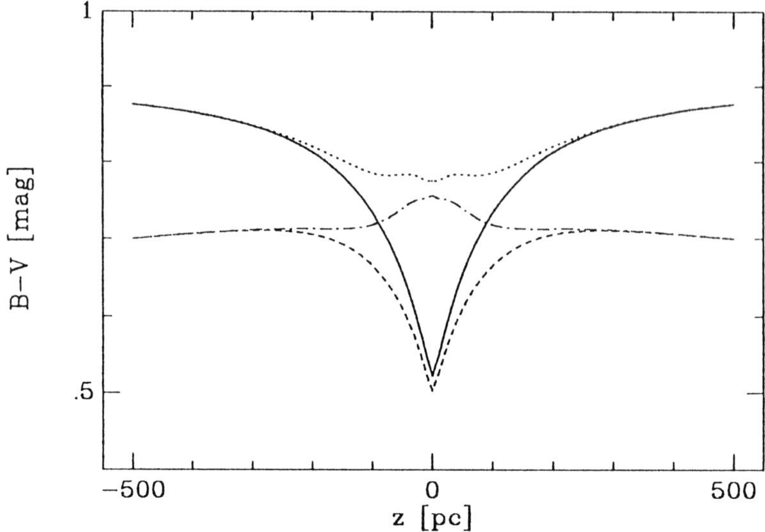

Fig. 5. Theoretical prediction for a colour profile in an edge-on galaxy, based on the diffusion theory of stellar orbits. The colour B-V (with a realistic zero point) is shown as a function of z. Full curve: no dust, constant metallicity; dashed curve: no dust, age-dependent metallicity; dotted curve: with dust, constant metallicity; dash-dotted curve: with dust and age-dependent metallicity.

rather difficult to disentangle the population effect $h_z(\tau)$ from other effects in the profiles of edge-on galaxies, especially from dust reddening. Nevertheless, it would be encouraging if the predicted luminosity and colour profiles can be brought into agreement with the observational data. This would strengthen our confidence that the increase of the scale height $h_z(\tau)$ with increasing age is a general phenomenon in disk galaxies, which may, of course, differ quantitatively from galaxy to galaxy, because of various heating mechanisms, star formation rates etc. .

4. Spiral Arms

There are two main dynamical processes which could cause spiral structure of grand design in a galactic disk: density waves and tidal interactions. Tidal interactions are certainly able to trigger spiral structure (e.g. Toomre and Toomre, 1972; Hearnquist, 1990). Such tidal arms are, however, transient phenomena. It is difficult to answer the question of whether or not long-living density waves occur in isolated galaxies because of some inherent instabilities of a galactic disk. C.C. Lin and his coworkers (Lin and Lowe, 1990; Bertin et al., 1989a,b) favour the existence of such quasi-stationary density waves, mainly on the basis of their analytical studies of the dynamics of galactic disks. Other theoreticians doubt the possibility of such global density waves, and are in favour of short-lived wavelets which are excited by 'swing amplification' (e.g. Toomre, 1981, 1990). Numerical simulations have difficulties in producing long-living density waves. Most simulations of isolated, non-barred galaxies produce wavelets and a spiral structure which varies rather rapidly in time (e.g. Sellwood, 1987). In barred galaxies or in galaxies with other oval distortions, it is much easier to excite spiral density waves of grand design by gravitational torques.

Whatever the dynamical cause of spiral structure is, it is very plausible that the higher density in the spiral arms, in contrast to the interarm region, triggers the formation of young stars in the spiral arms. In the frame-work of the density-wave theory, spiral shock-fronts are predicted in the gaseous interstellar medium or in the ensemble of interstellar clouds. Such shock fronts are normally also present in tidally-induced spiral arms. This triggering of star formation in spiral arms would explain the concentration of extremely young stars towards spiral arms. The initial dispersion of space velocities of young stars is already sufficient for the migration of these stars out of the spiral arms in which they were born. This produces an increase in the thickness of the spiral arms with increasing age of the stars used to define such arms. If the spiral structure is in addition a wave phenomenon, with a wave velocity different from that of the material, then the center of the spiral arm as defined by older stars, would also be displaced relative to that of newly born stars. This 'ageing' of spiral arms has been studied by numerical calculations of orbits of stars (e.g. Wielen 1978, 1979). The results are rather complicated, and it does not seem very promising to deduce the basic mechanism of the formation of spiral arms from a study of the spatial distributions and kinematics of stars of various ages.

References

Bertin, G., Lin, C.C., Lowe, S.A., Thurstans, R.P. (1989a), Astrophys.J. **338**, 78.

Bertin, G., Lin, C.C., Lowe, S.A., Thurstans, R.P. (1989b), Astrophys.J. **338**, 104.

Carlberg, R.G., Dawson, P.C., Hsu, T., Vandenberg, D.A. (1985), Astrophys.J. **294**, 674.

Carney, B.W., Latham, D.W., Laird, J.B. (1990), Astron.J. **99**, 572.

Fuchs, B., Wielen, R. (1987), in The Galaxy, G. Gilmore and B. Carswell (eds.), D. Reidel Publ.Comp., Dordrecht, p.375.

Gilmore, G., Reid, N. (1983), Mon.Not.R.Astron.Soc. **202**, 1025.

Gilmore, G., Wyse, R.F.G. Kuijken, K. (1989), Annual Rev.Astron.Astrophys. **27**, 555.

Gliese, W. (1969), Veroeff.Astron. Rechen-Inst.Heidelberg No. 22.

Hearnquist, L. (1990), in Dynamics and Interactions of Galaxies, R. Wielen (ed.), Springer-Verlag, Berlin, p. 108.

Ipser, J.R., Semenzato, R. (1985), Astron.Astrophys. **149**, 408.

Jahreiss, H., Wielen, R. (1983), in IAU-Colloquium No. 76, The Nearby Stars and the Stellar Luminosity Function, A.G.D. Philip and A.R. Upgren (eds.), L. Davis Press, Schenectady, New York, p. 277.

Jenkins, A., Binney, J. (1990), Mon.Not.R.Astron.Soc. **245**, 305.

Knude, J., Schnedler Nielsen, H., Winther, M. (1987), Astron.Astrophys. **179**, 115.

Lacey, C.G. (1984), Mon.Not.R.Astron.Soc. **208**, 687.

Lacey, C.G. Ostriker, J.P. (1985), Astrophys.J. **299**, 633.

Lewis, J.R., Freeman, K.C. (1989), Astron.J. **97**, 139.

Lin, C.C., Lowe, S.A. (1990), in Galactic Models, Proceedings of the 4. Florida Workshop on Non-linear Dynamics, J.R. Buchler, S.T. Gottesman, J.H. Hunter (eds.), Ann. New York Acad.Sci., Vol. 596, p. 80.

Mayor, M. (1974), Astron.Astrophys. **32**, 321.

Meusinger, H., Reimann, H.-G., Stecklum, B. (1991), Astron.Astrophys. **245**, 57.
Sellwood, J.A. (1987), in Proceedings of the Tenth Europ. Regional Astron. Meeting of the IAU, Vol. 4, J. Palous (ed.), Publ.Astron.Inst.Czech.Acad.Sci., No. 69, p. 249.
Stroemgren, B. (1987), in The Galaxy, G. Gilmore and B. Carswell (eds.), D. Reidel Publ.Comp., Dordrecht, p. 229.
Toomre, A. (1981), in The Structure and Evolution of Normal Galaxies, S.M. Fall and D. Lynden-Bell (eds.) Cambridge Univ. Press, Cambridge (UK), 1981, p. 111.
Toomre, A. (1990), in Dynamics and Interactions of Galaxies, R. Wielen (ed.), Springer-Verlag, Berlin, p. 292.
Toomre, A., Toomre, J. (1972), Astrophys.J. **178**, 623.
van der Kruit, P.C. (1986), Astron.Astrophys. **157**, 230.
Wainscoat, R.J., Freeman, K.C., Hyland, A.R. (1989), Astrophys. J. **337**, 163.
Wielen, R. (1974), Highlights of Astronomy **3**, 395.
Wielen, R. (1977), Astron.Astrophys.**60**, 263.
Wielen, R. (1978), in IAU Symposium No. 77, Structure and Properties of Nearby Galaxies, E.M. Berkhuijsen and R. Wielebinski (eds.), D. Reidel Publ.Comp., Dordrecht, p. 93.
Wielen, R. (1979), in IAU Symposium No. 84, The Large-Scale Characteristics of the Galaxy, W.B. Burton (ed.), D. Reidel Publ.Comp., Dordrecht, p. 133.
Wielen, R. (1982), Kinematics and Dynamics of the Galaxy, in Landolt-Boernstein, Group VI, Vol. 2, Subvol.2c, Chapter 8.4, H.H. Voigt and K. Schaifers (eds.), Springer-Verlag, Berlin, p. 208.
Wielen, R. (1985), in IAU Symposium No. 113, Dynamics of Star Clusters, J. Goodmann and P. Hut (eds.), D. Reidel Publ.Comp., Dordrecht, p. 449.
Wielen, R. (1987), in Proceedings of the Tenth Europ.Regional Astron. Meeting of the IAU, Vol. 4, J. Palous (ed.), Publ.Astron.Inst.Czech.Acad.Sci., No. 69, p. 157.
Wielen, R. (1988), in IAU Symposium No. 126, The Harlow-Shapley Symposium on Globular Cluster Systems in Galaxies, J.E. Grindlay and A.G.D. Philip (eds.), Kluwer Acad.Publ., Dordrecht, p. 393.
Wielen, R. (1991), in The Formation and Evolution of Star Clusters, K. Janes (ed.), Astron.Soc. of the Pacific, Conference Series, Vol. 13, p. 343.
Wielen, R., Fuchs, B. (1983), in Kinematics, Dynamics and Structure of the Milky Way, W.L.H. Shuter (ed.), D. Reidel Publ.Comp., Dordrecht, p. 81.
Wielen, R., Fuchs, B. (1985), in IAU Symposium No. 106, The Milky Way Galaxy, H. van der Woerden, R.J. Allen and W.B. Burton (eds.), D. Reidel Publ.Comp., Dordrecht, p. 481.
Wielen, R., Fuchs, B. (1988), in The Outer Galaxy, Proceedings of a symposium held in honor of Frank J. Kerr; L. Blitz and F.J. Lockmann (eds.), Lecture Notes in Physics, Vol. 306, Springer-Verlag, Berlin, p. 100.
Wielen, R., Fuchs, B.(1989), in Evolutionary Phenomena in Galaxies, J.E. Beckmann and B.E.J. Pagel (eds.), Cambridge Univ.Press, Cambridge (UK), p. 244.
Wielen, R., Fuchs, B. (1990), in Dynamics and Interactions of Galaxies, R. Wielen (ed.), Springer-Verlag, Berlin, p. 318.

Discussion

FREEMAN: I am curious to understand the difference between your age-velocity dispersion relation and that derived from the recent Stroemgren observations. Your highest velocity dispersion point lies midway between old disk and thick disk values, and your sample of 300 dwarfs might include about 10-15 thick disk stars. Is it possible that age errors from the Ca-emission technique could smooth the Stroemgren form of the σ-τ relation (flat σ-τ for old disk with $\sigma_W \simeq 20\,km/s$ and then $\sigma_W \simeq 40\,km/s$ for the oldest stars) into the form that you show ? How large do you think the Ca-age errors are ?

WIELEN: It is difficult to comment on the AVR based on 'recent Stroemgren observations' as long as neither the AVR nor the basic data and selection procedures have been published in detail or made available to us otherwise. General comments are given in Chapter 3.1 of the present paper. The absolute values of our Ca-ages depend on an assumed star formation rate (e.g. constant in time over 10^{10} years). The relative ages of old K and M dwarfs derived from the CaII emission intensities are probably more accurate than those derived for evolved stars by using isochrones based on stellar-evolution theory. For example, the group of stars with the lowest H and K emission intensities (-2 to -5) represents that 20 % of our sample with the highest ages (between 8 and 10 billion years). The main problem may be to select properly a representative sample of old disk stars.

NEMEC: Isn't the obvious explanation for why your age-velocity dispersion relationship differs from those of Carlberg et al.(1985) and Stroemgren (1987) that the three samples were constructed with different selection effects, causing the mixing proportions of stars from the different stellar populations to be different ?

WIELEN: Maybe. But even then the question remains which sample of stars represents most properly the old disk stars in which we are interested, when discussing diffusion mechanisms, for example.

OSTRIKER: One interesting feature of the massive black hole hypothesis for disc thickening is that it would definitely predict the existence of a thick disc. The break point found in the work I did with Lacey was at about 35 km/s in W.

WIELEN: In this case, the thick disk would not contain only old stars. It should be interesting to calculate the age distribution of stars at various heights z above the galactic plane, predicted from your scenario, and to try to test this prediction by observations.

SPERGEL: Two comments: (1) Numerical simulation done by Bill Bies, a Princeton undergraduate, using Matsuda's hydrocode finds that a triaxial bulge can excite believable spiral structure. (2) Massive black holes would have a dramatic effect on dwarf irregulars. Due to the high dark matter density and small velocity dispersions, MBHs would rapidly heat and disrupt the disks in these systems.

KURUCZ: It is easy to puff up a thin disk. Let there be 10000 globular clusters after the disk is formed. Their orbits carry them through the disk many times before they dissipate. Glancing collisions with disk stars heat the disk. The heating would decrease with radius and would decrease as a function of time as the clusters disintegrate.

WIELEN: This scenario is an interesting proposal. However, I think it meets quantitative difficulties: (a) Even if you put all the mass of the halo into permanent globular clusters of normal mass (say $2 \cdot 10^5$ solar masses each), the heating would be more than 10 times smaller then required by the local diffusion coefficient (derived from the observed AVR). (b) The main source of cluster disruption would be encounters among the globular clusters themselves. If one estimates in this scenario the number of clusters surviving over a Hubble time (by using procedures similar to those applied by Wielen (1985, 1987, 1988, 1991) for calculating the disruption of globular clusters by MBHs), the predicted number of surviving globular clusters is probably higher than we actually observe in our Galaxy.

MATEO: You mentioned that MBHs will destroy globular clusters; yet McClure and van den Bergh pointed out some time ago that the old open clusters are also the ones found furthest from the plane of the Galaxy. Assuming these clusters diffused to their current heights, how can MBHs account for this ?

WIELEN: It is true that massive black holes are also very efficient in destroying older open clusters (e.g. Wielen, 1985). However, distant encounters between MBHs and open clusters can accelerate open clusters (by the total gravitational forces of MBHs) without destroying them (by the tidal forces of MBHs). This would explain old open clusters at high distances z.

CARNEY: Would you expect such a large number of massive black holes to have a disruptive effect on very wide binaries in the disk and in the halo ? By wide, I mean separation of 0.1 pc and up, very loosely bound.

WIELEN: For binaries with separations of about 0.1 pc, the mean disruption time due to massive black holes is roughly of the order of 10^{10} years. Due to the stochastic nature of close encounters between the binaries and MBHs, a significant fraction of such binaries would survive over a much longer time, avoiding by chance close encounters with MBHs. Encounters with giant molecular clouds are probably more dangerous for wide binaries than those with MBHs.

ZINNECKER: Wouldn't it be better to test your predicted colour-height relation for edge-on spirals in the near-infrared (e.g. J-K), where dust extinction is greatly reduced while the sensitivity to metallicity is preserved ?

WIELEN: It is easy for us to use any filter you like. V and B-V, shown in Figs. 4 and 5, have been selected as examples only. We are already making predictions for a large number of colours. The aim is to find an optimum colour in which simultaneously the effect of different ages is maximized and the effect of dust is minimized.

THE STELLAR CONTENT OF SPIRAL ARMS

P.S. CONTI
Joint Institute for Laboratory Astrophysics
University of Colorado
Boulder, Colorado 80309-0440 USA

ABSTRACT. I will discuss the spiral arms of our Galaxy as indicated by the Wolf-Rayet star population and their connection to GMCs, the sizes and shapes of some nearby associations in the northern Milky Way, and the slope of the IMF for Cyg OB2 and several regions of the Magellanic Clouds.

1. Introduction

Our knowledge concerning the **stellar** content of spiral arms in our Galaxy is limited to O stars, supergiants, and those highly evolved objects of Wolf-Rayet (W-R) type. These stars should in principle define the spiral structure. By contrast, when we visually examine other galaxies the morphology we observe is usually **not** that of the individual hottest stars but rather the ionized gas (HII regions) excited by these objects along with the older supergiants (which have smaller bolometric corrections and thus are visually brighter). In this review I wish to touch briefly on several aspects of the title, posed as three questions: (1) Where **are** the spiral arms of our Milky Way? (2) What are the **sizes** and **extent of** stellar associations? (3) How **many** massive stars are found in associations (**IMF slope; upper, lower mass "cutoffs"**)? This discussion should be considered as a progress report.

2. Where are the Spiral Arms of the Galaxy?

The first identification of the spiral arms of our galaxy, as outlined by stars, was given nearly forty years ago by Morgan et al. (1953) who utilized two dozen northern OB stars with distances up to a few kpc. Humphreys (1970) presented a similar diagram for many supergiants (in both hemispheres) and the positions of multitudes of O stars and W-R stars have been considered by Garmany et al. (1982), and van der Hucht et al. (1988), respectively. Interspersed with these efforts have been radio observations of the cool HI gas and hot HII regions (a thorough discussion was given by Georgelin and Georgelin (1976). The cold gas and H_2 can be traced out by CO measurements at 2.6mm; a summary is provided by Combes (1991). Galactic structure determinations based upon stars has the advantage that the distances can be determined from well established calibrations of the spectral subtypes. Unfortunately, stars are difficult to observe at great distances, or through appreciable Galactic extinction and

B. Barbuy and A. Renzini (eds.), The Stellar Populations of Galaxies, 93–101.

stellar surveys are very incomplete. By contrast, HI and HII regions and CO clouds can be observed across the galaxy and the surveys are reasonably complete. They suffer from less certain distance estimates, being derived from a galactic rotation model with kinematic uncertainties (non-circular velocities).

I would like here to discuss a connection between the distribution of W-R stars and massive molecular clouds (GMCs) in the Galactic plane. W-R objects are the descendants of the most massive O stars (Conti et al. 1983; Humphreys et al. 1985) and should be excellent indicators of extreme Population I. These stars have very strong winds leading to a broad emission line spectrum (Abbott and Conti 1987). While we do not fully understand the physics of radiative transfer in dense moving media, the strong emission features can readily be recognized on objective prism plates, or with narrow filter techniques (Amrandroff and Massey 1985), and relatively faint objects identified and classified. Using "classic" spectroscopic techniques (calibration of Mv and $(b-v)_o$ with spectral subtypes), Conti and Vacca (1990) have determined distances (good to 50%) to most of the 160 (known) W-R stars of our Galaxy. For W-R stars within 4 kpc, with the most accurate distances, the "z" scale height is 45 pc, thus they are indeed extreme Population I. The numbers of known W-R stars in our Galaxy are, however, **seriously incomplete.**

Leo Bronfman first called my attention to the spatial correlation between the Galactic GMCs and W-R stars in the fourth quadrant. One would expect that the former objects are the birthplaces of many (all?) associations and massive stars. The latter, with (total) ages of only a few million years, cannot move far from their origins. I would like to present a **preliminary** plot here; a more complete discussion will be presented elsewhere (Conti et al. 1992). In Fig. 1 I show the distribution of massive GMCs inward from the sun (longitude only from 280 to 45 degrees - obtained from Bronfman) overlain on the W-R stars (from Conti and Vacca 1990). There is in the fourth quadrant a nice correlation between some of the GMCs and several W-R stars, especially at large distances of 10-12 kpc towards longitudes 315 and 340 degrees (which might be identified with the "Scutum-Crux" and "Norma arm" features, respectively, of Georgelin and Georgelin 1976). There appears to be a long trailing arm (Sag-Carina) near the sun towards 290 degrees (Shara et al. 1991 have detected a well-defined additional 13 W-R stars in just this direction, four at large distances).

There is a curious absence of **either** GMCs or W-R stars from 4-9 kpc away from the galactic center between 310 and 340 degrees; neither spiral arm as defined by these objects extends through this region. Furthermore, closer to the sun in these same directions and around through the galactic center to about 20 degrees longitude, there are considerable numbers of W-R stars and GMCs, with **no discernable delination of "arms"**. One has the impression that these relatively nearby extreme Pop I indicators have the distributon of a **bar** inward from the sun, aligned along an angle behind it. One is reminded of the arguments given by Blitz and Spergel (1991) that our Galaxy is a tri-axial ellipsoid with major axis aligned at 45 degrees with respect to the sun line to the Galactic center. Their Fig. 10 schematic representation is in reasonable agreement with the distribution shown in Fig. 1 (the more distant W-R stars towards the first quadrant are, of course, missing - hidden behind the extensive dust in these directions). More complete statistics on W-R stars would greatly help our understanding of Galactic structure.

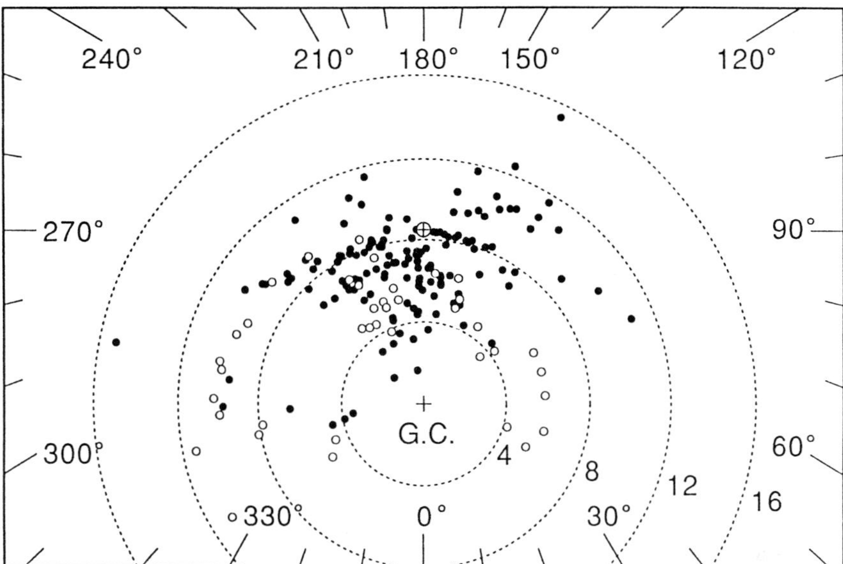

Figure 1. Galactic plane distribution of W-R stars (Conti and Vacca 1990). Filled circles, and GMCs (Bronfman, private communication). Open circles, with $M > 10^6\ M_\odot$. Is a "bar" inward from the sun present in these data?

3. What are the Sizes and Extent of Galactic Associations?

I would like to share with you some of the work my colleague Dr. Garmany and her associates are doing to answer these questions. Stellar associations are groupings of very young stars at comparable distances with similar kinematic properties. Proper motion information for O stars with distances of kpc is difficult to obtain; radial velocity data is often compromised by binary stars. A necessary requirement is that all stars belonging to an association be at the same distance. This step requires careful spectral classification and reliable photometry, especially for the Galaxy, to accurately derive the extinction and the distance for individual stars.

Garmany and Stencel (1991) have re-discussed the membership of a number of nearby associations in the northern sky using data in the literature and newly acquired spectra and photometry. Figure 2 displays a small region of their survey. The upper panel shows the IAU defined association boundaries (aligned with the cardinal directions!); the lower panel is a hand drawn attempt to encircle **only** those OB stars and supergiants that are at similar distances.

The relatively distant Cep OB3 association is about half the IAU size; the nearby Cep OB2 star group has a distinctly non-rectangular shape; Cep OB1 appears to be two subgroups. The physical sizes are about 50 - 100 pc when one takes account of the distances. Careful analysis like this must proceed over the entire Galactic plane before we can be completely confident of the stellar populations in various associations. In the Magellanic Clouds, association perimeters are more readily determined from just photometry since the distance ambiguity does not exist (e.g. Lucke 1972).

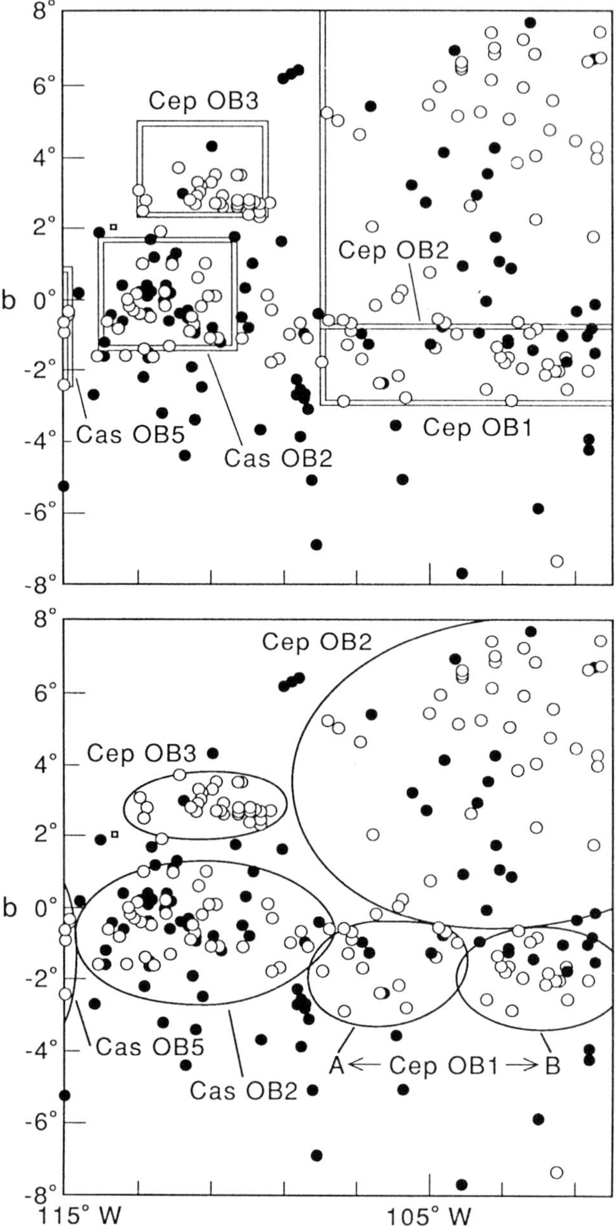

Figure 2. Spatial distribution of early type stars and supergiants. The open circles are stars at the same distances, thus defining the listed association; the filled circles are other stars. Upper panel: IAU defined boundaries; lower panel: hand drawn circumferences of associations (adapted from Garmany, private communication).

4. How Many Massive Stars are Found in Associations; IMF; Upper, Lower Mass Cutoffs?

Following Scalo (1986) one can define the slope of the Initial Mass Function as $\Gamma = d \log \xi(\log m)/d\log m$ where $\xi(\log m)$ is the mass function in units of number of stars born per unit logarithmic (base ten) mass interval per unit area. For a power law mass spectrum $f(m) = A m^{\gamma}$ and the index $\gamma = \Gamma - 1$. In Tinsley's (1980) notation, $\gamma = -(x+1)$ and $x = -\Gamma$; in these units the Salpeter (1955) slope is =-1.35. Garmany et al. (1982) utilized published photometry, spectral types and luminosity classes to place the OB stars within 2.5 kpc of the sun on a theoretical HR diagram constructed with mass tracks. It was then a simple procedure to count all the stars within various mass ranges and determine the IMF slope. Their result suggested that for massive stars $\Gamma = -1.3$ inward and $= -2.1$ outward of the solar circle. The statistics of this 5 kpc diameter region may not be complete (e.g. Humphreys and McElroy 1984) and a distinction between the inner and outer galaxy might not exist. Additional identifications of OB stars within this volume are currently underway by Garmany and associates and future results will be forthcoming.

I will now consider recent work on the massive star content of several well studied associations in the Magellanic Clouds, along with Cyg OB2, by Garmany and Massey and colleagues. Their procedure is to first acquire CCD UBV photometry to derive the reddening and Mv accurately. They then take spectra of the bluest and brightest stars so as to obtain accurate effective temperatures of the O stars (for which the photometry is degenerate - Massey 1985). For B0.5 and later type stars, the B-V colors are sufficient to determine temperatures. The stars are then located on Maeder and Meynet's (1988) evolution tracks and enumerated with respect to the masses. This leads directly to Γ with an estimated uncertainty of perhaps 0.2 in the slope.

Massey et al. (1989a) have studied LH 117 and 118 in the LMC. An analysis of NGC 346, the brightest HII region in the SMC, was made by Massey et al. (1989b). In that paper, they give a nice demonstration of how one would derive a different (steeper) slope for the IMF if one used **only** photometry for the O type stars. Parker et al. (1991) have analyzed two other associations in the LMC, LH 9 and LH 10. The latter is the next brightest HII region after 30 Dor. Massey and Thompson (1991) have discussed Cyg OB2. At the poster session of this meeting, Parker has given preliminary results for the 30 Dor region. Parker and Walborn (1991) have recently suggested that Γ varies within the 30 Dor complex and it may depend on sequential effects of star formation. All of the studies listed here have treated the comparable data in the same fashion; thus differences in Γ are undoubtedly significant.

Table 1. IMF slopes for various associations.

Association	Γ	galaxy
Galaxy field (inner)	-1.3	Milky Way
Galaxy field (outer)	-2.1	Milky Way
Cyg OB2	-1.0	Milky Way
LH9	-1.6	LMC
LH10	-1.0	LMC
LH117	-1.8	LMC
LH118	-1.8	LMC
30 Dor (mean)	-1.5	LMC
NGC 346	-1.8	SMC

There are substantial differences in the derived values of Γ for Cyg OB2, LH10 and portions of 30 Dor (Parker and Walborn 1991) compared to the other associations, which are similar to one another. Furthermore, the Γ are **not** a simple function of the metal abundances (which are lower in the LMC and SMC than in the solar vicinity). These conclusions are the results of careful investigations based upon **direct counts of stars** and would seem to be firmly established.

A perusal of the referenced papers reveals **no** evidence for **upper mass cutoffs** of star formation, beyond the statistics of each association. There is also **no** evidence from star counts bearing upon the existence of **low mass cutoffs**. One could imagine a scenario in which the formation of large numbers of massive stars, with their attendant radiation and winds, might snuff out formation of low mass stars. 30 Dor is a likely location and deep photometry of this region might reveal such a phenomenon if it exists.

Studies of the integrated properties of other galaxies have sometimes suggested the existence of upper or lower mass cut-offs. For example, Reike et al. (1980) suggest a general absence of low mass stars in the starburst region of M82; this result has been critcized by Lester et al. (1990). My reading of the controversy is that the jury is still out, but my instinct is that low mass cut-offs may occur under extreme conditions; the study of 30 Dor will tell the tale. As another example, Joseph (1991) suggests that the HeI 2.058 μm/Brackett γ line ratio observed in the starburst galaxy NGC 3256 implies no O stars earlier than type O8. While the argument seems sound, it appears to me that this may merely be an effect of evolution. While the age of the burst may well be 12 million years, as suggested by Joseph, it could well have more recently **turned off**. Then the absence of hotter O stars would merely be an effect of subsequent evolution away from the main sequence. This may also be the situation in NGC 1569 (Waller 1991), a starburst galaxy with bright blue stars (B supergiants?) and strong diffuse Hα but without evidence for high excitation from early O stars. The peculiar gas morphology might be due to numerous recent SN events, something that might be expected if the burst is over and the late type O stars are just now dying.

5. Some Future Perspectives

Population studies of relatively nearby giant HII regions such as 30 Dor can lead us to a better understanding of more distant starburst galaxies. When we investigate more distant objects, individual stars cannot be identified and we must depend on integrated spectra. It is thus important to couple our nearby stellar census with **spatially integrated** spectra of the same regions. In this connection, we (PSC, Mark Phillips and Bill Vacca) have obtained an 8'x8' minute moderate resolution spectrum of the 30 Dor region (by drifting slowly in RA and accumulating counts on the CCD). Estimating the number of Lyman continuum photons (NLyc) from the Hβ flux in the usual way, and taking one O7V star to provide 7 x 10^{48} Lyc/s, we predict the presence of about 100 O7V "equivalents." We believe this is reasonably close to the number of hot stars present but will have a better result when Parker's PhD census of 30 Dor is complete. (We also plan to obtain higher resolution spectra and do a "full blown" analysis of the entire emission line spectum).

Finally, let me turn to a new direction my work has been taking, that of W-R galaxies. These are a subset of starburst galaxies in which a broad 4686 HeII line, due to W-R stars, has been found in the integrated spectrum (Conti 1991). These galaxies also have a strong nebular emission line spectrum, due to the presence of many O stars. In Table 2 I abstract some of the hot star properties inferred for several of these objects, compared to better known nearby HII and GHII regions. The data in the last 3 rows is shown in the poster session here and will be presented in detail elsewhere as part of Bill's PhD thesis.

Table 2. Hot Star Properties of Recent Starbursts

Name	# O7V equivalents	Object
Orion	1 (!)	HII
NGC 346	5	HII SMC
30 Dor	100	GHII LMC
Mrk 1236	1600	W-R galaxy
IIZw40	10000	W-R galaxy
POX 4	30000	W-R galaxy

In the W-R galaxies, the presence of substantial numbers of W-R stars ensures that we are viewing a relatively recent burst of massive star formation. The IMF of these regions may be "flatter" than normal, but other complications (e.g. leakage of Lyman α from the galaxy) may play a role (Vacca and Conti 1992). NGC 346 and 30 Dor are nearby paridigms of the more distant galaxies with many O type and W-R stars and the detailed analyses of their integrated properties, compared to the individual stellar content, will help our understanding of recent starbursts.

ACKNOWLEDGMENTS

I appreciate continuing support by the NSF, most recently under grant AST90-15240. I would like to thank Leo Bronfman, Katy Garmany, Phil Massey, Joel Parker and Bill Vacca for continuing fun interactions and preprints.

REFERENCES

Abbott, D.C. and Conti, P.S. 1987 **ARAA** 25, 113.
Armandroff, T.E. and Massey, P. 1985 **ApJ** 291, 685.
Blitz, L. and Spergel, D.N. 1991 **ApJ** 370, 205.
Combes, F. 1991 **ARAA** 29, 195.
Conti, P.S. 1991 **ApJ** 377, 115.
Conti, P.S., Garmany, C.D., de Loore, C. and Vanbeveren, D. 1983 **ApJ** 274, 302.
Conti, P.S. and Vacca, W.D. 1990 **AJ** 100., 431.
Conti, P.S., Bronfman, L., May, J. and Vacca, W.D. 1992 in preperation.
Garmany, C.D., Conti, P.S., and Chiosi, C. 1982 **ApJ** 263, 777.
Garmany, C.D. and Stencel, R.E. 1991 **AA** in press.
Georgelin, Y.M. and Georgelin, Y.P. 1976 **AA** 49, 57.
Humphreys, R.M. 1970 **AJ** 75, 602.
Humphreys, R.M. and McElroy, D.B. 1984 **ApJ** 284, 565.
Humphreys, R.M., Nichols, M. and Massey, P. 1985 **AJ**, 90, 101.
Joseph, R. 1991 in "Massive Stars in Starbursts" Eds. C. Leitherer, N.R. Walborn, T.A. Heckman and C.A. Norman, p. 259.

Larson, R.B. 1987 in "Starbursts and Galaxy Evolution" Eds. T.X. Thuan, T. Montmerle, J. Tran Thanh Van, p. 467.
Lester, D.F., Carr, J.S., Joy, M and Gaffey, N. 1990 **ApJ** 352, 544.
Lucke, P.B. 1972 Ph.D. Thesis University of Washington.
Massey, P. 1985 **PASP** 97, 5.
Massey, P., Garmany, C.D., Silkey, M. and Degioia-Eastwood, K. 1989a **AJ** 97, 107.
Massey, P., Parker, J.W. and Garmany, C.D. 1989b **AJ** 98, 1305.
Massey, P. and Thompson, A.B. 1991 **AJ** in press.
Maeder, A. and Meynet, G. 1988 **AAS**, 76, 411.
Morgan, W.W., Whitford, A.E. and Code, A.D. 1953 **ApJ** 118, 318.
Parker, J.W. and Walborn, N.R. 1991 **ApJL** in press.
Reike, G.H., Lebofsky, M.J., Thompson, R.I., Low, F.J. and Tokunaga, A.T. 1980 **ApJ** 238, 24.
Salpeter, E.E. 1955 **ApJ** 121, 161.
Scalo, J.M. 1986 Fundamentals of Cosmic Physics 11, 1.
Shara, M.M., Moffat, A.F.J., Smith, L.J. and Potter, M. 1991 AJ 102, 716.
Tinsley, B.M. 1980 Fundamentals of Cosmic Physics 5, 287.
Vacca, W.D. and Conti, P.S. 1992 in preparation.
van der Hucht, K.A., Hidiyat, B., Admiranto, A.G., Supelli, K.R. and Doom, C. 1988 **AA** 199, 217.
Waller, W.H. 1991 **ApJ** 370, 144.

DISCUSSION

Question (V.H. Hughes) We have identified from the IRAS Point Source Catalogue about 220 HII regions (confidence level of 80%) by making use of the color-color diagrams. The HII regions are confined closely to the galactic plane (within 3°). Using radio observations, the HII regions are generally not resolved at 0.3" (maximum sizes of 0.1 pc) and they appear to contain hot stars of similar spectral type. The stellar distribution may be different from what is found in optically observed systems (Hughes and Macleod AJ, 97, 786, 1989).
Response (P.S.Conti) You seem to be suggesting that obscured HII regions might be smaller than optical ones and their stellar content might be different. While I can accept the former, due to their youth, the latter conclusions seems to me to be unlikely. Many HII and GHII regions in the Milky Way are highly obscured and their stellar content not yet studied due to dust extinction. Infra-red spectroscopy might help.
Question (C. Wilson) Two comments: First, I have recently re-identified the OB associations in the inner disk of M33 and MGC 6822 and the mean diameters are roughly 80 pc, very similar to the sizes from Garmany's new boundaries for Galactic associations. Second, the spatial distribution of Galactic W-R stars and molecular clouds more massive than 10^6 *M* looks very nice, but in the inner kpc of M33, which has recently been surveyed for GMCs, there is not more than one massive cloud, whereas there are 10-20 W-R stars identified.
Response (P.S.Conti) Thank you for reminding me to say that any correlation between W-R stars and GMCs cannot be exact because (1) many W-R stars are completely hidden behind dust in our galaxy, (2) some GMCs might not yet have formed massive stars, or their ages are still less than the formation time for W-R stars (a few million years) and (3) some GMCs may have

been entirely dissipated by the actions of their constituent stellar winds and radiation. Finally, some massive stars, including those of W-R type, do not seem to be associated with any other massive stars, or the spiral structure.

Question (H. Zinnecker) I would like to caution against the IMF slopes you have given for the LMC/SMC associations, mainly because the multiplicity of the massive stars is not resolved. If you placed the Trapezium cluster at the distance of the LMC, the separation of the four stars would be of the order of 0.1 arc second and unresolved in ground based CCD images.

Response (P.S.Conti) I agree binaries are a problem for the absolute numbers. However, I consider it unlikely that the binary fractions in the associations I discussed conspired to be different in just the right way so as to create IMF slope differences when in fact none existed. I need to remind you that this was a differential comparison.

Question (G. Meurer) You say there are about 100 O7V star equivalents in 30 Dor from the integrated spectrum. Do the number counts agree with this?

Response (P.S.Conti) I think they are in rough agreement but this awaits the census currently being carried out by Joel Parker for his PhD thesis.

MIXTURE MODELS FOR STUDYING STELLAR POPULATIONS: BIVARIATE RELATIONSHIPS AND POSTERIOR MIXING PROPORTIONS APPLICABLE TO THE THICK-DISK PROBLEM

JAMES M. NEMEC
Program in Astronomy
Washington State University
Pullman, WA 99164-3112 USA

AMANDA F. LINNELL NEMEC
International Statistics and Research Corp.
P.O. Box 496, Brentwood Bay
British Columbia, V0S 1A0 Canada

ABSTRACT. Finite mixture models provide a useful framework for analysing the overlapping spatial, kinematical, chemical, and age distributions of stellar populations. In this paper, the age-metallicity relationship is used to illustrate the properties of bivariate mixture distributions. The interpretation of bivariate scatter plots and 'binning' diagrams is discussed, and population membership is examined by computing posterior mixing proportions.

1. Introduction

It is generally recognized that in the solar neighbourhood there exist stars of all ages, up to $A{\sim}15$ Gyr. These stars are known to exhibit a diversity of chemical compositions, ranging from [Fe/H] $\sim$ 0.3 to -4.0 dex, and have space motions corresponding to U velocities (directed away from the Galactic center) and W velocities (directed toward the North Galactic Pole) between -200 and +200 km s^{-1}, and V velocities (in the direction of Galactic rotation) between -400 and +50 km s^{-1}. The observed *overall* distribution functions of the variables U and W are symmetric about zero (*i.e.*, the motions are symmetric about the local standard of rest). In contrast, the overall distribution functions for A, [Fe/H], and V are asymmetric. The asymmetry in the V distribution is the well-known asymmetric drift.

The notion that stars in the solar neighbourhood constitute a mixture of discrete stellar populations can be traced to the early ideas on stellar kinematics of Kapteyn, Schwarzschild, and Eddington. The modern concept of a stellar population originated with Baade's 1944 introduction of Populations I and II for modelling galaxies, and the five-component model for the Galaxy advocated at the 1957 Vatican conference. Within the last ten years, there has been considerable discussion about the number of distinct components that are needed to represent the chemical and dynamical history of our Galaxy (see Sandage 1987b), the proportion of stars belonging to each component, and the nature of the individual components. Much of this debate has been fuelled by various theories for the early evolutionary phases of our Galaxy (Eggen, Lynden-Bell & Sandage 1962, Searle and Zinn 1978, Norris & Ryan 1989).

According to the Bahcall & Soneira model (see Bahcall 1986), the solar neighbourhood can be represented by only two stellar populations - a disk (D) and a halo (H) - with 99.8% of the nearby stars brighter than M_V=16.5 belonging to the disk component, and only 0.2% belonging

B. Barbuy and A. Renzini (eds.), The Stellar Populations of Galaxies, 103–108.

to the halo, *i.e.*, D:H=500:1. Owing to uncertainty in the estimated mixing proportions, the ratio might actually be ~ 2-4 times greater than the quoted value. The Norris & Ryan model is another example of a two-component model, but in that case the two components comprise an extended disk component and a discrete halo. In contrast, proponents of the Gilmore-Reid-Wyse three-component model claim that the Galaxy has three distinct components - a flat thin disk (t), a spheroidal halo (h), and an intermediate thick disk (T). Estimated mixing proportions for this model range from a t:T:h ratio of 200:22:1 (Sandage 1987a) to a ratio of 1200:20:1 (Gilmore & Wyse 1985).

The ratios quoted above are estimates of the *true* or *absolute* mixing proportions of the components that make up the two- or three-component model. These ratios are obviously of interest since they have a direct bearing on questions related to the vertical acceleration, K_z, and dark matter in the solar neighbourhood. However, because complete samples are rarely available in practice, it is generally difficult to determine the true mixing proportions. For most samples, only the *apparent* mixing proportions (which depend on the survey limits, the location of the survey volume, and other imposed selection criteria) can be estimated directly.

Determination of the appropriate number of components and estimation of either the true or apparent mixing proportions are two essential steps in modelling galactic structure. The nature of the individual components is also an important consideration. This aspect of the model has raised questions about the shapes of the underlying within-component distribution functions, in particular whether there are correlations between age, kinematical and chemical variables, *e.g.*, metallicity and other gradients within the components. These questions must be answered if the 'best' fitting model is to be identified.

For simplicity, the discussion in this paper is limited to bivariate mixture models. After briefly reviewing the fundamental ideas on which finite mixture models are based (§2), we discuss bivariate scatter plots and 'binning' diagrams (§3), and bivariate posterior mixing proportions (§4). Such topics as parameter estimation, and the problem of the optimum number of components, were discussed in Nemec & Nemec (1991a, hereafter Paper I), and in Nemec & Nemec (1991b, hereafter Paper II), and will not be repeated here. Generalization of the bivariate mixture model to the multivariate case is straightforward. The reader is referred to Papers I and II, and the references therein, for a more detailed exposition of univariate and multivariate mixture models and their application to the study of stellar populations.

2. Bivariate Mixture Model for Age-[Fe/H] Relationship

To illustrate the application of mixture models to stellar population problems (in particular the thick-disk problem), we consider here an idealized age-metallicity relationship for the solar neighbourhood (*i.e.*, stars with distances nearer than about 500 pc), which will be assumed to comprise thin-disk, thick-disk, and halo stars. The basic equation for a three-component mixture model is

$$f(A,m) = \sum_{k=1}^{3} p_k f(A,m|k), \tag{1}$$

where $f(A,m)$ is the overall *joint* or *bivariate* probability density function of age (A) and metallicity (m = [Fe/H]), p_k is the mixing proportion for the kth population, and $f(A,m|k)$ is the corresponding joint density function of the kth component ($k = 1,2,3$). To obtain the univariate, or *marginal*, density function of an individual variable (in this case, A or m), the joint density function (Eqn. 1) is integrated over the other variable. The result is a univariate mixture distribution with the same mixing proportions as the joint distribution.

According to Eqn. 1, if the true mixing proportions are in the ratio t:T:h=200:20:1 (see above), then one would expect to find, in a random sample of 1000 nearby stars, ~905 thin-disk stars, ∼90 thick-disk stars, and ∼5 halo stars, each of which would have an age and a metallicity drawn from their respective within-component joint distribution functions of A and [Fe/H], *i.e.*, $f(A, m|k)$. For a sample of stars selected in a way that strongly favours nearby halo and thick-disk stars (e.g., the Sandage & Fouts 1987a and Carney *et al.* 1989 samples, which consist of high-proper-motion stars drawn from the Lowell proper motion survey), the true mixing proportions are replaced by sample-dependent apparent mixing proportions, *e.g.*, t:T:h=2:2:1.

At present, the forms of the within-component distribution functions are not well known. However, there is little evidence to suggest that a bivariate Gaussian distribution is not a good first order approximation. We will, therefore, assume that the within-component distributions are Gaussian. Thus,

$$f(A, m|k) = [2\pi\sigma_{mk}\sigma_{Ak}(1-\rho_k^2)^{1/2}]^{-1}\exp\{-Q_k\} \tag{2}$$

where

$$Q_k = \frac{1}{2(1-\rho_k^2)}\left[\frac{(m-\mu_{mk})^2}{\sigma_{mk}^2} + \frac{(A-\mu_{Ak})^2}{\sigma_{Ak}^2} - 2\rho_k\frac{(m-\mu_{mk})}{\sigma_{mk}}\frac{(A-\mu_{Ak})}{\sigma_{Ak}}\right],$$

and

$$\rho_k = \frac{\text{Cov}_k(A, m)}{\sigma_{Ak}\sigma_{mk}}$$

is the (Pearson) correlation between m and A for component k.

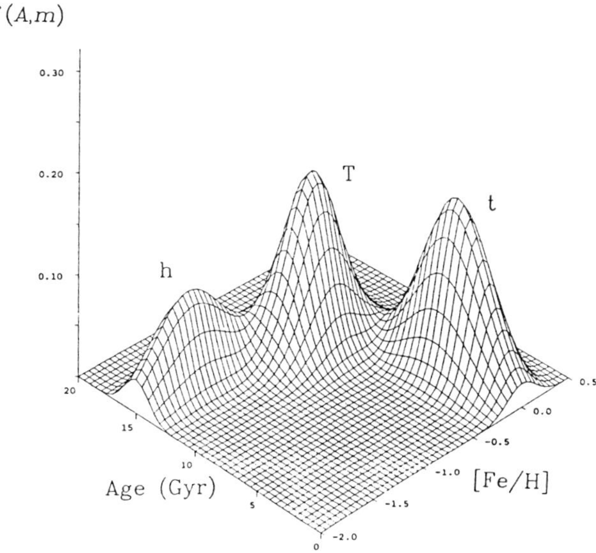

Figure 1. Overall bivariate distribution function corresponding to the A-[Fe/H] relationship for solar neighborhood stars.

Fig.1 illustrates a typical overall age-metallicity distribution that might arise under a Gaussian mixture model. The plot was generated by substituting a particular set of parameter values into Eqn. 2 and plotting the resultant mixture density function. For the (apparent) mixing proportions, we have substituted 20% halo, 40% thick disk, and 40% thin disk, which are consistent with the values that might be expected for high-proper-motion samples of the type mentioned above. For the means and standard deviations of the three components, we have adopted the following parameters: for the halo, a mean metallicity of μ_{mh}=−1.50 dex, with a standard deviation of σ_{mh}=0.35 dex, and a mean age of μ_{Ah}=15 Gyr, with a standard deviation of σ_{Ah}=1 Gyr; for the thick disk, a mean metallicity μ_{mT}=−0.50 dex, with a standard deviation of σ_{mT}=0.25 dex, and a mean age of μ_{AT}=13 Gyr, with a standard deviation of σ_{AT}=1.5 Gyr; and for the thin disk, a mean metallicity of μ_{mt}=0 dex, with a standard deviation of σ_{mt}=0.15 dex, and a mean age of μ_{At}=5 Gyr, with a standard deviation of σ_{At}=2.5 Gyr. These values were chosen to be representative of recent estimates (see Paper I). We have also assumed that m and A are statistically independent for each component, *i.e.*, $\rho_k = 0$ for $k = 1, 2, 3$.

The number of distinct peaks in Fig.1 is clearly equal to three. However, because the means for the individual components may be less well separated for some pairs of variables, the thick-disk might be less obvious in such bivariate plots as the Bottlinger U-V diagram, the V-m diagram, and the W-m diagram (see Nemec & Nemec 1992, in preparation). Of course, if these bivariate plots are compared with plots of the corresponding marginal distributions (cf. Fig. 1 of this paper and Fig. 1 of Paper I), it is evident that the ability to discriminate components improves as additional variables are taken into consideration.

In Fig. 1, we assumed that A and m are independently distributed for each component k, *i.e.*, there are no *within-component gradients.* To test the validity of this assumption, it is sufficient, in the Gaussian case, to test the null hypothesis that $\rho_k = 0$ for $k = 1, 2, 3$. Regardless of whether or not there are within-component gradients, it is evident that mixture distributions (Eqn. 1) do not generally factor into the corresponding marginals - a condition that is necessary for independence. Thus mixtures invariably exhibit some sort of *overall gradient.*

3. Bivariate Scatter Plots and Binning Diagrams

In the study of stellar populations, two types of bivariate diagrams are commonly used to display the data: (1) a scatter plot of the raw data, *i.e.*, y versus x for individual stars or star clusters; and when the sample is sufficiently large, (2) 'binning plots', in which the data are binned according to x and, for each bin, either the mean of the y-values, $\bar{y}(x)$, or the corresponding standard deviation, $s_y(x)$, is plotted against x (the mid-point of the x-values).

Scatter plots, and plots of $\bar{y}(x)$ versus x, which are simply smoothed scatter plots, are useful for depicting overall trends in the relationship between two variables. For example, the A-m scatter plots of Carlberg *et al.* (1985, Fig. 3a) and Schuster & Nissen (1989, Fig. 7), and the $\bar{A}(m)$-m binning diagrams of Twarog (1980, Figs. 1-3), Carlberg *et al.* (1985, Fig. 3b) and Schuster & Nissen (1989, Table 2), show that, for samples of solar neighbourhood stars, there is a tendency for older stars to be more metal poor. This type of non-linear trend is expected if the data are drawn from a mixture of stellar populations, each of which has a different mean age and mean metal abundance. In fact, if Eqn. 2 holds, the exact functional form of the relationship can be determined by computing the *conditional mean* of A given m (see Paper II).

Patterns of dispersion are often of as much interest as overall trends. For example, a scatter plot of W versus m typically shows no trend in the mean (*i.e.*, the points are scattered about the line $W = 0$) but has an obvious wedge-shaped appearance which is indicative of a heterogeneous variance. Such a pattern can readily be explained by a mixture, in which the constituent stellar populations have different W-dispersions (with $\mu_{Wk} = 0$ for all components k) and different mean metallicities (see Fig. 3 of Paper II). Although scatter plots are useful for identifying heterogeneity in the dispersion, they are not very effective for describing precisely how the dispersion varies. A binning diagram of $s_y(x)$ versus x is more useful for this purpose.

Some examples of this approach are the $s_U(m)$-m, $s_V(m)$-m, and $s_W(m)$-m values, which have been tabulated and plotted by Sandage & Fouts (1987), Carney, Latham & Laird (1989), Yoshii & Saio (1979), and Norris & Ryan (1989). Computation of the *conditional standard deviation* of y given x, under an appropriate mixture model (Eqn. 1 or 2), yields the relevant analytical expression for the dispersion.

The bivariate diagrams described above are generally thought to provide important clues about galactic evolution, such as evidence of a thick-disk component, information about within-component gradients, or features that might suggest the number of components. The great utility of mixture models is that by appealing to theory and to simulations much insight into the interpretation of these diagrams can be gained. For instance, simulated scatter plots of A versus m, and plots of the corresponding conditional mean, can be used to examine the effects of varying the mixing proportions (see Fig.6 of Paper II). Furthermore, if the properties of $s_W(m)$-m plots are understood (see Fig. 4 of Paper II), then it is clear that such diagrams cannot, in general, be used to determine the number of components in a mixture. They can, however, be useful for examining gradients within components, but only at the extremities of metallicity range where the influence of overlapping components is minimized, if not eliminated. Of course, the exact range over which this is achieved is difficult to determine and there is always the possibility of contamination from unidentified extreme components, *e.g.*, the presence of possible Pop. III stars, or metal-rich bulge stars. Because evidence for a gradient within a given component (*e.g.*, the Norris & Ryan extended disk) always depends on the underlying model (in particular, the assumed number of components - see Fig.3 of Paper II), any conclusions should be accepted with caution. Likewise, claims that one can, using such plots as $\bar{V}_{rot}$ versus m, distinguish between a 'continuum model' and a discrete component mixture model can be misleading, since what is often purported to support the continuum model is often equally explicable by a simple mixture model.

4. Bivariate Posterior Mixing Proportions: Population Membership

The mixing proportion p_k in Eqn. 1 is the unconditional (or overall) probability that a randomly selected star belongs to the kth stellar population. Once age and [Fe/H] measurements have been made, then the probability that the star belongs to a particular population should be revised to take into account this information, *e.g.*, if the star is very metal-poor then the probability that the star belongs to the halo should be adjusted upwards. The revised, or conditional, probabilities are known as *posterior* mixing proportions and will be denoted $p(k|A,m)$, for k=1,2, 3. These probabilities are functions of A and m and satisfy the constraint $\sum_{k=1}^{K} p(k|A,m) = 1$ for each (A,m).

Posterior mixing proportions are easily calculated using Bayes' Theorem. In the bivariate case of age and metallicity,

$$p(k|A,m) = \frac{p_k f(A,m|k)}{\sum\limits_{j=1}^{K} p_j f(A,m|j)}. \tag{3}$$

The right-hand side of this equation depends on the unknown model parameters, μ_{Ak}, σ_{Ak}, etc., which must be replaced by estimates before the posterior mixing proportions can be evaluated. Notice that even if A and m are independent for each component, *i.e.*, $f(A,m|k) = f(A|k)\ f(m|k)$, the posterior probability $p(k|A,m)$ does not factor into the univariate posterior functions.

Fig. 2 shows bivariate posterior mixing proportions $p(k|A,m)$ corresponding to the thick-disk and halo components of the model illustrated in Fig.1. The surfaces give membership probabilities for the thick disk and the halo components, for each value of A and m. In practice, such diagrams (computed and plotted for any number of components, and based on estimated parameters for a specific catalog of data) provide a quick visual estimate of the probability that a star belongs to a given stellar population.

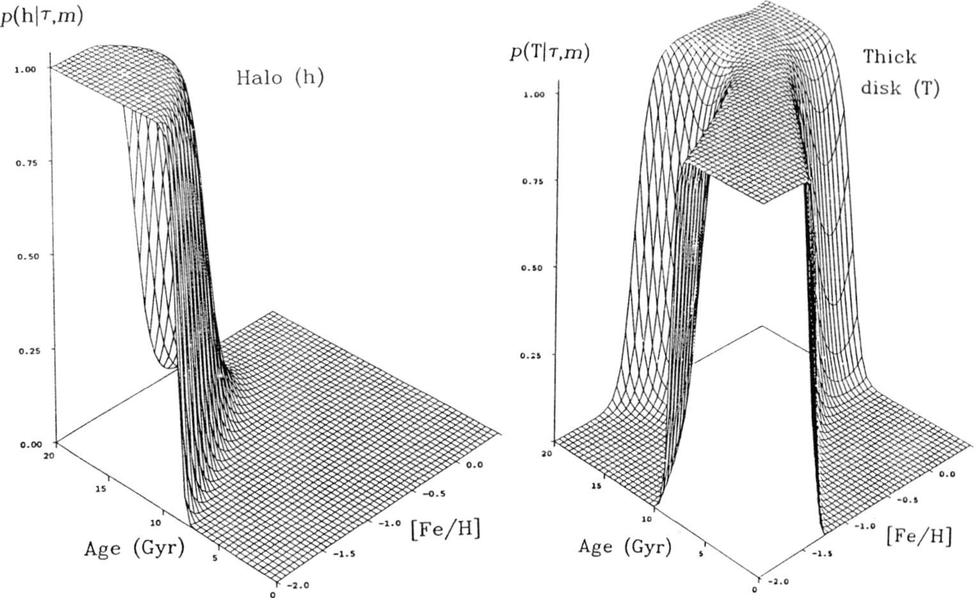

Figure 2. Posterior mixing proportions, conditional on age and metallicity, for the halo and thick-disk components of the three-component Gaussian model shown in Figure 1.

REFERENCES

Bahcall, J. 1986, ARA&A 24, 577.
Carlberg, R.G., Dawson, P.C., Hsu, T. & VandenBerg, D.A. 1985, ApJ, 294, 674.
Carney, B.W., Latham D.W. & Laird, J.B. 1989, AJ, 97, 423.
Eggen, O.J., Lynden-Bell, D. & Sandage, A.R. 1962, ApJ, 136, 748.
Gilmore, G. & Wyse, R. 1985, AJ, 90, 2015.
Nemec, J. & Nemec, A.F.L. 1991a, PASP, 103, 95 (Paper I).
Nemec, J. & Nemec, A.F.L. 1991b, ASP Conf. Ser., 13, 512 (Paper II).
Norris, J. 1986, ApJS, 61, 667.
Norris, J. & Ryan, S. 1989, ApJ, 340, 739.
Sandage, A. 1987a, AJ, 93, 610.
Sandage, A. 1987b, in The Galaxy, eds. G.Gilmore & B.Carswell, Dordrecht:Reidel, p.321.
Sandage, A. & Fouts, G. 1987, AJ, 93, 74.
Schuster, W.J. & Nissen, P.E. 1989, A&A, 222, 69.
Searle, L. & Zinn, R. 1978 ApJ, 225, 357.
Twarog, B. 1980, ApJ, 242, 242.
Yoshii, Y. & Saio, H. 1979, PASJ, 31, 339.

STELLAR EVOLUTION AND STELLAR POPULATIONS IN GALAXIES

André Maeder
Geneva Observatory
CH-1290 Sauverny, Switzerland

1 INTRODUCTION: FROM STAR PROPERTIES TO STELLAR POPULATIONS

Models of population synthesis are resting on an ensemble of other models, data and assumptions, in particular the models of stellar evolution, the data on initial compositions, the data on stellar spectra, the initial mass function (IMF), the star formation rate (SFR), the infall rate, etc... Here we shall concentrate on the properties of stellar models and their consequences for the global properties of stellar populations in galaxies.

The basic link between star properties and population synthesis is expressed in the so-called fuel consumption theorem (cf. Renzini and Buzzoni, 1986). It says that the contribution of stars in any post main sequence stage to the total luminosity of a star population is proportional to the amount of fuel burnt in the considered evolutionary stage. The demonstration of this theorem is rather straightforward. And although it is not used in recent population synthesis (cf. Guideroni and Rocca-Volmerange, 1987; Charlot and Bruzual, 1990), its didactical value is great. It shows how the integrated properties of star populations in galaxies directly depend on stellar internal properties. Indeed, the amount of fuel burnt in a given stage depends, in turn, on many physical assumptions in the models, such as the nuclear reaction rates, the opacities, convection and overshooting, mass loss, mixing processes, etc... Thanks to the above theorem, the effect of different model assumptions on population synthesis can be predicted, at least qualitatively.

2 THE NEED FOR A CHANGE IN ALL STELLAR MODELS

A few times per astronomer generation, stellar opacities are changing: 1955, 1970, 1977, 1991...This occurs just now again. The changes brought about by the new Livermore opacities (Iglesias and Rogers, 1991ab; Rogers and Iglesias, 1991) are quite large. The new opacities for metallicity $Z = 0.02$ are larger by a factor of 2–3 at log T = 5.5 compared to previous Los Alamos data. The results are well confirmed by independent works by Mihalas, Daeppen, Seaton and colleagues. Thus, it is not exaggerated to say that at the time of this writing, all existing models are obsolete and need to be reinvestigated.

B. Barbuy and A. Renzini (eds.), The Stellar Populations of Galaxies, 109–118.

New complete grids with the new opacities and other up–to–date ingredients are now undertaken at Geneva Observatory for the mass range 0.8 to 120 $M_\odot$ and $Z = 0.02, 0.004$ and 0.001. From the first results by G. Schaller, we can mention the following changes: a slight decrease in luminosity and Teff by a few 10^{-2} dex, with a corresponding need to increase the solar helium content to Y = 0.30–0.31, a negligible reduction of the overshooting parameter. However, the main effect concerns the occurrence, duration and extension of the blue loops in the He–burning stage and a considerable widening of the main sequence band in massive stars. The many consequences of the opacity enhancements are now examined and will be analysed in coming works.

Apart from opacities, a very critical point in model computations is the extent of mixing by convective cores, since this extent directly determines the amount of fuel burnt. In absence of any reliable theory about overshooting the amount d_{over} of overshoot beyond the classical Schwarzschild limit is determined on the basis of observations (cf. Maeder and Mermilliod, 1981; Maeder and Meynet, 1989). The proposed value is $d_{over}/H_p = 0.25$, i.e. a modest value, much smaller than the very large overshooting applied by Bertelli et al. (1985) in their models and related investigations. As seen above, the overshooting parameter may undergo a small change due to the use of the new opacities. There are great differences in the locations of the main sequence envelope for various sets of models: VandenBerg (VdB, 1985), Yale (Y; Green et al., 1987), Bertelli et al. (BBC, 1985), Maeder and Meynet (MM, 1989). The VdB and Y models do not include overshooting, while the other two sets include it with different rates. There are also large age differences in the models. In that respect we must also emphasize that the MM models in the range 1.3 to 2.5 $M_\odot$ need to be corrected for a sizeable error in the time scales; the corrected values are available on request. Despite that fact, due to the account of overshooting or not, there remain appreciable differences in the age estimates.

Several mass limits are particularly critical for the models of population synthesis (cf. Renzini and Buzzoni, 1986; Charlot and Bruzual, 1990). As their values are greatly influenced by overshooting, let us examine the most significant mass limits.

- M_{HeF} is the maximum initial mass for degenerate He Ignition. The classical value is about 2.2 $M_\odot$ (cf. Becker and Iben, 1980); with moderate overshooting, we find it around 1.85 $M_\odot$, while Bertelli et al. (1985) found it at 1.6 $M_\odot$. Stars with $M < M_{HeF}$ develop a well populated red giant branch.

- M_W is the maximum initial mass leading to the formation of white dwarfs. The best estimates are observational and based on the presence of white dwarfs in clusters (cf. Weidemann, 1990), which lead to a value of about 8 $M_\odot$.

- M_{up} is the minimum initial mass for central C–burning. Below M_{up} there is no hydrostatic C–burning, stars form a degenerate C,O core surrounded by He– and H–burning shells and evolve as AGB stars. M_{up} is about 8.95 $M_\odot$ in models without overshooting (cf. Becker and Iben, 1980), while with moderate overshooting it is around 6.6 $M_\odot$; Bertelli et al. (1985) found 5.2 $M_\odot$ with large overshooting.

- M_{EC} is the maximum initial mass for stars undergoing core collapse by electron capture. Its value is around 10.2 $M_\odot$ (cf. Nomoto, 1984). It corresponds to no significant change for visible stages in the HR diagram, but only for the final event. Schematically one has the following mass intervals for models without overshooting:

Red supergiants. Most nuclear burning is hydrostatic. Core collapse. Neutron star.

$M_{EC} = 10.2$ $M_\odot$

Bright giants. Hydrostatic C–burning. e^- capture. Neutron star.

$M_{up} = 8.95$ $M_\odot$

AGB stars. Degenerate core after He–burning. C–detonation. SN I 1/2. No remnant.

$M_W \simeq 8$ $M_\odot$

AGB stars. Degenerate core after He–burning. Complete loss of envelope to form a white dwarf.

$M_{HeF} = 2.2$ $M_\odot$

Red giant. He–flash, horizontal branch, AGB, white dwarf.

For models with overshooting, even moderate, the mass ranges are very different with large consequences for the mass interval of AGB stars, the appearance of a red giant branch and the nature of the final stages.

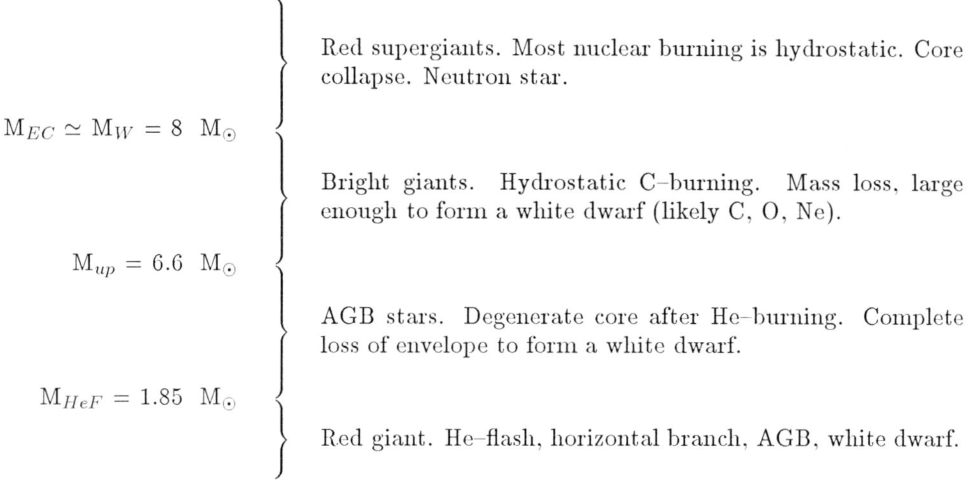
Red supergiants. Most nuclear burning is hydrostatic. Core collapse. Neutron star.
$M_{EC} \simeq M_W = 8$ $M_\odot$
Bright giants. Hydrostatic C–burning. Mass loss, large enough to form a white dwarf (likely C, O, Ne).
$M_{up} = 6.6$ $M_\odot$
AGB stars. Degenerate core after He–burning. Complete loss of envelope to form a white dwarf.
$M_{HeF} = 1.85$ $M_\odot$
Red giant. He–flash, horizontal branch, AGB, white dwarf.

Since AGB stars only appear below M_{up}, and RGB (red giant branch) stars below M_{HeF}, we notice that with overshooting AGB stars will turn on later during galactic evolution and the same for RGB stars. This has a large impact on the flux of galaxies, particularly in the near infrared. Also, with overshooting, the relative contribution of main sequence stars to the total flux is larger. Thus, overshooting is a critical problem for both stellar models and population synthesis.

3 MASSIVE STAR POPULATIONS IN GALAXIES

Due to their high luminosities, massive stars are observable in galaxies at large distances. As their lifetimes are short, they also are tracers of star formation. Particularly WR stars with their bright emission lines offer prominent signatures of starbursts in galaxies, even at very large distances. A most interesting case is that of the so-called WR galaxies, 40 of which are presently known (cf. Conti, 1991). Among the stellar properties showing very large differences from galaxy to galaxy and very large gradients in our Galaxy, the number frequencies of Wolf–Rayet stars present an extreme case. The number ratio of WR to O stars is larger by a factor of 10 in the Milky Way than in the Small Magellanic Cloud. Even more extreme, the number ratio WC/O of WC and O stars exhibits a difference of about two orders of a magnitude between the two galaxies. Some WR subtypes are totally missing in some galaxies, despite the fact that very young star populations are present. For example, there are no late WC stars in the LMC and SMC while there are many in the Milky Way.

The changes of Wolf–Rayet populations in galaxies with active star formation were first discovered in the pioneer work by Smith (1968, 1982). Further observational studies, particularly in the Magellanic Clouds by Breysacher (1981) and Azzopardi & Breysacher (1985), have confirmed the differences in the WR populations. Maeder et al. (1980) have proposed an explanation of the observed differences by a connection between the local metallicity Z and mass loss. The proposition by Maeder et al. was criticized by some authors who claimed that the galactic gradient in WR stars simply reflects the gradient in O stars. However, Meylan & Maeder (1983) showed that the galactic gradient of WR stars is steeper than the gradient of O stars. Thus the WR distribution in the Galaxy is not just a reflection of the O–star distribution and the WR gradient cannot be explained only by a change in the initial mass function.

Massive stars copiously evaporate during their evolution. Recent mass–loss rates for O–stars, supergiants and WR stars indicate that all stars with initial masses larger than 30 $M_\odot$ in Population I finish their life with final masses between 5 and 10 $M_\odot$ (cf. Maeder, 1990). Thus, it is not exaggerated to say that evaporation by stellar winds is the dominant factor in massive star evolution. In this context we immediately realize that a key point about massive star evolution in different galaxies is the relation between mass loss rates $\dot{M}$ and metallicity Z. If there were no such relation, massive star evolution would be about the same everywhere. However, recent models of stellar winds (cf. Kudritzki et al., 1987, 1991; Leitherer, 1991) have suggested the existence of a relation of the form $\dot{M} \sim Z^\alpha$, with $\alpha = 0.5 - 0.7$. In this way Z–effects enter massive star evolution. In the SMC, in blue compact galaxies or in elliptical galaxies at high redshift, Z is low and so are the mass–loss rates, while in the solar neighborhood Z is high and mass loss effects are substantial. This mass loss versus metallicity relation influences all the outputs of massive star evolution.

For Wolf–Rayet stars, which are essentially He–CO stars, the initial metallicity is not expected to have great consequences on the effective mass loss rates. Recent works have shown that the mass loss rates $\dot{M}$ of WR stars depend on their actual masses M with an exponent of about 2.5 (cf. Langer, 1989) when hydrogen is no longer present. Such a relation is also accounted for in recent models (cf. Maeder, 1990, 1991), which give the lifetimes t_{WR}

in the WR stage as a function of mass and Z. The clear trend is an increase of t_{WR} with initial mass and Z. Also, the minimum mass for WR formation is lower at higher Z. The lifetimes in the various phases can be used to derive relative number frequencies WR/O, WC/WR, WC/WN.

For a galaxy or a large galactic ring the assumption of a constant SFR over the last few 10^7y is reasonable. (For a single HII region, this would not be acceptable and aging effects are likely to intervene). For the IMF the standard Salpeter's law is taken. Table 1 shows the theoretical values of some number ratios as a function of metallicities. We notice the great increase of the relative numbers WR/O and WC/WN with the initial Z. Detailed comparisons with observations based on data by van der Hucht et al. (1988), Conti and Vacca (1991), Smith (1988), Arnault et al. (1989) show a very good agreement. This gives powerful support to the evolutionary models and to the idea that metallicity Z, through its effects on the mass loss rates $\dot{M}$ of O–stars and supergiants, is responsible for the enormous differences of the WR populations in galaxies with active star formation, which also has great consequences for the chemical evolution of galaxies, as shown below.

Table 1: Theoretical number ratios for WR stars

Z	WR/O	WC/WR	WC/WN
0.002	0.0032	0.057	0.061
0.005	0.0182	0.192	0.237
0.020	0.0752	0.640	1.784
0.040	0.1557	0.736	2.784

4 CHEMICAL EVOLUTION OF GALAXIES

The key effect in the interpretation of past chemical abundances rests on the fact that massive stars have small lifetimes. In the early phases of galactic evolution, only massive stars contribute to the chemical enrichment due to their short lifetimes. As time goes by, smaller stellar masses come into the game: firstly, only SNII contribute to the enrichment (mainly in O, Ne–Ca elements and r–process elements), then appears the production of the intermediate mass stars (with mainly C, N and s–process elements); they are followed by the contributions from supernovae SNIa (mainly Fe injection). Thus the ages at which stars of a given mass release their nucleosynthetic production are generally considered as the major effect regarding the changes of stellar yields as a function of time (cf. Matteucci, 1991; Truran, 1991). One should, however, also take into account the fact that the chemical yields are changing with the initial Z and may in this way influence the picture of chemical evolution of galaxies. The previous results on WR statistics and their differences in galaxies are relevant, since the WR stage is the last observable stage of massive stars before supernova explosion. Mass loss in massive stars acts as nucleosynthesis in two main ways: 1) The direct enrichment by stellar winds. 2) The difference in the remnant mass left at the time of core collapse. Figs. 1 and 2 from Maeder (1992) show the mass fractions of various elements ejected in the winds (hatched areas) and in the final stages for the models at $Z = 0.02$ and $Z = 0.002$.

From the figures, we can notice the following properties:

— For massive stars, the dominant effect is that of stellar winds which influence both the yield in the wind and in the supernovae. At high Z, large amounts of He and C are ejected before being further processed to heavy elements; this results in large He and C stellar yields and in a drastic reduction of the production of heavy elements, particularly of oxygen. At low Z values, especially at $Z = 0.002$, the situation is just opposite: we notice in Fig. 2, both in the winds and in SN ejecta, the huge yield in oxygen and the very small one in carbon.

— It is also visible from Figs. 1 and 2 that the final masses of initially massive stars with $Z \geq 0.02$ are considerably reduced by mass loss at the time of SN explosions. At $Z = 0.02$, all stars with initial mass larger than about 30 $M_{\odot}$ finish their life with small masses in the range of 5 to 10 $M_{\odot}$. For such masses, the nucleosynthetic yields in O, C and Z are derived from models by Woosley and Weaver (1986), Meynet (1990) and Thielemann et al. (1991), which all give very similar relations between these yields and the mass M_{CO} of the carbon–oxygen core. The small leftover masses contribute to a drastic reduction of the oxygen yields in high Z models.

— For low and intermediate mass stars, the stellar yields in helium are relatively small in higher Z models. The reason is the much thinner He–rich shell in higher Z models; also, in these models one has to substract a larger amount of initial helium.

— On the whole, the main result is that the nucleosynthesis of helium, carbon, oxygen, neon (a very special case) and heavy elements is a function strongly depending on both mass and Z.

Table 2 shows the corresponding net yields y_i in Z, He, C and O and the returned fraction R. All quantities have been integrated over a Salpeter's mass spectrum. We strongly emphasize that the exact values of the net yields very much depend on the masses adopted for the collapsed remnants (lower part of Figs. 1 and 2, based for lower mass stars as current data for white dwarfs, cf. Weidemann, 1990, and on a relation between residual baryon masses and M_{CO} for larger masses, cf. Woosley, 1986). It is not well known under which conditions the stellar cores collapse to black holes. This is nevertheless a very critical point because if a black hole is formed at SN explosion, all or most of the stellar mass may be taken by the collapsed remnant and does therefore not contribute to nucleosynthesis. This uncertainty affects all nucleosynthetic predictions, in the past as well as the present ones. In Table 2 the yields at low Z for stars with $M > 20$ $M_{\odot}$ are also given. These may be considered as representative of the yields for very early galactic evolution, since there only stars with lifetimes shorter than the galactic age may contribute to the chemical enrichments. Mass loss effects at $Z = 0.002$ are small and we expect that for lower Z values the mass loss rates are similarly negligible.

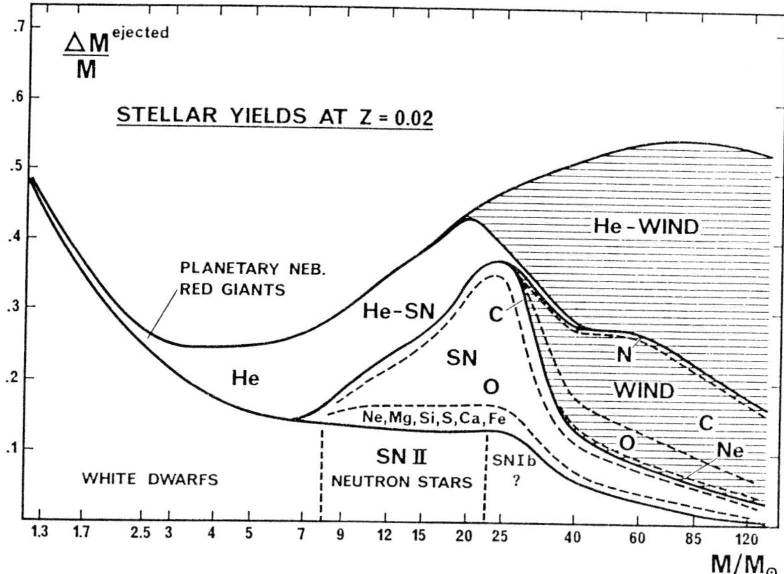

Figure 1 Mass fractions ejected as a function of initial masses for $Z = 0.02$.

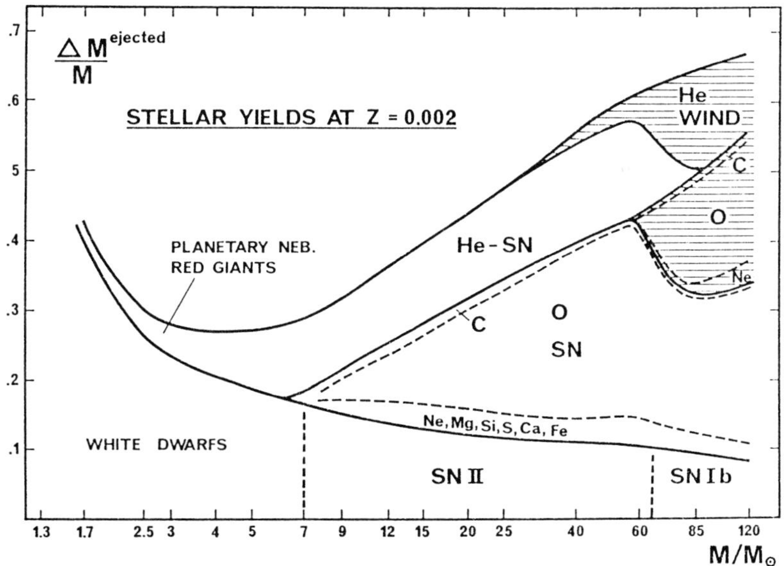

Figure 2 Mass fractions ejected as a function of initial masses for $Z = 0.002$.

Table 2: Net yields in heavy elements Z, in helium, carbon and oxygen. R is the returned fraction.

Z initial	Z	He	C	O	R returned
1–120 $M_\odot$					
0.002	0.0191	0.0181	0.00090	0.0150	0.7656
0.005	0.0191	0.0171	0.00240	0.0144	0.7728
0.020	0.0141	0.0201	0.00465	0.0076	0.7770
0.040	0.0111	0.0162	0.00401	0.0050	0.8036
20–120 $M_\odot$					
0.002	0.0325	0.0155	0.00091	0.0276	0.8937

From Table 2 we notice that the yields in Z and O strongly decrease during galactic evolution, in agreement with recent results by Josey and Tayler (1991). Several authors (cf. Twarog and Wheeler, 1982, 1987; Larson, 1986; Matteucci, 1986; Olive et al., 1987; Wheeler et al., 1989) have shown that with the best current estimates of the yields, IMF and SFR, the theory of galactic chemical evolution leads to a large overproduction of oxygen with respect to the observed abundance. The typical overproduction amounts to a factor 2.6 (cf. Matteucci, 1986). Some authors have proposed various manipulations of the IMF, of the upper and lower mass limits, etc...to solve the problem. However, as emphasized by Wheeler et al. (1989), the problem of the overproduction of oxygen is exacerbated if one simultaneously wants to account for the large O/Fe abundance ratios in halo stars: even more severe "actions" on the IMF and SFR are then required in the modellisations. In this context, the models given above may be welcome. Models with no or low mass loss, like at $Z = 0.002$, predict very high net yields y_O in oxygen, while at $Z = 0.002$ or 0.04 the values of y_O are a factor 4 to 6 smaller. Thus, the present models imply that the oxygen production is much smaller during a large part of galactic evolution than predicted by constant mass models. This is such as to considerably alleviate the well known problem of the overproduction of oxygen.

Metal deficient stars in the Galaxy generally show an excess in the abundance ratio $[O/Fe]$. This trend, firstly found by Conti et al. (1967), is well illustrated in various plots, such as the plot of $[O/Fe]$ vs. $[O/H]$ by Wheeler et al. (1989). Thus, the present stellar models also exhibit the same trend.

The ratios $(\Delta Y/\Delta O)$ and $(\Delta Y/\Delta Z)$ of the relative helium to oxygen or metal enrichments are parameters of major importance for galactic chemical evolution and cosmology. These ratios critically influence the estimate of the primordial helium originating from big bang nucleosynthesis (for recent references, see Audouze, 1987; Olive et al., 1990; Steigmann et al., 1991). Helium is generally best determined from extragalactic HII regions, which span a wide range of metallicities and therefore allow the determination of the helium content near zero metallicity (e.g. Peimbert and Torres Peimbert, 1974; Lequeux et al., 1979; Kunth and Sargent, 1983; Kunth, 1983; Pagel et al., 1986; Peimbert, 1986; Pagel, 1989, 1991).

The ratios of the yields from Table 2 are $y_Y/y_O = 3.26$, 1.21, 0.56 for the cases $Z = 0.04$, 0.002 and the case $Z = 0.002$, $M > 20\ M_\odot$ respectively. The ratios y_Y/y_Z are 1.46 at $Z = 0.04$ and 0.48 for $Z = 0.002$ with $M > 20\ M_\odot$. We notice that these theoretical values, although larger than most previous theoretical values, are smaller than some observed values, in particular that by Peimbert or by Pagel, who give a value of 3 or more. Among possible causes for the differences we must mention two main ones. Firstly, the contamination of HII regions by the winds of massive stars may be a problem. On the theoretical side, the uncertainty regarding the initial mass M_{BH} above which core collapse leads to a black hole is very critical. It is clear that if M_{BH} is equal to 60 $M_\odot$ or 25 $M_\odot$, the nucleosynthetic yields are very different, in particular the latter value leads to a much larger $\Delta Y/\Delta Z$ ratio. This is due to the fact that above M_{BH} the helium ejected in stellar winds contributes to galactic enrichment but not the heavy elements locked in the collapsed object. This problem was already discussed by Maeder (1984) and by Schild and Maeder (1985). From their data, a $dY/dZ = 3$ corresponds to a M_{BH} value around 20 $M_\odot$. However, the results about dY/dZ have ranged from 1 to 6 during the last decade, so it is perhaps premature to draw any firm conclusion about the value M_{BH}.

5 REFERENCES

Arnault P., Kunth D., Schild H., 1989, A&A **224**, 73

Audouze J., 1987, in *Observational Cosmology*, IAU Symposium 124, Eds. A. Hewitt et al., Reidel Publ. Co., p. 89

Azzopardi M., Breysacher J., 1985, A&A **149**, 213

Becker S.A., Iben I., 1980, ApJ. **237**, 111

Bertelli G., Bressan A., Chiosi C., 1985, A&A **150**, 33

Breysacher J., 1981, A&AS **43**, 203

Charlot S., Bruzual G.A., 1990, STSCI preprint no 452

Conti P.S., 1991, ApJ **377**, 115

Conti P.S., Greenstein J.L., Spinrad H., Wallerstein G., Vardya M.S., 1967, ApJ **148**, 105

Conti P.S., Vacca, 1991, Astron. J. in prep.

Green E.M., Demarque P., King C.R., 1987, *The revised Yale Isochrones and Luminosity Functions*, Yale University Obs., New Haven

Guideroni B., Rocca–Volmerange B., 1987, A&A **186**, 1

van der Hucht K.A., Hidayat B., Admiranto A.G., Supelli K.R., Doom C., 1988, A&A **199**, 217

Iglesias C.A., Rogers F.J., 1991a, ApJ **371**, 408

Iglesias C.A., Rogers F.J., 1991b, ApJ **371**, L73

Josey S., Tayler R.J., 1991, MNRAS in press

Kudritzki R.P., Pauldrach A., Puls J., 1987, A&A **173**, 293

Kudritzki R.P., Pauldrach A., Puls J., Voels S.R., 1991, in *The Magellanic Clouds*, IAU Symp. 148, Eds. R. Haynes, D. Milne, Kluwer Acad. Publ. p. 279

Kunth D., 1983, in *Primordial Helium*, ESO Workshop, Eds. P.A. Shaver et al., ESO Garching, p. 305

Kunth D., Sargent W.L.W., 1983, ApJ **273**, 81

Langer N., 1989, A&A **220**, 135

Larson R.B., 1986, MNRAS **218**, 409

Leitherer C., 1991, private communication

Lequeux J., Peimbert M., Rayo J.F., Serrano A., Torres–Peimbert, 1979, A&A **80**, 155

Maeder A., 1984, in *Stellar Nucleosynthesis*, Eds. C. Chiosi, A. Renzini, Reidel Publ. Co., p. 115

Maeder A., 1990, A&AS **84**, 139
Maeder A., 1991, A&A **242**, 91
Maeder A., 1992, A&A, in press
Maeder A., Lequeux J., Azzopardi M., 1980, A&A **90**, L17
Maeder A., Mermilliod J.C., 1981, A&A **93**, 136
Maeder A., Meynet G., 1989, A&A **210**, 155
Matteucci F., 1986, ApJ **305**, L81
Matteucci F., 1991, in *Chemistry in Space*, Erice School, Ed. M. Greenberg, Kluwer Acad. Publ., p. 1
Meylan G., Maeder A., 1983, A&A **124**, 84
Meynet G., 1990, Thesis University of Geneva
Nomoto K., 1984, in *Stellar Nucleosynthesis*, Eds. C. Chiosi, A. Renzini, Reidel Publ. Co., p. 205
Olive K.A., Schramm D.N., Steigman G., Walker T.P., 1990, Phys. Lett. B **236**, 454
Olive K.A., Thielemann F.-K., Truran J.W., 1987, ApJ **313**, 813
Pagel B.E.J., 1989, Rev. Mex. Astron. Astrofis. **18**, 153, 161
Pagel B.E.J., 1991, in *Dynamical and Chemical Evolution of Galaxies*, Elba Conference, Eds. J.J. Franco, F. Matteucci, Kluwer Acad. Press, in press
Pagel B.E.J., Terlevich R.J., Melnick J., 1986, PASP **98**, 1005
Peimbert M., 1986, PASP **98**, 1057
Peimbert M., Torres-Peimbert S., 1974, ApJ. **193**, 327
Renzini A., Buzzoni A., 1986, in *Spectral Evolution of Galaxies*, Eds. C. Chiosi, A. Renzini, Reidel Publ. Co., p. 195
Rogers F.J., Iglesias C.A., 1991, ApJ Suppl., in press
Schild R., Maeder A., 1985, A&A **143**, L7
Smith L.F., 1968, MNRAS **141**, 317
Smith L.F., 1982, IAU Symp. **99**, 597
Smith L.F., 1988, ApJ **327**, 128
Steigmann G., Gallagher J.S., Schramm D.N., 1991, Comment in Astrophys. and Space Sci., in press
Thielemann F.-K., Nomoto K., Shigeyama T., Tsujimoto T., Hashimoto M., 1991, in *Elements and the Cosmos*, Proc. of the 31st Herstmonceux Conf., Ed. R.J. Terlevich, Cambridge Univ. Press, in press
Truran J.W., 1991, in *Evolution of stars: the photospheric abundance connection*, IAU Symp. 145, Ed. G. Michaud, A. Tutukov, Kluwer Acad. Publ., p. 13
Twarog B.A., Wheeler J.C., 1982, ApJ **261**, 636
Twarog B.A., Wheeler J.C., 1987, ApJ **316**, 153
VandenBerg D.A., 1985, ApJ Suppl. **58**, 711
Weidemann V., 1990, Ann. Rev. Astr. Ap. **28**, 103
Woosley S.E., 1986, in *Nucleosynthesis and Chemical Evolution*, 16th Saas-Fee Course, Eds. B. Hauck et al., Geneva Observatory, p. 1
Woosley S.E., Weaver T.A., 1986, in *Radiation Hydrodynamics in Stars and Compact Objects*, IAU Colloquium no 89, Eds. D. Mihalas, K.-H. A. Winkler, Springer-Verlag, Lecture Notes in Physics **255**, 91
Wheeler J.C., Sneden C., Truran J.W. jr., 1989, Ann. Rev. Astron. Astrophys. **27**, 279

The Formation of Disks in Galaxies

G. HENSLER[1], A. BURKERT[3,4], J.W. TRURAN[3], H. DÜNHUBER[2], C. THEIS[1]

[1] Inst. of Theoretical Physics and Observatory, University of Kiel, Olshausenstr. 40, D-2300 Kiel 1, Fed. Rep. of Germany
[2] Max-Planck-Institut für Astrophysik, Karl-Schwarzschild-Str. 1, D-8046 Garching, Fed. Rep. of Germany
[3] Dept. of Astronomy, University of Illinois, 1011 West Springfield Ave., Urbana Champaign, IL 61801, USA
[4] University of California, Lick Observatory, Santa Cruz, CA 95064, USA

ABSTRACT. Different chemo-dynamical models dealing with the formation of stellar disks in the course of galaxy evolution are presented. One-dimensional models concerning the vertical stratification of disks are in strikingly good agreement with the solar vicinity. Two-dimensional investigations of the global evolution of disk galaxies show from preliminary results that the disk formation is delayed by means of an equatorial outflow of hot metal-enriched gas that stems from supernova ejecta in the star-forming central region.

1 General remarks on dissipative galaxy evolution

In order to understand central concentrations in galaxies, disky isophotes and metallicity gradients in giant ellipticals (gEs) as well as the mere existence of gas in all morphological types of galaxies, and most evidently the gaseous disks in spirals, **dissipation** must have played a dominant rôle during galactic evolution.

The stars formed during the galactic collapse decouple from the dissipative dynamical evolution of the gas, by this leading to the initial spheroidal stellar population, the halo. Its shape (and that of gEs) is supported by an anisoptropic velocity dispersion of the stars. A disk can only be formed if the remaining gas is enabled to settle down towards the equatorial plane according to its angular momentum. The amount of stellar matter that can be formed within this disk depends then only on the pre-history of star formation, in particular, on its dependence on the gas properties in the collapsing proto-galaxy and, by this, on the ratio of dissipative and collapse timescale to the star-formation timescale τ_{SF}.

2 Hydrodynamical models of disk galaxies

First simple hydrodynamical models of the evolution of disk galaxies had been calculated by *Larson (1975, Mon. Not. R. astr. Soc.* **173**, *671, 1976; Mon. Not. R. astr. Soc.* **176**, *31).* He obtained stellar systems that are comparable to disk galaxies even including a central bulge component, but he had to vary star-formation rate (SFR) and viscosity artificially. Nevertheless, kinematics, density, and metallicity of the stars in his models are showing too smooth gradients, not in agreement with the almost distinct stellar components like the observed disk and halo populations.

B. Barbuy and A. Renzini (eds.), The Stellar Populations of Galaxies, 119–122.

Burkert and Hensler (1987, Mon. Not. R. astr. Soc. **225**, *21p, 1988; Astr. Astrophys.* **199**, *131*: hereafter *BH88)* improved *Larson's* models by allowing for an anisotropic velocity dispersion of the stellar component. Although this improvement led to two distinct stellar populations because stars formed in the halo are prevented by their anisotropy to settle into the equatorial plane (like the gaseous component does) due to its isotropic pressure, they failed to produce a central bulge. This problem, however, can be avoided if the proto-galaxy is centrally condensed and non-rigidly rotating *(Dünhuber, 1990, Dipl. thesis, University of Munich)*.

All these above-mentioned models achieved a fast formation of the disk after already three free-fall times (τ_{ff}) and in most cases a too rapid consumption of gas due to a high SFR in contradiction to observationally derived SFRs within the disk *(Sandage, 1986, Astr. Astrophys.* **161**, *89)*. Although such a rapid collapse agrees with the first naive picture for our Galaxy proposed by *Eggen, Lynden-Bell, and Sandage (1962, Astrophys. J.* **136**, *748)*, it disagrees to that by *Yoshii and Saio (1979, Publ. Astr. Soc. Pac.* **31**, *339)*.

Detailed observations of the stellar disk population provide strong constraints on evolutionary models of the galactic disk and presumably on the proto-galactic conditions. In the solar vicinity the lack of metal-poor G-dwarfs, the age-metallicity relation, the metallicity-scaleheight-relation, the step-wise distributions of stellar kinematics and metallicity from halo to thin disk with the transition phase of a thick disk, age differences between halo and disk stars, and hints of a radial propagation of the onset of star formation within the disk itself serve as a challenge for our understanding of galactic evolution and as a sensitive probe for the quality of models. Several explanations have been proposed to solve some of these problems, however, applying non-dynamical studies only.

The above-mentioned purely dynamical models can also not account for all these problems, because they assume too unrealistic simplifications like a one-phase interstellar medium (ISM), instantaneous recycling, a dependence of the SFR on the total gas density, and the neglection of star-gas interactions. In particular, the G-dwarf problem, the thick-disk formation, and a time delay of the disk formation as well as the influence of superbubbles, galactic winds, and starburst epochs on the evolution of galaxies and their disks cannot be investigated by such a simple single-phase ISM description.

Supernovae (SNe) are violently acting upon the ISM by means of thermal (formation of hot bubbles, evaporation, condensation) and dynamical processes (sweep-up of surrounding material, formation of hot bubbles) and by metal enrichment due to stellar nucleosynthesis yields. In addition, the different gas phases need not show strongly coupled dynamics, but the SN-heated gas expands and can even be expelled from the galaxy, by this, carrying a non-negligible amount of metals away. While τ_{ff} and τ_{SF} determine the dynamical evolution of galaxies in the purely hydrodynamical one-phase description, in reality, the stellar lifetimes of massive stars inherently involve a heating timescale due to SN explosions. On the other hand, this requires to account for the cooling timescale from the hot to the cool star-forming phase. The latter one can exceedingly influence the collapse and galaxy evolution.

As demonstrated elsewhere *(Hensler and Burkert, 1990, in "Windows on Galaxies", eds. P. Fabbiano et al., Kluwer, p. 321)*, only the most sophisticated chemo-dynamical investigations taking a multi-phase ISM and the star-gas interactions with the metal dependence of processes into account are suitable for those problems. Tests have convincingly demonstrated that the neglection of a particular process has in most cases much stronger consequences for the evolution of the ISM than the acceptance of an uncertainty in its parametrization *(Hensler, 1988, Habil. thesis, University of Munich; Theis, 1990, Dipl.*

thesis, University of Munich; Hensler and Burkert, 1991, Astr. Astrophys. , submitted). The main features of the chemo-dynamical treatment are reviewed e.g. by *Hensler (1987, Mitt. Astron. Ges.* **70**, *141).*

3 Chemo-dynamical evolution of disk galaxies

Although, several observers have invoked the existence of different step-wise disk stratifications from halo to thick and thin disk populations, one of the major uncertainties at present and one of the most basically addressed questions of this conference concerns with the relative ages of these structures.

Two strategies for chemo-dynamical calculations of the galactic disk formation can now be followed:
1) A one-dimensional (1d) consideration of the vertical disk formation can study the disk stratification and the existence of different disk layers.
2) Two-dimensional (2d) global calculations of the evolution of disk galaxies can yield an insight in the evolutionary epochs of the different galactic components including the bulge and their interferences.

3.1 One-dimensional vertical formation of the galactic disk

As a first step towards getting an insight into the timescales of the disk settling 1d chemo-dynamical calculations have been performed by *Burkert, Truran, and Hensler (1991, Astrophys. J., in press*: hereafter *BTH)*. Starting with an initial stratification of a 10^6 K hot gaseous proto-disk with a typical column density for the solar vicinity of $\Sigma \approx 50 M_\odot pc^{-2}$, that is metal-enriched by population II stellar ejecta up to already 5% $Z_\odot$, the model is able to reproduce the different vertical disk layers in strikingly good agreement with observations.

The results of *BTH* are in particular briefly summarized here: After a cooling time of approximately 200 Myrs the star formation increases accordingly. Stars in the low metallicity range of $-1.5 \leq [Fe/H] \leq -0.5$ form only during the first 350 Myrs. In agreement with the thick disk values their vertical velocity dispersion amounts to $c_w \leq 40$ $km\ sec^{-1}$, providing a scaleheight of $h_z = 1.3$ kpc. While the star formation in the metal-poor gas ceases abruptly due to the gas consumption, metal-enriched SNII ejecta of the thick-disk stars provide a hot gas phase that starts after its cooling 250 Myrs after the onset of the thick-disk formation to form new stars, now with $-0.5 \leq [Fe/H] \leq 0.25$. This new stellar disk population represents the old thin disk and is shrinked within the next 4 Gyrs to only $h_z = 300$ pc with a smaller c_w of 20 $km\ sec^{-1}$. Meanwhile, the star formation decreases from $10^{-7.4}$ $M_\odot$ pc^{-2} yr^{-1} by one order of magnitude.

After 6 Gyrs 80 % of the gas is converted into stars and more metal-enriched gas of $[Fe/H] \geq 0.25$ can further concentrate to an accordingly young thin disk with $h_z \approx 50$ pc.

The final vertical stellar stratification agrees well with the exponential law derived by Sandage *(1987, Astron. J.* **93**,*H 610)*, if a stellar halo component is added accordingly.

If Σ is larger the thin disk forms faster. A radial column density gradient in a galactic disk would thus lead to a radial propagation of the thin disk formation. Furthermore, the model predicts a sharp edge of galactic disks for $\Sigma \leq 10 M_\odot pc^{-2}$. Due to the strong self-regulation effects that balance heating and cooling a thin disk cannot form. For the same reason the collapse of low-density galaxies can sometimes be prevented *(Theis, Burkert, Hensler, 1991, Astr. Astrophys. , submitted).*

3.2 Two-dimensional models

In order to date the onset of the thick-disk formation as well as the timescale for the formation of the halo and bulge, i.e. in order to study the evolution of galactic disks in the context of the whole galaxy evolution, global, at least, two-dimensional investigations have to be carried out. We have commenced the calculations with the same initial conditions as a typical model galaxy of *Larson* and *BH88*: The total mass is $10^{11} M_{\odot}$, however, now devided into two gas phases in pressure equilibrium, a hot one of 10^6 K and a cool cloudy one with 10^3 K and a volume filling factor of 0.1; the radius amounts to 30 kpc; the rigid rotation leads to a spin parameter $\lambda \approx 0.15$. Although the calculations have not yet be performed to more than 10 τ_{ff}, what amounts to approximately 2.7 Gyrs, the preliminary results by *Dünhuber, Theis, Burkert, Hensler (1990, Astron. Ges. Abstract Series* **4**, *61)* already reveal an impressing insight to the early stages of halo, bulge, and disk formation and will, therefore, be presented here.

The SFR drops from a high initial value caused by heating of the gas due to the collapse and in addition due to first stellar ejecta. Within the first 2 τ_{ff} the collapse of the proto-galactic cloud proceeds nearly spherical symmetric while it cools and the SFR increases again. However, because of the higher central gas condensation the SFR in the center exceeds that in the halo by far, i.e. more than 4 orders of magnitude. Then the still collapsing configuration flattens due to its angular momentum and the stellar distribution consists after $\approx 4\tau_{ff}$ of a prominent central dense almost spherical structure with an additional flat spheroid in the equatorial plane. Intimately coupled with the burst-like increase of the SFR in the center, a hot (several 10^6 K) central bubble of ejected stellar material forms and expands due to continuous feeding by SN explosions.

Differently from superbubble calculations *(Tomisaka and Ikeuchi, 1988, Astrophys. J.* **330**, *695)* but in agreement with investigations of the gas flow in gEs *(Yorke and Kunze, 1990, in "Physical Processes in Fragmentation and Star Formation", Capuzzo-Dolcetta et al. (eds.), p. 241)* this hot gas phase expands pressure-supported (due to further infalling cooler material) primarily in the galactic plane towards larger radii. By this it delays the formation of a stellar disk but forms a hot metal-enriched disky layer of a vertical height of 1-2 kpc. Because this global picture has not changed for at least almost three Gyrs, we suggest that the hot gas can produce a metal pre-enrichment of the disk material, by this providing the formation of a thick disk, an explanation of the G-dwarf problem, and the radial star-formation propagation. The star formation in the disk can further be fuelled by gas infalling from the halo, because at that time not more than 60% of the matter is transformed into stars. Further studies will clarify the evolutionary scenario and are in progress.

Acknowledgements: This work was partly (A.B.) supported by the NSF under grants AST 86-11500 and AST 89-17442 and the Theodor Lynen Foundation (A.B.) at the University of Illinois, and in part by the DFG under grants He1487/4-1 (H.D.) and He1487/5-1 (C.T.). The authors gratefully acknowledge travel support by the NATO under grant 0792/88.

Discussion

A. Serrano: *1) Can you form a bulge? 2) Can you explain galaxies without thick disks?*
G. Hensler: *to 1): The 2d chemo-dynamical evolutionary models show already after few τ_{ff} the formation of a clearly distinct central spheroid.*
to 2): No, not with our prescription of the star-formation. If such systems exist, specific processes have to be invoked that allow cooling of the gas but prevent star formation within the cool clouds.

TRENDS OF ELEMENT ABUNDANCES IN THE STARS OF OUR GALAXY

Spite M.
Observatoire de Paris Meudon
92195 Meudon Cedex

ABSTRACT. The analyses of element abundances in a variety of stars, formed at different times in our Galaxy, are presented in an effort to determine the distribution of elemental abundances at various stages in the evolution of the Galaxy.

1. Introduction. The different populations in the solar vicinity.

Our Sun is located in the external part of the disk of our Galaxy. Different investigations based on star counts (cf. in particular: Gilmore and Reid (1982), Gilmore and Wyse (1985), Sandage(1987) Gilmore et al. (1989))have shown that, in a cylinder perpendicular to the galactic plane above the Sun, it is possible to disentangle up to four different Populations: the halo , and the disk which is now divided in three different subsystems: the old thick disk, the old thin disk, and the young thin disk where star formation still occur.

Recently Burkert Truran and Hensler (1991) have proposed a chemodynamic model of the collapse of our Galaxy. They show with a minimum of hypotheses, that the gas collapsed into the equatorial plane in succeeding steps: when the gas collapses indeed, the star formation rate increases and provides sufficient energy to stop the free fall collapse. Later, when star formation slows down, due for example to the lack of gas, the collapse restarts. The different populations correspond to the different steps of this collapse. First the halo collapses on a rapid time scale (10^9years) in a "thick disk". During about another 10^9years stars are formed inside this "thick disk". Then it collapses into an "old thin disk" and during about 5×10^9 years stars are formed inside, and finally the old thin disk itself collapses into the young thin disk feature.
Moreover, if the material in the Galaxy, at each step of its formation, has been well mixed, a correlation between metallicity and age is expected . For example it can be admitted that all stars with [Fe/H]<-1 dex belong to the halo. The characteristics of the different subsystems are summarised in Table 1.

Since the Galactic plan belongs to all the subsystems, it is possible to find in this Galactic plane, stars formed at different steps of the life of the Galaxy. In the solar vicinity, these stars are close to us: it is possible to analyse in detail their chemical composition as a function of metallicity for a better understanding of the processes which determine the chemical evolution of the galaxies.

In section 2, recent spectroscopic studies of detailed abundance patterns in field Galactic stars located in the solar neighborhood (less than about 100pc from the Sun) are presented as a function of metallicity. In section 3 the composition trends in the field stars and in the cluster stars are compared.

B. Barbuy and A. Renzini (eds.), The Stellar Populations of Galaxies, 123–132.

Table 1. The different Populations in the Galaxy

	scale height (pc) (pc)	time scale of formation (years)	age of the stars* (years)	metallicity
galactic halo	3200	$>10^9$	15×10^9	[Fe/H]<-1.0
galactic disk				
thick disk	1000	0.5×10^9	14×10^9	–1<[Fe/H]<–0.5
old thin disk	270	5×10^9	14×10^9 to 9×10^9	–0.5<[Fe/H]<–0.3
young thin disk	50	...	$<9\times10^9$	[Fe/H]>–0.3

*Approximate values of the ages are determined assuming that the age of the halo stars is $\approx15\times10^9$years.

2- Composition trends in field stars

There have been several reviews of stellar abundance trends in recent years in articular Spite and Spite (1985), Gustafsson (1987), Lambert(1989), Wheeler et. al. (1989), Nissen, P.E.(1990), and recently a very interesting paper from Ryan et al. (1991) who observed extremely metal poor stars (eight of them with [Fe/H]<–3), but with a rather low S/N ratio. We will discuss here only some elements for which new measurements have been recently obtained.

2-1 THE VERY LIGHT ELEMENTS: Li, Be

These elements are very peculiar since they cannot be formed inside supernovae neither in normal hydrostatic nor explosive conditions: lithium and beryllium are formed by cosmic ray spallation (Meneguzzi et al. 1971), ^{7}Li (but not ^{9}Be) can be formed during the standard Big-Bang (SBB). Other possible sources of lithium are the novae, the red giants, and perhaps, the very massive stars.

The lithium abundance in stars of different populations has been extensively studied by different groups.In a main sequence star, lithium is destroyed as soon as the convective zone reaches the layers where the temperature is larger than 2×10^6K, but in the very young disk stars which have had no time to destroy their lithium, log N(Li)≈3 and log N(Be)≈1.3 (for log N(H)=12).

In the oldest halo dwarfs of our Galaxy the abundance of lithium is lower: log N(Li)=2.1±0.1 (Spite and Spite 1982, Spite et al. 1987, Rebolo et al. 1987, Hobbs and Pilachowski 1988, Hobbs & Thorburn 1991). In a large interval of metallicity -3<[Fe/H]<-1.2 there is NO correlation between lithium abundance and metallicity (Fig.1). Moreover it seems that the variation of the lithium abundance from star to star is within the error bars

The standard Big Bang produces a very small amount of Be. The observed beryllium is probably built by spallation (action of cosmic rays on CNO elements). In Fig.1 the abundance of beryllium is seen to track the iron abundance; the data are taken from Ryan et al. (1990), Gilmore et al. (1991) and Boesgaard (1990). It can be seen that, in the oldest stars, the abundance of beryllium is very low and, as a consequence the spallation processes has not been very efficient, not efficient enough to build a significant amount of lithium.

The constant lithium abundance (independent of metallicity and mass) in the halo stars, provides the primary basis for the view that the surface lithium found today in the halo dwarfs constitutes the nearly unaltered lithium fraction produced by the Big-bang. It must be clear however,that this is not a proof. Nevertheless, if we assume it, a consequence is that an important formation of lithium occured later in our Galaxy. The possible sites of this production could be the AGB phase of the intermediate mass red giants (Smith & Lambert 1990) and the novae (D'Antona & Matteucci 1991).

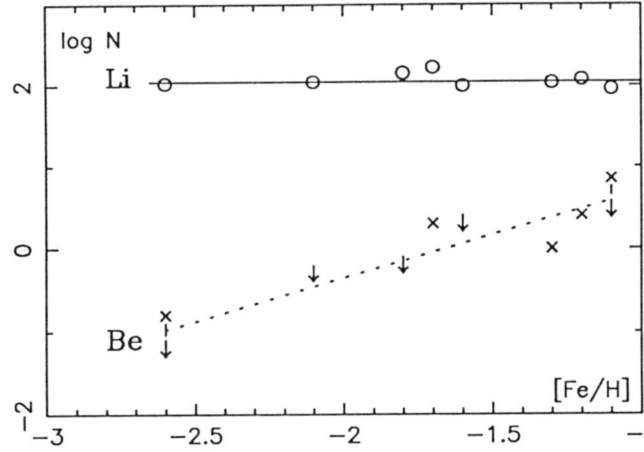

Fig.1 The abundances of both beryllium and lithium has been measured for 8 hot halo dwarfs. For these stars log N(Li) (open circles) and log N(Be) (crosses, or arrows for upper limits) have been plotted versus [Fe/H]. It can be seen that the abundance of lithium is independent of the metallicity and that, on the contrary, the abundance of beryllium seems to track the iron deficiency.

2-2 THE CNO GROUP.

These elements are of particular significance since after hydrogen and helium they are the most abundant elements in the Universe.

Carbon: The abundance of carbon in the atmosphere of a star may be altered at various stages of stellar evolution and especially at the red giant phase (dredge-ups of CN cycle fusion). Therefore in order to ascertain the trend of C/Fe as a function of metallicity, it is particularly important to concentrate on dwarf stars.

Laird (1985) has not observed any variation in [C/Fe] with metallicity,but Tomkin et al.(1986) and Carbon et al.(1987) asserted that a trend of increasing [C/Fe] ratios with decreasing metallicity existed. Wheeler et al.(1989) compiled all these data, trying to correct some systematic errors of the analyses. In this way the overall trend toward larger C/Fe ratio in the halo becomes more obvious . However, for most of these stars, the oxygen abundance is unknown, and the trend in [C/Fe] is affected by the unknown fraction of C combined in the CO molecule, and thus very dependent on the oxygen abundance assumptions. Ryan et al.(1991) measured the carbon abundance in 3 dwarfs with [Fe/H]<2.6, their data are consistent with the previous works and mildly support a larger value of C/Fe in the halo.

Oxygen: The trend of [O/Fe] with [Fe/H] can provide a check of the time scale of the formation of the Galaxy. The creation site of oxygen indeed, is restricted to very massive stars which have a very short life time; in contrast, iron can be created also during the evolution of lower mass stars which have a life time which spans the time of the Galaxy formation. Only the very massive stars had time to enrich the material which formed the halo stars.

All surveys of the oxygen abundances concluded that, in the thick disk and the halo, oxygen is overabundant relative to iron: cf. Clegg et.al (1981), Gratton and Ortolani (1986), Barbuy (1988), Magain (1988), Barbuy and Erderlyi-Mendes (1989), Abia and Rebolo (1989), Sneden et al. (1990). In the disk, the oxygen-to-iron ratio decreases slowly when [Fe/H] increases to reach the solar value. However the abundance values depend (Figure 2) on the type of oxygen lines used (forbidden, permitted, or lines of the OH molecule) and some more work has to be done in order to achieve a good agreement. Following Barbuy and Erderlyi-Mendes, [O/Fe] deduced from the forbidden lines, is about +0.4 dex in the halo; it is more than +1.0 dex following Abia and Rebolo who used the permitted triplet. Recently we could observe at the CFH telescope with a resolution of 80000, and a very high S/N ratio, the forbidden line at 630nm in some of the stars observed by Abia and Rebolo. We found a systematic difference of about 0.5 dex between the abundances deduced from the permitted lines by Abia and Rebolo and the abundances we deduced from the forbidden line with the same atmospheric model (Spite & Spite 1991). A similar difference had been found by Magain (1988) on one dwarf HD76932.

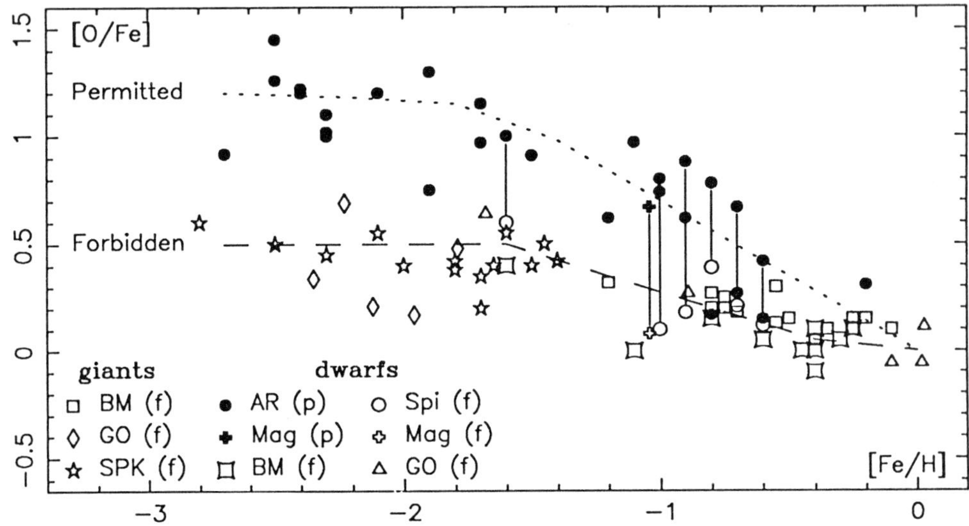

Fig.2 [O/Fe] versus [Fe/H] following Barbuy 1988 and Barbuy & Erderlyi-Mendes 1989 (BM), Gratton & Ortolani 1986 (GO), Abia & Rebolo 1989 (AR), Magain 1988 (Mag), Sneden et al. 1990 (SKP) and Spite & Spite (Spi). It seems that there is no systematic difference between dwarfs and giants but a systematic difference between the measurements based on the permitted lines and those based on the forbidden lines. (When the two systems have been measured in the same star the two measurements are joined by a straight line.)

It has been shown that the permitted lines are sensitive to non-LTE-effects. Recently Kiselman (1991) has studied the Non-LTE effects on oxygen abundance determinations; the correction computed for the permitted lines at 777nm, does not exceed 0.3dex. As a conclusion we would say that the overabundance of oxygen in halo stars is about +0.4dex, and that the large discrepancy between the permitted and the forbidden lines remains partly unexplained.

2-3 THE LIGHT METALS FROM Na to Ca ($11 \leq Z \leq 14$)

From the nucleosynthesis theory, these elements are formed in the same nucleosynthesis process but the formation rate of the odd light metals relative to the even metals can depend on the metallicity of the supernova. It is thus usual to consider the group of the odd Z elements (Na, Al) and to compare their abundance to the group of the even Z elements like Mg or Si.

The even Z elements: All recent studies of these elements have yielded similar results. The even Z elements are overabundant in the halo relative to iron: for exemple [Mg/Fe]≈+0.4dex up to [Fe/H]≈−3. In the disk, this ratio decreases slowly to reach the solar value. (Laird 1986, François, 1986a, 1987, 1988, Magain 1987, 1989, Gratton & Sneden 1987, 1988, 1991, Ryan et al. 1991). Following Molaro & Castelli (1990) and Molaro & Bonifacio (1990), the ratio [Mg/Fe] increases slowly in the extreme halo stars ([Fe/H]<-3). The mean behavior of [Mg/Fe] as a function of [Fe/H] is displayed in fig.3.
Similar enhancement of the ratio [M/Fe] in the halo are obtained for sulfur, and silicon (François, 1988, Gratton and Sneden 1991) , but Peterson *et al.* (1990) suspect a tendency for this enhancement to decrease with the increasing atomic number Z.

The odd Z elements: Aluminum and sodium are represented by only few unblended lines in the stellar spectra, and probably for this reason, the general trend of their abundances is up to now not completely clear.

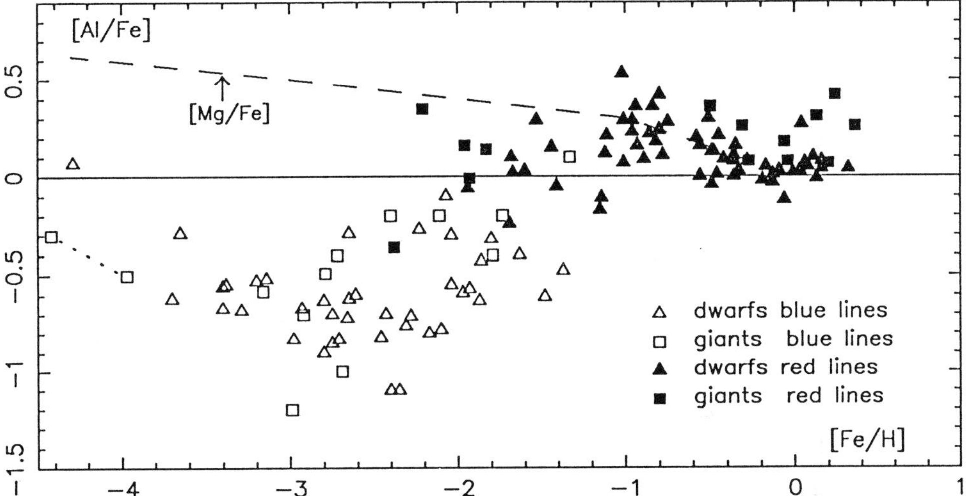

Fig.3 [Al/Fe] versus [Fe/H] for dwarf and giant stars. The filled symbols represent the aluminum abundance deduced from the "red" lines and the open ones, the abundance deduced from the "blue" resonance lines. In the region of the figure where blue or red lines have been used (-2.4>[Fe/H]>-1.3) it seems that there is a systematic difference between the two sets of measurements.
As a comparison, the dotted line represents the mean behavior of [Mg/Fe] versus [Fe/H].

-Na:
[Na/Fe] is constant up to [Fe/H]=-4.4: [Na/Fe]≈0 according Gratton and Sneden (1988), Peterson et al. (1990), and Molaro and Bonifacio (1990), who used the resonance sodium lines and [Na/Fe]}≈-0.2 according François (1986b) who used weak lines at about $\lambda \approx$ 616nm.
-Al:
To determine the abundance of aluminum, sometimes the "red" lines at λ 699.6nm and sometimes the "blue" resonance lines at λ 394.4nm are used (cf. François, 1986a, Magain 1987, Gratton & Sneden, 1988, Molaro & Bonifacio 1990, Ryan et al. 1991).

Clearly, the "blue" lines are used only if the metallicity of the star is so weak that the "red" lines disappear. Figure 3 shows the ratio [Al/Fe] as a function of [Fe/H] as computed by different authors. From the "red" lines: [Al/Fe] is almost constant. However, as soon as the "blue" aluminum lines are used the picture changes completely. It is apparent from Figure 3 that in the range -2.2<[Fe/H]<-1.4, the mean value of [Al/Fe] is found to be +0.1 if the "red" lines are used and -0.4 if the "blue" lines are used. To check this effect, it would be very interesting to measure the "blue" and the "red" lines in the same stars.

The odd-even effect Since Mg, Al and Na are supposed to be built in the same nuclear process, it is usual to compare directly the ratios [Al/Mg] or [Na/Mg] with [Mg/H] or [Fe/H]. The ratios [Na/Mg] and [Al/Mg] are independent of the metallicity in the disk, then these ratios decline with decreasing metallicity.

We should expect that when sodium and aluminum are directly compared to magnesium the scatter decreases, but this is not the case. It is likely that this scatter is mainly attributable to measurement errors as suggested in figure 3.

2-4 THE IRON PEAK ELEMENTS FROM Ca TO Ni ($20 \leq Z \leq 28$)

These elements are mainly produced by explosive Si burning during supernovae explosions. Recent nucleosynthesis models predict that the odd elements of the Fe group might have been underproduced in the massive supernovae which alone were able to enrich the material during the halo phase. The poor knowledge of the hyperfine structure which affects the lines of odd elements of the Fe group and the lack of reliable oscillator strengths for the relevant transitions explain that few reliable data existed. Recently Sneden and Crooker (1988), Gratton (1989), Peterson et al. (1990) and Gratton & Sneden (1991) have observed a sample of metal poor stars for a precise analysis of the relative abundances of these different elements, studying the best accessible transitions of these elements from high resolution high S/N spectra. Their results are summarized in fig.4 where are compared the pattern of these elements in the halo and in the disk.

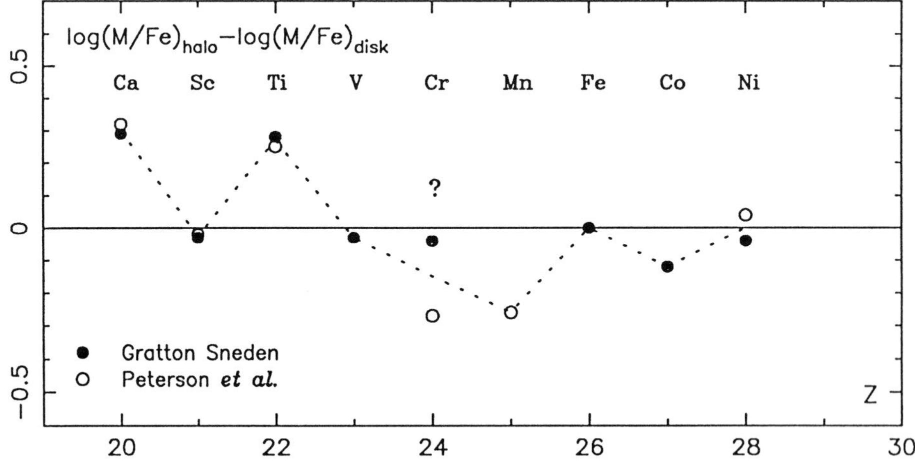

Fig.4 Difference between the relative abundances in the halo and in the disk stars for the iron peak elements. The abscissa is the atomic number Z.

Let us note that the overabundance of Scandium and Nickel found by Magain & Zhao (1989) and Luck & Bond (1983,1985) in the halo stars, has not been confirmed.

2-5 THE RARE EARTHS

The elements heavier than iron are synthesized by neutron capture reactions, in either the r (rapid) or the s (slow) process. From the measurements of the abundances of the different isotopes in the solar system we know for example that Ba and Sr, in the Sun, have been mainly formed by the "s" process and Europium by the "r" process.

In the halo stars, in the range $-2\leq[Fe/H]\leq-1$, these elements seems to be slightly less deficient than iron, then their abundance decreases steadily with [Fe/H] up to about [Fe/H]=–3. However the scatter is rather large, several halo stars are known for having a relatively high barium abundance as for example HD115444. This star has also a high Europium abundance. The scatter cannot be explained by measurements errors . In particular, Ryan et al. (1991) and Molaro & Bonifacio (1990) found a very large scatter in the strontium abundance of stars with $[Fe/H]\leq-3.4$. This intrinsic scatter suggests that the early Galaxy was not well mixed.

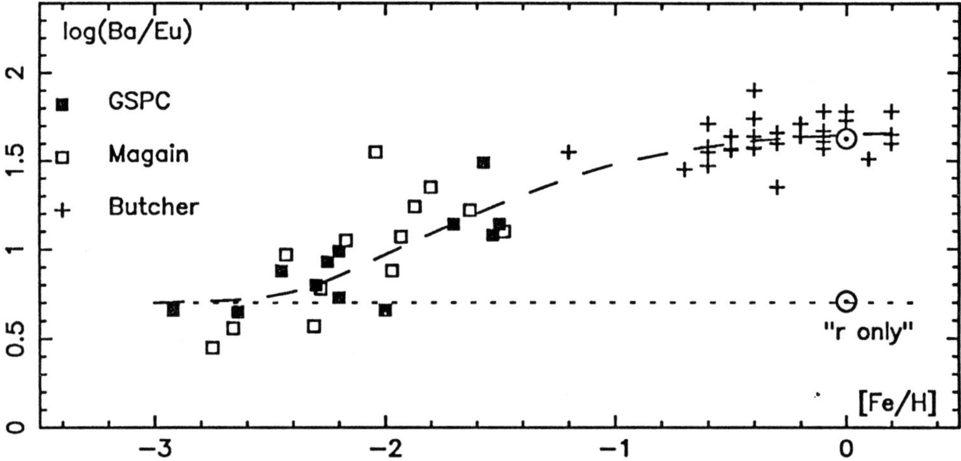

Fig 5. log (Ba/Eu) versus [Fe/H]. The dotted circles represent the solar value of Ba/Eu when either the total abundance of Ba and Eu or only the part formed by *r* process, are taken into account. The abundances have been taken from Butcher(1975), Gilroy et al.(1988) and Magain (1989).

Truran in 1981 remarked that the abundances of the different neutron capture elements in the halo stars strongly suggest a predominantly r-process origin. Gilroy et al. (1988) showed (fig.5) that, in the remote halo, the abundances of Barium and Europium are correlated; the ratio N(Ba)/N(Eu) is the same as the ratio, in the Solar system, of the only “r” process isotopes $N(Ba)_r/N(Eu)_r$ (Lambert 1989). The Ba/Eu behavior could be explained simply by asserting that at the beginning of the Galaxy both Eu AND Ba are pure “r” process elements (i.e. Ba is represented only by its “r” isotopes). Later the “s” process would begin to contribute substantially to the abundance pattern and the ratio [Ba/Eu] rises rapidly to reach the solar value.

However this explanation is not universally accepted: Magain (1989) and Ryan et al. (1991) proposed that at least part of the heavy elements are formed in the halo by the “s” process.

The "s" process is often called "secondary" because it is generally admitted that the first generation of stars (stars without metals), is not able to build the "s" process elements. In this theory, the Sr poor stars would be formed by the material ejected from this first generation of stars.

3. Chemical abundances in globular clusters

With the advent of the CCD detectors, precise measurements of abundances in the globular clusters became available. It is thus possible to compare the abundances in the field halo stars located in the solar vicinity, and in the halo clusters stars. However in the globular clusters only giants and supergiants can be observed. The abundance trends in the globular cluster stars have been recently reviewed by Wheeler et al. (1989) and Gratton (1990). Generally speaking, stars in globular clusters have the same abundance pattern as the field stars of the same metallicity. Possible differences are the abundances of oxygen and of the light metals Na and Al.

oxygen : Generally oxygen is overabundant in the cluster stars as well as in the metal deficient field stars [O/Fe]=0.5 (Pilachowski et al. 1983, Gratton and Ortolani 1989, Brown et al.1991b). However several cluster stars are suspected to be "oxygen poor" ($-0.3 \le [O/Fe] \le 0$). Following Brown et al. (1991a), the sum of the CNO nuclides is the same in the oxygen rich and in the oxygen poor stars; the deficiency of oxygen would be thus the signature of a mixing with CNO-cycled material.

the odd Z light metals: Although these elements are rather overdeficient in the halo field stars (fig 3), several cluster stars are known to have very strong aluminum and sodium lines.It has been shown that the strength of these lines is correlated with the strength of the CN band (e.g. Norris and Pilachowski 1985), but that the ratio [Na/Al] varies from star to star (François, 1991).
Gratton (1990) remarks that sodium overabundances (about +0.4 dex) are also observed in the field disk K supergiants (Smith and Lambert 1987) and suggests that these anomalies would be the consequence of severe non-LTE effects on the formation of the lines.
This explanation seems to be not completely satisfactory:

-A systematic overabundance of sodium (about +0.4 dex) is also observed in the field disk F supergiants and it has been shown that this overabundance cannot be explained by departure from LTE (Boyarchuk et al. 1988) .

-Moreover two stars in the same cluster, with about the same atmospheric parameters can have very different abundances of sodium (or aluminum).

The cause of the sodium/aluminum anomalies in the globular cluster stars is up to now not completely clear. The strong Na and/or Al stars in the globular clusters could be perhaps stars which have accreted matter processed in previously evolved AGB stars.

REFERENCES

Abia C., Rebolo R., 1989, ApJ 347, 186

Boesgaard A., 1990, Proceedings of the sixth Cambridge workshop "Cool stars, stellar systems, and the Sun" ed. G.Wallerstein, A.S.P. Conf. Serie 9, San Francisco, p.317

Brown J.A., Wallerstein G., Oke J.B., 1991a, AJ 101, 1693

Brown J.A., Wallerstein G., Cunha K., Smith V.V., 1991b, A&A, in press

Burkert A., Truran J.W., Hensler G., 1991, ApJ in press
Carbon D.F., Barbuy B., Kraft R.P., Friel E.D., Suntzeff N.B., 1987, PASP 99, 335
D'Antona F., Matteucci F., 1991, A&A 248, 62
François P., 1986a, A&A 160, 264
François P., 1986b, A&A 165, 183
François P., 1988, A&A 195, 226
François P., 1991, A&A 247, 56
Gilmore G., Reid N., 1982, MNRAS 202, 1025
Gilmore G., Wyse R.F.G, 1985, AJ 90, 2015
Gilmore G., Edvardson B., Nissen P.E., 1991, ApJ 378, 17
Gilmore G., Wyse R.F.G., Kuijken K., 1989, ARA&A 27, 555
Gilroy K.K., Sneden C., Pilachowski C.A., Cowan J.J., 1988, ApJ 327, 298
Gratton R.G., 1989, A&A 208, 171
Gratton, R.G., 1990, Memorie Soc. Astron. Ital. 61, 647
Gratton R.G., Ortolani S., 1989, A&A 211, 41
Gratton R.G., Sneden C., 1987, A&A 178, 179
Gratton R.G., Sneden C., 1988, A&A 204, 193
Gratton R.G., Sneden C., 1991, A&A 241, 501
Gustafsson B., 1987, "Stellar evolution and dynamics of the outer halo", M.Azzopardi and F.Matteucci, eds., (ESO-Garching) p.33
Hobbs L.M., Pilachowski C., 1988, ApJ 326, L23
Hobbs L.M., Thorburn J.A., 1991, ApJ 375, 116
Kiselman D., 1991, A&A 245, L9
Laird J.B., 1985, ApJ 289, 556
Lambert D.L., 1989, in "Cosmic abundances and dark matter" A.I.P. Conf. Proc. 183, (AIP, New York) p.168
Luck R.E., Bond H.E., 1983, ApJ 277, L75
Luck R.E., Bond H.E., 1985, ApJ 292, 559
Magain P., 1987, A&A 179, 176
Magain P., 1988, Proc. IAU Symp 132 "The impact of very high S/N spectroscopy on Stellar Physics", eds G.Cayrel and M.Spite, Kluwer, Dordrecht, p.485
Meneguzzi M., Audouze J., Reeves H., 1971, A&A 15, 337
Molaro P., Bonifacio P., 1990, A&A 236, L5
Molaro P., Castelli F., 1990, A&A 228, 426
Nissen P.E., 1990, Proceedings of the 31st Herstmonceux conference, "Elements and the Cosmos" ed. R.J. Terlevich, Cambridge University press. *in press*
Norris J., Pilachowski C., 1985, ApJ 299, 295
Peterson R.C., Kurucz R.L., Carney B.W., 1990, ApJ 350, 173
Pilachowski C., Sneden C., Wallerstein G., 1983, ApJS 52, 241
Ryan S.G., Bessell M.S., Sutherland R.S., Norris J.E., 1990, ApJ 348, L57
Ryan S.G., Norris J.E., Bessell M.S., 1991, AJ *in press*
Smith V., Lambert D.L., 1987, MNRAS 227, 563
Smith V., Lambert D.L., 1990, ApJ 361, L69
Sandage A., 1987, AJ 93, 610
Sneden C., Crocker D.A., 1988, ApJ 335, 406
Sneden, C., Kraft, R.P., Prosser, C., 1990, Proceedings of the sixth Cambridge workshop "Cool stars, stellar systems, and the Sun" ed. G.Wallerstein, A.S.P., San Francisco, p.369

Spite F., Spite M., 1985, ARA&A 23, 225
Spite M., Spite F., 1991, A&A *in press*
Tomkin J., Sneden C., Lambert D.L., 1986, ApJ 302, 415
Wheeler J.C., Sneden C., Truran J.W.Jr., 1989, ARA&A 27, 279
Zhao G., Magain P., 1989 in the Proceedings of Elba workshop "Chemical and dynamical evolution of Galaxies" F.Matteucci and J.Franco eds. *in press*

DISCUSSION

D. Johnson: Recent work by Smith and Lambert may solve the primordial Li abundance quandary. They targeted 5 maximum luminosity AGB stars in the LMC and 2 in the SMC. All seven show incredible Li enhancements. The conclusion is that these stars are producing Li by the Be-transportation mechanism. A simple closed box chemical evolution model shows that the yield from these stars can enrich the disk by a factor of 10 from the halo abundance. Of course with the metallicity differences between the clouds and our Galaxy, detection of such Li producing AGBs in the Milky Way is really needed.

R. de la Reza: Complementing the precedent speaker who mention the discovery of S stars in the Magellanic Clouds as a source of Li, I can say that we discovered other kind of sources of Li in our Galaxy that is Li-rich K giants, one of them having about 10 times larger abundance than the interstellar matter (see poster at this meeting).

M. Pinsonneault: The Spite Li plateau is a beautiful observation which places strong constraints on theory. Models with rotational mixing, however, can nearly uniformly deplete the halo plateau stars by a factor of up to 10. There are also many observations of high Li abundance (~ 3.5 to 3.6) in young stars, which indicates that the current Li abundance is higher than the "cosmic" value of 3.0. So it appears that we need both higher initial Li in halo stars and galactic enrichment

B.Pagel: I should like to report a new twist in the beryllium saga. Steigman & Walker point out that while Be production by cosmic ray spallation in the interstellar medium is proportional to CNO abundance, that of lithium is largely due to α,α collisions and therefore nearly independent of metallicity. Consequently, if Be in halo stars like HD140283 is due to spallation in the interstellar medium, it should be accompanied by a much larger amount of Li which could require significant corrections compared to the amount that is seen and usually ascribed to Standard Big Bang.

CHEMICAL EVOLUTION OF STELLAR POPULATIONS

B.E.J. PAGEL
NORDITA
Blegdamsvej 17
DK-2100 Copenhagen Ø
Denmark

ABSTRACT. Stars can now mostly be separated into disk and halo members when *both* Galactic rotation velocity and metallicity are considered. New O/Fe data give results similar to α/Fe data and enable the O or α-element distribution function for the halo to be fitted to Hartwick's modification of the Simple Model up to nearly solar abundances, implying strong overlap with the disk and no sharp break in [O/Fe] when [Fe/H] $\simeq -1$. The G-dwarf problem and large-scale radial abundance gradients are discussed; in both cases, there are deficiencies in the data and too many alternative hypotheses capable of explaining them.

1. Introduction

In this talk, I make some remarks on chemical evolution of stellar populations in our Galaxy, with a few references to those of nearby galaxies such as the Magellanic clouds. Topics include: ingredients of GCE models with comments thereon, the Simple Model, the systematics of oxygen and α-element abundances relative to iron, abundance distribution functions in the halo, bulge and disk, and abundance gradients in the Milky Way and some nearby spirals.

2. Ingredients of Galactic Chemical Evolution (GCE) Models

2.1. INITIAL CONDITIONS

The most natural initial abundances to assume are the pre-galactic ones, believed to result from the Big Bang (e.g. Boesgaard and Steigman 1985). Pagel et al. (1992) have re-worked the data on helium in extragalactic H II regions and its regression relations with oxygen (cf. Peimbert and Torres-Peimbert 1974, 1976; Lequeux et al. 1979) and nitrogen (cf. Pagel, Terlevich and Melnick 1986) with various improvements including deletion of objects with definite detections of broad WR features, some of which appear to be biased by local enhancements of helium and nitrogen from stellar winds, and find the same result from both regressions: $Y_p = 0.228 \pm 0.005$ (s.e.), which does not differ significantly from our earlier estimates (e.g. Pagel 1991). More surprising is that we still find a large value of the slope of the He$-$O regression, corresponding to a dY/dZ of at least 3 and probably 4, which is six or seven times the value recently predicted theoretically by Maeder (1991) for these low-metallicity objects (I Zw 18 to the LMC and NGC 5461 in M 101). Pre-galactic enrichment in carbon and heavier elements now looks unlikely in view of the metallicity distribution function of halo field stars (see below), but there are still interesting question marks hanging over beryllium (cf. M. Spite in this volume).

Invited talk at IAU Symposium No. 149, Angra dos Reis, Brazil, August 1991. B. Barbuy (ed.), Kluwer.

B. Barbuy and A. Renzini (eds.), The Stellar Populations of Galaxies, 133–142.

2.2. END-RESULTS OF STELLAR EVOLUTION

The general idea is clear: nuclear reactions in massive stars synthesize elements that are ejected into the interstellar medium (ISM) via supernova explosions and stellar winds, whereas small stars merely serve to lock up diffuse material. However, the details are far from clear, involving reaction rates, convection, mass loss, the range of masses within which stars ultimately explode, and effects of rotation and close binary evolution. Maeder (1991) has recently given theoretical estimates of yields as a function of initial mass and metallicity, assuming mass loss rates proportional to $Z^{0.5}$, but because of dY/dZ, I do not feel that this scheme is yet very close to the final answer.

2.3. INITIAL MASS FUNCTION

The IMF is another grey area. How does it go and does it vary in space or time in systematic ways that need to be taken into account in GCE theory? In our own neighbourhood, one can observe only the present-day luminosity function and the IMF depends on the past star formation rates assumed as well as on the mass–luminosity relation. In the Magellanic Clouds, steep power laws have sometimes been claimed, but recent results (Melnick 1987; Richtler, de Boer and Sagar 1991) are in agreement with Salpeter's law between 1 and 100 $M_{\odot}$, which is nominally flatter than the solar neighbourhood IMF after Scalo (1986), resulting in about three times as large a proportion of total mass in massive stars. Flattening of the slope towards low metallicities, once invoked to explain the appearance of increasing effective temperatures of ionising stars of extragalactic H II regions (Terlevich 1985), has gone away (McGaugh 1991) thanks to Maeder's (1990) models for the zero-age main sequence, but there are still cases including some blue compact galaxies and infra-red luminous starburst galaxies where "top-heavy" IMFs have been invoked for various reasons (Scalo 1990). This leads to the vexed question of bimodal star formation, which is still wide open.

Folding-in end products of stellar evolution with the IMF leads to two important quantities for GCE theory: the *returned fraction* and the *yield* of one or more elements. I refer to the yield derived in this way as the *true yield*, which is predicted on a theoretical basis to be comparable to solar abundance for carbon and heavier elements but has enormous uncertainties, as indicated above. One can define also an *effective yield*, deduced by fitting a GCE model to a set of observed abundance data. This again has considerable uncertainties, but these are probably less than those of the true yield.

2.4. STAR FORMATION RATES

The SFR is often parameterised as a power law in gas volume density or surface density. My prejudice is that only laws involving the first power make any physical sense, although the coefficient may vary as a function of ambient conditions. Two modifications of a simple linear law seem to be of particular interest:

2.4.1. *A Threshold Gas Surface Density* $\sim 8\ M_{\odot}\ pc^{-2}$ for all (or just for massive) star formation (e.g. Kennicutt 1989). As discussed by Phillipps, Edmunds and Davies (1990), this can lead to a correlation between surface brightness and mean metallicity related to radial abundance gradients, and possibly to changes in the abundance ratio of intermediate-mass star products like carbon to massive-star products like oxygen.
2.4.2. *Self-Regulating Star Formation.* This can lead to wild oscillations in SFR (Hensler and Burkert 1990; Arimoto 1989) that may be relevant to bursting modes of star formation seen, for example, in blue compact galaxies. In less extreme form, it can again help to account for radial abundance gradients (Dopita 1990; Matteucci et al. 1989; Parravano 1989; Phillipps and Edmunds 1991). Stochastic self-propagating star formation (Gerola and Seiden 1978; Dopita 1985) can perhaps also be regarded as a form of self-regulation.

2.5. THE GALACTIC CONTEXT

From the theoretical point of view, this includes different scenarios of galactic evolution – one-zone models or models with inflow and/or outflow affecting the outcome. From the observational point of view, we have stellar populations – the subject of this Symposium. Their definition has been greatly clarified in recent years by the beautiful work of Norris and Ryan (1989), Carney, Latham and Laird (CLL 1990) and Nissen and Schuster (1991) on proper motion stars, which give a kinematically biased sample of disk stars (favouring the "thick disk") but a relatively unbiased sample of halo stars. While there is substantial overlap between disk and halo (as between cats and dogs) in any one property – either kinematic or chemical, or even in age – the two-dimensional scatter diagram of Galactic rotational velocity versus [Fe/H] shown by CLL and by Nissen (1991) shows a marked diagonal gap (the "Nissen line") between halo and disk stars that enables nearly all of them to be classified unambiguously as belonging to one or the other. The distinction between these two populations is confirmed by their very different metallicity distribution functions, discussed below. Bulge stars, with a different distribution function yet again, are entitled on this basis to be treated as yet another separate population, with its own distinct evolutionary history.

More controversial is the idea that the thick disk constitutes a separate population as opposed to being just what is left over from the disk before it became thin. This idea seems to be gaining popularity, but I am not convinced that the evidence for discontinuity in metallicities from the rest of the disk – to the extent that it is statistically significant – does not result from selection biases.

3. Simple GCE Models

Owing to the numerous unknowns, I prefer to deal with very simple analytically soluble models with yields chosen at will to fit the data and simple prescriptions for SFR, gas flows and time delays. A good reference standard is still provided by the Simple Model with a capital S, which is a one-zone model treated in the instantaneous recycling approximation with constant yields for "primary" nucleosynthesis products from short-lived massive stars, e.g. oxygen and α-elements (maybe). This model leads to the characteristic abundance distribution function

$$\frac{ds}{d\ln z} \propto z\,e^{-z}; \qquad z \leq \ln\left(\frac{m}{g}\right) \tag{1a}$$

$$= 0; \qquad z > \ln\left(\frac{m}{g}\right), \tag{1b}$$

where s is the mass of stars with abundance $\leq z$ in units of the yield, g the mass of gas (and dust), and $m = g + s =$ the (constant) total mass of the system. Hartwick (1976) extended this model to include loss of diffuse material from the system at a rate proportional to the (net) SFR with a factor Λ. This leads to a similar distribution but with m no longer constant and with an effective yield a factor $(1 + \Lambda)$ less than the true yield.

Many models with SFR proportional to g predict a more or less linear increase in z with time; slightly higher powers of g lead to a flattening that is barely perceptible on a logarithmic scale. Figure 1 shows the prediction of such a linear model going through the Sun and assuming a Galactic age of 16 Gyrs compared to the data for a sample of field stars from the HD Catalogue by Gustafsson (in this volume) and collaborators. The arithmetic mean abundance at a given age is fairly well fitted by this linear model, but the scatter is large and well above likely experimental errors, and there are undoubtedly significant selection effects. Ages less than a few Gyrs may also need to be increased to allow for convective overshoot (Maeder and Meynet 1989).

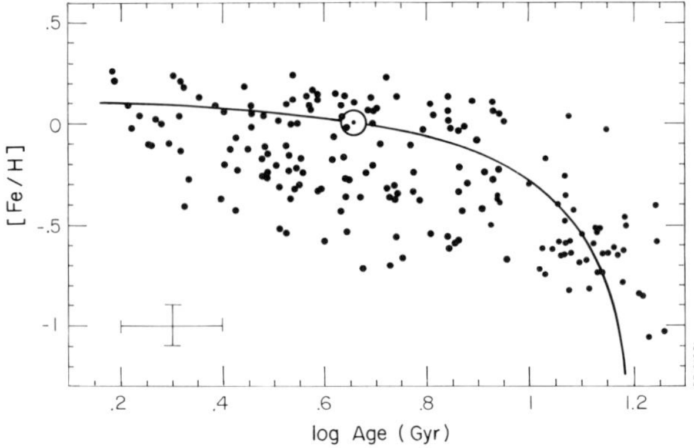

Figure 1. Age–metallicity relation for F stars in the HD catalogue by Gustafsson (these proceedings) and collaborators. The curve is based on a naive GCE model which assumes [Fe/H] = log{(16 – A)/11.4}, where A is the age in Gyr. The position of the Sun is shown, as is an estimated error bar of ± 0.1 dex in each coordinate.

4. α/Fe and O/Fe Relations

Aller and Greenstein (1960) discovered a so-called α-rich effect (cf. Wallerstein 1962) in metal-deficient stars, i.e. Mg, Si, Ca, S and Ti are deficient by smaller factors than iron in Galactic disk and halo stars for which [Fe(H] < 0. A corresponding effect was later discovered for oxygen in Arcturus (Gasson and Pagel 1966) and in a sample of red giants in general (Conti et al. 1967) using the forbidden [O I] line λ6300. Tinsley (1979) suggested that the effects result from a failure of the instantaneous recycling approximation for iron, to which a substantial contribution comes from Type I supernovae taking a significant amount of time (~1 Gyr) to complete their evolution, and this kind of model has been studied in detail by Matteucci et al. (e.g. Matteucci and François 1989) and also analytically (Pagel 1989a). In such models, halo stars are expected to have a plateau with $[\alpha/\mathrm{Fe}] = [\mathrm{O/Fe}] \simeq 0.5$, as is suggested by many observations, and then at some point in [Fe/H], depending on the time scale for SN I relative to the time scales for formation and enrichment of the halo and disk, [α/Fe] and [O/Fe] go down with increasing Fe/H to hit and pass through solar values. This subject has been well reviewed by Wheeler, Sneden and Truran (1989), who point out that [O/H] (rather than [Fe/H]) forms the best available "clock" to provide a (nonlinear) time-equivalent to measure progressive enrichment of the ISM. However, the existence of this plateau has been challenged by Abia and Rebolo (1989), who measured permitted, high-excitation O I lines in halo dwarf stars and found a continuous rise in [O/Fe] towards lower [Fe/H]. Furthermore, there could be a different sort of explanation for the [O/Fe] effect if the relative yields vary systematically as a function of metallicity (Maeder 1991) or something else.

Recent studies of [O I] (Nissen 1991) and OH (Bessell, Sutherland and Ruan 1991) have tended to cast doubt on the results derived from permitted oxygen lines at all metallicities; they could be incorrect partly because of non-LTE effects (Kiselman 1991), but mainly because these high-excitation lines are sensitive to small changes of temperature in relatively deep layers of stellar atmospheres, where the models are still uncertain. In contrast to the old picture accepted by Matteucci and François (1989), Pagel (1989a) and Gilmore, Wyse and Kuijken (1989), in which halo stars were thought to have a plateau with $[\mathrm{O/Fe}] \simeq 0.5$ extending to $[\mathrm{Fe/H}] \simeq -1$ (sometimes regarded as a transitional metallicity between halo and disk), followed by a roughly linear descent through the Sun, the new oxygen data of Bessell, Sutherland and Ruan, as well as the ensemble of α-element data presented by Andersen, Edvardsson, Gustafsson, Nissen and Lambert, and by Ryan, Norris and Bessell (1991) seem to be better represented by the relation

$$[\mathrm{O/Fe}] \simeq [\alpha/\mathrm{Fe}] \simeq 0.5; \; [\mathrm{Fe/H}] \leq -1.7 \tag{2a}$$

$$\simeq -0.3 \, [\mathrm{Fe/H}]; \; [\mathrm{Fe/H}] \geq -1.7, \tag{2b}$$

which means that halo stars with [Fe/H] > -1.7 are already affected by whatever it is that leads to the production of additional iron (relative to oxygen and α-elements) over and above the production from the first massive stars that presumably caused the initial enrichment of the halo. On the differential time-scale hypothesis, this suggests a longer time scale than 1 Gyr for formation of the halo (plausible on other grounds, such as age differences) and/or a contribution to iron from stars with larger masses than the standard SN Ia models. The hypothesis also implies that there should be cosmic scatter in the O, Fe relation, which may well be the case.

Can one discriminate between the above-mentioned differential time-scale hypothesis and the alternative suggestion that Fe/O production ratios increase with metallicity, either owing to mass loss and other factors in stellar evolution (Maeder 1991) or owing to changes in IMF? In principle we can, by comparing [Fe/O] in young metal-deficient stars of the Magellanic Clouds with the values in old metal-deficient stars of the Milky Way (Dopita 1990), or in old metal-rich stars of the Galactic Bulge with younger metal-rich stars of the solar neighbourhood (Matteucci and Brocato 1990; Barbuy and Grenon 1990; Matteucci 1991). Data available at present seem to marginally favour the differential time-scale hypothesis, but the full picture is not yet clear.

5. Metallicity Distribution Functions

The nice thing about the Simple Model with instantaneous recycling is that the abundance distribution function (equation 1) is independent of the star formation history. The price that one has to pay for this is that one needs to study an element to which the assumptions of instantaneous recycling and constant yield apply, and they clearly do not apply to iron (except along the [O/Fe], [α/Fe] plateau). They may apply well enough to oxygen and α-elements, so that it is their distribution function that should be compared to the prediction.

In fig. 2, I have taken the [Fe/H] distribution function for halo field stars lying below the gap near the Nissen line in the (V_{rot}, [Fe/H]) diagram of Carney, Latham and Laird (1990) and converted [Fe/H] into [O/H] using a relation similar to equation 2. It fits quite well on to Hartwick's modification of the Simple model with an effective yield $p/(1 + \Lambda) = 10^{-1.1} Z_{\odot}$, suggesting that 90% of the halo gas was lost in parallel with star formation if the true yield for oxygen and α-elements is $10^{-0.1} Z_{\odot}$, which fits the solar neighbourhood abundance distribution (Pagel 1989a).

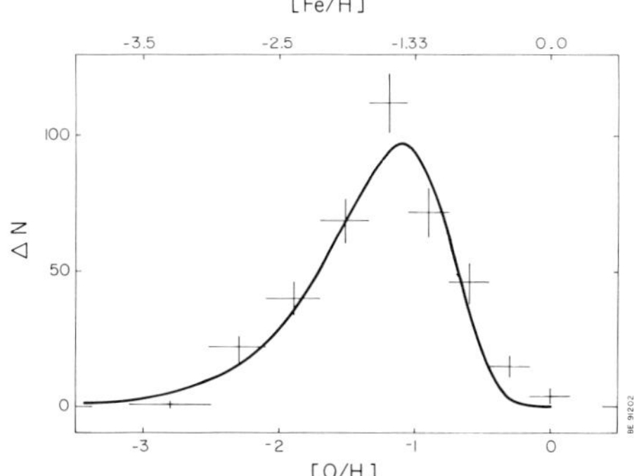

Figure 2. Distribution function of oxygen abundances in stars of the Galactic halo taken from CLL's (1990) plot of rotational velocity against [Fe/H] and assuming

$$[O/Fe] = 0.5; \ [Fe/H] \leq -2$$
$$= -0.25 \ [Fe/H]; \ [Fe/H] > -2.$$

ΔN is the number of stars in a bin of width 0.3 dex in [O/H]. Vertical error bars are $\pm\sqrt{\Delta N}$ and horizontal bars show the widths of the bins actually used. The curve is based on the Simple model, with an effective yield $10^{-1.1} Z_{\odot}$ for oxygen.

Of course, the true yield could have been higher with more than 90% mass loss, in which case there might have been enough material initially in the halo and bulge to form the disk. Alternatively, the halo distribution function can be fitted by a different model involving an inhomogeneously collapsing halo (Malinie, Hartmann and Mathews 1991). The most striking feature about this distribution function is the wide range in metallicities, corresponding to a Simple model that has gone to complete gas exhaustion (cf. Searle and Zinn 1978). As first noted by Beers, Preston and Shectman (1986) and Beers (1987), the Simple model works towards extremely low metallicities, indicating that there is no evidence for pre-galactic oxygen or metal production; but equally striking is the fact that high metallicities are also found overlapping very extensively with disk stars up to about solar abundance. Disk stars themselves go down as far as the peak of the halo distribution function at [Fe/H] = −1.6 (Morrison, Flynn and Freeman 1990), and there no longer seems to be a sharp halo–disk transition near [Fe/H] = −1, such as was deduced from the metallicity and spatial distributions of globular clusters (Zinn 1985; Pagel 1989b). How these fit into the new picture is not yet clear, owing to small-number statistics and some problems in their metallicity scale.

The Simple model also gives a good fit to the metallicity distribution in the Galactic Bulge (Rich 1988, 1990; Geisler and Friel 1990), but this time with a large effective yield of twice solar metal abundance (mainly, though not exclusively, iron), which may correspond to a still higher factor in oxygen abundance if the bulge stars are very old. This figure could perhaps represent the true yield in both the bulge and the halo, with applications to the G-dwarf problem in the solar neighbourhood, the problem consisting in the fact that the Simple model does not fit in this case (van den Bergh 1962; Schmidt 1963; Pagel and Patchett 1975); cf. fig. 3.

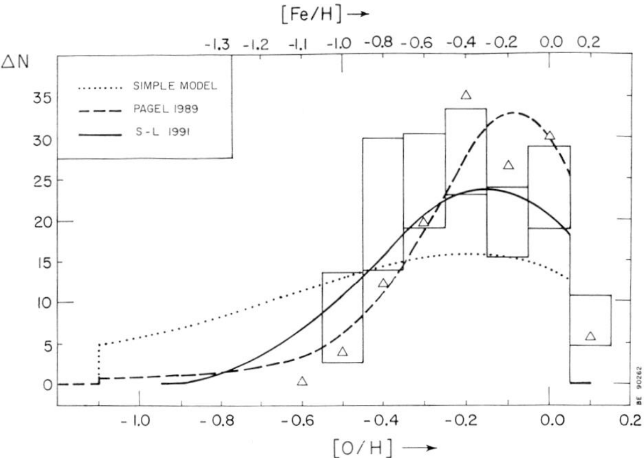

Figure 3. Distribution function of oxygen abundances for G-dwarf stars in a cylinder through the Sun perpendicular to the Galactic plane after Pagel (1989a,b: triangles) and Sommer-Larsen (1991: boxes). Curves represent a Simple model with modest prior enrichment from the halo and inflow models by Pagel (1989a) and Sommer-Larsen (1991), respectively. ΔN is the number of stars in a bin of width 0.1 dex in [O/H] and the heights of the boxes include both sampling errors and uncertainties in the correction from the solar neighbourhood to the solar cylinder.

Numerous (really too many) solutions to the G-dwarf problem have been proposed, including:

1. The oldest stars have either evolved away (Biermann and Biermann 1977; Bazan and Mathews 1990; but cf. Mould 1976) or moved away towards central parts of the Galaxy (Grenon 1989), so No Problem!

2. Larger yields at low metallicities owing to IMF (Schmidt 1963) or effects on stellar evolution (Maeder 1991). However, data on extragalactic H II regions rather suggest the opposite effect (Peimbert and Serrano 1982; Edmunds and Pagel 1984).

3. Prior enrichment from the halo and bulge (Ostriker and Thuan 1975; Binney and Tremaine 1987; Köppen and Arimoto 1991), viable if the true yield is large enough to accommodate the present halo : disk ratio deduced from star counts.

4. Inflow of unprocessed (or only slightly processed) material, corresponding to gradual formation of the disk (Larson 1976; Lynden-Bell 1975, 1991; Clayton 1985; Pagel 1989a; Sommer-Larsen 1991). This is still my favourite, partly because it makes dynamical sense and partly because it

readily accommodates the "metal-weak thick disk" (Morrison, Flynn and Freeman 1990). However, there should probably be a moratorium on discussions of the G-dwarf problem until there is an improved data base generally available.

6. Radial Abundance Gradients in Spiral Galaxies

Abundance trends between and across galaxies, deduced from observations of emission lines in H II regions, can provide a useful perspective supplementing the picture derived from individual stars in our Galaxy and members of the Local Group. Pagel (1981), Edmunds and Pagel (1984), and Axon et al. (1988) have given some comparisons between oxygen abundances in extragalactic H II regions and rough estimates of the gas fraction (a more detailed discussion is in preparation by Vila-Costas and Edmunds 1992), which suggest a crude fit to the Simple model prediction

$$z = \ln(m/g) \tag{3}$$

(Searle and Sargent 1972) with a rather low yield $p \simeq 0.2\, Z_{\odot}$ for oxygen in dwarf and Magellanic irregulars. There is a rather more impressive correlation between oxygen abundance and absolute luminosity of the parent galaxy (Skillman, Kennicutt and Hodge 1989) or maybe its mean surface brightness (Phillipps, Edmunds and Davies 1990). McCall (1982) and Edmunds and Pagel (1984) found a good correlation of oxygen abundance with total local mass surface density in different parts of spirals, while other authors have suggested a universal radial gradient in spirals of about -0.07 dex kpc^{-1} (Dufour et al. 1980; Belley and Roy 1991) similar to the one found for Milky Way H II regions (Shaver et al. 1983) and planetary nebulae (Maciel 1991). Other indicators like cepheids, open clusters and supergiants give a similar result in our Galaxy (see review by Pagel 1985), but there are also discrepant results from deep surveys of red giants (Neese and Yoss 1988, Lewis and Freeman 1989) and from B stars in young associations (Fitzsimmons et al. 1990), both of which suggest negligible gradients, apart from the young association Dolidze 25 at a galactocentric distance of 13 kpc (Lennon et al. 1990). These discrepancies are difficult to understand, but we believe that the existence of appreciable gradients in ISM abundances is well attested by the data on H II regions in external spirals (cf. Vila-Costas and Edmunds 1992).

Numerous causes for abundance gradients have been suggested, including:

1. Straightforward gas fraction effect in isolated concentric zones (Searle and Sargent 1972), driven by time-scale variations in SFR and/or a gas surface density threshold for star formation (Phillipps, Edmunds and Davies 1990 and references therein) and/or self-regulated star formation (Dopita 1985; Matteucci et al. 1989; Parravano 1989; Phillipps and Edmunds 1991), which two latter can also lead to the surface density effect. The Simple model can account for a part, but not for the whole, of the range of H II region abundances deduced from observation.

2. Varying true yields from a bimodal IMF (Güsten and Mezger 1982). This would not account for the existence of a shallower gradient in old stars compared to the ISM.

3. Ejection of hot gas in galactic winds, fountains or chimneys (Tinsley and Larson 1979; Vigroux, Chièze and Lazareff 1981; Arimoto and Yoshii 1987; Vader et al. 1988; Franx and Illingworth 1990; Lynden-Bell 1991). This is often assumed to account for low abundances in small galaxies with shallow potential wells and low escape velocities, and can be applied to isolated concentric zones to produce abundance gradients. The winds may involve the ISM in general or just hot, metal-enhanced material from supernovae; the latter hypothesis has some difficulties in accounting for the relative constancy of N/O in dwarf irregular and blue compact galaxies (Pagel et al. 1992).

The remaining hypotheses abandon the Simple one-zone model:

4. Inflow of unprocessed material from outside the disk (good for the G-dwarf problem), with radial variations in the inflow and star formation rates (Tinsley and Larson 1978; Chiosi 1980; Díaz

and Tosi 1984; Pagel 1989a; Sommer-Larsen 1991). The numerical model by Matteucci and François (1989) gives separate abundance gradients for different elements, taking into account non-instantaneous recycling, and explains some otherwise mysterious differential gradients that we seem to observe, in particular a flatter gradient in sulphur than in oxygen (Shaver et al. 1983; Vilchez et al. 1988; Díaz et al. 1991; Henry et al. 1992). It seems that sulphur (and presumably neon) may be better candidates for instantaneous recycling than oxygen itself! However, Garnett (1990) finds no significant change in S/O in extragalactic H II regions with abundances below solar.

Abundance gradients are not necessarily generated, but they can be substantially modified, by:

5. Radial gas flows arising either from a mismatch of angular momentum between infalling gas and the part of the disk on which it falls (Mayor and Vigroux 1981; Lacey and Fall 1985; Pitts and Tayler 1989) or from viscous transfer of angular momentum, which is also helpful in accounting for exponential disks (Sommer-Larsen and Yoshii 1989, 1990; Clarke 1989, 1991).

Playing with all these ideas is fun, but we do need better data, not only on abundances but also on ambient structural parameters such as H I and H_2 column densities, mass surface densities, disk–bulge deconvolution, and the nature and distribution of dark matter, in order to gain some more understanding of how abundance gradients arise.

References

Abia, C. and Rebolo, R. (1989), Astrophys. J. 347, 186.

Aller, L.H. and Greenstein, J.L. (1960), Astrophys. J. Suppl. 5, 139.

Arimoto, N. (1989), in "Evolutionary Phenomena in Galaxies", J. Beckman and B.E.J. Pagel (eds.), Cambridge University Press, p. 341.

Arimoto, N. and Yoshii, Y. (1987), Astr. Astrophys. 173, 23.

Axon, D.J., Staveley-Smith, L., Fosbury, R.A.E., Danziger, J., Boksenberg, A. and Davies, R.D. (1988), Mon. Not. R. astr. Soc. 231, 1077.

Barbuy, B. and Grenon, M. (1990), in "Bulges of Galaxies", B.J. Jarvis and D. Terndrup (eds.), Garching: ESO Conf. and Workshop Proceedings No. 35, p. 83.

Bazan, G. and Mathews, G.J. (1990), Astrophys. J. 354, 644.

Beers, T.C. (1987), in "Nearly Normal Galaxies, From the Planck Time to the Present", S.M. Faber (ed.), Springer, New York, p. 41.

Beers, T.C., Preston, G.W. and Shectman, S.A. (1986), Astr. J. 90, 2089.

Belley, J. and Roy, J.-R. (1991), Astrophys. J. Suppl., in press.

Bessell, M.S., Sutherland, R.S. and Ruan, K. (1991), preprint.

Biermann, P. and Biermann, L. (1977), Astr. Astrophys. 55, 63.

Binney, J. and Tremaine, S. (1987) "Galactic Dynamics", Princeton University Press.

Boesgaard, A.M. and Steigman, G. (1985), Ann. Rev. Astr. Astrophys. 23, 319.

Carney, B.W., Latham, D..W. and Laird, J.B. (1990), Astr. J. 99, 572.

Chiosi, C. (1980), Astr. Astrophys. 83, 206.

Clarke, C.J. (1989), Mon. Not. R. astr. Soc. 238, 283.

Clarke, C.J. (1991), Mon. Not. R. astr. Soc. 249, 704.

Clayton, D.D. (1985), in "Nucleosynthesis: Challenges and New Developments", W.D. Arnett and J.W. Truran (eds.) p. 65.

Conti, P.S., Greenstein, J.L., Spinrad, H.E., Wallerstein, G. and Vardya, M.S. (1967), Astrophys. J. 148, 105.

Díaz, A.I., Terlevich, E., Vilchez, J.M., Pagel, B.E.J. and Edmunds, M.G. (1991), Mon. Not. R. astr. Soc., in press.

Díaz, A.I. and Tosi, M. (1984), Mon. Not. R. astr. Soc. 208, 365.

Dopita, M.A. (1985), Astrophys. J. Lett. 295, L5.

Dopita, M.A. (1990), in "The Interstellar Medium in Galaxies", H.A. Thronson, Jr. and J.M. Shull (eds.), Kluwer Academic Publishers, Dordrecht, p. 437.

Dufour, R.J., Talbot, R.J., Jensen, E.P. and Shields, G. (1980), Astrophys. J. 236, 119.

Edmunds, M.G. and Pagel, B.E.J. (1984), Mon. Not. R. astr. Soc. 211, 507.

Fitzsimmons, A., Brown, P.J.F., Dufton, P.L. and Lennon, D.J. (1990), Astr. Astrophys. 232, 437.

Franx, M. and Illingworth, G. (1990), Astrophys. J. Lett. 349, L41.

Garnett, D. (1990), Astrophys. J. 363, 142.

Gasson, R.E.M. and Pagel, B.E.J. (1966), The Observatory 86, 196.

Geisler, D. and Friel, E.D. (1990), in "Bulges of Galaxies", B.J. Jarvis and D. Terndrup (eds.), Garching: ESO Conf. and Workshop Proc. No. 35, p. 77.

Gerola, H. and Seiden, P.E. (1978), Astrophys. J. 223, 129.

Gilmore, G., Wyse, R. and Kuijken, K. (1989), Ann. Rev. Astr. Astrophys. 27, 555.

Grenon, M. (1989), Astrophys. Sp. Sci. 156, 29.
Güsten, R. and Mezger, P.G. (1982), Vistas in Astr. 26, 159.
Hartwick, F.D.A. (1976), Astrophys. J. 209, 418.
Henry, R.B.C., Pagel, B.E.J., Lasseter, D.F. and Chincarini, G.L. (1992), submitted to M.N.R.A.S.
Hensler, G. and Burkert, A. (1990), Astrophys. Sp. Sci. 170, 231; 171, 149.
Kennicutt, R.C. (1989), Astrophys. J. 344, 685.
Kiselman, D. (1991), Astr. Astrophys. 245, L9.
Köppen, J. and Arimoto, N. (1991), in "Chemical and Dynamical Evolution of Galaxies", F. Ferrini, J. Franco and F. Matteucci (eds.), Pisa: Giardini Editore.
Lacey, C.G. and Fall, S.M. (1985), Astrophys. J. 290, 154.
Larson, R.B. (1976), Mon. Not. R. astr. Soc. 176, 31.
Lennon, D.J. Dufton, P.L., Fitzsimmons, A., Gehren, T. and Nissen, P.E. (1990), Astr. Astrophys. 240, 349.
Lequeux, J., Peimbert, M., Rayo, J.F., Serrano, A. and Torres-Peimbert, S. (1979), Astr. Astrophys. 80, 155.
Lewis, J.R. and Freeman, K.C. (1989), Astr. J. 97, 139.
Lynden-Bell, D. (1975), Vistas in Astr. 19, 299.
Lynden-Bell, D. (1991), in "Elements and the Cosmos" (31st Herstmonceux Conf.), M.G. Edmunds, B.E.J. Pagel and R.J. Terlevich (eds.), Cambridge University Press.
Maciel, W. (1991), in "Elements and the Cosmos", M.G. Edmunds, R.J. Terlevich and B.E.J. Pagel (eds.), Cambridge.
Maeder, A. (1990), Astr. Astrophys. Suppl. 84, 139.
Maeder, A. (1991), Astr. Astrophys., in press.
Maeder, A. and Meynet, G. (1989), Astr. Astrophys. 210, 155.
Malinie, G., Hartmann, D.H. and Mathews, G.J. (1991), Astrophys. J. 376, 520.
Matteucci, F. (1991), in "Morphological and Physical Classification of Galaxies", G. Longo (ed.), Kluwer.
Matteucci, F. and Brocato, E. (1990), Astrophys. J. 365, 539.
Matteucci, F. and François, P. (1989), Mon. Not. R. astr. Soc. 239, 885.
Matteucci, F., Franco, J., François, P. and Treyer, M.A. (1989), Rev. Mex. Astr. Astrofis. 18, 145.
Mayor, M. and Vigroux, L. (1981), Astr. Astrophys. 98, 1.
McCall, M.L. (1982), Thesis, University of Texas at Austin.
McGaugh, S.S. (1991), Astrophys. J., in press.
Melnick, J. (1987), in "Starbursts and Galaxy Evolution", T.X. Thuan, T. Montmerle and J.T.T. Van (eds.), Paris: Ed. Frontières, p. 215.
Morrison, H.L., Flynn, C. and Freeman, K.C. (1990), Astr. J. 100, 1191.
Mould, J.R. (1976), Mon. Not. R. astr. Soc. 117, 47 P.
Neese, C.L. and Yoss, K.M. (1988), Astr. J. 95, 463.
Nissen, P.E. (1991), in "Elements and the Cosmos" (31st Herstmonceux Conf.), M.G. Edmunds, B.E.J. Pagel and R.J. Terlevich (eds.) Cambridge University Press.
Nissen, P.E. and Schuster, W.J. (1991), Astr. Astrophys., in press.
Norris, J.E. and Ryan, S.G. (1989), Astrophys. J. 340, 739.
Ostriker, J.B. and Thuan, T.X. (1975), Astrophys. J. 202, 353.
Pagel, B.E.J. (1981), in "The Structure and Evolution of Normal Galaxies", S.M. Fall and D. Lynden-Bell (eds.), Cambridge University Press, p. 211.
Pagel, B.E.J. (1985), in "Production and Distribution of the CNO Elements", I.J. Danziger, F. Matteucci and K. Kjär (eds.), Garching: ESO, p. 155.
Pagel, B.E.J. (1989a), Rev. Mex. Astr. Astrofis. 18, 161.
Pagel, B.E.J. (1989b), in "Evolutionary Phenomena in Galaxies", J.E. Beckman and B.E.J. Pagel (eds.), Cambridge University Press, p. 368.
Pagel, B.E.J. (1991), Phys. Scripta T36, 7.
Pagel, B.E.J. and Patchett, B.E. (1975), Mon. Not. R. astr. Soc. 172, 13.
Pagel, B.E.J., Simonson, E.A., Terlevich, R.J. and Edmunds, M.G. (1992), Mon. Not. R. astr. Soc., in press.
Pagel, B.E.J., Terlevich, R.J. and Melnick, J. (1986), Pub. Astr. Soc. Pacific 98, 1005.
Parravano, A. (1989), Astrophys. J. 347, 812.
Peimbert, M. and Serrano, A. (1982) Mon. Not. R. astr. Soc. 198, 563.
Peimbert, M. and Torres-Peimbert, S. (1974), Astrophys. J. 193, 327.
Peimbert, M. and Torres-Peimbert, S. (1976), Astrophys. J. 203, 581.
Phillipps, S. and Edmunds, M.G. (1991), Mon. Not. R. astr. Soc. 251, 84.
Phillipps, S., Edmunds, M.G. and Davies, J.I. (1990), Mon. Not. R. astr. Soc. 244, 168.
Pitts, E. and Tayler, R.J. (1989), Mon. Not. R. astr. Soc. 240, 373.
Rich, R.M. (1988), Astr. J. 95, 828.
Rich, R.M. (1990), Astrophys. J. 362, 604.
Richtler, T., de Boer, K.S. and Sagar, R. (1991), ESO Messenger no. 64, p. 50.
Ryan, S., Norris, J. and Bessell, M.S. (1991), Astr. J. 102, 303.
Scalo, J.M. (1986), Fund. Cosm. Phys. 11, 1.
Scalo, J. (1990), in "Windows on Galaxies", A. Renzini, G. Fabbiano and J.S. Gallagher (eds.), Kluwer.
Schmidt, M. (1963), Astrophys. J. 137, 758.
Searle, L. and Sargent, W.L.W. (1972), Astrophys. J. 173, 25.

Searle, L. and Zinn, R. (1978), Astrophys. J. 225, 357.
Shaver, P.A., McGee, R.X., Danks, A.C. and Pottasch, S.R. (1983), Mon. Not. R. astr. Soc. 204, 53.
Skillman, E.D., Kennicutt, R.C. and Hodge, P.W. (1989), Astrophys. J. 347, 875.
Sommer-Larsen, J. (1991), Mon. Not. R. astr. Soc. 250, 356.
Sommer-Larsen, J. and Yoshii, Y. (1989), Mon. Not. R. astr. Soc. 238, 133.
Sommer-Larsen, J. and Yoshii, Y. (1990), Mon. Not. R. astr. Soc. 243, 468.
Terlevich, R. (1985), in "Star Forming Dwarf Galaxies", D. Kunth, T.X. Thuan and J.T.T. Van (eds.), Paris: Ed. Frontières, p. 395.
Tinsley, B.M. (1979), Astrophys. J. 229, 1046.
Tinsley, B.M. and Larson, R.B. (1978), Astrophys. J. 221, 554.
Tinsley, B.M. and Larson, R.B. (1979), Mon. Not. R. astr. Soc. 186, 503.
Vader, J.P., Vigroux, L., Lachièze-Rey, M. and Souviron, J. (1988), Astr. Astrophys. 203, 217.
van den Bergh, S. (1962), Astr. J. 67, 486.
Vigroux, L. Chièze, J.P. and Lazareff, B. (1981), Astr. Astrophys. 98, 119.
Vila-Costas, M.B. and Edmunds, M.G. (1992), in preparation.
Vilchez, J.M., Pagel, B.E.J., Díaz, A.I., Terlevich, E. and Edmunds, M.G. (1988), Mon. Not. R. astr. Soc. 235, 633.
Wallerstein, G. (1962), Astrophys. J. Suppl. 6, 407.
Wheeler, J.C., Sneden, C. and Truran, J.W. (1989), Ann. Rev. Astr. Astrophys. 27, 279.
Zinn, R. (1985), Astrophys. J. 293, 424.

DISCUSSION

A. Maeder: In order to derive the dY/dZ ratio from H II regions, you have with reason removed those regions showing WR signatures, since these H II regions are polluted by the winds of WR stars which enrich them in helium. I want to emphasize that the other H II regions may also be severely polluted for two reasons: (1) New helium is not only ejected in WR stars, but in all phases before; (2) the ages of WR stars are very small: after about 4–5 millions of years, they are turned off. Thus, you do not see them, but their pollution is there, contributing to increase the dY/dZ ratio. Thus, I would suggest that future analyses carefully consider this important effect before making comparisons with the models.

B. Pagel: A necessary (though not suffient) condition for local pollution by winds to be suspected is the appearance of "secondary" nitrogen, i.e. an excess of N/O above the value of about 0.035 that is typical of low-abundance extragalactic H II regions, the excess increasing with oxygen abundance. No such excess is noticeable in any but the three most oxygen-rich of the 19 objects that we have used, and the deletion of these does not affect the result significantly.

G. Meurer: How long do starbursts last – especially those in dwarf galaxies? I ask because the standard timescales of $10^6 - 10^7$ years are derived by assuming closed box models, and we know that dwarf galaxies not only can undergo stellar winds but are observed to do so (NGC 1705, Meurer et al. 1991, A.J. submitted). If you look at these starburst dwarf galaxies, we see strong emission lines but relatively red colours, even after correcting for underlying populations. If continuous star formation models are adopted, these bursts may have ages of the order of Gyrs.

B. Pagel: A possible way might be to study the strength of WR features relative to model predictions (Arnault, Kunth & Schild, AA 224, 73, 1989). The problem is that I do not really trust the model predictions.

H. Dottori: About the duration of the starbursts. I want to point out that the simultaneous presence of WR and red supergiant features in the spectrum of the jumbo H II region of NGC 3310 implies, within the framework of Maeder's models, that at least in this case the burst duration must be of the order of 10^7 years.

THE [O/Fe] RATIO IN HALO, DISK AND BULGE STARS

B. Barbuy
IAG-USP, Depto. Astronomia
C.P. 9638, São Paulo 01065, Brazil

1. Introduction

The oxygen-to-iron ratio in the different stellar populations is a tracer of the chemical enrichment by supernovae of types II and I: SNII/SNI along the Galaxy lifetime, given that the bulk of oxygen is produced and ejected by massive stars ($M > 10\ M_\odot$), whereas the bulk of iron is produced by SNI of intermediate masses. Otherwise, the O overabundances are generally accompanied by α-elements (^{24}Mg, ^{28}Si, ^{40}Ca, ^{48}Ti) overabundances (Wheeler et al, 1989).

The classical scenario of the time lag between the enrichment by massive stars, which can be considered to start immediately after the formation of the first stars, and that by the intermediate mass type I SNae, which starts to occur at about 10^8 years later, leads to the following characteristics for the different stellar populations: (a) [O/Fe] > 0 in the old stars; (b) [O/Fe] grows to the solar value from the old disk to the young stars.

2. Observational evidences

Only a few oxygen lines are available in stellar spectra, as reported by Lambert (1978). The OI triplets at λ 615.6, .7, .8 nm, λ 777.1, .4, .5 nm, λ 844.57, .64, .67 nm, λ 926.0, .2, .5 nm are commonly used to derive the oxygen abundances in dwarf stars, whereas the forbidden [OI] λ 557.7 nm, λ 630.03 nm and λ 636.38 nm are used in giants. The use of different lines to derive O abundances in giants and dwarfs, is due to the fact that the forbidden lines disappear in dwarfs, and more so in hotter ones, whereas the OI triplets disappear in giants.

2.1 Halo

Oxygen overabundances in the halo, first obtained by Conti et al. (1967), were confirmed by all subsequent work: Lambert et al. (1974), Sneden et al. (1979, SLW79), Clegg et al. (1981), Leep & Wallerstein (1981), Barbuy (1983), Luck & Bond (1985), Gratton & Ortolani (1986, GO86), Barbuy (1988, B88), Barbuy & Erdelyi-Mendes (1989, BE89), Abia & Rebolo (1989, AR89), Gratton (1990, G90), Sneden et al. (1990, SKP90), Spiesman & Wallerstein (1991, SW91), Bessell et al. (1991, BSR91), besides the preliminary work by Edvardsson et al. (1991), Tomkin et al. (1991, TLLS91), Spite & Spite (1991).

Disagreements occur however regarding the [O/Fe] absolute value in the halo.

B. Barbuy and A. Renzini (eds.), The Stellar Populations of Galaxies, 143–146.

A particularly striking discrepancy is seen between data by GO86, B88, SKP90, SW91, who used the [OI] lines in giants, and those by AR89 who used OI lines in dwarfs - see Figure 1. Two main evidences, shown in fig. 1, are to be pointed out: (1°) a disagreement between data by AR89 and results also derived using the OI triplet lines in dwarfs by SLW79, G90 and TLLS91; (2°) the results by G90 and SW91 based on [OI] lines in dwarfs, and those by BSR91 using OH bands in dwarfs, where [O/Fe] $\approx$ +0.4 and +0.5 respectively, eliminate the suggestion by AR89 of a conversion of a large fraction of O into N, occurring from the dwarf to the red giant stage. This possibility is largely unlikely, in any case, since overabundances of N in halo giants are only moderate (Kraft et al., 1982).

These evidences leave only two possibilities:

The OI 777 nm triplet give systematically higher oxygen abundances than the other lines: this possibility has been quantified by Kiselman (1991), through non-LTE calculations of the OI triplet for a grid of stellar parameters. The results (his fig. 3) is not able to reconcile results from the permitted oxygen lines with those from the forbidden lines. Kiselman (1991) does not succeed to reproduce even the solar lines adequately, and therefore recommends not to use such lines.

(2) Stellar parameters used are not appropriate, and the agreement of oxygen abundances derived from the forbidden and permitted lines is sensitive to the stellar parameters. One example of the influence of stellar parameters is illustrated by the case of the halo star HD 19445: adopting stellar parameters by Spite & Spite (1978, SS78) and by AR89, (θ_{eff}, log g, [M/H]) = (0.87, 4.0, -1.9) and (0.86, 4.5, -2.3), respectively, the result is that in order to bring the synthetic oxygen line computed with parameters by AR89 to the intensity of that for the SS78 parameters, an O overabundance of [O/Fe] = +0.55 is necessary.

Halo field versus globular clusters stars: The oxygen abundances in the halo field stars using the [OI] line result is an approximately constant value, cf. B88. In globular clusters giants, however, a variation seems to be present (Pilachowski et al., 1983; Brown et al., 1990). This may be a very important information regarding a possible self-enrichment of globular clusters, but further studies are still necessary for such inferences.

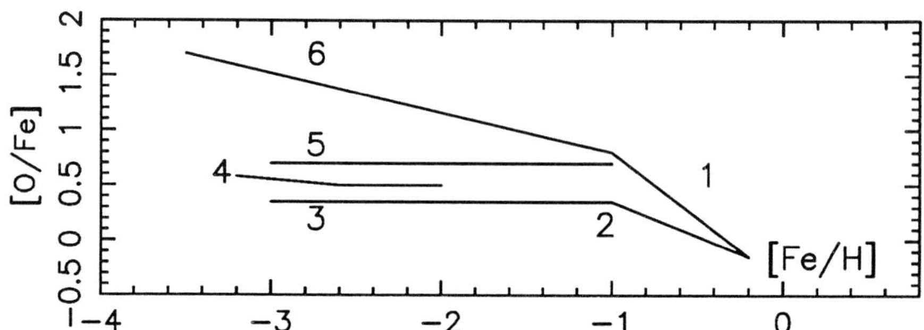

Figure 1 - Schematic behaviour of [O/Fe] vs. [Fe/H] by 1 Clegg et al, Edvardsson et al; 2 GO86,BE89; 3 GO86, B88, SKP90, SW91; 4 BSR91; 5 SLW79,G90,TLLS91; 6 AR89

2.2 Disk

The [O/Fe] data for the disk also show a disagreement between results from the forbidden and permitted lines, as can be seen in Fig. 1. It is clear that during the disk evolution the [O/Fe] drops from the halo value, gradually reaching the solar value at the solar metallicity. Interesting questions to ask are: (i) which is the metallicity corresponding to the transition halo-disk ? An interesting plot using [Ca/Fe] was presented by Nissen (1990), where a drop is seen at [Fe/H] $\approx$ -0.9, however the transition might occur at metallicities as low as [Fe/H] $\approx$ -1.6. (ii) Is there a spread of CNO/Fe abundances at this metallicity transition ? A mixture of populations (halo, thick disk, thin disk) might give this effect. No strong spread is seen, although some spread is detected at metallicities typical of the thick disk ([Fe/H] $\approx$ -0.9 to -0.6).

2.3 Bulge

The interest in the study of bulge stars resides, not only in the understanding of the chemical enrichment steps of our own Galaxy, but also in their probable similarity to the old galaxy populations of ellipticals and bulges of spirals.

The stellar content can be investigated in much greater detail than is possible by methods that seek to match the spectrum of the integrated light of unresolved systems. A star-by-star determination of their properties could yield information on the evolutionary history of such systems.

A major problem in the abundance determination of the bulge metal-rich stars is the uncertainty in their temperature. Besides the fact that relations temperature vs. colours for metal-rich giants are not precise, there is a strong reddening. It is therefore preferable to study the bulge clusters, given that the reddening can be roughly derived by comparing their colour-magnitude diagrams to that of 47 Tuc for example.

2.3.1 Individual stars in NGC 6553: NGC 6553 is the closest bulge cluster at a distance d $\approx$ 4.1 kpc, its brightest stars, of V $\approx$ 15, being observable at the 3.6m telescope using the Caspec spectrograph, at ESO.

Barbuy et al. (1991) have studied the star III-17, for which a metallicity of [M/H] $\approx$ -0.2 was found. Preliminary CNO abundances found are [C/Fe] = +0.1, [N/Fe] = +0.4, [O/Fe] = 0.0; the O abundance derived from the [OI]λ557.7 nm line, is imprecise, due to a defect in the spectrum in that region.

2.3.2 Bulge-like nearby metal-rich stars: A sample of nearby metal-rich stars was selected from the proper motion NLTT catalogue, combined with a study of their photometric metallicities and space velocities. The selected stars are candidates to be the local component of the bulge population, since about 4% of nearby stars correspond to an old disk population characterized by eccentricities $e < 0.5$, pericentric distances $R_p \approx 3.5$, apocentric distances $R_a < 11$ kpc, and high metallicities.

The CNO results for a dozen of these stars (Barbuy & Grenon, 1991) give: [C/Fe] $\approx$ 0.0, [N/Fe] $\approx$ 0.0 and [O/Fe] $\approx$ 0.0 to 0.2. The O abundance was derived for the forbidden plus permitted lines, and there was agreement.

As conclusion, these first [O/Fe] $\approx$ 0.0 for bulge stars lead to idea that the bulge seems to be somewhat different from the halo.

3. The "cosmic" oxygen abundance

O emission lines in nearby HII regions - such as the Orion nebula, are well-known to provide a O abundance lower than that of the Sun by about a factor 2. Luck &

Lambert (1985, LL85) also obtained an oxygen deficiency of 0.2 to 0.3 dex for intermediate mass supergiants, having evolved from main-sequence B stars, therefore in agreement with the HII regions value.

These deficiencies, together with the evidences for a He, C, N and O depletion in the solar wind, might indicate, as proposed by LL85, that the solar photospheric CNO abundances are enriched relative to the original solar nebula.

Abia, C., Rebolo, R.: 1989, ApJ 347, 186
Barbuy, B.: 1983, A&A 123, 1
Barbuy, B.: 1988, A&A 191, 121
Barbuy, B., Erdelyi-Mendes, M.: 1989, A&A 214, 239
Barbuy, B., Grenon, M.: 1991, A&A, to submit
Barbuy, B., Castro, S., Ortolani, S., Bica, E.: 1991, A&A, submitted
Bessell, M.S., Sutherland, R., Ruan, K.: 1991, preprint
Brown, J.A., Wallerstein, G., Oke, J.B.: 1990, AJ 100, 1561
Clegg, R.E.S., Lambert, D.L., Tomkin, J.: 1981, ApJ 250, 262
Edvardsson, B., Gustafsson, B., Lambert, D.L., Nissen, P., Tomkin, J., Andersen, J.: 1991, preprint
Gratton, R.: 1990, in IAU Symposium 145
Gratton, R., Ortolani, S.: 1986, A&A 169, 201
Kraft, R.P., Suntzeff, N.B., Langer, G.E., Carbon, D.F., Trefzger, C.F., Friel, E., Stone, R.P.S.: 1982, PASP 94, 55
Kiselman, D.: 1991, A&A Letters 245, L9
Lambert, D.L.: 1978, MNRAS 182, 279
Lambert, D.L., Sneden, C., Ries, L.M.: 1974, ApJ 188, 97
Leep, E.M., Wallerstein, G.: 1981, MNRAS 196, 543
Luck, R.E., Bond, H.: 1985, ApJ 292, 559
Luck, R.E., Lambert, D.L.: 1985, ApJ 298, 782
Nissen, P.E.: 1990, in *Elements and the Cosmos*, ed. R. Terlevich, Cambridge Univ. Press, in press
Ortolani, S., Barbuy, B., Bica, E.: 1990, A&A 236, 362
Pilachowski, C.A., Sneden, C., Wallerstein, G.: 1983, ApJS 52, 241
Sneden, C., Lambert, D.L., Whitaker, R.W.: 1979, ApJ 234, 964
Sneden, C., Kraft, R.P., Prosser, C.: 1990, in *VI Cambridge Workshop on Cool Stars, Stellar Systems and the Sun*, ed. G. Wallerstein, 369
Spiesman, W.J., Wallerstein, G.: 1991, preprint
Spite, F., Spite, M.: 1978, A&A 67, 23
Spite, F., Spite, M.: 1991, this symposium
Tomkin, J., Lemke, M., Lambert, D., Sneden, C.: 1991, in preparation
Wheeler, J.C., Sneden, C., Truran, J.W.: 1989, ARA&A 27, 279

B. Pagel: In my talk I suggested [O/Fe] starts to go down as [Fe/H] increases through -2; Serrano privately persuaded me that it should be -1.6. You say -1 to -0.9. It is important from the modelling point of view to fix this point if we can.

B. Barbuy: I have only shown the data on [Ca/Fe] by Nissen (1990) where a clear drop is seen at [Fe/H] $\approx$ -0.9.

J. Laird: I want to emphasize the possible danger of using only metallicity to separate stars. We may be, as Dr. Pagel said yesterday, mixing cats and dogs, that is, mixing stars of different populations. There may not exist a single unique relationship between [O/Fe] and [Fe/H]. It is important to consider the velocities of the stars being studied.

THE STELLAR POPULATIONS OF THE MAGELLANIC CLOUDS

Mario Mateo
The Observatories of the Carnegie Institution of Washington
813 Santa Barbara Street
Pasadena, CA 91101 USA

I. Introduction

In many ways, studies of the stellar populations in the Magellanic Clouds are more straightforward than in our own Galaxy because of our external vantage point of the Clouds. This allows us to study the global properties of their stellar populations, as well as the individual stars that comprise them. The resulting picture describing the stellar content of the Clouds is obviously complex, defying any attempt to present a complete description in a short review such as this. Thus, my goal is to provide a very broad overview of the global properties of the Magellanic Clouds, along with some slightly more detailed discussions on a few selected topics addressed by recent studies. Recent reviews of the global properties and stellar content of the Magellanic Clouds based on observations ranging from radio to x-ray wavelengths can be found in the Proceedings of IAU Symposium 148 (Haynes and Milne 1991) and in Westerlund (1990). More specialized reviews have been written recently by van den Bergh (1991), Freeman (1989), and Feast (1989), and by numerous authors in the proceedings of a recent European Colloquium on the Magellanic Clouds (de Boer, *et al.* 1989).

II. Global Properties

Some of the global properties of the Magellanic Clouds are summarized in Table 1, along with corresponding values for the Galaxy. It has long been appreciated – and is evident in Table 1 – that many of the global properties of the three galaxies vary smoothly along the sequence SMC ⇒ LMC ⇒ Galaxy (as represented by the solar neighborhood). For example, the mean present-day abundances of these galaxies increase along this sequence (Russell and Bessell 1989; Russell and Dopita 1989), while the gas-to-dust ratios decrease. Such trends make the Magellanic Clouds invaluable tools to study how the properties of stellar populations vary with global galaxian properties. As an illustration, Maeder (this volume) has shown that the changes in the relative numbers of C and N Wolf-Rayet stars and of the numbers of O stars in the three galaxies can be understood by the increased efficiency of stellar winds with increasing metallicity; similarly, the form of the chemical abundance enrichment histories of the three galaxies differ systematically (see §IV), perhaps related to their total masses or gas fractions.

A classic review of the global optical properties of the Magellanic Clouds can be found in de Vaucouleurs and Freeman (1972), while a more recent discussion of the large-scale photometric parameters of both galaxies was published by Bothun and Thompson (1988). Irwin (1991) discusses the structure of the Magellanic clouds based on deep star counts using Schmidt plates. While the LMC shows a generally regular outer structure, the SMC is clearly disturbed on the side facing the LMC, suggestive of a recent encounter between the two galaxies. The growing body of evidence of a significant line-of-sight depth in various parts of the SMC (nicely summarized by Hatzidimitriou and Hawkins 1989) supports this

B. Barbuy and A. Renzini (eds.), The Stellar Populations of Galaxies, 147–156.

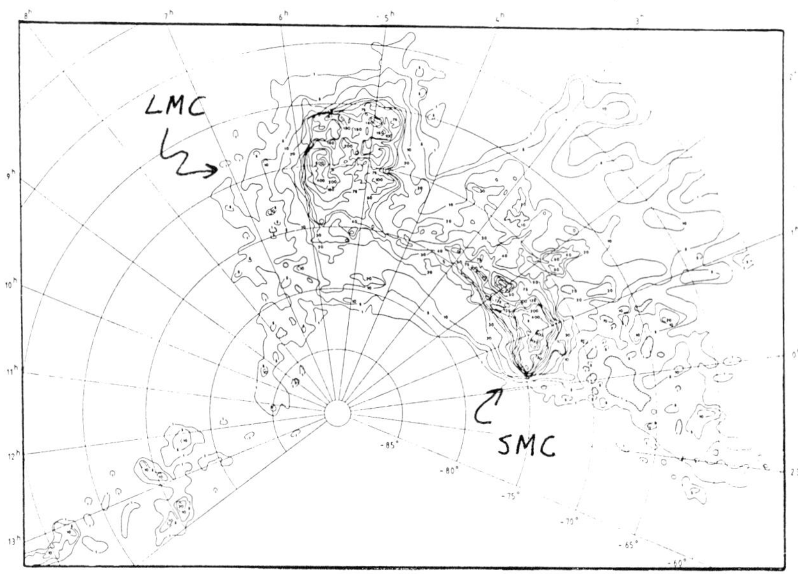

Figure 1 – An HI map of the Magellanic Clouds from Mathewson and Ford (1984). Note the complexity of the distribution of the gas over the entire Magellanic system.

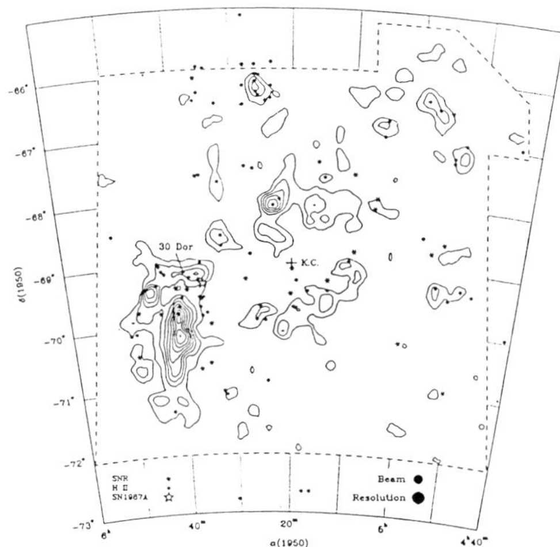

Figure 2 – The CO map of Cohen, *et al.* (1989). The gigantic CO cloud South of 30 Dor is a prominent feature.

Table 1

Global Properties of the Galaxy and Magellanic Clouds

	SMC	LMC	Galaxy
Distance (kpc)	58 ± 10	50 ± 5	...
$V_{0,tot}$	2.2	−0.1	...
$(B-V)_{0,tot}$	0.50	0.55	...
Luminosity ($L_\odot$)	4×10^8	2×10^9	2×10^{10}
Angular Size	14°	23° × 17°	...
Mass ($M_\odot$)	1×10^9	2×10^{10}	5×10^{11}
	($R < 3$ kpc)	($R < 6$ kpc)	($R < 50$ kpc)
Global M/L (Solar Units)	3	10	25
M_{HI} ($M_\odot$)	4×10^8	5×10^8	5×10^9
M_{HI}/M_{tot}	0.35	0.03	0.01
M_{molec} ($M_\odot$)	'low'	1.4×10^8	4×10^9
M_{molec}/M_{HI}	'low'	0.3	0.8
[Fe/H]$_0$	−0.7	−0.3	0.0

interpretation, as does the recent discovery of *young* stellar associations located between the Clouds (Irwin, *et al.* 1990; Grondin, *et al.* 1990). Radio astronomers have long appreciated that the Magellanic Clouds are the most prominent components of the single, much larger system shown (in part) in Figure 1 (from Mathewson and Ford 1984). How this complex structure has come about remains controversial. A review of the various possibilities is given by Wayte (1991) who favors the interpretation that the Magellanic Stream has been stripped via ram pressure from the Magellanic Clouds during their passage through the Galactic halo. In contrast, Murai and Fujimoto (1980) interpret the Stream as a tidal tail expelled from the Clouds during a recent near-collision of the two galaxies. Either way, it is clear that interactions – both with each other and with the Galaxy – have played important roles in the global evolution of the Clouds.

Recent studies have extended our view of the Magellanic Clouds to wavelengths other than the optical and radio. Notable among these is the CO survey by Cohen, *et al.* (1989) who mapped the inner 6° × 6° of the LMC during a multi-year survey from CTIO. The resulting map (Figure 2) reveals large CO clouds near a number of large star-forming regions, especially to the south of 30 Dor and near Shapley Constellation III north of the LMC Bar. A comparison of CO and H I maps in the vicinity of 30 Dor (Figures 1 and 2) with each other and with maps of the far-UV (Page and Carruthers 1981; Smith, *et al.* 1987) and Hα emission (Davies, *et al.* 1976) provides one of the clearest examples of how star formation propagates through massive gas complexes. It is not hard to guess that the next gigantic star formation region in the LMC will probably be located just south of the 30 Dor region where the largest CO cloud complex is observed.

III. The Populous Star Clusters of the Magellanic Clouds

The Magellanic Clouds possess rich populations of star clusters; just how rich can be seen from inspecting the map of the spatial distribution of the clusters in both Clouds (Irwin 1991). Many of these clusters are luminous objects that are morphologically similar to Galactic globular clusters. Some contain RR Lyr variables, possess blue horizontal branches, (*e.g.*, NGC 2257 and NGC 1841; Walker 1989, 1990) and have red integrated colors consistent with their identification as old stellar systems (van den Bergh 1981); in other words, these *are* globular clusters. However, most of the the bright Magellanic clusters are clearly very different. Their integrated spectra and colors imply that they contain stellar populations considerably younger than those found in true (*i.e.*, ancient) globular clusters (van den Bergh 1981). Detailed studies of the stellar content of these clusters have confirmed their relative youth compared to globular clusters (useful bibliographies of recent age determinations for Magellanic Cloud clusters can be found in Seggewiss and Richtler (1989) and Sagar and Pandey (1989)). Although the Clouds contain many of these luminous ($M_V \lesssim -8$) young clusters, there are few, if any, in the Galaxy. Are they simply globular clusters that have formed recently?

Table 2 lists the present-day integrated photometric properties and ages (based on analyses of color-magnitude diagrams) of five populous Magellanic Cloud clusters. Also listed are the predicted absolute visual magnitudes for these clusters for an age of 15 Gyr based *only* on the stellar evolutionary fading predicted by simple models for a Salpeter IMF slope (Elson, *et al.* 1987). Most of these clusters will have luminosities comparable to that of globular clusters in our Galaxy and M 31 ($\overline{M}_V = -7.1$, with an intrinsic dispersion of about 1 magnitude (van den Bergh 1985)). Alternatively, we can avoid using evolutionary models and compare clusters masses directly. As an example, consider NGC 1866. Lupton, *et al.* (1989) obtained velocities for 29 members of this cluster and derived an upper limit of $4 \times 10^5 M_\odot$ for its mass. Fischer, *et al.* (1991) analyzed more precise velocity measurements for 69 cluster members. Combined with a newly determined surface density profile spanning 3.5 decades of surface brightness, they derived a mass of $1.7 \pm 0.3 \times 10^5 M_\odot$ for NGC 1866. The mass of a typical Galactic globular cluster is about $1.2 \times 10^5 M_\odot$, assuming M/L = 2.

Although these comparisons suggest that *some* Magellanic Cloud clusters may be 'proto-globulars', it is important to recall that the overall luminosity function (LF) of clusters in the Clouds more closely resembles a power-law (similar to the LF of Galactic open clusters) than the Gaussian LF characteristic of globular clusters (Elson and Fall 1985). This suggests that *most* of the Magellanic Cloud clusters probably comprise a population similar to the Galactic open clusters. Nevertheless, assuming that some of the brightest blue clusters in the Clouds are in fact young globulars, their presence suggests that the Clouds today somehow resemble conditions in the early Galactic halo (Renzini 1991; but see Fujimoto and Noguchi 1990). Isolating these properties (low specific angular momentum? weak tidal fields?) may ultimately allow us to study how star formation proceeded during the early history of our Galaxy.

As emphasized by many authors (*e.g.*, Mould 1991, Renzini 1991), the star clusters of the Magellanic Clouds are invaluable laboratories for studies of the global properties of 'simple' stellar populations. Two examples can be cited from recent observational work. First, studies of the internal kinematics of a number of Magellanic Cloud clusters are now published or in progress, providing an opportunity to measure how mass-to-light ratios vary with age. Some results are shown in Figure 3 illustrating that the simple models mentioned

Table 2

Luminosities of Populous Cloud Clusters at Age = 15 Gyr

Cluster	Age (Myr)	M_V	M_V(15 Gyr)
NGC 2070 (LMC)	3	−10.9	−7.2
NGC 330 (SMC)	10	−9.9	−6.2
NGC 2100 (LMC)	10	−9.6	−5.9
NGC 1866 (LMC)	100	−9.5	−7.0
NGC 1978 (LMC)	2000	−8.5	−7.2

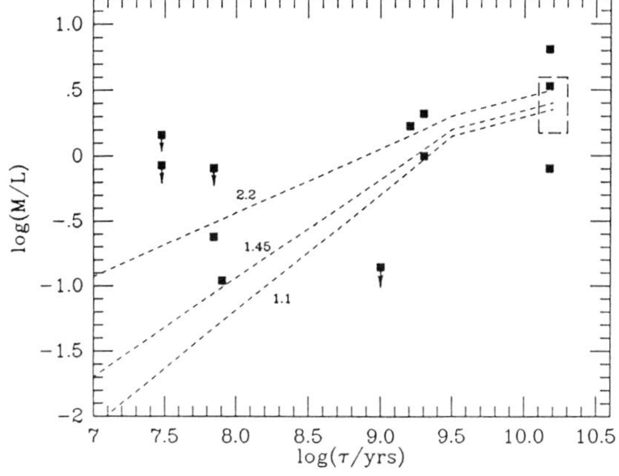

Figure 3 – The run of M/L vs. age for a number of Cloud clusters. For some, only upper limits are available. The curves refer to different fading rates calculated by Elson, *et al.* (1987) for the indicated IMF slopes.

above (Elson, *et al.* 1987) do in fact appear to describe the evolutionary fading of Cloud clusters reasonably well. Second, improved integrated photometry and spectroscopy of numerous Cloud clusters is now available in the UV (*e.g.*, Cassatella, *et al.* 1987) and IR (*e.g.*, Bica, *et al.* 1990). These data will provide important constraints on stellar synthesis models (*e.g.*, Chiosi, Bruzual, this volume; Barbaro and Olivi 1991) and should be of great interest to compare with direct CCD observations of the stellar content of populous Cloud clusters.

IV. The Age-Metallicity Relations and Star Formation Histories of the Magellanic Clouds

In a classic study, Butcher (1977) concluded that the bulk of the stellar population in the LMC is younger than about 3-4 Gyr based on the presence of a break in the main sequence luminosity function corresponding to the turnoff luminosity of an intermediate-age cluster. Subsequent studies using the same sort of analysis have confirmed this result for additional LMC fields (Stryker 1984; Hardy, *et al.* 1984). Most recently, Bertelli, *et al.* (1991) used synthetic color-magnitude (CM) diagrams based on the Padova stellar evolutionary models (Bertelli, *et al.* 1990) to estimate the star formation history in three LMC fields in greater detail than possible from a simple analysis of the luminosity functions. From the models, indices based on the ratios of the numbers of stars in strategically chosen regions of the CM diagram were identified as useful indicators of the age of a postulated 'burst' in star formation, the slope of the IMF, and the duration of enhanced star formation. For example, the ratio of the number of main sequence stars with $1.5 \leq M_V \leq 3$ divided by the number of subgiants in the same absolute magnitude interval is strongly affected by changes in the star formation rates at different epochs. Using three indices of this sort, Bertelli, *et al.* (1991) were able to isolate a consistent set of parameters describing the star formation history of the LMC fields. All three fields confirmed and extended the results of the earlier studies to a remarkable extent: the star formation rate throughout the LMC increased dramatically (by at least a factor of five) about 4 Gyr ago and has remained high ever since. Furthermore, the age of the 'burst' derived from the field stars is remarkably similar to that implied from the lack of LMC star clusters with ages (derived from CM diagrams) between about 4 and 10 Gyr (*e.g.*, Mateo 1988; Jensen, *et al.* 1988; Olszewski 1988). The latter is very likely not a selection effect since it is clearly possible to identify and study 7-10 Gyr old clusters in the SMC without difficulty (*e.g.*, Mateo, *et al.* 1986). Interestingly, an 'eyeball' application of the Bertelli, *et al.* (1991) analysis applied to the SMC field observed by Hardy and Durrand

(1984) implies that the bulk of the field stars in the SMC is substantially older than the LMC field stars, in agreement with the apparent age distribution of the SMC clusters (van den Bergh 1991).

Likewise, the chemical abundance patterns of the Clouds appear to be distinct from one another and from that of the Galaxy. These differences were apparent from the first spectroscopic studies of individual stars in Magellanic Cloud clusters (Cohen 1982; Cowley and Hartwick 1982), and have been subsequently confirmed by analyses of the photometric data of stars in clusters (*e.g.*, Da Costa 1991) and more extensive spectroscopy of cluster members (Olszewski, *et al.* 1991). The LMC appears to have been enriched in metals by over a factor of 30 during the epoch when the star formation rate was exceptionally low, while the SMC appears to have maintained a nearly constant abundance during the era when stars and clusters were being formed efficiently. The details hidden by this schematic description have only recently begun to be explored. For example, fine analyses of high dispersion spectra of Magellanic Cloud supergiants have provided information on the present-day abundances of individual elements (*e.g.*, Spite, *et al.* 1986; Russell and Bessell 1990; Spite and Spite 1990); previously, this sort of information could only be determined for emission nebulae in the Clouds (Dufour 1984). One particularly curious problem has emerged from these studies. Spite, *et al.* (1986) noted that the mean heavy-metal abundance of the SMC cluster NGC 330 is considerably lower than the mean abundance derived for luminous, young SMC field supergiants; Richtler, *et al.* (1989) and Reitermann, *et al.* (1990) reached a similar conclusion for the young LMC cluster NGC 1818. Given the complexities of these analyses of luminous supergiants (Russell and Bessell 1990) and their importance for our understanding of the chemical abundance histories of the Clouds, additional studies are clearly needed.

Two general conclusions seem to emerge from the discussion in the previous paragraphs. First, the star formation and chemical enrichment histories of the Magellanic Clouds are undoubtedly very different; this implies that tidal interactions have not been the dominant trigger of star formation activity in the Clouds *if* they have been bound during most of their lifetimes (Murai and Fujimoto 1980). Second, populous star clusters appear to be good tracers of the star formation histories of the Clouds, lending confidence to their use in more distant galaxies as probes of the overall stellar age distribution (*e.g.*, Schommer, *et al.* 1991). The same cannot be claimed regarding the use of clusters as reliable tracers of the chemical enrichment history of the Clouds (Richtler, *et al.* 1989). Precise abundance *and* age estimates of Magellanic Cloud field stars (*e.g.*, along the lines suggested by Mould 1991) are needed to compare with the cluster age-metallicity relations in both galaxies.

V. The Kinematics of the Stellar Populations of the Magellanic Clouds

Considerable information is available on the internal kinematics of the Magellanic Clouds (see Westerlund 1990). In this section, I will focus on recent work dealing with the LMC primarily because the kinematics of the SMC are so complex. For example, Martin, *et al.* (1989) identified no fewer than four distinct velocity groups in the SMC from an analysis of the radial velocities of objects associated with the young stellar population of that galaxy. In contrast, the kinematics of the older population of the SMC are consistent with a single spherical distribution (*e.g.*, Hardy, *et al.* 1989).

Although complex in their own right, the kinematics of the LMC appear considerably simpler by comparison. For example, Freeman, *et al.* (1983; FIO) analyzed radial velocity data for a large sample of populous LMC star clusters; they noted that (a) the kinematics of the youngest clusters were similar to that of the HI gas and consistent with both rotating in a flattened disk, (b) the intermediate-age and old clusters also rotate in a disk of comparable thickness as that defined by the youngest clusters, and (c) the old-cluster disk appeared to be tilted and offset (in systemic velocity) relative to the young-cluster disk. Four recent

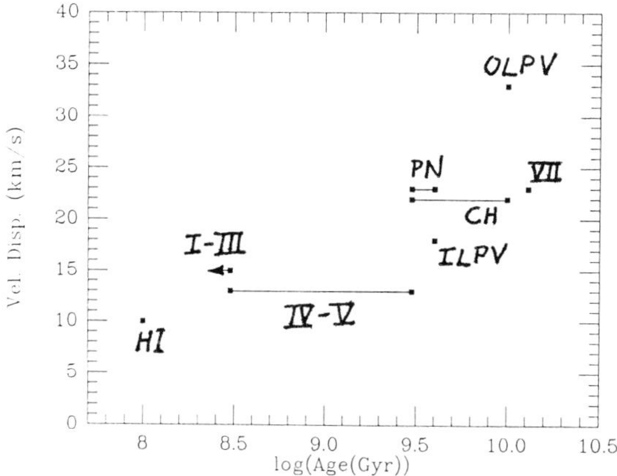

Figure 4 – The variation in the LOS velocity dispersion as a function of Age in the LMC. PN refers to planetary nebulae; OLPV and ILPV to old and intermediate-age LPVs, respectively; CH to CH stars; VII to the oldest LMC clusters; IV-V to intermediate-age clusters; I-III to young clusters; HI to neutral hydrogen.

studies of different stellar population tracers have extended and refined the FIO results:

1) Hartwick and Cowley (1988) measured velocities of 74 CH stars in the LMC. One subgroup of their sample has kinematics consistent with that of the HI, while the remaining stars (mostly located near the LMC center) exhibit much less rotation, and a higher line-of-sight (LOS) velocity dispersion. Interestingly, the two CH star subgroups have different systemic velocities, reminiscent of one of the more curious FIO results.

2) Hughes, *et al.* (1991) obtained radial velocities of 144 long-period variables (LPVs). Sixty-three are short-period LPVs (100-250 days) for which pulsational and evolutionary theory give ages of $\gtrsim$ 10 Gyr; the remaining 81 intermediate-period LPVs (225-450 days) have a mean age of 4 Gyr. The older LPVs showed little rotation and an LOS dispersion consistent with a spheroid axis ratio of $c/a \sim 0.3$. In contrast, the intermediate-age subgroup showed more rotation and a smaller LOS dispersion characteristic of a more flattened distribution.

3) Meatheringham, *et al.* (1988) studied a sample of 95 planetary nebulae in the LMC. By comparing their velocities with that of the HI (reanalyzed by them using the results of Rohlfs, *et al.* 1984), they found that the planetaries rotate with the HI gas but have a much higher LOS dispersion. Based on a simple diffusion argument, they concluded that the mean age of the nebulae in their sample is 3 Gyr.

4) In a recent paper, Schommer, *et al.* (1991) report velocities for 83 intermediate-age and old LMC clusters; most are located in the outer parts of the galaxy. They conclude that all of the clusters in their sample (which included most of the older FIO clusters) lie in a *single* disk; moreover, the systemic velocity offset noted by FIO was not confirmed, but ascribed to velocity errors in the earlier study. Thus, the most puzzling result of the FIO study appears to have been spurious; however, the conclusion that the oldest clusters lie in a flattened distribution, coplanar with the youngest clusters, was fully confirmed.

The LOS velocity dispersions from these studies are plotted as a function of age in Figure 4. There is a clear increase in the dispersion as a function of age as observed in the Galaxy (Wielen, Freeman, this volume). Is this increase due to heating by massive clouds or clusters in the LMC disk, or do the dispersions reflect conditions during different stages of the formation of the LMC? What is missing from Figure 4 is evidence for a true kinematic halo population with an LOS dispersion of about 50 km s^{-1} (Schommer, *et al.* 1991). Dispersions based on radial velocities of a large sample of unquestionably old objects (*e.g.*, RR Lyr stars) will clearly be needed to kinematically identify the halo of the LMC. Attempts to do this are in progress (Reid and Freedman, Freeman, private communications).

I thank M. Bessell, C. Chiosi, G. Da Costa, K. Freeman, S. Hughes, E. Olszewski, N. Reid, R. Schommer, W. Seggewiss, and S. van den Bergh for useful discussions on some of the topics covered in this review and for allowing me to quote results prior to publication. This work has been supported by a Hubble Fellowship through NASA grant # HF-1007.01-90A.

References

Barbaro, G. and Olivi, F. M. 1991, *A. J.*, **101**, 922.

Bertelli, G., Betto, R., Bressan, A., Chiosi, C., Nasi, E. and Vallenari, A. 1990, *Astr. Ap. Suppl.*, **85**, 845.

Bertelli, G., Mateo, M., Chiosi, G., and Bressan, A. 1991, *Ap. J.*, in press.

Bica, E., Alloin, D. and Santos, J. F. C. 1990, *Astr. Ap.*, **235**, 103.

Bothun, G. D. and Thompson, I. B. 1988, *A. J.*, **96**, 877.

Butcher, H. 1977, *Ap. J.*, **216**, 372.

Cassatella, A., Barbaro, J. and Geyer, E. H. 1987, *Ap. J. Suppl.*, **64**, 83.

Cohen, J. G. 1982, *Ap. J.*, **258**, 143.

Cohen, R. S., Dame, T. M., Garay, G., Montani, J., Rubio, M. and Thaddeus, P. 1989, *Ap. J. Lett.*, **331**, L95.

Cowley, A. P. and Hartwick, F. D. A. 1982, *Ap. J.*, **259**, 89.

Da Costa, G. S. 1991, in *The Magellanic Clouds*, eds. R. Haynes and D. Milne (Dordrecht: Reidel), p. 183.

Davies, R. D., Elliott, K. H. and Meaburn, J. 1976, *Mem. R. A. S.*, **81**, 89.

de Boer, K. S., Spite, F. and Stasińska, G. 1989, *Recent Developments of Magellanic Cloud Research*, (Paris: Obs. de Paris).

de Vaucouleurs, G. and Freeman, K. C. 1972, *Vistas in Astronomy*, **14**, 163.

Dufour, R. 1984, in *Structure and Evolution of the Magellanic Clouds*, eds. S. van den Bergh, and K. S. de Boer, (Dordrecht: Reidel), p. 353.

Elson, R. A. W. and Fall, S. M. 1985, *Pub. A. S. P.*, **97**, 692.

Elson, R. A. W., Fall, S. M. and Freeman, K. C. 1987, *Ap. J.*, **323**, 54.

Feast, M. W. 1989, in *The World of Galaxies*, eds. H. G. Corwin, and L. Bottinelli, (New York: Springer-Verlag), p. 118.

Fischer, P., Welch, P., Mateo, M., Côté, P. and Madore, B. 1991, *A. J.*, in press.

Freeman, K. C. 1989, in *The World of Galaxies*, eds. H. G. Corwin, and L. Bottinelli, (New York: Springer-Verlag), p. 99.

Freeman, K. C., Illingworth, G. and Oemler, A. 1983, *Ap. J.*, **272**, 488.

Fujimoto, M. and Noguchi, M. 1990, *P. A. S. Japan*, **42**, 505.

Grondin, L., Demers, S., Kunkel, W. E. and Irwin, M. J. 1990, *A. J.*, **100**, 663.

Hardy, E., Buonanno, R., Corsi, C. E., Janes, K. A., and Schommer, R. A. 1984, *Ap. J.*, **278**, 592.

Hardy, E. and Durrand, D. 1984, *Ap. J.*, **279**, 567.

Hardy, E., Suntzeff, N. B. and Azzopardi, M. 1989, *Ap. J.*, **344**, 210.

Hartwick, F. D. A., and Cowley, A. P. 1988, *Ap. J.*, **334**, 135.

Hatzidimitriou, D. and Hawkins, M. R. S. 1989, *M. N. R. A. S.*, **241**, 667.

Haynes, R. and Milne, D. 1991, *The Magellanic Clouds*, (Dordrecht: Reidel).

Hughes, S. M. G., Wood, P. R. and Reid, N. 1991, *A. J.*, **101**, 1304.

Irwin, M. J. 1991, in *The Magellanic Clouds*, eds. R. Haynes and D. Milne (Dordrecht: Reidel), p. 453.

Irwin, M. J., Demers, S. and Kunkel, W. E. 1990, *A. J.*, **99**, 191.

Jensen, J., Mould, J. and Reid, N. 1988, *Ap. J. Suppl.*, **67**, 77.

Lupton, R. H., Fall, S. M., Freeman, K. C., and Elson, R. A. W. 1989, **347**, 201.

Martin, N., Maurice, E. and Lequeux, J. 1989, *Astr. Ap.*, **215**, 219.

Mateo, M. 1988, in *Globular Cluster Systems in Galaxies*, eds. J. E. Grindlay and A. G. D. Philip, (Dordrecht: Reidel), p. 557.
Mateo, M., Hodge, P. and Schommer, R. A. 1986, *Ap. J.*, **311**, 113.
Mathewson, D. S. and Ford, V. L. 1984, in *Structure and Evolution of the Magellanic Clouds*, eds. S. van den Bergh, and K. S. de Boer, (Dordrecht: Reidel), p. 125.
Meatheringham, S. J., Dopita, M. A., Ford, H. C. and Webster, B. L. 1988, *Ap. J.*, **327**, 651.
Mould, J. R. 1991, in *The Magellanic Clouds*, eds. R. Haynes and D. Milne (Dordrecht: Reidel), p. 7.
Murai, T. and Fujimoto, M. 1980, *P. A. S. Japan*, **32**, 581.
Olszewski, E. W. 1988, in *Globular Cluster Systems in Galaxies*, eds. J. E. Grindlay and A. G. D. Philip, (Dordrecht: Reidel), p. 159.
Olszewski, E. W., Schommer, R. A., Suntzeff, N. B. and Harris, H. C. 1991, *A. J.*, **101**, 515.
Page, T. and Carruthers, G. R. 1981, *Ap. J.*, **248**, 908.
Reitermann, A., Baschek, B., Stahl, O. and Wolf, B. 1990, *Astr. Ap.*, **234**, 109.
Renzini, A. 1991, in *The Magellanic Clouds*, eds. R. Haynes and D. Milne (Dordrecht: Reidel), p. 165.
Richtler, T., Spite, M. and Spite, F. 1989, *Astr. Ap.*, **225**, 351.
Rohlfs, A., Kreitschmann, J., Siegman, B. C., and Feitzinger, J. V. 1984, *Astr. Ap.*, **137**, 343.
Russell, S. C. and Bessell, M. S. 1989, *Ap. J. Suppl.*, **70**, 865.
Russell, S. C. and Dopita, M. A. 1989, *Ap. J. Suppl.*, **74**, 93.
Sagar, R. and Pandey, A. K. 1989, *Astr. Ap. Suppl.*, **79**, 407.
Schommer, R. A., Christian, C. A., Caldwell, N., Bothun, G. D. and Huchra, J. 1991, *A. J.*, **101**, 873.
Schommer, R. A., Olszewski, E. W., Suntzeff, N. B. and Harris, H. C. 1991, *A. J.*, in press.
Seggewiss, W. and Richtler, T. 1989, in *Recent Developments of Magellanic Cloud Research*, eds. K. S. de Boer, F. Spite, and G. Stasińska, (Paris: Obs. de Paris), p. 45.
Smith, A. M., Cornett, R. M. and Hill, R. S. 1987, *Ap. J.*, **320**, 609.
Spite, M., Cayrel, R., Francois, P. and Spite, M. 1986, *Astr. Ap.*, **225**, 351.
Spite, M. and Spite, F. 1990, *Astr. Ap.*, **234**, 67.
Stryker, L. L. 1984, in *Structure and Evolution of the Magellanic Clouds*, eds. S. van den Bergh, and K. S. de Boer, (Dordrecht: Reidel), p. 79.
van den Bergh, S. 1981, *Astr. Ap. Suppl.*, **46**, 79.
van den Bergh, S. 1985, *Ap. J.*, **297**, 361.
van den Bergh, S. 1991, *Ap. J.*, **369**, 1.
Walker, A. R. 1989, *A. J.*, **98**, 2086.
Walker, A. R. 1990, *A. J.*, **100**, 1532.
Westerlund, B. E. 1990, *Astr. and Ap. Review*, **2**, 29.

B. Carney: There are two blue populous clusters I know of in the SMC: NGC 330 and NGC 346. The former's metallicity is very low according to echelle analyses by Francois and Monique Spite. The H II region surrounding NGC 346 is very deficient in C and N. Aside from the impact on the age-metallicity relation, could it be that metallicity is important in determining the mass of a cluster? Can you tell us something about the metallicities of the LMC blue populous clusters?

M. Mateo: I'll answer those in reverse order. As you know from your work on NGC 330, it's tough to get abundances from the CM diagrams of really young clusters; the high dispersion work is very important. The low abundances reported for some very young clusters (in both Clouds) make me nervous because the metallicities for somewhat older (about 1 Gyr; Olszewski, *et al.* 1991) clusters is considerably higher ([Fe/H] ~ -0.3). We really need to resolve this. As for the effect of metallicity on mass, I again refer to Olszewski, *et al.*; for clusters spanning a large range in luminosity (and presumably mass), the metallicities are similar. For [Fe/H] $\gtrsim -0.6$, there doesn't seem to be a strong correlation between [Fe/H] and cluster mass.

D. Hatzidimitriou: On the metallicity of NGC 330: M. Bessell presented some new results on the metal abundance of this cluster at the IAU General Assembly last week, showing a metallicity of [Fe/H] $= -0.75$. He claims that the significantly lower values given by, *e.g.*, Spite are due to problems with model atmospheres.

M. Mateo: And I believe he claims the reddening of the cluster is considerably different than what the Spite's assumed.

S. van den Bergh: I would like to quarrel a bit with your conclusion that the blue populous clusters in the LMC are young *globular* clusters. In a wide variety of environments (M 87, M 31, the Fornax dwarf) globular clusters have a Gaussian luminosity function. Data on the old (age $> 10^{10}$ yrs) clusters in the LMC are also consistent with such a luminosity function. It seems to me that the blue populous clusters in the LMC belong to a population with an *open* cluster luminosity function which is, however, somewhat enhanced in massive clusters compared to the open cluster luminosity function in the Galaxy.

M. Mateo: I really don't disagree. *Most* of the populous LMC clusters are in fact part of a population with a luminosity function like that of open clusters; only *a few* clusters appear to be luminous and massive enough that they will look like globular clusters when 15 Gyr old. Two excellent candidates are NGC 1866 and NGC 1978. If such objects make up only a small fraction of the total cluster population, they won't significantly perturb the luminosity function, especially if the open cluster population is rich as in the Clouds. They might account for the slight enhancement of the luminosity function you mentioned. Finally, since evolutionary fading is fastest at a young age, if there is a 2-3 Gyr range in the epoch of globular cluster formation, the resulting cluster LF won't necessarily look Gaussian at first (it may have a high-luminosity tail), even though after 15 Gyr it would.

J. Frogel: 1) I thought that earlier work showed that the integrated luminosities of the intermediate-age clusters would fade substantially if evolved to the age of Galactic globular clusters. 2) The estimation of a molecular mass based on CO observations always suffer from the converstion of H_2 mass via the Galactic ratio. If mean [Fe/H] of the LMC and SMC are low, then $M(H_2)$ will be significantly underestimated if based on a CO value.

M. Mateo: 1) The evolutionary fading was taken into account in my estimates of the cluster luminosities at an age of 15 Gyr. Most of the fading (if only stellar evolution is taken into account) occurs in the first 1 Gyr. I assumed a Salpeter slope ($x = 1.35$) for the IMF; the fading is much faster if x is smaller. Recent results suggest that the IMF slope in Cloud clusters is $\gtrsim$ 1.35, so the fading will be slightly less than I mentioned. The biggest uncertainty, really, is ignoring dynamical effects. 2) The CO-H_2 conversions I used in Table 1 come from Cohen, *et al.* (1989). They claim to take the metallicity deficiency of the LMC into account.

E. J. Alfaro: This comment is concerning your last conclusion. Some regions of our Galaxy show an age-metallicity relationship more similar to that obtained for the LMC than for the solar neighborhood. In particular, seven clusters studied by Geisler in the Galactic anticenter and the cluster IC 1311 (in the Cygnus region) observed by myself and colleagues display CMDs and locations in the age-metallicity diagram which indicate a clear similarity between these and the intermediate-age open clusters in the LMC.

M. Mateo: I agree. My conclusion refers to the age-metallicity relation for the solar neighborhood.

SURVEY OF STELLAR POPULATIONS IN THE SMC

D.HATZIDIMITRIOU

Anglo-Australian Observatory, P.O. Box 296, NSW 2121, Australia

and

L.T. GARDINER

Department of Astronomy, University of Edinburgh, Blackford Hill, Edinburgh EH9 3HJ, U.K.

Abstract. A stellar population survey covering a total of 130 square degrees in the outer parts (beyond 2 kpc from the optical centre) of the Small Magellanic Cloud is presented, based on colour-magnitude diagrams constructed from COSMOS measurements of a series of UKST photographic plates.

1. Introduction

As a dwarf irregular galaxy distinguished by its membership of an interacting system of galaxies, the Small Magellanic Cloud (SMC) can provide us with insights into the evolutionary development of dwarf irregular galaxies as well as the role of external dynamical interactions in stimulating star formation events.

The aim of the present survey of the stellar content of the outer regions of the SMC is to achieve a complete description of its star formation history from the age spectrum and spatial distribution of its stellar population.

The data set comprises colour (B$-$R) and magnitude (R) information for 1.1×10^6 stars in the outer parts of the SMC, covering a total area of 130 square degrees on the sky. The photometry is photographic from UKST plates digitised with the COSMOS automatic microdensitometer. The limiting magnitude of the survey is R=20mag, and the completeness better than 90% down to R=19.5mag. Details on the accuracy and completeness of the data-set and on the methodology of the analysis can be found in Hatzidimitriou *et al.* (1989) and Gardiner and Hatzidimitriou (1991; hereafter, GH). Colour-magnitude diagrams (CMD) were constructed in a grid, consisting of 0.87-square-degree cells. A complete series of these diagrams is published elsewhere (GH).

2. The Age Distribution in the Outer Parts of the SMC

a. Populations Younger than 2Gyr: In the following, age estimates based on the main-sequence are derived using the Revised Yale Isochrones. The main-sequence (MS) stars visible above the survey limit correspond to ages younger than $\simeq$2 Gyr, for a distance modulus of 18.8. Fig.1 shows the distribution of these younger populations over the area studied. The Wing region is the most conspicuous feature of this distribution. The younger population also appears to be more extended in the NE parts of the SMC (to the North of the Wing). However, some –but not all– of this effect is due to the shorter mean distance modulus in these areas (see Hatzidimitriou and Hawkins 1989).

B. Barbuy and A. Renzini (eds.), The Stellar Populations of Galaxies, 157–160.

Luminosity functions were also constructed for the MS stars (with a limiting magnitude of R=19.5 in order to avoid incompleteness problems) over extended circular annuli (for statistical reasons). A detailed analysis can be found in GH. Generally, a mixture of stellar generations with a median age increasing with distance from the SMC centre was found. Only two stellar generations could be identified as discrete SF events: the well-known very young population in the Wing (5×10^7yr, see also Irwin *et al.* 1990) and a $4 - 6 \times 10^8$yr population most conspicuous in the NE area (near the 'outer arm'). There is evidence that the mean age increases with increasing distance from the optical centre of the SMC (always assuming that the IMF remains the same at least during the last 2 Gyr).

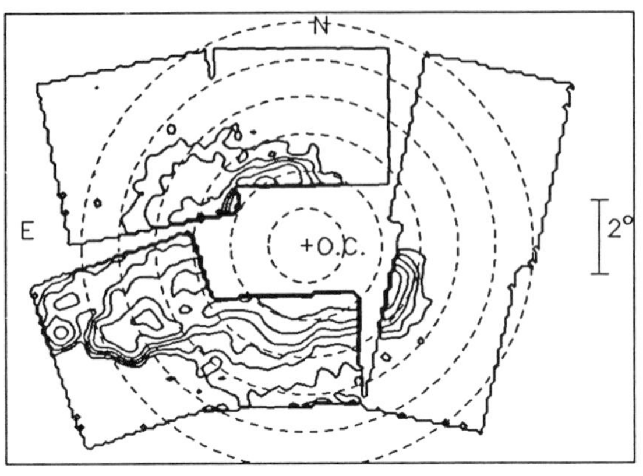

Figure 1. *Distribution of MS stars.*

b. Populations older than 2Gyr: The clump/red horizontal branch(RHB) is the most conspicuous feature on the CMD in the SMC field beyond 2kpc from the centre (except for the Wing region). Stars belonging to a wide range of ages (from $\simeq 10^8$yr to $\geq$10Gyr) can populate the RHB.

Comparison of the observed numbers of MS stars ($\tau < 1-2$Gyr) and of clump/RHB stars in conjunction with stellar evolutionary arguments indicates that the majority of the clump/RHB stars in the outer areas studied here are older than 1-2Gyr. Fig.2 shows a contour map of the surface number density of the clump/RHB stars. Although this distribution is much smoother than that of Fig.1, there are indications of asymmetries especially in the N and E areas.

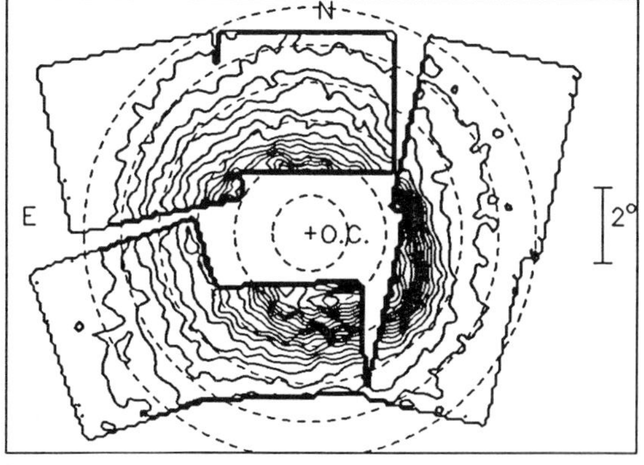

Figure 2. *Distribution of RHB stars.*

Using the colour difference between the clump/RHB and the red giant branch at the level of the RHB as a *median* age indicator (see Hatzidimitriou 1991 for the definition and calibration of the indicator), we find that beyond 2–2.5kpc from the optical centre, the SMC field has a *median* age of 10 ± 2Gyr.

There is a weak 'horizontal extension' of the red clump towards bluer colours (but to the red of the instability strip). This can be interpreted (using the above age indicator) as belonging to a population older by 2-3Gyr than the median population represented by the main clump. This population –if present– would account for $\simeq$7% of the mass of the total population. This estimate agrees with that by Frogel (1984) on the relative contribution of the mass of old stars (inferred by the number of RR-Lyrae stars) to the total mass of the SMC.

3. Kinematics of intermediate age stars in the outer parts

In the SMC, young stars and gas are known to have very complex kinematics (e.g. Torres and Carranza 1987), and they form four apparently distinct velocity groups. This kinematical behaviour along with (still controversial) claims for large line-of-sight depth in the SMC has led some authors to the conclusion that the SMC is in the process of 'irreversible disintegration' (Mathewson *et al.* 1988). A recent study based on RHB stars has shown that large line-of-sight depths exist among populations older than $\simeq$1-2Gyr (see Hatzidimitriou and Hawkins 1989; Gardiner and Hawkins 1991; Mateo & Hatzidimitriou, this volume).

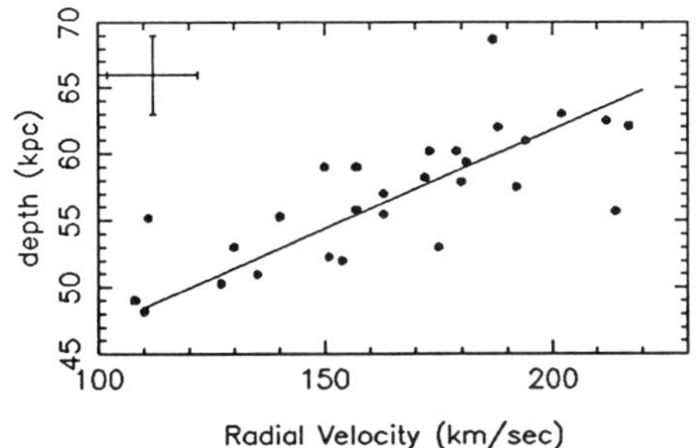

Figure 3. *Radial Velocity-Distance correlation for RHB stars in the NE.*

Radial velocities for a sample of RHB/clump stars in a 40arcmin field in one of the 'deep' regions in the NE (at 2.7kpc from the optical centre) were recently obtained using the fibre spectroscopic facility Autofib at the 3.9m AAT (Hatzidimitriou *et al.*, in prep.). Figure 3 shows the discovered correlation between radial velocity and line-of-sight distance for these stars. If this is due to simple streaming motion, the timescale of the motion is a few 10^8yr, which coincides with the recent encounter between the LMC and the SMC (Fujimoto and Murai 1984). This result, in conjunction with the large depth in extended areas in the N and E, confirms that the recent close encounter between the LMC and the SMC had a profound effect on the dynamical stability of the latter.

4. On the Star Formation History of the SMC

On the basis of the results briefly described in the previous paragraphs, the following comments can be made on the star formation (SF) history of the SMC and its possible connection with tidal interactions with the LMC and the Galaxy (details in GH):

(i) The major SF period in the SMC outer regions appears to have occurred $\sim$10 Gyr ago, although there is evidence for the presence of an older population amounting to $\sim$7% in mass.

(ii) There is a progressive aging of the stellar populations with distance from the optical centre. This effect may be connected to the density of the gas available for SF as a function of time and radius (e.g. critical-density SF scenario by Kennicutt 1989).

(iii) There is some evidence of a dynamical disturbance of the SMC in the Eastern and Northern outer regions (for both young and old populations), which appears to be connected with the most recent close encounter between the SMC and the LMC. However the effect of such interactions to the SF rate is difficult to assess. In the last 1-2 Gyr, there appear to be two distinct stellar generations which may be connected with dynamical events: the Wing and the 0.4-0.6Gyr population in the NE. In earlier periods the situation becomes very unclear. The age distribution of star clusters and of the general field populations in the two Clouds are significantly different (see also Mateo, this volume). Using the RHB dating method used above for published CMDs in the LMC, we find that the major SF event in the LMC occurred probably 2-3Gyr later than in the SMC (see also GH). It is therefore difficult to connect the SF histories of the two Clouds in this respect. However, we should keep in mind that the LMC and SMC were not necessarily always bound to each other and that the perigalactic distance may have been progressively larger (and hence the encounters less disruptive) in the past.

5. References

Frogel, J.A. 1984. *P.A.S.P.*, **96**, 856.
Fujimoto, M. and Murai, T. 1984. *IAU symp.No.108*, p.115, Reidel.
Gardiner, L.T. and Hawkins, M.R.S. 1991. *M.N.R.A.S.*, **251**, 174.
Gardiner, L.T. and Hatzidimitriou, D. 1991. *M.N.R.A.S.*, submitted.(GH)
Hatzidimitriou, D. 1991. *M.N.R.A.S.*, **251**, 545.
Hatzidimitriou, D., Hawkins, M. and Gyldenkerne, K. 1989. *M.N.R.A.S*, **241**, 645.
Hatzidimitriou, D. and Hawkins, M.R.S. 1989. *M.N.R.A.S*, **241**, 667.
Hatzidimitriou, D., Cannon, R.D., Hawkins, M.R.S., and Teo, A. *in prep.*
Irwin, M.J., Demers, S. and Kunkel, W.E. 1990. *Astron.J.*, **99**, 191.
Kennicutt, R.C. 1989. *Ap.J.*, **344**, 685.
Mathewson, D.S., Ford, V.L. and Visvanathan, N. 1988. *Ap.J.*, **333**, 617.
Torres, G. and Carranza, G.J. 1987. *M.N.R.A.S*, **226**, 513.

DISCUSSION

SERRANO: Can you tell the dependence of $\rho(r)$ in your old population?
HATZIDIMITRIOU: In the North the projected surface density profiles are relatively well represented by an exponential law $N \propto e^{-r/l}$, with $l \simeq 1.2$kpc (Gardiner and Hawkins 1991). However one should be aware of the very different distribution of populations of different ages.
ZINNECKER: I am a little confused about the evidence on interaction-triggered SF in the SMC. While you concluded that there is good evidence, Mario Mateo told us before that there is no such evidence. Could you clarify the situation?
HATZIDIMITRIOU: I agree with M.Mateo's conclusion that there is no evidence of major SF events triggered by tidal interactions in the MCs for at least the first ~10 Gyr of their lives. But there is evidence of both dynamical disturbance and of discrete SF events in the E and NE in the SMC outer parts. There is also evidence for an increase of the enrichment rate in the last 2 Gyr in the SMC (Da Costa 1991; IAU 148). All these effects could be due to the recent close encounter of the MCs with the Galaxy and of the LMC and SMC with each other.

STELLAR POPULATIONS IN M31 AND M33

S. VAN DEN BERGH
Dominion Astrophysical Observatory
National Research Council
5071 West Saanich Road
Victoria, B.C., V8X 4M6, Canada

C.J. PRITCHET
Department of Physics and Astronomy
University of Victoria
Box 3055
Victoria, B.C., V8W 3P6, Canada

1. Introduction

M31 and M33 are the two nearest extragalactic spirals. They are therefore particularly suitable for studies of stellar populations. From integrated photometry spiral galaxies are known to consist of four components: (1) a nucleus, (2) a nuclear bulge, (3) an (exponential) disk and (4) a halo. These stellar components are embedded within a massive invisible halo.

2. The Nuclei of M31 and M33

All spiral galaxies appear to contain nuclei while no irregular galaxy is known to contain a nucleus. The nucleus of M33 has B = 14.5 ± 0.1 (Nieto & Aurière 1982), which corresponds to $M_B \simeq -10.3$. The internal velocity dispersion of this nucleus is small ($\sigma \leq 30$ km s^{-1}), which indicates that its mass-to-light ratio must be low. The spectrum of the nucleus of M33 (van den Bergh 1976) is composite, with K/H + Hε yielding a late A spectral type, while CH/Hγ gives type F3 - F4. The observed spectrum and integrated colors of the nucleus of M33 might be produced by either (1) a young metal-rich population or (2) by an old stellar population that is *very* metal-poor. Van den Bergh's observations showed that λ 4325 of Fe I was stronger in M33 than in the spectra of very metal-poor globular clusters. The conclusion that the nucleus of the Triangulum nebula consists of young relatively metal-rich stars is strongly supported by recent near-infrared spectra (Schmidt, Bica & Alloin 1990). O'Connell (1983) obtains a mean nuclear star formation rate over the last 1 Gyr of $\sim 3 \times 10^{-4} M_{\odot}$ yr^{-1}. Gallagher, Goad & Mould (1982) estimate the mass of the nucleus of M33 to be $\sim 10^6 M_{\odot}$.

The nucleus of M31 has B = 13.6 ± 0.3, which corresponds to $M_B = -11.0$. This value is almost two orders of magnitudes brighter than an average globular cluster. Tremaine, Ostriker & Spitzer (1975) have suggested that the nucleus of the Andromeda nebula was formed from the

B. Barbuy and A. Renzini (eds.), The Stellar Populations of Galaxies, 161–168.

debris of globular clusters that had been dragged inwards by dynamical friction. However, the observation (van den Bergh 1969) that more than 97% of the globulars associated with the Andromeda nebula have integrated spectra which indicate that they are metal poorer than the nucleus of M31, militates against this suggestion. Probably most of the stars in nucleus formed from gas that had already been enriched in heavy elements before it flowed into the center of M31. Sandage et al. (1969) find that the inner ~ 40" of M31 exhibits an ultraviolet excess. This observation might possibly be accounted for by assuming that tidal friction dragged in some globular clusters containing blue horizontal branch stars. The large rotational velocities and velocity dispersion in the nucleus of M31 (Dressler & Richstone 1988, Kormendy 1988) requires a high central mass concentration of 10^7 - 10^8 $M_\odot$ if the nucleus is ellipsoidal, or of $10^{6.5}$ - 10^7 $M_\odot$ if the nucleus is a disk. The observed strengths of the H_2O and CO bands at 2.1 μm and 2.3 μm (Baldwin *et al.* 1973, Persson *et al.* 1980), and the low strength of the dwarf-sensitive Wing-Ford band (Whitford 1974) all indicate that the high mass-to-light ratio of the nucleus is *not* due to the presence of a dwarf-enriched lower main sequence. Jones, Alloin & Jones (1984) reached a similar conclusion from observations of the gravity sensitive Ca II triplet. Population modelling by Schmidt et al. (1989) suggest that the average logarithmic metallicity of giant stars in the semi-stellar nucleus of M31 is $[Z/Z_\odot] \approx + 0.6$.

3. The Nuclear Bulges of M31 and M33

The nuclear bulge of M31, which accounts for ~ 30% of the total visual light of the Andromeda nebula (de Vaucouleurs 1958), has an effective major axis of 17$'$.5 (3.7 kpc). Over the range 0$'$.1 < $\tilde{\omega}$ < 10' Kent (1983) finds that available photometry and the bulge rotation curve can be fit with a model having a mass-to-light ratio (in solar units) of $M/L_B \approx 3\ (\sigma/160)^2$, in which σ is measured in km s^{-1}. Population modelling by Schmidt et al. (1989) suggests that the average logarithmic metallicity of giant stars in the nuclear bulge of M31 is $[Z/Z_\odot] \approx + 0.3$. In this respect the stars in the bulge of the Andromeda nebula are similar to those in the bulge of the Galaxy (Whitford 1985), which are also found to be super metal-rich. According to Mould (1986) the brightest M31 bulge stars are about one magnitude more luminous in I than the brightest stars in the halo of the Andromeda nebula. IUE observations (Welch 1982), and direct imaging in U and B, show that the central bulge of M31 does not contain luminous metal-rich main sequence stars. In fact most population models for the bulge of the Andromeda nebula require no main sequence stars with spectral types earlier than G0V.

The existence of a tiny nuclear bulge in M33 remains controversial. Such a bulge was first reported by Patterson (1940) and confirmed by Boulesteix et al. (1979), who found it to have an effective radius of 2$'$.75 and a luminosity of ~ 1% of that of the exponential disk of the Triangulum nebula. More recently Kent (1987) has, however, suggested that the decomposition of the integrated light of M33 into disk and bulge components has been affected by spiral structure. The existence of the nuclear bulge of M33 therefore remains to be established with certainty.

4. The Disks of M31 and M33

According to Walterbos & Kennicutt (1988) the disk of M31 contributes 65% of the U light and 55% of the V luminosity of the Andromeda nebula. The disk scale-length of M31 decreases with increasing wavelength. It is 7.1 ± 0.4 kpc in U, 5.5 ± 0.3 kpc in R and 4.1 kpc in the K-band at 2.2 μm (Hiromoto *et al.* 1983). This suggests that the average age of stars in the disk of the Andromeda nebula decreases with radius. This conclusion is confirmed by Walterbos & Kennicutt (1988) who find that the disk of M31 becomes slightly bluer at large radii. The ring-shaped region of active star formation between 8 and 14 kpc from the nucleus is observed to be slightly bluer than are the zones on either side of this feature. The disk scale-length of M31 is comparable to, or slightly larger than, that of the Galaxy which lies in the range 3.5 - 5.5 kpc (Freeman 1987).

The distribution of late-type stars in the disk of M31 has been studied by Richer, Crabtree & Pritchet (1990). The carbon to late M star ratio in Baade's Field IV at 20 kpc from the nucleus was, perhaps surprisingly, found to be similar to that obtained by Richer & Crabtree (1985) in a field at only 11 kpc from the nucleus, in which the stellar metallicity is expected to be higher (Blair, Kirshner & Chevalier 1982).

Color-magnitude diagrams for disk stars, based on CCD images, are now available from the work of Crotts (1986), Hodge, Lee & Mateo (1988) and Hodge & Lee (1988). Unfortunately the latter two investigations do not reach deep enough to study the oldest population component of M31. Crott's data might perhaps be understood in terms of a model in which the dominant population of the outer disk consists of stars similar to, or slightly metal-richer than, those in the Galactic globular cluster 47 Tucanae, on which a lesser component resembling the intermediate - age Galactic cluster NGC 2158 is superimposed.

From digital stacking of Palomar Schmidt plates Innanen et al. (1982) found that the outermost part of the disk of M31 is warped. The fact that this warp is visible on both yellow and red-sensitive plates, shows that it is due to starlight, rather than to emission nebulosity. That the outermost "Population II suddenly swirls off to one side" was first noted by Baade (1963). It is of interest to note that the optical and radio (Newton & Emerson 1977, Cram, Roberts & Whitehurst 1980) images are warped in the same direction.

According to Kent (1987) the exponential disk of M33 has a scale-length of 9'.6 (2.2 kpc). Integrated photometry by de Vaucouleurs (1959) shows that the outer regions of the disk of the Triangulum nebula are slightly bluer (B - V = 0.50, U - B = -0.17) than are its inner regions (B - V = 0.59, U - B = -0.04). This effect might be due to a radial population gradient and/or to lower dustiness (Israel & Kennicutt 1980) of the outer metal-poor (Pagel & Edmunds 1981) regions of the disk.

From radial velocity observations Boulesteix & Monnet (1970) showed that the mass-to-light ratio of M33 increases by a factor of eight between 5' and 40' (1 - 9 kpc) from the nucleus. This observation constituted the first evidence for the existence of dark matter in the halo of the Triangulum nebula.

The resolution of the disk of M33 into stars was first achieved by the Earl of Rosse (1850). Lundmark (1921) found the brightest stars in the Triangulum nebula to have B $\simeq$ 15.7. Counts of early-type stars over the face of M33 have been published by Madore, van den Bergh & Rogstad (1974). Their data showed no simple relationship between the surface density of OB stars and that of neutral hydrogen gas. Reasons for this are probably that (1) a significant fraction of the gas in the central regions of M33 is in molecular form (Wilson et al. 1988), (2) the thickness of the M33 gas layer may increase with radial distance (as it does in the Galaxy), so that there is no one-to-one correspondence between the *surface* density of young luminous stars and the *space* density of HI, and (3) a minimum gas density may be required to trigger star formation (Kennicutt 1989).

Freedman (1985) observed that the upper ends of the M31, M33 and LMC luminosity functions are similar. Humphreys & Sandage (1980) find that the brightest blue and red supergiants in M33 have M_V = -9.4 and M_V = -8.15, respectively.

According to Walker (1964) the ratio of the number of blue to red supergiants in M33 varies with distance from the nucleus. This conclusion was subsequently confirmed by Humphreys & Sandage (1980) but not by Freedman (1985). Recent CCD observations by Wilson (1990) show no gradient in the ratio of red to blue supergiants in the inner 2 kpc of M33. Data on OB associations at larger radii (and lower metallicities) should be obtained to confirm this conclusion.

Since M33 has a radial abundance gradient (Vilchez et al. 1988, Zaritsky et al. 1989) one would expect the WC-to-WN ratio to decrease with increasing galactocentric distance. This expectation is confirmed by Massey & Conti (1983). According to Schild, Smith & Willis (1990) line-widths of early WC stars in M33 also appear to correlate with galactocentric distance.

5. The Halos of M31 and M33

Mould & Kristian (1986) have obtained I versus V - I color magnitude diagrams for halo fields in M31 and M33. Their data show that the halo of the Triangulum nebula has a red giant branch similar to those of metal-poor globular clusters, whereas the stars in the halo of the Andromeda nebula appear to exhibit a high mean metallicity and a large metallicity dispersion. The existence of old stars of low metallicity in the halo of M33 is confirmed by the discovery of 6 RR Lyrae stars by Pritchet & van den Bergh (Pritchet 1988). Further support for the existence of a halo population in M33 is provided by the important discovery (Schommer et al. 1991) that old red star clusters with B - V ≥ 0.6 have a significantly larger velocity dispersion than do younger blue clusters. In this respect the oldest clusters in M33 differ from those in the Large Magellanic Cloud which appear to belong to a (thick) disk population (Freeman, Illingworth & Oemler 1983). The fact that M33 contains a significant halo population, but little or no nuclear bulge suggests that *the halo and bulge constitute separate building blocks of galaxies i.e. the halo is not just a continuation of the nuclear bulge to large radii.* According to Harris (1991) the globulars in M33 have $< M_V >$ = -7.0 ± 0.2, with a dispersion of 1.2 mag. Taken at face value this result suggests that the luminosity function of M33 globular clusters is similar to that of globulars associated with the Milky Way. It would be important to obtain accurate reddening values for individual globular clusters in M33 to strengthen and confirm this conclusion.

Mould & Kristian (1986) have studied the color-magnitude diagram of a halo field that is located 40' (8.4 kpc) from the nucleus of M31. They find that the majority of stars in this region have metallicities between those of the Galactic globular clusters M92 and 47 Tuc. From a more detailed study of the same zone in the halo Pritchet & van den Bergh (1988) showed that the average metallicity in the inner halo of M31 is [Fe/H] ≃ -1.0, with a dispersion $\sigma_{[Fe/H]} \simeq 0.3$. This mean metallicity is slightly higher than the average metallicity of M31 globular clusters. For 150 clusters Huchra, Brodie & Kent (1991) obtain < [Fe/H] > = -1.21 ± 0.02. This value is significantly higher than < [Fe/H] > = -1.40 ± 0.01 for 121 Galactic globulars (Brodie & Huchra 1991). The observation that the M31 globulars are, in the mean, somewhat metal-richer than their Galactic counterparts is consistent with the well-established correlation between mean metallicity of globular cluster systems and the luminosity of their parent galaxies (van den Bergh 1975,

Mould, Oke & de Zeeuw 1991).

Huchra *et al.* (1991) find that all M31 globulars metal-richer than [Fe/H] = -0.5 are located within 10 kpc of the nucleus of that galaxy. However, no obvious correlation between metallicity and projected distance from the nucleus is seen for clusters with [Fe/H] < -0.5. Metal-rich clusters with [Fe/H] ≥ -0.8 in M31 appear to form a rotating disk that extends out to $\tilde{\omega} \simeq 5$ kpc. In the Milky Way all but one of the metal-rich disk (|Z| < 2 kpc) clusters are also located in a disk with $\tilde{\omega} \simeq 5$ kpc (Armandroff 1989). This result suggests that the metal-rich disk clusters in the Milky Way extend over a larger *fraction* of the optical disk than do those in the Andromeda nebula.

From Palomar 5-m spectra van den Bergh (1969) found that CN is stronger in M31 globular clusters than it is in Galactic globular clusters of similar metallicity. This conclusion has more recently been strengthened and confirmed by Burstein et al. (1984), Tripicco (1989), Brodie & Huchra (1990) and in the infrared by Davidge (1990). Burstein *et al.* have claimed that M31 globulars also exhibit significantly stronger Balmer lines than do Galactic globulars. However, CCD spectra by Tripicco, which cover the range λλ 3850 - 4200, appear to rule out a significant contribution of horizontal branch stars to the integrated spectra of M31 globulars.

For field stars in the inner halo of M31 Pritchet and van den Bergh (1988) conclude that the blue horizontal branch is probably weak. For 28 cluster-type variables in M31 these authors find $\langle P_{ab} \rangle = 0.55$ days, which indicates that the RR Lyrae stars in the inner halo of the Andromeda nebula belong to Oosterhoff's type I.

The fact that the inner halo of M31 is both relatively metal-rich ([Fe/H] ≃ -1.0), and rich in RR Lyrae stars, suggests that its stellar population is similar to that in the Galactic globular cluster NGC 6171. This conclusion is strengthened and confirmed by the location of the tip of the M31 giant branch in the V versus B-V color magnitude diagram (see Figure).

Observations of the halo planetary nebulae M31 - 290 and M31 - 372 (Henry 1990) yield logarithmic oxygen abundances of 8.54 and 8.05, on a scale where the logarithmic hydrogen abundance is 12. These observations provide direct evidence for a significant spread in the metallicity of stars in the halo of the Andromeda nebula.

Using the 3.6-m CFH telescope we have observed a number of areas in the halo of M31. These fields are located along both the minor axis (from 40' to 5° from the nucleus), and at an angle of 30° to the major axis (out to 2° from the nucleus). All fields were observed with an RCA CCD at the prime focus of CFHT (field 2' x 3'). Exposure times were typically 45 min to 1 hour through B and V filters.

Data reduction techniques were standard. We used DAOPHOT (Stetson 1987) to measure magnitudes, and calibrated each field using Landolt (1983) standard stars. The limiting magnitude of the data (S/N = 4) appears to be about V = 24.5 and B = 25. Here we briefly discuss the data for the innermost three fields. The outer fields are much sparser, due to the steep stellar density gradient in the halo of M31. A detailed analysis of the CMD's in these outer fields will have to await completeness and error analysis using DAOPHOT add-star experiments.

The color-magnitude diagrams for our fields are shown in Figure 1. The fields shown are M0 (40' from nucleus, originally studied by Pritchet and van den Bergh 1987), M1 (1° from nucleus along the minor axis), M2 (1°.5 from the nucleus along the minor axis) and E1 (40' along major axis and 40' along the minor axis from the nucleus. The figures also show fiducial sequences for some Galactic globular clusters with a range of metallicity, adjusted to the reddening and absorption of M31. The clusters shown are M92 [Fe/H = -2.24], M5 [Fe/H = -1.40], and 47 Tuc [Fe/H = -0.71] (Suntzeff, Kinman & Kraft 1991).

The fields M0, M1 and E1 show a prominent giant branch that reaches up to between V = 22.5

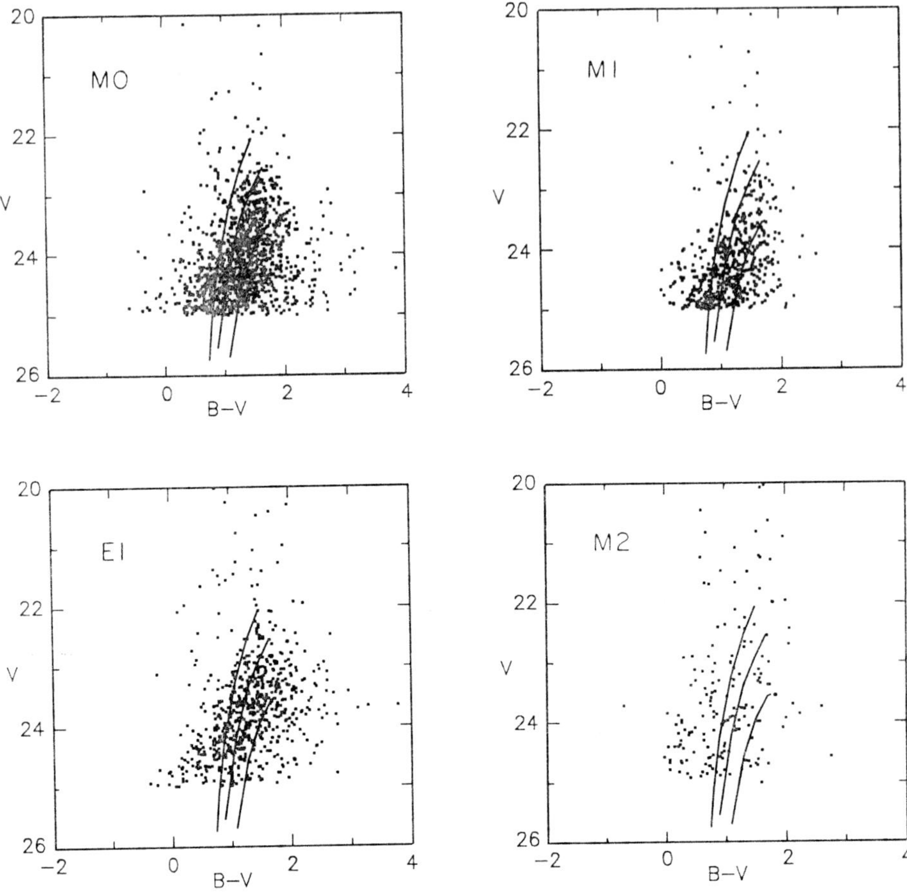

Figure 1 Color-magnitude diagrams for four M31 halo fields. M0, M1 and M2 are located along the minor axis at distances of 0°.67, 1°.0 and 1°.5 from the nucleus, respectively. Field E1 is located 40' along the major axis and 40' along the minor axis. Due to a steep stellar density gradient most stars in field M2 are probably Galactic foreground objects. No obvious metallicity differences are seen between fields M0, M1 and E1. The fiducial sequences shown are those of M92, M5 and 47 Tuc shifted by $E_{B-V} = 0.08$ and $A_V = 0.24$ mag.

and V = 23, which is intermediate in color between those of the giant branches of 47 Tuc and M5 - in good agreement with the results of Pritchet and van den Bergh (1987). None of the CMD's reach deep enough to show the horizontal branch. The principal conclusion from overlaying these CMD's, and also from comparing histograms of the color distribution of stars between V = 23 and 24, is that *the median B-V color of the halo giant branch in the three inner halo fields does not vary by more than ± 0.1 mag peak-to-peak in B-V.* Adopting d(B-V)/d[Fe/H] ≈ 0.5 at [Fe/H] = -1 (e.g. Pritchet & van den Bergh 1987), this maximum range in B-V corresponds to ± 0.2 in [Fe/H]. This indicates that, within the inner halo of M31, there is not a significant gradient in metallicity either perpendicular or parallel to the disk. The range in color is consistent with the observation that the V magnitude of the giant branch tip is the same for all three inner halo fields to within ± 0.2 mag peak-to-peak, correspoding to a metallicity difference of ± 0.3 in [Fe/H] (Pritchet & van den Bergh 1987).

Further results for these and other M31 fields are in preparation (Pritchet & van den Bergh 1992).

6. References

Armandroff, T.E. 1989, *AJ*, 97, 375

Baade, W. 1963, *Evolution of Stars and Galaxies* (Cambridge:Harvard Univ. Press), p. 73

Baldwin, J.R., Danziger, I.J., Frogel, A., & Persson, S.E. 1973, *Ap Letters*, **14**, 1

Blair, W.P., Kirshner, R.P., & Chevalier, R.A. 1982, *ApJ*, **254**, 50

Boulesteix, J., Colin, J., Athanassoula, E., & Monnet, G. 1979, in *Photometry, Kinematics and Dynamics of Galaxies*, ed. D.S. Evans (U. Texas:Austin), p. 271

Boulesteix, J., & Monnet, G. 1970, *A&A*, **9**, 350

Brodie, J.P., & Huchra, J.P. 1990, *ApJ*, **362**, 503

Brodie, J.P., & Huchra, J.P. 1991, *ApJ*, in press

Burstein, D., Faber, S.M., Gaskell, C.M., & Krumm, N. 1984, *ApJ*, **287**, 586

Cram, T.R., Roberts, M.S., & Whitehurst, R.N. 1980, *A&AS*, **40**, 215

Crotts, A.P.S. 1986, *AJ*, **92**, 292

Davidge, T.J. 1990, *ApJ*, **351**, L37

de Vaucouleurs, G. 1958, *ApJ*, **128**, 465

de Vaucouleurs, G. 1959, *ApJ*, **130**, 728

Dressler, A., & Richstone, D.O. 1988, *ApJ*, **324**, 701

Freeman, K.C. 1987, *ARAA*, **25**, 603

Freedman, W.L. 1985, *ApJ*, **299**, 74

Freeman, K.C., Illingworth, G., & Oemler, A. 1983, *ApJ*, **272**, 488

Gallagher, J.S., Goad, J.W., & Mould, J. 1982, *ApJ*, **263**, 101

Harris, W.E. 1991, *ARAA*, **29**, in press

Henry, R.B.C. 1990, *ApJS*, **356**, 229

Hiromoto, N., Maihara, T., Oda, N. & Okuda, H. 1983, *PASJ*, **35**, 413

Hodge, P. & Lee, M.G. 1988, *ApJ*, **329**, 651

Hodge, P.W., Lee, M.G. & Mateo, M. 1988, *ApJ*, **324**, 172

Huchra, J.P., Brodie, J.P., & Kent, S.M. 1991, *ApJ*, **370**, 495

Humphreys, R.M., & Sandage, A. 1980, *ApJS*, **44**, 319

Innanen, K.A., Kamper, K.W., Papp, K.A., & van den Bergh, S. 1982, *ApJ*, **254**, 515

Israel, F.P., & Kennicutt, R.C. 1980, *Ap. Letters*, **21**, 1

Jones, J.E., Alloin, D.M., & Jones, B.J.T. 1984, *ApJ*, **283**, 457

Kennicutt, R.C. 1989, *ApJ*, **344**, 685

Kent, S.M. 1983, *ApJ*, **266**, 562

Kent, S.M. 1987, *AJ*, **94**, 306

Kormendy, J. 1988, *ApJ*, **325**, 128

Landolt, A.U. 1983, *AJ*, **88**, 439

Lundmark, K. 1921, *PASP*, **33**, 324

Madore, B.F., van den Bergh, S. & Rogstad, D.H. 1974, *ApJ*, **191**, 317

Massey, P., & Conti, P.S. 1983, *ApJ*, **273**, 576

Mould, J.R. 1986, in *Stellar Populations*, eds. C. Norman, A. Renzini & M. Tosi (Cambridge:Cambridge Univ. Press), p. 9

Mould, J., & Kristian, J. 1986, *ApJ*, **305**, 591

Mould, J.R., Oke, J.B., & de Zeeuw, P.T. 1991, *AJ*, in press

Newton, K., & Emerson, D.T. 1977, *MNRAS*, **181**, 573

Nieto, J.L., & Aurière, M. 1982, *A&A. Ap.*, 108, 334

O'Connell, R.W. 1983, *ApJ*, 267, 80

Pagel, B.E.J., & Edmunds, M.G. 1981, *ARAA*, 19, 77

Patterson, F.S. 1940, *Harv. Bull.*, **No. 914**, 9

Persson, S.E., Cohen, J.G., Sellgren, K., Mould, J., & Frogel, J.A. 1980, *ApJ*, **240**, 779

Pritchet, C.J. 1988, in *The Extragalactic Distance Scale*, eds. S. van den Bergh & C.J. Pritchet (BYU Press:Provo), p. 59

Pritchet, C.J., & van den Bergh, S. 1987, *ApJ*, **316**, 517

Pritchet, C.J., & van den Bergh, S. 1988, *ApJ*, **331**, 135

Pritchet, C.J., & van den Bergh, S. 1992, in preparation

Richer, H.B. & Crabtree, D.R. 1985, *ApJ*, **298**, L13

Richer, H.B., Crabtree, D.R., & Pritchet, C.J. 1990, *ApJ*, **355**, 448

Rosse, W. Parsons Earl of 1850, *Phil. Trans. R. Soc.*, **140**, 499

Sandage, A.R., Becklin, E.E., & Neugebauer, G. 1969, *ApJ*, **157**, 55

Schild, H., Smith, L.J., & Willis, A.J. 1990, *A&A*, **237**, 169

Schmidt, A.A., Bica, E., & Alloin, D. 1990, *MNRAS*, **243**, 620

Schmidt, A., Bica, E., Alloin, D., & Dottori, H. 1989, *Ap. Space Sci.*, **157**, 79

Schommer, R.A., Christian, C.A., Caldwell, N., Bothun, G.D., & Huchra, J. 1991, *AJ*, **101**, 873

Stetson, P.B. 1987, PASP, 99, 191

Suntzeff, N.B., Kinman, T.D., & Kraft, R.P. 1991, *ApJ*, **367**, 528

Tremaine, S.D., Ostriker, J.P., & Spitzer, L. 1975, *ApJ*, **196**, 407

Tripicco, M.J. 1989, *AJ*, **97**, 735

van den Bergh, S. 1975, 1969, *ApJS*, **19**, 145

van den Bergh, S. 1975, *ARAA*, **13**, 217

van den Bergh, S. 1976, *ApJ*, **203**, 764

Vílchez, J.M., Pagel, B.E.J., Diaz, A.I., Terlevich, E., & Edmunds, M.G. 1988, *MNRAS*, **235**, 633

Walker, M.F. 1964, *AJ*, **69**, 744

Walterbos, R.A.M. & Kennicutt, R.C. 1988, *A&A*, **198**, 61

Welch, G.A. 1982, *ApJ*, **259**, 77

Whitford, A.E. 1974, in *IAU Symposium No. 58, The Formation and Dynamics of Galaxies*, ed. J.R. Shakeshaft (Dordrecht:Reidel), p. 169

Whitford, A.E. 1985, *PASP*, **97**, 205

Wilson, C.D. 1990, Caltech PhD dissertation

Wilson, C.D., Scoville, N., Freedman, W.L., Madore, B.F., & Sanders, D.B. 1988, *ApJ*, **333**, 611

Zartisky, D., Elston, R., & Hill, J.M. 1989, *AJ*, **97**, 97

THE DWARF ELLIPTICAL GALAXIES OF THE LOCAL GROUP & THE STELLAR POPULATIONS AND AGE OF M32

Wendy L. Freedman
Carnegie Observatories
813 Santa Barbara Street
Pasadena, CA 91101 USA

I. Introduction

An active debate continues over whether elliptical galaxies are primarily old stellar systems or whether they have had major star formation events in the recent past. Not only is this question of interest with regard to understanding the stellar populations and star formation history of nearby systems, but the resolution of this issue influences the interpretation of the spectra of high-redshift galaxies and has profound consequences for our understanding of galaxy, and therefore ultimately, of cosmological evolution.

Our lack of understanding of the stellar make-up in elliptical galaxies has persisted for some time because there are no giant elliptical galaxies near enough to allow the study of their stellar populations directly. Most information on the stellar populations of elliptical galaxies rely on the interpretation of integrated light. However, direct information on the bright stellar content of low-luminosity elliptical galaxies *can* be obtained from a study of the Local Group dwarf ellipticals. The nearby Andromeda galaxy, M31 has four low-luminosity elliptical companions: M32, NGC 205, NGC 185 and NGC 147, the subjects of this review.

This review will begin with a broad summary of population characteristics of dwarf elliptical galaxies (dE's), it will briefly summarize what is known about the stellar populations of the four Andromeda companions, and then discuss the specific case of M32 in detail. M32, the highest surface brightness Andromeda companion, has characteristics very simliar to the giant ellipticals, and has therefore been the focus of much of the controversy surrounding the issue of the ages of elliptical galaxies. Studies of its integrated light, in combination with new studies of its brightest resolved giants, particularly in the near-infrared, may help to resolve many of the outstanding questions regarding the stellar populations in elliptical galaxies.

II. General Properties

All four dwarf elliptical companions to Andromeda were first resolved by Baade (1944). M32, a compact dwarf, is the closest companion, lying 24 arcmin away from the M31 nucleus. The next nearest companion is NGC 205, well-known to have undergone very recent star formation as evidenced by a small population of young OB stars. Seven degrees away lie the low-luminosity pair of dwarf ellipticals, NGC 147 and NGC 185, themselves separated by 58 arcmin. A few bright blue stars in NGC 185 have also been noted (Baade 1951). Properties of the Andromeda companions have previously been reviewed by van den Bergh (1975) and Hodge (1989).

Relation to Giant Ellipticals

The relation of the low-luminosity dwarf ellipticals to giant ellipticals has been dis-

B. Barbuy and A. Renzini (eds.), The Stellar Populations of Galaxies, 169–179.

cussed extensively in the literature. For example, Wirth and Gallagher (1984) and Kormendy (1985) conclude that the more luminous ellipticals and compact ellipticals (for example, M32) are physically distinct from the diffuse (low-surface-brightness) dwarfs. Alternatively, Sandage *et al.* (1985) have argued that many observed global properties (for example, the luminosity and effective surface brightness) show a continuity between the massive ellipticals and the low-surface-brightness ellipticals. Very recent results on the kinematics of dwarf ellipticals (*e.g.* Bender and Nieto 1990; Bender, Paquet and Nieto 1991; Carter and Sadler 1990; Held, Mould, and de Zeeuw 1990) have found that contrary to earlier expectations (Davies *et al.* 1983), these galaxies are ***not*** rotationally flattened, and are thus they are similar in this respect to the the giant ellipticals.

Color-Magnitude Diagrams

With the availability of CCD detectors, the stellar content of all four Andromeda dE's has been subject to recent investigations. *I* versus *(V-I)* color magnitude diagrams for the four companions are presented in Figure 1.

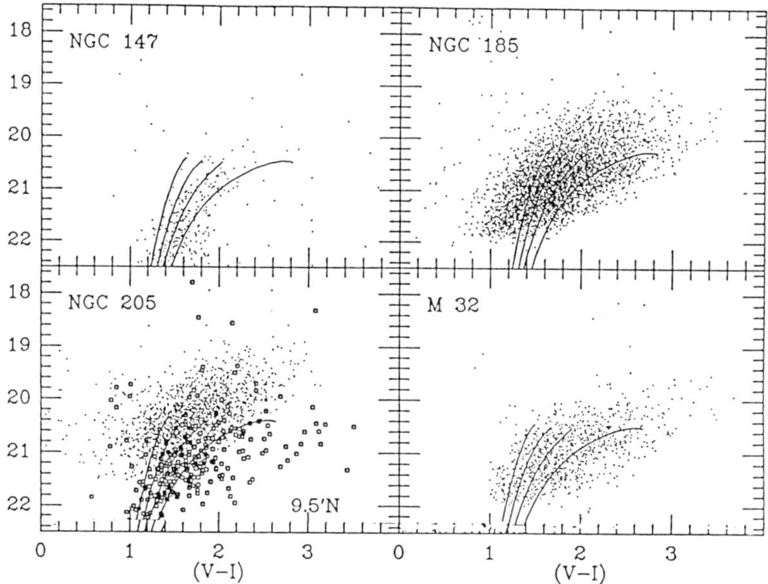

Fig. 1 – *I* versus $(V\text{-}I)_0$ color-magnitude diagrams.

The data for NGC 147 are from Mould, Kristian and da Costa (1983); those for NGC 185 are from Lee, Freedman and Madore (1992a); open circles for NGC 205 are from Mould, Kristian and da Costa (1984) and small dots are from Lee, Freedman and Madore (1992b); finally the data for M32 are from Freedman (1989). Globular cluster sequences for M15, NGC 6752, NGC 1851 and 47 Tuc from Da Costa and Armandroff (1990) have been shifted to the appropriate distance modulus for each galaxy. These sequences correspond to [Fe/H] values of −2.2, −1.6, −1.3, and −0.71, respectively. The distance moduli and the mean metallicity of the giant branch stars and its dispersion are summarized for all four galaxies in Table 1. Lower limits are given in the cases where there is incompleteness in the *V* data. Note that in the case of NGC 205, the data of Lee, Freedman and Madore were obtained in the central regions of NGC 205, and crowding effects result in a much brighter limiting magnitude than for the case of the outer Mould, Kristian and Da Costa field.

TABLE 1

Galaxy	μ_0	<Z> $\pm$ $1\sigma_z$
M32	24.2 $\pm$ 0.3	>-0.7 $\pm$ 0.3
N205	24.3 $\pm$ 0.2	>-0.9 $\pm$ 0.4
N147	24.0 $\pm$ 0.15	-0.9 $\pm$ 0.3
N185	23.9 $\pm$ 0.2	>-1.3 $\pm$ 0.3

Ultraviolet observations

Large elliptical galaxies form a fairly narrow sequence in a plane defined by Mg_2 versus the ultraviolet (1550 – V) color (Burstein *et al.* 1988). M32 lies on the low-luminosity end of that sequence, but NGC 205 has an extremely blue UV color and is displaced from the sequence by about 4 magnitudes. These ultraviolet studies indicate that if elliptical galaxies are undergoing very recent star formation (as is the case for NGC 205), such galaxies may be identified on the basis of their ultraviolet properties. For example, both NGC 205 and NGC 185 have very flat UV spectra shortward of 3000 Angstroms (Buson, Bertola and Burstein 1990). In contrast the UV spectrum of M32 falls rather sharply in the wavelength range from 3000 to 2000 Angstroms.

III. Individual Galaxies

NGC 205

The stellar content of NGC 205 was first discussed by Baade (1951) who noted that the dominant population of this galaxy was made up of red giant stars, while the presence of OB supergiants indicated a very recent epoch of star formation. In addition to very young supergiants, NGC 205 also contains a population of luminous red stars (Gallagher and Mould 1981) and luminous carbon stars (Richer *et al.* 1984), which are likely stars of intermediate age (*i.e.*, several Gyr). From Figure 1 it can be seen that the brightest red giants in NGC 205 are about 0.8 mag brighter than those in M32, for example. The mean metallicity of this system falls between that of M32 and NGC 185, and there is a significant dispersion about this mean (see Table 1). Several prominent dust patches are visible in NGC 205, the neutral hydrogen has been mapped and appears to be distributed in a rotating disk, very near to the dust and blue stars (Johnson and Gottesman 1983). CO has been searched for in NGC 205, but more sensitive measurements are needed for a positive detection (Sage and Wrobel 1989). The origin of the gas in this galaxy is still unclear; the proximity of NGC 205 to M31 leaves open the option that the gas has been stripped from M31. Hodge (1973) noted an isophote twist of about 30 degrees toward M31, most likely due to a tidal interaction with M31 (*e.g.*, see discussion and Fig. 7 of Kormendy 1982). The galaxy has a color gradient in the sense that the inner regions are significantly bluer than the outer regions (*e.g.*, Hodge 1973). As noted in the previous section, the ultraviolet colors of NGC 205 are very blue relative to elliptical galaxies in general. All of these data suggest a complex history of star formation in NGC 205, with star formation continuing to the present day.

NGC 185

The dominant population of resolved stars in this galaxy is again composed of red giant stars, although Baade (1951) noted the presence of a small number of blue supergiants. The luminous red stars in NGC 185 resemble those in NGC 205, the brightest giants in this case being approximately 0.6 mag brighter at I than those in M32. Recently Saha and Hoessel (1990) have detected about 150 RR Lyraes in this galaxy, signalling the presence of a *bona fide* old ($>10^{10}$ year old) population. NGC 185 contains 6 known globular clusters (*e.g.*, Harris 1990). The mean metallicity of the globular clusters has been measured by Da Costa and Mould (1988) who find [Fe/H] = -1.65 ± 0.25, somewhat lower than (but consistent to within the errors) the mean metallicity of -1.3 ± 0.3 for individual giant stars measured by Lee, Freedman and Madore (1992a). The central regions of NGC 185 are also bluer than its outer regions (Hodge 1963; Price 1985). NGC 185 also has two rather conspicuous dust patches, it contains approximately 10^5 $M_\odot$ of both neutral hydrogen (Johnson and Gottesman 1983) and molecular gas (Wiklind and Rydbeck 1986), and it has blue ultraviolet colors. Much like NGC 205, NGC 185 appears to have continued forming stars for an extended period of time. However, NGC 185 is much further away from M31 than NGC 205, and the origin of its interstellar medium remains to be explained.

NGC 147

In contrast, NGC 147 displays no evidence for recent star formation. The galaxy has not been detected in neutral hydrogen, no prominent dust patches are obvious, no blue supergiants are present, and as can be seen from Figure 1, unlike NGC 185, NGC 205 and M32, very few stars are seen above the luminosity defined by those in Galactic globular clusters. From these data Mould, Kristian and Da Costa (1983) place a limit of 10% on the possible contribution from a population of intermediate age. Despite the lack of recent activity, the mean metallicity of giants in NGC 147 is similar to those in NGC 205: -0.9 ± 0.3 according to the calibration of Da Costa and Armandroff (1990). Saha, Hoessel and Mossman (1990) have recently detected an old population of RR Lyraes in this galaxy.

M32

No young blue stars are observed in M32 (although it should be noted that the center of this galaxy cannot be resolved as in the case of NGC 205 or NGC 185). No dust patches are evident. The most prominent characteristic of the color-magnitude diagram is the presence of a population of red giant branch stars, the brightest of which resolve at an I magnitude of about 19.5 mag. Although it is very close to M31, Kent (1987) finds no truncation in its surface brightness profile which might indicate a significant tidal interaction. M32 has the highest surface brightness of the four Andromeda companions, and it is the most metal rich. Freedman (1989) places a lower limit of [M/H] = -0.5 dex (based on the revised Yale isochrones) on the metallicity of the giant stars in a field 2 arcmin south of the M32 nucleus, and studies of nuclear integrated spectra suggest a metallicity of about solar (*e.g.*, O'Connell 1985). M32 is the only Andromeda companion with no globular clusters of its own.

Because of its high surface brightness in comparison to the other Andromeda dE's, the similarity of many of its structural properties relative to giant ellipticals (gE's), and because of its proximity (in contrast to the gE's), M32 has acted as a focal point for stellar population studies. It is everyone's "stellar population standard galaxy" for population synthesis studies. It must be borne in mind that it is not yet established whether M32 is typical of elliptical galaxies in general, or merely the "M32 prototype" (that is, a low luminosity elliptical galaxy which is a companion to a luminous spiral galaxy). However, it is clear that if we do not understand M32, it is doubtful that we can understand the stellar populations in more distant galaxies. For example, current population synthesis techniques applied to M32 *and* giant elliptical galaxies lead to a similar end result (see discussion below).

Integrated Studies

As distinct from spiral galaxies, elliptical galaxies have only trace amounts of gas and dust, and the colors of nearby ellipticals are red and exhibit very little scatter. Moreover, the Hubble diagram (magnitude versus redshift) also has very little scatter. These observations have given rise to the classical picture in which elliptical galaxies are viewed as old systems evolving passively with little or no recent star formation.

Opponents to this classical picture (*e.g.*, O'Connell 1980; Pickles 1985) argue that the integrated spectra of the nuclei of M32 and giant elliptical galaxies show evidence for substantial recent star formation activity, while at the same time showing very little dispersion in abundance. Alternatively, it has been argued (*e.g.*, Renzini 1986; Renzini and Buzzoni 1986; Frogel 1988) that if stellar population models adequately allow for a dispersion in metallicity within ellipticals, an older age follows.

In actual fact, both groups are highlighting the same underlying problem discussed extensively many times before (Renzini and Buzzoni 1986; O'Connell 1986); namely, that from integrated spectra alone, it is not possible to measure ***directly*** the underlying age ***and*** abundance distributions of a composite stellar population. Rather, in practice, a temperature is measured, and disentangling the effects of age and metallicity is non-trivial. Extracting the age and metallicity then requires either (1) ***a priori*** knowledge of the star formation history or abundance distribution, or (2) an assumption about the dispersion in one of the parameters. The absence of knowledge of (1) has resulted in the dichotomy discussed above: two different groups arguing either for a low metallicity dispersion, or alternatively for a small age dispersion. The one robust conclusion to emerge from all these studies is that early-type galaxies are not simple stellar populations with only a single-age, single-metallicity population.

As discussed elsewhere in this volume by E. Bica (and references therein), and of interest in the context of M32, are recent population synthesis studies which make use of star clusters as templates. The abundance distribution inferred from these models for M32 agrees with the inferred abundance spread (based on the observed color spread) of giant stars by Freedman (1989). At the same time, in agreement with previous studies of the integrated nuclear spectrum of M32, they argue for a presence of an intermediate age component, although somewhat less substantial than concluded in earlier studies. However, as noted by Schmidt *et al.* (1991), detailed simulations and tests of their method indicate that one of the most difficult cases to model are those where a model galaxy evolves to solar metallicity "*in the presence of old and intermediate-age populations*" [*i.e.*, precisely the type of model believed to apply to M32]. Even when constraints (evolutionary scenarios) are imposed on the models (Bica, Alloin and Schmidt 1990), their two best solutions yield very different relative contributions of the intermediate-age component of 30% and 70% respectively. The final adopted solutions are still constrained more by assumptions than by the input data.

Given the notorious difficulty in putting together a complete library (containing stars and/or clusters representative in age and metallicity of stars in elliptical galaxies), and given the non-uniqueness of population synthesis, it is perhaps not surprising that the results of such studies are still very controversial, and that the issue of the age spread within M32 and other elliptical galaxies is not yet resolved.

Color Gradients

In the case of M32, spectral synthesis studies to date have been confined to the nucleus of the galaxy, whereas crowding and confusion effects have limited color-magnitude diagrams to fields off the nucleus. The lack of positional overlap in the two types of studies leads to the question of whether the stellar populations being sampled in the inner and outer areas might be systematically different. One possible manifestation of differences in stellar population might be a color gradient as a function of radius.

An early UBV photoelectric study of an on- and off-nuclear region in M32 by Sandage, Becklin and Neugebauer ([SBN], 1969) suggested that a color gradient might be present across M32. However, a later compilation of a more extensive set of data (including those of SBN) covering 100 arcsec in extent, failed to reveal a significant gradient in *(U-B)* or (B−V) (Sharov and Lyuti (SL) 1983, 1988 and references therein). Readers can judge for themselves whether Freedman (1989) has misquoted these results as claimed by O'Connell (1990). New results obtained at high resolution (Michard and Nieto [MN] 1991; Lauer, private communication) find no significant gradient even in the innermost regions. In Figure 2, *UBVI* data obtained at the Palomar 1.5m in August 1990 also show no evidence for a significant color gradient. In addition Peletier (1991, private communication) has completed both an optical and near-infrared study of M32 and again finds no significant gradients.

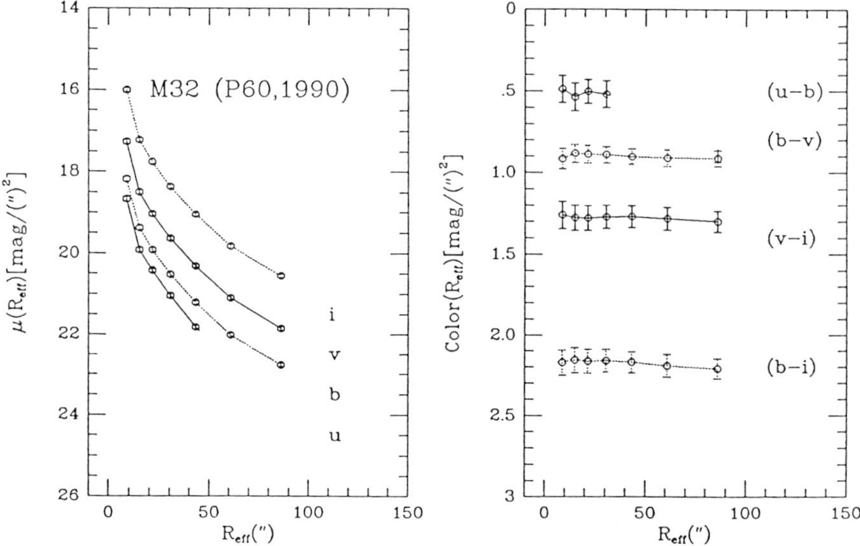

Fig. 2 – Surface photometry and color gradients in M32 based on Palomar 1.5m data. The calibration of the UBV photometry is from SL and MN. The *I* data are uncalibrated.

An interesting (and as yet unexplained) new observation reported by O'Connell (this conference) is a large ultraviolet color gradient in M32. The ultraviolet color gets bluer with increasing radius (which is contrary to the sense of a decreasing metallicity gradient) and is in the opposite sense to that observed in M31, M81 and NGC 1399.

Spectral Gradients

In his extensive population study, Rose (1985) found no evidence for spectral line gradients in M32. Both Cohen (1979) and Davidge (1991) found absorption features which weaken with radius; however, in detail the agreement of individual features is very poor. Faber (this conference) reports that no significant spectral gradients are present in M32.

Direct Studies of Individual Stars in M32

Freedman (1989) resolved individual giant stars in M32 and found (1) evidence for a spread in metallicity of the giants, and (2) a population of stars more luminous than the tip of the first red giant branch. A lower limit to the mean metallicity from the mean (V−I) color is −0.5 dex with a 2σ spread of 0.6 dex. As discussed by Freedman (1989) this dispersion is larger than can be accounted for by photometric error alone. Furthermore, the observed spread is a lower limit only, since the *V*-band frame does not extend faint enough to measure extremely red and metal-rich stars, if they are indeed present in M32. These results appear to rule out population synthesis models which assume a single metallicity population. Spectroscopic followup for stars in M32 is planned.

As can be seen in Table 1, this inferred abundance spread is found in all four of the Andromeda dE's. It is interesting to note that abundance spreads of similar magnitude were predicted by Searle (1979) for the halos of giant galaxies in addition to dwarf galaxies. His prediction (σ = 0.45 dex [FWHM=1.05]) was based on simple considerations of mass loss in a stellar system undergoing chemical enrichment.

In order to study this population of red giant stars in more detail, new near-infrared data were obtained at the Palomar 5m. A region in M32 covering 90" x 60" was mosaiced in *JHK*. An interesting new result to emerge from this study is that stars up to 2 bolometric magnitudes above the tip of the first red giant branch are present in M32. Many of these stars are not present in Figure 1 because they are so red that they were simply undetected on the *V* CCD frame. These stars are illustrated in a K_0 versus $(J-K)_0$ color magnitude diagram (Figure 3). Positions of the giant branch loci for several Galactic globular clusters are drawn to illustrate the extent to which the luminosities of the giant stars in M32 exceed those of first red giant branch stars. A comparison field taken at the same isophotal level in M31, but removed from the body of M32 indicates that the bulk of this population does indeed belong to M32 itself, and is not simply contamination from the disk of M31.

The presence of such bright red stars in consistent with the recent 2μm spectroscopy of Davidge (1990). Davidge concludes that a component more luminous and cooler than the first red giant branch is needed to fit the near-infrared spectrum of the M32 nucleus.

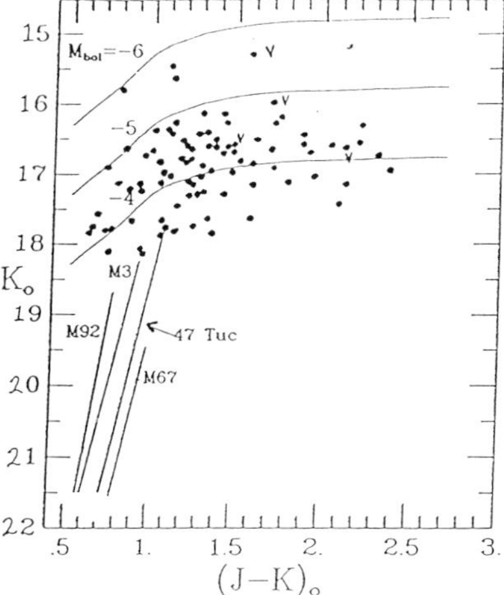

Fig. 3 – K_0 versus *(J-K)*$_0$ color-magnitude diagram for M32. The giant branch loci are from Cohen, Frogel, and Persson (1978). For the purposes of this plot, distance moduli of 14.2 and 24.2 mag have been assumed for the Galactic Bulge and M32, respectively.

The brightest stars in M32 have $M_{bol} = -5.5$ mag. In contrast, the brightest stars in the Galactic Bulge have a sharp cutoff at $M_{bol} = -4.2$ mag (Frogel and Whitford 1987). The sample of stars observed in the near-infrared by Frogel and Whitford were selected on the basis of the objective-prism survey carried out in the Bulge by Blanco, McCarthy and Blanco ([BMB], 1984). It is entirely possible that the bolometrically most luminous stars found at K in M32 would have been missed in the optical BMB survey.

The existence of a population of bright stars in the bulge of M31 similar to those in M32 has been reported by Rich (this conference) and Rich and Mould (1991), but Davies, Frogel and Terndrup (1991) suggest that these stars result from contamination by the disk of M31. However, the latter study assumes the maximum disk contribution consistent with the photometry of Kent (1987). Further studies already in progress by these groups should soon resolve this issue.

There are several possible explanations for this newly-discovered bright population in M32: If the stars belong to an *old* population then:

a) the stars may be similar to the most luminous long-period variable stars (LPVs) observed in Galactic globular clusters. Frogel and Elias (1988) have shown that a population of long-period variable stars having bolometric magnitudes in excess of the core-helium-flash luminosity exists in the most metal-rich clusters. A population of this type could be present in M32 since its mean metallicity is comparable to, or slightly greater than, the most metal-rich Galactic globular clusters. Such stars are also observed in the Galactic Bulge. Frogel *et al.* (1990) find that the mean bolometric luminosity of LPVs in the Bulge and Galactic globular clusters is –4.2 mag in both cases.
b) as suggested by Renzini and Greggio (1990) the stars may be a result of a population of binary stars which have merged to produce stars with sufficient mass and fuel in their envelopes to achieve higher luminosities as they evolve up the asymptotic giant branch.
c) the stars belong to a super-metal-rich tail of the metallicity distribution of M32. For example, Frogel and Whitford (1987) have argued that the most luminous stars in the Galactic Bulge may be explained as a result of the fact that the main-sequence lifetimes of metal-rich stars are longer (*e.g.*, VandenBerg and Laskerides 1987). Thus stars with a higher metallicity enter the giant phase at a later age. Old, super-metal-rich stars may therefore have slightly higher-mass progenitors than those of metal-poor giants, and again have additional fuel which allows them to evolve up the asymptotic giant branch at a higher luminosity.

Alternatively, d) the stars may be evidence for a *younger* population: that is an extended asymptotic giant branch population as observed in Searle-Wilkinson-Bagnuolo (1980) clusters of intermediate age in the Magellanic Clouds (Frogel, Mould and Blanco 1990).

Can the observed population of bolometrically luminous stars in M32 then be simply explained by a population of old stars similar to those observed in the Galactic globulars? This question can be addressed by the near-infrared and optical data already obtained, and the answer is a very clear no. For 3 years I have been monitoring the M32 giants for variability. A search for variable star candidates has been successful (there are about 350 candidates based on 4 epochs of data). Several of the brightest variable candidates are indicated in Figure 3. Given the areal coverage of the search, the numbers of variable candidates in M32 are consistent with that predicted by scaling the numbers of long-period variables observed in Galactic globular clusters to the luminosity of M32. However, the fraction of variable star candidates that have been detected at K is *less than 10% of the total.* That is, most of the stars which have been observed in the near-IR are, in fact, non-variable. These numbers are consistent with the conclusion of Impey *et al.* (1986) that <15% of the 1-2μm radiation comes from asymptotic giant branch stars. More stars with such cool photospheres would result in a larger observed dispersion in the near-infrared colors.

The numbers of stars also appear to be larger than can be accounted for by a population of binaries (Renzini, private communication). Possibility c) (very high metallicity) is unlikely since the luminosity of stars in M32 exceed those observed in the Galactic Bulge by 1 magnitude. In this case, a population of stars must exist in M32 with a metallicity greater than that observed in the Galactic Bulge. The mean metallicity of giants in the Galactic Bulge is about twice solar (Rich 1988), significantly larger than the mean metallicity of M32 stars.

Spectra of the bolometrically most luminous stars in M32 are being obtained in an effort to distinguish between alternatives c) and d). In addition, it will be interesting to determine the period distribution of the long-period variable stars (LPVs) for a comparison with the Galactic Bulge. The period distribution of the Galactic Bulge stars peaks at about 400 days (Whitelock, Feast and Catchpole 1991), and contrary to earlier reports (lacking direct period determinations), there are no LPV's in the Bulge with periods of 1500 days. It had been argued (*e.g.,* Harmon and Gilmore 1988) that the presence of >1000 day long-period variables was a signature of a population of a few billion years in the Bulge. The period distribution of LPVs in M32 will provide a further constraint on the age distribution of stars in this galaxy.

Undoubtedly some of the brightest stars in M32 belong to an old population resembling the long-period variables in Galactic globular clusters. However, most of these luminous red stars are perhaps simplest understood if they belong to a population of intermediate age (5-10) Gyr stars. The existence of an age spread for stars in M32 of course cannot be used to infer the existence of an age spread for more distant ellipticals. But given that population synthesis techniques applied to elliptical galaxies yield results similar to those for M32, these results underscore the difficulty and complexity inherent in the interpretation of the integrated spectra of both nearby and high-redshift ellipticals.

I thank Barry Madore, Alvio Renzini, Jay Frogel, Bob O'Connell, and Myung Gyoon Lee for interesting discussions about many of the issues raised in this review, and Leonard Searle and Allan Sandage for comments on an earlier draft of this paper. Particular thanks go to Myung Gyoon Lee for allowing me to use his modified version of DAOPHOT for surface photometry, for instruction in its use, and for producing Figures 1 and 2. Many thanks also to Lori Clampitt for her help in identifying long-period variable candidates in M32. I have profitted greatly from the use of the NASA/IPAC extragalactic database (NED). Finally, I thank the AAS for their travel support to this meeting.

References

Baade, W. 1944, *Ap. J.*, **100**, 79.
Baade, W. 1951, *Publ. Obs. Univ. of Michigan*, No. 10, 7.
Bender, R. and Nieto, J.-L. 1990, *Astr. Ap.*, **239**, 97.
Bender, R., Paquet, A. and Nieto, J.-L. 1991, *Astr. Ap.*, **246**, 349.
Bica, E., Alloin,D. and Schmidt, A. A. 1990, *Astr. Ap.*, **228**, 23.
Blanco, V. M., McCarthy, M. F., and Blanco, B. M. 1980 *Ap. J.*, **242**, 938.
Burstein, D. Bertola, F., Buson, L. M., Faber, S. M., and Lauer, T. R. 1988, *Ap. J.*, **328**, 440.
Buson, L. M., Bertola, F., and Burstein, D. 1990, in *Windows on Galaxies*, eds. G. Fabbiano *et al.*, (Netherlands: Kluwer), p. 51
Carter, D., and Sadler, E. M. 1990, *M. N. R. A. S.*, **245**, 12p.
Cohen, J. 1979, *Ap. J.*, **228**, 405.
Cohen, J. G., Frogel, J. A., and Persson, S. E. *Ap. J.*, **222**, 165.

Da Costa, G. S. and Armandroff, T. E. 1990, *A. J.*, **100**, 162.
Da Costa, G. S. and Mould, J. R. 1988, *Ap. J.*, **334**, 159.
Davidge, T. J. 1990, *A. J.*, **99**, 561.
Davidge, T. J. 1991, *A. J.*, **101**, 884.
Davies, R. L., Frogel, J. A., and Terndrup, D. M., 1991 *A. J.*, in press.
Davies, R. L., Efstathiou, G., Fall, S. M., Illingworth, G. D., Schechter, P. 1983 *Ap. J.*, **266**, 41.
Freedman, W. L. 1989, *A. J.*, **98**, 1285.
Frogel, J. A. 1988, *Ann. Rev. Astr. Ap.*, **26**, 51.
Frogel, J. A., and Elias, J. H. 1988, *Ap. J.*, **324**, 823.
Frogel, J. A., Mould, J. R., and Blanco, V. M. 1990, *Ap. J.*, **352**, 96.
Frogel, J. A., and Whitford, A. E. 1987, *Ap. J.*, **320**, 199.
Frogel, J. A., Terndrup, D. M., Blanco, V. M., and Whitford, A. E. 1990, *Ap. J.*, **353**, 494.
Gallagher, J. S. and Mould, J. R. 1981, *Ap. J. (Letters)*, **244**, L3.
Harmon, R., and Gilmore, G. 1989, *M. N. R. A. S.*, **235**, 1025.
Harris 1991, *Ann. Rev. Astr. Ap.*, **29**, 543.
Held, E. V., Mould, J. R., and de Zeeuw, P. T. 1990 *A. J.*, **100**, 415.
Hodge, P. W. 1963, *A. J.*, **68**, 691.
Hodge, P. W. 1973, *Ap. J.*, **182**, 671.
Hodge, P. W. 1989, *Ann. Rev. Astr. Ap.*, **27**, 139.
Impey, C. D., Wynn-Williams, C. G., and Becklin, E. E. 1986, *Ap. J.*, **309**, 572.
Johnson, D. W. and Gottesman, S. T. 1983, *Ap. J.*, **275**, 549.
Kent, S. 1987, *A. J.*, **94**, 306.
Kormendy, J. 1982, in *Morphology and Dynamics of Galaxies*, eds. L. Martinet, and M. Mayor, (Geneva: Geneva Observatory), p. 115
Kormendy, J. 1985, *Ap. J. (Letters)*, **292**, L9.
Lee, M. G., Freedman, W. L. and Madore, B. F. 1992a, *A. J.*, in preparation .
Lee, M. G., Freedman, W. L. and Madore, B. F. 1992b, *A. J.*, in preparation .
Michard, R. and Nieto, J.-L. 1991, *Astr. Ap.*, **243**, L17.
Mould, J. R., Kristian, J., and da Costa, G. S. 1983, *Ap. J.*, **270**, 471.
Mould, J. R., Kristian, J., and da Costa, G. S. 1984, *Ap. J.*, **278**, 581.
O'Connell, R. W. 1980, *Ap. J.*, **236**, 430.
O'Connell, R. W. 1986, in *Spectral Evolution of Galaxies*, ed. C. Chiosi and A. Renzini, (Dordrecht: Reidel), p. 321.
O'Connell, R. W. 1990, in *Bulges of Galaxies*, eds. B. J. Jarvis and D. M. Terndrup, (Garching: ESO), p. 187.
Pickles, A. J. 1985, *Ap. J. Suppl.*, **59**, 33.
Price, J. S. 1985, *Ap. J.*, **297**, 652.
Renzini, A. and Buzzoni, A. 1986, in *Spectral Evolution of Galaxies*, ed. C. Chiosi and A. Renzini, (Dordrecht: Reidel), p. 135.
Renzini, A. 1986, in *Stellar Populations*, ed. C. A. Norman, A. Renzini, and M. Tosi, (Cambridge: Cambridge University Press), p. 213.
Renzini and Greggio 1990, in *Bulges of Galaxies*, eds. B. J. Jarvis and D. M. Terndrup, (Garching: ESO), p. 47.
Rich, R. M. 1988, *A. J.*, **95**, 828.
Rich, R. M. and Mould, J. R. 1991, *A. J.*, **101**, 1286.
Richer *et al.* 1984, *Ap. J.*, **287**, 138.
Rose 1985, *A. J.*, **90**, 1927.
Sage, L. J., and Wrobel, J. M. 1989, *Ap. J.*, **344**, 204.
Saha, A., and Hoessel, J. G. 1990, *A. J.*, **99**, 97.
Saha, A., Hoessel, J. G. and Mossman, A. E. 1990, *A. J.*, **100**, 108.

Sandage, A., Binggeli, B., and Tammann, G. A. 1985, *A. J.*, **90**, 1759.
Sandage, A. R., Becklin, E. E. and Neugebauer, G., 1969 *Ap. J.*, **157**, 55.
Schmidt, A. A., Copetti, M. V. F., Alloin, D. and Jablonka, P. 1991, *M. N. R. A. S.*, **249**, 766.
Searle, L. 1979, "Les Elements et leurs Isotopes Dans l'Univers",(Liege: Institut d'Astrophysique, p. 437.
Searle, L., Wilkinson, A. and Bagnuolo, W. 1980 *A. J.*, **239**, 803.
Sharov, A. S. and Lyuti, V. M. 1983, *Sov. Astron.*, **27**, 1.
Sharov, A. S. and Lyuti, V. M. 1988, *Sov. Astron.*, **65**, 469.
VandenBerg, D. A. and Laskarides, P. G. 1987, *Ap. J. Suppl.*, **64**, 103.
van den Bergh, S. 1975, *Ann. Rev. Astr. Ap.*, **13**, 217.
Whitelock, P., Feast, M., and Catchpole, R. 1991, *M. N. R. A. S.*, **248**, 276.
Wiklind, T. and Rydbeck, G. 1986, *Astr. Ap.*, **164**, L22.
Wirth, A. and Gallagher, J. S. 1984, *Ap. J.*, **282**, 85.

P. te Lintel Hekkart: Did IRAS detect M32, and was the FIR flux consistent with the number of LPVs found?

W. Freedman: Yes, M32 was detected by IRAS, as were NGC 147 and NGC 185 (*e.g.,* Jura *et al.* 1987, *Ap. J.*,**213**,L11). The latter two galaxies were found to have anomalously low 12μm emission relative to other ellipticals. M32 was also measured at 2 and 10 μm by Impey *et al.* (1986) who found that the 10 μm emission cannot be accounted for by photospheric emission from giant stars alone. Both the 2 and 10 μm emission in M32 have the same spatial distribution as the stars. As mentioned in the talk, this flux is consistent with the emission from LPVs as found for example in the Galactic Bulge. And the upper limit from the Impey *et al.* study of a 15% contribution from luminous LPVs is consistent with the fraction of LPVs found in M32.

G. Da Costa: Can you tell whether or not the very young stars in the center of NGC 205 come from a single burst of short duration or from a more extended or continuous star formation?

W. Freedman: I have not undertaken any modelling of the type discussed by Monica Tosi (this volume). But my suspicion is that ultimately there will still be a problem of non-uniqueness; that is several models would be consistent with the available data.

A. Renzini: I've two questions: 1) on the basis of the metallicity distributions inferred from the RGB color distribution, did you try to infer an average turnoff color for M32, assuming it is as old as globulars? And 2) did you try to estimate the frequency *per unit sampled luminosity* of the stars brighter than, say, M_{bol} = -4.5 respectively in the Bulge and in M32?

W. Freedman: In answer to your first question, no I myself have not inferred an average turnoff color, but as you yourself have stressed many times, an older turnoff age will result when a metallicity spread is taken into account. But the difficulty of course is measuring the true underlying distributions in *both* age and metallicity. The existence of a metallicity spread indicates that models with a small adopted abundance dispersion are in error, but it cannot be used to *rule out* the possibility that the population may also have a spread in age. The answer to your second question is yes, I calculated the fraction of the total light contributed by these bright giants, and it amounts to approximately 10%.

INTERMEDIATE AGE POPULATIONS

J.R. MOULD
Palomar Observatory, California Institute of Technology
Pasadena, California 91125, U.S.A.

ABSTRACT. Our understanding of asymptotic giant branch evolution has reached the point where the timescales are well enough known that realistic luminosity functions can be computed. The Mira and OH-IR star phases of stellar evolution truncate the AGB well before Reimers' Law mass loss exhausts the envelope. These luminosity functions allow us to compute the luminosity variance, which may be sensitive to the presence of intermediate age populations.

1. Introduction

In a distant stellar system an intermediate age population manifests itself most clearly through its asymptotic giant branch properties. Globular clusters have an AGB which stops at or around the luminosity of the red giant branch tip. This termination of the AGB is due to mass loss. Very young populations don't have an AGB at all, because $9M_{\odot}$ (or $5M_{\odot}$ models with convective overshooting) ignite their carbon cores non-degenerately. But between the limits of low mass stars and massive stars, the brightest stars in stellar populations are AGB stars. Here I plan to discuss progress in understanding the AGB in the last several years. Iben and Renzini (1983) remains the baseline review article in this area.

2. AGB Luminosity Functions in Magellanic Cloud Clusters

A census of stars in Cloud clusters has been published by Frogel, Mould and Blanco (1990), building on earlier work by Aaronson and Mould (1985) and Lloyd-Evans (1980). Figure 1 shows the luminosity functions accumulated in Searle, Wilkinson and Bagnuolo (1980) type. Type is a measure of age. Note the following things in Figure 1.

1. The AGB tip increases in luminosity in younger clusters, but does not reach the high luminosities anticipated with Reimers' (1975) mass loss rate with scaling parameter $\eta = 1/3$ (see Figure 7 of Iben and Renzini for those predictions).

2. The luminosity at which thermal pulses begin, as indicated by the presence of carbon and S stars, increases in luminosity in younger clusters. The parameterization by Lattanzio (1991) predicts this trend satisfactorily, and is used here in preference to the interpolation in Figure 7 of Iben and Renzini.

3. The following simple estimates of lifetime on the AGB give approximate (factor of two) agreement with the numbers of AGB stars per unit luminosity of the population in Figure 1. For thermal pulsing AGB stars: 1.3 Myrs/mag (Renzini 1978). For early

B. Barbuy and A. Renzini (eds.), The Stellar Populations of Galaxies, 181–190.

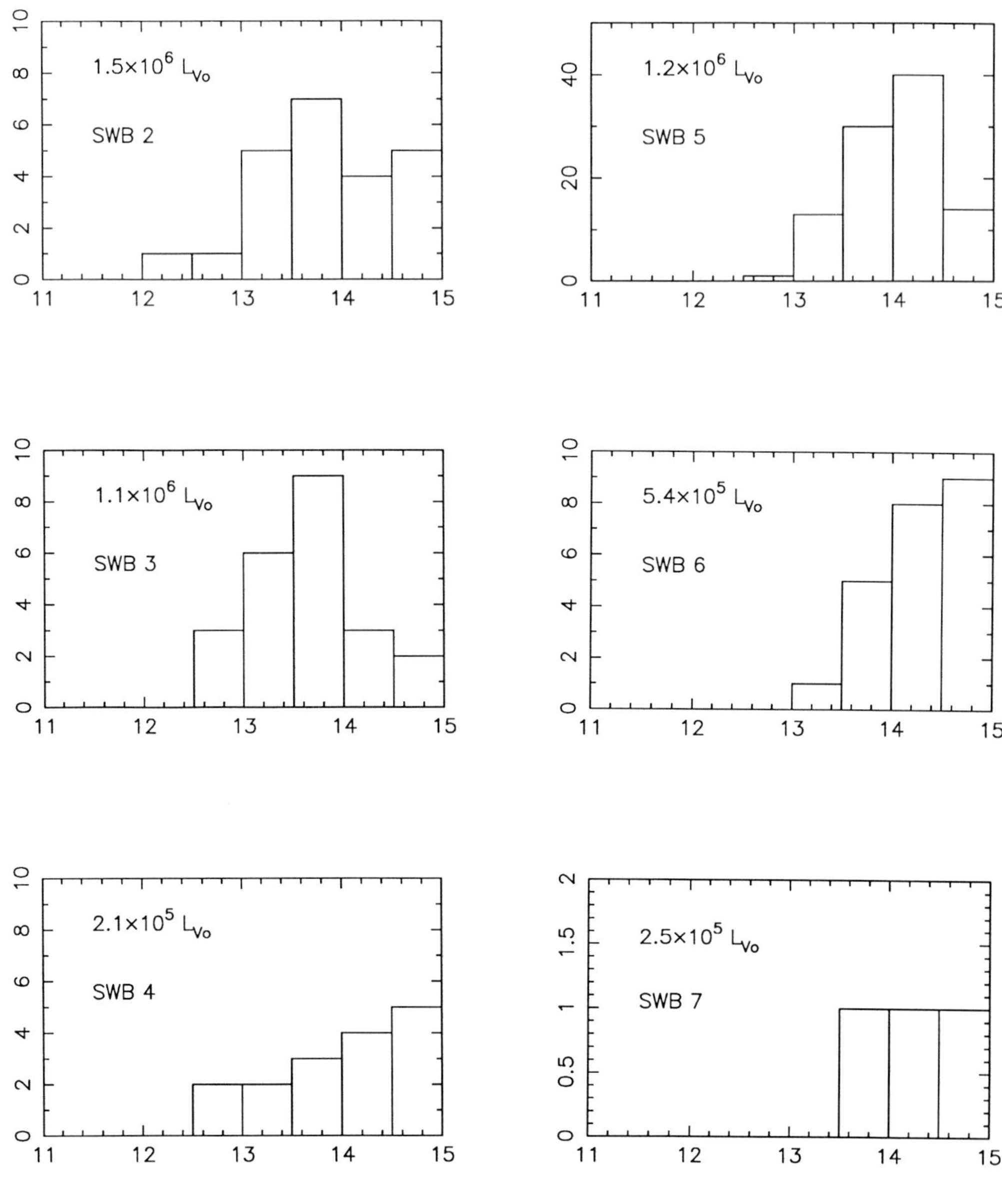

Figure 1. AGB luminosity functions for accumulated Magellanic Cloud clusters of different SWB type. The horizontal axis is apparent bolometric magnitude in the LMC. The total visual luminosity of the clusters in each panel is indicated. Source: Frogel *et al.* (1990).

AGB stars (before thermal pulses begin): 2.1 Myrs/mag (estimated here from the results of Lattanzio 1986).

3. The AGB Luminosity Function in Reid's LMC Field

Reid and Mould (1984) published an AGB luminosity function for a 15 square degree region of the Large Magellanic Cloud by selecting stars from Schmidt plates with V–I > 1.6 and $m_{bol} < 15$. This is shown in the upper part of Figure 2. There is some incompleteness at the faint end of this luminosity function, and some contamination by core helium burning massive stars (which are presumably much rarer than the 1–$2M_{\odot}$ stars dominating this distribution). The lower part of Figure 2 shows the Long Period Variable stars (LPVs) in this field studied by Reid, Glass, and Catchpole (1988). The factor of ten scale change between the top and bottom of Figure 2 emphasizes the small fraction of AGB stars that are Miras, and the correspondingly short lifetimes of AGB stars once they have become pulsationally unstable and entered the Mira phase. Still rarer are the "cocoon stars" (Reid 1991), optically unidentified IRAS sources in the same field. There is undoubtedly some incompleteness at the faint end of the cocoon star luminosity function. These luminosity functions are consistent with the notion that, once an AGB star's envelope has become unstable, its evolution is no longer controlled by burning on the inside, but rather by mass loss on the outside. Compared with the 10^6 year lifetimes associated with less advanced AGB stars, 10^5 year lifetimes should be associated with intermediate age Miras, and 10^4 years with cocoon stars.

Population synthesis models that include these evolutionary stages have recently been presented by Charlot and Bruzual (1990), based on a semi-empirical study of OH-IR stars by Bedijn (1988). These models predict a smaller AGB contribution to the total light of a simple stellar population of age 10^8 years, than did the Renzini and Buzzoni (1986) models based on Reimers' mass loss rates. Furthermore, they have been shown to fit the observations of Frogel, Mould and Blanco in this respect.

Simple population synthesis models can incorporate AGB evolution by including the following in their summation over all evolutionary phases:

1. Lattanzio's (1991) expressions for the core mass at the first thermal pulse,
2. Weidemann's (1987) initial-final mass relation,
3. the core-mass / luminosity relation of Lattanzio (1986) or Boothroyd and Sackmann (1988),
4. lifetimes of 1.3 Myr/mag (TPAGB) and 2.1 Myr/mag (E-AGB).

As an illustration, consider 7 epochs of star formation in the LMC, each a simple stellar population, corresponding to SWB types I–VII. If we employ the physical parameters (ages and metallicities) given for these populations by Frogel, Mould and Blanco, together with equation (14) of Renzini and Buzzoni (1986), and if we suppose the LMC has a star formation history such that there is 10^6 $L_{V\odot}$ in each of the 7 populations, we obtain a luminosity function shown in the top part of Figure 3. If we want to emphasize recent star formation and the peak in the star formation history ~2 Gyr ago, which almost everyone seems to see in the LMC, we can double the type I contribution and quadruple the type V contribution. We then obtain the luminosity function shown in the bottom part of Figure 3. The latter is approaching the shape of the observed luminosity function in Figure 2.

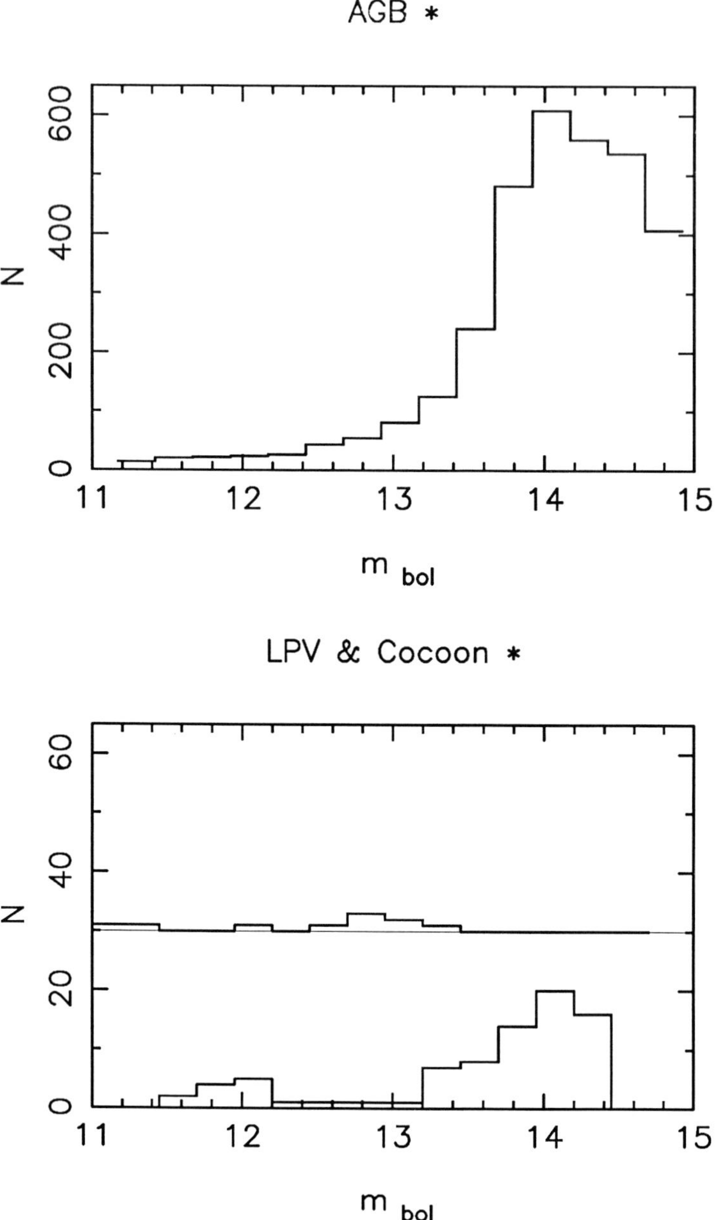

Figure 2. (above) Colour selected AGB luminosity function for Reid's LMC field. (below) LPV and "cocoon star" luminosity functions in this field. The IRAS sources are in the top part of the lower panel; the much more numerous LPVs in the bottom part.

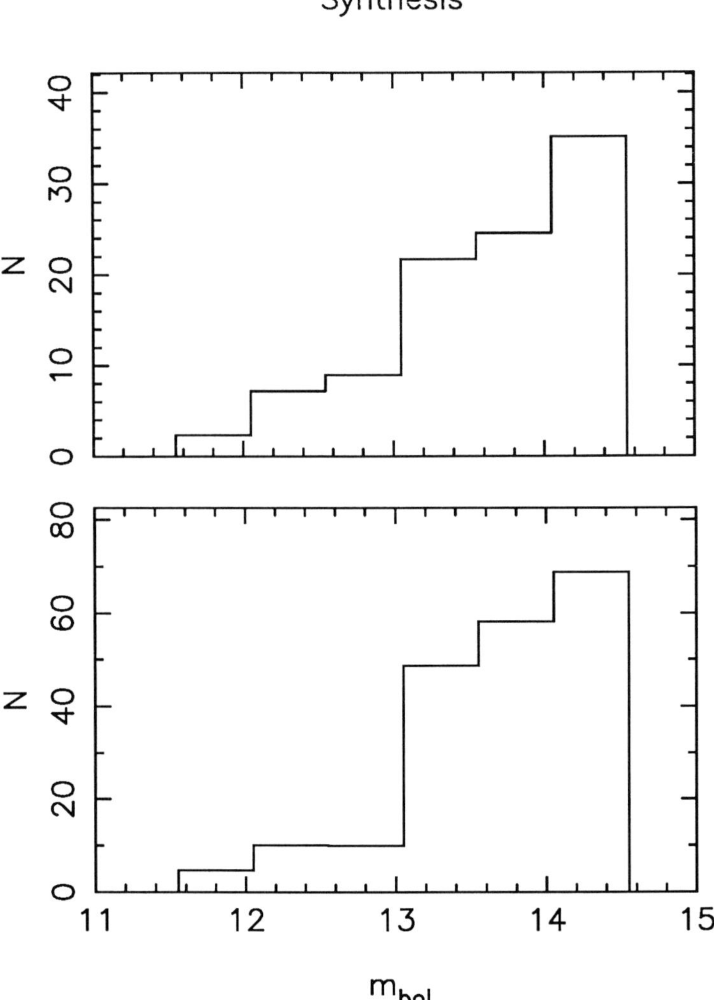

Figure 3. Synthesized AGB luminosity functions for the LMC. Top: assuming a luminosity of 10^6 $L_{V\odot}$ for all SWB types. Bottom: with enhanced type I and V populations.

4. Intermediate Age Populations in Ellipticals ?

Our desire to understand the stellar population of elliptical galaxies has tended to have a cosmological motivation. The traditional substitutes for inaccessible gE galaxies have been the bulge of M31 and its dE companion, M32. At this meeting R.M. Rich and W. Freedman have shown fascinating new evidence that both these objects have unexpectedly bright (M_{bol} = −5) AGB stars. Some of these stars are likely to be long period variables, and some are so red that measurement of their light curves will be required to establish $< M_{bol} >$. In respect of the bulge of M31 Rich and Mould (1991) considered four possible sources of these AGB stars, and it is worthwhile to review them here:

1. contamination by the disk of M31. Separation of the two components is generally an imprecise exercise in surface photometry, and M31 is no exception. Our best estimates were that the disk of M31 could account for between 17 and 56% of the observed bright AGB stars. But this fraction was not so well determined that we could confidently rule out the hypothesis that all the bright AGB stars actually belong to the disk. Study of fields closer to the center of M31 is feasible, and should resolve this problem, possibly soon. Presumably, the contamination hypothesis does not explain the observation of bright AGB stars in M32.

2. The bright AGB stars are super-metal-rich stars. Metal rich globular clusters have red horizontal branches and thus relatively thick post core helium burning envelopes, which can carry them to $< M_{bol} > = -4.5$ on the AGB (see Frogel and Elias 1983, 1988). Compare this with −3.6 for metal poor globulars. So, if we extrapolate this trend, what should we expect ? The answer is unclear, particularly since super-metal-rich compositions are now thought to exhibit blue horizontal branches (Pinsonneault *et al.* , this volume). Super-metal-rich composition is less likely to explain the bright AGB stars in M32, although Bica *et al.* (this volume) argue that the upper limit of the metallicity distribution in the M32 remains undetected.

3. The bright AGB stars are a 2–5 Gyr population. In the case of M32 this is exactly what population synthesis has been saying for years (see O'Connell 1986). Figure 1 would suggest approximately 4 AGB stars per magnitude at m_{bol} = 19.5 for a simple SWB type V population of 10^5 $L_{V\odot}$

4. Bright AGB stars result from stellar mergers. There are, of course, no bright AGB stars in globular clusters, even where blue stragglers are present, but this is probably just a result of small number statistics and the ratio of lifetimes of these phases (at least 10^3). Dwarf spheroidal galaxies may permit a better test of this hypothesis, and in the case of Carina it seems that an intermediate age population is required to explain simultaneously the bright AGB and the main sequence luminosity function (Mighell 1990). The dynamical conditions in globular clusters dwarf spheroidals, bulges, and ellipticals are very different, and it is quite a challenge to any theory of stellar mergers to populate the HR diagram appropriately and retain some correspondence with observations.

5. The effect of bright AGB stars on the luminosity variance statistic

If the traditional motivation for the study of old stellar populations is cosmological, a new twist has appeared recently in the discovery by Tonry and Schneider (1988) that a statistic of the luminosity function is measureable in a partially resolved population. That statistic is the variance of the luminosity function, and it is observable, when other noise sources

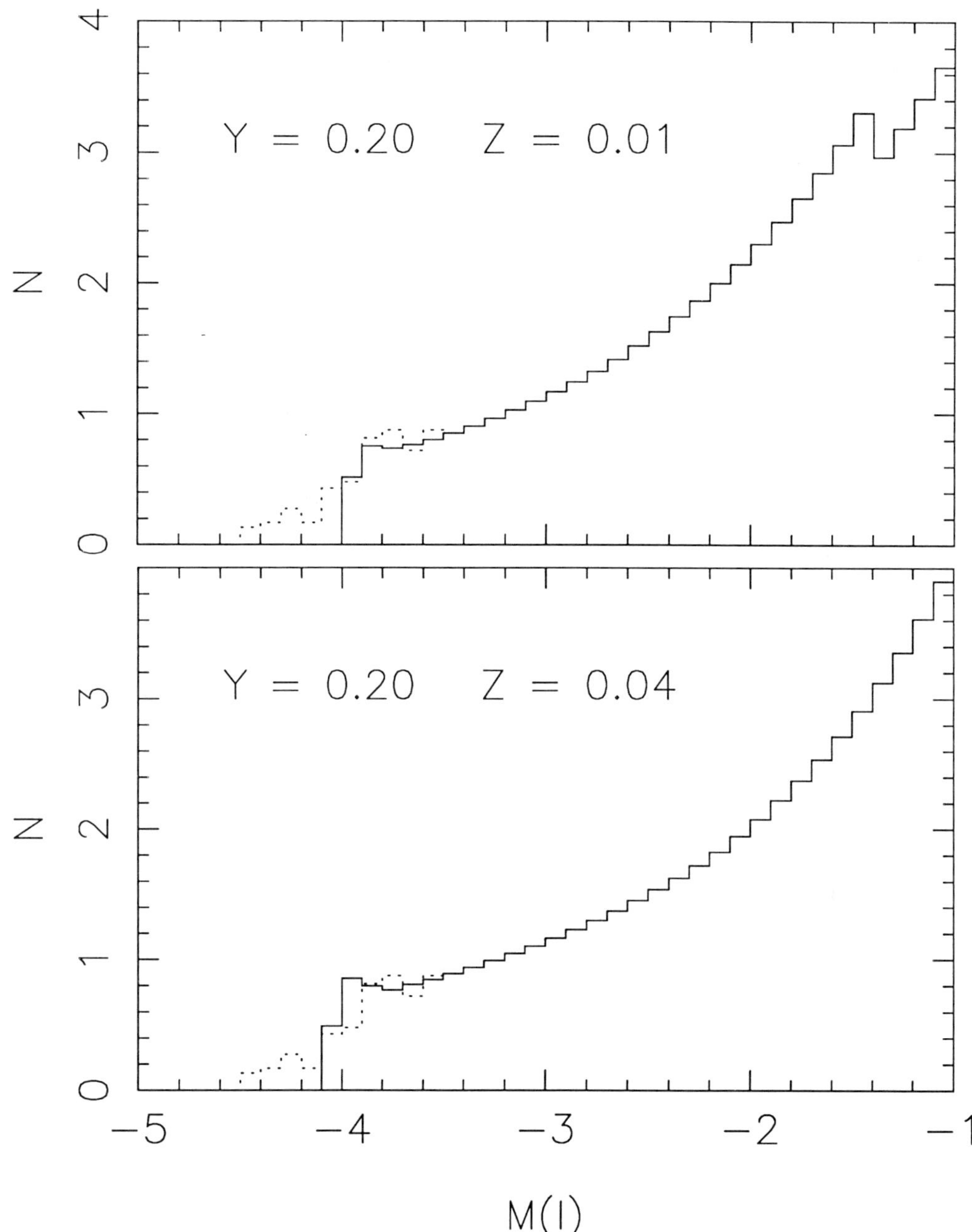

Figure 4. Revised Yale Luminosity Functions for $M_I < -1$ (solid lines). Observed M32 luminosity function (Freedman 1989) for $M_I < -3.5$ (dashed line).

are appropriately small and distinct, as the variation in the number of stars per pixel in the sample that one draws when imaging a stellar population. The luminosity variance, $\overline{m}_I$, is a very tight function of the color of elliptical galaxies (Tonry 1991), which may appear as something of an embarrassment for the theory of stellar populations, as the sense of the color dependence is the opposite of the prediction (Tonry, Ajhar and Luppino 1990), if metallicity variation is the source of the color range. Fortunately, calculations confirm Tonry's suggestion that the origin of this discrepancy is bolometric corrections to cool stars. If we substitute the quadratic relation of Bessell and Wood (1984) between BC_I and V–I for the linear relation adopted in the Revised Yale Isochrones and Luminosity Functions (Green, Demarque and King 1987), we obtain the results shown in the first 3 rows of Table 1: super-metal-rich stellar populations of age 17 Gyr are indeed expected to have a lower luminosity variance at I than their less metal rich counterparts. The observed difference in $\overline{m}_I$ between M31 and M32 is 0.35 ± 0.08 mag, a large fraction of the 0.42 mag difference between predictions for Z = 0.04 and Z = 0.01 in row (3) of Table 1. I must caution that these calculations should not be used to calibrate $\overline{m}_I$, as Wood and Bessell's bolometric corrections are only valid for the average composition of the stars they observed. Moreover, the treatment of the post giant branch evolution adopted here is rudimentary at best. A helium abundance Y = 0.2 and Salpeter initial mass function were employed. Mean Kron-Cousins V–I colors of the model populations are given in the second part of the table.

Table 1: Calculated Luminosity Variance

$\overline{m}_I$	Z = 0.01	Z = 0.04	
Yale	–1.85	–2.10	(1)
Yale + AGB	–2.07	–2.26	(2)
Modified BC	–1.84	–1.42	(3)
V–I	Z = 0.01	Z = 0.04	
Yale	1.12	1.40	(1)
Yale + AGB	1.15	1.43	(2)

A more interesting exercise in the present context is to add in the bright AGB stars seen in M32 to the theoretical luminosity functions. This is shown in Figure 4, in which the dashed curve represents the luminosity function of M32 observed by Freedman (1989), normalized to the Yale luminosity functions where they overlap. Table 1 shows that stars brighter than the termination of the Yale giant branches cause a 0.2 mag increase in the luminosity variance. This is not surprising, as these AGB stars are the brightest stars in the galaxy, and therefore most effective in increasing the variance. We conclude that, whatever their source (and each of the four items listed in the previous section is a candidate), bright AGB stars must be factored into calculations of luminosity variance. If bright AGB stars occur in some elliptical galaxies and not in others, the luminosity variance will reflect this phenomenon.

6. References

Aaronson, M. and Mould, J. 1985, *Astrophys.J.*, **288**, 551.

Bedijn, P. 1988, *Astron & Astrophys.*, **205**, 105.

Bessell, M. and Wood, P. 1984, *Pub. Astron. Soc. Pacific*, **96**, 247.

Boothroyd, A. and Sackmann, I.-J. 1989, *Astrophys.J.*, **328**, 641.

Charlot, S. and Bruzual, G. 1990, *Astrophys.J.*, **367**, 126.

Freedman, W. 1989, *Astron.J.*, **98**, 1285.

Frogel, J. and Elias, J.1983, *Astrophys.J.*, **272**, 167.

Frogel, J. and Elias, J.1988, *Astrophys.J.*, **324**, 823.

Frogel, J., Mould, J. and Blanco, V. 1990, *Astrophys.J.*, **352**, 96.

Green, E., Demarque, P. and King, C. 1987, *The Revised Yale Isochrones and Luminosity Functions*, (Yale University Observatory).

Iben, I. and Renzini, A. 1983, *Ann. Rev. Astron. Astrophys.*, **21**, 271.

Lattanzio, J. 1986, *Astrophys.J.*, **311**, 708.

Lattanzio, J. 1991, *Astrophys.J. Suppl.*, **76**, 215.

Lloyd-Evans, T. 1980, *M.N.R.A.S.*, **193**, 97.

Mighell, K.1990, *Astron & Astrophys.Suppl.*, **82**, 1.

O'Connell, R. 1986 in *Stellar Populations*, eds. C. Norman, A. Renzini and M. Tosi, (Cambridge), p.167.

Reid, I.N. 1991, *Ap.J.*, in press.

Reid, I.N., Glass, I. and Catchpole, R. 1988, *M.N.R.A.S.*, **232**, 53.

Reid, I.N. and Mould, J. 1984, *Astrophys.J.*, **284**, 98.

Reimers, D. 1975, in *Problems in Stellar Atmospheres and Envelopes*, ed. B. Bascheck, W. Kegel & G. Traving (Berlin: Springer-Verlag), p.229.

Renzini, A. 1978 in *Advanced Stages in Stellar Evolution*, eds. P. Bouvier and A. Maeder, (Sauverny: Geneva Observatory), p.149.

Renzini, A. and Buzzoni, A. 1986, in *Spectral Evolution of Galaxies*, eds. C. Chiosi & A. Renzini, (Dordrecht: Reidel), p.195.

Rich, R.M. and Mould, J. 1991, *Astron.J.*, **101**, 1286.

Searle, L., Wilkinson, A. and Bagnuolo, W. 1980, *Astrophys.J.*, **239**, 803.

Tonry, J. 1991, *Astrophys.J. Letters*, **373**, L1.

Tonry, J., Ajhar, E. and Luppino, G.1990, *Astron.J.*, **100**, 1416.

Tonry, J. and Schneider, D. 1988, *Astron.J.*, **96**, 807.

Weidemann, V. 1987, *Astron & Astrophys.*, **188**, 74.

DISCUSSION

RENZINI: For over ten years I've been troubled with the problem of the missing bright AGB stars, and was unable to find a sensible solution. For example, I was not satified with *ad hoc* appeals to extra mass loss and the like. Having said this, I would like also to say that the solution to the problem may finally have been found. In a recent paper in A. & A. Blocker and Schonberner report evolutionary calculations of a thermally pulsing $7M_{\odot}$ star, in which the mixing length parameter is set in such a way for the so-called "envelope-burning" process to occur. With great surprise the models *do not* follow the famous core-mass luminosity relation, but evolve quickly to very high luminosities, spending a nearly ten times shorter time between M_{bol} = −6 and −7 than models assuming the canonical core-mass luminosity relation would predict.

MOULD: All the progress in AGB theory in the last few years has come about through the explicit calculation of evolutionary models rather than trying to guess the results for one set of parameters from those for another, and this is no exception. Observationally and theoretically, the high mass end of AGB evolution is now the frontier. The acceleration of high mass AGB stars up the AGB after ignition of envelope burning overcomes the objection that Reid and I made to this process, namely that there is a deficiency of high luminosity M stars as well as C stars. There are also the Lithium observations of Lambert and Smith in luminous AGB stars to support the reality of envelope burning.

Most of the AGB stars in the LMC, however, are older than 10^8 years, of course, and of too low a mass for envelope burning. For these stars I'm confident that the processes I've described of early termination of AGB evolution following the onset of Mira pulsation are required to explain the luminosity distribution.

DWARF SPHEROIDAL GALAXIES

G. S. DA COSTA[1]
Anglo-Australian Observatory
P.O. Box 296
Epping, NSW 2121
Australia

ABSTRACT. The properties of dwarf spheroidals are reviewed with emphasis on the newly discovered Sextans system, as well as on the star formation histories, the dark matter content and the abundance - luminosity relation of these galaxies. The relation of dwarf spheroidals to other dwarf galaxies is also discussed.

1. Introduction

The term *dwarf spheroidal* (hereafter dSph) is conventionally applied to the 8 low luminosity galaxies that are companions to the Milky Way (Carina, Draco, Fornax, Leo I, Leo II, Sculptor, Sextans and Ursa Minor) and to the 3 apparently similar systems that are companions to M31 (And I, And II and And III). Although often thought of in the past as merely large, low density globular clusters, detailed studies over the past ten years or so have revealed that the dSph galaxies possess a more diverse set of properties and contain more complex stellar populations than the globular cluster analogy would predict. Indeed an alternative definition of the term "dwarf spheroidal galaxy" might now be "*a low luminosity* ($M_V > -14$) *non-nucleated dwarf elliptical galaxy*" and as such, they are worthy of continued attention. In particular, since the individual stars in dSph galaxies can be resolved, their study has, and will continue to, contribute to the understanding of the origin and evolution of dwarf galaxies in general.

In the next section the properties of the recently discovered Sextans dSph are briefly discussed. Then follow sections dealing with the star formation histories, the mass-to-light ratios and the abundance-luminosity relation for dSph galaxies, while the final section contains some questions and speculations regarding the relation of dSphs to other dwarf galaxies.

2. Sextans

Unlike the discovery of the other 10 dSphs, the Sextans dSph was not found by visual inspection of photographic plates. Instead it was found as an excess of faint stellar images

[1]The Brazilian or Portuguese spelling of my surname, used at the conference, is "da Costa" but my Anglo-Australian ancestors altered the spelling to "Da Costa" and that is what I prefer people to use.

B. Barbuy and A. Renzini (eds.), The Stellar Populations of Galaxies, 191–200.

above the background on an APM scan of a UK Schmidt telescope plate. The discovery paper (Irwin *et al.* 1990) listed the following properties: an estimated distance of 85 kpc, an ellipticity of 0.4, and a density profile that fits a King model with a core radius of approximately 0.4 kpc and a limiting radius of ~2.2 kpc. Further, the photographic c-m diagram indicated a predominantly red horizontal branch but did not reveal any obvious population of bright very red stars. Subsequent to the discovery paper, two further studies have appeared, each based on observations made in the first season (1990) of Sextans observing. These are Da Costa *et al.* (1991) in which low signal-to-noise spectroscopy of a sample of 6 Sextans red giants was reported, and Mateo *et al.* (1991) which presented a CCD based c-m diagram. The former gave a first abundance estimate ([Fe/H] = -1.7 ± 0.25 dex) and a determination of the Sextans radial velocity (230 ± 6 km/s) while the latter, *inter alia*, confirmed the distance and the horizontal branch morphology suggested by the photographic results. The CCD c-m diagram also implied that Sextans was likely to be rich in RR Lyrae variables and based on deeper imaging of a smaller field, which revealed the main sequence turnoff, that the bulk of the stellar population of Sextans was comparable in age to the galactic globular clusters; *i.e.* no substantial younger component was present. However, the c-m diagram did identify the presence of a blue straggler population.

The second season of Sextans observing (1991) has now concluded and not surprisingly, the quality and scope of the studies of this dSph have improved significantly. For example, Mateo *et al.* (this volume) have used a CCD survey to establish the presence of a number of RR Lyrae variables as well as 3 anomalous cepheids in a single field near the center of the galaxy. Similarly, Suntzeff *et al.* (1991) have used fiber spectroscopy to observe a sample of approximately 30 Sextans red giant members. Based on a comparison of line strengths with those of globular cluster stars, these authors derive a mean abundance for Sextans of [Fe/H] = -2.05 ± 0.05 dex with a likely intrinsic abundance range of approximately -1.7 to -2.5 dex, corresponding to a σ([Fe/H] ≈ 0.15 dex. From the same set of observations Suntzeff *et al.* (1991) also estimate the velocity dispersion of Sextans as 5.6 ± 2.1 km/s. This value is lower than the 10 km/s dispersion found for Draco and Ursa Minor (see Pryor 1991 for a summary) but, given the low central density of Sextans, there is little doubt that this dispersion implies an M/L value that significantly exceeds the M/L for globular clusters. Hence it seems likely that Sextans also contains dark matter in the same way as do the other galactic dSphs studied in detail.

To summarize then, Sextans is a "conventional" dSph; *i.e.* its properties are very similar to those of the other galactic dSphs. The only unconfirmed characteristic at the present time is the identification of a small population of upper-AGB carbon stars, but given their presence in most if not all the galactic dSphs (see Azzopardi, this volume) and in And II (Aaronson *et al.* 1985), it would be surprising if Sextans lacked them.

3. Star Formation Histories

Perhaps the most interesting result to come from recent studies of dwarf spheroidal galaxies is the indication that all dSph show, to a greater or lesser extent, evidence for *star formation over extended periods*. This result was unexpected given that dSphs show no signs of current or recent star formation and have no detectable HI (*e.g.* Knapp *et al.* 1978, Mould *et al.* 1990). Nevertheless the stellar populations of dwarf spheroidals can be thought of as being made up of two basic components. These are an *old metal-poor population* similar to that of globular clusters, though in dSphs this old population invariably shows an intrinsic abundance range, and an *intermediate-age population, i.e.* a population of stars whose ages range from approximately 2 to 10 Gyr. The existence of this latter population is revealed by, for example, the presence of carbon stars on the AGB with luminosities above that of the first giant branch tip, and by the presence of main sequence stars whose luminosities exceed that of the turnoff for an old population. *The ratio of these two population types however, varies considerably from dSph to dSph.* For example, Ursa Minor consists almost completely of old stars (Olszewski & Aaronson 1985) while the stellar population of Carina is dominated by stars of intermediate-age (Mould & Aaronson 1983) with the old population being only a minor contributor to the total (Saha *et al.* 1986). It should also be noted that this characteristic of multiple age populations is not restricted to the galactic dSphs; the presence of upper-AGB carbon stars in And II (Aaronson *et al.* 1985) shows that this galaxy has an intermediate-age population also. The c-m diagram study of the giant branch of And I (Mould & Kristian 1990) however, did not reveal any candidate upper-AGB stars. These authors place a 2σ upper limit of 20% on the fraction of intermediate-age population in this dSph, but this limit is not inconsistent with the presence of a small intermediate-age component as found for example in Sculptor (Aaronson & Mould 1985).

The population mix in Carina has been recently placed on a more quantitative footing through the work of Mighell (Mighell 1990, Mighell & Butcher 1991). In these papers, estimates of the fraction of the old component comes from analysis of a deep color-magnitude diagram while estimates of the age and age-range come from a detailed luminosity function study. In the c-m diagram work (Mighell 1990), it is shown that the spread in V-R color fainter than $V \approx 23$ is significantly larger than that expected for a single age population. The cause of the color spread is identified as subgiants from an older population, which are redder than the dominant intermediate-age main sequence stars. Analysis of color distribution histograms then suggests that the old component comprises 17 ± 4 per cent of the total, a fraction in accord with earlier estimates (old ~ 30%) based on carbon star numbers (Aaronson & Mould 1985). In addition however, Mighell's analysis suggests that the old population was probably created in a single "burst " ($\tau < 3$ Gyr) and that the rate of star formation between the "bursts" that created the old (age > 13 Gyr) and intermediate-age (age ~ 7 Gyr) populations was probably negligible, otherwise the observed asymmetry in the color distribution histograms would not be as marked.

In the luminosity function study (Mighell & Butcher 1991), an observed luminosity function containing more than 2500 Carina stars with $1 < M_V < 5$, *i.e.* a magnitude interval ranging from just below the horizontal branch to somewhat fainter than the turnoff, was compared with the predictions of a number of theoretical models. The models were based on the luminosity functions contained in the Revised Yale Isochrones (Green *et al.* 1987) together with various assumptions for the initial epoch and duration of the star formation "burst" that produced the majority of Carina's stars. Despite a broad exploration of parameter space, only a small number of models gave acceptable fits to the observed luminosity function. These successful models all imply that the bulk of the star formation in Carina began approximately 9 Gyr ago, reached a peak between perhaps 6 and 8 Gyrs and had ceased by 5 Gyr.

As a consequence of this work Carina emerges as a dwarf galaxy which apparently has had the following star formation history: an initial epoch of star formation in which relatively few stars were formed, a significant quiescent interval, then a period of active star formation which lasted for at least 2 billion years, and then finally the cessation of all star formation approximately 5 Gyr ago. A star formation history of this type raises a number of questions! These include "What regulates the star formation?" or equivalently, "Why did Carina wait almost half the age of the Universe before forming significant numbers of stars?". A further point that should not be overlooked in this context is that the duration of the major episode of star formation in Carina was apparently much longer than the lifetime of massive stars. How then did Carina, a low density system, retain the gas necessary for such an extended period of star formation against the energy input from supernovae? This latter question is probably answered by noting that the rate of star formation implied is actually quite low (comparable to or less than the rate in the solar neighborhood, consequently the term "burst" does not apply here). Thus the rate of energy input from supernovae into what was presumably a large gas mass was no doubt insufficient to quench the star formation for a significant interval. The fact that Carina is "gas-free" at the present time however, indicates that the energy input did eventually become sufficient to drive the remaining gas from the system and stop the star formation.

This section can be summarized then by noting that the *star formation histories of dwarf spheroidal galaxies are clearly complex and varied.* The inevitable conclusion that follows from these results is that the star formation histories of the more luminous dE galaxies are therefore also likely to be complex and varied. Thus evidence for the existence of intermediate-age populations in dE galaxies (*e.g.* Gregg, this volume) should not come as a surprise.

4. Dark Matter

The subject of dark matter in dSph galaxies has recently been thoroughly reviewed by Pryor (1991) and his article contains the most up-to-date results for Draco and Ursa Minor. Both these systems have velocity dispersions of the order of 10 km/s and central M/L values of approximately 100 in solar units, indicating that they are dominated by dark matter. While work on these systems and others such as Carina and Sextans continues, recent new results on Sculptor and Fornax are of some interest. For both these systems, the samples of stars with accurate velocities are now sufficiently large that the form of the velocity dispersion profile, as distinct from just its central value, can now be investigated for the first time. The dispersion profile is a vital quantity if the distribution of the dark matter in these dSphs is to be constrained.

For Fornax, Mateo *et al.* (1991) have used the Las Campanas Observatory 2.5m echelle spectrograph to obtain accurate velocities for approximately 40 stars in two fields; one near the center and one situated approximately 1.5 core radii from the center, along the major axis. From these observations they find a central velocity dispersion of $\sim 10 \pm 2$ km/s (Paltoglou & Freeman 1991 report a similar number). This value however, is made rather more uncertain than it might otherwise be by the low systemic velocity of Fornax which makes it difficult to unambiguously assign membership to individual stars with velocities far from the mean; the low S/N of the high dispersion spectra prevent the application of any line strength based membership criterion. Nevertheless, combined with the most recent values for the structural parameters (Eskridge 1988b) this dispersion together with the observed central surface brightness leads to a central mass-to-(visual) light ratio of $\sim 11 \pm 5$ in solar units indicating a substantial dark matter component in this dSph. It should perhaps also be emphasized however, that, leaving aside questions concerning membership of individual stars and their effect on the observed dispersion, it is an unfortunate but true situation that the uncertainties in the distance of Fornax and in its central surface brightness and core length scale make a large contribution to the uncertainty in the M/L value. This is illustrated by the results of Paltoglou & Freeman (1991) who, although finding a similar velocity dispersion, derive a significantly lower central M/L value through the use of a different distance and different structural parameters.

Mateo *et al.* (1991) also find no evidence for any rotation about the minor axis in Fornax, a result consistent with Paltoglou & Freeman's determination of a rotation velocity of 3.5 ± 2.0 km/s at a radial distance of one core radius. Paltoglou & Freeman further demonstrate that this rotation velocity is inconsistent with the hypothesis that Fornax has evolved from a dwarf irregular (*i.e.* a system with a rotating disk). Perhaps more significantly though, the Mateo *et al.* results do not indicate any decrease in velocity dispersion outside the core; indeed within the error bars, the central and 1.5 core radius dispersions are the same. Further, the 1.5 core radius observed dispersion lies ~1.5σ above the prediction of the King model that best fits the surface density data. While these results are at most suggestive since, for example, the velocity dispersion profile of an isotropic King model is not likely to be a good representation of the dispersion profile of anisotropic flattened system even if the mass is distributed in the same way as the light, they may represent the first indication that the dark matter in this dSph has a more extended distribution than that of the stars.

The results of Armandroff & Da Costa (1992) for Sculptor are similar. Using velocities determined at CTIO for a sample of 32 stars spread from the center to radial distances of approximately two core radii, these authors find a central velocity dispersion of 7.0 ± 1.2 km/s and a central mass-to-light ratio of 8.2 ± 3.7 in solar units, using the most recent values for the structural parameters (Eskridge 1988a). As for Fornax, the largest contributor to the uncertainty in the M/L value is not the dispersion but is, in this case, uncertainty in the central surface brightness. The sample shows no evidence for rotation about the minor axis with an upper limit on the rotation velocity of this dSph being $V_{rot} < 2$ km/s at one core radius. Unlike Fornax however, as illustrated in Fig. 1 the velocity dispersion observations are consistent with either the flat dispersion profile expected if the dark matter is spatially much more extended than the stars, or the dispersion profile of the King model that fits best the stellar surface density profile.

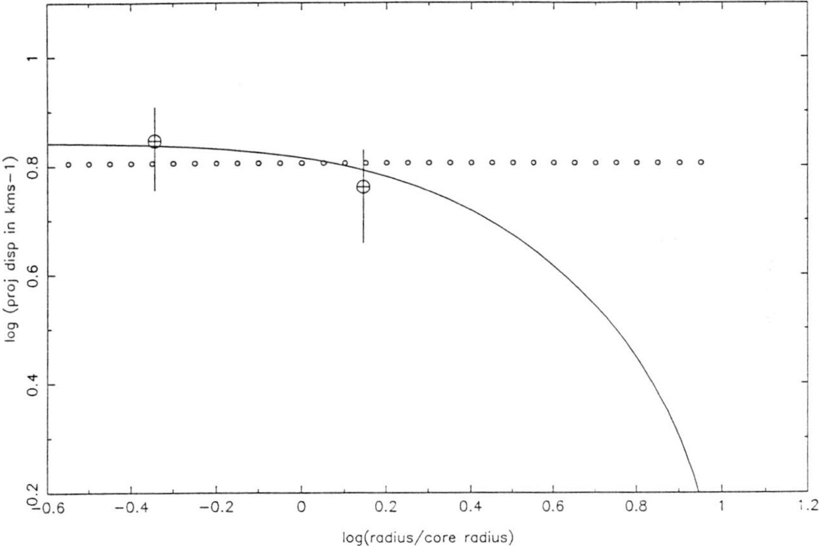

Figure 1. Velocity dispersion versus radius for the Sculptor dSph from Armandroff & Da Costa (1992). The open circles are the observed points with the inner sample comprising 18 stars and the outer 14 stars. The solid curve is the dispersion profile of the King (1966) model that fits the surface density data for this dSph. The dotted line is a schematic representation of what the dispersion profile might be if the dark matter is spatially much more extended than the stars.

In both cases these first results on velocity dispersion profiles are tantalizing, but it is clear from Fig. 1 for example, that observations of stars still further from the centers of the galaxies are required before any meaningful constraints can be placed on the distribution of the dark matter. Fortunately, both Sculptor and Fornax are sufficiently rich that enough members with magnitudes brighter than the observational limit are likely to exist at large radial distances. Identifying such stars however, is a formidable task in itself even with the advent of fast CCD cameras capable of surveying large areas (square degrees!) of sky in relatively small amounts of telescope time.

5. Luminosity - Abundance Relation

Since this topic was last reviewed (*e.g.* Da Costa 1988), the publication of a number of new results necessitates a re-evaluation of the dSph luminosity - abundance relation. These new results include improved determinations of density profiles (*e.g.* Eskridge 1988a,b) that allow better (but still uncertain) estimates of absolute magnitudes, and additional abundance estimates from both spectroscopic (*e.g.* Armandroff & Da Costa 1991, Lenhert *et al.* 1992) and color-magnitude diagram studies (*e.g.* Reid & Mould 1991). Further, the addition of Sextans (see above) and And I (abundance from Mould & Kristian 1990, absolute magnitude from Caldwell *et al.* 1992) means that there are now 9 dSphs with both abundance and luminosity estimates. These data (see Caldwell *et al.* 1992 for a compilation of dSph abundance estimates and for new estimates of the absolute magnitudes of 5 of the 8 galactic dSphs as well as determinations of the luminosities of the three M31 companions; luminosities of the remaining 3 galactic dSphs come from Webbink 1985) reveal a relation that has more apparent scatter than earlier versions. Given the likely ±0.5 mag errors in the absolute magnitude estimates of most of the galactic dSphs (for which surface brightnesses are rarely directly measured, rather they are usually inferred from surface densities using a luminosity function) and the typical ±0.2 - 0.3 dex errors in the mean abundance values, it is difficult to estimate the size of any *intrinsic* dispersion in this relation. Plausible estimates are perhaps ±0.5 - 0.7 mag at constant [Fe/H] or ±0.2 dex at constant M_V. This is similar to what is seen for brighter dEs (*e.g.* Brodie & Huchra 1991). With the addition of the dE galaxies NGC 147 and NGC 205, whose abundances have been determined in the same way as for (at least some of) the dSph (Mould *et al.* 1983,1984), the relation becomes better defined. A unweighted least squares fit to these 11 dwarf galaxies yields the relation $Z \propto L^{0.36\pm0.07}$ or $L \propto Z^{2.8\pm0.5}$ which, coincidentally or not, is close to the relation predicted by the scaling law relations of Dekel & Silk (1986) in which dwarf galaxies form and lose gas inside a dominant extended dark matter halo.

The extent to which this dwarf galaxy abundance - luminosity relation matches the equivalent relation for the more luminous E galaxies is not easily assessed because of the uncertainty in converting measured line strength indices to abundance for high abundance objects (*e.g.* Sadler this volume; Faber *et al.* this volume). Nevertheless extrapolating the dwarf galaxy relation to higher luminosities yields solar abundance at $M_V \approx -22.5$ mag. In contrast, most calibrations of the line strengths of E galaxies indicate solar abundances for less luminous E galaxies. This difference would then appear to indicate that there is a steepening of the abundance luminosity relation in moving from dEs to ellipticals, but since the dEs appear to be morphologically different from the ellipticals and the lower luminosity compact ellipticals (*e.g.* Wirth & Gallagher 1984, Kormendy 1985), it may simply be evidence that the two types of galaxies follow intrinsically different luminosity - abundance relations. This point is also made by Caldwell *et al.* (1992); see also Bender (this volume). On the other hand, the dwarf galaxy luminosity - abundance relation is defined solely by Local Group galaxies and one may question its universality. However, the publication of abundances and absolute magnitudes for a large sample of dwarf galaxies outside the Local Group by Brodie & Huchra (1991) allows this question to be investigated. The Brodie & Huchra (1991) abundances have been derived from line strength indices calibrated using integrated spectra of globular clusters and should therefore be directly comparable with the available Local Group data. The results of this comparison are shown in Fig. 2; the M_B values listed by Brodie and Huchra (1991) have been converted to M_V by using the relation between $(B-V)_0$ and [Fe/H] defined by the galactic globular clusters. While the Brodie & Huchra (1991) data have large uncertainties, typically ~0.6 dex in abundance and ~0.5 mag in luminosity, resulting in a large apparent scatter, it is evident that these galaxies do generally follow the luminosity - abundance relation defined by the Local Group dwarfs. However, better data are required before the status of the few outliers, and thus the extent of intrinsic scatter in this relation, can be assessed.

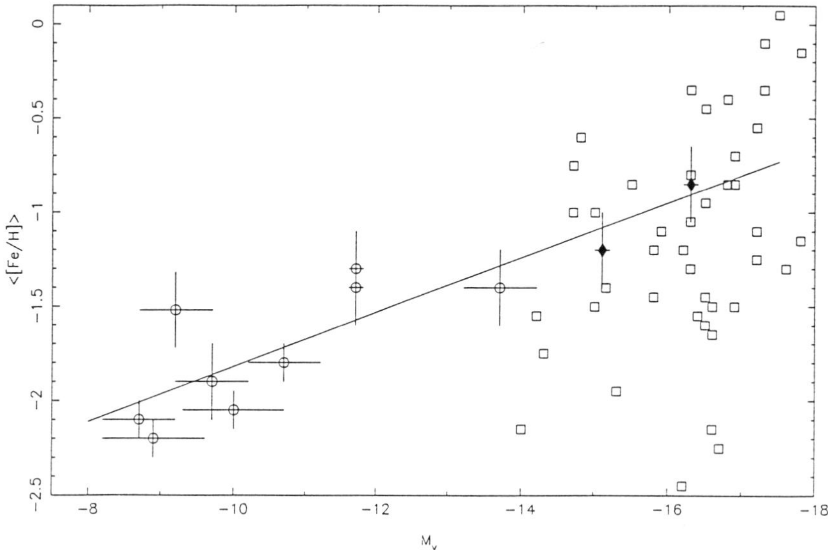

Figure 2. The abundance - luminosity relation for dSph and dE galaxies. The open circles with error bars are the dSph galaxies (8 Galactic dSph plus And I) while the filled diamonds are the dE galaxies NGC 147 and NGC 205. The open squares are dwarf galaxies from Brodie & Huchra (1991). The straight line is an unweighted least squares fit to the data for the 9 dSphs and NGC 147 and NGC 205.

6. Dwarf Spheroidals in the General Context of Dwarf Galaxies

Dwarf spheroidal galaxies are characterized by the absence of HI and by the presence of only old and intermediate-age stellar populations. Dwarf irregular galaxies, on the other hand, are characterized by the presence of large amounts of HI and by indications of current or recent star formation. Are these categories of dwarf galaxy disjoint or do transition objects exist? The recent results of van de Rydt *et al.* (1991), following on earlier work of Ortolani & Gratton (1988), suggest that the Phoenix dwarf galaxy may be such a transition object. This galaxy is a Local Group member that lies at a distance of some 400 kpc from the Milky Way. The van de Rydt *et al.* study shows that Phoenix consists mostly of an old metal-poor population which nevertheless possesses an intrinsic abundance dispersion. There is also a population of red stars with luminosities beyond the giant branch tip, one of which has been spectroscopically confirmed as a carbon star, thus indicating the presence of an intermediate-age population in this galaxy. These properties are all reminiscent of those of dSphs and in fact ,with $M_V \approx -10$ and <[Fe/H]> ≈ -2.0 dex, Phoenix even lies on the abundance - luminosity relation of Fig. 2. However, Phoenix also contains an "association" of young blue stars whose age is estimated as less than 150 million years (van de Rydt *et al.* 1991) and further, HI was recently detected in this galaxy (Carignan *et al.* 1991). Both these latter properties are characteristic of a dwarf irregular. Thus it does appear that Phoenix is a dwarf galaxy whose properties span the gap between the dIrrs and the dSphs, the transition nature being further emphasized by the low (0.1) value of M_{HI}/L_V for Phoenix compared to typical values (~2) for dIrrs (Carignan *et al.* 1991). The Local Group member LGS3 may be a similar transition type dwarf since its properties are also similar to those of a dwarf spheroidal except for the presence of a moderate amount of HI (Cook & Olszewski 1989, Sargent & Lo 1985). It is then interesting to speculate as to what extent transition dwarfs like Phoenix and LGS3 have been able to retain modest amounts of HI gas because they lie far from any "large" galaxy such as M31 or the Milky Way; certainly with the exception of the Magellanic Clouds, all the companions to M31 and the Milky Way are gas poor (though the dEs NGC 185 and NGC 205 are not gas free).

The current catalog of dwarf spheroidals includes only objects that are companions to either our galaxy or to M31. Is this a selection effect or do isolated dwarf spheroidals exist? The recent discovery of a faint dwarf galaxy in the constellation of Tucana (Lavery 1990) may provide the first indication that isolated dwarf spheroidals exist. This galaxy presents a smooth appearance on optical images, *i.e.* there are no obvious HII regions or associations. It is noticeably flattened with $e \approx 0.5$ and has a diameter of approximately 1.2 kpc if its distance is of the order of 800 kpc (Lavery & Mighell 1991). It resolves into stars on deep images and the brightest stars are red. Preliminary color-magnitude diagrams (Lavery & Mighell 1991, Suntzeff & Seitzer 1991) suggest a distance modulus of approximately 24.5 making it a likely Local Group member with an absolute magnitude of approximately $M_V \approx -9.5$. No HI was detected in an initial search (Lavery & Mighell 1991). All these properties are reminiscent of the dSph companions to M31 and it thus seems quite probable that Tucana is indeed an isolated dSph galaxy. It is then of more than passing interest to note that Tucana was discovered entirely by chance(!) and to speculate how many similar galaxies are lurking both within the Local Group and at larger distances. The low surface brightness and small apparent size of such galaxies makes a directed search for such objects a daunting task!

7. Summary

The dwarf spheroidal companions to the Milky Way and M31, together with galaxies like Phoenix and Tucana are examples of what are probably the most common type of galaxy in the Universe. They differ from more luminous dEs only in intrinsic scale (mass,size); their properties are otherwise similar. Hence, because the individual stars can be studied in these galaxies, inferences can be made that are otherwise hard to draw when only integrated-light indices are available. For this reason dwarf spheroidals are worthy of continued attention. In particular, the following questions deserve serious study. *What governs the Star Formation History in dwarf spheroidal galaxies*, or equivalently, why does the ratio of old to intermediate-age populations vary so widely from dSph to dSph? *What role do the dark matter halos play in the evolution of dSphs? To what extent is proximity to a large "parent" galaxy important in the evolution of dSphs?* If these questions can be answered for the nearby dSph systems, then it will represent considerable progress towards answering similar questions for more distant and more luminous, but presumably otherwise quite similar, dwarf galaxies.

I would like to acknowledge the generosity of my colleagues, especially Dr. Nick Suntzeff, in supplying preprints and descriptions of work-in-progress that provided much of the material on which this review is based.

8. References

Aaronson, M., and Mould, J. 1985, ApJ, 290, 191
Aaronson, M., Gordon, G., Mould, J., Olszewski, E., and Suntzeff, N. 1985, ApJL, 296, L7
Armandroff, T.E., and Da Costa, G.S. 1991, AJ, 101, 1329
Armandroff, T.E., and Da Costa, G.S. 1992, in preparation
Brodie, J.P., and Huchra, J.P. 1991, ApJ, 379, 157
Caldwell, N., Armandroff, T.E., Seitzer, P., and Da Costa, G.S. 1992, AJ, in press
Carignan, C., Demers, S., and Côté, S. 1991, ApJL, 381, L13
Cook, K.H., and Olszewski, E. 1989, BAAS, 21, 775
Da Costa, G.S. 1988, in IAU Symp. 126, The Harlow Shapley Symposium on Globular Cluster Systems in Galaxies, edited by J. Grindlay and A.G.D. Philip (Kluwer, Dordrecht), p. 217
Da Costa, G.S., Hatzidimitriou, D., Irwin, M.J., and McMahon, R.G. 1991, MNRAS, 249, 473
Dekel, A., and Silk, J. 1986, ApJ, 303, 39
Eskridge, P.B. 1988a, AJ, 95, 1706
Eskridge, P.B. 1988b, AJ, 96, 1352
Green, E.M., Demarque, P., and King, C.R. 1987, The Revised Yale Isochrones and Luminosity Functions (Yale Univ. Obs., New Haven)
Irwin, M.J., Bunclark, P.S., Bridgeland, M.T., and McMahon, R.G. 1990, MNRAS, 244, 16P
King, I.R. 1966, AJ, 71, 64
Knapp, G., Kerr, F., and Bowers, P. 1978, AJ, 83, 360
Kormendy, J. 1985, ApJ, 295, 73
Lavery, R.J. 1990, IAU Circ. 5139
Lavery, R.J., and Mighell, K.B. 1991, AJ, submitted
Lenhert, M.D., Bell, R.A., Hesser, J.E., and Oke, J.B. 1992, ApJ, in press
Mateo, M., Nemec, J., Irwin, M., and McMahon, R. 1991, AJ, 101, 892
Mateo, M., Olszewski, E., Welch, D.L., Fischer, P., and Kunkel, W. 1991, AJ, 102, 914
Mighell, K.J. 1990, A&AS, 82, 1
Mighell, K.J., and Butcher, H.R. 1991, A&A, in press
Mould, J., and Aaronson, M. 1983, ApJ, 273, 530
Mould, J.R., Bothun, G.D., Hall, P.J., Staveley-Smith, L., and Wright, A.E. 1990, ApJL, 362, L55
Mould, J., and Kristian, J. 1990, ApJ, 354, 438
Mould, J.R., Kristian, J., and Da Costa, G.S. 1983, ApJ, 270, 471
Mould, J., Kristian, J., and Da Costa, G.S. 1984, ApJ, 278, 575
Olszewski, E.W., and Aaronson, M. 1985, 90, 2221
Ortolani, S., and Gratton, R.G. 1988, PASP, 100, 1405

Paltoglou, G., and Freeman, K.C. 1991, in preparation
Pryor, C. 1991, in Morphological and Physical Classification of Galaxies, edited by G. Busarello *et al.* (Kluwer, Dordrecht), in press
Reid, N., and Mould, J. 1991, AJ, 101, 1299
Saha, A., Monet, D.G., and Seitzer, P. 1986, AJ, 92, 302
Sargent, W.L.W., and Lo, K.-Y. 1985, in Star Forming Dwarf Galaxies, edited by D. Kunth *et al.* (Frontières, Gif sur Yvette), p. 253
Suntzeff, N.B., and Seitzer, P. 1991, in preparation
Suntzeff, N.B., Mateo, M., Terndrup, D., Weller, W., and Olszewski, E. 1991, in preparation
van de Rydt, F., Demers, S., and Kunkel, W.E. 1991, AJ, 102, 130
Webbink, R.F. 1985, in IAU Symp. 113, Dynamics of Star Clusters, edited by J. Goodman and P. Hut (Reidel, Dordrecht), p. 541
Wirth, A., and Gallagher, J.S. 1984, ApJ, 282, 85

DISCUSSION

Mateo: Just a comment. Sextans *has* been surveyed for variables as described in our poster. We find a total of 42 variables of which 3 are anomalous cepheids.

Da Costa: Obviously I was on the beach when I should have been looking at the posters. The written version will contain mention of your work.

Silk: The extended period of star formation in dwarf spheroidals such as Carina may be understood as follows. Depletion of the gas via a supernova-driven wind requires both that the energy input be sufficient to drive a wind, a condition easily satisfied, and that the radiative cooling timescale be longer than a dynamical timescale. The latter condition occurs only once the gas supply is sufficiently depleted by star formation. Hence the star formation efficiency determines the duration of the star formation phase, and so this can be long-lived.

Da Costa: This may be the case but remember that the metal abundance of the system is low (< 1/30 solar) - that may make the cooling less efficient. On the other hand, the rate of star formation implied is not high so perhaps you are right.

Gregg: Just as dSph galaxies are not the same kind of object as globular clusters, dSphs are also very different from the compact M32-like ellipticals. The existence of intermediate-age populations in dSphs and in some nucleated dEs should therefore *not* be interpreted as supporting evidence for intermediate-age populations in objects like M32.

Da Costa: I agree. Dwarf spheroidals and dwarf ellipticals form a different sequence from the compact ellipticals and so, yes, it is less certain that the results on the presence of intermediate-age populations can be applied to these galaxies.

Schommer: When I looked at the distributions of velocities in dSphs, I found all of them consistent with a gaussian except your earlier Sculptor data. These velocities show a minor skew. Have you looked at the actual distribution with your new data?

Da Costa: The distribution of the 32 velocities in the current sample is consistent with a gaussian.

SURVEYING CARBON STARS IN THE DWARF SPHEROIDAL GALAXIES

M. AZZOPARDI[1], J. LEQUEUX[2]
[1]*Observatoire de Marseille, 2, Place Le Verrier*
13248 Marseille Cedex 4, France
[2]*Observatoire de Meudon*
92195 Meudon Cedex , France

ABSTRACT. We have completed a blue-green Grism survey for carbon stars in the seven "classical" dwarf spheroidal galaxies. The results for six of them have been published already, while the results for the Fornax galaxy are presented here: this galaxy contains 77 carbon stars, including 30 new objects. The bolometric luminosity function of Fornax is intermediate between that of the Small and of the Large Magellanic Cloud. We will also compare these carbon stars with the fainter objects in the galactic bulge and in the galactic halo and discuss their properties in relation with the metallicity and history of star formation in the parent galaxies.

1. Introduction

Carbon stars are a very late stage in the evolution of intermediate-mass stars. Unfortunately their formation is still not well understood (Iben and Renzini 1983) although progress has been done recently (see e.g. Boothroyd and Sackman 1988 and Lattanzio 1989). There seems to exist at least three kinds of carbon stars with different origins: i) the "normal" ones which are Asymptotic Giant Branch (AGB) stars enriched in carbon brought to the surface by dredge-ups driven by helium-burning thermal pulses; ii) fainter and bluer ones which have been discovered in the galactic bulge and whose origin is not understood (Westerlund et al. 1991) and iii) dwarf carbon stars presumably resulting from mass transfer from a more massive companion at the time it was a carbon star (Dearborn et al. 1986). Observationally, there seems to be continuity between classes i) and ii), the brightest bulge carbon stars being not strikingly different from the faintest observed SMC carbon stars.

Although we are not yet able to make full use carbon stars for studying the properties of the stellar populations of which they are a part, their potential is very great as they are relatively bright and easily recognizable while their number relative to other stars is obviously very sensitive to metallicity and age. Consequently, a number of astronomers have made systematic searches for carbon stars in our Galaxy and in nearby galaxies.

2. Finding relatively blue carbon stars

We have been surveying carbon stars in the Magellanic Clouds, the dwarf spheroidal galaxies and the galactic bulge using very-low resolution field spectroscopy in the green. The spectral range (4350-5400 A) allows to recognize carbon stars from their deep absorption bands of the C_2 molecule (the Swan bands) with bandheads at 4737 and 5165 A. The observations were made at the prime foci of the ESO and of the CFH 3.6 m telescopes, equipped with wide-field correctors and a Grism or a Grens with respectively 2200 and 2000 A/mm dispersion (see Breysacher and Lequeux 1983 for more information on the technique we used). Comparison with a similar, more classical technique using the CN bands in the near-infrared shows that our

B. Barbuy and A. Renzini (eds.), The Stellar Populations of Galaxies, 201–205.

technique is roughly equivalent for the red, bright carbon stars and more powerful for the fainter and bluer carbon stars (see McCarthy 1987 and Blanco and McCarthy 1990).

Subsequent spectroscopy is very useful for confirming the nature of the carbon star candidates found in the surveys. We have made medium-resolution spectroscopy of many of the candidates in the Fornax, Sculptor, Carina and Leo I galaxies using the ESO 3.6 m telescope with the Boller and Chivens spectrograph (long slit mode and Optopus multifiber device) or EFOSC and using the NTT with EFOSC 2. Infrared (IR) photometry has been made at the ESO 3.6 m telescope for a number of the Fornax stars. This, added to IR photometry obtained by other authors and after correcting for extinction given in the literature, allows to calculate the bolometric magnitude using a formula for the bolometric correction to the K magnitude given by Wood et al. (1983): $BC_K = 0.55 + 2.65\ (J\text{-}K)_0 - 0.67\ (J\text{-}K)_0^2$,
J and K being the magnitudes in the Johnson system (Johnson et al. 1966). Then the absolute bolometric magnitude can be calculated using distances from the literature.

3. The number and luminosity functions of carbon stars in the dwarf spheroidal galaxies

The results for Sculptor, Carina, Leo I, Leo II, Draco and Ursa Minor have been presented by Azzopardi et al. (1985, 1986). These surveys should be essentially complete. Preliminary results for the Fornax dwarf spheroidal galaxy are presented here. Fornax contains 77 carbon stars: 30 new ones and 47 already known (Azzopardi, Lequeux and Muratorio 1992). We have added 13 carbon stars in the areas searched for by Blanco and McCarthy (Frogel et al. 1982) and by Westerlund et al. (1987) using near-infrared Grism technique. 20 carbon stars have been found outside these areas, 3 of which, belonging to the list of the very red giants discovered by Demers and Kunkel (1979) in the Fornax galaxy, were previously classified as carbon stars by Aaronson and Mould (1980) and Lundgren (1990). Note that the stars designated as "Ctm" by Aaronson and Mould (1980) and by other authors have not been considered as carbon stars. Although medium-resolution spectroscopy has not been secured yet for all stars, the list should be essentially complete at least for "normal" carbon stars (stars of the other categories are too faint to have been detected). Table 1 summarizes the results for the carbon stars in the dwarf spheroidals; for comparison it also contains estimates for the Magellanic Clouds (field carbon stars, the census in the clusters being certainly incomplete), the galactic bulge and the galactic halo.

Table 1. Census of field carbon stars in local group galaxies

Galaxy	Candidates (our surveys)	Candidates (previous)	Confirmed by spectroscopy	Estimated total nb.	References
Fornax	77	47	53	77	1, 2, 3, 4, 5, 6
Leo I	19	1	18	19:	7, 8, 9
Carina	11	6	9	11	7, 8, 10, 11
Sculptor	8	3	8	8	3, 4, 7, 8
Leo II	7	4	5	7	7, 9
Draco	4	3	3	4	8, 12
Ursa Minor	1	1	1	1	8, 9
Sextans	0	0	0	0	13
LMC	-	849	?	11000	14, 15
SMC	1707	860	?	3100	14, 16, 17
Gal. Bulge	34	0	34	-	18
Gal. halo	-	10	10	-	19, 20, 21

Note that we do not take into account the CH star K = SOC 215 in Ursa Minor (Zinn 1981).

References to Table 1: (1) this paper and Azzopardi et al. 1992; (2) Aaronson and Mould 1980; (3) Frogel et al. 1982; (4) Richer and Westerlund 1983; (5) Westerlund et al. 1987; (6) Lundgren 1990; (7) Azzopardi et al. 1985; (8) Azzopardi et al. 1986; (9) Aaronson et al. 1983; (10) Cannon et al. 1981; (11) Mould et al. 1982; (12) Aaronson et al. 1982; (13) Irwin et al. 1990; (14) Blanco and McCarthy 1983; (15) Blanco and McCarthy 1990; (16) Rebeirot et al. 1992; (17) Azzopardi and Rebeirot 1991; (18) Azzopardi et al. 1991; (19) Margon et al. 1984; (20) Mould et al. 1985; (21) Green et al. 1991.

As discussed by Richer and Westerlund (1983) and Azzopardi, Lequeux and Westerlund (1985) there is some correlation between the number of carbon stars per unit luminosity and the absolute luminosity of the parent galaxy, and also a trend with metallicity, the number of carbon stars per unit luminosity increasing with decreasing absolute luminosity and decreasing abundance. The new data do not change these conclusions. However their interpretation is difficult because the galaxies we consider presumably have different histories of star formation, and also because there is a correlation between absolute luminosity and metallicity which does not allow to disantangle the different possible effects.

Table 2 presents the bolometric luminosity functions for the galaxies or parts of galaxies in our sample, together with the adopted extinctions and distance moduli. The bolometric magnitudes have been calculated in an homogeneous way as explained in Section 2. Most of the samples are incomplete in the sense that all discovered carbon stars do not have the IR photometry necessary to calculate the bolometric magnitudes, but we do not think that the results are strongly biased by this incompleteness.

Table 2. Bolometric luminosity functions for dwarf spheroidal galaxies and other systems

System	n (C stars)	Completeness %	E(B-V)	$(m-M)_0$	<-M_{bol}>	(-M_{bol}) range
Fornax	25	32	0.02	21.0	4.6	3.7-5.6
Leo I	5	26	0.00	21.7	4.3	4.1-4.5
Carina	8	73	0.06	19.7	3.9	2.9-4.6
Sculptor	2	25	0.02	19.5	3.2	3.0-3.4
Leo II	5	71	0.00	21.7	4.0	3.7-4.4
Draco	3	75	0.03	19.4	3.1	2.8-3.5
Ursa Minor	1	100	0.02	19.3	2.9	-
LMC	74	< 1	0.05	18.5	5.1	3.7-6.4
SMC	112	< 4	0.03	18.8	4.4	2.7-5.6
Gal. Halo	7	?	0.00	-	3.2:	1.8-5.2:
Gal. Bulge	34	?	var.	14.5	1.2	-0.2-2.8

Note that the results from the galactic halo carbon stars are extremely uncertain: the distances are very poorly known, and a large fraction of these objects may be dwarfs (Green et al. 1991).

4. Discussion and conclusions

From Table 2 one can draw the following conclusions:

i) The bolometric luminosity functions of the carbon stars in the dwarf spheroidal galaxies differ somewhat from each other: those in Fornax appear to be brighter than those in Carina, Leo I and Leo II which themselves are brighter than those in the other dwarf spheroidals. One must be careful in interpreting the numbers literally, because of uncertainties in the distances and of small-number statistics. There is no obvious correlation with metallicity, that itself is poorly known, and presents a large range of values inside at least several galaxies. There is a correlation with the absolute magnitude of the galaxy, which may reflect to some extent a small-number statistical effect similar to that discussed by Schild and Maeder (1983) for the brightest stars in galaxies,

ii) The bolometric luminosity function of the Fornax galaxy is intermediate between those of the SMC and the LMC. This result cannot be appreciably biased by incompleteness, small-number statistics, etc.... It is very probably related to a different history of star formation in all three galaxies, rather than to a metallicity effect, since the metallicity of Fornax is smaller than those in both the LMC and the SMC. This property as well as property (i) deserves further studies,

(iii) The luminous carbon stars predicted by the "classical" theory of carbon star formation are not found in any of the systems we have studied. This is a major problem, already known for a long time (see e.g. Iben and Renzini 1983). There are ways out e.g. through mass loss and convective overshooting that we cannot discuss here,

(iv) We have not (yet?) found in any other system the equivalent of the low-luminosity carbon stars that we discovered in the galactic bulge. These objects are very rare in the bulge (one for several hundreds of M stars) and they are very faint, so their detection even in the nearest galaxies requires very deep low-resolution field spectroscopy (or alternatively, but less securely, deep narrow-band filter imaging). We are starting such a program in the Magellanic Clouds,

v) The status of the halo carbon stars is uncertain. Current theories open the possibility that 0.8 $M_\odot$ stars of low metallicities can become carbon stars. Very recently, Tsuji et al. (1991) have found that some halo CH stars have very high $^{12}C/^{13}C$ ratios, suggesting that they have been produced by the third dredge-up; however they state that these stars may be (like the halo dwarf carbon stars) binary objects which have been enriched in ^{12}C by mass transfer from a normal carbon star in the past or by some other process. More work is necessary before we understand the halo carbon stars. Globular clusters are good places for looking systematically for new candidates and we are undertaking this program.

5. References

Aaronson, M. and Mould, J. (1980) ApJ 240, 804.
Aaronson, M., Liebert, J. and Stocke, J. (1982) ApJ 254, 507.
Aaronson, M., Olszewski, E.W. and Hodge, P.W. (1983) ApJ 267, 271.
Azzopardi, M., Lequeux, J. and Muratorio, G. (1992) in preparation.
Azzopardi, M., Lequeux, J. and Westerlund, B.E. (1985) A & A 144, 388.
Azzopardi, M., Lequeux, J. and Westerlund, B.E. (1986) A & A 161, 232.
Azzopardi, M., Lequeux, J., Rebeirot, E. and Westerlund, B.E. (1991) A & AS 88, 265.
Azzopardi, M. and Rebeirot, E. (1991) IAU Symp. 148, in R. Haynes, D. Milne (eds.), The Magellanic Clouds, Kluwer Academic Publishers, Dordrecht, pp. 71-76.
Blanco, V.M. and McCarthy, M.F. (1983) AJ 88, 1442.
Blanco, V.M. and McCarthy, M.F. (1990) AJ 100, 674.
Boothroyd, A.I. and Sackmann, I.G. (1988) ApJ 328, 671.
Breysacher, J. and Lequeux, J. (1983) The Messenger 33, 21.
Cannon, R.D., Niss, B. and Norgaard-Nielsen, H.U.(1981) MNRAS 196, 1P
Dearborn, D.S.P., Liebert, J., Aaronson, M., Dahn, C.C., Harrington, R., Mould, J. and Greenstein, J.L. (1986) ApJ 300, 314.
Demers, S. and Kunkel, W.E. (1979) PASP 91, 761.
Green, P.J., Margon, B. and MacConnell, D.J. (1991) ApJ, Letter in Press.
Frogel, J.A., Blanco, V.M., McCarthy, M.F. and Cohen, J.G. (1982) ApJ 252, 133.
Iben, I. Jr. and Renzini, A. (1983) Ann. Rev. A & A 21, 271
Irwin, M.J., Bunclark, P.S.n, Bridgeland, M.T. and McMahon, R.G. (1990) MNRAS 244, 16P.
Johnson, H.L., Mitchell, R.I., Iriarte, B. and Wisniewski, W.Z. (1966) Com. Lun. Plan. Lab. 4, 99.
Lattanzio, J.C. (1989) ApJ 344, L25.
Lundgren, K. (1990) A & A 233, 21.
McCarthy, M.F. (1987) ESO conf. & Workshop Proc. No. 27, M. Azzopardi, F. Matteuci (eds.) p. 203.
Margon, B., Aaronson, M., Liebert, J. and Monet, D. (1984) AJ 89, 274.
Mould, J.R., Cannon, R.D., Aaronson, M. and Frogel, J.A. (1982) ApJ 254, 500.
Mould, J.R., Schneider, D.P., Gordon, G.A., Aaronson, M. and Liebert, J.W. (1985) PASP 97, 130.
Rebeirot, E., Azzopardi, M. and Westerlund, B.E. (1992) in preparation.
Richer, H.B. and Westerlund, B.E. (1983) ApJ 264, 114.
Schild, H. and Maeder, A. (1983) A & A 127, 238.
Tsuji, T., Iye, M., Tomioka, S., Okada, T. and Sato, H. (1991) A & A, Letter in press.
Westerlund, B.E., Edvardsson, B. and Lundgren, K. (1987) A & A 178, 41.
Westerlund, B.E., Lequeux, J., Azzopardi, M. and Rebeirot, E. (1991) A & A 244, 367.
Wood, P.R., Bessell, M.S. and Fox, M.W. (1983) ApJ 272, 99.
Zinn, R. (1981) ApJ 251, 52.

Discussion:

Schommer: I think the first carbon star in a dwarf spheroidal was discovered by Canterna and Schnickehooper in 1978, based on Washington photometry. Zinn showed a spectrum, confirming it to be a CH star, and Bessell and Norris called attention to the possible importance of carbon stars in dwarf spheroidal galaxies.

Da Costa: I was interested to see that there appears to be a good correlation between the luminosity of the carbon stars in the dwarf spheroidals and the age of the intermediate-age population. Fornax has the youngest intermediate-age stars and brightest carbon stars, opposite of the case for Sculptor and Draco. Difference between LMC and SMC could then be the result of different mean age for LMC and SMC as suggested by Mateo and others.

Nemec: Why are the N(C,L) for Draco and Carina excessively high, whereas that for Ursa Minor appears to be normal ? Do you think that the differences are intrinsic or due to selection effects ?

Azzopardi: Indeed the number of carbon stars per unit luminosity of the parent galaxies is larger by about a factor 4 in Carina and Draco. We do not believe that this can be due to biases in the carbon star searches. However the integrated luminosities of the dwarf spheroidal galaxies are so poorly known that the reality of the difference can be questioned.

STELLAR POPULATIONS IN DWARF IRREGULAR GALAXIES OF THE LOCAL GROUP

M.Tosi[1], L.Greggio[2], P.Focardi[2], G.Marconi[2]
[1] *Osservatorio Astronomico, Via Zamboni 33, Bologna, Italy*
[2] *Astronomy Dept., Bologna University, Via Zamboni 33, Bologna, Italy*

Given the uncertainties in the derivation of the current and past SFR in galaxies of any kind, the SF regime in irregulars is subject of wide debate. Since the most direct information on the stellar populations and relative histories in any system can be derived from their CMD, we have undertaken a project for accurately studying the CMDs and luminosity functions (LF) of nearby, well resolved, irregulars. The method proceeds on two tracks: a) we have developed a numerical code for Montecarlo simulations of CMDs and b) we have taken deep and accurate CCD photometry of several galaxies and derived the corresponding CMDs and LFs. The comparison of the observational data with the corresponding theoretical expectations provides several constraints on the SF history and on the IMF of the analysed objects. We do not pretend to reach unique conclusions on the evolution of irregular galaxies, but we can sensibly reduce the range of possible interpretations.

All the irregulars in our sample were selected from the DDO Catalogue and were supposed to belong to the Local Group. Detailed descriptions of the data acquisition and reduction and of the simulation code can be found in Tosi et al. (1991). In the following we briefly summarize the results relative to Sextans B, NGC 3109 and DDO 210.

Fig.1a shows the CMD of one of the two regions observed in Sextans B. This diagram contains 819 objects and its morphology is typical of all irregulars: a large dispersion of the data points, a bright blue plume, a number of bright red stars significantly lower than the corresponding number of bright blue stars. The synthetic diagram shown in Fig.1b is one of those in better agreement with the data of Fig.1a (Tosi et al. 1991). It assumes a distance modulus $(m\text{-}M)_o=25.6$ and is based on evolutionary tracks with large overshooting and low metallicity, Z=0.001 (Bertelli et al. 1986 and Greggio 1984). The adopted IMF has an exponent -2.6 (slightly steeper than Salpeter's -2.35), and the SF in the last one billion years has proceeded in two separate episodes of activity. The first from 1 Gyr to 1.5×10^8 yr ago, at the moderate rate of $2\times10^{-3}M_\odot yr^{-1}$, and the second from 1.3×10^8 yr to 3×10^6 yr ago, at an even lower rate of $1\times10^{-3}M_\odot yr^{-1}$. The cessation time of 3 million years ago may be not significant, but if the SF activity is allowed to reach the present time, bright blue stars not observed in the actual galaxy inevitably appear in the synthetic diagram.

The synthetic diagram of Fig.1b takes into account all the photometric errors, including the possible stellar blend of objects which happen to be too close in the projected plane of the CCD frames to be distinguishable from each other. This blend leads to assign spurios magnitudes and colours to the detected objects and is responsible for a large fraction of the spread in the stellar distribution on the CMD. If stellar blend is not included the same synthetic diagram becomes that of Fig.1c where the various evolutionary phases are much more separated from each other. In particular, the blue plume is now splitted into two parts: objects at the left of the vertical gap are main sequence (MS) stars, objects at its right are evolved stars at the hot edge of the blue loop evolutionary phase. This implies that the observational blue plume is populated by both MS and evolved stars. This finding

B. Barbuy and A. Renzini (eds.), The Stellar Populations of Galaxies, 207–210.

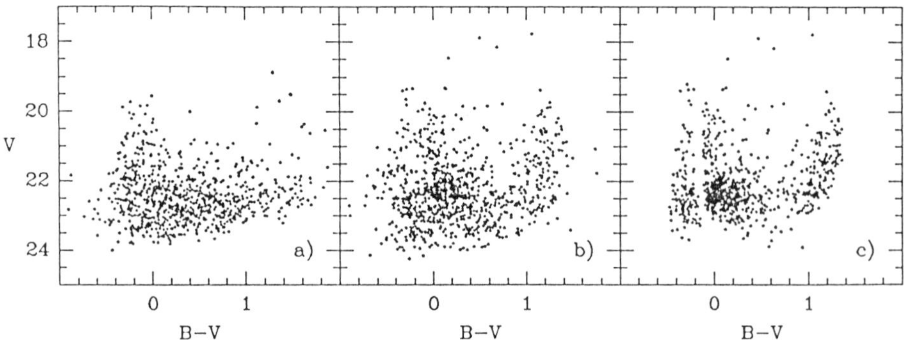

Fig.1. C-M diagrams for one region of Sextans B: a) observational, b) synthetic, c) synthetic without taking stellar blend into account.

represents a serious warning for people deriving the LF of the blue plume to infer the LF of the MS and the corresponding IMF: a safe criterion for MS selection must always be applied to avoid misleading conclusions.

The global and MS LFs of the above synthetic models are both in good agreement with the corresponding observational LFs.

From the various simulations performed for the two observed fields that cover all Sextans B, we infer (Tosi et al. 1991) that this galaxy can be treated as a single homogeneous body because the two regions contain roughly the same stellar populations, with the same low metallicity, the same IMF slightly steeper than Salpeter's and a *gasping* regime of SF.

Fig.2a shows the CMD derived from our observations of the central region of NGC 3109. It contains 1019 stars and its overall morphology is similar to that of Sextans B. According to our analysis, which is still in a preliminary stage, the best synthetic diagram for this region is shown in Fig.2b. It is based on the same tracks as that of Fig.1b and assumes a distance modulus $(m\text{-}M)_o=25.7$. The SF has again proceeded in two episodes, the most recent one stopped 7×10^6 yr ago. In this case we are confident that there is no on-going SF activity, otherwise there would be too many non observed bright MS stars. The IMF is incredibly flat, with an exponent of -1.2 which reminds the value suggested by Melnick (1987) for galaxies of such low metallicity. If stellar blending is not taken into account (Fig.2c) the blue plume again splits into two parts: the MS, on the left, fainter by almost one mag than the hot edge of the blue loops, a feature which can be recognized in Fig.2a as well. The predicted and observed LFs are in very good agreement.

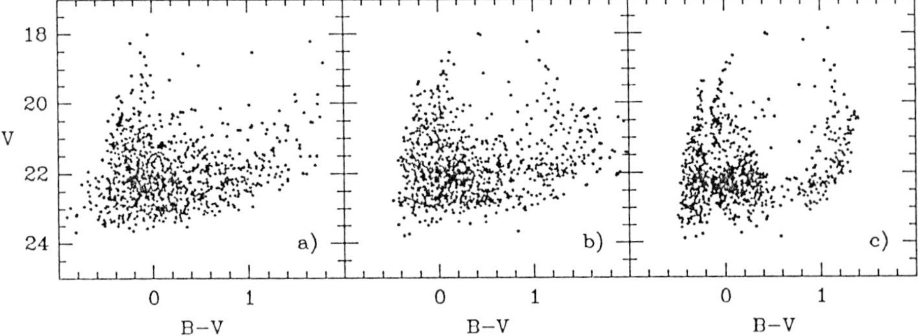

Fig.2. C-M diagrams for one region of NGC 3109: a) observational, b) synthetic, c) synthetic without taking stellar blend into account.

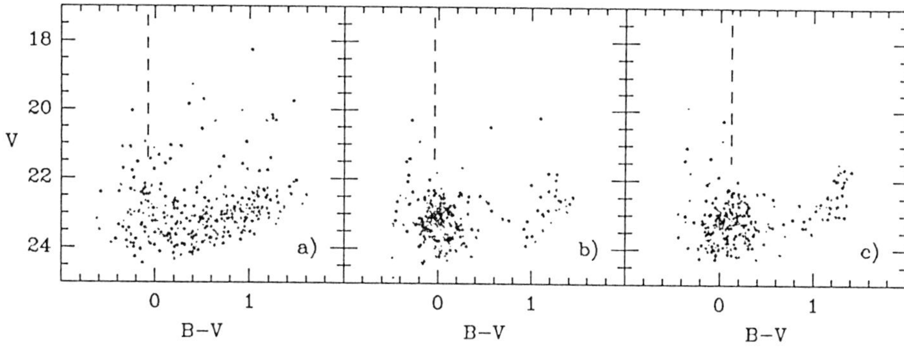

Fig.3. C-M diagrams for one region of DDO 210: a) observational, b) synthetic with $(m\text{-}M)_o=28$, c) synthetic with $(m\text{-}M)_o=27$.

DDO 210 is a highly contaminated galaxy only recently resolved in its stellar content (Marconi et al. 1990). The CMD of its central region is shown in Fig.3a. Photometric data obtained for a nearby external field allow us to infer that the stars brighter than $V\simeq21$ are most probably foreground stars, that at least half of the objects with $21\leq V\leq22$ can be members of DDO 210 (including all those bluer than $B\text{-}V\simeq0.2$), and that several background galaxies contaminate the diagram below $V\simeq21$. When only probable members are considered, the usual blue plume of irregular galaxies can be better distinguished in the diagram, the location of its main body being indicated by a vertical dashed line in Fig.3a.

In spite of the low number of objects and the high degree of contamination, some interesting indications can be derived from our analysis, especially on the distance modulus to this galaxy, which has never been properly evaluated. Figs 3b and 3c show the best synthetic diagrams obtained with $(m\text{-}M)_o=28$ and $(m\text{-}M)_o=27$, respectively. In both cases two episodes of SF have been considered: one, old and rather intense, has produced almost all the stars in the diagram; the second, very recent and very inefficient, is necessary to provide the few blue and bright stars of DDO 210. The comparison of the predicted with the observed B-V location of the blue plume shows that the model assuming $(m\text{-}M)_o=28$ (Fig.3b) is more consistent with the data. The shorter distance modulus, infact, implies smaller masses for the stars populating the blue plume, which are characterized by redder colours. This result is clearly independent of the adopted IMF and SF regime.

More conclusive results on NGC 3109 and DDO 210 will be presented in Greggio et al. (1991). We anticipate that in NGC 3109 different regions contain different stellar populations, as already found by Ferraro et al. (1989) for WLM, another Local Group irregular.

Our analysis performed so far suggests the following scenario:
a) The SF regime in the last one billion years in dwarf irregular galaxies seems to have proceeded in long periods of moderate activity interrupted by short quiescent intervals. This regime is clearly different from a bursting mode, where the single SF episodes are short and intense and separated by long quiescent intervals. From our simulations we exclude a time-decreasing SFR, as it would predict too many unobserved stars in late evolutionary phases, while the possibility of a constant SFR cannot be ruled out although it tends to predicts too many bright MS stars.
b) In spite of the modest SF activity proposed above, if the same *gasping* regime is extrapolated to all the galaxy lifetime the metallicity resulting from the corresponding stellar nucleosynthesis is much larger than observed in these galaxies. Diluting mechanisms able to reduce the metal content without altering the stellar content are then required. For a number of Blue Compacts (i.e. in a bursting regime of SF) Matteucci and Tosi (1985) found that galactic winds powered by SN explosions can make the proper effect; we then suggest that this mechanisms may be the way out of the problem for dwarf irregulars as well.

c) Strong deviations from a Salpeter's IMF do not seem to be required, except perhaps for one region of NGC 3109. This case, however, may depend on the adopted set of tracks. In this context we emphasize the need for homogeneous sets of stellar evolutionary models with different metallicities and overshooting parameters, which are still lacking in the literature.

Finally, we recall the *caveat* concerning the derivation of MS luminosity function: the observed blue plumes of irregulars can be largely populated by evolved stars and cannot thus be taken as representative of the pure MS.

References

Bertelli G., Bressan A., Chiosi C., Angerer K. 1986, *Astron.Astrophys.Suppl.Ser.* **66**, 191.
Ferraro, F.R., Fusi Pecci, F., Tosi, M., Buonanno, R. 1989, *M.N.R.A.S.* **241**, 433
Greggio L., 1984 in *Observational Tests of the Stellar Evolution Theory*, A.Maeder and A.Renzini eds (Dordrecht:Reidel), p. 329.
Greggio, L., Marconi, G., Focardi, P., Tosi, M. 1991, in preparation
Marconi, G., Focardi, P., Greggio, L., Tosi, M. 1990, *Astrophys.J.* **360**, L39
Matteucci, F. & Tosi, M. 1985, *M.N.R.A.S.* **217**, 391
Melnick J. 1987, in *Stellar Evolution and Dynamics in the Outer Halo of the Galaxy*, M. Azzopardi and F. Matteucci eds (ESO Garching FRG), p.589.
Stetson, P.B. 1987, *Pub.A.S.P.* **99**, 191
Tosi, M., Greggio, L., Marconi, G., Focardi, P. 1991, *Astron.J.* **102**, 951

Discussion

Hensler. 1) If the SF has been ceased several Myrs ago, one would expect the according SNII rate to be observed. Does a hot SN-driven gas phase exists ? Does the SN rate agree well with your model ? 2) How is the WR phase taken into account ?
Tosi. 1) No SN has been detected in the 3 irregulars discussed here, which seems consistent with the low number of massive progenitors available in the models. 2) The WR phase is taken into account implicitely when computed in the adopted stellar evolution tracks.
Serrano. Did you take into account that some of the stars are binaries ?
Tosi. In the simulations shown here binaries are not included. I have included them in the simulations for galactic open clusters and I believe that they would not make much difference in the case of our dispersed CMDs.
Faber. Have the vertical gaps ever been seen in the CMDs of nearer clusters where the errors are small, and, if not, what does that mean about the accuracy of the tracks ?
Tosi. The vertical gap has never been observed. Star clusters, though, are populated by a low number of stars, so that their CMDs are dominated by stochastic effects.
Schommer. N.Caldwell and I have surveyed DDO 210 for variables, but found none. The galaxy is well resolved, and seems to have a well developed red giant branch. I think its distance might be closer to $(m\text{-}M)_o \simeq 26$. Then, it becomes an interesting transition object, like Phoenix.
Tosi. If $(m\text{-}M)_o = 26$, all the stars of DDO 210 should be evolved off the MS and the turn-off definitely below the frame limit. But what could the blue bright stars of the diagram be?
Bruzual. What is the IMF upper mass limit in your simulations ? Have you found any evidence for dust in these galaxies ?
Tosi. The limit is 100 $M_\odot$. We have not checked the dust content.
Meurer. 1) Three of your galaxies have a definite tip to the blue plume which you interpret as a cessation time for the most recent SF event. Can this also be interpreted as an upper limit to the IMF ? 2) Have you done simulations of the CMDs you would get with observations done on nights of excellent seeing (say $0.3''$) with appropriate equipment ?
Tosi. 1) In order to interpret the MS tip only in terms of IMF cutoff (i.e. with on-going SF) we should assume an upper mass limit as low as 20 $M_\odot$, otherwise the blue plume would be too bright. But too many bright evolved stars would be predicted that are not observed. 2) No: the simulations are performed according to the real observational conditions.

BLUE LOW SURFACE–BRIGHTNESS GALAXIES

N. Bergvall
J. Rönnback
Astronomiska observatoriet
Box 515
S–751 20 Uppsala
Sweden

ABSTRACT: The first results from an analysis of CCD photometry of blue low surface–brightness galaxies are presented. The galaxies have a wide range of morphologies, luminosities and ages – all appear to be older than 2 Gyr. We also discuss the results from spectroscopy of ESO 146–G14, a large disk galaxy with low chemical abundances.

1. Introduction

Low surface–brightness galaxies (LSBGs) have come to be one of the more interesting topics in extragalactic astronomy today. Deep searches for LSBGs in clusters of galaxies [e.g. 1] have been initiated in an effort to track the faint tail of the luminosity function and to study the distribution and nature of the dark matter component of the universe. LSBGs also play an important role for the understanding of the chemical evolution of galaxies.

Different explanations for the low surface brightness have been suggested: a) The galaxies are faded remnants of starforming galaxies which have consumed the gas or lost it due to wind stripping b) The star formation rate is low because the surface gas density is close to the threshold for star formation [e.g. 2] c) The formation of the first stellar generation is just commencing – i.e. the galaxies are truly young. We wanted to investigate the last possibility, at the same time contributing to the understanding of the chemical evolution of galaxies in general.

It is obvious that the detection of young galaxies at low redshifts would have important consequences. It would give us a unique opportunity to investigate the conditions for galaxy formation, early stellar evolution and the primordial abundances in the interstellar gas. Galaxy formation theories predict [3] that star formation in protodisks occasionally may be delayed until present times if they are subject to slow contraction, thus maintaining the gas surface density below the critical threshold. A sudden very large increase in the star formation rate (SFR) can then be initiated [4] as the gas density increases due to contraction, gravitational interaction, accretion of gas or smaller galaxies. A galaxy fitting into this scenario could be the isolated HI cloud in Virgo [5], although the true age of the optical counterpart is still a matter of controversy. We recently started a pilot study of blue LSBGs (BLSBGs) in the optical and in the HI 21–cm line. Here we will discuss some of the first results from the optical study.

B. Barbuy and A. Renzini (eds.), The Stellar Populations of Galaxies, 211–214.

2. Observations

When selecting the galaxies of our sample [6] we searched the L–V ESO/Uppsala catalogue [7] for extremely blue galaxies, having $|B-R|<0^m.5$. This should guarantee a low age of the luminous stellar component and would definitely exclude normal LSB spirals and irregulars. To secure age homogeneity, we also demanded the gradient in B–R to be low. Finally, only galaxies with a surface magnitude $\mu(B,eff) > 23^m.5$ were accepted. This left us with about 60 candidates. We are aware that the uncertainties in the catalogued magnitudes and colours will have the consequence that some redder galaxies will be scattered into our sampling area. Only broadband CCD photometry can give us a definite indication as to whether these galaxies have extreme properties or not. So far, we have obtained b,V and "Gunn i" images at ESO and NOT and HI observations at Westerbork of about 50% of the sample. Spectra have been obtained of a few galaxies. The data will be used to study the stellar content and distribution, the chemical abundances, the fractional HI mass and the spatial distribution of the galaxies.

3. Results.

3.1 MORPHOLOGIES, SIZES AND LUMINOSITIES

Morphologically the sample is quite heterogeneous. Three main morphological types can be recognized – disks, irregular galaxies with prominent HII regions and amorphous galaxies. A small part of the sample seem to be nearby dwarfs.The disk galaxies often exhibit warped structures and no or inconspicuous bulges. In contrast to the irregulars, the amorphous galaxies, despite their blue colours, have a smooth appearance and do not seem to contain any bright HII regions.This could be due to a recent gas heating and expulsion by winds from massive stars. These are also the galaxies which tend to increase in number as one approaches the limiting surface magnitude of the ESO Schmidt plates. About a dozen measured velocities yield dimensions and luminosities in the ranges 10–50 kpc and $M_B = -15$ to -19 (H_0=50 kms^{-1}Mpc^{-1}).

3.2 COLOURS AND AGES

Relevant information about the initial mass function (IMF) and ages of the dominant stellar generations can be obtained from optical/infrared spectroscopy or from a combination of spectroscopy and broadband photometry. Since spectroscopy is very difficult to obtain for the majority of the objects we have to rely on information from the broadband CCD photometry. These data can be compared with predictions from spectral evolutionary models under different assumptions about the IMF and the temporal variation of the star formation rate (SFR).

Fig. 1 shows the colour properties of galaxies for which we have reliable photometry. The diagram also includes the predicted evolutionary tracks for two different star formation histories. We assumed a Salpeter mass function, a mass range $0.1<M<100\ M_\odot$ and a metallicity of 5% solar. Nebular emission was included. When we calculated the synthetic colours we took into account both the actual filter transmission profiles and the response curve of the CCD used for the observations. Therefore our b–V is about $0^m.1$ redder than the corresponding Johnson/Cousins B–V for our programme galaxies. If we keep this in mind we see that the mean B–V of our BLSBGs are about $0^m.1$ bluer than the bluest of the previous samples of LSB galaxies discussed in the literature [e.g. 8]. Still we note that none of the galaxies observed so far is extremely blue. As a whole, the sample does not agree with a young stellar population. Even if we account for a small amount of reddening, the stellar population appears to be > 2 Gyr old for the majority of the galaxies. Some exceptions may be found at the lower left envelope of the distribution. One of the galaxies in this region is ESO 146–G14, one of the few for which we have obtained a spectrum.

This large (≈30 kpc) disk galaxy has an unusually low oxygen abundance, about 4% of the solar value. With M_B=–17 it thus breaks the metallicity–luminosity relation [9] that holds for other low luminosity galaxies. The oxygen abundance, as derived from the empirical relations discussed by [10], is nearly constant across the disk, indicating strong mixing or infall of processed gas. The rotation curve shows signs of mass outflows from the star forming regions but this result probably needs to be checked [11] with spectroscopy at higher dispersion. The stellar absorption spectrum shows prominent Balmer lines across the galaxy, including the bulge. From model comparisons [12,13] we find that the colours (after extinction corrections), the spectrum and the low abundances are all consistent with an age of 4–6 Gyr, corresponding to z ≈ 0.4, assuming Ω=1 and a normal IMF.

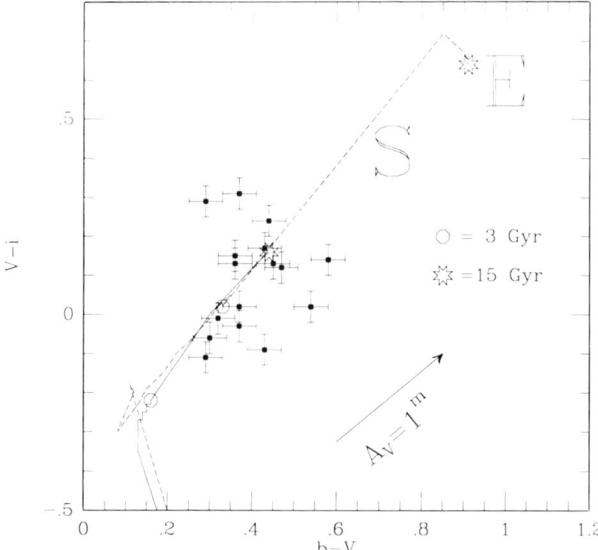

Figure 1. Colour–colour diagram of b–V versus V–i for BLSBGs. Also marked in the figure is the position of normal spiral galaxies (S) and elliptical galaxies (E). Evolutionary tracks up to an age of 15 Gyr are presented for a Salpeter mass function and two different star formation histories, constant SFR (—) and a short burst (– – –). The evolution up to 3 Gyr is from our models, assuming 5% solar metallicity. The later stages are from [14]. The reddening vector and the mean errors in the observations are indicated.

References

1. Davies, J.I., Phillipps, S., Disney, M.J. (1989) Mon. Not. R. Astr. Soc. 244, 385

2. van der Hulst, J.M., Skillman, E.D., Kennicutt, R.C., Bothun, G.D. (1987) Astr. Astrophys 177, 63

3. Silk, J., Szalay, A.S. (1987) Astrophys J. 323, L107

4. Struck–Marcell, C., Scalo, J.M. (1987) Astrophys. J. Suppl. Ser. 64, 39

5. Salzer, J., Alighieri, S., Matteucchi, F., Giovanelli, R., Haynes, M.P. (1991) Astron. J. 101, 1258

6. Bergvall, N., Rönnback, J. (1990) Proc. Nordic Baltic Meeting, ed. C–I. Lagerkvist, D. Kiselman, M. Lindgren, Uppsala, 71

7. Lauberts, A., Valentijn, E.A. (1989) "The Surface Photometry Catalogue of The ESO/Uppsala Galaxies", European Southern Observatory, Munich

8. Schombert, J.M., Bothun, G.D., Impey, C.D., Mundy, L.G. (1990) Astron. J. 100, 1523

9. Skillman, E.D., Kennicutt, R.C., Hodge, P.W. (1989) Astrophys. J. 347, 875
10. Skillman, E.D. (1989) Astrophys. J. 347, 883
11. Schweizer, F. private communication
12. Bergvall, N. (1991) in preparation
13. Olofsson, K. (1991) private communication
14. Arimoto, N., Yoshii, Y. (1986) Astron. Astrophys. 164, 260

Discussion:

H. Ferguson: Your velocity diagram for ESO 146–G14 showed no sign of coherent rotation. Is this unusual for a disk of its scale length?

Bergvall: Since the spectra are of low dispersion, our data for the stellar absorption features are afflicted with rather large uncertainties and do not exclude a rotation of the order of 200 kms^{-1}.

G. Hensler: 2 comments: Firstly, Theis, Burkert and Hensler (1991, subm. to Astron. Astrophys.) demonstrate by means of chemo–dynamical evolutionary models of galaxies that all cosmological 1σ and 3σ density fluctuations with masses of 10^9 to $10^{12}\,M_\odot$ are evolving into the $\tilde{\rho}$–M (mean density–mass) region where dwarf and giant spheroidal systems are located from observations. Exceptionally, only galaxies that start with $10^{11}\,M_\odot$ from 1σ fluctuations are not able to collapse thermally but are self–regulated by their star formation, i.e. stars produce hot gas that prevents the collapse and controls by this the subsequent star formation and the cooling timescale determines the galactic evolution. Such a galaxy remains at approximately $\tilde{\rho}\approx10^{-3}\,M_\odot pc^{-3}$ and looses mass continuously. Secondly, these low surface–brightness galaxies would be disrupted in clusters of galaxies due to encounters and are, therefore, observationally found only in the field.

Bergvall: Our data indicate that the BLSBGs indeed tend to avoid regions of high galaxy density.

J. Frogel (Question to Bergvall and Tosi): Can there be an evolutionary link between the dIrr and blue low surface–brightness galaxies?

M. Tosi: Yes, in my opinion these galaxies may well be related to the irregulars of our sample. Clearly our selection of only galaxies in the Local Group prevents any suggestion about clustering.

Bergvall: Our sample seems to contain galaxies of widely different sizes and luminosities, some of which I am sure are related to dIrrs. The difference lies in the surface brightness, reflecting either different ages or different SFR.

H. Dottori: Have you determined the equivalent width of the H lines in ESO 146–G14?

Bergvall: The W(Hα) in emission is between 10 and 350 Ångström. If you refer to the absorption lines we find that W(Hδ) is 5–8 Å in all parts of the galaxy. For a stellar population with an exponentially fading star formation rate, the measured equivalent widths of the absorption lines, in conjunction with the measured W(Hα) in emission, imply an intermediate age.

STAR AND STAR CLUSTER SPECTRAL LIBRARIES

E. Bica
Instituto de Física-UFRGS, Porto Alegre, RS, Brazil

ABSTRACT. This paper reviews spectral libraries of stars and star clusters, together with their applications to population synthesis. The problem of abundance calibrations for metal rich populations is also addressed, in particular index definitions and non-solar CNO/Fe ratios. A stellar population data bank would be important to accelerate progress in the field and would optimize the use of future telescope time.

1. INTRODUCTION

Stellar libraries have been collected since quite long for the analysis of composite stellar populations. An early example is de Vaucouleurs and de Vaucouleurs' (1959) synthesis of an LMC bar integrated spectrum. In the 60's and early 70's the libraries consisted of photographic spectra (e.g. Spinrad 1962) and photoelectric photometry like Faber's (1973) 10 colour system. The natural evolution of these techniques with the development of linear detector arrays was spectrophotometry. In the early studies and in the subsequent ones with low resolution scanners it was clear the intention by the authors to visualise the models they were computing (e.g. O'Connell, 1976). Detailed spectral visualisations became possible by means of observations carried out with high resolution scanners and/or CCD detectors (e.g. Pickles 1985; Bica 1988, hereafter B88).

2. STELLAR LIBRARIES

So far the stellar libraries which have been used more often are Gunn and Stryker's (1983, hereafter GS83) and Jacoby *et al.*'s (1984, hereafter JHC84). The main advantage in GS83 is the wide spectral coverage (3130< λ <10800 Å), but the resolution is low (20 Å in the blue and 40 Å in the near-infrared). JHC84 spans 3510< λ <7427 Å at a resolution of $\approx$4.5 Å. The libraries contain respectively 175 and 161 entries, which consist mostly of solar neighbourhood stars of spectral types O to M and luminosity classes from V to I. Pickles (1985) presented a library of 200 stars at $\approx$15 Åresolution, in the range 3600< λ <10000 Å ; this library was later complemented with Baade window giants (Pickles and van den Kruit 1990). Faber *et al.* (1985) studied 110 stars at 9 Å resolution in the range 4000< λ <6200 Å. They are K giants and subgiants, as well as some giants in metal poor globular clusters. Alloin and Bica (1989 and references therein) studied the near-infrared NaI and CaII lines at 3 Å resolution in F to M stars of luminosity classes V to I. These spectra are corrected for earth atmosphere absorptions, and are complementary to JHC84 in wavelength range. Rose (1985) collected a high resolution blue- violet library, in a stellar population study based on central depth of lines.

Many other stellar libraries and data sets exist in the literature, but it would be impossible to mention all of them here. As an illustration of spectra dedicated to particular types of stars, I mention Mould and Aaronson's (1980) library of carbon and other late

B. Barbuy and A. Renzini (eds.), The Stellar Populations of Galaxies, 215–224.

type stars and Melnick's (1985) one of early type stars in 30 Doradus.

2.1 INFRARED AND ULTRAVIOLET

Detector developments in the infrared have made possible the acquisition of high quality spectra. Frogel *et al.* (1991) studied 18 solar neighbourhood and 14 Baade window M giants in the range $1< \lambda <2.5$ μm, which were complemented with visible spectra. The IUE data bank contains most types of stars, in the range $1000< \lambda <3000$ Åat an average resolution of ≈7 Å.

2.2 A DATA BANK FOR STELLAR POPULATIONS

Now it has been ≈20 years of linear detector spectra of stars and star clusters (sect. 3). In addition, an enormous amount of galaxy spectra have been accumulated in the literature in the form of individual objects or sets. Examples of large sets are the blue to red spectra of 455 Ellipticals by Faber *et al.* (1989), 320 galaxies of various types (Véron-Cetty and Véron 1986) and 161 spirals (Keel *et al.* 1985). Bica and Alloin (1987a, 1987b; hereafter BA87a and so on) have collected spectra for 170 galaxies in the interval $3700< \lambda <10000$ Å , which were recently complemented with the near-ultraviolet, in view of connecting template populations to IUE spectra. It would be important to store all these spectra in a stellar population data bank, which could be operated by existing facilities like the CDS and the NASA-IPAC. This would certainly accelerate progress in the field and would optimize future telescope time allocation, by improving S/N ratio and complementing wavelength ranges for the same objects, and by observing new ones. The stored data should be as much calibrated as possible by the authors: absolute flux calibrations are not essential, relative ones are ideal, but only wavelength calibrated spectra are useful too.

2.3 MODEL LIBRARIES

Spectral models based on stellar atmospheres may turn out to be the only way to have some stars with special characteristics in libraries, like low luminosity metal poor stars. Kurucz (1979) presented a model library which included ≈1 million atomic lines, but the absence of molecular lines precluded the generation of realistic models for late spectral types. Only recently molecular lines have been included in detail, together with atomic lines (e.g. Erdelyi Mendes and Barbuy 1991). Kurucz (this meeting) has presented a new series of models with a comprehensive set of molecular and atomic data. Much work is still necessary in spectral models in order to fit in detail observed spectra, in particular for late type and/or metal rich stars, and as well as for objects with non-solar CNO/Fe ratio.

2.4 POPULATION SYNTHESIS WITH STELLAR LIBRARIES

Population synthesis using stellar libraries has been often applied to nuclei (e.g. Pickles 1985) and other subsystems in galaxies (Gregg 1989), allowing one to obtain information such as fractions of different stellar types and age components. Applications of stellar libraries to star cluster integrated spectra have not been exploited as much. Clusters present an additional synthesis constraint with respect to galaxies, which is the statiscally complete parts of the observed HR diagram. Santos Jr. *et al.* (1990) have studied the rich Galactic open cluster M11 and now we have applied the same method to the moderately metal rich globular cluster 47 Tuc in the range 3100< λ <9800Å. It is possible to infer on the low main sequence IMF: the near-infrared is essential for this purpose and the infrared range should be even more discriminating. The 47 Tuc minus model residuals in the MgI 5175Å region point to a higher heavy element abundance in the model, as expected because it uses solar neighbourhood stars from GS83's library. However 47 Tuc presents a *stronger* blue-violet blanketing which, according to a synthesis with laboratory molecular patterns, it is very possibly caused by molecules involving C,N and O (Santos Jr. et al., this meeting). This would suggest that non-solar CNO/Fe ratios occur in the Halo/Bulge transition, similarly to those detected for O/Fe in halo giants (Barbuy 1988).

Stellar libraries have also been used for spectral visualisation of evolutionary synthesis models (Bruzual 1983; Guiderdoni and Rocca-Volmerange 1987).

3. STAR CLUSTER LIBRARIES

Early studies of cluster integrated spectra were based on photographic material (e.g. van den Bergh 1969). During the 80's many observations based on modern detectors were published: Burstein *et al.* (1984) studied M31 and Galactic globular clusters (GGC) in the range 3900< λ <6200 Å; Rabin (1982) analysed intermediate and old age Magellanic Cloud clusters (MCC), GGCs and the Galactic open cluster (GOC) NGC2243 in the range 3800< λ <6200 Å ; Rose (1985) collected high dispersion spectra of GGCs and the GOC M67 spanning 3800< λ <5200 Å, whereas Tripicco (1989) performed a similar analysis for M31 clusters. Huchra *et al.* (1991 and references therein) have derived velocities and metallicities for M31 and GGCs. All such studies concentrated efforts on the visible part of the spectrum, in particular the blue region. BA86a and BA87b have studied GGCs, GOCs and MCCs: in addition to the visible range, the near-infrared one has been observed and recently the data set has been extended to the near-ultraviolet.

3.1 POPULATION SYNTHESIS BASED ON A CLUSTER LIBRARY

B88 presented a population synthesis method of galaxy nuclei using the base of star cluster spectra in BA86a and BA87b. The synthesis derives ages and metallicities of population components. Other parameters controlling the star formation are inlaid in the cluster spectra, which considerably simplifies the analysis. A *grid* of cluster equivalent widths and underlying continua as a function of age and metallicity (BA86b, BA87b) is used in the computations, whereas the cluster spectra are employed for visualisation purposes. Two algorithms were developed for the solution of the inverse problem: a) combinations of grid elements (B88); (b) multi-minimization which basically merges the advantages of the combination method with those of classical minimization algorithms (Schmidt *et al.* 1989). In addition to Shapley-Ames galaxy nuclei (B88), the synthesis has been applied to the M31, M32 and NGC205 nuclei (Bica *et al.* 1990a), M33 and other blue nuclei (Schmidt *et al.*

1990), and galaxies in the distant cluster Abell 370 (Jablonka *et al.* 1990). The subtraction of an appropiate population model has allowed one to study in detail emission components, in some cases to an unprecedent luminosity level (Bonatto *et al.* 1989; Storchi-Bergmann *et al.* 1991).

Metallicity and age scales are not yet settled, but the synthesis results are essentially independent of these uncertainties because they connect directly observables in the galaxy spectra to those in the cluster grid.

3.2 ULTRAVIOLET STUDIES

The cluster library, combined to IUE data, was used to probe the three possible explanations for the UV turnup in giant E galaxies (BA88 and references therein). No new synthesis was performed, we simply checked in the UV the behaviour of the synthesis components previously obtained from the visible/near-infrared ranges in B88. a) A young burst of star formation reproduces the turnup and is compatible with the synthesis; b) horizontal branch stars associated to the old metal poor components have a negligible UV contribution, and even if one exagerates such amount, it does not fit the shape of the UV turnup unless a very anomalous HB is present; c) post AGB stars associated to the old metal rich components (the dominant ones in the giant E synthesis) explain the turnup, if ≈5 of such UV bright stars were detected in very metal rich globular clusters like NGC6528 and NGC6553.

The near-ultraviolet range is very important not only for the possibility of connections to UV satellite observations, but also for its spectral features. A new library of star clusters and galaxies has been collected in this range (Bica et al. 1991a). The 3360 Å NH band shares the effect of metallic feature wavelength dilution (BA86a, BA87b), as one approaches younger ages. Index definitions for the balmer jump and the 4000 Å break throughout the literature are basically the same. We illustrate in Fig. 1 their behaviour in globular clusters of various metallicities. For the metal rich group G1 the feature is purely 4000 Å break, whereas for the metal poor group G5 it is mostly balmer jump from blue horizontal branch stars.

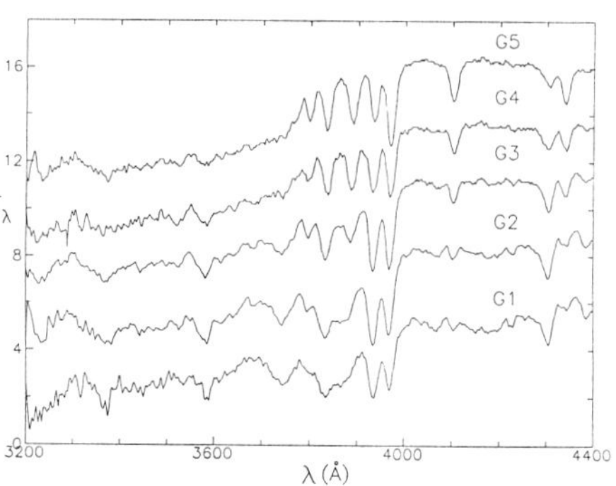

Fig. 1

3.3 CLUSTER SPECTRA AND UBV PHOTOMETRY IN THE LMC

The spectral evolution of LMC star clusters younger than 500 Myr presents two phases where the flux from red stars is enhanced in the red/near-infrared ranges (Bica et al. 1990b, BA87b, BA86a): a) the red supergiant phase (RSG) at t≈10 Myr and b) there is evidence of an effect involving M type AGB stars at t≈100 Myr. Recently we have enlarged the sample of LMC clusters with UBV photometry to 624 entries. We have detected the Helium-flash gap (Fig. 2) at t≈600 Myr (Bica it et al. 1991b), which denotes the first appearance of red giant branch stars (Sweigart *et al.* 1991, and references therein). Fig. 2 presents many other features which are worth commenting. The youngest HII regions are not in the upper left corner of the (U-B) vs (B-V) diagram, mainly because of the way emission lines evolve in the different filters. The massive clusters containing RSGs are concentrated in a clump, whereas many small ones jump to red (B-V) colours because of stochastic effects. The RSG phase has been confirmed in models of integrated colour evolution using two different sources for massive star tracks (Arimoto and Bica 1989; Girardi and Bica in this meeting). There is a marginal evidence for a gap near t≈20 Myr in Fig. 2 which could be associated to the AGB phase transition. It should be recalled that the age ticks are from a scale without overshooting.

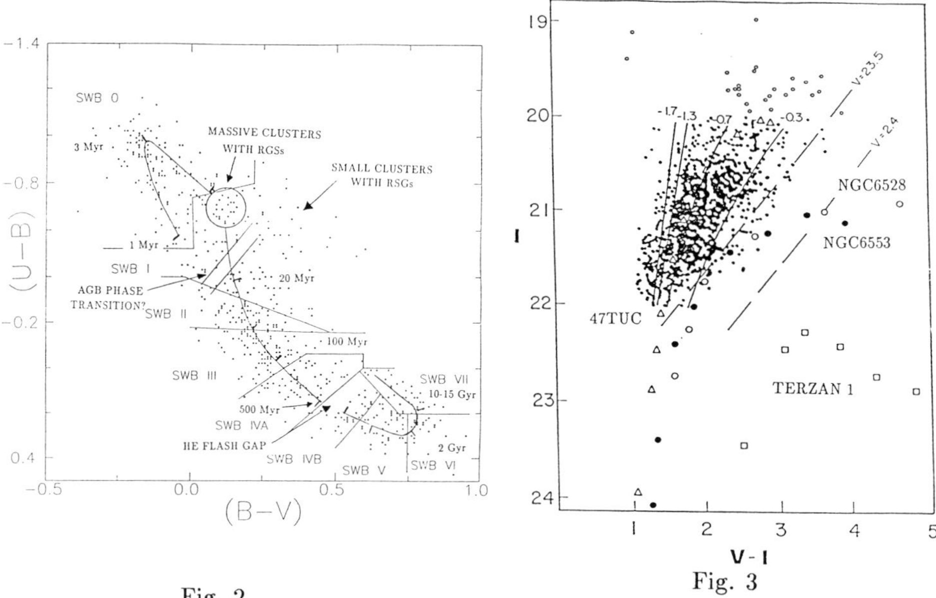

Fig. 2

Fig. 3

3.4 CLUSTER LIBRARY AND CMDs OF LOCAL GROUP GALAXIES

The bulge globular clusters NGC6528 and NGC6553 present very strong-lined spectra, comparable to those in many giant galaxy nuclei (BA86a, BA87b). They are clearly more metallic than clusters like 47 Tuc, which is usually taken as prototype of metal rich ones in the literature. Ortolani *et al.* (1990) have started a systematic colour magnitude diagram (CMD) survey of bulge clusters, under excellent seeing conditions, because of crowded fields. Bica *et al.* (1991c) have gathered such clusters in the absolute I vs (V-I) diagram and compared them to CMDs of local group galaxies in the literature. It is clear from this study that metallicity histograms based on the red giant branch (RGB) width should not be performed at constant luminosity level because metal rich RGBs change their morphology and become fainter, owing to blanketing effects. We illustrate in Fig. 3 the cluster sequences on the M32 CMD by Freedman (1989). There is an inclined observational cutoff through the diagram because the plate limits are different. Metal rich populations such as that in NGC6553 have not been attained in M32's CMD. According to the M32 synthesis in Bica *et al.* (1990), this component is necessary because those corresponding to the observed part of the CMD are not strong-lined enough to account for the M32 spectrum. Freedman pointed out that the average metallicity she was deriving was possibly a lower limit. Our synthesis intermediate age component is compatible with Freedman's detection of AGB stars, whose luminosity points to an age of ≈5 Gyr. It should be pointed out that the intermediate age giant branch certainly overlaps with part of the old age RGB. The synthesis metallicity dispersion corresponding to the observed part of the CMD is also compatible with Freedman's study: small contributions are observed for [Z/Z⊙]<-1, in particular if one considers that the stars close to the plate limit have larger photometric errors. The synthesis shows a metallicity dispersion, not only for a similarly small amount of metal poor components, but also because the solar and the [Z/Z⊙]=-0.5 components are spectroscopically very different. Population synthesis in a wider spectral range and a deeper CMD are necessary to shed more light on the M32 question.

3.5 M31 CLUSTERS

We have recently observed 2 open and 7 globular clusters in M31 (Jablonka *et al.* 1991). The wide spectral range allows one to infer confidently on age, metallicity and reddening, based on our previous grid and spectral cluster data. In particular we find evidence that the cluster Mayall IV (G219) is not a classical metal poor globular cluster as previously classified. Its properties resemble those of an intermediate age cluster of [Z/Z⊙]≈-1. Previous studies were performed in the blue-visual region, where metallic features appear to be considerably diluted by the age effect, which is not the case in the near- infrared. An inspection of the spectra in Burstein *et al.* (1984), which reach the red range, shows that indeed M IV is considerably bluer than globular clusters of [Z/Z⊙]≈-2.

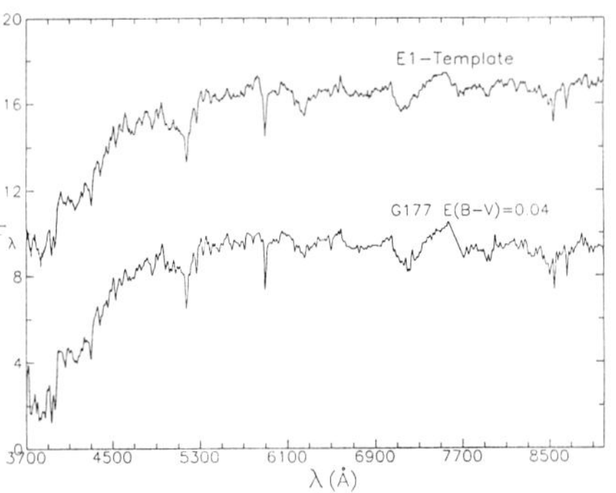

Fig. 4

In the inner bulge of M31, G170 is a cluster as strong-lined as NGC6553 in our Galaxy; G177 and G158 are even more metallic. G177 is compared in Fig. 4 to an average of the strongest-lined E galaxies in B88. The absorption features are comparable. This definetely throws down a dogma in the stellar population literature, which used to state that star clusters do not overlap with giant galaxies in spectral properties. This wrong notion dissiminated because in previous comparisons of clusters and galaxies the "metal rich" clusters were similar to 47 Tuc.

3.6 CNO ENHANCEMENT IN BULGES?

In addition to the super metal rich cluster G177 in the central bulge of M31, we have studied G158, which is as strong-lined as G177 (Fig. 5a). However they differ in the sense that G177 has a stronger blue-violet blanketing, as illustrated by the difference spectrum in Fig. 5b, where the residuals between 47 Tuc and its model built with solar neighbourhood disc stars (section 2.4) are also shown. The blue-violet residuals are very similar, although in one case we are dealing with super metal rich clusters while in the other with sub-solar and solar heavy element abundances. The fact that G158 behaves like disc stars led us to suspect that it is an inner disc cluster, whereas G177 is a genuine inner bulge cluster (Bica *et al.* 1991d). According to the molecular synthesis in sect. 2.4 the blue-violet blanketing arises from molecules involving CNO elements. These evidences point to a non-solar CNO/Fe ratio in bulges.

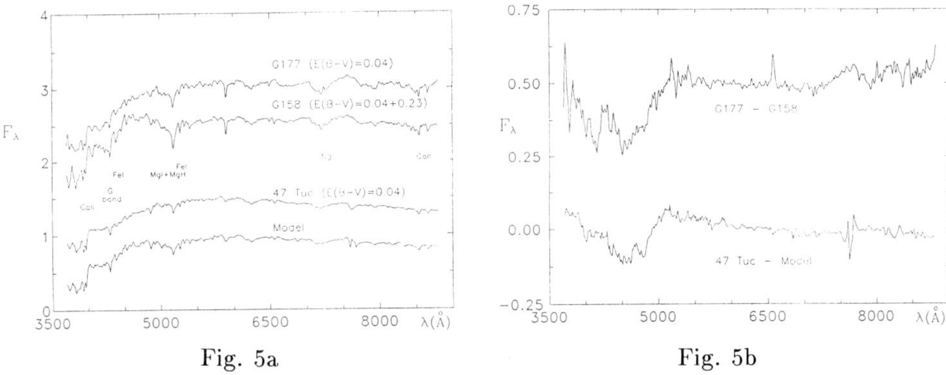

Fig. 5a Fig. 5b

3.7 THE CALIBRATION PROBLEM OF METAL RICH POPULATIONS

Recently new results on the metallicity calibration of metal rich populations have been presented. According to Zinn and West's (1984) scale, NGC6553 has [Fe/H]=-0.29, slightly lower than that derived by Barbuy *et al.* (1991) from high dispersion analysis of the giant III-17. The star shows a Nitrogen enhancement.

Brodie and Huchra (1990) have derived a spectral metallicity calibration for extra-galactic globular clusters, which was tied to Zinn and West's scale. The application of this scale to G158 and G177 led to [Fe/H]=-0.26 and -0.15 respectively (Huchra *et al.* 1991). This would imply that the average [Fe/H] for the strongest-lined E galaxies is essentially solar (Fig. 4). In this scenario, the blue-violet blanketing excesses should be due to higher than solar CNO/Fe ratios. However in detail there are still discrepancies; a comparison of G158 to a solar abundance old cluster model (Fig. 5a) shows that G158 cannot have [Fe/H]=-0.26, because line strengths are at least a factor two stronger in G158.

The problem of scales appears to be not just one of ranking and selecting cluster calibrators. The way that indices are defined might play an important role too. The continuum tracings for equivalent widths in BA86a are sensitive to a global blanketing in the region 4050< λ <4500 Å, which is almost comparable to the C_2 and MgH absorptions around 5100 Å. In the Lick system the best metallicity indices are MgH and Mg2 which have flux side bands in high continuum zones. Mgb is not as good, certainly because its sidebands are located within the MgH absorption. Brodie and Huchra have used six indices as primary calibrators. Four of them, i.e. the CNB, G, Fe52 and Δ , have sidebands in absorbed regions. Some of these indices might saturate, whereas the tracings would still show differences among metal rich populations. It would be important to compare in detail index behaviour among different systems.

The metallicity scale used in B88 relies on an average of cluster values from Zinn's third calibration (Zinn and West, 1984) and from Bica and Pastoriza (1983). The latter calibration is similar to Zinn's first one (high values). It might turn out that Zinn's first calibration is closer to a Z_{CNO} scale, while the third one to a Z_{Fe} scale.

4. CONCLUSIONS

A data bank for spectra of stars, star clusters and galaxies would accelerate progress in stellar population studies, and would optimize the use of telescope time. Much of the blanketing excess in bulge populations with respect to the solar spectrum appears to be caused by molecules involving CNO elements. The problem of line index definitions is of major importance for the abundance calibration of metal rich populations.

ACKNOWLEDGMENT: I thank the State of Rio Grande do Sul Science Foundation FAPERGS for a grant which made possible UFRGS group members to attend this meeting.

REFERENCES

Alloin, D., Bica, E. 1989, A&A, 217, 57
Arimoto, N., Bica, E. 1989, A&A, 222, 89
Barbuy, B. 1988, A&A, 191, 121
Barbuy, B., Castro, S., Ortolani, S., and Bica, E. 1991, A&A, submitted
Bica, E. 1988, A&A, 195, 76
Bica, E., Alloin, D. 1986a, A&A, 162, 21
Bica, E., Alloin, D. 1986b, A&AS, 66, 171
Bica, E., Alloin, D. 1987a, A&AS, 70, 281
Bica, E., Alloin, D. 1987b, A&A, 186, 49
Bica, E., Alloin, D. 1988, A&A, 192, 98
Bica, E., Alloin, D., and Schmidt, A. 1990a, A&A, 228, 23
Bica, E., Alloin, D., and Santos Jr., J. F. C. 1990b, A&A, 235, 103
Bica, E., Alloin, D., and Schmitt, H., 1991a, in preparation
Bica, E., Claría, J. J., Dottori, H., Santos Jr., J. F. C., and Piatti, A. 1991b, ApJ Letters, in press
Bica, E., Barbuy, B., Ortolani, S. 1991c, ApJ Letters, in press
Bica, E., Jablonka, P., Santos Jr., J. F. C., Alloin, D., Dottori, H. 1991d, A&A, in press
Bica, E., Pastoriza, M. 1983, ApSS, 91, 99

Bonatto, Ch., Bica, E., and Alloin, D. 1989, A&A, 226, 23
Brodie, J. P., Huchra, J. P. 1990, ApJ, 362, 503
Bruzual, G. 1983, ApJ, 273, 105
Burstein, D., Faber, S. M., Gaskell, C. M., and Krumm, N. 1984, ApJ, 287, 586
Erdelyi-Mendes, M., Barbuy, B. 1991, A&A, 241, 176
Faber, S. M. 1973, ApJ, 179, 731
Faber, S. M., Friel, E. D., Burstein, D., and Gaskell, C. M. 1984, ApJS, 57, 711
Faber, S. M., Wegner, G., Burstein, D., Davies, R. L., Dressler, A., Lynden- Bell, D., and Terlevich, R. J. 1989, ApJS, 69, 763
Freedman, W. L., 1989, AJ, 98, 1285
Frogel, J. A., Tendrup, D. M., and Whitford, A. E. 1991, ApJ, in press
Gregg, M. D. 1989, ApJ, 337, 45
Guiderdoni, B., Rocca-Volmerange, B. 1987, A&A, 186, 1
Gunn, J. E., and Stryker, L. L. 1983, ApJS, 52, 121
Huchra, J. P., Brodie, J. P., and Stephen, M. K. 1991, ApJ, 370, 495
Jablonka, P., Alloin, D., and Bica, E. 1990, A&A, 235, 22
Jablonka, P., Alloin, D., and Bica, E. 1991, A&A, submitted
Jacoby, G. H., Hunter, D. A., and Christian, C. A. 1984, ApJS, 56, 257
Keel, W. C., Kennicutt, R. C., Hummel, E., and van der Hulst, J. M. 1985, AJ, 90, 708
Kurucz, R. L. 1979, ApJS, 40, 1
Melnick, J. 1985, A&A, 153, 235
Mould, J., Aaronson, M. 1980, ApJ, 240, 464
O'Connell, R. W. 1976, ApJ, 206, 370
Ortolani, S., Barbuy, B., and Bica, E. 1990, A&A, 236, 362
Pickles, A. J. 1985, ApJ, 296, 340
Pickles, A. J., van der Kruit, P. C. 1990, A&AS, 84, 421
Rabin, D. 1982, ApJ, 261, 85
Rose, J. A. 1985, AJ, 90, 1927
Santos Jr., J. F. C., Bica, E., and Dottori, H. 1990, PASP, 102, 454
Schmidt, A., Bica, E., and Dottori, H. 1989, MNRAS, 238, 925
Schmidt, A., Bica, E., and Alloin, D. 1990, MNRAS, 243, 620
Spinrad, H. 1962, ApJ, 135, 715
Storchi-Bergmann, T., Bica, E., and Pastoriza, M. 1990, MNRAS, 245, 749
Sweigart, A. V., Greggio, L., and Renzini, A. 1990, ApJ, 364, 527
Tripicco, M. J. 1989, AJ, 97, 735
de Vaucouleurs, G., de Vaucouleurs, A. 1959, PASP, 71
van den Bergh, S. 1969, ApJS, 19, 145
Véron-Cetty, M. P., Véron, P. 1986, A&AS, 66, 335
Zinn, R., West, M. 1984, ApJS, 55, 45

DISCUSSION

VAN DEN BERGH: The metal-poor cluster Mayall IV is located in the *outer* halo of M31. How would you account for the existence of a young population component in such an object?

BICA: M IV might have been formed in an interaction of M31 with a companion, or else it suggests an usual scenario for galaxy formation with late gas clouds. In our Galaxy the cluster Ruprecht 106 poses a similar problem (Buonanno *et al.* 1991, AJ, 100, 1811). M IV might as well have originated from a gas cloud accreted by M31. The occurence of an intermediate age cluster in the outer halo is not more defying to canonical models than that of the metal rich globular M II (G1), which is more distant in the M31 halo. In Christian *et al.*'s CMD (1991, AJ, 101, 848), M IV shows two RGB sequences (their Fig. 3). They interpreted one as a metal poor classical globular and the other as a metal rich field contamination. An intermediate age RGB might occupy the same locus as the former. Another possibility is a merger, as might be the case of ω Cen.

Rocca-Volmerange: We compared the template globular cluster G2 from B88 with our synthetic stellar population model using Yale tracks. The comparison gives an excellent fit at the same age 17 Gyrs for both. But if the evolutionary tracks change I assume that your time evolution scale has also to change.

BICA: Ages attributed to clusters through different tracks may change, but the synthesis results with the cluster grid will remain essentially unchanged because it is an inverse problem of feature equivalent widths in a galaxy against those in the cluster grid. The fractions will not change, only the ages attributed to them. The same holds true for eventual changes in metallicity calibrations.

MOULD: You mentioned that you see evidence for the AGB phase transition in Magellanic Cloud clusters. What turn-off mass would you associate with this phase transition, and is this consistent with the massive AGB star evolution discussed by Renzini yesterday?

BICA: There is a clear gap denoting the RGB phase transition in SWB IV clusters; for the AGB phase transition there is only marginal evidence of a gap in the (U-B) vs (B-V) diagram in SWB I, close to the borderline with SWB II clusters (Fig. 2). We find spectroscopically an enhancement of M star features in clusters of age $\approx$100 Myr (SWB III) like NGC1866 (Bica *et al.* 1990b). In the CMD of Galactic open clusters of this age there is evidence for an extended AGB of M stars (Bica et al. 1990, Rev. Mexicana Å, 21, 202), which might be the peak flux contribution of massive AGB stars. According to Renzini the turn-off mass of clusters in the AGB phase transition should be at $\approx$5 M$\odot$ and for the RGB one at $\approx$2 M$\odot$. The observational evidences are basically consistent the massive AGB star evolution discussed by Renzini.

MODEL ATMOSPHERES FOR POPULATION SYNTHESIS

Robert L. KURUCZ
Harvard-Smithsonian Center for Astrophysics
60 Garden Street
Cambridge, MA 02138
U.S.A.

ABSTRACT. I have used my newly calculated iron group line list together with my earlier atomic and molecular line data, 58,000,000 lines total, to compute new opacities for the temperature range 2000K to 200000K. Calculations have been completed at the San Diego Supercomputer Center for 56 temperatures, for 21 pressures, for microturbulent velocities 0, 1, 2, 4, and 8 km/s, for 3,500,000 wavelength points divided into 1221 intervals from 10 to 10000 nm, for scaled solar abundances [+1.0], [+0.5], [+0.3], [+0.2], [+0.1], [+0.0], [-0.1], [-0.2], [-0.3], [-0.5], [-1.0], [-1.5], [-2.0], [-2.5], [-3.0], [-3.5], [-4.0], [-4.5], and [-5.0]. I have rewritten my model atmosphere program to use the new line opacities, additional continuous opacities, and an approximate treatment of convective overshooting. The opacity calculation was checked by computing a new theoretical solar model that matches the observed irradiance. Thus far I have completed a grid of 7000 model atmospheres at 2 km/s for all the abundances, for the temperature range 3500K to 50000K, and for log g from 0.0 to 5.0. This grid will allow a consistent theoretical treatment of photometry from K stars to B stars. Fluxes are tabulated from .09 to 160 micrometers. Preliminary results are reported for many photometric systems. Work is underway on grids for other microturbulent velocities. Microturbulent velocity strongly affects the interpretation of Cepheid and RR Lyrae photometry. The models, fluxes, and colors are available on magnetic tape and will also be distributed on CD-ROMs.

1. Old models

I start by listing the shortcomings in my old models as I did at an earlier meeting (Kurucz 1987) but this time I can report that I have corrected most of them.

My old models (Kurucz 1979a,b) were produced as long as 18 years ago on computers that are primitive by today's standards. The number of optical depth layers was limited by small memories and slow processors. Now I can compute with many more layers and go to shallower optical depths. This greatly improves the numerical accuracy of the calculated radiation field at wavelengths that have very high or rapidly varying opacity. Fortunately, such wavelengths do not very much affect the structure of a model. There was also a limit to the number of frequencies that I could afford to compute. Now I use up to 1221 wavelength intervals from 9 to 160,000 nm. I can now use 1 nm resolution in the ultraviolet for better comparison to satellite observations.

My old models for F and G stars are systematically in error and predict color indices that are off by as much as 0.05 mag. I assumed that the error was caused by problems in the mixing length treatment of convection and by the omission of molecular line opacity in the coolest models. My theoretical model for the solar photosphere has several per cent error in the flux in the red and, of course, cannot reproduce the molecular features in the ultraviolet. Improvements in my treatment of convection, even going so far as having hot and cold streams, have reduced the error somewhat for the hotter convective models. I have

B. Barbuy and A. Renzini (eds.), The Stellar Populations of Galaxies, 225–232.

also added approximate overshooting and have tried various increases in the mixing-length to scale-height ratio. I now use 1.25. However, I am now convinced that most of the error comes from missing line opacity, including missing atomic line opacity, which turns out to be significant at all effective temperatures. I hope that Nordlund and others will be able to produce models with realistic convection cells to take care of the convection physics, so now I am concentrating on the opacity. I discuss molecular and atomic opacity in the next section.

We do not know much about microturbulent velocity. It has been decreasing as a function of time as the models have improved. In the sun it is depth-dependent varying from about 0.5 to 1.8 km/s. The models assume a constant value. Twenty five years ago, I arbitrarily chose 2 km/s as a nice round number. In some stars it can be much larger or much smaller. Opacity, radiative acceleration, and model structure vary considerably with the microturbulent velocity. It may be that 1 or 1.5 km/s is a good choice for high gravity models. George Michaud tells me that the existence of some types of diffusion implies microturbulent velocities less than 100 m/s. Microturbulent velocity may vary strongly with phase and with depth in pulsating stars, thereby strongly affecting the atmospheric structure and colors. I plan to investigate this through spectrum synthesis. My new grids of models have microturbulent velocity as a parameter, so it must be specified when choosing a model.

The helium abundance is another arbitrary number. I chose 10% by number. Others use a 10% He/H ratio. I have switched to a smaller value because I think it more probable. Small errors in the helium abundance produce errors in the density, electron number, and opacity and consequently produce systematic errors in the derived stellar parameters. The "solar" metal abundances have also changed with time and are not yet final. I now use Anders and Grevesse (1989) abundances.

My old low gravity models have systematic errors because of non-LTE and sphericity effects. So do my new models.

2. New lines and new opacities

I reported on my line and opacity calculations at a NATO workshop in Trieste (Kurucz 1991). The details of my line lists and the opacities can be found in that paper. Here I will give only a brief outline.

My earlier model calculations used the distribution-function line opacity computed by Kurucz (1979ab) from the line data of Kurucz and Peytremann (1975). We had computed gf values for 1.7 million atomic lines for sequences up through nickel using scaled-Thomas-Fermi-Dirac wavefunctions and eigenvectors determined from least squares Slater parameter fits to the observed energy levels. That line list has provided the basic data and has since been combined with a list of additional lines, corrections, and deletions with the help of Barbara Bell and Terry Varner at the Center for Astrophysics. The line data are being continually, but slowly, improved. We collect all published data on gf values and include them in the line list whenever they appear to be more reliable than the current data. I have also completely recomputed Fe II (Kurucz 1981).

After the Kurucz-Peytremann calculations were published, I started work on line lists for diatomic molecules beginning with H_2, CO (Kurucz 1977), and SiO (Kurucz 1980). Next, Lucio Rossi of the Istituto Astrofisica Spaziale in Frascati, John Dragon of Los Alamos, and I computed line lists for electronic transitions of CH, NH, OH, MgH, SiH, CN, C_2, and TiO. In addition to lines between known levels, these lists include lines whose wavelengths are predicted and are not good enough for detailed spectrum comparisons but are quite adequate for statistical opacities. Work is continuing on other molecules and molecular ions, and on the vibration-rotation spectra. I also have data for terrestrial atmospheric molecules.

In 1983 I recomputed the opacities using the additional atomic and molecular data described above which totalled 17,000,000 lines. These opacities were used to produce improved empirical solar models (Avrett, Kurucz, and Loeser 1984), but were found to still not have enough lines. For example, there were several regions between 200 and 350 nm where the predicted solar intensities are several times higher than observed, say, 85% blocking instead of the 95% observed. The integrated flux error of these regions is several per cent of the total. In a flux constant theoretical model this error is balanced by a flux error in the red. The model thus predicts the wrong colors. In detailed ultraviolet spectrum calculations, half the intermediate strength and weak lines are missing. After many experiments, I determined that this discrepancy is caused by missing iron group atomic lines that go to excited configurations that have not been observed in the laboratory. Most laboratory work has been done with emission sources that cannot strongly populate these configurations. Stars, however, show these lines in absorption without difficulty. Including these additional lines produces a dramatic increase in opacity, both in the sun and in hotter stars. A stars have the same lines as the sun but more flux in the ultraviolet to block. In B stars and in O stars there are large effects from third and higher iron group ions. Envelope opacities that are used in interior and pulsation models are also strongly affected.

I was granted a large amount of computer time at the San Diego Supercomputer Center by NSF to carry out new calculations. To compute the iron group line lists I determined eigenvectors by combining least squares fits for levels that have been observed with computed Hartree-Fock integrals (scaled) for higher configurations including as many configurations as I can fit into a Cray. All configuration interactions are included. My computer programs have evolved from Cowan's (1968) programs. Transition integrals are computed with scaled-Thomas-Fermi-Dirac wavefunctions and the whole transition array is produced for each ion. The forbidden transitions can be computed as well. Radiative, Stark, and van der Waals damping constants and Lande g values are automatically produced for each line. The first nine ions of Ca through Ni produced 42,000,000 lines. I will recompute the energy levels and line lists when new analyses become available and I will make the predictions available to laboratory spectroscopists. I plan to do the heavier and lighter elements as a background project.

The models I can compute now are not valid for M stars. I eventually need line lists for the triatomic molecules, but I hope that other people will do the work before I have to learn the physics. I am working on the low temperature bands now, however, for atmospheric transmission.

In late 1988 I used the line data described above to compute new solar abundance opacity tables for use in my modelling. The calculations involved 58,000,000 lines, 3,500,000 wavelength points, 56 temperatures from 2000K to 200000K, 21 log pressures from -2 to +8, and 5 microturbulent velocities 0, 1, 2, 4, 8 km/s, and took a large amount of computer time. The opacity is tabulated both as 12-step distribution functions for intervals on the order of 1 to 10 nm, and as opacity sampling where, simply, every hundredth wavelength point in the calculation was saved. There are actually two sets of distribution functions, a higher resolution version with 1212 "little" intervals, and a lower resolution version with 328 "big" intervals. The "little" wavelength intervals are nominally 1 nm in the ultraviolet and 2 nm in the visible. The opacities were tested by computing a solar model as described below.

Since the beginning of 1990 I have been able to take tremendous advantage of the new Cray YMP at the San Diego Supercomputer Center. In a few months I finished more than I had expected to do in two years. I computed opacities ranging from 0.00001 solar to 10 times solar, enough to compute model atmospheres ranging from the oldest Population II stars to high abundance Am and Ap stars. The hardest part was transmitting the results (200 tapes) back to Cambridge over Internet, but even that usually worked quite well. The exact abundances are [+1.0], [+0.5], [+0.3], [+0.2], [+0.1], [+0.0], [-0.1], [-0.2], [-0.3], [-0.5], [-1.0], [-1.5], [-2.0], [-2.5], [-3.0], [-3.5], [-4.0], [-4.5], [-5.0], and [+0.0, no He]. The final files for each abundance require two 6250 bpi VAX backup tapes. I distribute copies of the tapes on request. I hope to produce CD-ROMs of these opacities that can be read on any workstation with a CD reader.

I am open to suggestions for computing opacities for other abundance mixes. I plan to compute Population II opacities with [+0.4] enhanced alpha process elements, some Am and Ap mixes, and C/O variations.

3. New models

I have rewritten my model atmosphere program to use the new line opacities, additional continuous opacities, and an approximate treatment of convective overshooting. The opacity calculation was checked by computing a small grid of solar models with various microturbulent velocities and mixing-length-to-scale-height ratios. I adopted a solar model shown in Figure 1 that matches the observed irradiance (Neckel and Labs 1984; Labs et al. 1987) with Vturb = 1.5 km/s and l/H = 1.25. I am confident that I have solved the missing opacity problem. I then computed (on Vaxstations) a grid of 400 solar abundance, 2 km/s models covering the effective temperature range from K stars to O stars. The models are listed in Table I. Models cooler than 9000K are convective with l/H = 1.25. The range of this grid should allow photometric calibrations consistent for both cool and hot stars. Note that since triatomic molecules are not included, the models cannot be used for M stars. I have distributed tapes of the models and the flux predicted from each model at 1221 wavelengths in the range .01 to 160 micrometers. The range is enough to treat ionization in H II regions and to calibrate the infrared.

I have computed preliminary UBV and uvby colors and bolometric corrections for the solar abundance grid described above following Buser and Kurucz (1978) and Relyea and Kurucz (1978) including the typographic correction from Lester et al (1986). The colors were normalized by finding the model in the grid that best interpolates the spectrophotometry of Vega (Hayes and Latham 1975; Tug, White, and Lockwood 1977), Figure 2, and that best matches the Balmer line profiles (Peterson 1969) and then by forcing the computed colors for that model to match the observed colors. That model has Teff = 9400K and log g = 3.95 just as in the earlier grid (Kurucz 1979a). However, remember that the helium abundance, metal abundances, line opacity, and resolution are different. Models with lower microturbulent velocity are not observationally distinguishable in the

Table I. Solar abundance models, 2 km/s microturbulent velocity

Teff	log g 0.0	0.5	1.0	1.5	2.0	2.5	3.0	3.5	4.0	4.5	5.0	Teff
3500	X	X	X	X	X	X	X	X	Z	Z	Z	3500
3750	X	X	X	X	X	X	X	X	Z	Z	Z	3750
4000	X	X	X	X	X	X	X	X	X	X	X	4000
4250	X	X	X	X	X	X	X	X	X	X	X	4250
4500	X	X	X	X	X	X	X	X	X	X	X	4500
4750	X	X	X	X	X	X	X	X	X	X	X	4750
5000	X	X	X	X	X	X	X	X	X	X	X	5000
5250	X	X	X	X	X	X	X	X	X	X	X	5250
5500	X	X	X	X	X	X	X	X	X	X	X	5500
5750	X	X	X	X	X	X	X	X	X	X	X	5750
6000	X	X	X	X	X	X	X	X	X	X	X	6000
6250	B	X	X	X	X	X	X	X	X	X	X	6250
6500	B	X	X	X	X	X	X	X	X	X	X	6500
6750	B	X	X	X	X	X	X	X	X	X	X	6750
7000	B	X	X	X	X	X	X	X	X	X	X	7000
7250	B	X	X	X	X	X	X	X	X	X	X	7250
7500	B	X	X	X	X	X	X	X	X	X	X	7500
7750		B	X	X	X	X	X	X	X	X	X	7750
8000		B	X	X	X	X	X	X	X	X	X	8000
8250		B	X	X	X	X	X	X	X	X	X	8250
8500		B	X	X	X	X	X	X	X	X	X	8500
8750			B	X	X	X	X	X	X	X	X	8750
9000			B	X	X	X	X	X	X	X	X	9000
9250				B	X	X	X	X	X	X	X	9250
9500				B	X	X	X	X	X	X	X	9500
9750				B	X	X	X	X	X	X	X	9750
10000				B	X	X	X	X	X	X	X	10000
10500				B	X	X	X	X	X	X	X	10500
11000					B	X	X	X	X	X	X	11000
11500					B	X	X	X	X	X	X	11500
12000					B	X	X	X	X	X	X	12000
12500					B	X	X	X	X	X	X	12500
13000					B	X	X	X	X	X	X	13000
14000				B	X	X	X	X	X	X	X	14000
15000					B	X	X	X	X	X	X	15000
16000					B	X	X	X	X	X	X	16000
17000					B	X	X	X	X	X	X	17000
18000					B	X	X	X	X	X	X	18000
19000					B	X	X	X	X	X	X	19000
20000						B	X	X	X	X	X	20000
21000						B	X	X	X	X	X	21000
22000						B	X	X	X	X	X	22000
23000						B	X	X	X	X	X	23000
24000						B	X	X	X	X	X	24000
25000						B	X	X	X	X	X	25000
26000						B	X	X	X	X	X	26000
27000							B	X	X	X	X	27000
28000							B	X	X	X	X	28000
29000							B	X	X	X	X	29000
30000							B	X	X	X	X	30000
31000							B	X	X	X	X	31000
32000								B	X	X	X	32000
33000								B	X	X	X	33000
34000								B	X	X	X	34000
35000								B	X	X	X	35000
37500									B	X	X	37500
40000									B	X	X	40000
42500										B	X	42500
45000										B	X	45000
47500										B	X	47500
50000										B	X	50000

X = converged
B = blows up from radiative acceleration
Z = converged, not realistic because no water opacity

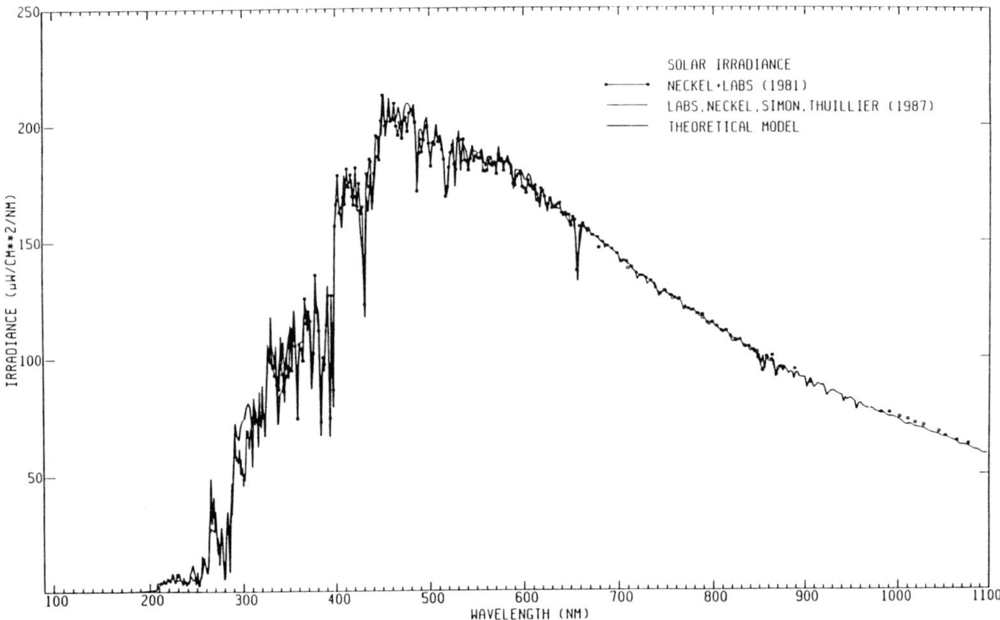

Figure 1. Predicted solar irradiance compared to observed.

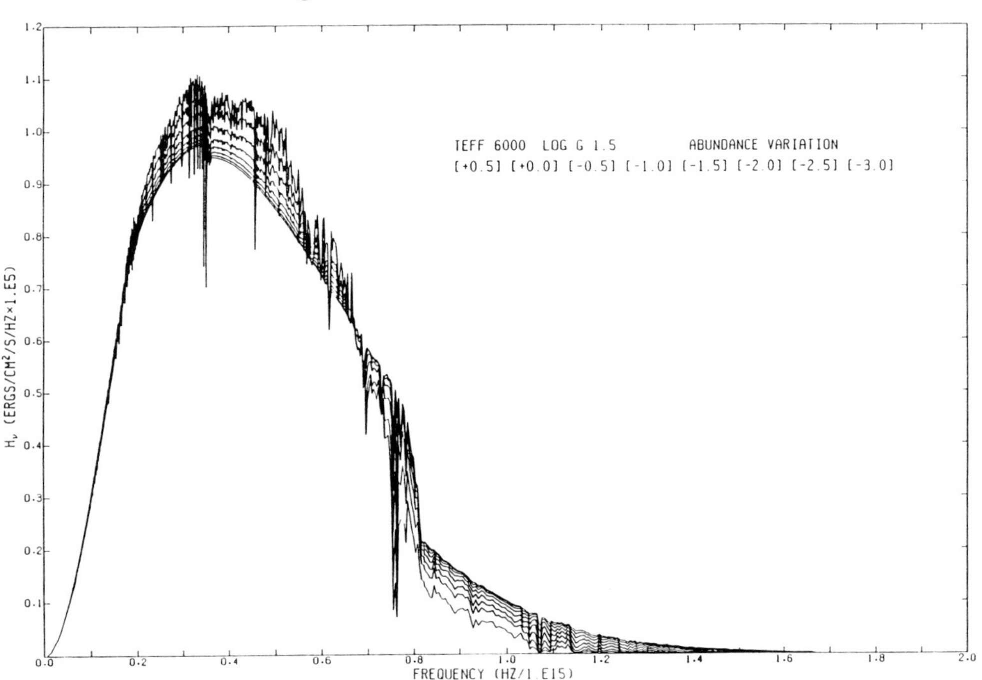

Figure 3. The change in flux distribution for a G giant as the abundance increases from extreme Pop II to extreme Pop I. At the 850 nm maximum the flux increases by 15% (.15 mag). In the Balmer continuum the flux decreases by more than a factor of 2 (1 mag).

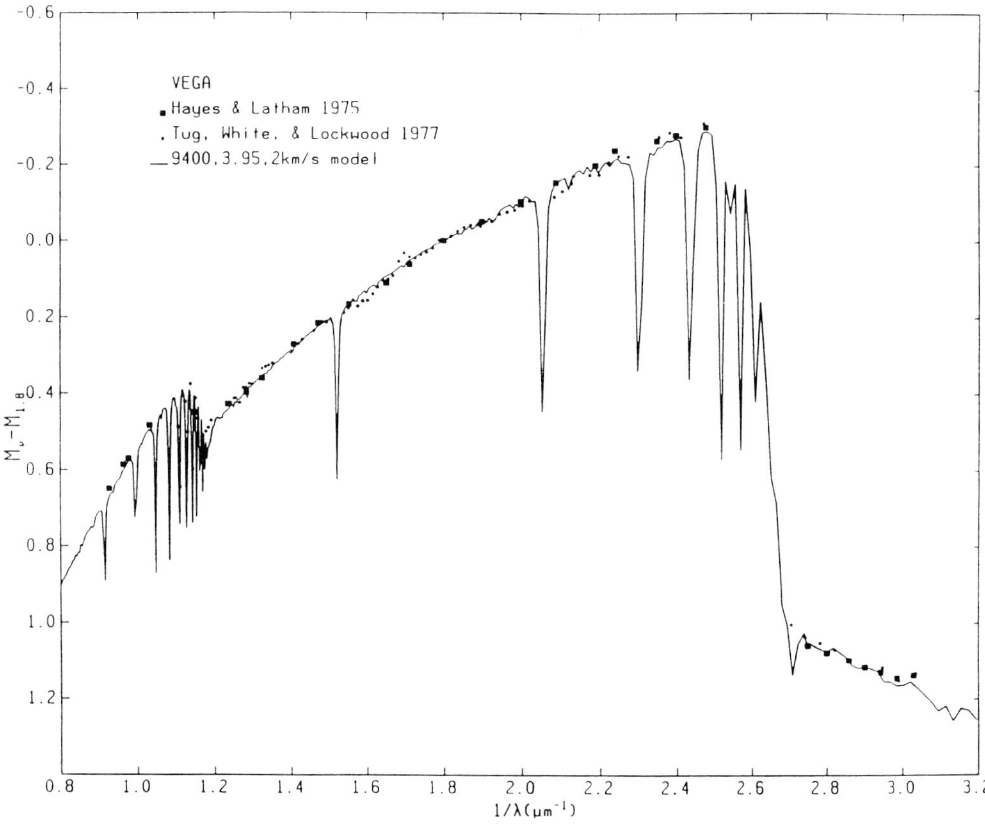

Figure 2. Observed and predicted spectrophotometry relative to 1.8 inverse micrometers (555.6 nm) for Vega. The computed slopes of the Balmer, Paschen, and Brackett continua and the Balmer discontinuity agree with observations.

visible. This model is not a physical model for Vega because Vega does not have solar abundances (for example, see Adelman and Gulliver 1990), but even if the physical parameters are somewhat off, the temperature-pressure structure must be correct in the continuum and Balmer-line-wing-forming layers. A correct model will have a similar structure but perhaps for a somewhat different effective temperature and gravity.

I have begun to compute grids of models on the Cray using the solar abundance, 2 km/s grid for starting models. Thus far I have run approximately 7000 models for all abundances at 2 km/s. Just checking the output is a tremendous amount of work for me. I have to individually rerun mistakes and failures. I will compute the fluxes, predicted photometry, Balmer line profiles, and limb-darkening for each model. At the present time I can supply magnetic tapes with the models for the various abundances. I will also publish the data on CD-ROMs. Most users will be able to find what they need by simple interpolation. For any model I expect to be able to compute a complete, full-resolution spectrum that can be compared to high resolution observations, or degraded to low resolution, say, 0.1 nm.

Figure 3 shows results from these calculations for G giants, a sample of what is now available for population synthesis of galaxies. It is clear that there are strong abundance effects throughout the spectrum. In addition to the variation in the spectral features, there is strong redistribution of energy from the ultraviolet to the red and infrared as the abundance increases. It is clearly impossible to do any type of evolutionary population synthesis based on observed Population I spectra. These models are still in need of improvement, but I hope they allow a significant advance in population synthesis.

ACKNOWLEDGEMENTS. This work is supported in part by NASA grants NSG-7054, NAG5-824, and NAGW-1486, and has been supported in part by NSF grant AST85-18900. The most important contribution to this work is a large grant of Cray computer time at the San Diego Supercomputer Center.

REFERENCES

Adelman, S.J. and Gulliver, A.F.: 1990, *Astrophys. J.* **348**, 712.

Anders, E. and Grevesse, N.: 1989, *Geochimica et Cosmochimica Acta* **53**, 197.

Avrett, E.H., Kurucz, R.L., and Loeser, R.: 1984, *Bull. Amer. Astron. Soc.* **16**, 450.

Buser, R. and Kurucz, R.L.: 1978, *Astron. Astrophys.* **70**, 555.

Cowan, R.D.: 1968, *J. Opt. Soc. Am.* **58**, 808.

Kurucz, R.L.: 1977, *SAO Spec. Rep.* No. 374, 170 pp.

Kurucz, R.L.: 1979a, *Astrophys. J. Supp. Ser.* **40**, 1.

Kurucz, R.L.: 1979b, *Dudley Observatory Report* No. 14, ed. A.G. Davis Philip, p. 363.

Kurucz, R.L.: 1980, *Bull. Amer. Astron. Soc.* **11**, 710.

Kurucz, R.L.: 1981, *SAO Spec. Rep.* No. 390, 319 pp.

Kurucz, R.L.: 1987, *IAU Colloquium No. 95, The Second Conference on Faint Blue Stars*, eds. A.G. Davis Philip, D.S. Hayes, and J.W. Liebert, L. Davis Press, Schenectady, 129.

Kurucz, R.L.: 1991, *Stellar Atmospheres: Beyond Classical Models* ed. by L. Crivellari, I. Hubeny, and D.G. Hummer, NATO ASI Series, Kluwer, Dordrecht, 440.

Kurucz, R.L., and Peytremann, E: 1975, *SAO Spec. Rep.* No. 362, 1219 pp.

Labs, D., Neckel, H., Simon, P.C., Thuillier, G.: 1987, *Solar Phys.* **107**, 203.

Lester, J.B., Gray, R.O., and Kurucz, R.L.: 1986, *Astrophys. J. Supp. Ser.* **61**, 509.

Neckel, H. and Labs, D.: 1984, *Solar Phys.* **90**, 205.

Peterson, D.M.: 1979, Personal communication.

Relyea, L.J. and Kurucz, R.L.: 1978, *Astrophys. J. Supp. Ser.* **37**, 45.

Tug, H., White, N,M., and Lockwood, G.W.: 1977, *Astron. Astrophys.* **61**, 679.

DISCUSSION

ROCCA-VOLMERANGE: For infrared studies it would be necessary to extend your models down to 2000K. Do you think that is possible?

KURUCZ: I do not have line opacity for any triatomic molecules so you would have to ask Gustafsson, or Jorgensen, or Johnson for those models. If the abundances are low so that triatomic opacity is not important, I should be able to compute down to 3000K. At lower effective temperatures the surface temperature is below 2000K which is the lower limit of my tabulated opacities.

GUSTAFSSON: Would you comment on your present recipe for convection and on what effects it has on colours and spectra, as compared with the standard local mixing-length theory?

KURUCZ: There have been several small changes over the years that add up to 0.05 - 0.06 mag in the colors for F and G stars. Lester et al. at Toronto tried several fixes. One I adopted is computing opacities for hot and cold elements and then averaging instead of computing opacity for the average temperature. I recently put in an approximate overshooting by assuming that the center of a bubble stops at the top of the convection zone so that there is convective flux one bubble radius above the convection zone. That flux is found by computing the convective flux in the normal way and then smoothing it over a bubble diameter.

ULTRAVIOLET OBSERVATIONS OF STELLAR POPULATIONS IN GLOBULAR CLUSTERS AND GALAXIES

Robert W. O'Connell
Astronomy Department
University of Virginia
Charlottesville, VA 22903 USA

ABSTRACT. We review some of the contributions made by vacuum ultraviolet observations to our understanding of stellar populations, with emphasis on recent results from the HST and the Astro-1 mission.

1. Introduction

The vacuum ultraviolet is a valuable new window on stellar populations, whose potential is comparable to that of the infrared window so fruitfully exploited over the last 20 years. Access to the UV has been more limited because of the need to place instruments above the atmosphere and their consequent small apertures or limited lifetimes. To date, there have been nine major UV observatories, in terms of number of returned observations: OAO-2, *Copernicus*, ANS, TD-1, IUE, SCAP/FOCA, HST, ROSAT/WFC, and Astro-1; data from the last four of these will reach publication only over the next several years.

Thanks mainly to the remarkable longevity of IUE, a total of about 100,000 UV observations of all types have been made. This is an impressive number, given the technical difficulties. However, in the context of this symposium, the available UV data has somewhat limited scope. First, about 98% of the observations are of individual Galactic stars, not composite systems such as clusters or galaxies. Further, only about 1% is derived from imaging devices or large telescopes capable of reaching objects fainter than the IUE limiting magnitude of $m_\lambda(2000\ \text{Å}) \sim 14$. [1] Finally, only TD-1 made an *all-sky survey*, and this had a limiting magnitude $m_\lambda \sim 9$—comparable to that of the Henry Draper Catalog. Consequently, we know as much about the UV sky today as we knew about the optical sky in 1900! This could quickly change during the 1990's, given favorable funding winds, and we look forward to rich data returns from EUVE, FOCA, Astro-2, the second-generation HST, Spectrum-UV, Lyman/FUSE, and perhaps a deep survey experiment.

The main advantage of the UV for stellar populations problems is its high sensitivity to stellar temperature, which offers greatly improved temperature resolution for a given photometric precision. This also permits the isolation of hot objects from the often dominant cool star background (e.g. in the crowded cores of globular clusters). Further, the UV contains much information on interstellar gas and dust. Finally, the sky background in the UV

[1] Magnitudes are quoted in the monochromatic system, defined as $m_\lambda = -2.5 \log F_\lambda - 21.1$, where $[F_\lambda] = \text{erg s}^{-1}\ \text{cm}^{-2}\ \text{Å}^{-1}$. UV colors used here, such as (1500–2500), are based on such magnitudes.

B. Barbuy and A. Renzini (eds.), The Stellar Populations of Galaxies, 233–243.

is significantly fainter than at longer wavelengths (Henry 1991), providing opportunities for study of very low surface brightness regions.

In this review, I try to give some sense of what UV experiments earlier than 1990 have contributed to our understanding of stellar populations but will concentrate on more recent results from the HST and the Ultraviolet Imaging Telescope (UIT) experiment on the Astro-1 mission. Earlier UV imaging results are covered in more detail in O'Connell (1991); the most comprehensive review of UV spectroscopy is Kondo (1987).

2. The Ultraviolet Imaging Telescope

The Astro-1 Spacelab mission flew for nine days in December 1990. Four experiments were carried: the Broad Band X-ray Telescope, the Hopkins UV Telescope, the Wisconsin UV Photo-Polarimeter Experiment, and the UIT. Despite a number of equipment difficulties, the mission obtained about 40% of its planned observations and reached an efficiency of 80% of the planned exposure time per orbit just prior to its termination due to poor weather at the landing site. As of this writing the UV instruments are scheduled for reflight in 1994. There will be a guest observer program open to the community on Astro-2.

The UIT was designed by a team at Goddard Space Flight Center under the direction of T. P. Stecher. It is a 38-cm telescope with a field of view $40'$ in diameter. It carries two image-tube cameras and a total of 11 filters plus a full-field grating. The combination of photocathode and filter response strongly rejects long-wavelength light, so "red leaks" are negligible for most objects. Most observations discussed here were made with the broad-band "FUV" filter (centroid 1500 Å, width 350 Å) or "NUV" filter (2500 Å, 1150 Å). Data is recorded on film and digitized to a 2048^2 pixel format. The stellar limit for a 10-min exposure is $m_\lambda(\mathrm{UV}) \sim 19.5$, or V $\sim$ 23 for a hot, unreddened source.

UIT performed well during the mission. Image FWHM's, attributable mainly to small guidance errors, averaged $3''$. A total of 821 of the 2050 planned exposures were made. About half of the exposures have been reduced to date, but only about 50 have received a first-order scientific analysis. Flux calibration is still preliminary.

3. Globular Clusters

Pioneering photometry by OAO-2 and ANS (Welch and Code 1980, van Albada *et al.* 1981) revealed that there is a large range in the integrated UV energy distributions of globular clusters, amounting to over 3 mags in (1500–3300). This was shown to reflect the population and temperature distribution of hot horizontal branch (HB) stars. The bluest clusters are not the most metal poor but are rather "second parameter" clusters whose HB's are bluer than normal for their metal abundance. IUE has obtained the integrated UV spectra of many clusters and some spectra for individual bright stars (reviewed in Castellani and Cassatella 1987). Methods of using integrated spectra to estimate relevant HB parameters (age, abundance, mass loss, rotation, etc.) are still under development (e.g. Nesci 1983, Caloi *et al.* 1985).

Since the hot HB produces no more than ~10% of the integrated optical light (e.g. Buzzoni 1989), the UV will be the only practical means of studying the HB in distant, unresolved clusters. Low S/N spectra are already available for clusters in the Magellanic Clouds (Cassatella *et al.* 1987) and M31 (e.g. Cowley and Burstein 1988, Crotts *et al.* 1990). UV data may settle controversies over the age spread of M31 clusters, but only when HST-quality spectra are obtained. UV observations of more distant cluster systems will be important to place Local Group results in a larger context.

UV imaging can provide a rapid census of the hot star population of resolved clusters. A 30-minute exposure in a broad band with a UIT-sized telescope could detect hot stars up to 9 magnitudes below the HB in a nearby cluster such as M5 (O'Connell 1991). Combined with the natural rejection of cool stars, this sensitivity implies that UV imaging can yield *complete samples* of hot populations. This includes the rapidly-evolving, and therefore rare, types of post-giant branch stars whose evolutionary status is not yet understood (Vauclair and Liebert 1987, Greggio and Renzini 1990). To date, only two large-sample HB studies, based on UV imaging of bright clusters with rockets or balloons, have been published (Bohlin *et al.* 1985, Laget *et al.* 1991).

An excellent example of the ability of UV imaging to penetrate even the crowded cores of clusters is the recent discovery of blue stragglers in the core of 47 Tuc (Paresce *et al.* 1991). Paresce *et al.* obtained HST/FOC images (field size 22″ square) in far-UV bands and found about 20 bright blue objects. In a UV color-magnitude diagram (CMD) these are $\sim$ 1 mag bluer than the warmest HB stars and have locations on the extrapolated main sequence of the cluster. Their density, about 0.05 arcsec^{-2}, is higher than for similar objects in the outer regions of the cluster, leading Paresce *et al.* to conclude they are binaries, formed by interactions, which have been concentrated in the core by mass segregation. It is evident that UV observations will strongly complement the rapidly growing body of ground-based data on modifications of cluster populations through interactions (e.g. Nemec and Cohen 1989, Djorgovski *et al.* 1991).

During Astro-1, UIT obtained good imaging of ω Cen, M79, and NGC 1851. Short exposures of three other clusters have yet to be evaluated. In a 1100 sec exposure of M79 we find a UV-bright center produced either by 2–3 very bright stars or the combined background. Our shorter exposures should permit resolution of individual central sources. We have obtained a UV CMD for 100 objects, which is shown in Fig. 1 together with a ZAHB locus based on Sweigert's (1987) models. Most of the objects cluster around the expected HB locus. (Note that the HB is not actually "horizontal" in this diagram.) We have not yet attempted to use the fit to estimate mean mass or abundance parameters. Interestingly, there are a number of hot objects at or above the extreme tip of the ZAHB locus. Eight have $T_e \sim$ 30,000–150,000. These objects would fall near V $\sim$ 20 and were not present in optical-band CMD's. If they are ZAHB or evolved HB stars, they have envelope masses $\lesssim$ 0.01 $M_\odot$. These are so small that they are not likely to proceed to the normal AGB phase but will rather follow exotic CMD paths involving large T and L excursions (Caloi 1989). Two stars in Fig. 1 are $\sim$ 2 mags brighter than the HB (and several yet brighter may be present in the core). They could be the later stages of such "extreme HB stars" or perhaps post-AGB objects moving blueward from the AGB or down the remnant cooling sequence (e.g. Schönberner 1983).

We observed ω Cen in daylight; our 5 min FUV exposure was excellent, but our NUV frame was contaminated with skyglow. On the FUV frame we detect 1360 stars with $T_e \gtrsim 10,000$ and $m_\lambda(1500) < 19$. The core is fully resolved; the hot objects have an irregular distribution with little central concentration. Using the catalog of Dickens *et al.* (1988), we have constructed a FUV,V CMD for objects in common at $r > 3'$ (about 20% of our sample). As expected, most lie near the HB. However, a significant number scatter up to 3 mags above the HB. We estimate that about 60 stars in the full sample lie above the HB. More such objects would be expected than in M79 since ω Cen is a factor of 10 more luminous. We plan to extend our FUV/optical CMD coverage using new U-band CCD images.

The three brightest objects within $r < 12'$ fall at $m_\lambda(1500) \sim$ 10.2–12.0, implying absolute

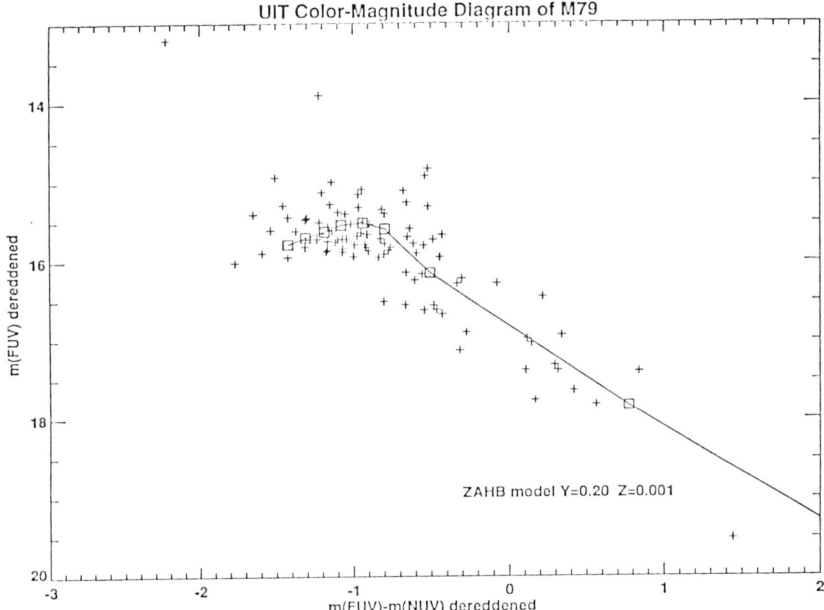

FIGURE 1: Ultraviolet color-magnitude diagram for $r > 40''$ in M79. Magnitudes and colors, defined as in §1, are corrected for extinction. ZAHB models from Sweigart (1987) are plotted; the hottest of these has an envelope mass of 0.01 $M_\odot$.

λ1500 magnitudes (corrected for extinction) of –3 to –5. Only the brightest of these (ROA 5701) had been recognized from the ground or had an IUE spectrum available (Cacciari *et al.* 1984). We recently obtained IUE spectra of the two new identifications. All three objects have similar UV SED's, corresponding to $T_e \sim 25,000$, and B-type absorption lines. We believe all are cluster members. Again, they could be P-AGB objects. Interestingly, one of the new identifications falls within the $41''$ error box for Einstein IPC source C, which previously had no known optical/UV counterpart.

A UIT-class instrument could quickly survey the brightest 50 globular clusters for such rare, hot objects, creating a sample suitable for making statistical inferences concerning their evolution. Another area for future UV exploration is the white dwarf luminosity function and its implication for remnant cooling physics.

4. Star-Forming Galaxies

The UV has already made considerable contributions to our understanding of systems containing massive OB stars, particularly in the form of IUE spectra. However, recent reviews have covered this area (Kondo 1987, O'Connell 1990a) concerning both populations and the ISM, so I will only briefly discuss some newer developments here.

The far-UV ($\sim$ 2000 Å) continuum measures the mean star formation rate over the last $\sim$ 100 Myr, about 50 times longer than the ionizing continuum ($\lambda < 912$ Å), and hence provides a more representative picture than do emission lines. A compilation by Buat *et al.* (1989) of integrated UV data for spirals, based mainly on SCAP balloon observations, shows that the star formation rate per unit area on this timescale is strongly correlated with the *total* (neutral + molecular) hydrogen surface density but with neither phase separately. They derive a power-law relation with exponent 1.6.

Because UV colors are a strong function of age, the appearance of galaxies depends on the observing wavelength. This is an important issue for assessing the evolutionary state of high redshift systems, which are often imaged in the rest-frame UV and at low S/N. Bohlin *et al.* (1990) recently simulated the appearance of high redshift galaxies based on rocket UV images and illustrated the remarkable "morphological transformations" which can occur: spirals ⇒ E/S0; Sb ⇒ Sc; barred ⇒ unbarred; interacting ⇒ single; single ⇒ multiple.

Recent HST/GHRS spectra have again demonstrated the value of the UV for determining the abundances of hot stars (which cannot be easily done at longer wavelengths owing to the paucity of useful lines). Heap *et al.* (1991) obtained excellent spectra of Melnick 42 in the 30 Dor complex of the LMC. They deduce $Z \sim Z_{\odot}/4$, $T_e \sim 42,500$, $L_{bol} \sim 2.5 \times 10^6$ L$_{\odot}$, $M \sim 100$ M$_{\odot}$, and a terminal wind velocity of $\sim$ 3000 km s^{-1}.

UIT imaged a number of Local Group and nearby star-forming galaxies. Since unreddened OB stars are $\sim$ 4 mags brighter per unit wavelength in the UV than at V, such images are valuable for tracing star formation histories, investigating the IMF, and identifying important spectroscopic targets. For instance, a 6-min FUV daylight exposure of the 30 Dor complex registered $\gtrsim$ 2000 hot objects fainter than the IUE limit. A 10-min NUV exposure of the disk of M31 is shown in Fig. 2; it includes the massive OB association NGC 206, a number of other OB complexes, and the companion elliptical galaxy M32. We also obtained a good image of the UV-bright plume along the minor axis of M82, presumably a product of forward-scattering by ejected dust grains (Courvoisier *et al.* 1990), and of the cooling flow system NGC 1275. Our preliminary photometry of 1275 indicates that if the IMF is normal, star formation is not occurring now but terminated $\sim$ 50–150 Myr ago.

5. Elliptical Galaxies and Spiral Bulges

The old, high metallicity populations of ellipticals and S0-Sb bulges are not rich in hot HB stars nor, usually, massive OB stars. Nonetheless, the UV can provide important insights here. Two areas are undergoing rapid development.

First, one can use integrated near-UV ($\lambda\lambda$ 2000-3500) spectra to determine their *main sequence turnoff* characteristics, which is crucial to age-dating old populations. It is difficult to extract information on turnoff stars in the optical/IR because their light is blended with cooler dwarfs, subgiants, and giants, which produce over 70% of the light longward of V. By contrast, the turnoff provides $\gtrsim$ 70% of the light in the NUV. As a bonus, the NUV contains many strong absorption lines (see, for example, Burstein *et al.* 1988), which will ultimately yield improved abundance determinations. Because of these advantages, the NUV promises the best near-term improvements in age-dating techniques, which have been a subject of controversy (O'Connell 1986, Renzini 1986).

To extract MSTO information from the NUV, one must first remove the residual effects of the cooler stars plus those of hot HB, UVX (see below), and massive MS populations. This requires long-baseline observations, preferably covering at least 1200–6000 Å. Early

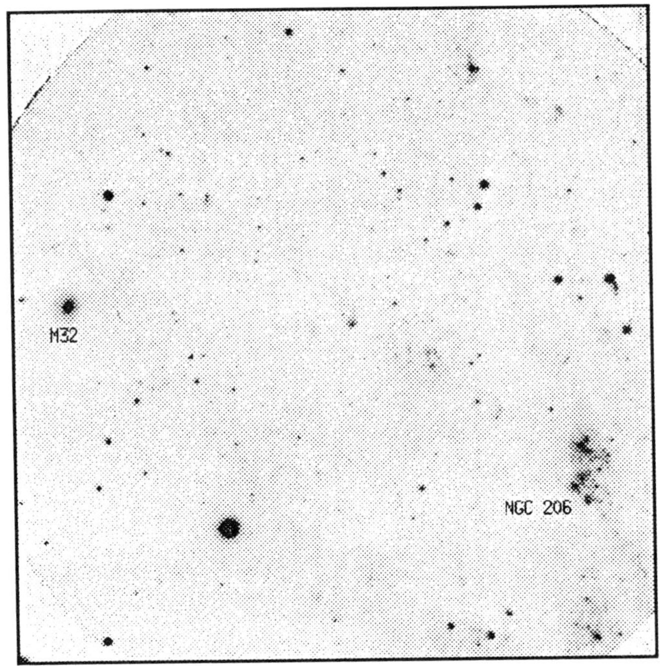

FIGURE 2: UIT image taken with the broad-band NUV filter of a field in M31's disk SW of the bulge. The region shown is 32′ on a side. The OB association NGC 206 and companion E galaxy M32 are marked. The edge of the circular UIT FOV is visible in the corners.

explorations of the problem, based on both "evolutionary" and "optimizing" synthesis techniques, included Wu *et al.* (1980), Gunn *et al.* (1981), and Bruzual (1983); more recent and detailed models include Rocca-Volmerange and Guiderdoni (1988), Buzzoni (1989), Barbaro and Olivi (1989), and Magris and Bruzual (this conference).

M32, which has assumed a key role at this conference as a testing-ground of population modeling, is an important case in point. IUE spectra of the central $\sim 15''$ of M32 indicate a significant contribution by F5-9 stars, presumably (though not definitely) on the main sequence (Bruzual 1983, Burstein *et al.* 1984, Rocca-Volmerange and Guiderdoni 1987, Kjærgaard 1987). An example of the good fits possible with such models is shown in O'Connell (1990b), where a 5 Gyr-old, solar abundance model is compared to the best available IUE spectrum (Burstein *et al.* 1988). Older, lower-abundance models would contain MSTO stars of similar temperatures. However, my preliminary tests of such models do not yield good NUV fits, and nearly all other studies of the optical/IR spectrum of M32 are consistent with $Z \sim Z_{\odot}$ and a relatively youthful age of 5–8 Gyr for the MSTO (O'Connell 1986, Boulade *et al.* 1988, Bica *et al.* 1990, Davidge 1990). Further refinement of NUV techniques should permit improved age and abundance resolution. The existing results, however, have long seemed entirely consistent with other evidence that elliptical populations are often the product of chaotic and episodic processes extending over a long period. Indeed, that is one of the main themes emerging at this conference.

The UV "rising branch" or excess ("UVX") is a distinct phenomenon affecting the far-UV spectra ($\lambda \lesssim 2000$ Å) of ellipticals and large bulges. It was discovered by OAO-2 (Code 1969) and is manifested by a sharp rise in the energy distribution below 2000 Å, which has a slope equivalent to $T_e \sim 20,000$. Since this is far hotter than the MSTO of old populations ($T_e \lesssim 7,000$), the upturn was a considerable surprise. More recent surveys, especially with IUE, have established that the (1500–V) color of ellipticals varies widely and becomes bluer as metal abundance increases (e.g. Burstein *et al.* 1988). Note that the sense of this relation is *reversed* with respect to the familiar dependence of UBV colors on abundance in old populations.

Since two recent reviews have thoroughly covered both observational and interpretational issues (Burstein *et al.* 1988, Greggio and Renzini 1990), I will not dwell on earlier results. The two main proposals for the source of the UVX are: *(i) massive OB stars*, formed during the past $\sim$ 50 Myr presumably from gas lost during giant branch evolution; and *(ii) low-mass, post-giant branch stars* in advanced evolutionary phases, the descendents of the objects making up the dominant old population. For some time the favored low-mass candidate has been "post-AGB" stars (Schönberner 1983), though it has been recently recognized that these are only one of a plethora of such candidates, others of which (e.g. hot HB, post-EAGB) may fit the data better (see Greggio and Renzini 1990 for details).

The balance of the data prior to 1990 supported the low-mass star interpretation, and the Astro-1 mission has provided yet stronger evidence that massive stars are not the dominant factor in the UVX phenomenon. HUT obtained high S/N spectra covering $\lambda\lambda$ 950–1800 for NGC 1399, the brightest E in the Fornax cluster. The continuum shape and the absence of C IV absorption indicate that massive stars hotter than B0 are absent. The detailed analysis, described by Ferguson at this conference, favors extreme HB stars as the source of the UVX.

UIT images of four objects (the bulges of the Sb spirals M31 and M81, and the E galaxies M32 and NGC 1399) likewise yield little evidence for massive stars. There is none of the structure characteristic of star-forming regions, despite the fact that such are readily visible in the surrounding disks of M31 and M81. All four objects show smooth de Vaucouleurs-type surface brightness profiles in the NUV and FUV to below the level of the night sky. Further, no resolved stars are present in M32 or the M31 bulge. In the bulge, the limit is $m_\lambda(2500) \sim 18.4$ at $r \sim 30''$. This implies that single stars hotter than B1 V or B8 Ia are absent; limits for blended images are cooler. HST imaging would be required to search for individual low-mass objects; PAGB or P-EAGB types would appear in the FUV at $\sim$ 21–24 mag in M31's bulge with a surface density of $\sim$ 0.1–1 arcsec^{-2}.

We have also derived (1500–2500) colors for these four systems. Results are shown in Fig. 3. We caution that these are based on preliminary calibrations and are subject to revision over the next several months. Individual point sources and regions contaminated by disk light (e.g. near M32) have been masked out of the photometry.

The behavior of these systems in the UV is dramatically different from that in the optical/IR, where they exhibit striking homogeneity. First, there are gross differences in their central colors, amounting to a $\sim$ 3 mag range in (15–25) between the bluest system (NGC 1399) and the reddest (M32). Our central colors are in agreement with those based on IUE spectra for objects in common with Burstein *et al.* (1988). Second, all four systems have (15–25) color gradients which are very strong ($\sim$ 1 mag over the observed regions)

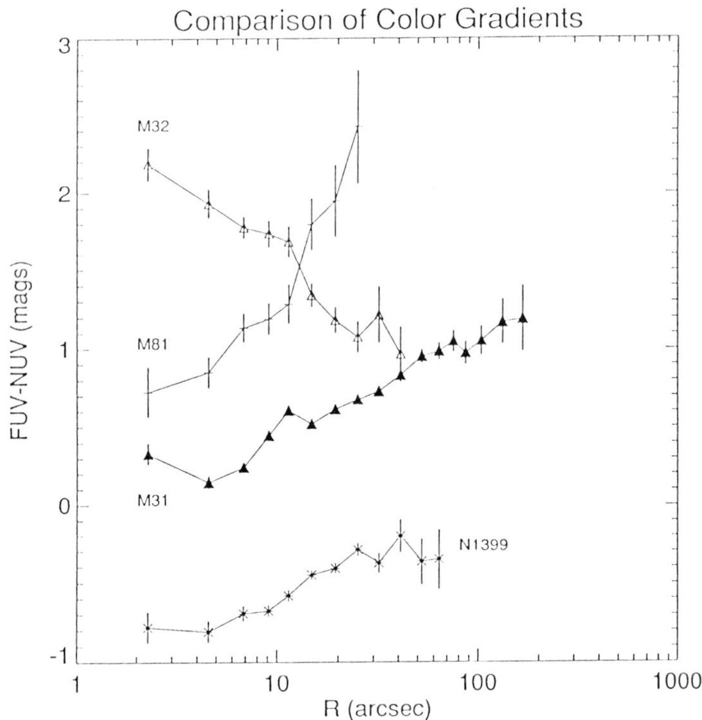

FIGURE 3: UV (1500–2500) colors as a function of radius for four old populations. The plots are truncated where the photometry becomes unreliable. For orientation, normal O7 and A5 main sequence stars would fall at colors of −1.3 and +2.2, respectively; the bluest and reddest globular clusters fall at 0.0 and +2.1.

by comparison to optical/IR gradients ($\lesssim$ few 0.1 mags). The color gradients are smooth from the cores outward. In all objects except M32 the colors become *redder* outward; in M32 they become *bluer*. There were earlier indications of color gradients in IUE spectra for the M31 nucleus (Welch 1982, Deharveng *et al.* 1982), but this is the first evidence for extended gradients of large amplitude.

It is not likely that dust is responsible for the gradients. First, the gradients are similar in the Sb's and the E galaxy NGC 1399 despite a likely major difference in dust content. Second, for the Galactic UV reddening law, the color excess E(15–25) $\sim$ E(B–V) $\lesssim$ 0.1. Instead, it is probable that we are seeing the response of the hot, low mass evolutionary phases which produce the UVX to a decline in metal abundance with radius. The sense observed in the systems other than M32 is the same as the overall UVX/abundance effect found for galaxy centers by Burstein *et al.* (1988). It is consistent with the expectation that giant branch mass loss increases with metal abundance, producing an increase in the net UV luminosity of low-mass remnants (Greggio and Renzini 1990). However, the non-overlap of the NGC 1399 and M31 color profiles in Fig. 3, despite similar optical spectra and line strengths, suggests that the UVX may be sensitive to parameters in addition to metal abundance. A more complex behavior has also been hinted at by the facts that the UVX seems stronger in "boxy" than in "disky" E's (Longo *et al.* 1989) and that S0's have a steeper NUV-luminosity relation than do E's (Smith and Cornett 1982, Kodaira *et al.* 1990).

The UV color gradient in M32 is not explainable as a simple metal abundance effect if the abundances are roughly constant or declining outward, as suggested by optical colors. Because M32 has the lowest mean abundance of the objects studied, its behavior could reflect a phase transition in the dominant type of UVX star. Alternatively, an age gradient, in which the center is younger, would have the correct sense to explain the colors, since it becomes more difficult to produce hot, long-lived post-GB remnants if the MSTO mass is larger (Greggio and Renzini 1990).

6. Conclusion

Important new insights into both young and old stellar populations can be obtained with UV observations. This review has concentrated mainly on systems in or near the Local Group, but rapid progress in covering a much larger volume is possible now that UV telescopes with faint thresholds are available. Ultimately, some of the most interesting applications of UV diagnostics will be to the evolutionary histories of very distant systems in earlier stages of evolution.

ACKNOWLEDGEMENTS: The UIT project began in 1978. The twelve-year effort leading to its first mission was directed by PI Ted Stecher with Co-I's Ralph Bohlin, Mort Roberts, Andy Smith, and myself; Harvey Butcher served as a Co-I before moving to Holland. Ron Parise of CSC Corp. devoted many years to making the mission a success as a Payload Specialist. We gratefully acknowledge the numberless contributions made at Goddard by Ted Gull, Susan Neff, the engineering group led by Gerry Baker, and the STX Corp.; MSFC, KSC, and JPL; and the support of the three other Astro science teams.

REFERENCES

Barbaro, G., and Olivi, F.M. 1989, *Ap. J.*, **337**, 125.
Bica, E., Alloin, D., and Schmidt, A. 1990, *Astr. Ap.*, **228**, 23.
Bohlin, R.C., Cornett, R.H., Hill, J.K., Smith, A.M., and Stecher, T.P. 1985, *Ap. J.*, **292**, 687.
Bohlin, R.C., *et al.* 1991, *Ap. J.*, **368**, 12.
Boulade, O., Rose, J.A., and Vigroux, L., 1988, *A.J.*, **96**, 1319.
Bruzual, G.A. 1983, *Ap. J.*, **273**, 105.
Buat, V., Deharveng, J.M., and Donas, J. 1989, *Astr. Ap.*, **223**, 42.
Burstein, D., Bertola, F., Buson, L.M., Faber, S.M., and Lauer, T.R. 1988, *Ap. J.*, **328**, 440.
Burstein, D., Faber, S.M., Gaskell, C.M., and Krumm, N. 1984, *Ap. J.*, **287**, 586.
Buzzoni, A. 1989, *Ap. J. Suppl.*, **71**, 817.
Cacciari, C., Caloi, V., Castellani, V., and Fusi Pecci, F. 1984, *Astr. Ap.*, **139**, 285.
Caloi, V., Castellani, V., Nesci, R., and Rossi, L. 1985, *Astron. Astrophys. Suppl.*, **59**, 505.
Cassatella, A., Barbero, J., and Geyer, E.H. 1987, *Ap. J. Suppl.*, **64**, 83.
Castellani, V., and Cassatella, A. 1987, in *Exploring the Universe with the IUE Satellite*, ed. Y. Kondo (Dordrecht: Reidel), p. 637.
Cowley, A.P., and Burstein, D. 1988, *A.J.*, **95**, 1071.
Crotts, A.P., Kron, R.G., Cacciari, C., and Fusi-Pecci, F. 1990, *A.J.*, **100**, 141.
Courvoisier, T.J.-L., Reichen, M., Blecha, A., Golay, M., and Huguenin, D. 1990, *Astr. Ap.*, **238**, 63.
Davidge, T.J. 1990, *A.J.*, **99**, 561.
Deharveng, J.M., Joubert, M., Monnet, G., and Donas, J. 1982, *Astr. Ap.*, **106**, 16.
Djorgovski, S., Piotto, G., Phinney, E.S., and Chernoff, D.F. 1991, *Ap. J. (Letters)*, in press.
Greggio, L., and Renzini, A. 1990, *Ap.J*, **364**, 35.
Gunn, J.E., Stryker, L.L., and Tinsley, B.M. 1981, *Ap. J.*, **239**, 48.
Heap, S.R. *et al.* 1991, *Ap. J. (Letters)*, in press.
Kjærgaard, P. 1987, *Astr. Ap.*, **176**, 210.
Kodaira, K., Watanade, T., Onaka, T., and Tanaka, W. 1990, *Ap. J.*, **363**, 422.

Kondo, Y. 1987, *Exploring the Universe with the IUE Satellite* (Dordrecht: Reidel).
Laget, M., Burgarella, D., Milliard, B., and Donas, J. 1991, *Advances in Space Research* (COSPAR XXVIII), in press.
Longo, G., Capaccioli, M., Bender, R., and Busarello, G. 1989, *Astr. Ap.*, **225**, L17.
Nemec, J.M., and Cohen, J.G. 1989, *Ap. J.*, **336**, 780.
Nesci, R. 1983, *Astr. Ap.*, **121**, 226.
O'Connell, R.W. 1986, in *Stellar Populations*, eds. C.A. Norman, A. Renzini, and M. Tosi (Cambridge University Press), p. 167
O'Connell, R.W. 1990a, in *Windows on Galaxies*, eds. G. Fabbiano, J.S. Gallagher, and A. Renzini (Dordrecht: Kluwer), p. 39.
O'Connell, R.W. 1990b, in *Bulges of Galaxies*, eds. B.J. Jarvis and D.M. Terndrup (Garching: ESO), p. 187.
O'Connell, R.W. 1991, *Advances in Space Research* (COSPAR XXVIII), in press.
Paresce, F. *et al.* 1991, *Nature*, **352**, 297.
Renzini, A. 1986, in *Stellar Populations*, eds. C.A. Norman, A. Renzini, and M. Tosi (Cambridge University Press), p. 213.
Rocca-Volmerange, B., and Guiderdoni, B. 1987, *Astr. Ap.*, **175**, 15.
Rocca-Volmerange, B., and Guiderdoni, B. 1988, *Astr. Ap. Suppl.*, **75**, 93.
Smith, A.M., and Cornett, R.H. 1982, *Ap. J.*, **320**, 609.
Sweigart, A.V. 1987, *Ap. J. Suppl.*, **65**, 95.
van Albada, T.S., de Boer, K.S., and Dickens, R.J. 1981, *M.N.R.A.S.*, **195**, 591.
Vauclair, G., and Liebert, J. 1987, in *Exploring the Universe with the IUE Satellite*, ed. Y. Kondo (Dordrecht: Reidel), p. 355.
Welch, G.A. 1982, *Ap. J.*, **259**, 77.
Welch, G.A., and Code, A.D. 1980, *Ap. J.*, **236**, 798.
Wu, C.-C., Faber, S.M., Gallagher, J.S., Peck, M., and Tinsley, B.M. 1980, *Ap. J.*, **237**, 290.

DISCUSSION

LEE: One comment regarding the ω Cen CM diagram. You showed that there is a large scatter in luminosity. I think this is mostly due to the following effects: (1) internal variations in [Fe/H] in ω Cen and the HB luminosity dependence on [Fe/H]; and (2) evolution off the ZAHB, which itself produces a large range in luminosity. One or two stars may be post-AGB, but most in your sample are indeed HB stars.

O'CONNELL: I agree that such effects must be present, and we are in the process of examining the location and frequency of the brighter objects with respect to post-ZAHB evolutionary tracks. We do find objects up to 3–4 mags above the ZAHB, however, and our first impression is that not all of them can be evolved HB stars.

CARRASCO: I suggest that the hot UV bright stars you detect in globular clusters correspond to the same population that exists in the disk and that spectroscopically mimics OB stars of extreme Pop I. These are the so-called "runaway" stars. It has been found that they share the same kinematics as planetary nebulae, indicating that this UV bright population is indeed an evolved one.

O'CONNELL: I can't comment on their relation to the runaway OB stars, but it is certainly true that the IUE spectra of the bright objects in ω Cen look at first glance like those of main sequence B stars. We will make a more detailed comparison soon.

CACCIARI: You mentioned that ages can be derived from UV colors. I would like to make a comment on that. This method, which was applied by Dickens and van Albada about 10 years ago using ANS data for Galactic globular clusters, may be biased by the presence of UV bright stars, such as post-AGB stars. For example, the PAGB star VZ1128 in M3 produces about 20% of the total light at 1500 Å; the same type of star in 47 Tuc would produce about 90% of the 1500 Å light.

O'CONNELL: Yes, it is essential to remove the contaminating effects of very hot stars, as well as the influence of cooler dwarfs and giants. In the case of galaxies, one must remove the long-wavelength tail of the UVX stars. This can be done with some confidence if one has long baseline spectra available. For the hot components, that means data with reasonable S/N down to 1200 Å.

RICH: Have you compared your UV surface photometry for the M31 bulge with the V-band photometry of Kent? Is the UV light profile like the visual or is it more concentrated?

O'CONNELL: The near-UV profile appears to be very similar to the optical-band profile; the far-UV surface brightness declines faster with r. The (NUV–optical) colors will be affected by UVX contamination as well as by blanketing, and we have not yet tried to disentangle these.

FERGUSON: (1) Can you set any interesting limits on the number of young stars in NGC 1399 from the smoothness of the surface brightness profile? (2) Do any of the UIT exposures of globular clusters go deep enough to pick up the white dwarf cooling sequence?

O'CONNELL: (1) There is possibly some low-level structure in the NGC 1399 images, and we are re-digitizing our data with finer microdensitometer resolution. We think that we may be detecting the brighter globular clusters. (2) Unfortunately, our exposures were too short. We would have needed at least 30-minute exposures, and preferably stacked images from several orbits. Our primary candidate for such studies, 47 Tuc, was also not well placed for long nighttime exposures.

PELETIER: A large fraction of the galaxies in the paper by Burstein *et al.* (1988) showing UV upturns contained active galactic nuclei (e.g. M87, NGC 6166). Is is possible that the "reversed" gradients you find are signs of induced star formation in the inner regions?

O'CONNELL: We don't think so, since in M31, M32, and M81 there is no sign of individual OB stars, associations, or clusters. Also, the objects in the Burstein *et al.* sample which you mention have UV spectra which are significantly *flatter* than those of objects like NGC 1399 or M31. This is probably consistent with the broader range in temperature expected from the upper main sequence in a star-forming system. The steeper spectrum in NGC 1399 probably indicates that low-mass post giant branch stars dominate the UV.

ELLIS: Milliard *et al.* (preprint) present galaxy counts to $m_\lambda(2000) = 19.5$. I find their counts to be above a simple no-evolution expectation, indicating recent changes in the bulk star formation rate. However, their image quality is not as good as the UIT's, and their star/galaxy separation is done in a complex, uncertain way. Will UIT be able to produce UV counts significantly deep to address this?

O'CONNELL: We certainly hope so. We made a number of deep exposures in high latitude fields centered on quasars with this kind of problem in mind.

THE STELLAR CONTENT OF THE BULGES OF SPIRAL GALAXIES: WHAT DO WE KNOW AND HOW DO WE KNOW IT?

JAY A. FROGEL
Astronomy Department
The Ohio State University
174 West 18th Avenue
Columbus, Ohio 43210

ABSTRACT. For a true picture of the stellar content of the bulges of spiral galaxies it is necessary to combine spectroscopic and photometric observations from the ultraviolet to the infrared. In early-type spirals one generally finds a metal-rich, old population with no excess of low mass stars. In later type systems there is an increasing contribution from young stars. The population in the Galactic bulge is representative and does not contain a significant number of stars with large IR excesses.

1. Introduction

There is a poem which I am sure most of us have read as children about three blind men and an elephant. It tells the story three sightless people trying to describe an elephant based on what they can feel. Unfortunately, the poet allowed each of the three people to feel only one part of the elephant. So, we have the first person describing the elephant as snake-like as he could only feel the trunk. The second person thought that elephants must be like trees since he was feeling one of the elephant's legs. The third person concluded that elephants are hairy beasts – he was feeling the tail.

A not too dissimilar situation exists when it comes to describing spiral galaxies: conclusions about their stellar content by observers who concentrate primarily on one region of the electromagnetic spectrum will, not unexpectedly, be focused on those stars that are the main contributors to the region of the spectrum under observation. Ultraviolet observers will emphasize stars that may be metal poor or young, or a combination of the two. K and M dwarfs will make their presence felt most strongly in the red spectral region. Observations in the thermal infrared will be unduly influenced by the presence of IR/OH stars and other objects with significant amounts of mass loss. Although each of these galaxy observers will, like the elephant investigators, be partially correct, all of them will be completely wrong in trying to generalize their limited findings. An elephant certainly has all of the traits attributed to it by its three examiners but altogether can be characterized by none of them. The same is true for a spiral galaxy: it will contain some of each of the types of stars deduced by its examiners, but none of these types can be said to give an overall picture.

When multi-wavelength observations of spiral galaxies are combined, it becomes obvious that the stellar populations of these galaxies are highly composite. There are late-type dwarfs and giants, metal poor and metal rich stars, old and young ones, and a smattering of stars with infrared excess emission. The relative numbers of each of these types of stars varies in a more or

B. Barbuy and A. Renzini (eds.), The Stellar Populations of Galaxies, 245–254.

less systematic way along the Hubble sequence. In fact, the delineation of this variation and its interpretation is of fundamental importance in understanding the evolution of spiral galaxies. In this brief review I will examine the compositeness of the stellar population of spiral galaxies and bring to your attention a number of interesting recent pieces of research. These represent some, but certainly not all of the important findings on the topic that have appeared since the thorough review by O'Connell (1983).

2. Issues of Technique and Input for Stellar Synthesis Models

Let me first call your attention to a recently completed Ph.D. thesis by David Silva from the University of Michigan, now a post-doctoral fellow at Kitt Peak. He is the latest of several researchers in the field to emphasize the need for improved accuracy of the observations that go into synthesis models rather than achieving a yet finer grid to cover the phase space of the models. A number of well chosen examples serve as clear illustrations of his point. One of these is a comparison of the stellar libraries of Pickels and Jacoby and their collaborators. Particularly below 4500 Å there are major differences in the spectrophotometry of stars of ostensibly the same spectral type drawn from the two libraries. Whether one set of spectra or the other is used in models has a major effect on the conclusions drawn concerning the presence of a young stellar population. Presumedly, these differences in the spectrophotometry arise from calibration differences in the original observations.

Silva also points out that in linear evolutionary population synthesis, if the entire energy distribution is used, the continuum information will dominate over the line information. This can lead to significant degeneracy in the solutions and underscores the need to get the correct luminosity function. Two examples that he gives are the high degree of correlation between the energy distributions of K2V and K2III stars (r=0.998) and between the energy distributions of weak and strong line K giants (r=0.997). The result is that in evolutionary population synthesis, degenerate relations will often be selected rather than physically correct ones. The implication is that one should not just rely on a single goodness of fit parameter; careful examination of the residuals between model and observations is required. Silva also emphasizes the importance of broad band colors from the ultraviolet to the infrared. As I have emphasized elsewhere (Frogel 1985, 1988), a comparison of observed and predicted colors provides a simple but key first test of a model and can set parameter space boundary conditions for luminosity functions.

Next, I went to mention the work of E. Bica, D. Alloin, and their collaborators. At meetings on topics similar to the theme of this one, we have often been admonished by A. Renzini that we are neglecting one of the best tools for population synthesis, namely clusters. Although there have been limited attempts to use clusters as building blocks, Bica and Alloin are the first to make a major effort in this direction. Their data base for population synthesis consists of integrated colors and spectra of star clusters in the Magellanic Clouds and the Milky Way from 3700 to 8000 Å region at 11 Å resolution. They use absorption features and the overall shape of the continuous energy distribution to match clusters to galaxies. Their cluster data define two sequences: changing [Fe/H] at constant age (Galactic globulars); nearly constant [Fe/H] with variable age (Magellanic Clouds). Many of their most important results are discussed in other papers presented at this conference. For spiral bulges, their results are in general agreement with the earlier work of Turnrose (1976). For the late-type spirals, they agree with my own, considerably less detailed, analysis of UBVJHK colors (Frogel 1985) of the nuclei of the spirals, also based on integrated cluster photometry.

A potential shortcoming in the work of Bica and his collaborators is an incomplete sampling of the necessary parameter space of temperature, metallicity, and age for the stars that go into synthesis models of galaxies. This should not be confused with Silva's claim, mentioned above, that better sampling of parameter space is not necessarily needed. He was referring to the fineness of the grid. Here we are talking about undefined boundaries. The problem is the following. Whitford and Rich (1983) and Rich (1988) have shown that the K giants in Baade's Window in the bulge of the Milky Way, have a mean [Fe/H] of twice solar. The cluster data base use by Bica *et al.* has no high [Fe/H] globular clusters nor does it have high metallicity analogues of the intermediate age clusters in the LMC.

While some of the oldest open clusters in the Galaxy have the potential to remedy the latter shortcoming, none of these open clusters comes even close to the richness needed in insure proper sampling of the relatively short-lived but major contributors to the bolometric luminosity of such stellar assemblages, namely luminous late-type stars on the asymptotic giant branch (cf. Houdashelt *et al.* 1991). On the other hand, it may be possible to alleviate the problem of missing old, high [Fe/H] stars. As reported at this conference, Ortolani, Bica, and their collaborators are obtaining new optical data for several bulge globular clusters in the Milky Way. These are the most metal rich clusters known in the Galaxy (Armandroff and Zinn 1988). An effort is also underway to obtain infrared photometry for these objects (Frogel, Terndrup, and Armandroff, in preparation). Infrared data is crucial for a determination of bolometric luminosity and temperature of the giant stars. In addition new data on metal rich globular clusters in M31 (Jablonka, private communication) confirms the extrapolations used by Bica *et al.* to compensate for a somewhat incomplete data base. Finally, one could treat the stellar population of the Galactic bugle as a cluster, albeit one of a unique type, and include it in parallel with the other clusters in a library. We (Terndrup and myself) are pursuing such an approach.

To conclude this section, I would like to remind you of something obvious but that is often overlooked: the problem of radial gradients. Even at the relatively close distance of the Virgo Cluster, spectrophotometric observations with a slit size as small as 2" still corresponds to ~45" in M31 or 1.3° in the bulge of our own galaxy. We know that there are strong metallicity gradients in the Galactic bulge on this scale. Therefore, in interpreting observations with even relatively high spatial resolution in extragalactic systems, we must always bear in mind that not only will there be a range in age and metallicity (not to mention velocity dispersion and possibly IMF), that characterizes the stars at every position in a galaxy, but the observed distribution functions will be affected by strong spatial gradients.

3. The Composite Population of Spiral Bulges

3.1. DWARFS VERSUS GIANTS

The Na I doublet near 8200 Å is an often used spectral feature in attempts to determine the relative contribution of dwarf and giant stars in a stellar population since it is quite strong in dwarfs but weak in giants. Unfortunately, contamination by Telluric features has made interpretation of observations of Na I difficult and controversial. In addition, it now appears that these observations are affected by other stellar absorption features (Alloin and Bica 1989; Xu, Veron-Cetty, and Veron 1989). Blended with the Na I absorption are metallic lines in K giants and TiO absorption at 8198.5 and 8205 Å in late-type giants. Xu *et al.* conclude that in luminous galaxies the observed Na I feature is actually a blend of all three components, each contributing about equally. Alloin and Bica find that in the center of M31 the enhanced absorption in the 8200 Å region is due to the 8205 Å TiO feature rather than Na I. They also demonstrate that in giant stars Na I is weakly correlated with [Fe/H] because of the blends with other stellar features. The correlation is in the sense that Na I absorption gets stronger in cooler stars. Thus it would appear that claims based on Na I feature for a dwarf enriched population in the integrated light of many galaxies are exaggerated.

In a recent preprint, Delisle and Hardy present a study of the spatial variation in the nuclei of early-type spirals and E galaxies of the Ca II triplet, TiO bands, and the Na I doublet. Their conclusions as to the contamination of the Na I absorption by other stellar absorption features is in agreement with those of Alloin and Bica and of Xu *et al.* They also find that Na I correlates more strongly with global parameters likes M_V and the Mg_2 lines rather than with details of the initial mass function deduced from models. This finding, too, is consistent with the importance of the contaminating features rather than just a simple dwarf - giant effect on the observed Na I strength.

There is a well defined decrease in the strength of the Na I blend with increasing central distance in half of the galaxies in the Delisle and Hardy sample. Since the lines that blend with the Na I feature are expected to be metallicity sensitive, the observed radial dependence, as well as the global correlation mention above, probably reflects an overall correlation with metallicity. The CaII triplet shows similar but less well defined dependencies. An interesting finding of Delisle and Hardy is that one TiO band increases with increasing radial distance. Could this be related to the fact that in a metal rich population the brightest stars in M_{bol}, which also happen to be the reddest, are relatively quite faint in the *I* band (see the contribution by Ortolani, Barbuy, and Bica elsewhere in the volume)? If a declining value of [Fe/H] is associated with increasing radial distance, than as the distance increases, so would the relative contribution to the I band of the reddest stars thus causing an apparent increase in the strength of some TiO bands. The behavior of TiO bands may also be complicated if one encounters a significant population of luminous AGB stars in regions that have experienced star formation in the past few billion years.

The Wing-Ford band of FeH can provide another important constraint on the relative numbers of M giants and dwarfs as it is very sensitive to the presence of dwarfs later than M4 (Whitford 1977). Unfortunately, it is in an even more difficult part of the spectrum to observe than the Na I doublet. Hardy and Couture (1988) reported the first clear-cut detection of this band in several E galaxies. They showed that measurements of the FeH band are probably affected, and may even be dominated by the (2-3) band of the δ system of TiO. Hardy and Couture's results "very likely exclude dwarf-enriched models" for the E galaxies they observed. Whitford (1977) was unable to detect the band in any of the galaxies he observed.

To sum up, recent observations of spectral features that are most sensitive to the presence of late type dwarfs in the integrated light of galaxies give no convincing evidence for the presence of an excess number of such stars in the systems studied to data. A luminosity function similar to the Salpeter one is adequate, in most cases, to understand the spectral features and continuous energy distributions of these galaxies.

3.2. OLD VERSUS YOUNG STARS

What observational criteria can be used to distinguish between age differences and metallicity differences? The fact that observations of galaxies are averages over volumes that probably encompass significant radial variations in these two physical parameters greatly complicates the problem. This issue is particularly critical when very distant galaxies are observed. It is also significant when observing spiral bulges as, except for edge-on systems, there will always be a contribution from the part of the disk that lies along the line of sight to the bulge.

If young stars are present in a galaxy they will be obvious at blue wavelengths and have little or no effect in the red. Therefore, the detection and analysis of a young stellar component can best be done by the measurement of spectral features and colors in the blue. Determination of the main sequence turnoff is particularly important. Examples of how a young stellar component can be detected and distinguished from a blue but old and metal poor component are given by Frogel (1985) and Bica and Alloin (1987).

In contrast to the blue, the red and near-infrared spectral regions are dominated by cool giants from a relatively old stellar population. Colors and spectral features in the red and near-ir will be sensitive primarily to abundance variations and secondarily to details of the AGB and to the ratio of dwarf to giant stars. The presence of supergiants can introduce some complications, but these are easy to sort out. If used in combination, the blue and red spectral region together can set strong constraints on the age and metallicity of a stellar population. An example of the use of just broad band colors to constrain isochrone synthesis models is provided by Charlot and Bruzual's (1991) recent work.

In another example Bica and Alloin (1987) compare blue and red spectral features in the same set of galaxies. In late type spirals the blue features, e.g. Ca II H&K lines and CN absorption, are considerably weakened compared with the red features. If blue features alone had been observed, one might draw the conclusion that the late-type spirals are metal weak. But Bica and Alloin argue convincingly that the weakness of the blue features is due almost exclusively to *dilution* by the continuum light of young, hot stars. Central to their reasoning is the fact that the red spectral features in the same galaxies show a much smaller change with galaxy type. Bica and Alloin's conclusions for the nuclei of late-type spirals are in accord with the earlier work of O'Connell (1982) and Frogel (1985) who based their analysis on lower resolution spectral data or just broad-band colors. In the latter work, Frogel showed that the UBV colors are decoupled from the JHK ones in a sample of about 20 late-type spirals. This suggests that the current level of star forming activity varies considerably from one galaxy to the next, even at the same Hubble type. An obvious cause of this would be if periods of particularly active star formation were discrete and took place on a time scale significantly less than the age of the galaxy.

3.3. OBSERVED TRENDS WITH HUBBLE TYPE

In the previous two subsections I have reported that recent studies of the stellar content of spiral bulges lean heavily towards three conclusions. In varying degrees, these conclusions require that observations be made over a wide enough baseline in order to sort out the effects of age from metallicity. In the 1990s, it has finally become clear that "wide enough baseline" means at least from the blue cutoff of the atmosphere to the near infrared. These three general conclusions concerning the stellar content of spiral bulges are:

1) they can be highly composite;
2) the initial mass function of the stars appears normal, i.e. there is no evidence for an enhancement in the numbers of late-type dwarfs;
3) the available observations of spiral bulges point to the presence of a young population in some of them and a metal poor population in (almost) none of them.

How do these conclusion vary with Hubble type? An answer may be found in a recent, comprehensive survey of stellar populations in the bulges of spiral galaxies by Bica and Alloin (1987) and Bica (1988). I will briefly summarize their findings relevant to the question.

Earlier and more luminous spirals are generally very similar to E and S0 galaxies. Bica and Alloin find little or no evidence for blue HB stars in these systems, implying that there is no significant metal poor population in the bulges of early-type spirals. Later type spirals and the less luminous ones of a given type have more prominent contributions from young stars, but the change in the mean [Fe/H] is small. Their findings for the three main classes of spirals are:

> Sa: Spirals of this class are very homogeneous. They are characterized by a red continuum and strong absorption features. There could be a small contribution from a 5 Gyr old population.
>
> Sb: The higher luminosity Sbs are red and strong lined, similar to ellipticals and Sas. Lower luminosity Sbs are also red, but the absorption spectrum is not as strong lined. Occasionally one finds an Sb with a blue nucleus, e.g. N4569, probably due to a recent period of particularly active star formation.
>
> Sc: As a group Scs are much less homogeneous than Sbs or Sas. The lower luminosity ones have blue to very blue energy distributions implying a significant population of young stars. In the bluest group, a component of the stellar population with an age of $\leq 3\times10^8$ yrs contributes nearly 90% of light at 4000 Å and nearly 60% at 9000 Å.

4. How Similar are Spiral Bulges and Elliptical Galaxies?

It has been known for a long time that the colors and spectral absorption features of elliptical galaxies and the bulges of early type spirals are quite similar. Morgan's (1956) qualitative spectroscopy of the integrated light from Baade's Window in the bulge of the Milky Way and Whitford's (1978) seminal study of the same field raised the possibility that delineation of the properties of the stars in Baade's Window would be valuable in furthering our understanding of the stellar content not just of spiral bulges, but of elliptical galaxies as well (Frogel and Whitford 1987; Frogel 1985)

Recent studies already mentioned provide further evidence for a strong similarity in the stellar content of E galaxies and the bulges of *early-type* spirals. For example Delisle and Hardy (1991) find that to within the observational uncertainties, the mean strengths of red stellar absorption features in early-type spiral bulges and E galaxies are indistinguishable. Bica, Arimoto, and Alloin (1988) combine evolutionary models, e.g. those described by Arimoto and Yoshii (1987) and cluster synthesis technique of Bica and Alloin (1987) to show that spiral nuclei and elliptical nuclei of comparable total mass have had very similar star forming histories and, consequentially, similar present day stellar populations.

What about late-type spirals? In so far as their old stars are concerned, they appear to be quite similar to E type galaxies (e.g. Frogel 1985), although it is difficult to distinguish between populations that are 5 to 10 Gyr old and those that are 10 to 15 Gyr old. Overall, though, there are significant differences between the bulges of late-type spirals and typical E galaxies (Alloin 1973; Turnrose 1976; O'Connell 1976 and 1982; Frogel 1985; the papers of Bica and collaborators referred to earlier). These differences may best be summarized by saying the late-type spirals have energy distributions that are significantly bluer than those of the early type spirals and E galaxies. This blue energy distribution arises from a population of young stars mixed with the old population. The relative proportion of young and old stars varies strongly from galaxy to galaxy. In fact, it is misleading to speak of an "old" and a "young" population in these galaxies since star formation has probably never really stopped in many of them.

There are some striking differences between spiral bulges and E galaxies. The nuclei of many spirals have strong emission line spectra indicative of H II regions as well as non-thermal emission. While as many as half of E and S0 galaxies have detectable [N II] 6584 emission (Phillips *et al.* 1986), this emission tends to be quite weak and difficult to detect in contrast to that found in the spiral nuclei. The emission in spirals may more properly be associated with the inner part of a disk rather than the bulge. Gas flowing inwards through the disk or along a bar could fuel the star formation and non-thermal activity observed in spiral nuclei. Perhaps the most fundamental difference between spiral bulges and ellipticals, and the whose origin is the least understood, is the fact that spiral bulges are rotationally supported whereas E's are, for the most part, pressure supported.

5. Evolutionary Synthesis Models

Stellar synthesis models tell us primarily about spiral bulges as they are today, not their evolutionary history. We need to account for the observed distributions of metallicity and age of the stars as well as their kinematics, Sandage's (1986) "three dominant parameters in the population concept". The recent work of Arimoto and his collaborators (e.g. Arimoto and Yoshii 1987; Koppen and Arimoto 1990) has done much to help us understand the evolution of stellar properties. Their underlying premise is that rapid star formation drives chemical enrichment of the interstellar medium in spiral bulges until a supernova driven wind expels the remaining gas. The effects of such a wind are discussed in detail by Vader (1987).

Koppen and Arimoto (1990) use the above scheme to explain a number of the global characteristics of spiral galaxies that vary systematically along the Hubble sequence. They consider a simple 3 component model: 1) a bulge with relatively low angular momentum that collapses in a free-fall time (~0.1 Gyr); a SN driven wind turns on after ~1 Gyr to terminate star formation; 2) a disk with high angular momentum, a long collapse time scale, and slow star formation rate (SFR) time scale; 3) a halo in which there is no star formation. The halo gas is enriched by material blown out of bulge by SN, and subsequently falls back into disk and enriches it in turn.

For a metallicity distribution, N(Z), they use data on the Galactic bulge from Rich (1988) and published data on G dwarfs to fix parameters for one galaxy - the Milky Way. The peak of N(Z) gives the yield, y, within the bulge almost independent of other parameters. The width of N(Z) gives the gas depletion rate times the age. The metallicity distribution for G dwarfs sets an upper limit on the time scale for a SN driven wind from the bulge. They can then explain the correlation with Hubble type of the mean metallicity of spiral disks and the gas content of the disks for Sa through Sc spirals.

Koppen and Arimoto construct a model sequence in which the bulge to disk ratio, B/D, and the SFR in the disk vary monotonically with Hubble type. The B/D variation in terms of relative integrated luminosity, is consistent with observations. From Sa through Scd Koppen and Arimoto's model sequence has the following properties: The SFR per unit mass in the disk declines by only factor of 3 from the early to the late type spirals. At the same time, the mean metallicity of the bulge relative to that of the disk increases from unity to a factor of 4 while the ratio of bulge mass to disk mass declines from 2.5 to less than 0.1. As Koppen and Arimoto point out, the higher SFR in the disks of early type spirals means that the enrichment rate is higher than in the disks of later type spirals. This results in the smaller contrast in mean metallicity between bulge and disk for earlier type systems than later type ones. In the bulge component of their models the initial accretion of gas and the SFR are independent of the B/D ratio. The result of this is that the mean metallicity of the bulge does not change much over the spiral sequence. On the other hand, pollution of the disk by bulge gas is greatest in early type spirals because the mass of the bulge is greatest so the largest amount of matter is lost. They also require that the yield for the bulge be 4 times greater than that for the disk in order to account for differences in N(Z) between the two components distribution, similar to Pagel's (1987) result. This difference in yield can be accounted for by an increase in the slope of the IMF in the disk by only 20%, 1.3 to 1.55. Such a small difference in slope between disk and bulge is not surprising as the time scales for the SFR in the disk and bulge are probably driven by different processes, e.g. spiral density waves in the disk, and cloud-cloud collisions in the bulge.

There are two obvious properties of spiral galaxies that have not been taken into account in the above model. First of all, spirals show a big range in properties even at the same Hubble type. Secondly, the range in metallicity within a given spiral disk is greater even than the range from one galaxy to another. Nevertheless, this model provides a simple but enlightening approach to understanding the global properties of the stellar content of spiral bulges.

6. Are the Bulges of the Galaxy or of M31 Representative?

Whitford's (1978) study of the integrated light in Baade's Window provided quantitative evidence that the stellar content of the Galactic Bulge is similar to that of early-type galaxies, both ellipticals and spirals. Use of Galactic bulge M giants in stellar synthesis models has resulted in significantly improved fits in the red and near-infrared to the integrated light of E galaxies (Frogel and Whitford 1987; Frogel 1988; Terndrup *et al.* 1990). In brief, evidence in favor of the stellar content of the Galactic bulge being typical of other early type systems is accumulating. Nothing has yet been found that seriously challenges this result.

M31 is the nearest large spiral to the Milky Way. Individual stars in its bulge and disk can be studied without recourse to observations from space. Thus it is important to determine as quantitatively as possible whether or not the stellar population of its bulge is similar to that of our own Galaxy. In resolving this issue we are also addressing the representative nature of the Galactic bulge. All recent studies conclude that M31's bulge in integrated light is typical of other early-type systems. Delisle and Hardy (1991) find absorption features in the bulge of M31 to be indistinguishable from other early type spiral bulges or from E galaxies. In central 4" of M31, Alloin and Bica (1989) find that the enhanced absorption near the 8200 NaI doublet is due to a feature at 8205, probably TiO. Thus there is no evidence for dwarf enhancement, consistent with observations of other most other galaxies. Bica, Alloin, and Schmidt (1990) find that both the bulge and the semi-stellar nucleus of M31 are dominated by the old, metal-rich stellar component. They estimate that at most 10-20% of the flux can come from an intermediate age component. Again, these are limits characteristic of E galaxies and early-type spirals.

Recent remarkable advances in the technology of infrared arrays has finally allowed us to image the bulge of M31 at 2.2 μm and compare the physical characteristics of M31's luminous M giants with those in the Galactic bulge (Rich *et al.* 1989). Results from two programs have recently been published (Rich and Mould 1991; Davies *et al.* 1991), and the conclusions and not nearly as clear cut as in the case of the integrated light of the two bulges. Bolometric luminosity functions for two fields in M31 are similar to the luminosity function for Baade's Window except that for M31 the brightest stars are 0.5 to 1.0 magnitude brighter than those in Baade's Window. Although this result could be taken as evidence for a component to the stellar population with an age several Gyr younger than that of the Galactic bulge, Davies *et al.* argue that these bright stars are from M31's disk that lies along our line of sight but behind its bulge. Observations of M32 reported at this conference by W. Freedman show a population of luminous stars in that galaxy as well. They cannot be explained away by a disk. Some resolution of the problems raised by these observations of M31 and M32 is expected by the time these conference proceedings appear in press.

The Galactic Bulge and IR/OH Sources

From balloon and other data, emission at ~2.2 μm from the Galaxy shows a well defined disk and bulge structure, typical of other spiral galaxies. For example, Melnick *et al.* (1987) find for $15° \leq l \leq 60°$ the FWHM of 2.4 μm emission is ~5°. For the central 10°, on the other hand, the FWHM is nearly twice this amount. Emission at 2.4 μm is primarily from photospheres of K and M giants (Frogel 1988).

IRAS has detected numerous stellar sources in the bulge, many of which are classified as OH/IR stars on the basis of their colors and luminosities (Whitelock, *et al.* 1991; van der Veen and Habing 1990). Are there sufficient numbers of these sources to constitute an important component of the population and make a significant contribution to the integrated light of the bulge at 12μm? Stellar emission at this wavelength will be dominated by circumstellar dust. Rowan-Robinson and Chester (1987) make corrections for the severe extinction for the IRAS sources, particularly at low latitudes, calculate the effects of sources with low optical depths in their dust shells, and predict $L = 1.7 \times 10^8 L_{sun}$ for the total luminosity. They also predict what would be seen in the central 4' of M31 and find consistency with Soifer *et al.*'s (1986) value - 4.9Jy at 12μm. This is about 2 times greater than expected from purely photospheric emission but in close agreement with the excess predicted by Frogel and Whitford for Baade's Window from ordinary M giants alone. So the stars contributing to excess in both galaxies are probably similar.

There is a more direct and less model dependent way to investigate the influence of OH/IR stars on the integrated light from the bulge. That is to map out the spatially integrated 12 μm flux measured by IRAS. This values are derived from the total flux detected by IRAS, thus removing limitations associated with source confusion and the detection limit for faint sources. For example, the untreated IRAS data revealed a FWHM of ≤1° within 6° of the Galactic center (Gautier *et al.* 1984). I have completed an analysis of the preliminary, prerelease version of the "Super Sky Flux" maps from IRAS. These represent the best attempt to date to give maps of the sky at all IRAS wavelengths with all known instrumental effects plus the zodiacal light contribution removed. Within the statistical uncertainties I find that the integrated 12 μm flux between $b = \pm 2$ to $\pm 12°$ is no greater than that predicted from models based on optically identified M giants alone (Blanco *et al.* 1984; Frogel and Whitford 1987; Frogel *et al.* 1990). The conclusion, then, is that the OH/IR stars are not a significant component of the bulge population, even at 12μm.

My own research has been supported in part by NASA grant NAG 5-1367 through their ADP program.

References

Alloin, D. 1973, A&A, 27, 433
Alloin, D., and Bica, E. 1989, A&A, 217, 57
Arimoto, N., and Yoshii, Y. 1987, A&A, 173, 23
Armandroff, T. E., and Zinn, R. 1988, AJ, 96, 92
Bica, E. 1988, A&A, 195, 88
Bica and Alloin 1987, A&AS, 70, 281
Bica, E., Alloin, D., and Schmidt, A. A. 1990, A&A, 228, 23
Bica, E., Arimoto, N., and Alloin D. 1988, A&A, 202, 8
Blanco, V. M., McCarthy, M. F., and Blanco, B. M. 1984, AJ, 89, 636
Charlot, S., and Bruzual A., G. 1991, ApJ, in press.
Davies, R. L., Frogel, J. A., and Terndrup, D. M. 1991, AJ, in press.
Delisle, S., and Hardy, E. 1991, AJ, in press.
Frogel, J. A. 1985, ApJ, 298, 528
Frogel, J. A. 1988, ARA&A, 26, 51
Frogel, J. A., Terndrup, D., Blanco, V. M., and Whitford, A. E. 1990, ApJ, 353, 494
Frogel, J. A., and Whitford, A. E. 1987, ApJ, 320, 199
Gautier, T. N., Hauser, M. G., Beichman, C. A., Low, F. J., Neugebauer, G. et al. 1984, ApJL, 278, L57
Hardy, E., and Couture, J. 1988, ApJL, 325, L29
Houdashelt, M. L., Frogel, J. A., and Cohen, J. G. 1991, AJ, in press
Koppen, J., and Arimoto, N. 1990, A&A, 240, 22
Melnick, G. J., Fazio, G. G., Koch, D. G., Rieke, G. H., Young, E. T., et al. 1987, in The Galactic Center, ed. D. C. Backer, AIP Conf. Proc. No. 155, p. 157
Morgan, W. W. 1956, PASP, 68, 509
O'Connell, R. W. 1976, ApJ, 206, 370
O'Connell, R. W. 1982, ApJ, 257, 89
O'Connell, R. W. 1983, in Highlights of Astronomy, ed. R. M. West (Dordrecht: Reidel), 147
Pagel, B.E.J., 1987, in The Galaxy, eds. G. Gilmore, & R. Carswell (Dordrecth: Reidel), 341
Phillips, M. M., Jenkins, C. R., Dopita, M. A., Sadler, E. M., and Binette, L. 1986, AJ, 91, 1062
Rich, R. M. 1988, AJ, 95, 828
Rich, R. M., and Mould, J. 1991, AJ, 101, 1286
Rich, R. M., Mould, J., Picard, A., Frogel, J. A., and Davies, R. 1989, ApJ, 341, L51
Rowan-Robinson, M., and Chester, T. 1987, ApJ, 313, 413
Sandage, A. 1986, ARA&A, 24, 421
Soifer, B. T., Rice, W. L., Mould, J. R., Gillett, F. C., Rowan-Robinson, M., and Habing, H. J. 1986, ApJ, 304, 651
Terndrup, D. M., Frogel, J. A., and Whitford, A. E. 1990, ApJ, 357, 453
Turnrose, B. E. 1976, ApJ, 210, 33
Vader, J. P. 1987, ApJ, 317, 128
van der Veen, W. E. C. J., and Habing, H. J. 1990, A&A, 231, 404
Whitelock, P., Feast, M., and Catchpole, R. 1991, MNRAS, 248, 276
Whitford, A. E. 1977, ApJ, 211, 527
Whitford, A. E. 1978, ApJ, 226, 777
Whitford, A. E., and Rich, R. M. 1983, ApJ, 274, 723
Xu, Z., Veron-Cetty, M.-P., and Veron, P. 1989, A&A, 211, L12

ABSORPTION-LINE SPECTRA OF ELLIPTICAL GALAXIES AND THEIR RELATION TO ELLIPTICAL FORMATION

S. M. FABER, GUY WORTHEY, and J. JESÚS GONZALEZ
UCO/Lick Observatories
Board of Studies in Astronomy and Astrophysics
University of California
Santa Cruz, CA 95064
USA

ABSTRACT. Recent spectral data strongly suggest that elliptical galaxies represent at least a two- and probably a three-parameter sequence: light-element abundance, iron-peak abundance, and probably age. These data are discussed in the context of new ideas about elliptical formation, including hierarchical clustering, merger-caused starbursts, and a variable IMF. Non-solar abundance ratios set important constraints on elliptical formation.

1. INTRODUCTION

The accuracy with which we need to model elliptical galaxy populations is conditioned by the use to which this knowledge will be put. Some applications are basically qualitative, others highly quantitative. An example of the latter is the use of models in cosmology to predict the statistical properties of populations of galaxies at large lookback times. The accuracy needed in this case is discouragingly high, roughly ±20% in age or ±10% in luminosity, if useful limits are to be set.

These two sorts of applications put vastly different demands on population models. In some ways, as we shall see, the quantitative situation now looks bleaker than before, and there is much work needed in stellar evolution and spectral synthesis before old stellar populations can be a truly quantitative tool. On the other hand, new data have yielded important qualitative insights to elliptical formation. This paper describes recent results on the spectra of elliptical galaxies and places them in the context of current thinking on how ellipticals formed.

2. NEW VIEWS ON ELLIPTICAL GALAXY FORMATION

The classical view of elliptical (and spheroid) formation held that these objects formed stars efficiently in the early universe and that their stellar populations can be modeled as an old, nearly coeval, single burst (Baade 1941; Morgan 1959; Eggen, Lynden-Bell, and Sandage 1962; Tinsley 1968; and Larson 1974a). A proto-elliptical was assumed to be a spherically symmetric, collapsing gas cloud, with successive generations of metal-enriched stars formed closer to the center, the observed metallicity gradient being built up by inflow of more enriched

B. Barbuy and A. Renzini (eds.), The Stellar Populations of Galaxies, 255–265.

gas from the outer to the inner zones. Any later star formation was assumed to be describable by a simple, smooth mathematical function, such as $\exp[-t/\tau]$.

Over the past decade, evidence has been accumulating that this picture is too simple. We highlight briefly some of the major new ideas (and some speculations) that we think should be part of a more realistic picture of elliptical galaxy formation:

1.) *Hierarchical clustering:* Numerous evidence suggests that structure in the universe formed hierarchically through the merger of small bits and pieces to create larger units. Protogalaxies therefore had a clumpy substructure, with essentially no spherical symmetry. This hierarchical merging lasted many billions of years and continues on galaxy-sized scales at a low level even today. Galaxies are still colliding and merging to make yet larger galaxies. One therefore cannot speak, as we did earlier, of 'the epoch of galaxy formation'. Galaxy formation is a long-drawn-out process that continues to the present.

2.) *Mergers and starbursts:* Many gas-rich merging galaxies are sites of intense star formation. These starbursts appear to be stimulated by the collision of gas clouds in the two galaxies, and the remnant galaxies in many cases are expected to evolve in several billion years to look like normal ellipticals (*e.g.*, Toomre 1977; Schweizer 1982, 1987, Lutz 1991). We believe that the star formation and morphology of many nearby colliding galaxies provide a model - albeit at reduced intensity - for how stars actually formed in elliptical protogalaxies at high redshift. These nearby systems are an important window on the past.

3.) *Episodic star formation:* Merger-induced starbursts lead naturally to the notion that star formation in ellipticals is episodic, with long periods of quiescence punctuated by intense bursts. Since the entire process took billions of years, the stellar populations of spheroids encompass a wide range of ages, but the birthrate function was highly non-uniform. The usual, smoothly declining mathematical approximations do not apply.

4.) *The importance of metallicity:* There is a general lack of appreciation among cosmologists of how sensitively the evolution rates of old stellar populations depend on composition. Solar-abundance models are widely used for distant galaxies despite the fact that their spectra are potentially badly wrong. Observable parameters that correlate with age are colors, 4000 Å break, and Balmer line strength, all of which depend mainly on turnoff color. However, this is a strong function of Z as well as age: an error of 0.5 dex in Z translates to a factor of three in turnoff age, a huge error in cosmological terms. Getting evolutionary rates accurate to 20% requires metallicities accurate to 0.1 dex, which is currently quite beyond reach. The real situation is worse because of probable non-solar abundance ratios in giant E's (see below), necessitating models that vary these elements separately if high accuracy is to be achieved.

5.) *Bimodal star formation and a variable IMF:* This final notion is quite speculative. We noted above how rapid star formation is common in colliding galaxies and how it contrasts strongly with the sluggish, inefficient star formation in isolated galaxies. We suggest that this duality may be related to the theory of *bimodal star formation,* developed to account for star births in the Milky Way (see review by Shu et al. 1987). According to this theory, one mode is 'low-mass' or 'spontaneous' star formation, which is quiescent, continuous, widespread, and regulated by *internal* cloud parameters such as ionization and magnetic field. Opposed to this is 'high-mass' or 'induced' star formation, which is bursty, localized, and triggered by *external* factors such as passage through a spiral arm.

We speculate that the starbursts seen in merging galaxies are essentially just more intense versions of the high-mass process going on in spiral arms. Silk notes (this volume) how it is plausible theoretically that high-velocity cloud-cloud shocks could favor the formation of high-mass stars. The cloud-cloud shock environment in mergers may be similar to, yet more extreme, than the pressure jump across a spiral arm. Many authors (see review by Rieke 1988) have noted how the large mass requirements of certain starbursts could be eased if the initial mass function (IMF) were weighted toward high-mass stars. Moreover, it would not be surprising in this picture to find close ties between a galaxy's dynamics (especially the velocity dispersion) and such IMF-related quantities as mean metal abundance and certain abundance ratios. These trends are in fact observed and are discussed in more detail below.

3. THE LICK GALAXY PROJECT

Most of our results come from the library of galaxy and stellar spectra collected at Lick Observatory from 1972 to 1985. The spectra were taken with the Image-Dissector Scanner (IDS) on the Shane 3 m telescope and cover approximately 4000 Å to 6200 Å with 10 Å resolution (Faber et al. 1985, Burstein et al. 1984). The ultimate goal is to understand the compositions of elliptical galaxies well enough to use them as cosmological tools. The visible portion of the spectrum was selected because it is less sensitive than the blue to turnoff temperature and less sensitive than the red to the details of M giant evolution. The G- and K-type spectra of stars that dominate the visual region are amenable to calibration both theoretically and empirically, using samples of local stars.

Integrated line-strengths are calculated in two steps (Worthey, this volume). We utilize a set of empirical fitting functions (Gorgas et al. 1991) that model 11 strong lines as a function of stellar temperature ($V-K$), metallicity (taken to be [Fe/H]), and surface gravity. These functions were derived by fitting to a sample of ~300 field and cluster G and K stars. The fitting functions are coupled to published evolutionary tracks and isochrones (VandenBerg and Bell 1985, VandenBerg 1985, VandenBerg and Laskarides 1987, Green et al. 1987, Siedel et al. 1987, Swiegart 1987, Lattanzio 1991). The models are preliminary, lacking an accurate treatment of the M-giant tip and utilizing a provisional color-temperature calibration (Green 1988) and flux distributions (Bell and Gustafsson 1989). However, the basic conclusions are qualitative and should not be affected by these uncertainties.

Fitting functions are a novel approach in that they smooth over a stellar library in a controllable way and even allow limited creation (via interpolation) of certain stars that do not exist in the library (*e.g.*, very old super-metal-rich stars). However, there is an important limitation in that the fitting functions incorporate implicitly the *same abundance ratios* that are present in the calibrating stars. Since our calibrators are local, their abundances presumably reflect general solar-neighborhood values (*e.g.*, Wheeler et al. 1989). These ratios (and trends in these ratios) cannot be altered; they are locked into the mathematical coefficients. Our models are therefore potentially inconsistent, as they combine published evolutionary tracks for solar abundance ratios with fitting functions based on stars with ratios that may vary. This proves to be a crucial limitation in what follows.

4. IRON VS. MAGNESIUM

Our most important result concerns the strength of Fe vs. Mg in elliptical galaxies. Fe5270 vs. Mg_2 is shown for some well-observed ellipticals in Fig. 1a. An important feature is the large scatter. This is confirmed by residuals in other Fe lines and also by independent observations. We return to this point below.

A second feature is the rather shallow slope in Fe relative to Mg_2. Two squares indicate averages for typical ellipticals of high and low luminosity (termed 'giant' and 'compact' respectively). Mg_2 is 60% larger in giant E's, whereas Fe5270 is only 20% larger. This shallow slope is a distinct break from the globular clusters (Burstein et al. 1985), in which Fe5270 is linearly proportional to Mg_2 up to nearly the compact E level. The same break is seen also in Fe5335.

An Fe deficit (or Mg excess) is also indicated by the models shown in Fig. 1a. The models match globular clusters rather well up through nearly solar metallicity. The model lines also pass through the compact-E points, showing solar abundance ratios for these objects. The giant E's, however, show a systematic Fe deficit relative to Mg_2; their Fe5270 strengths rise only about half as fast as the models predict. A similar result was also noted by Peletier (1989). The star-cluster data of Bica and Alloin (1986) for our Galaxy, if extrapolated, also predict stronger Fe in giant E's than is seen.

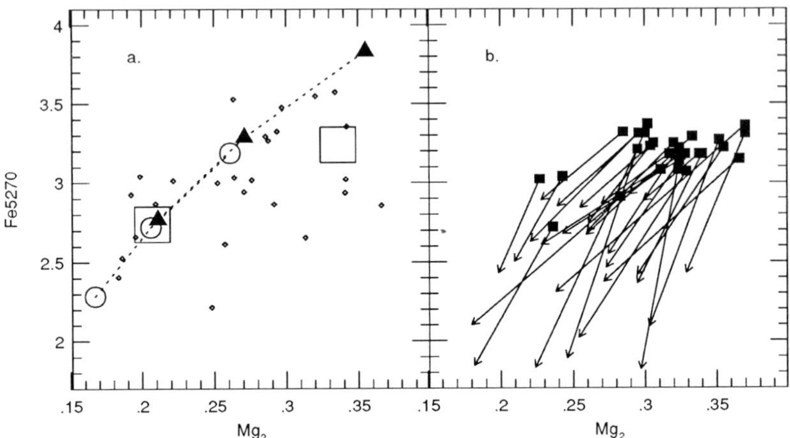

Figure 1a. Fe5270 vs. Mg_2 for Lick elliptical nuclei (small dots). Median loci for compact and giant E's are shown as large squares. Burst models for ages of 6 and 18 Gyr are shown as lines with symbols at [Fe/H] = -0.25, 0.0, and +0.25 dex. The triangles represent the older sequence. Figure 1b. CCD gradient data. Values for the central 5" of each galaxy are shown as black squares, and the averages of the furthest radii (typically 15-20") as the tips of the attached arrows.

The models in Fig. 1a show that the Fe/Mg_2 *ratio* is insensitive to both age and overall Z. Separate tests indicate that IMF is also unimportant. Since populations add vectorially in Fig. 1, the Fe discrepancy cannot be cured by invoking spreads in any of these quantities.

The only remaining option, it seems, is that giant E's have higher [Mg/Fe] than the most metal-rich stars of the solar neighborhood. Measuring this excess accurately is impossible, as we lack models with the proper abundance ratios. However, we can perhaps get a rough estimate from Fig. 1a. Hβ evidence (coming up) suggests that the average compact E is rather young, about 6 Gyr. Both Fe and Mg then agree in indicating a slight underabundance of $\sim$-0.1 dex relative to solar. This agrees well with the empirical fact that the spectra of compact E's can be closely matched using solar-neighborhood stars (Spinrad and Taylor 1968, Faber 1973, O'Connell 1980).

The giant E's are different. Regardless of whether an old or young age is chosen, the abundance from Mg_2 comes out 0.2-0.3 dex higher than from Fe5270. Interestingly, if an old age of 15 Gyr is selected, Fe is up by only +0.05 dex (and even Mg_2 is up by only +0.25 dex). These numbers might change with proper models, but it is possible that the absolute overabundance of Fe in giant E's is actually rather small.

These derived abundances are very sensitive to age, however; both Fe and Mg increase by +0.2-0.3 dex for an age increase of from 6 to 15 Gyr. Therefore, if these abundances were input back into evolving models, there would still be an uncertainty of a factor of 1.6-2 in the predicted rate of spectral evolution. Thus, we have not significantly sharpened our hold on metallicity by studying lines even in this (most favorable) part of the spectrum. The age-metallicity degeneracy can only be attacked, it seems, by modeling the near-UV spectrum between 2400 Å and 3200 Å (Fanelli et al. 1990), or perhaps from exquisitely accurate measurements of the Balmer line absorption (see below).

In an independent study, Gonzalez (1991) has measured Mg_2 and Fe gradients in 28 (mostly cluster) ellipticals using a high-accuracy CCD detector. His results are shown in Fig. 1b. Note that the trends *within* galaxies are steeper than the relation linking the nuclei. The same difference is seen in Fe5335 and was also detected by Efstathiou and Gorgas (1985), Peletier (1989), and by Gorgas et al. (1991) for a smaller sample of galaxies (but *cf.* Sadler and Davies 1991 for a different view).

These slopes help us to understand the scatter for the nuclei in Fig. 1a. Within a galaxy, Fe and Mg evidently vary in rough proportion; however, the *global average* of Mg to Fe is clearly varying from galaxy to galaxy. The collective Fe-Mg relationship for all populations is thus two-dimensional, with each galaxy following its own enrichment line. The precise nuclear location along this line depends on the maximum degree of nuclear enrichment achieved and on the spatial resolution of the observations.

Since Fe and Mg are made by two different types of supernovae - Type Ia vs. Type II - a change in their ratio could signal important differences in the nucleosynthetic enrichment process. Excess Mg could indicate an overabundance of *all* Type II-light elements in giant E's, relative to Fe. This possibility can be assessed by comparing the predicted impacts of the two element groups on the HR diagram and on the spectrum (Renzini 1977). The Fe-peak controls the temperature of the giant branch and thus the contribution of the coolest M giants. Fe should be tied to TiO strength, infrared spectral features, and M_I, the mean I-band brightness on the giant branch (Tonry et al. 1990). The rather small variations seen in these quantities are consistent with their being controlled by Fe.

Conversely, the impact of the Type II light elements should be be large and pervasive. Since O opacity largely controls the turnoff temperature, Type II abundances should couple closely to the blue-visual colors and Hβ line strength. Large variations are also expected in individual light-element features such as Mg b, MgH, NaD, and Na 8190 (Faber 1973, Bica and Alloin 1986, Deslisle and Hardy 1991). Production of C and N is less clear, but the strong CN excess suggests that these light elements, too, are also enhanced. The only known contradictions are the G-band (CH) and Ca (H-K and the IR triplet), but these are features that show little variation even in stars where Mg is strong (Faber et al. 1985, Díaz et al. 1989). Further checks are needed, but it appears plausible that the major driver of spectral variations in elliptical galaxies comes mainly from the light elements, not the Fe-peak.

5. Hβ VS. IRON AND MAGNESIUM: A THREE PARAMETER SCENARIO?

Hβ line strength may signal yet a third parameter in the system. Hβ is a difficult line because it is often filled in by emission. However, we have found that [O III] 5007 Å is a reliable flag for Hβ in nuclear spectra, and Gonzalez's long-slit data show this line clearly. Fig. 2 shows Hβ vs. Mg_2 for Gonzalez's nuclei, pruned of all suspected emission. The relation is exceptionally narrow, unlike Fe versus Mg_2. On it are superimposed Worthey's models. They overlap the galaxies but have a shallower slope at fixed age. The steepness of the galaxies reflects the fact that compact E's have high Hβ. Taken at face value, this suggests an age trend among Gonzalez's galaxies, with the compact E's some 5-6 Gyr old and the giant E's much older (these are mean ages only and are not meant to rule out that there could be a large age range within each type of system). Whether this result is general or merely reflects small-number statistics in Gonzalez's sample is not yet clear. However, it is interesting that the age for M32 estimated this way agrees well with detailed population synthesis (Faber 1973, O'Connell 1980).

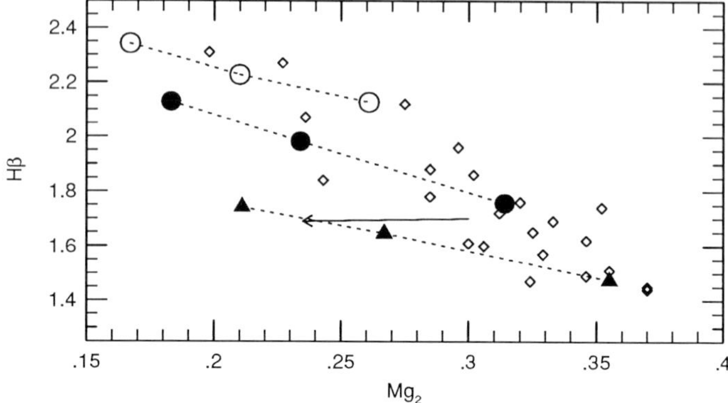

Figure 2. Hβ vs. Mg$_2$ for a sample of 28 nuclei without Hβ emission. The arrow shows a mean radial gradient. Model predictions appear as in Fig. 1a, except that a 9-Gyr locus is added as the middle line.

A mean radial $H\beta$ gradient is also indicated in Fig. 2. For clarity, we plot only an average of the best-measured 28 galaxies. The average gradient is flat, as was found also by Efstathiou and Gorgas (1985) and Gorgas et al. (1990). This flatness is surprising because of the steep gradients in both Fe and Mg$_2$. Worthey's models predict that, if all radii of a galaxy have the same age and differ only in Z, Hβ should *increase* slightly outward, not remain flat (see Figure 2).

These results are still preliminary, and the discrepancy may not be real. However, the implications are important and are worth a brief mention. To counteract metallicity, there would have to be a third parameter operating, most plausibly a gradient in the mean age. The sense is such that the centers of galaxies would have to be younger on average than the outer parts, by roughly one-quarter or one-third the total age.

This is interesting precisely because it agrees with the hierarchical merging picture. In this picture, a starburst is induced over a large volume while at peak intensity, but leftover gas eventually finds its way to the center, to be consumed there in the final stages of star formation. Actual starbursts follow this pattern. Young ones such as NGC 7252 show new stars over a large area (Schweizer 1982), while older bursts are strongly concentrated to the nuclei. Well known examples include NGC 205, NGC 1275, NGC 5102, NGC 2681, and M82. Among the Lick sample we have off-nuclear spectra of 8 early-type galaxies that show excess nuclear Hβ; in all of them, this excess declines away from the nucleus. The conclusion seems clear that the infusion of young stars from starbursts is relatively largest in the central regions. In merging and hierarchical clustering, then, elliptical galaxies form 'from the outside in', just as they do in the classic spherical collapse picture.

Growing evidence suggests that subtle age differences may in fact be common in ellipticals. Schweizer et al. (1991) showed that a sample of disturbed (mostly field) ellipticals tended to have systematically weak metal lines and strong Hβ for their luminosities, as though their spectra were diluted by young stars. Gregg (this volume) finds that the same galaxies yield systematically positive peculiar motions from the $D_n - \Sigma$ relation, due, he thinks, to enhanced surface-brightness. De Carvalho and Djorgovski (this volume) find a similar effect in field ellipticals from Faber et al. (1990), whereas cluster galaxies appear to be more homogeneous and dimmer. Gonzalez's very narrow Hβ relation in Fig. 2 is consistent with these findings since his galaxies are mainly cluster members.

The evidence thus suggests that star formation continued for a long time in ellipticals but cut off sooner in clusters than it did in the field. An earlier cutoff in clusters is plausible if only moderate-velocity galaxy collisions are effective in producing mergers and starbursts. At the high encounter velocities in large clusters, two colliding systems will not coalesce, and shocked gas will be heated to escape velocity and will not form stars. Thus we have a natural cutoff in the age of stellar populations located in dense environments.

6. THREE WAYS TO VARY [Mg/Fe]

Three possible scenarios might account for variations in [Mg/Fe]. All utilize the fact that Mg is produced by Type II supernovae, whereas the bulk of Fe comes from Type Ia.

6.1. Different Star Formation Timescales

This first suggestion exploits the fact that Type II SNe come from massive stars with lifetimes shorter than 10^8 yr, while Type Ia SNe result from binary mass transfer, which sets in more slowly and lasts for Gyr. There results a well known time delay between the production of the light elements and the bulk of the Fe-peak, with the consequence that the first generations of stars have enhanced [Mg/Fe]. As applied to giant E's, however, this argument appears to have the wrong sign: star formation would need to occur faster to produce a higher yield of Mg, yet the evidence suggests that these systems actually took longer to accumulate. Longer timescales are suggested by the fact that the giants are larger objects (and hence in a hierarchy would have to form later) and also by their lower densities and longer dynamical timescales (a factor of 3 longer compared to compact E's [*e.g.* Blumenthal et al. 1984]). Giant E's are therefore required to accumulate their gas gradually, not making stars, and then make them all suddenly in a final burst. This scenario seems forced.

On the other hand, the argument can be turned around to exploit the observed fact that, for whatever reason, stellar populations in compact E's appear to be *younger* than those in giant E's. Smaller Es agglomerated their material quickly, so this argument goes, but turned it into stars more slowly. Perhaps slower star formation could be related to the lower cloud-cloud collision velocities in these systems.

6.2. Selective Loss of Mg vs. Fe

Modulating net yield by allowing metals to escape is the classic way to account for the higher metallicities of giant E's (Larson 1974a). The same mechanism has also been invoked to explain radial abundance gradients within galaxies (Franx and Illingworth 1990). To explain differences in Mg/Fe, we would need to appeal to the selective loss (or retention) of light vs. Fe-peak elements in galaxies of different sizes at different times.

To match the observations requires either that giant ellipticals lose more Fe or that compact ellipticals lose more Mg. The second choice is preferable, since it is hard to see how large galaxies with their deeper potential wells could lose more of anything. Preferential loss of light elements could have occurred early in compact ellipticals provided supernovae-driven winds were most intense at that time. However, in standard wind scenarios (Arimoto and Yoshii 1987, Matteucci and Tornambè 1987), the onset of winds is irreversible, so the loss of Fe always *exceeds* the loss of Mg. Winds do not appear to be the answer unless they can be shown to be more effective in giants than in compacts, which seems unlikely.

6.3. Selective Production of Mg vs. Fe: Variable IMF

The final mechanism posits a difference in the IMF, with giant E's having more very massive stars, more Type II SNe, and more light element production. No conclusive evidence exists at this time to prefer this view over the others; however, it has no known drawbacks and fits rather neatly with the concept of bimodal star formation mentioned above. High-mass star formation, we recall, is triggered by external conditions such as passage through a spiral arm. A comparable external variable in ellipticals could be the ambient effective gas pressure due to cloud-cloud collisions. Since this scales roughly as the dispersion σ_{cl}^2, the effective ambient pressure in merging galaxies would be about 100 times higher than in spiral arms. Even among E's, there is a difference of a factor of $\sim$10 between compacts and giants. If higher σ_{cl} favored more massive stars, the higher abundance of light elements in giant galaxies would be naturally explained.

The variable IMF theory also helps to explain another observational fact - the very tight correlation between Mg_2 and σ. Throughout this paper we have spoken loosely of 'giant' versus 'compact' E galaxies, tacitly implying that abundance scales best with galaxy mass. In fact, Mg_2 scales much more closely with *velocity dispersion* than it does with luminosity or any other parameter, in a relation that is closely obeyed over an astonishingly wide range of velocity dispersions (Bender et al., this volume). One explanation is that metal abundance is related to the local escape velocity, and thus to σ (Franx and Illingworth 1990). However, this idea is perhaps less attractive now that Fe is known to decouple from Mg - why should one element follow the escape velocity but not the other?

On the other hand, a tight Mg_2-σ correspondence is a natural consequence of the variable IMF scenario. High σ translates directly to high Mg_2 through the strength of the upper IMF, while Fe, being produced by lower-mass stars, is at best loosely correlated. In fact, an intimate relation between σ and *all* Type-II light elements is expected in this scenario, in agreement with the data.

Each theory makes distinct predictions, and it should be possible to choose among them by measuring accurate line strengths in the outer parts of ellipticals of different sizes and types. It is already clear, for example, that current standard enrichment models fail badly in predicting universally *more* Fe in giant E's rather than less (Matteucci and Tornambè 1987). As this result is generic under the simplest assumptions, the line strengths are clearly telling us something very fundamental and unexpected about elliptical formation. However, before accurate comparisons with models are possible, we must first learn how to translate line strengths reliably into age and element abundances. Before that can be done, 1) stellar evolutionary models must be broadened to include variable element ratios, and 2) spectral synthesis must be perfected to provide spectra of stars with non-solar abundance ratios. Our net reliance on theory will be vastly increased, and an extensive testing program will have to be undertaken before the results can be believed.

7. CONCLUSION

Although quantitative goals remain elusive, our qualitative understanding of ellipticals seems to be growing rapidly. IRAS and other facilities have shown conclusively that galaxy collisions trigger intense bursts of star formation. Many merging galaxies are today generating dynamically hot stellar populations that will evolve to look like ellipticals eventually. Some mergers are even forming proto-globular clusters (Lutz 1991, Holtzman et al. 1991).

This observational evidence is consistent with our growing theoretical understanding of merging and how it operates via hierarchical clustering to make structure in the universe (see review by Frenk 1988). Merging on galaxy-sized scales has passed its peak, and merging galaxies today are the last dregs undergoing this process. Though systematically less gas rich than earlier collisions, today's merging galaxies are nevertheless crucial because they show us, quite faithfully, what proto-spheroid formation must have looked like at high redshift. We see a close facsimile of it taking place now in the Galactic neighborhood.

Many of the early objections to formation of ellipticals by mergers (*e.g.*, Ostriker 1980) have been blunted by the realization that mergers often contain gas and that gaseous dissipation plays a major role in determining the final structure of the remnant. Thus, ten years ago, we witnessed a classic 'nature vs. nurture' debate on galaxy formation, whether protogalaxies were basically isolated structures, or whether their fate was determined mainly by interaction with the environment (*c.f.* White 1982). The realization that structure forms hierarchically and with gaseous dissipation is steadily erasing that distinction. On the one hand, the classic protogalaxy picture now takes clumpy substructure for granted, while, on the other, mergers now routinely include gaseous dissipation and star formation. Ironically, the two formerly opposing theories have themselves now merged!

The spectral data reported here are an important part of this picture. The knowledge that there are (probably) three parameters is crucial to making models. Moreover, all three parameters are intimately tied to exactly how and when these galaxies made stars. Merging predicts that in some galaxies there should exist subtle correlations between dynamical irregularities and spectral properties, depending on the time since the last major interaction. These correlations are now being discovered (Schweizer et al. 1991; Bender et al., this volume). The challenge in future is to make all this information quantitative.

In conclusion, we offer a caveat. Our whole approach here rests implicitly on the assumption that $H\beta$ is produced mainly by the turnoff stars and thus is simply related to age. Were it to be demonstrated that some other population, such as horizontal branch stars, contributes heavily to $H\beta$ (*e.g.*, Burstein et al. 1985), any simple three-parameter calibration involving Fe, Mg, and age would be threatened. The nearby galaxies M31 and M32 are the only stellar laboratories in which this possibility can be tested (using a repaired HST). Until this important loose end has been tied up, the basic calibration for old stellar populations will remain somewhat shaky.

References

Arimoto, N., and Yoshii, Y. 1987, Astron Astrophys, 173,23

Baade, W. 1941, ApJ, 100,137

Brocato, E., Matteucci, F., Mazzitelli, I., and Tornambè, A. 1990, ApJ, 349,458

Bica, E. and Alloin, D. 1986, Astron. Astrophys. 162,21

Blumenthal, G.R., Faber, S.M., Primack, J.R., and Rees, M.J., 1984, Nature, 311, p.517

Burstein, D., Faber, S.M., Gaskell, C.M., Krumm, N. 1984, ApJ, 287,586

Deslisle, J., and Hardy, E., 1991, A.J., in press

Díaz, A.I., Terlevich, E., Terlevich, R. 1989, MNRAS, 239,325

Efstathiou, G., and Gorgas, J. 1985, MNRAS, 215,37p

Eggen, O.J., Lynden-Bell, D., and Sandage, A. 1962, ApJ, 136,735

Faber, S.M., 1973, ApJ 179,731

Faber, S.M., Friel, E.D., Burstein, D., and Gaskell, C.M. 1985, ApJ Suppl, 57,711

Faber, S.M., Wegner, G., Burstein, D., Davies, R.L., Dressler, A., Lynden-Bell, D., Terlevich, R.J. 1989, ApJ Suppl, 69,763

Fannelli, M.N., O'Connell, R.W., Burstein, D., and Wu, C. 1990, ApJ, 364,272

Franx, M., and Illingworth, G., 1990, ApJ, 359,L41

Frenk, C.S., 1988, in eds. J. Audouze, M.C. Pelletan, and A. Szalay, IAU Symposium No. 130, *Large-Scale Structures of the Universe*, Kluwer Academic Publishers, Dordrecht, p.259

Gonzalez, J.J, 1991, Ph. D. Thesis, University of California, Santa Cruz

Gorgas, J., Efstathiou, G., and Aragón Salamanca, A. 1990, MNRAS, 245,217

Gorgas, J., Faber, S.M., Burstein, D., Gonzalez, J.J., Courteau, S., Prosser, C. 1991, ApJ Suppl, submitted

Green, E. 1988, in ed. A.G. Davis Philip, *Calibration of Stellar Ages*, L. Davis Press, Schenectady, NY, p.81

Green, E.M., Demarque, P., King, C.R. 1987, *The Revised Yale Isochrones and Luminosity Functions*, Yale University Observatory, New Haven

Holtzman, J.A., Shaya, E.J., Faber, S.M., Lauer, T.R., Lynds, C.R., Baum, W.A., Groth, E.J., Hester, J.J., Hunter, D.A., Light, R.M., O'Neil, E.J., and Westphal, J.A. 1991, preprint

Larson, R.B. 1974a, MNRAS, 166,585

Larson, R.B. 1974b, MNRAS, 169,229

Larson, R.B., and Tinsley, B. 1978, ApJ 219,46

Larson, R.B., 1986, MNRAS, 218,409

Lattanzio, J.C., 1991, ApJ Suppl, 76,215

Lutz, D. 1991, Astron Astrophys, 245,31

Matteucci, F., and Tornambè, A. 1987, Astron Astrophys, 185,51

Morgan, W.W. 1959, P.A.S.P. 71,92

O'Connell, R.W., 1980, ApJ 236,430

Ostriker, J.P., 1980, *Comments on Astrophysics*, 8,177

Peletier, R. 1989, Ph. D. Thesis, Rijksuniversiteit Groningen

Renzini, A. 1977, in eds. P. Bouvier and A. Maeder, *Advanced Stages in Stellar Evolution*, Saas-Fee Lectures 1977, Geneva Observatory, Switzerland, p.166

Rieke, G.H. 1988, in eds. R.E. Pudritz and M. Fich, *Galactic and Extragalactic Star Formation*, Kluwer Academic Publishers, Dordrecht, p. 561

Sadler, E.M., and Davies, R.L., 1991, preprint

Schweizer, F. 1982, ApJ, 252,455

Schweizer, F. 1987, in ed. S.M. Faber, *Nearly Normal Galaxies*, Springer - Verlag, New York, p.18

Schweizer, F., Seitzer, P., Faber, S.M., Burstein, D., Dalle Ore, C.M., and Gonzalez, J.J., 1990, ApJ, 364,L33

Seidel, E., Demarque, P., and Weinberg, D. 1987, ApJ Suppl, 63,917

Shu, F.H., Adams, F.C., and Lizano, S. 1987, Ann Rev Astron Astrophys, 25,23

Spinrad, H., and Taylor, B.J. 1968, ApJ Suppl, 22,445

Sweigart, A.V., 1987, ApJ Suppl 65,95

Tinsley, B.M. 1968, ApJ, 151,547

Tonry, J.L., Ajhar, E.A., and Luppino, G.A., 1990, AJ, 100,1416

Toomre, A. 1977, in eds. B.M. Tinsley and R.B. Larson, *The Evolution of Galaxies and Stellar Populations*, Yale University Press, New Haven, p.401

VandenBerg, D.A., 1985, ApJ Suppl 58,711

VandenBerg, D.A., and Bell, R.A., 1985, ApJ Suppl 58, 561

VandenBerg, D.A., and Laskarides, P.G., 1987, ApJ Suppl 64,103

Wheeler, J.C., Sneden, C., and Truran, J.W.Jr. 1989, Ann Rev Astron Astrophys, 27,279

White, S.D.M. 1982, in eds. L. Martinet and M. Mayor, *Morphology and Dynamics of Galaxies*, Saas-Fee Lectures 1982, Geneva Observatory, Switzerland, p.289

Discussion

J.P.Ostriker: A cautionary note on mergers. The thinness and coldness of spiral disks implies that recent merger activity is minimal. In recent calculations we found at Princeton that not more than 4% of the mass has been added (interior to our orbit) in the last 5 billion years.

Faber: Yes, I agree with that in the case of spirals. Even for ellipticals, I think it will turn out that a small percentage of them have undergone an interaction so recently as to sizeably perturb the colors and surface brightness. The real question for the typical elliptical is whether, via very careful observations of the line spectra or the dynamics, we can uncover evidence of even more ancient mergers. I think the answer with respect to the dynamics is incontrovertibly yes, and the line spectra are clearly tending in the same direction also.

D. Spergel: I want to reiterate your point about the role of magnetic fields. If magnetic fields grew by a dynamo mechanism, then the B field in the early universe was much weaker than today. Magnetic fields are believed to play an important role in transferring angular momentum from protostars to the molecular cloud. Without strong magnetic fields, old stars are much more likely to be rapidly rotating and much more likely to be in close binaries. These two effects need to be considered in models of stellar evolution.

Faber: Yes, I am particularly troubled by the role that binaries might play in massive star star evolution. Several people at this conference have privately noted that most Wolf-Rayet stars are close binaries. Might this fact influence the mass-loss process and hence the net yield of the various elements? Perhaps we should be thinking more closely about all these effects.

S.G. Djorgovski: A question and a comment: have you tried to divide your sample into the field and rich-cluster ellipticals and compared them? My comment is on M32. There is now a growing body of evidence that there are color and population gradients near the centers of highly concentrated globular clusters (bluer in the middle), which seem to involve changes in the horizontal branch and may be caused by dynamical processes. It is possible that similar phenomena may occur in high-density ellipticals like M32 and cause the 'inverse' UV color gradients, as seen in the ASTRO data.

Faber: On point one: yes, we did, and we found no net offset in predicted peculiar velocities between the field and clusters. However, we did not further subdivide the *field* sample according to environmental density, and this might reveal the effect that you and de Carvalho have found. On point two: that is an interesting suggestion that might perhaps be checked with the new HST images of M32 that have just been obtained.

Pagel: You mentioned the possibility that the universe could be only 10 Gyrs old. Recent development in Th/Nd/En chronology fully support ages up to 15-20 Gyrs deduced form HR diagrams in the conventional way.

1. EVIDENCE FOR MERGER SIGNATURES IN THE LINE-STRENGTH GRADIENTS OF ELLIPTICAL GALAXIES

2. A UNIVERSAL RELATION BETWEEN METALLICITY AND POTENTIAL DEPTH FOR ALL HOT STELLAR SYSTEMS?

Ralf Bender[1]
Landessternwarte Königstuhl
D-6900 Heidelberg, Germany

ABSTRACT In the *first* part of this paper it is shown that those giant ellipticals which presumably formed in a violent merging process (as inferred from their peculiar core kinematics) reflect the details of the merging in their Mg_2-index profiles. In the *second* part it is demonstrated that, independent from other structural parameters, central Mg_2 index and central velocity dispersion σ_o (presumably representing the mean depth of the potential well) follow a unique relationship for all types of hot stellar systems (including diffuse dwarf ellipticals, compact ellipticals, bulges and giant ellipticals).

1. Evidence for merger signatures in the line-strength gradients of elliptical galaxies

INTRODUCTION: In recent years it became evident that probably up to 1/3 of luminous ellipticals contain cores which are kinematically de-coupled from the main-bodies of the galaxies (for a review, see [1]). This de-coupling can plausibly be only achieved if these ellipticals coalesced from constituents which had transformed most of their mass into stars already, i.e. from galaxy-like objects. In the merging process the stars violently relaxed and formed an anisotropic main body, while some fraction of the gas settled in the core of the merger remnant and formed stars in a starburst (a process which is presumably still observed today in some IRAS galaxies). Because stars and gas reacted differently in the merging process the final mean angular momentum vectors of gas and stars can differ considerably. This is most likely the origin of the peculiar cores observed in lumimous ellipticals. Some of these objects may still form today (e.g. [2]) but the majority of them is presumably quite old, especially when located in clusters of galaxies.

The basic question now is whether the details of the merging process not only

[1] Visiting Astronomer, German-Spanish Astronomical Center, Calar Alto, operated by the Max-Planck-Insitut für Astronomie, Heidelberg, jointly with the Spanish National Commission for Astronomy.

B. Barbuy and A. Renzini (eds.), The Stellar Populations of Galaxies, 267–270.

show in the kinematics but also in the metallicity/stellar population profiles of these galaxies. Previous studies [3] based on color index profiles could not uncover any indications but were limited by the difficulty of seeing corrections and sky subtraction on small-field CCDs. These latter problems are much less severe in case of line strength indices which can be derived from a single longslit spectrum and no comparison between different exposures (as in the case of color studies) is necessary. Indeed, in all the results shown below, it was found that sky subtraction errors are much smaller than the errors caused by photon noise. Only in the inner parts the gradients may have been somewhat weakend due to seeing effects but this does not affect the conclusions. The data shown here were obtained with the 3.5 m telescope on Calar Alto, Spain during several observing runs between 1987 and 1989.

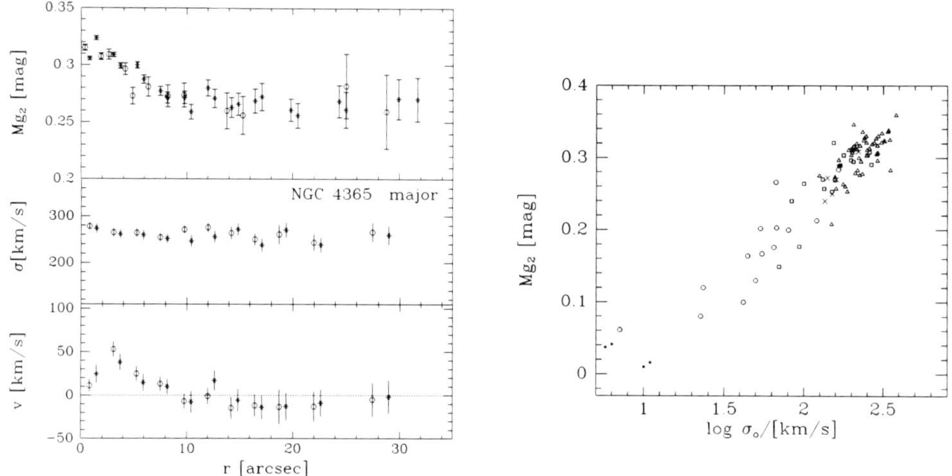

Figure 1 (left): Rotation velocity v, dispersion σ and Mg_2-index against radius r in NGC 4365. In the v and σ plots different symbols refer to different sides from the center along the *major* axis of the galaxy. In the Mg_2 plot, stars refer to the *major* axis, circles to the *minor* axis.
Figure 2 (right): Central Mg_2 against central σ_o for diffuse and compact low luminosity ellipticals (circles), ellipticals of intermediate luminosity (squares), luminous ellipticals (triangles), bulges (crosses) and dwarf spheroidals (dots).

RESULTS. Four ellipticals (NGC 4365, 4406, 4494, 5322) known to have peculiar core kinematics were analysed with respect to their line-strength profiles in Mg_2, <Fe> and H_β (line-strength indices in accordance with [4]. In *all* of them it was found that the Mg_2-index profile exhibits a discontinuous change of slope at roughly the same radius at which the peculiar core kinematics starts to show up. One example is presented in Fig.1 (NGC 4365). Outside the kinematic transition radius between core and main body (at $\approx$ 8 arcsec in NGC 4365) the Mg_2 gradient is rather weak, while inside of this radius the Mg_2-index rises steeply. The other indices (not shown here) show similar behaviour, but relative change with respect to MG_2 is not the same in all galaxies. This may indicate that core and mainbody not only have different metallicities but also that other parameters (like age, or the ratio of iron peak to light elements) are different (see also Faber, this conference). In a first simplified interpretation the Mg_2 profile can be understood in terms of a superposition of a lower metallicity main body having a only shallow gradient in Mg_2 and a higher metallicity core region which shows de-coupled kinematics and formed as a consequence of the merger event. The high metallicity of the core can be explained by the reasonable assumption that it formed from pre-enriched material and that, because of the depth of the potential well, the metals produced during the central starburst were well confined to the central region and could not escape from the galaxy. Furthermore, the central starburst may have produced stars with an IMF different from the one which was typical for the main body stars and this may possibly cause a radially changing [Mg/Fe]. The in some objects observable low Mg_2 gradient in the main body may be the residual of the Mg_2 gradients in the progenitor objects, these are expected to be weakened but not erased by merging (the more tightly bound parts of the progenitors are also more tightly bound in the merger product, see e.g. [5]).

2. A universal relation between metallicity and potential depth for all types of hot stellar systems?

It is well known that the metallicity of hot stellar systems correlates with their mass (e.g. [6], [7]). As realized by [8] there is however considerable scatter around the mean relation with the residuals from the Mg_2-luminosity correlation being correlated with the residuals from the mean velocity-dispersion luminosity relation. Since at a given luminosity higher velocity dispersion indicates higher surface brightness (because of the virial theorem), this in turn implies that at a given luminosity objects of higher surface brightness also have higher metallicity (see the papers by Gregg and Prugniel, this conference). Therefore, surface brightness is a second parameter in determining the metallicity of hot stellar systems. This can also be interpreted in the sense that metallicity is more closely correlated with the mean depth of the potential well (which roughly scales with the velocity dispersion) than with the mass of the object.

It is crucial for our understanding of the star formation history in hot stellar systems whether the metallicity velocity-dispersion relationship is different for different types of hot stellar systems or whether a universal relation exists. In order to investigate this question diffuse and compact dwarf ellipticals (dE's, cE's) in the Virgo cluster and the three brightest dE companions of M 31 (NGC 147, 185, 205) were observed with high spectral resolution at the Calar Alto 2.2m and 3.5m telescopes. Spatially resolved kinematics was derived (see [9]) and Mg_2 line-strengths were obtained and calibrated to the Faber/Burstein system. Furthermore, data of bulges of S0 galaxies (from the data set of Paquet et al., this conference) and of bright ellipticals with high accuracy kinematic data (see [1]) were compiled and added to the sample. Finally, the Mg_2 values of 5 dwarf spheroidal companions were estimated on the basis of their metallicity derived from cm-diagrams ([7]) and using the globular cluster based Mg_2-[Fe/H] calibration of [4]. In this way it was possible to derive for a large variety of hot stellar systems (dwarf spheroidals, diffuse and compact dwarf ellipticals, bulges and luminous ellipticals) a standardized data set covering 15 mag in luminosity, 10 mag/arcsec^2 in surface brightness, a factor 30 in velocity dispersion, and about a factor 100 in metallicity.

As shown in Fig.2, all the various types of hot stellar systems, independent from their large variety in structure and kinematics, follow a single, apparently universal relation between Mg_2 index and velocity dispersion. This is most remarkable since, e.g., giant ellipticals have been influenced by merging (see above) and faint ellipticals most likely suffered from wind-driven mass-loss (see e.g. [10]). Several important conclusions can be drawn on the basis of Fig.2:

(1) As the Mg_2-index is moderately age dependent, the age spread at a given velocity dispersion (or depth of the potential well) cannot be very large. There is however an indication that some part of the scatter in the Mg_2-σ_o-relation is indeed due to the presence of intermediate age stars (in case of luminous E's this has been shown by [11]; in case of dE's this is indicated by the presence of dust and gas).

(2) *The metallicity spread at a given potential depth/velocity dispersion cannot be large* either. This requires that the various formation/evolution processes (wind-driven mass-loss, merging) influence σ_o and Mg_2 in a similar way.

(3) Since the [Mg/Fe] ratio most likely depends on metallicity, Mg_2 can only be representative for the abundance of light elements. The [Fe/H]-σ_o-relation may flatten significantly at high velocity dispersions (see also part 1 of this paper).

References:

[1] Bender, R.: 1990, in *Dynamics and Interactions of Galaxies*, ed. R. Wielen, Springer Verlag Heidelberg, p. 232

[2] Schweizer, F.: 1990, in *Dynamics and Interactions of Galaxies*, ed. R. Wielen, Springer Verlag Heidelberg, p. 60

[3] Franx, M., Illingworth, G: 1990, ApJ. Lett. 359, L41

[4] Burstein, D., et al.: 1984, ApJ. 287, 586

[5] White, S.D.M.: 1979, MNRAS 189, 831

[6] Faber, S.M.: 1973, ApJ. 179, 731

[7] Aaronson, M., Mould, J.: 1985, ApJ. 290, 191

[8] Terlevich, R., et al.: 1981, MNRAS 196, 381

[9] Bender, R., et al.: 1991, A&A, 246, 349

[10] Dekel, A., Silk, J.: 1986, ApJ. 303, 39

[11] Schweizer, F., et al.: 1990, ApJ. 364, L33

Questions:

Renzini: How do you fit the Mg_2-σ relation with the merging scenario?

Bender: There is probably no simple answer to that question but it may all work out well if merging is accompanied by dissipation and star formation, as I tried to indicate in the first part of my talk.

Young Giant Elliptical Galaxies: where are they?

ROBERTO TERLEVICH

Royal Greenwich Observatory. Madingley Road, Cambridge, CB3 0EZ U.K.

Abstract.
Recent work have raised the intriguing possibility that the activity seen in most active galactic nuclei (AGN) is powered solely by young stars and supernova remnants in a burst of star formation at the time when the metal rich core of the spheroid of a normal, albeit young galaxy, was formed. The predicted emitted multifrequency spectrum, line width, variability and luminosity function of the young cores of ellipticals, are indistiguishable from those observed in Quasars. Only a small fraction ($\sim$5 %) of the total mass of elliptical galaxies, the core mass, is needed to explain the observed luminosities and luminosity function of Quasars at z$\gtrsim$2.0.

Key words: Young galaxies - Elliptical galaxies - Active Galactic Nuclei

1. Introduction

There are several lines of evidence suggesting that giant elliptical galaxies are among the oldest stellar systems: their colours and line strengths similar to those of galactic giants, the continuity of the colours and stellar absorption line strengths with those of galactic globular clusters and their high metallicity coupled with a very low content of cold gas and dust (Faber 1977). Recent work based on the extremely small scatter in the colour-magnitude or colour-diameter relations for the Coma and Virgo clusters (Bower *et al.* 1992), provides direct evidence that cluster ellipticals have probably completed star formation by redshift $z > 2$. If this is true, a first rank elliptical with $M = 5 \times 10^{12}\,M_\odot$forming in about $2Gyr$ will have an *average* star formation rate $2500\,M_\odot yr^{-1}$, thus reaching absolute magnitudes of -27 in the blue band (H_0=50 $km\,s^{-1}\,Mpc^{-1}$, q_0=0.5 and Λ=0, are used throughout this paper). If there is a peak in the star formation rate as suggested by dissipative galaxy formation models (Larson 1974,Carlberg1984), the maximum luminosity could be much higher. A short peak in the star formation rate can be also triggered by tidal encounters or mergers. Thus, it is possible that during a short period, first rank ellipticals may have reached absolute magnitudes $\sim -28, -29$ comparable with that of the most luminous Quasars. A similar result is obtained by considering the luminosity in massive stars needed to produce by redshift 2 the metals we observe today in first rank ellipticals (Cowie 1988).

Where are these bright blue galaxies?

In this paper I explore the possibility that many (most?) of the high redshift blue emission line objects discovered in wide field optical surveys and classified as Quasars due to the presence of broad emission lines in their spectrum, are in fact young ellipticals forming their metal rich core.

2. The Starburst model for AGN

The possibility that a starburst can power the most extreme forms of activity that are seen in Quasars and luminous Seyfert nuclei has been proposed several times in the past (Shklovskii, 1960; Field, 1964; Mc Crea, 1976), but was abandoned because it failed to explain satisfactorily the observed variability of luminous Quasars, their

B. Barbuy and A. Renzini (eds.), The Stellar Populations of Galaxies, 271–276.

radio emission, unresolved images, the presence of broad permitted and narrow forbidden emission lines and their intensity ratios.

More recently Terlevich and Melnick (1985; hereafter TM85) started a systematic study of Starburst in high metallicity environment. In this model (Terlevich *et al.* 1987), nuclear activity is the *direct* consequence of the evolution of a massive young cluster of coeval stars in the high metal abundance environment of the nuclear region of early type galaxies (Pagel and Edmunds 1981; Díaz *et al.* 1985).

The phenomenology of narrow line AGN (Seyfert 2 and LINERS) and its relation to nuclear starbursts has been analyzed by TM85 who further predicted an evolutionary sequence which follows the evolution of a coeval nuclear cluster. For luminous objects the nuclear emission line regions evolve from normal HII regions to type 2 Seyferts and later to LINERS with transitions after about 3 Myrs and 5 Myrs, respectively. The subsequent evolution of these young clusters into the supernova phase and the development of the broad line region (BLR), has been described by Terlevich *et al.* (1987) and Terlevich (1989,1990a,b). During this supernova or Quasar phase, most of the bolometric luminosity is emitted by the young stars, while the broad permitted emission lines and their variability are due to supernova (SN) and supernova remnant (SNR) activity (Terlevich and Melnick 1985, 1988; Terlevich 1989; Filippenko 1989;Terlevich 1990b; Terlevich *et al.* 1992).Heckman (1991) and Filippenko (1992) excellent reviews of the Starburst model for AGN provide also good discussions of some potential problems. Terlevich (1992) addressed some of the problems raised by Heckman (1991).

3. The evolution of nuclear star clusters

The early evolution of a massive metal rich star cluster presents four different phases. The appearance of the first extreme Wolf-Rayet or WARMERS (TM85) marks the beginning of the Seyfert phase and the end of the HII region phase. The explosion of the first SN of type Ib corresponds to the onset of non-thermal radio emission while the first explosions of type II SN lead to the formation of the BLR.

Phase 1. (From 0 to 3 Myr). During this phase the photoionization is dominated by hot main sequence stars and the nuclear spectrum is typical of a low excitation, high metallicity, HII region. Computations of the ionizing continuum and emission-line spectra are given by García Vargas and Díaz (1992).

Phase 2. (From 3 to 4 Myr). The most massive stars in the cluster ($M \geq 40\text{-}60\,M_\odot$) become extreme WC or WO Wolf-Rayet stars and reach the *warmer* phase.

A key aspect is that stellar mass loss rates increases with increasing metal abundance in massive stars. Thus, very massive stars formed in metal rich environments end their lives as bare C-O cores with effective teperatures reaching $T_{eff} \sim 2\times10^5$K and luminosities reaching 4×10^6 $L_\odot$. Models show that after 3 Myr, the ionizing continuum of a young cluster is nearly a power law of index $\alpha\sim -1.5$ (where $f_\nu \alpha \nu^\alpha$), with an exponential cutoff at about 30 Ryd (TM85, Cid Fernandez *et al.* 1992). The emission line spectrum of the ionized surrounding gas is that of type 2 Seyferts. Furthermore, the "blue featureless continuum" and the associated big UV bump observed in most Seyferts may be the reddened spectrum of the ionizing cluster (Terlevich 1990a, Cid Fernandez and Terlevich 1992). Strong support for this interpretation comes from the discovery that the strength of infrared CaII triplet absorption lines in Seyfert 2 nuclei with weak or absent optical stellar absortion lines, is as strong as, or stronger than, those in normal galactic nuclei, as is expected from a young cluster containing red supergiants (Terlevich *et al.* 1990).

Large amounts of dust are synthesized just before the Wolf-Rayet stage of the most massive stars, so we expect to see an extremely reddened high excitation Seyfert type 2 nucleus at this stage.

Phase 3. (From 4 to 8 Myr). The first SNIb with massive progenitors ($M \geq 40\,M_{\odot}$) appear in the cluster and with them copious amounts of non-thermal radio emission from the SNR. These SN are optically dim (due to the lack of an extended envelope to thermalize the energy) and probably very luminous in radio frequencies (Weiler and Sramek, 1988). During this phase, the ionization sources are main sequence hot stars, *warmers*, and SN; the emitted spectrum resembles that of a Seyfert 2 (or a LINER for the less luminous objects).

Due either to dynamical friction or initial conditions, the most massive stars in nearby young clusters and giant HII regions populate the inner core of the cluster (Terlevich 1987), while the less massive stars tend to live in a more extended region. The collective action of stellar winds and SN explosions from these very massive stars creates a hot cavity in the interstellar medium (ISM). This hot "superbubble" will expand along the steepest density gradient (along the poles in the case of a disk). Material flowing along this axis, "superwinds", will give rise to elongated, mildly collimated radio structures (Heckman, Armus and Miley 1990). The combined ejecta of several SN may in some cases reach the outer parts and shock the ISM at large distances from the nucleus. Ionizing radiation will also escape along this tunnel and photoionize the shocked ISM at large distances from the nuclei giving rise to optical filaments correlated in position with the radio ejecta. Ionization cones, such as those seen by Tadhunter and Tsvetanov (1989) and Pogge (1988a,b), should be relatively common in this phase.

Phase 4. (from 8 to 60 Myr). This is the type II SN phase. The remnants of the metal rich massive stars ($M \sim 8 - 25\ M_{\odot}$) are presumably very luminous at radio, IR, optical, UV and X-ray wavelengths because their kinetic energy is rapidly thermalized by dense circumstellar material around the metal rich progenitor red giant stars. These remnants *are* the BLR.

The BLR is fully developed during this phase and the optical spectrum is dominated by broad and variable permitted lines. Most of the ionization comes from the strong UV and X-ray emission from the high velocity SNR shocks. Variability is due to SN flashes, to cooling instabilities in the expanding SNR shells and to the luminosity curve of the remnants. Most of the initial dust is evaporated by the first few SN.

Recent observations show that the spectrum of at least some luminous SN exploding in HII regions have a striking resemblance to that of the BLR of Seyfert galaxies (Filippenko 1989) and, conversely, that the flares of some Seyfert galaxies have the luminosity, life-time and spectral signatures of type II SN (Terlevich and Melnick, 1988). The fundamental difference between "Seyfert-like" and normal type II SN can be understood if the former are associated with shocks that, after leaving the envelope of the star, expand into a region of *high circumstellar gas densities.* Theoretical computations of the evolution of SNR in dense environments show that after sweeping up a small amount of gas these remnants become radiative and deposit most of their energy in very short time scales thus reaching very high luminosities. Because of the large shock velocities, most of the energy is radiated in the extreme UV and X-ray region of the spectrum (*e.g.* Shull, 1980; Terlevich *et al.* 1992).

4. Properties of the young cores in elliptical galaxies

There is mounting observational evidence that large spheroid formation is more or less complete by z=2 to 2.5 (Cowie, 1988 and references therein). I will assume that giant elliptical galaxies formed their metal rich cores in a core crossing time at the epoch of formation of the large spheroids, $2 < z_{form} \lesssim 10$. The size of the core of bright elliptical galaxies with $M_B = -23$ is $r_{c,app} = 1000$ pc while an elliptical with luminosity L* has $r_{c,app} \sim 200$pc (Kormendy, 1987). The core contains 1/20 of the total mass of the elliptical galaxy and its crossing time is, even for the largest galaxies, less than 2.5×10^7 years. A typical L* elliptical has a core crossing time of about 1.4×10^6 years (Terlevich, 1992). This time scale is smaller than the life time of the most massive stars, thus, star formation can be synchronized over the whole core. Clusters of about this size, and decoupled either photometrically or kinematically, have recently been found in many nearby elliptical galaxies (Franx and Illingworth, 1988; Bender, 1988).

The luminosity function of the population of young cores will be that of present day old cores shifted to higher luminosities by the change in the M/L ratio of the stellar population and with a co-moving volume density weighted by the short lifetime of the event. The luminosity function of elliptical galaxies (Tammann *et al.* , 1979) has a maximum value of about $2 \times 10^{-4} Mpc^{-3} mag^{-1}$ for low luminosity ellipticals and the absolute blue magnitude corresponding to L^* is M_B = -21.0. The reduction of the co-moving density associated with the short life-time of the young core can be estimated from the ratio of the age of the universe at z=2 (about 2.4×10^9 years) to the total lifetime of the burst at the end of the type II SN phase, 6×10^7 years. Thus for young cores the maximum density will be at $z = 2$, $(1/40) \times 2 \times 10^{-4} Mpc^{-3} mag^{-1} = 5 \times 10^{-6} Mpc^{-3} mag^{-1}$, very similar to the maximum density of Quasars at the same redshift (Boyle, 1991; Terlevich, 1992). The luminosity function of present day cores is that of present day ellipticals shifted to lower luminosities by the ratio of core luminosity to total galactic luminosity, $2.5 log(1/20) = 3.25$ mag. The typical L_B/M ratio for a young stellar cluster with type II SN activity is about 50 (Terlevich 1990).The present value of L_B/M for cores in a luminous elliptical is 1/20. Thus, the blue luminosity of young cores will be about 7.5 mag brighter than that of old ones. For a typical L^* elliptical, the young core will have $M_B = -25.2$ and $R_{eff} = 200$pc and, at a redshift of 2, its effective radius will be 0.05 arcsec with apparent blue magnitude, including K correction, of $B = 20.0$. Its density will be about $3 \times 10^{-6} Mpc^{-3} mag^{-1}$. For a bright elliptical, $M_B = -23$, the young core will have $M_B = -27.2$ and $R_{eff} =$

1000pc and, at a redshift of 2, its effective radius will be 0.25 arcsec with apparent blue magnitude, including K correction, of $B = 18.0$. Its density will be about $4 \times 10^{-8} Mpc^{-3} mag^{-1}$. The predicted parameters of young cores are in very good agreement with the observed ones for Quasars (Terlevich, 1992). The young cores of the brightest ellipticals will reach $M_B = -28.5$, similar to the most luminous Quasars.

5. Discussion

The possibility that the bulk of the stellar population we see today in the cores of elliptical galaxies was generated in a burst of star formation is supported by the short crossing time of the core. Also, dissipational galaxy formation models suggest that, at the end of the formation of a luminous elliptical, a short lived peak in the star formation occurs (Larson, 1974; Carlberg, 1984) in the central parts of the galaxy. This peak in the star formation rate should not be an exclusive property of dissipational models of galaxy formation. It can be generated in almost any galaxy formation scenario by the late infall of enriched gas into the central regions of the elliptical. There the metal rich gas can sit until star formation is triggered either by tidal interactions or by the generation of a bar instability when the gas density reaches some critical value (Efstathiou and Silk, 1983).

The predicted young core luminosity function is an excellent match to the observed luminosity function for Quasars in the redshift range from 2.0 to 2.9 (Boyle, 1991; Terlevich 1992; Terlevich and Boyle, in preparation). Thus, *the young cores of ellipticals containing only 5 % of the total galactic mass are capable of producing the luminosity of even the most luminous Quasars.* The upper limit on the size of the active region is compatible with their lack of resolution. The emitted multifrequecy spectrum is very similar to the average AGN spectrum (Terlevich 1990)

The suggestion is that perhaps most of the high redshift emission line objects found in optical surveys and classified as Quasars due to the presence of broad emission lines in their spectrum, are in fact young spheroids forming their metal rich core.

Acknowledgements

It is a great pleasure to thank my collaborators: Brian Boyle, Angeles Díaz , José Franco, Jorge Melnick, Guillermo Tenorio-Tagle and Elena Terlevich with whom most of the work presented in this paper was done. Travel grants by the Royal Society and the IAU are thanked.

6. References

Bender, R., 1988. *Astr. Astrophys.* , **202,** L5

Bower, R.G. Lucey, J.R. & Ellis, R.S., 1992. *Mon. Not. R. astr. Soc.* , in press

Boyle, B.J., 1991. *in: Texas ESO/CERN Meeting on Relativistic Astrophysics, Brighton, U.K.*

Carlberg, R.G., 1984. *Astrophys. J.* , **286,** 403.

Cid Fernandez Jr., R. & Terlevich, R., 1992. *in: Relationship Between Active Galactic Nuclei and Starburst Galaxies, ed. Filippenko, A.V.*

Cid Fernandez Jr,. R., Dottori, H.A., Gruenwald, R.B. & Viegas, S.M., 1992. *Mon. Not. R. astr. Soc.* , in press

Cowie, L.L., 1988. *in: The Post-recombination Universe*Eds. Kaiser, N. & Lasenby, A.N.

Efsthatiou,G. & Silk, J., 1983. *Fund. of Cosmic Physics*, **9,** 1.

Faber, S.M., 1977. *in: The evolution of galaxies and stellar populations, Eds. Tinsley, B.M. & Larson, R.B.*

Field, G.B., 1964. *Astrophys. J.* , **140,** 1434.

Filippenko, A., 1989. *Astron. J.* , **97,** 726.

Filippenko, A.I., 1992. *in: Relationship Between Active Galactic Nuclei and Starburst Galaxies, ed. Filippenko, A.V.*

Franx, M. & Illingworth, G., 1988. *Astrophys. J.* , **327,** L55.

García Vargas, M.L. & Díaz, A.I., 1992. *in: Relationship Between Active Galactic Nuclei and Starburst Galaxies, ed. Filippenko, A.V.*

Heckman, T.M., 1991. *in: Massive Stars in Starbursts, Eds. Walborn, N. & Leitherer, C.*.

Heckman, T.M.,Armus, L, & Miley, G.K., 1991. *Astrophys. J. Suppl.* , **74,** 833.

Kerr, F.J., 1957. *Astron. J.* , **62,** 93.

Kormendy, J., 1987. *in: Structure and Dynamics of Elliptical Galaxies, Ed. De Zeeuw, T.*

Larson, R.B., 1974. *Mon. Not. R. astr. Soc.* , **166,** 585.

McCrea, W.H., 1976. *in: The Galaxy and the Local Group, Eds. Dickens, R.J. & Perry, J.E. (RGO Bull 182).*

Pagel, B.E.J. and Edmunds, M., 1981. *Ann. Rev. Astr. Astrophys.* , **19,** 77.

Pogge, R.W., 1988a. *Astrophys. J.* , **328,** 519.

Pogge, R.W., 1988b. *Astrophys. J.* , **332,** 702.

Shklovskii, I.S., 1960. *Soviet Astronomy*, **4,** 885.

Shull, J.M., 1980. *Astrophys. J.* , **237,** 769.

Tadhunter, C. & Tsvetanov, Z., 1989. *Nature*, **341,** 422.

Tammann, G.A., Yahil, A. & Sandage, A., 1979. *Astrophys. J.* , **234,** 775.

Terlevich, E. , 1987. *Mon. Not. R. astr. Soc.* , **224,** 193.

Terlevich, R., 1989. *in: Evolutionary Phenomena in Galaxies, Eds. Beckman, J.E. & Pagel, B.E.J., Cambridge Univ. Press*

Terlevich, R., 1990a. *in: Windows on Galaxies, Eds. Fabbiano, G., Gallagher J. & Renzini, A., Kluwer: Dordrecht*

Terlevich, R., 1990b. *in: Structure and Dynamics of the Interstellar Medium, Eds. Tenorio-Tagle, G., Moles, M. & Melnick, J., Springer-Verlag: Berlin*

Terlevich, R. and Melnick, J., 1985. *Mon. Not. R. astr. Soc.* , **213,** 841; (TM85).

Terlevich, R. and Melnick, J., 1988. *Nature*, **333,** 239.

Terlevich, R., Melnick, J. and Moles, M., 1987. *in: Observational Evidence for Activity in Galaxies, Eds. Khachikyan, E.Ye., Fricke, K.J. & Melnick, J., Reidel:Dordrecht*

Terlevich, R., 1992. *in: Relationship Between Active Galactic Nuclei and Starburst Galaxies, ed. Filippenko, A.V.*

Terlevich, R., Tenorio-Tagle, G. ,Franco, J. and Melnick, J., 1992. *Mon. Not. R. astr. Soc. in press.*

Weiler, K.W. & Sramek, R.A., 1988. *Ann. Rev. Astr. Astrophys.* , **26,** 295.

Metallicity of Unresolved Stellar Populations

M.G.EDMUNDS
Department of Physics and Astronomy,
University of Wales College of Cardiff,
P.O.Box 913,
CARDIFF CF1 3TH,
U.K.

The determination of the metallicity of an unresolved population is not easy. An obvious problem is that stars of the same mass, but different metallicities, evolve at different rates and have different luminosities. For example, metal poor red giants are considerably brighter in the optical region than corresponding metal rich stars. The "mean" abundance for a stellar population therfore depends on whether the mean is taken with respect to *mass*, *star number* or *luminosity*. It is inevitable that observational work weights by luminosity, while chemical evolution models usually weight by mass. The two weightings can give rather different means, which may also depend on the "metallicity structure" - i.e. the relative numbers of stars of different metallicities, and rough calculations show that metallicity indicators like colours could give 0.3 dex lower abundance than the mass-weighted mean, if uniform (i.e. single metallicity population) calibrators have been used.

A modern, accurate calibration of the popular Mg_2 index has yet to be published, but a rough estimate of its behaviour is shown in Figure 1, based on Edmunds (1991). It shows that the indicator is quite a good one below solar abundances, but that above solar its behaviour is probably very non-linear with metallicity, and should not yet be trusted.

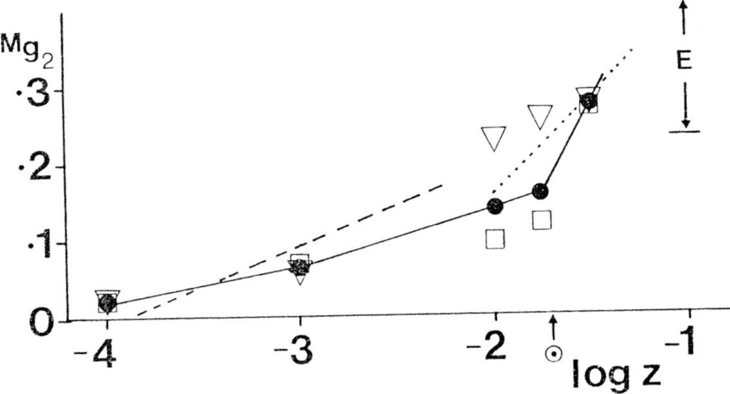

Figure 1. A rough calibration of the Mg_2 index for uniform metallicity, 15 Gyr old stellar populations. The open triangles are red giant indices, the open squares dwarfs, and the filled circles represent the combined population. The dashed and dotted lines are the empirical calibrations of Brodie and Huchra (1990) and Gorgas *et al* (1990).

B. Barbuy and A. Renzini (eds.), The Stellar Populations of Galaxies, 277–280.

An added complication with this index is the uncertainty over what to take as a [Mg/Fe] ratio. Although the [α element/Fe] ratio has sometimes been regarded (like O/Fe) as an age indicator (e.g. Matteucci and Brocato 1990), it can be shown (Edmunds *et al* 1991) that a quasi-secondary behaviour of iron, with the iron yield depending on overall metallicity, could also explain its behavior as observed in Galactic stars. It remains unclear what ratio should be adopted for elliptical galaxy simulations, but detailed analysis of stars in the Galactic bulge - which appear to reach up to metallicities of at least two or three times solar - should help to identify whether [Mg/Fe] $= 0.3$ in old, metal rich populations (as predicted by "age" models) or [Mg/Fe] ≤ 0 (as predicted by a simple "quasi-secondary" model)

The Yield Problem

Assuming that colour and Mg_2 abundance indicators are giving something like the true abundance in ellipticals, the implication would be that, at least in the centres of large ellipticals, the mean abundance in the stars may reach up to $z \sim 0.06$, i.e. about three times solar. This presents a problem, since although gas abundances can build up to this kind of value quite easily by recycling, it is much more difficult to produce such a high mean abundance in a stellar population. Indeed, as shown in Edmunds (1989), although already a well-known result, a region of a galaxy which suffers outflow, *unenriched* inflow, or no gas flow, can never have a mean abundance which exceeds the "true" yield. By true yield is meant the amount of heavy elements released per mass of interstellar material processed into long-lived stars or remnants. The problem is that irregular galaxies appear to show a yield of order 0.006, and for spiral galaxies there is some evidence (Vila and Edmunds 1992) of a metallicity-dependent yield which, however, could not give rise to a mean population metallicity of more than about 0.004 on the basis of a "simple" chemical evolution model. These values are an order of magnitude below what is required, even ignoring the probable extra factor of two from metallicity/luminosity effects in the abundance indicators. There are possible ways out. Enriched inflow is a possibility, where the central regions form out of gas which already contains considerable heavy element abundance. As an example of an elementary model of this type, Figure 2 shows the (mass-weighted!) mean abundance for a spherical model galaxy with star formation rate proportional to the gas density, and a constant velocity inflow of gas. The "true" yield is p, and the curves are numbered with the parameter $\mathcal{R}$ which represents flow timescale/star formation timescale for the model. For clarity, the $\mathcal{R} = 30$ model is shown dashed. The initial radius of the galaxy is taken to be unity, and the curves show the mean stellar abundance after gas exhaustion. This is one of a series of elementary models considered by Edmunds and Greenhow (1992), a paper which attempts to find general constraints on the effects of flows in more realistic galaxy models than was possible in Edmunds (1989). As is apparent, and known from the "concentration" model of Lynden-Bell (1975) and many numerical models, it is possible for the mean stellar abundance to reach two or three times the yield, at least near the centre. But this may not be enough, and it is quite difficult to fine-tune models to reach higher values. It is tempting to speculate that the true yield may actually vary, and that it this variation (due perhaps to stellar physics or imf variations) which produces the rather high values of mean stellar abundances which are apparently seen in the centres of galaxies.

An interesting property of the elementary inflow model of Figure 2 is its metallicity structure near the centre, which is shown in figure 3. This shows that for large values of the

$\mathcal{R}$ parameter (i.e. slow flow), the metallicity distribution can mimic the structure that a "simple" closed -box model would have with a true yield twice that actually present. This might be a good model for the Galactic bulge, as observed by Rich (1990).

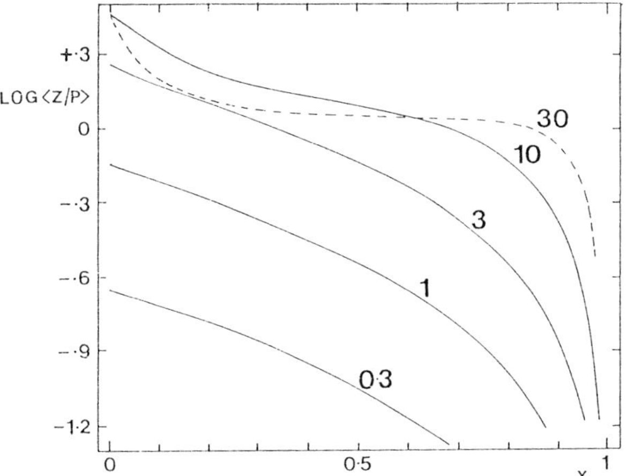

Figure 2. Mass-weighted mean stellar abundance as a function of radius x in a spherical inflow model (see text for details).

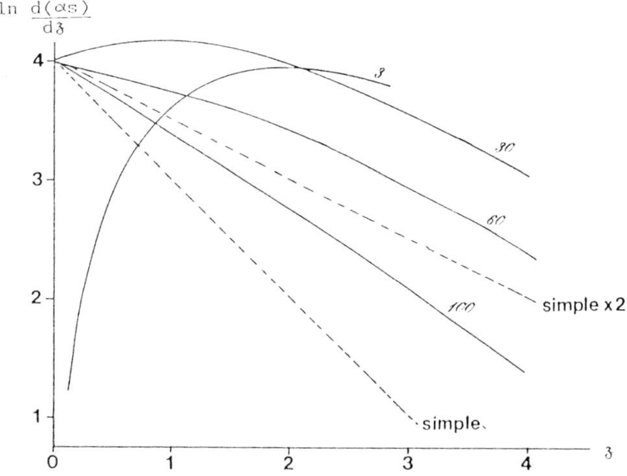

Figure 3. Metallicity structure of the spherical model of Figure 3 at radius x = 0.05. The curves are marked with the value of the $\mathcal{R}$ parameter, and the dashed lines show the metallicity structure of "simple" closed box models with yields of p and 2p.

The Importance of Surface Density

As a final provocative comment, I would like to re-emphasise the possiblity that the *surface density* or *surface brightness* of a system may be an important parameter in its chemical evolution. For spiral (Vila and Edmunds 1992), irregular (Phillipps, Edmunds and Davies 1990) and elliptical (Edmunds and Phillipps 1989) galaxies there does seem to be a link between surface density or surface brightness and metallicity, which may be more fundamental than absolute magnitude/metallicity relations. The latter may be dominated by sample selection effects, but not everyone will agree. Perhaps *both* luminosity and surface density/brightness are important. At least it is a problem that can be addressed by observation of larger data sets.

References

Brodie, J.P. and Huchra, J.P. 1990, Astrophys.J.,**362**,503.
Edmunds M.G. 1990, MNRAS,**246**,678.
Edmunds, M.G. 1991 in *Galaxies and the Cosmos* ed. R.J. Terlevich and M.G. Edmunds, Cambridge Univ. Press, *in press*.
Edmunds, M.G., Greenhow, R.G., Johnson, D., Kluckers, V. and Vila, B.M. 1991, MNRAS,**251**,33p.
Edmunds, M.G. and Greenhow, R.G. 1992, MNRAS, *submitted.*
Edmunds, M.G. and Phillipps, S. 1989, MNRAS,**241**,9p.
Gorgas, J., Efstathiou, G. and Aragon-Salamanca, A. 1990, MNRAS,**245**,217.
Lynden-Bell, D. 1975, Vistas in Astron.,**19**,299.
Matteucci, F. and Brocata, E. 1990, Astrophys.J.,**365**,539.
Rich, M. 1991, Astrophys.J.,**362**,604.
Phillipps, S., Edmunds, M.G. and Davies, J.I. 1990, MNRAS,**244**,168.
Vila, B.M. and Edmunds, M.G. 1992, MNRAS, *submitted.*

Discussion

G. Worthey: Just a word of comfort: you are exactly right about Mg_2. Models by Peletier and theoretical ones by Mould, as well as detailed ones by myself, confirm the steepening of the Mg_2/z relation near solar abundances.

R.Bender: Surface brightness alone is not sufficient to parameterise metallicity. From what I showed in my talk, there is a mass-term which combines with surface brightness in a way that Mg_2 seems to be correlated directly with velocity dispersion, with no correlation between residual scatter and surface brightness.

S.Faber: If you confine attention to just the bright ellipticals, have you tried plotting abundance vs log σ , as opposed to surface brightness? Bender's results suggest that surface brightness does not work nearly so well, and here the selection effects you referred to are small.

M.Edmunds: I havn't looked at correlations with log σ , and for ellipticals things seem to depend on whether the plot is of a *colour* based abundance indicator or Mg_2.

M.Bershady: In the magnitude limited redshift survey of Koo and Kron, I have found a peculiar population of low redshift, compact blue galaxies ($z \leq 0.1$). These objects are intrinsically faint and blue in optical and near-infrared, similar to, and bluer than, NGC 4449. Yet in optical-near-IR colour magnitude diagrams, these faint galaxies are too red, and display an "earlier" type C-M relation. It is possible that these objects are another example of a "2nd parameter" in the metallicity-absolute magnitude relation, namely compactness. So here surface mass density may be more relevant than surface brightness.

Stellar Populations and Peculiar Velocities of Elliptical Galaxies

MICHAEL D. GREGG
Mt. Stromlo and Siding Spring Observatories
Private Bag, Weston Creek P.O.
A.C.T. 2611, Australia

Abstract.
The estimated distances of elliptical galaxies, as determined by the $D_n - \sigma$ technique, can be affected by stellar population variations and perhaps also dynamical effects due to mergers, leading to spuriously large positive peculiar velocities.

1. Measuring Peculiar Velocities

When extragalactic astronomers speak about peculiar velocities, they usually mean the difference between a galaxy's measured radial velocity and the velocity it would have if it were at rest with respect to the general Hubble expansion or $V_{pec} = V_{rad} - V_{dist}$. A radial-velocity-independent distance estimate is needed to determine V_{dist}. Presently, the best distance estimators are the infrared Tully-Fisher relation (Aaronson, *et al.* 1979) for spirals and the $D_n - \sigma$ method (Dressler *et al.* 1987) for ellipticals. In determinations of cluster distances, both have an advertised accuracy of 10-20% (Dressler *et al.* 1987; Aaronson *et al.* 1989).

The $D_n - \sigma$ relation relies on the correlation between the central velocity dispersion σ and a photometric diameter D_n, defined as the diameter of a circular aperture within which the blue surface brightness is 20.75 mag/arcsec^2. The velocity dispersion is nearly independent of distance while D_n simply scales as distance^{-1}. The displacement of a galaxy in the D_n direction from a fiducial $D_n - \sigma$ relation (usually the Coma cluster) is a measure of its relative distance.

Using the $D_n - \sigma$ distance estimator, Faber *et al.* 1989 determined peculiar velocities for $\sim$ 400 ellipticals and found evidence for a large scale streaming motion through our part of the universe, of amplitude 600 km s^{-1} toward l=307°, b=9°, in very rough agreement with the microwave background dipole. This result has been generally supported by other studies, whether involving ellipticals (Lucey & Carter 1988) or spirals (Aaronson *et al.* 1989; Mathewson *et al.* 1991). Lynden-Bell *et al.* (1988) attribute the streaming motion to the gravitational influence of a "Great Attractor" located at a distance corresponding to $\sim$ 4500 km s^{-1}.

2. Fine Structure Results of Schweizer *et al.*

In a seemingly unrelated study, Schweizer *et al.* (1990, S90) studied the "fine structure" of elliptical galaxies and its correlation with various line strength residuals. Fine structure is any measurable deviation from elliptical isophotes, such as shells, jets, boxy isophotes, or X-structure. All of these features are interpreted by S90 as phenomena resulting from mergers. The amount of fine structure is parametrized in a semi-quantitative way in S90 with an index Σ which ranges from 0 to 7.6 in their sample of 36 ellipticals (see S90 or Schweizer & Seitzer, this conference, for details). Larger Σ presumably implies a more recent or more traumatic merger event.

B. Barbuy and A. Renzini (eds.), The Stellar Populations of Galaxies, 281–285.

S90 show that in their sample of ellipticals, Σ is correlated with the residuals of Mg_2, $H\beta$, and CN formed from the mean relation with absolute magnitude defined by a large sample of $\sim$ 400 ellipticals (Faber *et al.* 1991). Galaxies with larger Σ have more negative residuals of Mg_2 and CN and positive residuals of $H\beta$. They interpret this as differences in the mean ages of the ellipticals due to merger induced star formation, though except for a time dependence of some of the ingredients of the Σ index and the likelihood of star formation accompanying mergers, the correlations could be explained equally well by metallicity variations. Whether

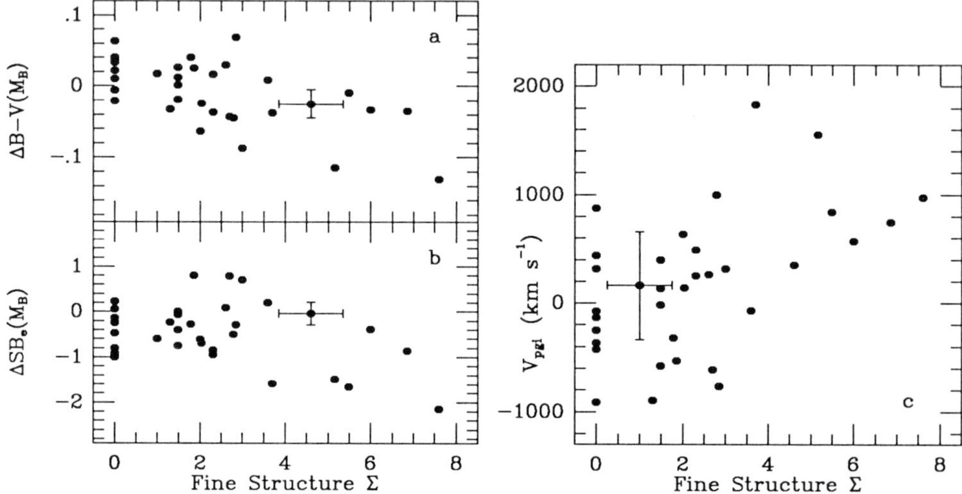

Fig. 1. Color and surface brightness residuals and peculiar velocity trends with the fine structure parameter Σ.

age or metallicity is responsible for the correlations, there should also exist trends with color and perhaps surface brightness. Using the data of Burstein *et al.* 1988, mean linear relations of $B - V$ and effective surface brightness, SB_e, versus absolute blue magnitude were defined. The residuals $\Delta(B - V)$ and ΔSB_e from these mean relations have been determined for 35 of the fine structure ellipticals. Figures 1a & b show the resulting correlations between the residuals and Σ. There is a clear trend for high Σ galaxies to be bluer, with somewhat higher surface brightness.

3. Effects on Distance Estimates

The determination of accurate distances to elliptical galaxies using the $D_n - \sigma$ relation requires a universal and tight correlation between D_n and $\log(\sigma)$. Yet the fine structure results paint a picture of common galaxy mergers and associated star formation, which must alter the relation between central velocity dispersion and D_n through mass-to-light differences due to mean age differences of the galaxies. Young stars in the high Σ galaxies lower M/L which inflates D_n, in turn causing an underestimate of the distance. This should result in systematically large positive

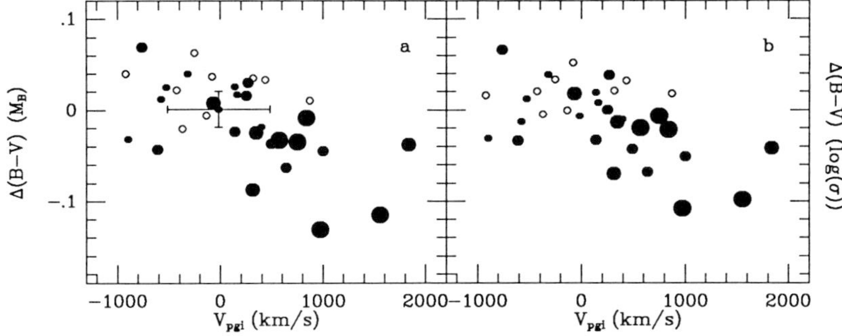

Fig. 2. Residuals of B − V from the mean relations with M_B and $\log(\sigma)$ for the fine structure galaxies, symbols scaled to Σ. See text for details.

peculiar velocities for the high Σ objects. To search for this effect, define a measure of peculiar velocity $V_{pgi} = V_{cmb} - R_e$ where V_{cmb} is the velocity of the group or cluster of which the individual galaxy is a member with respect to the microwave background (from Faber *et al.* 1987) and R_e is the *individual* $D_n - \sigma$ distance estimate, expressed as a velocity (computed using data in Burstein *et al.* 1988, Davies *et al.* 1988, and equation 3.1 of Lynden-Bell *et al.* 1988). This definition of peculiar velocity maximizes sensitivity to stellar population effects. Figure 1c shows the correlation obtained between Σ and V_{pgi}. The distances to the high Σ galaxies have been underestimated either because of stellar population variations or dynamical effects which alter the relation between D_n and σ.

Quantities more directly influenced by stellar populations than Σ provide a more sensitive test for effects on distance estimates. Figure 2a shows the trend of V_{pgi} with $\Delta(B-V)$, as defined above. Open circles are galaxies with $\Sigma = 0$; filled circles are scaled roughly with Σ. Objects with zero fine structure show no trend whatsoever with V_{pgi} while the other objects exhibit a clear effect, with the largest Σ galaxies tending to be at larger negative residuals and more positive peculiar velocities. Similar trends are found for surface brightness and Mg_2. The M_B used to generate these residuals are based on $D_n - \sigma$ distances, thus any distance errors will result in *positive* correlations in these diagrams, opposite to the trends seen. Because these plots show that the $D_n - \sigma$ distances *are* in error, the real effect is *larger* than demonstrated here because the distance errors work to weaken the correlations.

The results of Figure 2a are not dependent on using M_B as the independent variable to generate the residuals. In Figure 2b, residuals from the $\log(\sigma) - (B-V)$ relation are plotted against V_{pgi} for the fine structure galaxies, symbols the same as for Figure 2a. Here too, distance errors work against the trend seen. The correlations in Figure 2b are not as strong as in Figure 2a. This is due possibly to an effect demonstrated in the N-body simulations of Balcells & Quinn which show that a merger involving a small galaxy and a large elliptical can produce a dip in the central velocity dispersion of $\sim 40\,\mathrm{km\,s^{-1}}$. So even though an underestimate of

$\log(\sigma)$ will cause an underestimate of distance of the same magnitude as the stellar population inflation of D_n, it may not be manifested in Figure 2b because a decrease in color may be partly compensated for by a decrease in $\log(\sigma)$. That is, mergers cause galaxies to move along lines roughly parallel to the "unperturbed" $\log(\sigma) - (B - V)$ relation. That $\log(\sigma)$ is systematically too low for the high Σ objects can be demonstrated by calculating the residuals of $\log(\sigma)$ on M_B for the fine structure sample. All but 1 of the 9 objects with $\Sigma > 3$ have negative $\log(\sigma)$ residuals.

4. Discussion

How much of the large scale streaming motion can be explained away by stellar population effects? Little or none, probably, because both the spirals and the CMB dipole show rough qualitative agreement with the ellipticals. Faber *et al.* 1988 have also wisely bullet-proofed themselves against such effects by adopting the median $D_n - \sigma$ distance to groups and clusters, rather than relying on averages, though there may still be problems if too few objects in a group are measured or have young populations. However, taking stellar population effects into account, if only as a way of eliminating galaxies from a data set, will certainly result in smoother velocity fields which may facilitate modeling and help resolve some of the outstanding disagreements among the various large scale flow studies. One might be able to explain some of the rather large discrepancies between $D_n - \sigma$ and Tully-Fisher estimated distances to various groups and clusters (Mould *et al.* 1991).

Finally, the sensitivity of $D_n - \sigma$ to young stellar populations can be reduced by using R, I, or infrared bandpasses instead of B to define a fiducial diameter to correlate with $\log(\sigma)$.

References

Aaronson, M., Huchra, J., & Mould, J. 1979, ApJ, 229, 1

Aaronson, M., Bothun, G., Cornell, M.E., Dawe, J.A., Dickens, R.J., Hall, P.J., Han Ming Sheng, Huchra, J.P., Lucey, J.R., Mould, J., Murray, J.D., Schommer, R.A., & Wright, A.E. 1989, ApJ, 338, 654

Balcells, M. & Quinn, P.J. 1990, ApJ, 361, 381

Burstein, D., Davies, R.L., Dressler, A., Faber, S.M., Stone, R.S.P., Lynden-Bell, D., Terlevich, R.J. & Wegner, G. 1987, ApJS, 64, 601

Davies, R.L., Burstein, D., Dressler, A., Faber, S.M., Lynden-Bell, D., Terlevich, R.J. & Wegner, G. 1987, ApJS, 64, 581

Dressler, A., Lynden-Bell, D., Burstein, D., Davies, R.L., Faber, S.M., Wegner, G., & Terlevich, R. 1987, ApJ, 313, 42

Faber, S.M., Burstein, D., Dalle Ore, C., & Gonzalez, J. 1991, in prep.

Faber, S.M., Wegner, G., Burstein, D., Davies, R.L., Dressler, A., Lynden-Bell, D., & Terlevich, R.J. 1989, ApJS, 69, 763

Lucey, J.R., & Carter, D. 1988, MNRAS, 235, 1177.

Lynden-Bell, D., Faber, S.M., Burstein, D., Davies, R.L., Dressler, A., Terlevich, R.J. & Wegner, G. 1988, ApJ, 326, 19

Mathewson, D.S., Ford, V.L., & Buchhorn, M. 1991, ApJL, in press

Mould, J.R.,Stavely-Smith, L., Schommer, R.A., Bothun, G.D., Hall, P.J., Ming Sheng Han, Huchra, J.P., Roth, J., Walsh, W., & Wright, A.E. 1991, ApJ, in press

Schweizer, F., Seitzer, P., Faber, S.M., Burstein, D., Dalle Ore, C.M., & Gonzales, J.J. 1990, ApJL, 364, 33 (S90)

DISCUSSION

Terlevich: In a recent paper with J. Lucey and D. Carter we examined the properties of the fundamental plane (FP) with new improved data in Coma cluster including objects up to $5°$ away from the core, to check for environmental effects. The main conclusion is that the FP is very thin, all the scatter is consistent with observational errors. Furthermore, the FP is the same at all radii, *i.e.*, is independent of local density that changes by x100.

Gregg: Yes, the Coma cluster ellipticals do appear to be very well behaved in all this.

Faber: I just wish to point out that we have made an extensive test of the variation of fundamental-plane zero point for the whole sample in Burstein, Faber, & Dressler (*Ap. J.* 1989). There, we plotted V_{pec} (group) vs. richness and included also field galaxies. There was no correlation with R or offset for field galaxies. This suggests there is no wholesale problem with the 7S distances, which is what you said.

Djorgovski: The effects you discussed add to the growing body of evidence that differences in formation histories (which may vary systematically with the environment) can modify galaxian properties in a way which would masquerade as false peculiar velocities. One particularly instructive example is a direct comparison of the peculiar velocities for clusters, measured with the Tully-Fisher relation and with the $D_n - \sigma$ relation (*c.f.* Mould *et al. Ap. J.*, in press): there is no correlation whatsoever.

Gregg: I have seen the Mould *et al.* results. I would say there is some correlation present, but it is very poor. This is an intriguing state of affairs and should be worrisome to those who measure peculiar velocities.

THE STELLAR POPULATION IN GALAXIES OBSERVED ON THE ASTROPHYSICAL STATION "ASTRON"

I.Pronik, N.Merkulova, L.Metik, V.Pronik
Crimean Astrophysical Observatory, USSR

25 star systems were observed in 1983-1987 with the 80cm telescope on the board of astrophysical station "Astron". Entrance diaphragms were 60" and 10". Spectrometer with the exit slit 28 A has measured fluxes in erg/sec cm^2 A in the region 1600-3500 AA in regime of narrow band photometry. Exposures in every point were 2-5 min for bright objects and 10-30 min for week ones.

Spectral data of 18 extragalactic objects were analysed to investigate stellar population. Entrance diaphragm 60" permit to measure the fluxes of 0.2-20 kpc central regions (mainly 1-6 kpc). H=75 km/sec Mpc. Continuum energy distribution and absorption features in spectra of galaxies were considered.

The brightest absorption features in spectra of some galaxies were near λ2350 A and λ2800 A . Main contributors of λ2800 A feature are lines MgI,2852 A and MgII,2796, 2803 AA. Intensity index $C_{2800}=lg\ F_{cont}/F_{2800}$ calculated for all objects, shows the correlation with the spectral type of the galaxy obtained by van den Bergh (1960).The maximum values of C_{2800} correspond to *g-k* spectral types galaxies. Fig.1 shows large scattering of maximal values of C_{2800} but in Fig.2 these indexes are in good correlation with the dimension of central region of observed galaxies.When dimension of the observed region decreased,the values of C_{2800} increased: in accordance with the optical results for equivalent width W_λ (*MgII,5150* A) (Faber et al,1977;Faber,1983).

The main contributors of λ2350 A feature can be absorption lines FeII,2388, 2390, 2413 AA and NiII,2296, 2316, 2394 AA,known for A-F stars spectra (Cianni et al, 1984; Jamar,Macau,1974) and dust absorption, having maximum near λ2175 A (Aiello et al,1988).But brightness of λ2350 A feature is correlated with that of λ2800 A feature,caused mainly by stellar population (Fig.3).The comparison of the spectra obtained for E and Sa galaxies with that of α Car (F0 Ib-II) permit us to suppose that UV spectra of these galaxies are caused by A-F stars,having bright λ2350 A and λ2800 A absorption features.

B. Barbuy and A. Renzini (eds.), The Stellar Populations of Galaxies, 287–290.

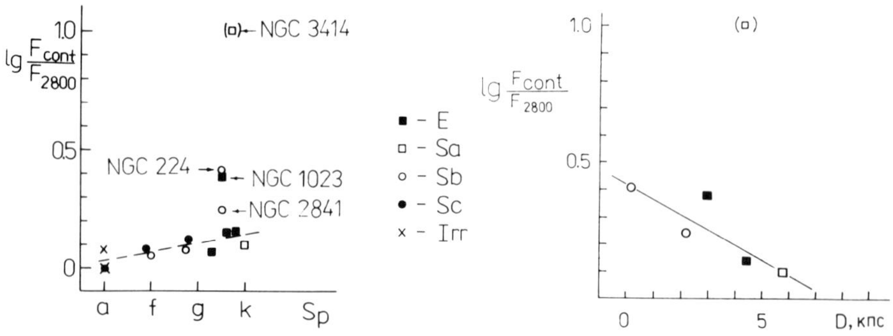

Fig.1. Comparison of λ2800 A absorption index with the spectral type of the galaxy. In brackets- weak value of flux.

Fig.2. Comparison of λ2800 A absorption index with the dimension of central region of galaxy,corresponding to 60" entrance diaphragm.

UV-continuum distribution in spectra of observed galaxies was compared with that of other UV explorers,using different diaphragms: IUE (10"x20"), ANS (2'.5 x 2'.5), OAO (10'). In the most cases (NGC 4486, E1; NGC 1023, E7; NGC 5194,Sc; NGC 1569,Irr, et al) energy distribution in continuum is not depend on diameter of diaphragm.

Two spectral indexes - red and blue - were calculated to describe the continuum of galaxies. Fig 4 showes that the spectral index $lg\ F_{3400}/F_{2800}$ is connected with the optical spectral type of the galaxy. So we suppose that this index is caused synonymously by stellar population of the galaxy. It distinguished the maximum for E,Sa(*g-k*) galaxies. But Sc, Irr galaxies,having *a-f* spectral types,show almost flat continuum in the region 3400-2800 AA.

Behaviour of continuum in the region 2200-1800 AA does not connect simultaneously with the optical spectral type of

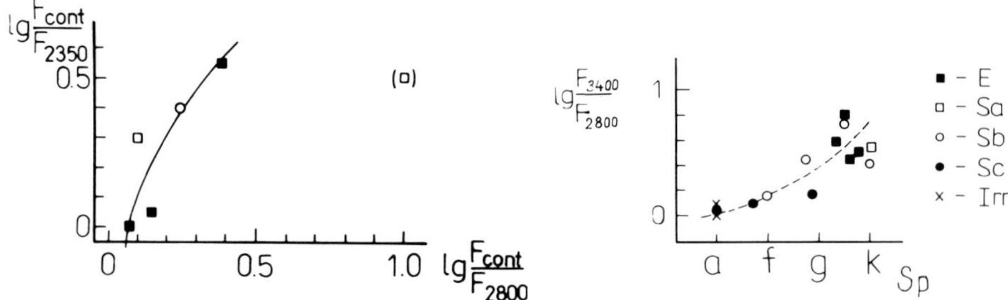

Fig.3. Comparison of λ2350 A and λ2800 A indexes for early types galaxies.

Fig.4. Dependence between spectral indexes and spectral types of galaxies.

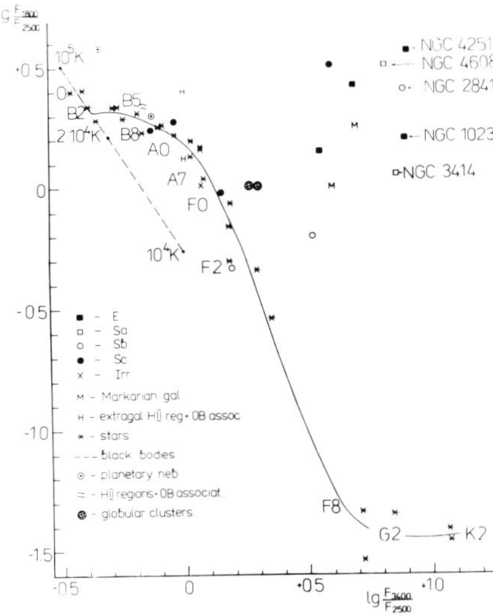

Fig.5. Two-colour diagram of central regions of 14 normal galaxies,2 Markarian galaxies,2 HII regions,containing O-B associations in spiral branches of S galaxies and of our Galaxy objects.

the galaxy and its morphological type,too. Flux increasing from λ2200 A to λ1800 A is observed as for Sc,Irr (*a-f*) galaxies so for E,Sa (*g-k*) ones. Maximum inclination of continuum in the region 1800-2200 AA is observed for E,Sa galaxies.This result was improved by two-colour diagram method.

The two-colour diagram method requires the knowledge of dust influence on colour indexes.Such diagram plotted for 65 O-A5 stars of our Galaxy,having different values of dust absorption, shows that when dust absorption increases, stars shift on two-colour diagram from left to right and down.Fig.5 gives two-colour diagram for 14 normal galaxies (E-Irr), two HII regions,containing O-B stars in S galaxies, two Markarian galaxies and objects in our Galaxy: 2 globular clusters (NGC 6397 and NGC 7078),60 unreddened O9-K2 stars,8 planetary nebulae and HII regions with the O-B associations and black bodies (10^4K≤T≤10^5K). One can see that extragalactic objects are divided in two sequences - left and right.

Left sequence consists of Sc,Irr galaxies and HII regions,containing OB associations.It coincides with the main-sequence B-F stars. The bluest galaxies of these types are located near B8 stars of main sequence. We suppose that UV emission of left sequence galaxies is caused by groups of early types, having normal luminosity function like that of O-B associations and A-B star clusters : population I stars. Radiation of its red stars was not detected in the region 1800-3500 AA.

Right sequence of galaxies on Fig.5 represents the central regions of E, Sa, Sb galaxies. Its both indexes $lg\ F_{1800}/F_{2500}$ and $lg\ F_{3400}/F_{2800}$ reach higher values than those of left sequence of galaxies. Extremely high values of

$lg\ F_{1800}/F_{2500}$ index of E galaxies are more than those of hottest stars and cannot be a result of dust influence, but only a result of stellar population. Luminosity function of right sequence objects has the gap between hot and cold stars, like that of population II stars. But it differs from the luminosity function of globular clusters of our Galaxy (see Fig.5) by high number of extremely cold stars.

Extreme right objects on diagram from top to bottom are: NGC 4251, NGC 4608, NGC 2841, NGC 1023 and NGC 3414. Index of absorption feature λ2800 A increases in this direction,showing that the main stellar population of horizontal branch of HR-diagram changes from hottest to moderate hot stars in this direction. Optical spectral classes of these galaxies are G5-K and spectral indexes $lg\ F_{3400}/F_{2800}$ are equal to those of G2-G3 stars. Main-sequence-turnoff in this case is near G3-G5 stars. Such systems are not younger than $10x10^9$ years. Model calculations give the similar ages. Dimensions of the central regions of investigated galaxies are mainly 1-6 kpc. They can evolve as closed systems.Models of such systems were calculated by Brusual (1983) and Burstein et al (1988). According to these models systems, having $lg\ F_{1800}/F_{2500} = 0.41$ (extreme right galaxies on Fig.5), are $16x10^9$ years old.

In the middle part of two-colour diagram there are three galaxies:NGC 4486,Mrk 573 and Mrk 800. In the spectra of these galaxies absorption features are weak or absent evidencing that their luminosity function in UV region caused not only by A-F, but by other early type stars,too. Their red and blue colour indexes are not so high as those of extreme right galaxies on Fig.5. We suppose that modern star formation influences on UV light of these galaxies. Galaxies,having regions of modern star formations, are known as "active". As a rule, Markarian galaxies belong to this type of objects. D.Burstein et al (1988) showed that NGC 4486 have regions of recent star formation, too, as a result of gas accretion from Virgo cluster of galaxies, nucleus of which is NGC 4486.

Full text will be published in Izv.Krimsk.Astrophys.Obs.

REFERENCES

Aiello,S.,Barsella,B.,Chlewicki,G.,Greenberg,J.,Patriachi,P., Perinotto,M.(1988),*Astron.Astrophys.*,V.**73**,P.195.

Bergh van den,S.(1960),*Publ.David Dunlap Obs.*,V.**11**,No 6.

Brusual,G.A.(1983),*Astrophys.J.*,V.**273**,P.105.

Burstein,D.,Bertola,F.,Buson,L.,Faber,S.,Lauer,T.,(1988), *Astrophys.J.*,V.**328**,P.440.

Ciani,A.,D'Odorico,S.,Benvenuti,P.(1984),*Astron.Astrophys.*, V.**137**,P.223.

Faber,S.(1983),*Highlights of Astronomy*,V.**6**,P.165.

Faber,S.,Burstein,D.,Dressler,A.(1977),*Astron.J.*,V.**82**,P.941.

Jamar,C.,Macau,D.(1974),*Astron.Astrophys.*,V.**33**,P.87.

The Hot Stellar Component in Elliptical Galaxies and Spiral Bulges

HENRY C. FERGUSON
University of Cambridge, Institute of Astronomy
and
ARTHUR F. DAVIDSEN and GERARD A. KRISS
Center for Astrophysical Sciences, The Johns Hopkins University

Abstract. Suggestions for the source of the ultraviolet flux in elliptical galaxies and spiral bulges include young stars, post-asymptotic-giant-branch (PAGB) stars, hot-horizontal-branch stars, and accreting white dwarfs. Each candidate has different implications for the spectral evolution of galaxies. We review current understanding of the origin of the far-UV flux, with emphasis on recent results from the Hopkins Ultraviolet Telescope (HUT), flown on the Astro-1 space-shuttle mission in December 1990.

The origin of the far-ultraviolet light in elliptical galaxies and spiral bulges poses one of the most serious challenges to our understanding of the evolution of the stellar populations in these systems. UV observations from the *OAO-2, IUE*, and *ANS* satellites have shown that these systems typically have an upturn in their spectra shortward of 2000 Å, with flux increasing to shorter wavelengths (see Burstein *et al.* 1988 and references therein). The galaxy surface-brightness profiles are similar, but not identical, in the optical and UV, but the strength of the upturn varies from galaxy to galaxy. Galaxies with similar optical colors can differ by up to 2.5 magnitudes at 1500 Å. Galaxies with high metallicities tend to have bluer 1550 Å$-V$ colors than those with low metallicities.

Possible contributors to the UV flux include 1) young hot stars, presumably forming as a minority population from the reprocessed gas shed from the old population; 2) post-asymptotic-giant-branch (PAGB) stars similar to the central stars of planetary nebulae; 3) hot horizontal branch (HB) stars, either from the metal-rich population that dominates the optical light, or from a minority metal-poor population or 4) accreting white dwarfs. These possibilities and variations upon them have recently been reviewed in great detail by Greggio and Renzini (1990).

From its inception in 1978, the Hopkins Ultraviolet Telescope (HUT) project (Davidsen *et al.* 1991) has had the solution, or at least the illumination, of this problem as one of its primary scientific goals. A fast focal ratio ($f/2$) and large apertures were incorporated in HUT to maximize its capability for measuring the far-UV spectra of giant ellipticals. HUT produces one-dimensional spectra from the galactic Lyman-limit to 1860 Å with a resolution of ~ 3 Å. During the Astro-1 space shuttle mission in December 1990, we used HUT to obtain long exposures of NGC 1399 and the bulge of M31, and shorter exposures of several other galaxies and globular clusters. Most of the galaxies were observed through a $9'' \times 116''$ aperture centered on the nucleus. First results for NGC 1399 are described by Ferguson *et al.* (1991) . Here we summarize these results, and present a comparison of the NGC 1399 spectrum to other spectra obtained.

NGC 1399 has the strongest UV upturn of the "quiescent" ellipticals analyzed by Burstein *et al.* (1988). The HUT spectrum of NGC 1399, binned over 10 Å, is shown in Fig. 1. To set limits on the amount of present-day star formation, we have constructed synthetic spectra using an "isochrone synthesis" technique similar to

B. Barbuy and A. Renzini (eds.), The Stellar Populations of Galaxies, 291–295.

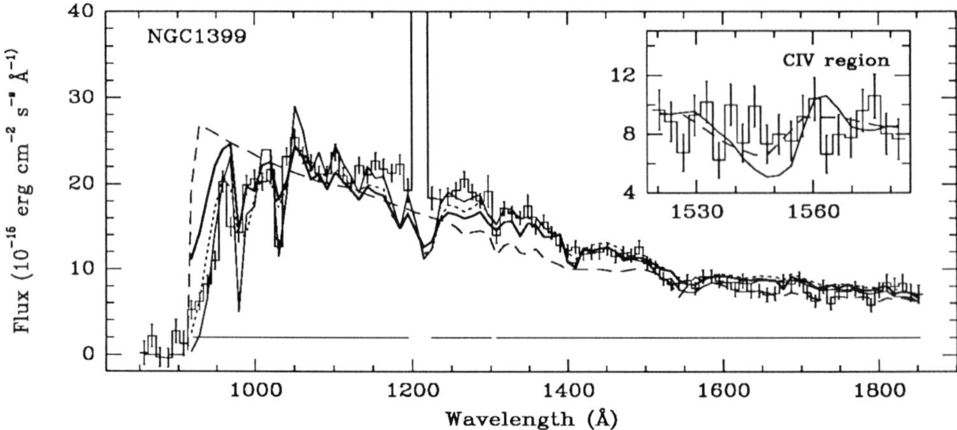

Fig. 1. HUT spectrum of NGC 1399. The histogram in the main figure is the flux-calibrated HUT spectrum, binned at 10 Å intervals. The solid line shows the best-fitting solar-metallicity Kurucz (1991) model atmosphere. The model parameters are $T_{\rm eff} = 24000$ K, and $log(g) = 4.0$. Regions used in the fit are indicated by the horizontal line at the bottom of the plot; regions contaminated by strong airglow lines were excluded. The dashed line shows the RG hot-E model at an age of 13 Gyr. The dotted line shows a model of constant star formation over 10^9 yr with an IMF slope of $x = -0.1$ and an upper mass cutoff of $M_{upper} = 20M_\odot$. Finally, the thick solid line shows a synthetic spectrum of a population of $0.546M_\odot$ PAGB stars. The inset shows a comparison of the observed C IV profile to that expected from a population of young stars. The histogram shows the data. The solid line is for a Salpeter IMF from 0.85 to $119M_\odot$ forming at a constant rate over 10^8 yr. The dashed line is the RG hot-E model (IMF slope x=1.7 with an upper mass limit of $80M_\odot$; exponentially declining SFR with timescale $\tau = 2.7$ Gyr).

that of Charlot and Bruzual (1990). Evolutionary tracks for high-mass stars from Maeder and Meynet (1988) (hereafter MM) were used, with bolometric corrections taken from Humphreys and McElroy (1984) and Flower (1977). Constraints on the young star population can be set by considering the entire spectral-energy distribution (SED), or just the region near C IV $\lambda 1550$, the strongest absorption line in the hot stars that should be contributing. Comparison with the SED has the danger that it must rely on stellar atmosphere models (Kurucz 1991) that may not match the metallicity of the population, and that are in any case poorly tested below Lyα. Furthermore, the uncertainties in extinction and the absolute HUT calibration are greater shortward of the IUE spectral range. On the other hand, the comparison near C IV, while mostly free from uncertainties in extinction and calibration, is hampered by low signal-to-noise and the possibility that the IUE spectral library (Heck *et al.* 1984) used to construct the synthetic spectrum does not match the metallicity of the stellar population.

Figure 1 (inset) shows the region around C IV compared to the C IV profile expected from a population of stars forming at a constant rate over 10^8 yr with a Salpeter IMF from $0.85M_\odot$ to $119M_\odot$. This model is excluded at the 99.9% confidence level by a χ^2 comparison to the data. A 2σ upper limit is that stars from this model can be responsible for no more than 50% of the 1550 Å flux (less if the component producing the rest of the light also has intrinsic C IV absorption). The dashed line shows the RG hot-E model, which is excluded at the 95% confidence level.

Comparison of the NGC 1399 SED to individual Kurucz (1991) solar-metallicity models suggests that stars cooler than $\sim$ 25000 K are the dominant contributors to the UV upturn. The best-fit model is shown as a solid line in Fig. 1. This temperature limit, set mostly by the turnover below 1000 Å, is consistent with the absence of C IV absorption. Stars hotter than $\sim$ 28000 K would probably have detectable C IV, especially if they are metal rich. The shape of the SED limits the number of massive stars that can be evolving at present along the main sequence in NGC 1399. The dashed line in Fig. 1 shows that the RG hot-E model far overshoots the continuum below 1000 Å. A lower upper-mass cutoff and flatter IMF slope are required to fit the HUT spectrum. The best-fit constant-star-formation model (shown as a dotted line in Fig 1) has $M_{upper} = 20 M_{\odot}$ and an IMF slope $x = -0.1$; however, such a sharp upper-mass cutoff on an IMF that is weighted toward massive stars seems implausible.

The turnover near the Lyman limit also imposes a serious constraint on PAGB models, since these stars are typically much hotter then 25000 K. To test the PAGB hypothesis, we constructed synthetic spectra of single-mass populations of stars evolving along the tracks given by Schönberner (1987). Solar-metallicity model atmospheres from Kurucz (1991) and Clegg & Middlemass (1987) were used to build the spectrum in the HUT bandpass. The resulting (best-fit) model for 0.546 $M_{\odot}$ stars is shown as a heavy line in Fig. 1. The match to the data becomes progressively worse for higher mass PAGB stars. If PAGB stars produce the UV upturn, then either they do not evolve along the Schönberner tracks, or the model atmospheres used to construct the synthetic spectrum are inappropriate. Once again, the lack of detectable C IV bolsters our confidence in the constraints derived from the SED; high-temperature metal-rich PAGB stars with enough of an envelope left to produce the Lyman lines and C III λ1175 (easily seen in the raw data) would probably produce detectable C IV. The most consistent interpretation of the data is that cooler stars are responsible for the UV upturn in this galaxy.

Further constraints on the UV stellar population may come from comparing the SED's of different galaxies. We have made a first cut at this for the low S/N spectra obtained of the centers of NGC 1316, and M 81, and the high-quality spectrum of M 31. In M 81, the continuum flux drops gradually from 1200 Å to the Lyman limit, probably due to internal extinction. The NGC 1316 SED is quite similar to NGC 1399's, but a quantitative comparison will require careful modelling of the contribution from scattered geocoronal Lyα. Figure 2 shows the ratio of the M 31 flux to the NGC 1399 flux as a function of wavelength. M 31 has significantly more flux than NGC 1399 near the Lyman limit. It is unlikely that errors in extinction can explain this difference, unless the extinction curve is radically different from that measured in our own galaxy. We conclude that the stars producing the strong UV upturn in NGC 1399 are not just "more of the same" type of stars that produce the upturn in M 31. Hotter stars are allowed in M 31, and, while we have not yet fit models in detail, it is possible that low-mass PAGB stars could contribute significantly. The population producing the *excess* flux in NGC 1399 must be significantly

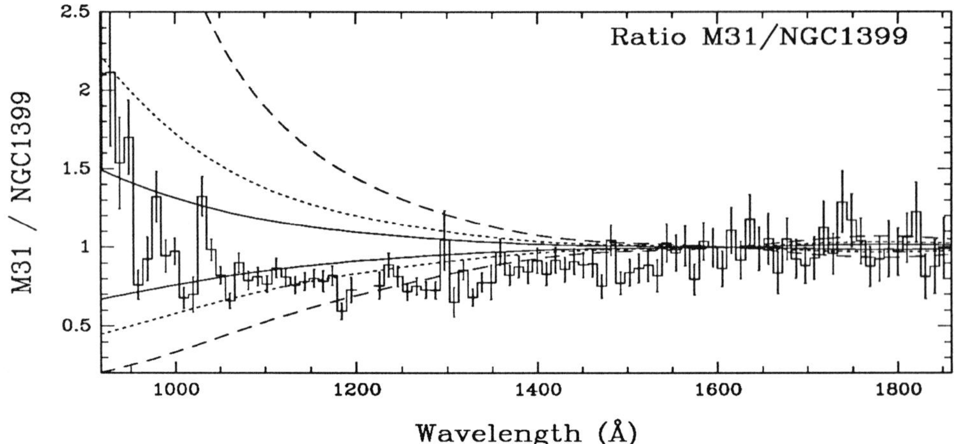

Fig. 2. The ratio of the M 31 spectrum to NGC 1399, normalized such that the mean is 1.0 between 1500 and 1700 Å. The M 31 spectrum has been corrected for galactic extinction, but not for internal extinction or contamination by disk light (both of which are probably small effects). The curves that bend upwards at short wavelengths show the deviation expected from a constant value of 1.0 if NGC 1399 were subject to extinction following the galactic curve of Longo *et al.* (1989). The solid, dotted, and dashed lines correspond to reddenings of $E(B-V) = 0.05, 0.1$, and 0.2 mag, respectively. The curves that bend downwards show the ratio expected if M 31 were subject to additional extinction. The fact that none of the extinction curves match the shape of the M 31/NGC 1399 ratio suggests the excess flux near the Lyman limit in M31 is due to hotter stars, rather than errors in the extinction correction.

cooler, and is unlikely to be PAGB stars or young stars, for the reasons discussed above.

Acknowledgements

The HUT project is supported by NASA contract NAS5-27000 to the Johns Hopkins University.

References

Burstein, D., Bertola, F., Buson, L. M., Faber, S. M., & Lauer, T. R. 1988, ApJ, 328, 440
Charlot, S. & Bruzual, G. A. 1990, STScI preprint no. 452
Clegg, R. E. S. & Middlemass, D. 1987, MNRAS., 228, 759
Davidsen, A. F., et al. 1991, In preparation
Ferguson, H. C., et al. 1991, ApJ, submitted
Flower, P. J. 1977, A&A, 54, 31
Greggio, L. & Renzini, A. 1990, ApJ, 364, 35
Guiderdoni, B. & Rocca-Volmerange, B. 1987, A&A, 186, 1
Heck, A., Egret, D., Jaschek, M., & Jaschek, C. 1984, A&AS, 57, 213
Humphreys, R. M. & McElroy, D. B. 1984, ApJ, 284, 565
Kurucz, R. L. 1991, CfA preprint no. 3181
Longo, R., Stalio, R., Polidan, R. S., & Rossi, L. 1989, ApJ, 339, 474
Maeder, A. & Meynet, G. 1988, A&AS, 76, 411
Schönberner, D. 1987, in Planetary Nebulae, ed. S. Torres-Peimbert (London: IAU), 340

Discussion

Gregg: NGC 1316 is known to have a lot of dust, so how do you explain the similarity of its UV spectrum and that of NGC 1399?
Ferguson: NGC 1399 is a magnitude fainter in B than NGC 1316, but its 1400 Å flux is about factor of 4.5 brighter, so I think NGC 1316 probably is affected by extinction. The fact that the UV spectral energy distributions are similar suggests that the intrinsic slope of the upturn may be steeper in NGC 1316, but I should caution that the spectral-energy-distribution I showed was based on a very low S/N spectrum and the airglow contribution has not yet been carefully modeled.
Whitelock: I don't think you should reject the accreting WD hypothesis on the basis of supernova rates. Accreting symbiotic WD's within the galaxy undergo periodic shell flashes and eject much of the material they have accreted. They may therefore not be SN progenitors.
Pinsonnealt: We have computed evolutionary models of metal-rich HB stars. Their high metallicity and helium accelerates H-burning, causing them to skip the AGB phase and become hot stars on the He-burning main sequence. Because these stars are both fainter ($\sim 100 L_{\odot}$) and long lived ($\sim 10^7$ years) their properties differ from the post-AGB models you have discussed. These results depend on age, metallicity and $\Delta Y/\Delta Z$, and will also affect the integrated colors by removing the AGB light.

GALAXIES AT INTERMEDIATE REDSHIFTS

RICHARD S. ELLIS
Physics Department, Durham University
Durham DH1 3LE, England

ABSTRACT: Faint object spectroscopy and deep infrared imaging is providing exciting opportunities to understand the astrophysical nature of normal galaxies up to ≃8-10 Gyr ago (H_o=50). Deep photometry of 10 distant clusters with 0.5< z <1 shows a systematic bluing of *red members* with look-back time, in good agreement with the view that the bulk of the early type population must be genuinely old (z_f >2) and remarkably homogeneous. Statistically complete redshift surveys of the highly abundant population of faint blue field galaxies, however, indicate sizeable changes must have occurred in the galaxy luminosity functions as recently as z≃0.5. Possible explanations include widespread merging of star-forming sub-units of present day galaxies, or an entirely new population of sources whose present-day remnants are intrinsically faint. The contrast between the long term evolution of the spheroidal population and that seen in the faint counts is striking. I finally discuss some new opportunities to identifying field galaxies with z>1 using gravitational lensing and QSO absorber identifications.

1. INTRODUCTION

My aim is to bridge the gap between the detailed studies of local stellar populations presented elsewhere in this symposium and the truly high redshift objects (QSOs and radio galaxies) whose present-day counterparts remain uncertain. We wish to use redshifted sources to deduce the evolutionary properties of *normal* galaxies studied locally. Since such galaxies cannot be readily identified beyond redshifts $z \simeq 1$ with 4-m class telescopes, I will define the 'intermediate redshift' range to be 0.1< z <1.

In examining the evolutionary properties of moderate z galaxies, we must recognise the severe limitations of the unresolved data compared to the detailed studies of nearby stellar systems. We need to ask simpler questions and aim for *statistical* conclusions. We must also realise that our selection procedures may play a critical role in determining the visibility of various kinds of sources. It is all too easy to compare a faint dataset with a bright counterpart and falsely attach differences that are purely induced by selection to some evolutionary process. Furthermore, whilst there is a temptation to construct detailed theoretical models of the data based on preconceived ideas of how local galaxies might have evolved (e.g. as self-contained systems parameterised by mass functions, star formation histories and luminosity functions), it is hardly surprising, given the uncertainties, that self-consistent solutions usually emerge. There have been enough surprises in recent faint galaxy data to warn us against such a detailed approach. At this stage, it is preferable to adopt an empirical analysis of the data with the minimum number of parameters.

B. Barbuy and A. Renzini (eds.), The Stellar Populations of Galaxies, 297–309.

I will focus on two topics which serve to highlight very contrasting views of the star formation history of normal galaxies. Firstly, local and distant *cluster* data reveal a remarkable homogeneity and synchronised evolution for luminous early-type galaxies. To first order, ellipticals galaxies appear to be passively evolving old systems. In the low density *field* however, a plethora of faint blue galaxies are found whose redshifts are surprisingly local, indicating changes in the galaxy luminosity function on much shorter timescales.

2. STAR FORMATION HISTORY OF LOCAL ELLIPTICALS

We begin with a simple question: did luminous elliptical galaxies form the bulk of their stars in a single burst at a well-defined epoch (i.e. coeval origin) and, if so, when was this epoch of principal star formation?

As a local benchmark, consider the colours of ellipticals in the contrasting environments of the Coma and Virgo clusters. Earlier studies of the homogeneity of the colour-luminosity (c-L) relation gave conflicting results. Sandage & Visvanathan (1977) claimed their universal $u-V$ relations demonstrated a simple evolutionary picture for early-types with little environmental dependance. On the other hand, Aaronson *et al* (1981) found the relative distance modulus of Coma and Virgo as determined from the $u-V$ c-L relation differed significantly from that using $V-K$. The implication was that Virgo ellipticals are redder in $V-K$ by as much as 1.0±0.3 magnitudes. Aaronson *et al* suggested that an asymptotic giant branch contribution in Virgo arising from recent star formation might account for the result.

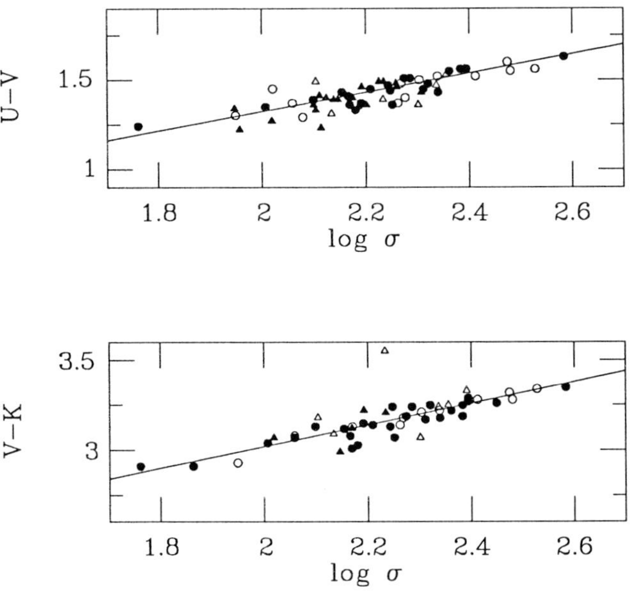

Figure 1: $U-V$ and $V-K$ vs velocity dispersion σ, for early-type galaxies in the Virgo (open symbols) and Coma (filled symbols) clusters from the survey of Bower, Lucey & Ellis (1991). Ellipticals are denoted by circles, S0s by triangles. The solid line is the best fit relation for the combined dataset.

We have recently completed a precision photometric survey of those 21 Virgo E/S0s and 50 Coma E/S0s with reliable velocity dispersions (σ) to re-examine this question (Bower, Lucey and Ellis 1991). By using colour - σ relations we can work independently of distance. The $U-V$ and $V-K$ vs σ relations are reproduced in Figure 1 for the ellipticals and S0s; colours refer to a $5h^{-1}$ kpc aperture and have been corrected for redshift and reddening effects. Adopting a common slope for each colour-σ relation, no significant colour offset is found between Virgo and Coma ($< 0.^m01$). Bower, Lucey & Ellis claim Aaronson *et al*'s result probably arose from their use of inhomogeneous photometry together with a systematic error in the K photometry of their faintest Coma galaxies.

More interestingly, however, is the remarkably small rms scatter found about these relations for the ellipticals. In Coma, the scatter, $\delta_{(U-V)}=0.^m035$, is consistent with observational error. Since the bulk of the U light arises from main sequence stars (c.f Buzzoni 1989), the homogeneity in colour can be converted into physical constraints on the age of single burst ellipticals in different environments. The time evolution of $U-V$ is given straightforwardly for any mass function from the main sequence lifetimes. Adopting a scatter of $0.^m04$, we can write:

$$\frac{\partial(U-V)}{\partial t} \leq \frac{0.^m04}{\beta(t_H - t_f)}$$

where t_H is the Hubble time, t_F is the look-back time to the initial burst, and β is a 'synchronicity' parameter. β=1 would correspond to no coordination of galaxy formation within the time interval $t_H - t_F$, $\beta < 1$ to coordinated formation. If there is no coordination between forming galaxies ($\beta \simeq 1$), ages of >13 Gyr are derived. Even if β=0.3, ages exceed 10 Gyr. Ages of less than 6-7 Gyr are only possible if $\beta \simeq 0.1$, i.e. remarkably synchronous formation. These conclusions are not sensitive to the adopted slope of the initial mass function. Whilst the age *scales* are metallicity-dependent, ellipticals appear to be a homogeneous population across different environments, and whose main star formation probably occurred before z$\simeq$2.

3. EVOLUTION IN DISTANT CLUSTER GALAXIES

Do we see the systematic colour evolution expected if the local ellipticals were produced by single bursts prior to z$\simeq$2? By z=0.7, we are probing look-back times of $\simeq$4h^{-1} Gyr where significant changes might be detected. Clusters of galaxies beyond z=0.5 have been compiled by Gunn *et al* (1986) and Couch *et al* (1991), and Aragón-Salamanca and I have recently completed an optical and infrared photometric survey of 10 such clusters whose results can be compared directly with those of Figure 1 (for a preliminary discussion see Aragón-Salamanca 1991).

Galaxies in each cluster were selected from deep K images since calculations (Aragón-Salamanca 1991) show that optical selection beyond $z \simeq 0.5$ would lead to an artificially-increased proportion of blue galaxies. At z=0.75, for example, the R band samples rest-frame U, which is sensitive to young hot stars (§2) whereas K samples rest wavelengths above 1μm. Figure 2 shows the most distant cluster in the sample. To a strict K magnitude limit, $V-K$ and $I-K$ aperture colours are determined for typically $\simeq$20 galaxies per cluster of which $\simeq$16 are expected to be members. Field contamination becomes more serious for the more distant clusters and for bluer colours, but to the precision required, it is sufficient to subtract appropriate K-limited field colour distributions from Cowie *et al* (1991)'s VIK deep survey.

The clusters group naturally in 3 narrow redshift intervals at $\overline{z_i}$=0.56, 0.70 and 0.86. Accordingly, we reduce the individual galaxy colours to those appropriate at a fixed luminosity (i.e. corrected for the c-L slope) and make minor corrections to bring the cluster to the closest of the 3 mean redshifts. The present-day 'no-evolution' prediction at that $\overline{z_i}$ is made by interpolating between appropriate rest-frame colours in local Virgo+Coma+field datasets.

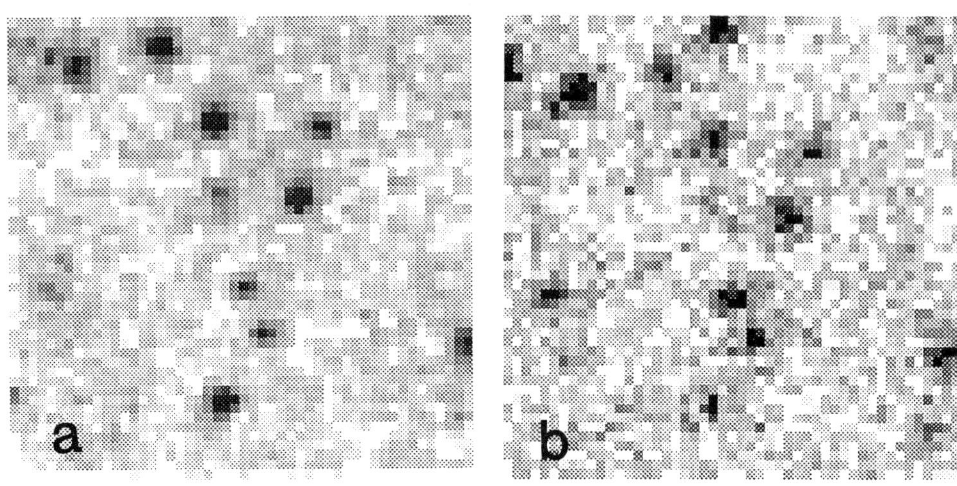

Figure 2: (a) K image of the distant cluster 1603+4329, z=0.92 secured with the IRCAM InSb array on the 3.8m UK Infrared Telescope, (b) corresponding I image using a EEV CCD on the 2.5m Isaac Newton telescope.

The $V-K$ colour distributions, before and after field subtraction, are shown for the 3 $\overline{z_i}$ samples in Figure 3; $V-K$=0 is defined to be the present-day expectation at that redshift. Not only does the mean colour move systematically blueward with redshift, but there is a complete absence of galaxies with present day colours beyond z≃0.7, i.e. we see clear evolution of the ridge-line in the c-L diagram. This observations is distinct from an increasing fraction of blue galaxies discovered in lower z clusters by Butcher & Oemler (1978). Unlike the B-O effect, imprecise field subtraction *cannot* explain the paucity of red cluster members.

A similar evolutionary phenomenon was reported by Dressler & Gunn (1990) for a similar sample using the spectroscopic discontinuity at 4000 Å , D_{4000}. Their sample is optically-selected, however, and therefore less reliable for quantitative work because of the 'UV bias' discussed above. However, it is satisfying to note an excellent one-to-one correlation between our $V-I$ colours and Dressler & Gunn's D_{4000} where common data has been compared.

The colour evolution expected by z≃1 in the single burst case (e.g. Bruzual's (1983) c-models with a burst of 1 Gyr) depends on the epoch of the burst, z_F, and the cosmological parameters, H_o and q_o. For H_o=50 and q_o=0.5, the observed evolution of $\Delta(V-K) \simeq 1.^m4 \pm 0.2$ by z≃0.9 suggests a high $z_F \geq$5-10, although formally $z_F \simeq 2$ is allowed if q_o is small. Of course, if the star formation era is more extended than 1 Gyr, z_F must generally be pushed higher to maintain the same evolutionary trend. The available data on elliptical galaxies at $z \simeq 0$ and to $z \simeq 1$ are *both* consistent with the simple picture of a single burst of star formation at $z > 2$ and the subsequent passive evolution of that stellar population.

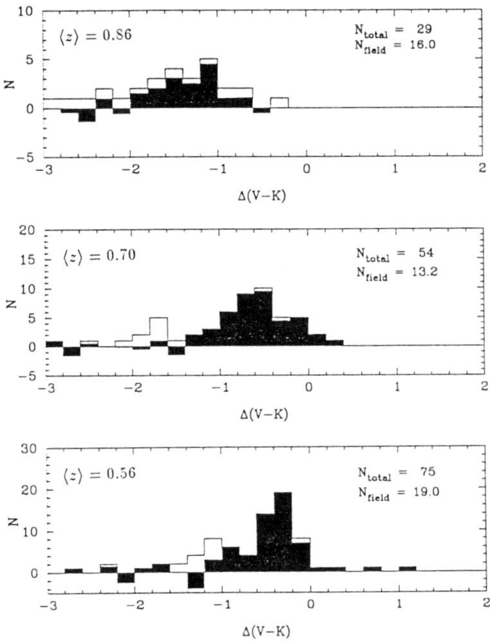

Figure 3: $V - K$ distributions for distant cluster galaxies, with $\Delta(V - K)=0$ defined for a non-evolving present day elliptical. Each panel shows the colour distribution for 2 or more clusters reduced to the indicated common redshift. Shading shows the effect of subtracting appropriately-scaled field data from Cowie et al (1991).

4. RECENT STAR FORMATION IN ELLIPTICALS?

The single burst hypothesis for spheroidal populations has been challenged by evidence for recent (<3-5 Gyr) star formation in certain ellipticals and bulge populations (e.g. O'Connell 1980, Pickles 1985). For example, Bruzual (1984) showed that whilst a 12 Gyr c-model evolutionary model reproduces the observed spectral energy distribution (SED) for $\lambda \geq 3200$ Å , there is flux shortfall at shorter wavelengths. The 'extra' component could be due to recent star formation or a missing evolved component in the models; the dilemma has been reviewed by Burstein *et al* (1988), Greggio & Renzini (1990). Longward of 2000 Å the mean SED is fairly well-determined amongst ellipticals, so any recent star formation would be a fairly widespread phenomenon. A 12 Gyr μ=0.8 declining star formation model fits the data but would indicate a greater amount of colour evolution by $z \simeq 1$ than observed.

On the other hand, Rose (1985) and Bower *et al* (1990) have demonstrated how diagnostic Sr II, Fe I and Hδ absorption lines can determine the mean surface gravity of a stellar population *independently of metallicity and reddening.* By studying representative ellipticals in various environments (rich clusters, groups and isolated fields), they find ellipticals in low density environments contain a stellar component $\simeq$ 6-7 Gyr younger than those found in the rich cluster counterparts. The separation of Virgo and Coma ellipticals on Rose's Sr/Fe/Hδ plane (Figure 4) demonstrates this but is in marked conflict with the identical integrated $U - V$ galaxy colours discussed earlier. It may be that the intermediate age component is strictly confined to the cores of the ellipticals where the spectral signal is sampled, i.e. a strong age gradient exists. Alternatively there could a metallicity difference between the two samples which compensates for the age difference in the integrated colours. Neither explanation is particularly convincing, however.

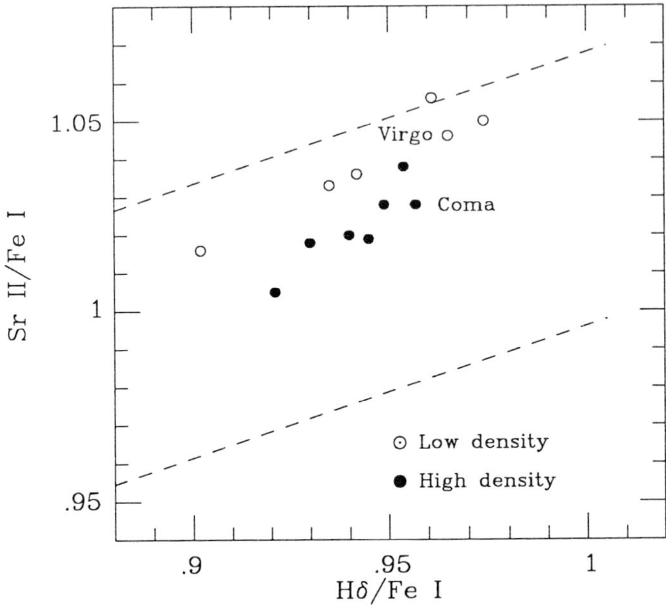

Figure 4: Sr II-Fe I-Hδ plane for ellipticals in various environments using preliminary data from Rose et al (1991) strengthening the environmental dependence of recent star formation claimed by Bower et al (1990). Dashed lines indicate the sequence for dwarf (upper) and giant (lower) Galactic stars.

Further evidence for recent activity in cluster galaxies includes the many manifestations of the Butcher-Oemler effect reviewed by Dressler (1984) and MacLaren *et al* (1988). There are blue members undergoing a short-term burst of star formation (Dressler & Gunn 1983); red members revealing an excess of UV flux (Ellis *et al* 1986) or post-starburst spectral features (Couch & Sharples 1988), and possibly red galaxies with an infrared excess attributable to an AGB population (Aragón-Salamanca *et al* 1991). Modest bursts of star formation draw our attention to what appear to be dramatic evolutionary differences at z$\simeq$0.2–0.5. A secondary burst at z$\simeq$0.3, whose strength is $\simeq$10% of that at z_F, would explain the moderate z cluster data without destroying the small $U-V$ scatter seen in local systems (Figure 1). Thus, these phenomena, whilst important in their own right, should not deter from understanding the overall trends defined by the high z cluster data. Additionally, in the local data, continuing star formation appears to be confined to galaxies in the lower density systems.

5. EVOLUTION IN FIELD GALAXIES

The gradual evolution witnessed for the spheroidal population in clusters contrasts with events occurring in the field at surprisingly recent times. Many reviews (e.g. Ellis 1990) have addressed the significant excess population of faint galaxies whose mean colour becomes bluer with increasing apparent magnitude. At b$_J$ ≃24, the observed surface density of galaxies in randomly-chosen fields exceeds the 'no evolution' prediction (based on local luminosity functions) by a factor ≃6 and the median colour of $b_J - R \simeq 0.6$ (Tyson 1988) corresponds to that of the bluest Sm/Irr galaxies observed today (irrespective of redshift since the optical SED is flat). The predominantly blue faint sky undoubtedly indicates much star formation and perhaps a significant contribution to the present-day metallicity (Cowie 1988).

Whilst it is difficult to constrain the redshifts of the faint blue galaxies from photometry, the absence of any U band drop-out for the faintest detected in b_J and R suggests virtually all have z<3 if the Lyman limit is as strong as seen in the QSO population (Guhathakurta *et al* 1991). Regardless of any evolution, the sheer number of faint galaxies within such a volume then presents a challenge to the inflationary (Ω=1, Λ=0) cosmological model if numbers are conserved (Yoshii & Takahara 1988). Resolving the faint galaxy problem is thus an outstanding question of modern cosmology.

Great strides in this area have been made in the last 5 years via multiple object spectroscopy on 4-m telescopes. Statistically-complete redshift surveys have been published by Broadhurst *et al* (1988, 20< b_J <21.5), Colless *et al* (1990, 21< b_J <22.5) on the AAT and deeper surveys are in press to b$_J$=24 (Colless *et al* 1991, Cowie *et al* 1991). Figure 5 shows the available redshift-magnitude data from these surveys. I have also included an unpublished intermediate depth (17< b_J <20.5) survey by Broadhurst & Ellis which aids greatly in delineating the general trend.

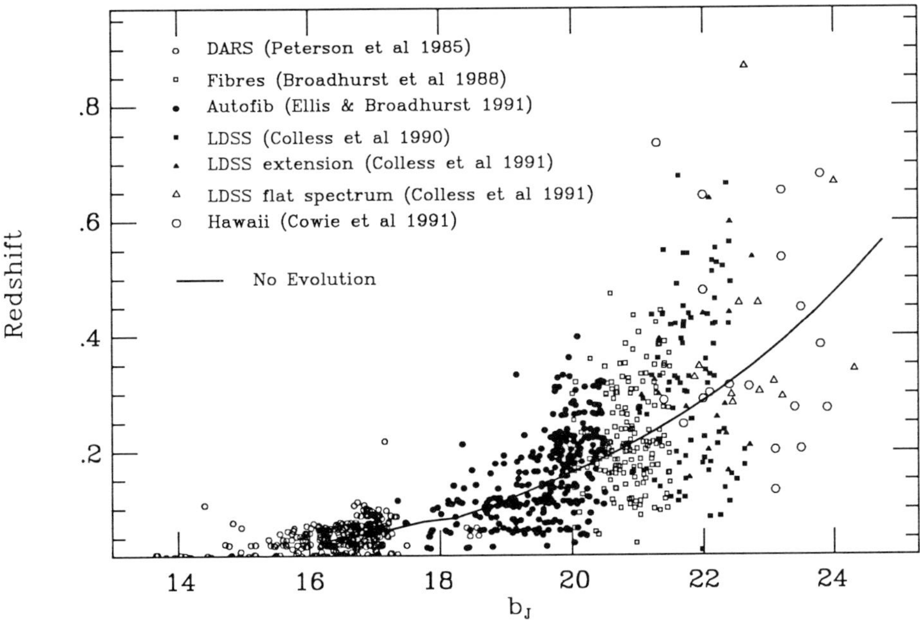

Figure 5: Redshift vs. b_J magnitude for recently completed faint galaxy surveys (see legend for details). The solid line represents the mean redshift at each b_J magnitude in the no evolution case; this evidently fits the data rather well despite the excess number seen in the counts beyond b_J ≃21.

Broadhurst *et al* (1988) demonstrated that the faint excess cannot reasonably arise from an underestimated faint end slope of the galaxy luminosity function since the mean redshift, $\bar{z}$, of their survey to b_J =21.5 is too high. Likewise Colless *et al* (1990) eliminated a high z tail in their distribution to b_J=22.5 as a possible source of the extra blue galaxies. Recently, Colless *et al* (1991) reduced the incompleteness of a portion of their earlier survey via additional observations (included in Fig. 5) from 19% to 7% with no change to their $N(z)$ distribution despite extending their [O II] spectral sensitivity to z=1. Although 7% of the b_J <22.5 population could still strictly lie beyond z$\simeq$1, $N(z)$ would now have to be bimodal. For the monotonic $N(z)$ decline expected in most models, the likely evolution in the bright end of the luminosity function since z$\simeq$1 cannot be significant. The Hawaii survey (Cowie *et al* 1991, 22.5< b_J <24) contains 13 redshifts deeper than Colless *et al*'s (1990) sample but $\bar{z}$ is barely increased providing valuable confirmation of the trend established in the AAT surveys (see Fig. 5).

The most significant new result, perhaps, is provided by Colless *et al*'s (1991) survey of objects with 22< R <23 whose $b_J - R$ and $R - I$ colours indicate the source to be very blue or *flat-spectrum* (f_ν ~const.). Tyson (1988) and Cowie & Lilly (1990) argued that beyond b_J $\simeq$23-24, a dominant population of much bluer objects are found that may differ from the z$\simeq$0.3 sources identified in brighter surveys. Of 8 flat spectrum sources surveyed with LDSS at the AAT, 7 redshifts yield $\bar{z}$ =0.43. Continuity of redshift and colour trends with b_J is a very important feature. Broadhurst, Ellis & Glazebrook (1991) demonstrate, by measuring [O II] equivalent widths for most of the sample plotted in Figure 5, that the *slope*, $\gamma = d \log N / dm$, of the galaxy counts is distinctly steeper for the strong star forming population over 17< b_J <22.5, than for the quiescent population which matches the no-evolution prediction. This demonstrates rather convincingly that it is these star forming galaxies, examples of which occur at apparent magnitudes at bright as b_J $\simeq$20-21, that provide the count excess.

6. PUZZLES AND POSSIBLE SOLUTIONS

The field surveys provide three puzzles. Firstly, increased star formation in any subset of the current luminosity function (LF) should imply a greater average luminosity for those objects (even if temporary) and hence an increased depth in any magnitude-limited survey. Yet the redshift distribution is similar to the no-evolution prediction. In short, there is an apparent increase in the comoving volume density of luminous galaxies with redshift.

Secondly, whilst the integrated number of faint galaxies cannot be reconciled with the present day space density and the volume bound by z <3 for the inflationary Ω=1, Cowie's (1991) K counts present no such dilemma and indicate no large excess over no evolution to K=22.

Finally, the absence of *any* high z luminous precursors to normal L* galaxies will soon become an embarrassment. The apparent magnitude at which one expects to see beyond z$\simeq$1 depends on the evolutionary model. For example, Tinsley's (1977) pioneering work predicted primaeval galaxies at z_F $\simeq$2 undergoing strong initial bursts would appear at B$\simeq$20. Likewise, Koo's (1990) mild evolutionary models predicts some z>1 flat-spectrum objects would be seen with b_J $\simeq$23. The absence of *any* high z objects other than QSOs and radio galaxies to b_J=24 is beginning to place a strong constraint on the likelihood of any luminous phase for normal sources unless this occurred at very high redshift.

Two explanations have been offered for the above paradoxes (earlier suggestions of a $\Lambda \neq 0$ cosmology are now inconsistent with the K counts). Broadhurst, Ellis & Glazebrook (1991) consider the faint blue galaxies to be precursors of the normal population and suggest they represent a gas-rich merging population whose star formation rate (SFR) declines with time. Looking back, a typical galaxy breaks into fragments whose optical luminosity is governed by the increasing SFR, but whose K luminosity declines in proportion to the decreasing mass. They show how it is possible to match both the b_J and K counts and the available redshift data in a simple model. Merging has the important effect of maintaining the average blue luminosity despite the smaller mass, and hence the increasing numbers are not accompanied by a high z tail. The model also explains the absence of high redshift L* galaxies in a natural way and predicts that a K-selected sample would yield a redshift distribution whose $\bar{z}$ is *less* than the no evolution value. Self-similar merging is a feature of standard cold dark matter models explaining many present day observations (e.g Frenk *et al* 1990) . Additionally, IRAS has shown merging occurs in low density environments (Soifer *et al* 1986) at a rate that appears to increase with redshift (Saunders *et al* 1990). The main difficulty is in reconciling the abundance of present day spirals with fragile disks which would be affected merging that extends to recent epochs (Ostriker 1990).

Alternatively, Babul & Rees (1991, see also discussion in Cowie 1991, Cowie *et al* 1991) invoke an entirely new population of star forming dwarfs which undergo spectacular star formation at modest $z(\simeq 1)$ but then decay beyond detection (or possibly to dE galaxies) by the present epoch. Detailed predictions are not possible because the new population is not physically constrained before or after its luminous phase. Although an *ad hoc* solution, we can place some constraints on an *additional* population from the counts alone. If the dwarfs are added to an unchanging mass function of normal galaxies, the $\times 2$ excess number over no evolution at $b_J \simeq 22.5$ and the absence of any excess at $K \simeq 20$ enables us to conclude that the average dwarf must have $b_J - K < 2.5$. But the median $b_J - K$ colour at K=20 is $\simeq 5$ (Cowie *et al* 1991). Thus it is difficult to reconcile the photometric data with a 2-component population. The merger model provides an elegant solution to this remarkably simple problem. Instead of *adding* sources to the no evolution prediction, merging and star formation are in effect *transforming* the counts. In K there is a very rapid redistribution to fainter magnitudes since the luminosity fades more quickly with redshift: the excess seen in b_J only appears at K magnitudes well beyond the current counting limits.

The two hypotheses are radically different in their implications. The star forming mergers picture implies the optically faint sky is telling us about remarkably recent dynamical evolution of galaxies like our own. The dwarf picture would, however, make it much harder to determine the history of L* galaxies since the faint optical sky rapidly becomes confused by local events. Regardless of which is correct, however, it is clear there must have been significant changes in the field galaxy luminosity function since z$\simeq$1.

7. OTHER PROBES OF HIGH REDSHIFT GALAXIES

In view of the above, it is important to consider alternative ways of finding high redshift field galaxies. If the merger picture is correct, one expects a paucity of luminous galaxies, especially at infrared wavelengths. Two methods appear promising: studies of gravitationally-lensed arcs and QSO absorption lines.

Soucail *et al* (1988) demonstrated beautifully how the giant arc in Abell 370 (z=0.37) is the single gravitationally-lensed image of a background z=0.72 galaxy. Since an arc image is produced only because a field galaxy happens to lie behind an unrelated foreground cluster, one might take such arcs as a serendipitous field sample. With the Toulouse group, Aragón-Salamanca, Smail & I have determined $B-R$ and $R-K$ colours for the sample of known arcs with spectroscopic data. Those with $z_{arc} < 1$ appear to be normal spirals, whereas those with $z_{arc} > 1$ are extraordinarily blue. The drawback of the lensing picture is that it tells us little about the *luminosity* of the source. Additionally, surface brightness dimming could play nasty tricks. At large z we might be restricted to examining high surface brightness *portions* of galaxies which would reflect an untypically high SFR: such blue colours might not that surprising.

More promising perhaps is the examination of matter at high redshift via its absorbing effect on distant QSOs. Bergeron & Boissé (1991) claim that the bulk of z_{abs} <1 Mg II systems are luminous galaxies with strong star formation. With Bergeron, we are searching for CIV absorbers with 1.2< z_{abs} <2 via deep *RIK* imaging, paying particular attention to absorbing systems *clustered* in redshift in order to maximise detections. An L* galaxy at z$\simeq$1.5 has K$\simeq$19-20 regardless of type and should be readily visible in a modest integration on UKIRT. Preliminary results indicate that those candidate objects lying close to the QSO are *fainter* than L^* in K even at the absorber redshift (and thus considerably fainter if foreground). Many of these candidates have colours typical of flat-spectrum galaxies. The preliminary results agree well with the star-forming mergers hypothesis.

8. CONCLUSIONS

There are no convincing conclusions. Our view of the intermediate redshift Universe is in turmoil. On the one hand, cluster galaxies are evolving passively on timescales of 5-10 Gyr as predicted by the most simple evolutionary models we can imagine. On the other hand, the field galaxy luminosity function has been transformed dramnatically in the past 3-5 Gyr. A consistent picture must weave together the old spheroidal population with a remarkably recent era of disk and metal formation. The past few years has demonstrated well how difficult observations acquired with large telescopes and good instrumentation have defined the problems and how theory has squirmed to accommodate the results. To me this seems an ideal state of affairs.

Acknowledgements

I thank my colleagues Alfonso Aragón-Salamanca, Tom Broadhurst, Matthew Colless, Karl Glazebrook and Jacqueline Bergeron for allowing me to quote results determined with their help. I also thank the organisers for their generous travel funds which enabled me to participate in this enjoyable meeting.

REFERENCES:

Aaronson, M, Persson, S E & Frogel, J A 1981 *Ap. J.*, **245**, 18.
Aragón-Salamanca, A 1991, Ph.D. thesis, University of Durham.
Aragón-Salamanca, A, Ellis, R S & Sharples, R M 1991, *MNRAS*, **248**, 128.
Babul, A & Rees, M J 1991, preprint.
Bergeron, J & Boissé P 1991 *Astron. Astr.*, **243**, 344.
Bower, R G, Ellis, R S, Rose, J & Sharples, R M 1991, *Astr. J.*, **99**, 530.
Bower, R G, Lucey, J R & Ellis, R S 1991 *MNRAS*, in press.
Broadhurst, T J, Ellis, R S & Shanks, T 1988 *MNRAS*, **235**, 827.
Broadhurst, T J, Ellis, R S & Glazebrook, K 1991 *Nature* in press.
Bruzual, G 1983 *Ap. J.*, **273**, 105.
Bruzual, G 1984 in *Evolution of Galaxies*, ed. Gondhalekhar, P, RAL Publications, RAL 84-003.
Burstein, D, Bertola, F, Buson, L M, Faber, S M & Lauer, T *Ap. J.*, **328**, 440.
Butcher, H & Oemler, A 1978 *Ap. J.*, **219**, 18.
Buzzoni, A 1989 *Ap. J. Suppl.*, **71**, 817.
Colless, M M, Ellis, R S, Taylor, K & Hook, R N 1990 *MNRAS*, **244**, 408.
Colless, M M, Ellis, R S, Broadhurst, T J, Taylor, K & Peterson, B A 1991, in preparation.

Couch, W J & Sharples, R M 1988 *MNRAS*, **299**, 423.
Couch, W J, Ellis, R S, Malin, D F & MacLaren, I 1991 *MNRAS*, **249**, 606.
Cowie, L L in *The Post-Recombination Universe*, ed. Kaiser, N & Lasenby, A, Kluwer, p1.
Cowie, L L 1991 in *Observational Tests of Cosmological Inflation*, eds. Shanks, T *et al*, Kluwer, p25.
Cowie, L L & Lilly, S J 1990 in *Evolution in the Universe of Galaxies: Hubble Centennial Symposium*, ASP Conference Series, ed. Kron, R, p212.
Cowie, L L, Gardner, J P, Hu, E M, Wainscoat, R J & Hodapp, K W 1991a *Ap. J.*, in press.
Cowie, L L, Songaila, A & Hu, E 1991b *Nature*, submitted.
Dressler, A 1984 *Ann. Rev. Ast. Astr.*, **22**, 185.
Dressler, A & Gunn, J E 1983 *Ap. J.*, **263**, 533.
Dressler, A & Gunn, J E 1990 in *Evolution in the Universe of Galaxies: Hubble Centennial Symposium*, ASP Conference Series, ed. Kron, R, p200.
Ellis, R S 1990 in *Evolution in the Universe of Galaxies: Hubble Centennial Symposium*, ASP Conference Series, ed. Kron, R, p248.
Ellis, R S, Couch, W J, MacLaren, I & Koo, D C 1985 *MNRAS*, **217**, 239.
Frenk, C S, White, S D M, Efstathiou, G & Davis, M 1990 *Ap. J.*, **351**, 10.
Greggio, L & Renzini, A 1990 *Ap. J.*, **364**, 35.
Guhathakurta, P, Tyson, A J & Majewski, S 1991 *Ap. J.*, **357**, L9.
Gunn, J E, Hoessel, J & Oke, J B 1986 *Ap. J.*, **306**, 30.
Koo, D C 1990, in *Evolution in the Universe of Galaxies: Hubble Centennial Symposium*, ASP Conference Series, ed. Kron, R, p268.
MacLaren, I, Ellis, R S & Couch, W J 1989 *MNRAS.*, **230**, 249.
O'Connell, R W 1980 *Ap. J.*, **236**, 340.
Ostriker, J P 1990 in *Evolution in the Universe of Galaxies: Hubble Centennial Symposium*, ASP Conference Series, ed. Kron, R, p25.
Pickles, A J 1985 *Ap. J.*, **296**, 340.
Rose, J 1985 *Astr. J.*, **90**, 1927.
Rose, J, Bower, R G, Ellis, R S, Sharples, R M, Caldwell, N M & Teague, P 1991, in preparation.
Sandage, A & Visvanathan, N 1977 *Ap. J.*, **216**, 214.
Saunders, W, Rowan-Robinson, M, Lawrence, A, Efstathiou, G P, Kaiser, N, Ellis, R S & Frenk, C S 1990 *MNRAS*, **242**, 318.
Soifer, B T, Sanders, D B, Neugebauer, G, Danielson, G E, Lonsdale, C J, Madore, B F & Persson, S E 1986 *Ap. J.*, **303**, L41.
Soucail, G, Mellier, Y, Fort, B, Matthex, G & Cailloux, M 1988 *Astron. Astr.*, **191**, L19.
Tinsley, B M 1977 *Ap. J.*, **211**, 621 (erratum **216**, 349).
Tyson, A J 1988 *Astr. J.*, **96**, 1.
Yoshii, Y & Takahara, F 1988 *Ap. J.*, **326**, 1.

DISCUSSION:

O'CONNELL: The idea that secondary star formation in ellipticals depends on the environment, and is therefore likely to be more common in the Local Group and Virgo than in dense clusters, makes good sense. Most of the evidence for late star formation is also confined to the *centers* of nearby galaxies, as pointed out by Sandy Faber earlier, so there is not necessarily any inconsistency with the homogeneity of large aperture colors which you find.
ELLIS: Agreed.

ELSTON: Since your faint galaxies are resolved, what are their morphologies and surface brightnesses? Surface brightness would increase by 2-3 magnitudes from evolutionary considerations, but fade by $(1+z)^4$. What would a present-day blue galaxy look like?
ELLIS: Good resolution (0.8 arcsec FWHM) NTT imagery of the [O II]-strong galaxies at z$\simeq$0.3 failed to reveal any obvious pecularities in terms of surface brightness or morpholgy. The typical star-forming galaxies are neither compact objects nor low surface brightness fuzzballs and do not show a strong excess of closeby companions, although the sample surveyed so far is small. The blue light appears to be shared by the bulk of the galaxy rather than, say, being confined to the nuclear region. However, better resolution data is needed to make progress. These objects are not faint and 0.3-0.4 arcsec FWHM would be sufficient to say a lot more.
SILK: The faint blue galaxies seen in the very deep counts are dwarfs, hence their relics, if they represent a new population that subsequently faded, would be metal poor and blue, rather than red and with low surface brightness. Also, if the triggering of the star formation bursts in galaxies is associated with denser regions, such as groups and clusters, one should search in those regions (rather than in the field) for the low surface brightness relics.
ELLIS: I think you have to be careful what you mean by 'blue'. A low luminosity dwarf elliptical is bluer than a giant elliptical but still much redder than the bulk of the faint population. The K counts eliminate such a red population unless they have very low surface brightnesses. Secondly, if the bursts were more common in groups/clusters, the [O II]-selected galaxies would be distributed differently to the remainder. With $\simeq$1000 faint redshifts in hand, no obvious separation is yet evident.
MELNICK: If the blue population you find in deep redshift surveys consists of dwarf galaxies, you'd expect to find them significantly less clustered than if their absolute magnitudes were $\simeq$ -18 as in your merger hypothesis. Have you looked at the clustering (e.g. $\xi(r)$) of these galaxies?
ELLIS: Spatially, there is no difference (see answer to Silk), but preliminary evidence by other groups to much fainter limits suggests the faint blue galaxies are less strongly clustered, in terms of their angular correlation function $w(\theta)$, than would be expected for normal galaxies at the observed surface density. However, the angular measurements are only currently available on small separations (<1 Mpc) and the comparison is notoriously difficult to make. With large format CCD data becoming available, it will be interesting to see more extensive results.
OSTRIKER: You need a candidate set of objects to be the numerous blue galaxies seen in the counts. Lyman limit systems seen in absorption between us and high redshifts QSOs contain metal lines and presumably star formation. If these numerous objects fade to dwarf spheroidals, could they be your blue systems at an intermediate epoch?
ELLIS: Transforming a blue galaxy close to L*, as required by the redshift surveys, to a M_B=-12 dSph would be a dramatic change, and one would have little time in which to manoevre. I should stress that these blue galaxies are bursting at remarkably recent lookback times, may be directly related to the IRAS phenomenon and thus I prefer to consider them as precursors of normal galaxies rather than to hide them away in uncharted regions of the luminosity function.
BERSHADY: Is the redshift distribution the same for the high and low [O II] equivalent width galaxies?
ELLIS: The distributions do not differ to any great extent. The proportion with strong [O II] increases with *apparent magnitude* but they populate a wide range in redshift and hence luminosity. We are just getting to the point where it will be possible to track the luminosity functions of the burst/non-burst components independently with redshift. If the burst+merger model is correct we would expect a steepening of the faint end slope with redshift at the expense of the non-burst LF.

RENZINI: You said the Strontium method is independent of the metallicity, but do you think it is also independent of the Strontium abundance?
ELLIS: I don't know. We checked the integrated spectrum of the globular 47 Tuc and M32 and the surface gravities derived by Rose's Strontium method is in good agreement with those implied by the turn offs seen in the resolved colour magnitude diagrams. I'd be surprised if such a clear distinction between rich cluster ellipticals and those in low density environments arises by Strontium variations, but I guess it can't be excluded at this stage.
KURUCZ: If you evolve a large elliptical galaxy backward, there is a point where it becomes transparent because the stars we see as giants are back on the main sequence. At that point, all the dwarf companions of a galaxy are visible, whereas today they are hidden or they cannot be seen because they are overwhelmed by the brightness of the large galaxy.
ELLIS: Interesting idea, but note that when a giant elliptical has all its stars on the main sequence it would become extraordinarily blue and luminous. No such objects are seen (excluding Terlevich's interesting suggestion concerning QSOs) either because this epoch lies at redshifts where the Lyman limit has passed through the optical, or because dust obscures such objects. In both cases, you would never see behind such an object and in any case, you would not resolve companions as independent objects. As I explained to others above, the extra blue objects are not strongly correlated in position with the quiescent population.

EVOLUTIONARY POPULATION SYNTHESIS

GUSTAVO BRUZUAL A.
Centro de Investigaciones de Astronomía, CIDA
Apartado Postal 264
Mérida 5101-A
Venezuela

ABSTRACT. A brief and non-exhaustive review of the main research papers on Evolutionary Population Synthesis is presented. The degree up to which these studies obey well known astrophysical constrains, such as the Fuel Consumption Theorem, is analyzed. A summary of the most significant results from the Isochrone Synthesis Spectral Evolution Models of Bruzual and Charlot (1992) concludes this presentation.

1. Introduction

Among the different techniques used in spectral synthesis, evolutionary population synthesis (EPS) has become a standard technique to study the spectrophotometric evolution of galaxies. The primary goal of EPS is to compute the time-dependent distribution of stars of various masses in the HR diagram for a given stellar initial mass function (IMF) and star formation rate (SFR). The evolving integrated spectrum of the stellar population for the assumed IMF and SFR is then derived by adding up the spectra of the individual stars, weighting each spectrum by the number of stars of each type. Without pretending to be an exhaustive review of the literature on EPS, the remaining of this section summarizes the main contributions on the subject.

The pioneer work of Tinsley (1972), followed up by Tinsley and Gunn (1976a, b) and Tinsley (1978) established the foundations of EPS and defined the general properties (IMF, SFR, age) of the EPS models that reproduced the photometric properties of various kinds of stellar systems. During the same period Searle, Sargent and Bagnuolo (1973) and Huchra (1977) developed parallel codes to study the photometric properties of star clusters and galaxies of different morphological types. Bruzual (1981, 1983) and Gunn, Stryker, and Tinsley (1981) were the first to use stellar spectrophotometric data in EPS. The prediction of detailed galaxy spectra allowed us to derive firmer conclusions on galaxy evolution than photometric studies had permitted in the past. The magnitude and color evolutionary corrections up to redshifts of cosmological interest derived from these spectrophotometric codes were more realistic than the ones derived previously from interpolated broad band fluxes. More *ad hoc* and less detailed models were constructed by Arimoto and Yoshii (1986, 1987). The evolution of these models in the UV range does not follow directly from stellar evolution, as in the previously mentioned cases, but is arbitrarily scaled according to the flux in the U band. Guiderdoni and Rocca-Volmerange (1987) introduced some refinements into EPS, such as including nebular emission and reddening by dust grains in their synthetic spectra. Their coverage of the HR diagram was more complete than in the work of previous authors.

A large amount of work on EPS has been performed by the *Italian School*. A careful set of spectral evolutionary models for elliptical galaxies has been developed by Barbaro and Olivi

B. Barbuy and A. Renzini (eds.), The Stellar Populations of Galaxies, 311–320.

(1989) in a paper that has not received due attention from other workers on the subject. The assemble of evolutionary tracks used in this paper includes all phases of stellar evolution, and an attempt is made to follow the chemical enrichment of successive generations of stars. The stellar spectra are obtained from Kurucz's (1979) set of model atmospheres. Barbaro and Olivi conclude that the UV flux in elliptical galaxies is produced by PAGB stars. A careful study of the most likely mass and chemical content of the PAGB stars producing this UV flux has been presented by Bertelli *et al.* (1989). They conclude that the final mass *vs.* metallicity relation suggested by the PAGB model of the UV emission in ellipticals is indeed compatible with stellar evolution results, provided that the correct dependence of this mass on both chemical composition and mass loss is taken into account. Buzzoni (1989) has developed an independent set of EPS models along the same lines of Barbaro and Olivi, using different evolutionary tracks and supplementing Kurucz's model atmospheres with Bell and Gustafsson (1978) models for cool stars. This code has been used by Buzzoni *et al.* to derive a metallicity scale for elliptical galaxies via a calibration of the Mg_2 index.

In parallel to the spectrophotometric models described above, there has been considerable effort invested in building purely photometric models. These models are very valuable for detailed studies of the population content of nearby stellar systems. Battinelli and Capuzzo-Dolcetta (1989) have built EPS models to study the dependence of the *UBV* fluxes and colors on time and metallicity. They discuss in detail several alternatives to explain the gap observed in the color distribution of the MC clusters, favoring a solution in terms of the high transit velocity shown by their synthetic models in the HR diagram around the $B-V$ color corresponding to the fast reddening of the clump of core helium-burning stars. Barbero *et al.* (1990) in an independent study of the emission from LMC clusters in selected far-UV bands suggest that the gap in the observed colors, at least for the LMC, is caused by the lack of clusters in the range of ages between 0.2 and 1 Gyr. Models that follow a detailed treatment of the chemical and photometric evolution in the *UBVR* bands of elliptical galaxies have been developed by Brocato *et al.* (1990). These authors conclude that the increasing UV flux with increasing metallicity observed in ellipticals is explained by the different behavior of stars of different metal content during the post-HB phases. Mazzei *et al.* (1991) built a detailed photometric model in the *UBVRIJKLMN* bands for the evolution of disc galaxies based on the evolutionary tracks of Bertelli *et al.* (1990). This model, which incorporates diffuse dust emission, is applied successfully to our own galaxy.

Charlot and Bruzual (1991, hereafter CB91) assembled a library of stellar evolutionary tracks for solar metallicity, largely based on Maeder and Meynet (1989), suited for an isochrone synthesis approach. At any age an isochrone is derived by interpolation of the evolutionary tracks in the HR diagram. By construction, all evolutionary stages are included in the isochrone. This insures a smooth time dependence of all the properties of the synthetic population. The photometric models of CB91 were shown to reproduce well the observations of nearby stellar populations in the *UBVRIJKL* bands. Bruzual and Charlot (1992, hereafter BC92) have extended the models of CB91 by adding a spectrophotometric stellar library which covers the range from the far UV to the far IR and is essentially based on observational data for solar metallicity stars. The evolutionary tracks used by BC92 contain some improvements over those of CB91. The main sequence lifetime of the stars with masses between 1.3 $M_\odot$ and 2.5 $M_\odot$ have been revised as indicated by Maeder and Meynet (1991) and Bertelli *et al.* (1991). The revised lifetimes range from about 90% (2.5 $M_\odot$) to about 50% (1.3 $M_\odot$) of the original values. Following the prescriptions of Magris and Bruzual (1992, hereafter MB92) the treatment of the post-AGB evolution of low and intermediate-mass stars was refined. The final tracks include the evolution of all stars initially less massive than 7 $M_\odot$ as planetary nebulae, bare PN nuclei, and through the white-dwarf cooling sequence. In section **3** I will discuss some of the results of CB91, BC92, and MB92. A detailed comparison of these results with previous work is beyond the scope of this short presentation (see BC92 for details).

2. The Fuel Consumption Theorem and Evolutionary Population Synthesis

Early EPS models were severely criticized by Renzini (1981) and Renzini and Buzzoni (1983, 1986, hereafter RB86) on the grounds that these models *more or less dramatically violate the prescriptions of the Fuel Consumption Theorem,* (FCT) *and should therefore be viewed with extreme caution even when they apparently provide satisfactory fits to the observations.* The FCT states that "the contribution of stars in any given post-main sequence stage to the integrated bolometric luminosity of a simple stellar population (SSP) is directly proportional to the amount of fuel burned during that stage". Figure 5 of RB86 (hereafter RB86-F5) shows the fractional contribution of the various evolutionary stages to the integrated bolometric luminosity of a SSP as a function of the age of the population, derived by these authors from the FCT for specific assumptions about stellar evolution.

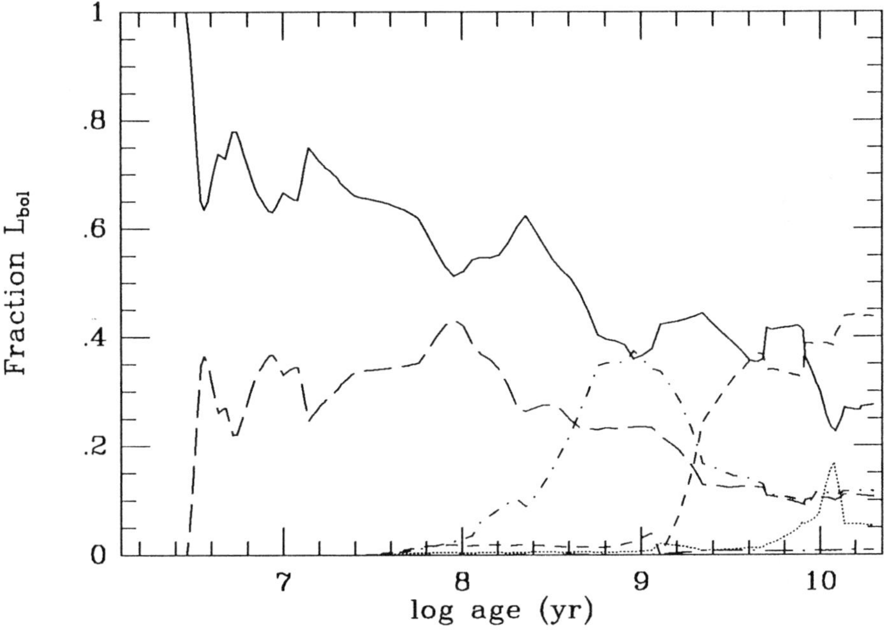

Figure 1. Fractional contribution of the various evolutionary stages to the integrated bolometric luminosity of a SSP as a function of time according to BC92. The different curves correspond to the following stages: MS (*solid line*), SGB (*dotted line*), RGB (*dashed line*), CHeB (*long-dashed line*), AGB (*dot and short-dashed line*), and PAGB (*dot and long-dashed line*). The Salpeter (1955) IMF was assumed.

Renzini's criticism is certainly valid for the early work on EPS by Tinsley (1972, 1978), Tinsley and Gunn (1976a, b), Bruzual (1981, 1983), Gunn, Stryker, and Tinsley (1981), Arimoto and Yoshii (1986, 1987), and Guiderdoni and Rocca-Volmerange (1987). Most of the papers attributed above to the *Italian School* do make an attempt to obey the prescriptions of the FCT and, in particular, to follow literally RB86-F5 (see also Wyse 1985). Unfortunately, the exact shape of the lines shown in RB86-F5 depends on the set of evolutionary tracks used in the computation. CB91 and BC92 have shown that their isochrone synthesis algorithm do strictly obey the FCT. However, the evolutionary tracks used by BC92 are independent from those used by RB86, and the fractional contributions to the total light derived by BC92 shown in Figure 1 do differ from those of RB86-F5.

The fluctuations seen in some of the curves in Figure 1 are intrinsic to the stellar tracks, since the contribution of a given evolutionary phase does not always vary smoothly with decreasing ZAMS mass. A striking feature is that MS stars dominate the integrated light of the burst population from early ages on until about 8 Gyr, contrary to the behavior shown in RB86-F5. This predominance of MS stars is essentially due to the inclusion of convective overshooting in the Maeder and Meynet (1989) stellar tracks. Convective overshooting lengthens the MS lifetime of stars more massive than 1.1 $M_{\odot}$. Since lower-mass stars do not develop a convective core on the MS, RGB stars dominate the bolometric light at later ages.

Figure 2 shows the fractional contribution of AGB stars to the integrated bolometric luminosity of a SSP according to BC92. The agreement of the model and the data is quite good, especially since the youngest clusters may be contaminated by M supergiants mistaken for AGB stars (Frogel *et al.* 1990). The contribution of the AGB stars becomes significant at much earlier ages in RB86 than in CB91 and BC92. The reason for this discrepancy is the assumed upper-mass limit for stars to go through the AGB phase. In the Maeder and Meynet (1989) tracks used by CB91 and BC92, stars with $m \leq 7M_{\odot}$ go through the double-shell burning phase that characterizes the AGB, whereas this limit is higher for the tracks of RB86.

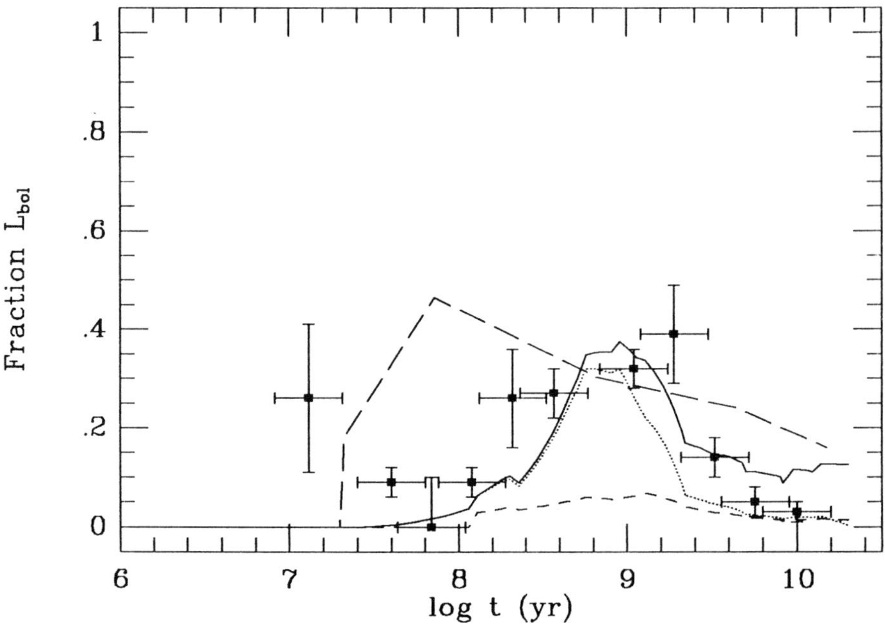

Figure 2. Fraction of the integrated bolometric light of a SSP with the Salpeter (1955) IMF accounted for: AGB stars brighter than M_{bol} = -3.6 (*dotted line*), all AGB stars (*solid line*), and TP-AGB stars (*short-dashed line*), according to BC92. The data points are the contribution observed in the MC clusters by Frogel *et al.* (1990). The long-dashed line is the contribution predicted by RB86-F5.

3. Isochrone Synthesis Spectral Evolution Models

BC92 have used their isochrone synthesis models to derive a best-fitting spectrum for several well-observed nearby galaxies. The selection of this best fit is done objectively, once a model with a given IMF and SFR is specified, and refers only to the age at which the model resembles most closely the observed spectrum. No attempt was made by BC92 to derive the optimal IMF or SFR for a given galaxy.

Figure 3 shows a comparison of the predicted spectral energy distribution of a 1 Gyr burst model, in which star formation takes place during the first Gyr of the life of the galaxy, with that of an *average* elliptical galaxy, at the best-fitting age of 13 Gyr. For details about this average spectrum see BC92. The fit is excellent over the whole spectral range, and no systematic trend is observed in the behavior of the residuals with wavelength. The main stellar absorption features in the observed spectrum and the amplitude of the 4000 Å break are well reproduced by the model. The quality of the fit in the UV, optical, and the near-IR, constitutes a major improvement over earlier population synthesis models (Bruzual 1983, Guiderdoni and Rocca-Volmerange 1987).

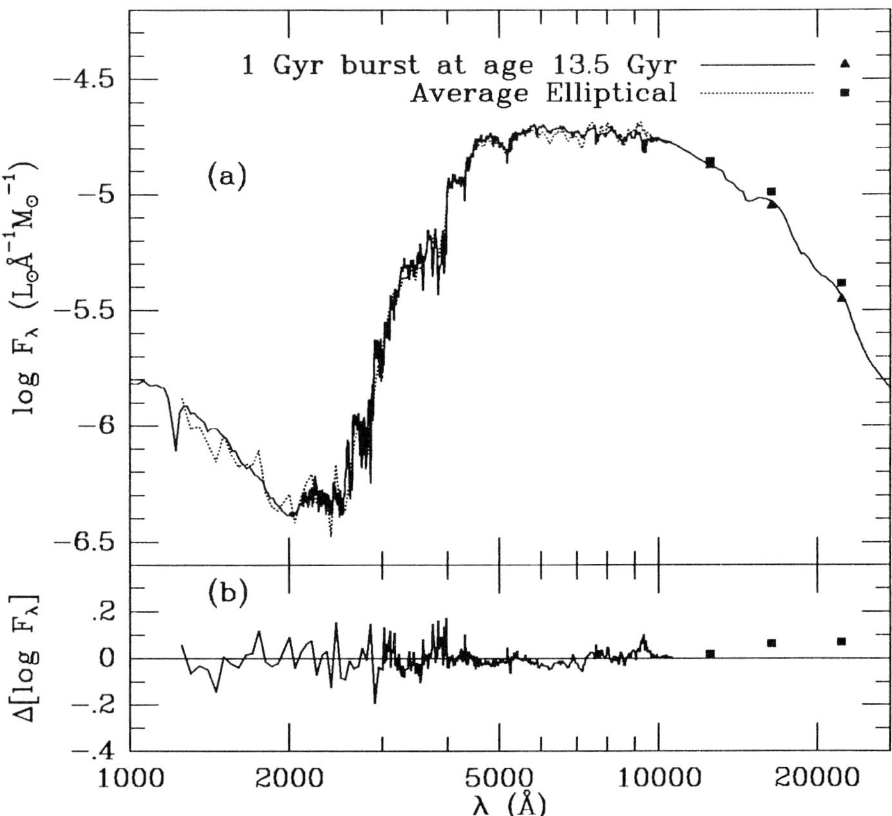

Figure 3. *(a)* Comparison of the 1 Gyr burst model of BC92 at age 13 Gyr (*thin line and triangles*) with the *average* observed spectrum of quiescent elliptical galaxies (*thick line and squares*). The squares and the triangles correspond to the fluxes at the effective wavelengths of the J, H, and K bands. The Salpeter (1955) IMF was assumed. *(b)* Residual of the comparison in panel *(a)*.

The UV-rising branch of quiescent elliptical galaxies (Burstein *et al.* 1988) is thus explained by BC92 as a natural consequence of late stellar evolution, superseding the conclusions of models without post-AGB stars that required recent star formation (Bruzual 1983, Guiderdoni and Rocca-Volmerange 1987), but in agreement with Barbaro and Olivi (1989).

Figure 4 compares a model with constant SFR with the observed spectral energy distribution of the irregular galaxy NGC 4449, assembled from *IUE*, optical, and near-IR data by Ellis and Bruzual (1983, unpublished). The best-fitting age in this case is 1 Gyr. The model does not include nebular emission by gas and hence it cannot reproduce the observed emission lines. The residuals between model and data are very small, except blueward of 1500 Å, where the model shows an excess of UV radiation. Reducing the proportion of massive stars in the IMF decreases this excess, but the 4000 Å break then rises in the model, and makes the fit worse in the optical. Alternatively, extinction by internal dust in this galaxy may reduce the ultraviolet flux.

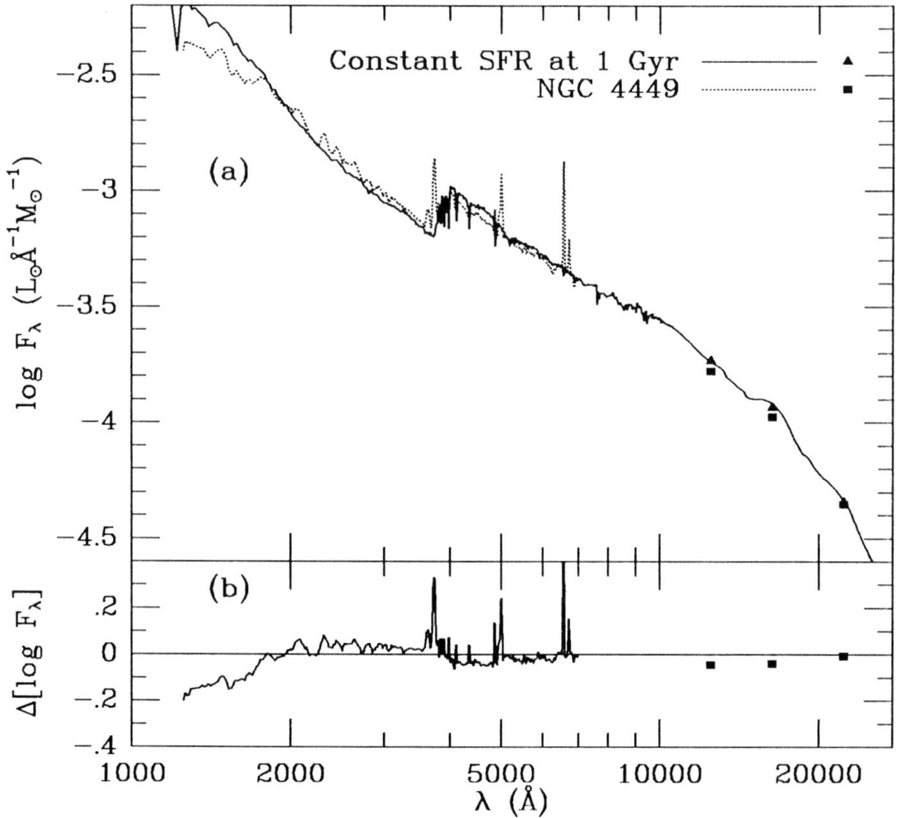

Figure 4. *(a)* Comparison of the constant SFR model of BC92 at age 1 Gyr (*dotted line and triangles*) with the observed spectrum of the irregular galaxy NGC 4449 (*thick line and squares*). The squares and the triangles correspond to the fluxes at the effective wavelengths of the J, H, and K bands. The Salpeter IMF was assumed. *(b)* Residuals of the comparison in panel *(a)*.

The spectra of galaxies of other morphological types are explained by BC92 with exponentially decreasing SFR models with characteristic time τ. The model fits are similar to the ones shown in Figures 3 and 4, except that the age inferred for the Coleman *et al.* (1980) Sdm (constant SFR, age = 2.5 Gyr) is older than the one found for NGC 4449 due to the redder spectrum of the former. Models with intermediate time scales of star formation and the Salpeter IMF can reproduce reasonably well the spectra of Sbc (τ = 4, age = 15 Gyr) and Scd (τ = 7, age = 14.5 Gyr) galaxies in the Coleman *et al.* (1980) sample.

The significance of the best-fitting ages is weak in the case of the average elliptical, whose spectrum can be fitted almost equally well by the 1 Gyr burst model at ages in the range 11 to 16 Gyr. The age for NGC 4449 is more strongly constrained by the rapid increase of the 4000 Å break as old stars accumulate. The ages and τ derived for the Sbc and Scd galaxies are not rigidly constrained by the observed spectra. Similar fits are obtained with shorter or longer τ's at younger or older model ages, respectively.

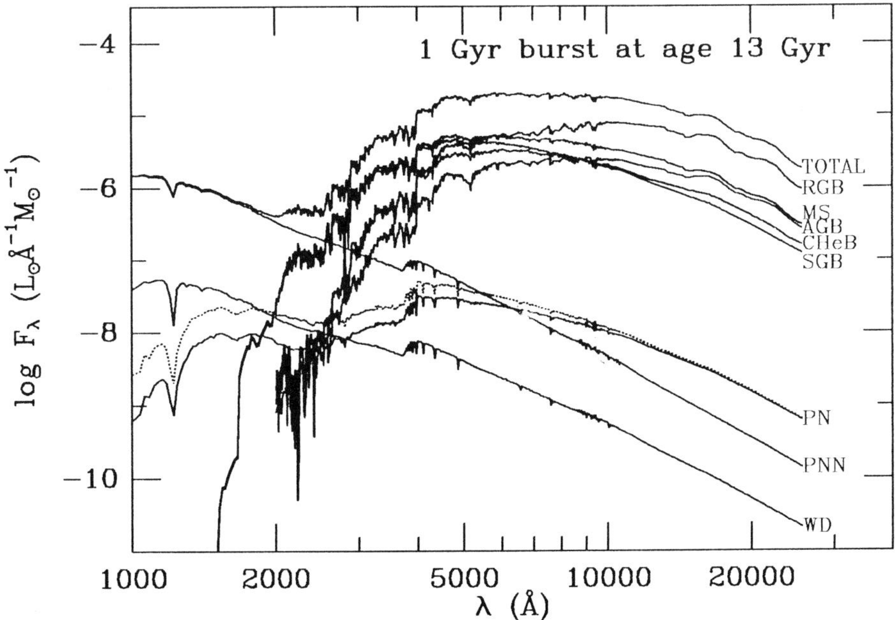

Figure 5. Contribution of stars in different groups to the total spectral energy distribution of a 1 Gyr burst model at 13 Gyr for the Salpeter IMF (PNN = bare PN nucleus). The dotted line next to the PN contribution corresponds to the case where extinction of the core radiation by the surrounding nebula is ignored. The vertical scale refers to a total mass in stars of 1 $M_\odot$.

Figure 5 shows the contribution to the total spectral energy distribution of the 1 Gyr burst model, at 13 Gyr, of different stellar groups. The spectrum blueward of 2000 Å is entirely dominated by low-mass PAGB stars whose envelopes have dissipated. MS and SGB stars account for most of the light between 3000 Å and 4000 Å, and the RGB produce more than half of the light at longer wavelengths. The exact shape of the spectrum from the optical to the near-IR, however, is a subtle combination of various nearly equivalent contributions. Note that PN have a negligible contribution to the integrated spectrum, even if one ignores the absorption by the envelope of the core radiation (*dotted line*).

MB92 have analyzed the UV upturn seen in elliptical galaxies using the evolutionary models of BC92. Figure 6 shows the average spectral energy distributions for 3 representative groups of galaxies with different levels of UV flux from the Burstein *et al.* (1988) sample. Galaxies with UV flux level like N4649 (top frame in Figure 6) are modeled by MB92 with a 14 Gyr old stellar population (consistent with Bertelli *et al.* 1989) in which the SFR corresponds to an initial burst of 100 $M_\odot\ yr^{-1}$ lasting for 1 Gyr plus a *residual* continuing star formation of 0.03 $M_\odot\ yr^{-1}$. Normal PAGB stars, included in the BC92 stellar library, contribute mainly at $\lambda \leq 2200\ \mathring{A}$, and cannot reproduce the flux level observed in these systems in the range 2200-2600 $\mathring{A}$. Other possible source of this excess UV flux besides MS stars have been explored by Greggio and Renzini (1990). On the contrary, the spectra of galaxies with an intermediate value of (1550–V) (like N4472, middle frame) are well fitted by a 1 Gyr burst BC92 model seen at 13 Gyr (Figure 3). In this case the UV flux is produced by normal PAGB stars, resulting from the evolution of a quiescent stellar population.

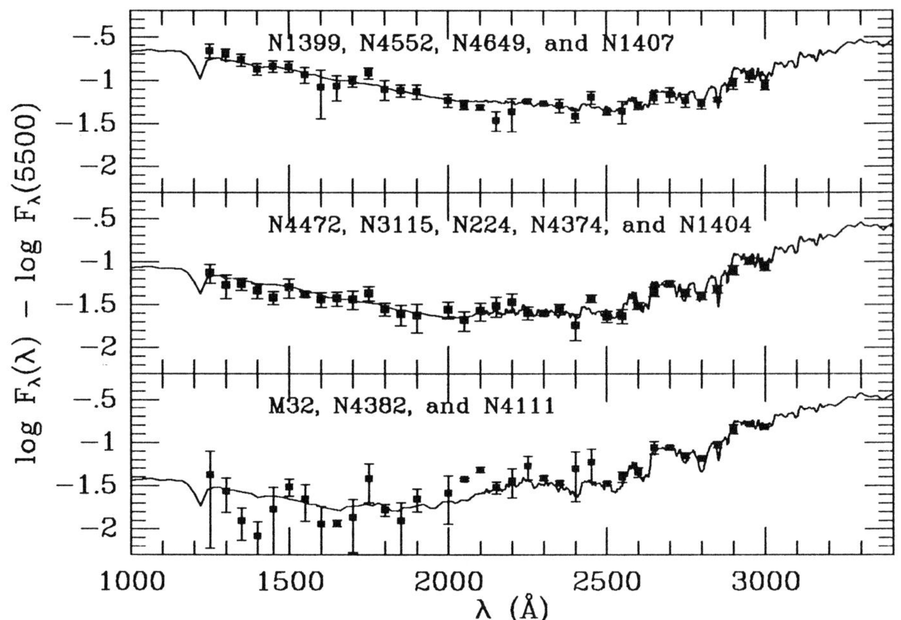

Figure 6. The data points represent the average spectral energy distributions for 3 representative groups of elliptical galaxies defined by Burstein *et al.* (1988). The error bars represent the dispersion around the mean flux. The lines correspond to the models by MB92 described in the text. The Salpeter IMF was used in the models.

The spectra of galaxies like M32 (bottom frame) can be understood if the stellar system underwent two events of star formation: an initial 1 Gyr burst, and a second burst of the same duration at an approximate age of 6 Gyr. Each of these bursts encompasses half of the galaxy mass seen through the *IUE* aperture. This result is in accord with O'Connell (1986), and Bertelli *et al.* (1989), but was questioned by Greggio and Renzini (1990).

Conclusions

Considerable progress has been achieved in recent years in the field of EPS. Complete libraries of evolutionary tracks and spectra that had been incorporated, together with increasingly sophisticated synthesis algorithms, have allowed several authors to develop EPS codes that overcome most of the problems present in early models. In particular, the isochrone synthesis models of BC92 can reproduce reasonably well the observed spectral or photometric properties from the UV to the near-IR of nearby galaxies of various Hubble types, in which young massive stars, old low-mass stars, or mixtures of stars of all masses and ages produce the light. These models constitute therefore a reliable mean to investigate the spectral evolution of any stellar population. The possibility that some of the other evolved hot star candidates (Greggio and Renzini 1990) contribute significantly to the UV flux is not excluded, but for many ellipticals, the normal evolution of stars beyond the AGB produces enough flux to account for their UV rising branch. Spectrophotometric models which also follow the chemical evolution of the stellar population (e.g. Barbaro and Olivi 1989, Brocato *et al.* 1990) still need further development.

References

Arimoto, N., and Yoshii, Y. 1986, A&A, 164, 260

______. 1987, A&A, 173, 23

Barbaro, G., and Olivi, F. M. 1989, ApJ, 337, 125

Barbero, J., Brocato, E., Cassatella, A., Castellani, V., and Geyer, E. H. 1990, 351, 98

Battinelli, P., and Capuzzo-Dolcetta, R. 1989, ApJ, 347, 794

Bell, R. A., and Gustafsson, B. 1978, A&AS, 34, 229

Bertelli, G., Bressan, A., and Chiosi, C. 1991, ApJ, submitted

Bertelli, G., Chiosi, C., and Bertola, F. 1989, ApJ, 339, 889

Brocato, E., Matteucci, F., Mazzitelli, I., and Tornambé, A. 1990, ApJ, 349, 458

Bruzual A., G. 1981, Ph.D. thesis, University of California, Berkeley

______. 1983, ApJ, 273, 105

Bruzual A., G., and Charlot, S. 1992, ApJ, in press (BC92)

Burstein, D., Bertola, F., Buson, L. M., Faber, S. M., and Lauer, T. R. 1988, ApJ, 328, 440

Buzzoni, A. 1989, ApJS, 71, 817

Buzzoni, A., Gariboldi, G., and Mantegazza, L. 1991, ApJ, submitted

Charlot, S., and Bruzual A., G. 1991, ApJ, 367, 126 (CB91)

Coleman, G. D., Wu, C. C., and Weedman, D. W. 1980, ApJS, 43, 393

Frogel, J. A., Mould, J., and Blanco, V. M. 1990, ApJ, 352, 96

Greggio, L. and Renzini, A. 1990, ApJ, 364, 35

Guiderdoni, B., and Rocca-Volmerange, B. 1987, A&A, 186, 1

Gunn, J. E., Stryker, L. L., and Tinsley, B. M. 1981, ApJ, 249, 48

Huchra, J. 1977, ApJ, 217, 928

Maeder, A. and Meynet, G. 1989, A&A, 210, 155

______. 1991, private communication

Kurucz, R. L. 1979, ApJS, 40, 1

Magris C., G. , and Bruzual A., G. 1992, ApJ, submitted (MB92)

Mazzei, P., Xu, C., and De Zotti, G. 1991, A&A, submitted

O'Connell, R. W. 1986, in Stellar Populations, eds. Norman, Renzini, and Tosi (Cambridge), p 167

Renzini, A. 1981, AnnPhysFr, 6, 87

Renzini, A., and Buzzoni, A. 1983, MemSAIt, 54, 739

______. 1986, in Spectral Evolution of Galaxies, eds. C. Chiosi and A. Renzini (Reidel), p 195 (RB86)

Salpeter, E. E. 1955, ApJ, 121, 161

Searle, L., Sargent, W. L. W., and Bagnuolo, W. G. 1973, ApJ, 179, 427

Tinsley, B. M. 1972, A&A, 20, 383

______. 1978, ApJ, 222, 14

Tinsley, B. M., and Gunn, J. E. 1976a, ApJ, 203, 52

______. 1976b, ApJ, 206, 525

Wyse, R. F. G. 1985, ApJ, 299, 593

Discussion

Rocca-Volmerange: There is a real danger to use, as you did, two different types of evolutionary tracks in population synthesis models. Discontinuities in evolutionary time, bolometric magnitude, and effective temperature are observed between the 1 $M_\odot$ evolutionary track from Maeder and Meynet (1989) and the 0.9 $M_\odot$ track from Chiosi's group.

Bruzual: This is certainly true and I have pointed that out myself in several occasions (Bruzual 1981, 1983). Charlot and myself were extremely careful to insure that these discontinuities did not affect the final results of our isochrone synthesis models. The revision of the MS lifetime of the stars with masses between 1.3 $M_\odot$ and 2.5 $M_\odot$ indicated by Maeder and Meynet (1991) and Bertelli *et al.* (1991) tends to iron out the discontinuities that you mentioned.

Rocca-Volmerange: The observed dispersion in $V-K$ of the MC clusters is likely explained by the stellar evolution of cold stars. Why does your sequence $V-K$ *vs.* age appear so smooth?

Bruzual: The models that I have discussed include the smooth evolution of the stellar population, and hence the predicted color *vs.* age lines show a smooth behavior. No attempt has been made to incorporate into the isochrone synthesis models the stochastic effects produced by fluctuations in the number of giant stars sampled by the aperture of the telescope, or fluctuations due to the relatively small numbers of stars present in clusters, etc. An approach to the inclusion of these effects in EPS models is discussed by Bertelli *et al.* (1991).

Vanbeveren: A stellar population may contain a significant fraction of interacting binaries. Since you do not include the evolution of close binaries in your models, how reliable then is a comparison between observations and theoretical models?

Bruzual: As far as I know no EPS model includes the evolution of close binary systems. I still think that the comparison of the model spectra and the observations has some significance in as much as it refers to the accuracy with which the observed spectral features are reproduced by the models. As long as the effect of close binary evolution is to populate parts of the HR diagram that would otherwise be void of stars, then it is dangerous to derive conclusions from current EPS models about the SFR, IMF, or age of the dominant population in a stellar system.

Ostriker: As you noted binaries are omitted. In globular clusters the interacting (WD + MS) systems contribute, I believe, a significant amount at UV wavelengths. Did you compare your UV spectra with those observed from globular clusters to see if close interacting binaries make a significant contribution?

Bruzual: No, I have not performed this comparison. I think that this is a very good suggestion and I will try to do this comparison in the near future.

Torres-Peimbert: How sensitive are your color predictions for evolved galaxies to the duration of the star formation burst?

Bruzual: Very insensitive. A few Gyrs after star formation ends, the colors for different models are independent of the assumed duration of the star formation burst.

Serrano: I believe that your solutions are not unique. It will be interesting to display iso-χ^2 contours in parameter space. I expect the solutions not to be very sensitive to changes of parameters along the metallicity axis.

Bruzual: This is correct and we are working in that direction. I have done some work in collaboration with D. C. Koo on the prediction of the number counts and color distributions of faint galaxies in which we do compute iso-χ^2 contours. This will be extended to spectral synthesis in the future.

MODELS OF POPULATION SYNTHESIS

Cesare Chiosi, Gianpaolo Bertelli, Alessandro Bressan
Department of Astronomy and Astronomical Observatory
Vicolo dell'Osservatorio 5
35122 Padova Italy

Abstract. We present models of chemical and photometric evolution of galaxies. We show that the onset of the AGB phase at a few 10^8yr causes a sudden change in the infrared colours of the galaxy. However, this powerful indicator of the time of galaxy formation is wiped out by cosmological effects.

1. Introduction

The analysis of the spectrophotometric properties of star clusters and galaxies, with the aid of the standard technique of population synthesis, provides important clues on the primary questions related to the epoch of formation and the evolution of stellar aggregates of various complexity. Existing models of population synthesis can be subdivided in two groups depending on which way they are constrained to the stellar input (O'Connell 1986). In "Evolutionary Population Syntheses" (EPS), a particular theoretical scenario is used to populate the HR diagram and to compute the integrated properties, i.e. magnitudes, colors, and spectral energy distribution (SED) assuming a limited number of parameters (in addition to those already adopted for the underlying stellar tracks). In "Optimizing Models" (OM), the observed SED is matched by summing up SED's taken from a library of stars or clusters. The elemental SED's are suitably weighted by the matching procedure. The OM method allows for the inclusion of stellar types whose evolution is not yet fully clarified, but it heavily depends on the completeness of the libraries themselves. The EPS method may easily include the effect of different chemical compositions, but the quality of the final result depends on the accuracy and homogeneity of the input stellar models.
In this paper we describe a new method for EPS, which is based upon modern stellar models and isochrones (Alongi et al. 1991). First, we show how the isochrones can be used to construct integrated magnitudes and colors in broad band photometry for a single stellar population (SSP). Second, adopting a model of galactic chemical evolution, we follow the variation of magnitudes and colors of the model galaxy both as a function of time in its rest frame and of the red-shift. This analysis made with the broad band colors can be easily extended to narrow band SED.

2. Data Bases of Stellar Models

A good data base of stellar models must include modern physical input, extend to the latest evolutionary phases, e.g. AGB, PN, WD, cover large intervals of stellar masses and chemical compositions. Because of free parameters adopted in stellar model calculations (mixing length in the atmosphere, mass loss, mixing inside the core, etc...), the evolutionary tracks and lifetimes must be tested against color-magnitude diagrams (CMD) and luminosity functions (LF's) of template star clusters. In doing this, particular care must be payed to the transformation from luminosities and T_{eff} to magnitudes and colors.
Most libraries of stellar models currently in use are either incomplete or heterogeneous (in the sense that models from different sources are sampled). This often gives rise to

B. Barbuy and A. Renzini (eds.), The Stellar Populations of Galaxies, 321–324.

spurious results as pointed out by Charlot and Bruzual (1990). Among the available data bases, the Yale tracks and isochrones are old and limited, while Bertelli's et al. (1986) data base in spite of its homogeneity and completeness up to the AGB phase for a wide range of initial masses and chemical compositions, was computed with old opacities. The models of Maeder and Meynet (1988, 1990) were computed with the most updated input physics, including the opacities from the Los Alamos Library (LAL), extended to massive stars, but did not include the late evolutionary phases (HB and beyond) of low mass stars. In addition to this, as noticed by Alongi et al (1991) and Stothers (1991), the core He-burning lifetime of these models are a factor of 2 to 3 longer than the values given by many other authors. Furthermore, Bertelli et al (1991) have clearly demonstrated that the core H-burning lifetimes for stars in the mass range 1.1 $M_\odot$ to 2 $M_\odot$ given by Maeder and Meynet (1988, 1990) are wrong by a large factor compared to the straightforward estimate obtained from the ratio of the available fuel to the luminosity of their own models. The most recent data base of evolutionary tracks was computed by Alongi et al (1991). These models and accompanying isochrones are briefly discussed by Bressan et al in this volume. The wide mass interval (0.6 $M_\odot$ to 100 $M_\odot$), the inclusion of late evolutionary phases (AGB), and the different initial chemical compositions (Z=0.02, Y=0.28; Z=0.008, Y=0.25; Z=0.001, Y=0.23; Z=0.1, Y=0.35) make these stellar models suitable for EPS studies. The evolutionary tracks by Alongi et al (1991) were tested, by means of the synthetic CMD technique, against old galactic open clusters, intermediate age clusters in the LMC, and young luminous stars both in the Galaxy and LMC. Finally, the effects of the new opacities (OPAL and OP projects) were also investigated albeit in a preliminary fashion. Since the enhancement (a factor of 3 for the solar abundance) found in the new opacities with respect to LAL occurs in the external layers (Log T $\simeq$ 5.3), the effects on the stellar models are small at least for the adopted chemical compositions. Work is in progress to include in the data base the PN and WD phases for intermediate and low mass stars, and the carbon burning phase for massive stars.

3. The Single Stellar Population

The integrated monochromatic flux from the stellar population of a galaxy of age T is defined as

$$F_\lambda(T) = \int_0^T \int_{M_L}^{M_U} S(m,t,Z)\, f_\lambda(m,\tau,Z)\, dt\, dm \tag{1}$$

where S(m,t,Z) denotes the stellar birth-rate and $f_\lambda(m,\tau,Z)$ the monochromatic flux of a star of mass m, metallicity Z(t) and age τ=T-t. Separating S(m,t,Z) in the product of a function of time $\Psi(t,Z)$ (otherwise called the star formation rate SFR) and the initial mass function $\phi(m)$, the above integral becomes

$$F_\lambda(T) = \int_0^T \Psi(t,Z)\, f_\lambda(\tau,Z)\, dt \tag{2}$$

where $f_\lambda(\tau,Z) = \int_{M_L}^{M_U} \phi(m)\, f_\lambda(m,\tau,Z)\, dm$ is the integrated monochromatic flux of a SSP with age τ and metallicity Z.

Although the algorithm is of general validity, in the following we present results limited to broad band magnitudes and colors. Extension to narrow band photometry to calculate SED's is underway. Integrated Johnson-Cousins U,B,V,R,I,J,K,L,M,N colours and magnitudes of SSP's are computed following the method described by Chiosi et al. (1988), and Bertelli et al. (1990). We adopt the Salpeter initial mass function with the normalization condition $\int_{M_L}^{M_U} \psi(m)\, dm = 1$.

Figure 1 (top panel) shows the temporal evolution of the integrated (B-V) and (V-K) colors for a SSP with assigned composition. The trend of the (B-V) color confirms what already pointed out by Chiosi et al (1988 and references therein). i.e. the lack of sudden changes in the color at the onset of the AGB and RGB phases, in other words the lack of the phase transitions advocated by Renzini & Buzzoni (1986). The trend of the theoretical (B-V) color well agrees with the observational data for LMC clusters (see Chiosi et al 1988). Since these new models of SSP extend to massive stars, another interesting feature is visible, i.e. the red peak in the colors in coincidence with the red supergiant phase of massive stars at about 10^7 yr. Remarkably, the AGB phase transition, that is invisible in (B-V), clearly shows up in the infrared colors (V-J), (V-K), (V-L), (V-M), (V-N), where a discontinuity of about one magnitude reveals the transition from massive to intermediate mass stars. See the case of the (V-K) color. The age at which the discontinuity occurs increases with the adopted amount of core overshoot and it is about 10^8 yr for mild overshoot. This feature is a good indicator of the SSP age. On the contrary, there is almost no sign of the transition to low mass stars (appearance of the RGB phase). There, the infrared colours run almost flat and consequently they become poor age indicators. Our data compare well with the infrared data for LMC clusters (Persson et al. 1983), considering that the finite number of bright red stars introduces significant dispersion in the observed colours (Chiosi et al 1988). Finally, while intermediate age star clusters suggest a metallicity $Z_{\odot}/2$, the oldest ones indicate Z=0.0004.

4. The Photometric Evolution of Galaxies

Once the integrated colors and magnitudes of SSP's are known as a function of time and chemical composition, and the initial mass function $\phi(m))$ is assigned, the integrated properties of a galaxy at the age T can be easily obtained from eq. (2), provided that the past history of $\Psi(t, Z)$ and metallicity Z(t) are known. Instead of assigning conventional analytical expressions to these quantities, we made use of the inflow model of galactic chemical evolution. developed by Chiosi (1981). In this model, all relevant physical quantities, i. e. gas mass, star mass, total mass, SFR, and metallicity are ultimately driven by the rate of mass accretion. This is expressed by $\dot{M}(t) = A \times e^{-\frac{t}{\tau}}$ where A is a normalization constant, τ is the timescale of the mass accretion process, and $M(t)$ is the current total mass of the system. The constant A is fixed by imposing that $M(t)$ is equal to the total mass $M(T_{GAL})$ at the galaxian age T_{GAl}. The time scale τ is a parameter. In Chiosi's (1981) model the rate of star formation is given by $\Psi(t, Z) = \nu \; [\frac{M(T_{GAL})}{M(t)}]^{k-1} \; [\frac{M_g(t)}{M(T_{GAL})}]^k$, where ν is an adjustable parameter, $M_g(t)$ is the current mass density of gas, and k comes from having included the Schmidt(1959) law in the derivation of the above SFR. For τ ($\rightarrow 0$) this law reduces to an initial burst of star formation, while for $\tau \rightarrow \infty$ it corresponds to an ever continuing star formation activity. In between there is an ample possibility for SFR's that started small, grew to a maximum, and then declined. Chiosi's (1981) model also allows for more complex SFRs, e.g. sporadic bursts or bursts superposed to a continuous star formation. The time scale τ and the coefficient ν are constrained by imposing that gas content, SFR, and metallicity match the observed present day values in a real galaxy (or part if it). Finally, the galaxian age T_{GAL} is fixed by adopting a particular model for the universe and a value for the red-shift of galaxy formation (H_0, q_0 and z_{for} in Friedmann universe).

5. The Color Evolution of Galaxies in Their Rest Frame

The EPS's were computed for a three groups of models characterized by different values of τ (0.1, 1, and 5 Gyr) at given values of q_0 and z_{for} hence age T_{GAL}. In their rest frame, all the model galaxies show a remarkable evolution of the magnitudes and colors in the various passbands. More precisely, unless τ is a significant fraction of the present galaxian age, in the past the magnitudes rose to a peak value about 3 to 4 mag brighter than seen at the present time. The peak tends to widen and appear later as τ increases. The variation in the infrared colors due to the onset of the AGB phase in the stellar component is still well visible (Figure 1, bottom panel) but instead of being sharply confined in time as it was for the SSP, now it spreads over a time scale of about 0.5 Gyr, being centered at at about 3×10^8 yr, almost independently of τ.

6. The Color Evolution of Galaxies at High Red-Shifts

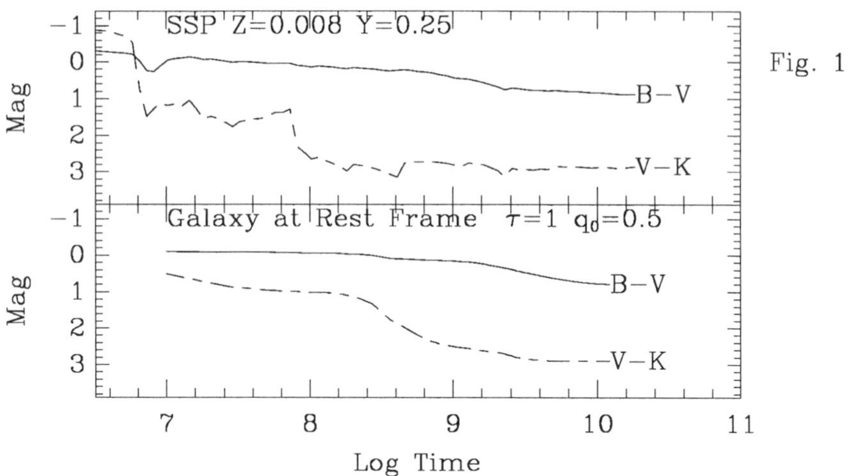

Fig. 1

Since the transition in the infrared colors is in principle a powerful indicator of the age of galaxy formation (let us remind that the transition in the rest frame occurs soon after the first generations of stars are borne, or in other words the galaxy has formed) we calculated the apparent magnitudes and colors as a function of the red-shift Z. To this aim, we derived the K- and E- corrections (Guiderdoni & Rocca-Volmerange 1987) as follows. Two ultraviolet colors were added to the available broad band colors and an analytical fit was applied to derive and to red-shift the SED. We are well aware that this is a crude approximation of the much more sophisticated models in which detailed SED's are used (see Guiderdoni & Rocca-Volmerange 1987). Nonetheless, in spite of the simple approach, we are able to show that the infrared colors lose memory of the phase transition that was visible in the SSP and whole galaxy at the rest frame. The cosmological effects wipe out this otherwise interesting age indicator.

REFERENCES

Alongi M., Bertelli G., Bressan A., Chiosi C., Fagotto F., Greggio L., Nasi E. 1991a. A&A submitted

Bertelli G., Bressan A., Chiosi C. 1991. ApJ submitted

Bertelli G., Bressan A., Chiosi C., Angerer K.. 1986. A&AS 66, 191

Bertelli G., Betto R., Bressan A., Chiosi C., Nasi E., Vallenari A. 1990. A &AS 85, 845

Buser R.& Kurucz R. L. 1989. Private communication

Charlot S., Bruzual G. 1990. ApJ 367, 126

Chiosi C., Bertelli G., Bressan, A. 1988. A&A 196, 84

Chiosi C. 1981. A&A 83, 206

Guiderdoni B., Rocca-Volmerange B. 1987. A&A 186, 1

Maeder A. 1990. A&A 84, 139

Maeder A., Meynet G., 1988. A&A 76, 411

O'Connel R. 1986. in *Spectral Evolution of Galaxies*, eds. C. Chiosi, A. Renzini, Reidel, Dordecht, p. 195

Persson S.E., Aaronson M., Cohen J.G., Frogel J.A., Matthews K. 1983, ApJ, 266, 105

Renzini A., Buzzoni A. 1986. in *Spectral Evolution of Galaxies*, eds. C. Chiosi, A. Renzini, Reidel, Dordecht, p. 195

Schmidt M. 1959. ApJ 129, 243

Stothers R. 1991. ApJ submitted

THE AGE LADDER FROM LOW- TO HIGH-REDSHIFT POPULATIONS

ALVIO RENZINI
Astronomy Department,
University of Bologna,
CP 596 I-40100 Bologna, Italy

ABSTRACT. The determination of the age of stellar populations in progressively more distant stellar systems requires successive calibrations which in some way remind the procedures followed in determining astronomical distances. A few exemplifications of this *Age Ladder* are schematically reviewed, such as the determination of the age of galactic globular clusters, of the youngest population in bulges and ellipticals, and of the most distant, high redshift radiogalaxies.

1. INTRODUCTION: GETTING AGES AND AGE ERRORS

The *Distance Ladder*, from nearby trig parallax stars to quasars, through e.g. galactic star clusters, Cepheids, supergiants, HII-regions, supernovae, rotational velocity and velocity dispersion of galaxies, first ranked cluster galaxies, and the Hubble law, is rather familiar to all of us. How errors propagate from primary, to secondary, to tertiary distance indicators is well known in principle (if not always in practice), and a wise empirical approach has been generally followed, as opposed to distance determination methods making extensive use of theoretical models that may introduce systematic errors which are difficult to quantify. But once we know the distance to galaxies, and thus their size and luminosity, in order to make astrophysics and cosmology, to understand how galaxies have formed and evolved, we need also to know the age (distribution) of their stellar populations. Ages, unfortunately, are more difficult to get than distances. The use of stellar evolution models is unavoidable, and this has far too often resulted in less rigorous measuring procedures, in which observational facts and theory are intermingled in far from optimal combinations, often losing track of how errors propagate and pile up, rather than choose those operational sequences that minimize the final error in age. I believe that the same rigor commonly requested to methods of galactic and extragalactic distance determinations should also be asked to dating methods, and we should speak of an *Age Ladder*, as that particular sequence of operations that starting from the derivation of precision ages for galactic open and globular clusters (GC), goes on making use of such ages to calibrate so-for-saying *secondary* and then *tertiary* age indicators. In this brief review I will restrict to just a few concrete examples that allow to exemplify a method which can be of more general validity.

2. THE TURNOFF LUMINOSITY AND THE AGE OF GLOBULAR CLUSTERS

The output of stellar model calculations are evolutionary sequences which can be used to construct the theoretical *isochrones*. By comparing isochrones to cluster color-magnitude diagrams (CMD) one can derive cluster ages, and this comparison can be made following several different procedures. However, not only the result, but also its uncertainty (i.e. the age error) together with the possibility itself of quantifying such uncertainty may depend on the particular procedure. In this respect, the

B. Barbuy and A. Renzini (eds.), The Stellar Populations of Galaxies, 325–336.

best procedure remains the classical one (e.g. Sandage 1970) in which the single detail of the isochrones that is used is the time-dependence of the luminosity of the main sequence turnoff:

$$\mathrm{Log}\, t_9 \simeq -0.41 + 0.37\, M_{\mathrm{V}}^{\mathrm{TO}} - 0.43\, Y - 0.13\, [\mathrm{Fe/H}], \tag{1}$$

where t_9 is the cluster age in Gyr units, $M_{\mathrm{V}}^{\mathrm{TO}}$ the absolute visual magnitude of the main sequence turnoff (TO), Y the helium abundance, and [Fe/H] the metallicity in standard notations*. In turn, $M_{\mathrm{V}}^{\mathrm{TO}} = V^{\mathrm{TO}} - \mathrm{mod}$, where V^{TO} (the TO apparent magnitude) is the directly *observable* quantity, and mod is the cluster distance modulus. Equation (1) then allows to estimate the relative importance of the uncetainty in each of the four input quantities (namely: V^{TO}, mod, Y, and [Fe/H]) in determining the total uncertainty of the age determination.

Clearly the *great villain* is the error in the distance of the clusters. Here I have estimated that current distances are typically affected by a 1/4 magnitude error in the modulus – $\sigma(\mathrm{mod}) \simeq 0.^{\mathrm{m}}25$ – which immediately translates into a $\sim$ 22% error in the derived cluster age ($\sim$ 3 Gyr for an age of 15 Gyr). All other IQs convey substantially smaller errors. The high photometric accuracy of CCDs now allows to determine a cluster's V^{TO} with an accuracy perhaps better than $0^{\mathrm{m}}.1$, which translates into a $\sim$ 9% error in age. The helium abundance is very well known, from either the R method, primordial nucleosynthesis, or empirical determinations of the *pregalactic* abundance, which all indicate $Y = 0.23 - 0.24$. Anyway, a ± 0.02 uncertainty in Y gives a negligible 2% error in age. I assume the metal content of the best studied clusters to be uncertain by perhaps 0.3 dex, which translates into a $\sim$ 9% uncertainty in age. There may be a problem with the *composition* of metallicity (e.g. [O/Fe], see the contributions of Barbuy and Carney), but there is no room to discuss it here. Clearly the *great villain* is the error in the distance of the clusters, and the question *"How good are globular cluster ages?"* immediately becomes: *"How good are globular cluster distances?"*

All in all, just three main standard candles have been used (or are being considered) for the determination of GC accurate distances, namely: RR Lyraes, subdwarfs, and white dwarfs. Pros and cons of each method are discussed elsewhere in some detail (Renzini 1991), suffice here a few schematic considerations.

RR Lyrae are rather bright, but rare objects, and even the nearest one is too distant for a trigonometric parallax to be obtained with an interesting accuracy. The calibration of this standard candle then relies on indirect physical or astrophysical methods, each making extensive use of pulsation, evolution, and stellar atmosphere models in some combination. Unfortunately, the more theory is used, the less we are able to quantify errors. Ultimately, even if we reach self-consistent distances (and then ages), we are left with the doubt that the models may have been slightly inaccurate in some way, thus introducing an unknown bias. A linear relation is usually assumed between the absolute magnitude of RR Lyraes and their metallicity: $M_{\mathrm{V}}^{\mathrm{RR}} = a[\mathrm{Fe/H}] + b$,, and the question becomes *"how can we directly measure the slope (a) and the zero point (b)?"*. A calibration of the zero point with $0^{\mathrm{m}}.1$ accuracy would need trig paralllaxes with $\sim$ 0.1 m.a.s. accuracy for a suitable sample of RR Lyrae stars, something we will not reach soon. The perspective looks more favorable for the slope a, which also plays an important role in the dating process. As Carney has reviewed at this meeting, some of the indirect methods give $a \simeq 0.35$, which would imply virtually coeval globular clusters in our Galaxy, while other methods suggest $a \simeq 0.2$ which may imply a large trend of the cluster age with metallicity if not compensated by a suitable trend in [O/Fe]. Therefore, disentangling among much different scenarios for the formation of the Galaxy ultimately relies on measuring the slope with better than $\sim 0^{\mathrm{m}}.05$ accuracy per dex in [Fe/H]. Obtaining CMDs for the globular clusters in M31 would provide a direct determination of the slope, as such clusters are virtually all at the same diatance, and span the full range of metallicity from $[\mathrm{Fe/H}] \simeq -2$ to ~ 0. HST was supposed to provide the necessary data, but now we will have to wait for the recovery of its full capabilities.

* This relation has been derived by Buonanno *et al.* (1989) from the isochrones of VandenBerg and Bell (1985).

Subdwarfs distances are currently based on five such stars with reasonably well known trig parallax (see van Altena *et al.* 1988). The average error in the modulus of the five calibrators is $<\sigma(\text{mod})>= 0.^{\text{m}}15$, which is rather good. However, the MS location is rather sensitive to [Fe/H], and distance determinations by this method make use of the metallicity of the subdwarfs and of the cluster, which are both subject to errors. Unfortunately the five subdwarfs span a very narrow range in [Fe/H], and to extend the calibration to other metallicities one has to rely on theoretical ZAMS models (or on the main sequence of the Hyades cluster) and proceed to interpolations and extrapolations. When paying attention to the propagation of the errors, one finds that $\sigma(\text{mod}) \simeq \sigma([\text{Fe/H}])$, and therefore the relative error in age is nearly equal to the error in [Fe/H]: i.e. a 0.3 dex error in the cluster [Fe/H] propagates into an error of ~30% in age. All in all, while future efforts (and specially the Hipparcos mission) can improve the trig parallax determinations for very many subdwarfs, the problem with the uncertainty in metallicity may remain.

The basic idea of using white dwarf as standard candles is very simple: to fit the WD cooling sequence of a globular cluster to either the appropriate theoretical or empirical WD cooling sequence (Renzini 1991). The procedure is analogous to the classical main sequence fitting to the local subdwarfs, but with some non-trivial advantages: the method does not involve metallicity determinations which inevitably bring along their uncertainties, and there is no mixing length calibration involved. In fact, WDs have virtually metal free atmospheres, coming either in the DA or non-DA varieties (nearly pure hydrogen or pure helium, respectively), and their radius is insensitive to the mixing length. In the most straightforward case of the fitting to an empirical cooling sequence one has only to apply a small correction taking into account the mass difference between the cluster WDs and the local calibrators. Moreover, WDs are locally much more abundant than subdwarfs, and therefore an accurate trig parallax can be obtained for a much larger sample of calibrators. All in all – with HST working at nominal performance – the distance modulus of a cluster should be obtained with an accuracy better than $\sim 0.^{\text{m}}1$, which translates into a better then 10% accuracy in age. Unfortunately, WDs are very faint, and the useful ones are fainter than $V \simeq 24$ even in the closest globular cluster. The spherical aberration syndrome affecting HST makes presently impossible to reach such faint oblects in the crowded field of the clusters, and again, we may have to wait for the deployment of WFPC II, unless adaptive optics on large ground based telescopes comes first. But I believe that, ultimately, white dwarfs will prove the best standard candles, able to improve the calibration of all other calibrators (RR Lyraes, subdwarfs, etc.).

3. THE TURNOFF COLOR AND THE AGE OF STARS IN ELLIPTICALS

The turnoff luminosity is not the only time-dependent feature of theoretical isochrones. Turnoff colors (or the whole shape of the isochrone from the MS to the base of the RGB) also depend on age, and in principle may be used to estimate cluster ages. Actually, this has been a very popular approach to GC dating, a technique also known as *isochrone fitting.* Having thoretical isochrones in one hand, and observed cluster loci in the other, it may be difficult to resist the temptation of overlapping isochrones to data points, pick up the single isochrone which *best fits* the data, and read out its age. Yet, this temptation should be resisted. Unlike luminosities, the temperatures (colors) of theoretical isochrones are in fact seriously affected by our current, still very rough way of parameterizing the efficiency of the convective energy transfer, which is perhaps the most rudimentary of all ingredients entering in the construction of stellar models. Thus, the *shape* of the isochrones, as well as the location of the lower MS, both depend on the so-called mixing-length parameter $\alpha = \ell/H_{\text{P}}$, which is not a priori known, and which requires an independent calibration. Indeed, when using uncalibrated theoretical isochrones to get GC distances and ages we would build up distance and age ladders on the most uncertain aspect of theoretical models: certainly not a cleaver way of proceeding. Moreover, part of this limitation survives even after the calibration has been applied, as the available calibrators are very few: basically, just the sun and the five best subdwarfs already encountered in the previous section (see VandenBerg 1990). In other words, we cannot be sure that the value of α required to fit e.g. the solar radius, also gives the correct radius (i.e. temperature, color) for turnoff stars in a GC.

Should we conclude that fitting isochrones to cluster CMDs is a futile exercise? In spite of the remarkably good match often reached, my answer is *"yes"*, if what matters is the determination of accurate GC ages. Indeed, if we have a direct access to TO luminosities, and we have to rely on subdwarfs for the calibration, why bring in all the uncertainties introduced by the theory of convection, that we cannot even quantify? However, for another purpose, matching isochrones to cluster loci is an indispensable step along the age ladder. Having determined in our best way GC distances and ages following the methods described in §2, then the calibration of α via accurate isochrone fitting provides us with a new tool, a clock that we can use for dating stellar systems that we cannot resolve into individual stars, but for which we have access only to the integrated light. This is the population synthesis technique, often applied to dating the stellar populations of elliptical galaxies, in which the correct sequence of operations is: 1) obtain GC distances using suitable standard candles, 2) get their ages, 3) calibrate α via isochrone fitting, 4) construct synthetic spectral energy distributions (SED) using calibrated isochrones, 5) estimate the age of ellipticals by comparing synthetic and observed SEDs. Far too often some of these operations have been omitted, or used in a different order, e.g. getting GC ages via isochrone fitting, or estimating the age of ellipticals using synthetic SEDs based on uncalibrated isochrones.

It is worth emphasizing that the calibration of the mixing-length parameter α is not such a trivial exercise, after all. Suffice to compare the results of different calibrations, such as Fig. 1 in Demarque *et al.* (1988), or Fig. 3 in VandenBerg (1990), which show that even if two calibrations coincide near the MS, they can dramatically differ in the TO region, to the extent that for fixed TO color the inferred age may differ by a factor of two. Therefore, before venturing into cosmological applications of population synthesis it is better to make sure that isochrones provide a nice match from the ZAMS all the way to the tip of the RGB, such as in the case e.g. of the calibrated isochrones of Straniero and Chieffi (1991).

Convection is not the only concern when using the TO color as an age indicator. As frequently emphasized, metallicity is another serious problem (see e.g. O'Connell 1980; Renzini 1986). Suppose we have been able to determine with extreme precision the TO color of a population, from either its CMD or its SED: to get the age we then need an independent estimate of the metallicity of the population. With little algebra, from Straniero and Chieffi (1991) one gets:

$$\left(\frac{\partial \mathrm{Log}\, t}{\partial [\mathrm{Fe/H}]}\right)_{(B-V)^{\mathrm{TO}}} = -0.88 - 0.35[\mathrm{Fe/H}], \qquad (2)$$

i.e. $d\mathrm{Log}\, t \simeq d[\mathrm{Fe/H}]$ for near solar metallicity, and a 0.3 dex error in the estimated metallicity translates into a factor of two error in age. A comparison with eq. (1) immediately reveals how much more vulnerable to metallicity errors are ages estimated from TO colors, compared to ages derived from the TO luminosity. In the case of colors the error is up to ten times larger than in the latter case. This argument not only provides further strength to the case against the use of the isochrone fitting method to date GCs, but it reveals also one of the most disturbing intrinsic weaknesses of the population synthesis approach, a drawback that simply cannot be eliminated: the interesting quantity to determine is age, but the SED of a population is more sensitive to metallicity (and to its detailed distribution) than it is to age.

The case of the nearby dwarf elliptical M32 provides the best illustration of the problem. It does not matter here that M32 may not be a typical elliptical, that its evolutionary history may have been much different from that of giant ellipticals. What matters is that we have excellent spectroscopic data for it, photometry of individual stars is now possible, and a long sought test of the population synthesis results is becoming feasible. Over the years the population synthesis method has indicated for M32 1) near solar metallicity ($[\mathrm{Fe/H}] \simeq 0$), 2) no metallicity dispersion, and thus 3) the presence of a dominating intermediate age ($\sim$ 5 Gyr old) component (see O'Connell 1986, and references therein). However, thanks to the superior imaging capabilities of the CFH telescope on Mauna Kea, Freedman (1989) has been able to resolve M32 into individual RGB and AGB stars and found $<[\mathrm{Fe/H}]> \simeq -0.5$, or slightly larger (see also her contribution to these proceedings), and a 1σ metallicity dispersion of ~ 0.5 dex. Given the reduction by a factor of 3 in average metallicity, I would maintain that adopting the *observed* metallicity distribution, and applying the *same* population synthesis methods which gave 5 Gyr, thanks to eq. (2) one should now obtain ~ 15 Gyr, i.e. no need for an intermediate age component. After this discussion two questions remain. Why O'Connell's 5 Gyr, [Fe/H] = 0 synthetic spectrum fits so well M32, while we know from Freedman that there is a large metallicity dispersion, and the average is below solar? Why Bica *et al.* (1990) cluster-melange method prefers a small metallicity dispersion and a large spread in age instead of the contrary? Answering these questions will help to understand first how our population synthesis tools work, and then how to most effectively apply them to other galaxies.

Finally, I would like to call attention to another relevant aspect. To my understanding, all evolutionary population synthesis models constructed so far have assumed a RGB made of K-type giants. Yet, the upper RGB of near-solar metallicity GCs in the Galactic Bulge is instead populated by M-type giants (Ortolani *et al.* 1990, see also this volume). In super metal rich (SMR) populations – such as those dominating in giant ellipticals – an even larger fraction of the RGB and AGB is likely to be populated by M giants. This means that perhaps $\sim 30\%$ or more of the total bolometric light may come from M giants, thus considerably decreasing the RGB and AGB contributions in the visual, compared to models in which the giant branch is supposed to be made of K giants. Curiously enough, turning K giants into M giants will make the synthetic $B-V$ *bluer* (and the $V-K$ redder), thus increasing the estimated age by an amount which remains to be properly evaluated.

4. THE AGB TIP AND THE AGE OF YOUNGEST STARS IN BULGES

Besides the TO region, other parts of the CMD are sensitive to age. For example, the color and morphology of the horizontal branch (HB) and the luminosity extension of the AGB are both sensitive to age, and can be used in age estimates. Unfortunately, both depend also on the amount of mass loss during the red giant phases, which is not known with the necessary precision from either observations or theory. This implies that the HB morphology and the AGB extension cannot be used for *absolute* age determinations, but relative age estimates are possible once mass loss is properly calibrated using populations of independently known age. Again, we find here another methodological analogy with the distance ladder: the dermination of the age of GCs via eq. (1) allows a calibration of mass loss once we demand that theoretical HBs and AGBs fit those observed in Galactic GCs. Having done so, it becomes possible to use the HB and/or the AGB of other clusters or populations to infer their age difference with respect to the calibrating clusters. The HB as a clock is discussed by others at this meeting (e.g. Y.-W. Lee and Suntzeff in connection with *Second Parameter* effects, while for a possible use on moderate redshift galaxies see Greggio and Renzini, 1990). In this section I will then concentrate on the AGB.

In Galactic GCs the AGB barely extends above the RGB tip, which implies that mass loss has to prevent the stellar core from growing beyond $0.54 - 0.55 M_\odot$ (Renzini and Fusi Pecci 1988). Several clusters of the Magellanic Clouds contain instead AGB stars which are much brighter than the RGB tip, and since Mould and Aaronson (1979) this is currently interpreted as an age effect. Fig. 1 shows theoretical RGB and AGB luminosity functions (LF) appropriate to stars less massive than $\sim 2M_\odot$. In these AGB models the core is allowed to grow up to $\sim 0.6M_\odot$, and therefore the composite RGB+AGB luminosity function is one appropriate to an intermediate age (few Gyr) stellar population. Note that the LF drops by a factor of ~ 4 at the tip of the RGB, as the RGB evolution is suddenly terminated by the onset of the core helium flash. As age is increased, the main effect is merely a decrease of the luminosity of the AGB tip, until in a very old (~ 15 Gyr) population the AGB tip gets very close or even fainter than the RGB tip whose luminosity is virtually independent of age. One can conclude that in a composite population consisting of an old plus an intemediate age component the LF must exhibit a drop by *more* than a factor of 4 at the luminosity corresponding to the tip of the RGB. Moreover, only having identified this feature in the LF one can draw the bottomline above which stars can certainly be ascribed to the AGB. More quantitatively, the expected RGB drop in the LF is given by $\sim 4(L_{\rm OLD} + L_{\rm IA})/L_{\rm IA}$, where $L_{\rm OLD}$ and $L_{\rm IA}$ are

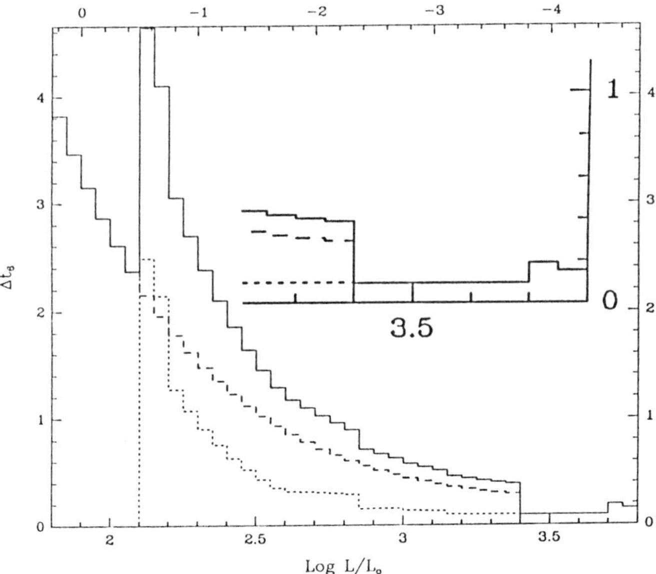

Fig. 1. The RGB (dashed line), AGB (dotted line), and cumulative (RGB+AGB, solid line) luminosity function for a population older than $\sim$ *1 Gyr, with* $(Z,\ Y) = (10^{-3},\ 0.3)$. *The vertical scale is in* 10^6 *yr per luminosity bin* $\Delta \mathrm{Log}\, L = 0.05$, *while the upper scale gives the bolometric magnitude. The inlet shows a magnification of the top end of the LF, showing the drop associated to the RGB tip which for* $Z = 10^{-3}$ *is at* $\mathrm{Log}\, L/L_\odot\ = 3.4$ $(M_{\mathrm{bol}} \simeq -3.8)$. *The RGB tip get* ~ 0.25 *mag brighter every dex increase in metallicity. The RGB and AGB models are respectively from Sweigart and Gross (1978) and Gingold (1974).*

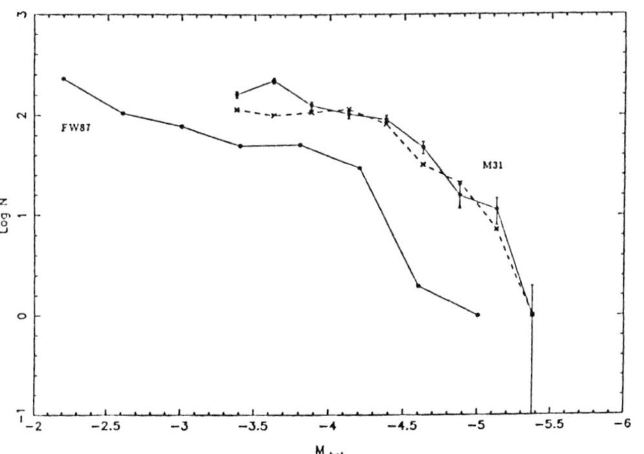

Fig. 2. The luminosity functions of the bulge of M31 (Mould and Rich 1991), and of Baade's window in the Galactic Bulge (Frogel and Whitford 1987, labelled FW). Note the absence in the LF of M31 of any feature resembling the RGB drop in Fig. 1. In the Galactic Bulge the only drop is at $M_{\mathrm{bol}} \simeq -4.2$, *suggesting this feature being the RGB drop itself.*

respectively the total luminosity of the old and of the intermediate age components.

Fig. 2 shows the LF of the Galactic Bulge from Frogel and Whitford (1987), compared to the one that Mould and Rich (1991) have obtained for the bulge of M31. From the lack of a drop off at $M_{\mathrm{bol}} \simeq -4.2$ Mould and Rich infer (among other possibilities) that the bulge of M31 may contain an intermediate-age component that according to Frogel and Whitford is not present in the Galactic Bulge. Yet, the LF of M31 does not show any sign of the expected RGB tip drop off, which in a SMR population should be found around $M_{\mathrm{bol}} \simeq -4$: something must have washed it out. I have no explanation for this, but certainly we need to uderstand why there is no RGB drop off before using the LF in Fig. 2 to make age inferences. Yet, this is still not enough. We need also to know what is the AGB tip luminosity in a SMR, 15 Gyr old population, so as to set the bottomline above which stars can be ascribed to an intermediate age component. Unfortunately, the composition dependence of the luminosity of the AGB tip has not been sufficiently explored, and recent calculations (e.g. Lattanzio 1991) don't extend beyond solar. Bulges instead contain SMR stars (see Rich, this volume), and for a moderate galactic helium enrichment ($\Delta Y/\Delta Z \simeq 3$) such stars would also be strongly enriched in helium (i.e. $Y \gg Y_\odot$, note that bulge stars would have really extreme composition if $\Delta Y/\Delta Z \simeq 6$, as advocated by Pagel at this meeting). In conclusion, while waiting for more extended evolutionary calculations we are left with the unanswered question: *"Are large AGB luminosities in bulges a result of young ages or of large $Z+Y$?"* To answer we need a suitable calibration of both giant branch tips as a function of composition, in particular in the SMR regime.

Even after the identification of the RGB tip and of the old age AGB tip, an additional complication remain. Blue stragglers (BS) are now found abundantly in globular clusters, and almost certainly BSs are the result of the merging of two main sequence stars in a close binary. Following merging BS evolve as single star with mass up to two times the turnoff mass of the parent population, and therefore when on the AGB they climb to much higher luminosities than single stars do. It follows that a BS progeny could be mistaken for an intermediate age component. Scaling from the frequency of BSs in M3, Renzini and Greggio (1990) estimate the number of BS progeny stars in a population to be $N_{\rm j}^{\rm BSP} \simeq 6\times 10^{-13} L_{\rm T} t_{\rm j}$, where $L_{\rm T}$ is the total luminosity of the population, and $t_{\rm j}$ is the duration of the $\rm j^{th}$ evolutionary phase. From this relation, adopting $10^{10}L_\odot$ for the luminosity of the Bulge, and 2×10^6 yr for the duration of the TP-AGB we obtain $N_{\rm TP-AGB}^{\rm BSP} \simeq 6\times 10^{-13}\times 10^{10}\times 2\times 10^6 = 12,000$ stars in the whole Bulge, a number that can be appropriately scaled to the actual luminosity sampled by a CCD frame or IR array. This is not a completely negligible contribution, and the presence of an intermediate age component in bulges (and in M32, see Freedman's contribution) would be confirmed only if the actual number of AGB stars brighter than the (still poorly known) AGB limit of an old SMR populations were to exceed this estimate, i.e. ~ 1 star every $10^6 L_\odot$. Conversely, in a few Gyr single age population the expected frequency of TP-AGB stars is about 30 times larger: the actual counts will decide.

5. THE AGB, THE RGB, AND THE AGE OF HIGH-z RADIOGALAXIES

Dating stellar populations in nearby elliptical galaxies being a rather cumbersome affair (see §2), the existence of *features* in the time evolution of the SED would certainly help to pinpoint ages more precisely. The rapid development at fairly precise ages of the AGB and later of the RGB (the so-called phase transitions) represent two such features (Renzini and Buzzoni 1983, 1986), and soon it has been attempted to use them in dating high redshift galaxies (e.g. Wyse 1985). Here I would like to report on recent progress along these lines.

In drawing their Fig. 5 showing the relative contribution to the integrated light of a population, Renzini and Buzzoni made use of the best stellar models available in the early '80s, without any attempt at fudging with parameters, but making clear that a calibration of mixing length, mass loss, and overshooting was necessary before applying their theoretical predictions to galaxies. Also, the natural calibrators were indicated in the GCs of the Magellanic Clouds, which span the interesting range in age, and contrary to open clusters are populous enough for the purpose. As pointed out by RB (and later emphasized by many other investigators) their Fig. 5 gives a gross overestimate of the AGB contribution for clusters younger than a few 10^8 yr, such as NGC 1866 that *uncalibrated* theory predicted to have some 10 bright AGB stars, while it has none. Indeed, the problem of the missing bright AGB stars in Magellanic fields and clusters has been around for over 10 years (see e.g. Frogel *et al.* 1990, and references therein), and a number of possible solutions have been proposed. I should say that I have never been satisfied with any of such solutions of the *"AGB mystery"*, including those that I have entertained myself. Now, finally, the real solution may have been found. Blöcker and Schönberner (1991) have recently computed TP-AGB models for a $7M_\odot$ star, with a mixing-length parameter $\alpha = 2$. As expected (see Renzini and Voli 1981), their models experience strong *envelope burning* process, but most surprizingly their luminosity evolution is much faster than predicted by the famous Paczyński's core mass-luminosity relation. The models spend between $M_{bol} = -6$ and -7 a time which is nearly 10 times shorter than was the case for models assuming Paczyński's relation, such as those of Renzini and Voli. These calculations suggest that, climbing quickly to very high luminosity, the more massive AGB stars which experience envelope burning will rapidly run into severe mass loss, thus leaving the AGB and evolving towards their final white dwarf configuration. The final result is a drastic reduction of the TP-AGB lifetime and of the amount of fuel which is burned during the TP-AGB, i.e. of the TP-AGB contribution to the total light of a population. Both results now offer what appears to be an excellent opportunity to finally explain the AGB luminosity function of MC clusters and fields (see Mould's contribution), while the high lithium abundance in bright AGB stars provides independent evidence for the envelope burning process being indeed active for $M_{bol} \lesssim -6$ (Smith and Lambert 1989). It is worth emphasizing that Blöcker and Schönberner's important results undermines two generally entertained dogmas: the universality of the core mass-luminosity relation, and the insensitivity of the stellar luminosity to the mixing-length parameter. *A posteriori*, it looks quite natural that both are invalid in presence of envelope burning, and – after all – it is somewhat reassuring for theory that the solution to the AGB mystery can evenually be found in just properly calibrating the envelope mixing-length parameter α!

Fig. 3 is an update of the old RB's Fig. 5. The input models are the same, apart from the use of Sweigart *et al.* (1989) models to nicely delineate the RGB phase transition at $t \simeq 5 \times 10^8$ yr*. Furthermore, a sharp switch on of envelope burning is assumed in stars more massive than $3M_\odot$ (the value appropriate for $\alpha = 2$ and $Z = 0.01$, see Renzini and Voli 1981), thus reducing by a factor of 10 the fuel consumption during the TP-AGB. Admittedly, this is still no more than an educated guess. Note that what was meant by AGB phase transition now splits into two evolutionary bifurcations: the first one still at $M \simeq 9M_\odot$ corresponds to the appearance of stars with degenerate C-O cores (i.e. the AGB transition in the *old* sense), the second one at $M = M_{HBB}$ (in the notations of Renzini and Voli 1981) corresponds to the rather sharp separation between stars with and without envelope burning. The mass (age) at this separation is a function of α, and decreases from $M_{HBB} \simeq 9M_\odot$ for $\alpha \lesssim 1$, to $\sim 3M_\odot$ for $\alpha = 2$, as assumed in drawing Fig. 3. Thus, for $\alpha = 2$ the first transition at $t \simeq 2 \times 10^7$ yr is not associated with the appearance of a major AGB contribution to the total light (see Fig. 3), while such a contribution is substantially delayed to $t \simeq 2 \times 10^8$ yr, and gets close to the RGB phase transition which rapidly brings the RGB contribution over 20%. In this way, the AGB+RGB contribution rapidly climbs from a fairly small fraction for $t \lesssim 2 \times 10^8$ yr, to over 50% at $t \simeq 10^9$ yr. It is particularly attractive to ascribe to this combined AGB+RGB development the very rapid jump of $V - K$ from ~ 1 to ~ 3.5 exhibited by SWB type 4 clusters of the Magellanic Clouds (see Fig. 3 in Renzini 1991). No such a major effect is predicted on the optical colors, since AGB and RGB stars radiate mostly in the near-IR.

What are the perspectives of detecting this ~ 2.5 mag color jump in young ($\lesssim 10^9$ yr), high redshift galaxies? Could the *"Red Bump"* in the SED of some 1 Jansky radiogalaxy be ascribed to the AGB or RGB phase transitions? Lilly (1988) has derived an age of $\gtrsim 10^9$ yr for one radiogalaxy at $z = 3.4$, and such an old age at such a large redshift would push the epoch of galaxy formation at $z > 10$, at odd with the standard CMD scenario. Lilly's determination relies on models which assume a 1 Gyr duration of the star formation process (Bruzual 1983), and in which the red giant component is not adequately treated. Either of these two aspects may have affected the result. More recently Chambers and Charlot (1990) have re-analyzed the problem, concluding that the galaxy with the highest redshift in their sample ($z = 3.8$) has an age of only $\sim 3.3 \times 10^8$ yr, thus somehow releaving the pressure on CDM. What is new in Chambers and Charlot's approach compared to that of Lilly? First they have assumed a shorter duration for the star formation process (i.e. $\sim 10^8$ rather than 10^9 yr), second, they have attempted a more sophisticated approach to the population synthesis, in particular improving the treatment of the red giant evolutionary phases, and assuming

* Sweigart *et al.* models have been implemented with a few HB and Early-AGB sequences kindly provided by Chieffi and Straniero.

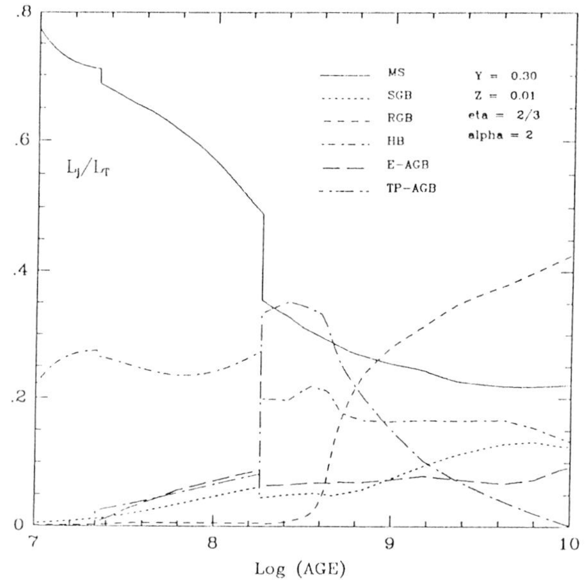

Fig. 3. The time evolution of the relative contributions L_j/L_T of stars in the various evolutionary stages to the integrated bolometric light of an evolving stellar population. Composition, mass loss, and mixing-length parameters (η and α) are indicated. No overshooting is assumed. The contributions of the Early- and Thermally Pulsing-AGB phases are plotted separately.

an important AGB contribution to appear at $t = 3 \times 10^8$ yr. At first sight I was striken by the derived age being only 10% older than the assumed AGB phase transition, is this transition the origin of the *red bump*? Probably not. A more meditated analysis of the result reveals in fact that Lilly's *red bump* isn't really very red, corresponding to a rest frame wavelength of only $\sim$ 5000 Å, where late type (M or C) AGB stars should contribute little flux. For this reason I am now inclined to believe that the younger age follows from the fact that Chambers and Charlot have assumed a 10 times shorter e-folding time for star formation (compared to Lilly). If this is the true reason, then it implies that the observed SED only constrains the bulk of star formation to have been completed some 10^8 yr ago, but tells very little on the age of the galaxy, i.e. on the beginning of the star formation, and therefore on the epoch of galaxy formation. It follows that other ways have to be envisaged to date high redshift galaxies, and the combined AGB+RGB phase transitions – with the associated 2.5 mag jump in $V - K$ – may help in this respect. Still, at $z = 4$ the restframe $V - K$ would be observed as $K - 10\mu$, and it may take a long time before we will be able to detect high redshift radiogalaxies at 10μ.

6. CONCLUSIONS

Having discussed four dating methods along the age ladder, I would like to conclude with a few schematic sentences:

• Current GC dating methods give ages of 13–15±3 Gyr. If $H_o = 100$, then we need $\Lambda \neq 0$, with $\Lambda + \Omega = 1$ to keep the inflationary scenario. For this among other reasons $\Lambda \neq 0$ is not regarded now as ugly as it used to be years ago (see e.g. Weinberg 1989; Peebles 1991).

• The question whether ellipticals contain an intermediate age component remains unsettled by population synthesis methods. Numerical experiments with synthetic populations would be very valuable in assessing the age vs. metallicity *resolution* of such methods.

• My prejudice is that bulges are (on average) older than halos (Renzini and Greggio 1990). To test this idea it would be interesting to compare the HB luminosity vs. HB color clocks, along the lines presented by Suntzeff and Y.-W. Lee at this meeting. Also, counts to luminosity ratios are needed to assess whether the brightest stars in bulges belong to an intermediate age AGB, or what.

• It may take some time before we are able to date very high-z radiogalaxies, but the mere existence of $z = 5$ quasars supports the notion of $z_{GF} \gtrsim 10$ (see Turner 1991).

I am indebted to Laura Greggio and to Michele Guastamacchia for the plot of Fig. 1 and Fig. 3, respectively, and to Sandro Chieffi and Oscar Straniero for having expressly computed and provided several evolutionary sequences incorporated into Fig. 3.

REFERENCES

Bica, E., Alloin, D., Schmidt, A.A. 1990, *Astr. Ap.*, **228**, 23

Blöcker, T., Schönberner, D. 1991, *Astr. Ap.*, **244, L43**

Boesgaard, A.M., Steigman, G. 1985, *Ann. Rev. Astr. Ap.,* **23**, 319

Bruzual, G. 1983, *Ap. J.*, **273**, 105

Buonanno, R., Corsi, C.E., Fusi Pecci, F. 1989, *Astr. Ap.*, **216**, 80

Chambers, K.C., Charlot, S. 1990, *Ap. J. (Letters),* **348**, L1

Demarque, P., Guenther, D.B., King, C.R., Green, E.M. 1988, in *Calibration of Stellar Ages*, ed. A.G.D. Philip (Schenectady: L. Davis), p. 101

Freedman, W. 1989, *A. J.*, **94**, 1285

Frogel, J.A., Mould, J.R., Blanco, V.M. 1990, *Ap. J.*, **352**, 96

Frogel, J.A., Whitford, A.E. 1987, *Ap. J.*, **320**, 199

Gingold, R.A. 1974, *Ap. J.*, **193**, 177

Greggio, L., Renzini, A. 1990, *Ap. J.*, **364**, 35

Lattanzio, J.C. 1991, *Ap. J. Suppl.*, **76**, 215

Lilly, S.J. 1988, *Ap. J.*, **333**, 161

Mould, J.R., Aaronson, M. 1979, *Ap. J.*, **232**, 421

O'Connell, R.W. 1980, *Ap. J.*, **236**, 430

O'Connell, R.W. 1986, in *Stellar Populations*, ed. C. Norman, A. Renzini, M. Tosi (Cambridge Univ. Press), p. 167

Ortolani, S., Barbuy, B., Bica, E. 1990, *Astr. Ap.*, **236**, 362

Peebles 1991, in *Observational Tests of Inflation*, ed. T. Banday, T. Shanks (Dordrecht: Kluwer), in press

Renzini, A. 1986, in *Stellar Populations*, ed. C. Norman, A. Renzini, M. Tosi (Cambridge Univ. Press), p. 213; 1991, in *Observational Tests of Inflation*, ed. T. Banday, T. Shanks (Dordrecht: Kluwer), in press

Renzini, A., Buzzoni, A. 1983, *Mem. S.A.It.* **54**, 739; 1986, in *Spectral Evolution of Galaxies*, ed. C. Chiosi and A. Renzini (Dordrecht: Reidel), p. 195

Renzini, A., Greggio, L. 1990, in *Bulges of Galaxies*, ed. B.J. Jarvis, D.M. Terndrup (Garching: ESO), p. 47

Renzini, A., Fusi Pecci, F. 1988, *Ann. Rev. Astr. Ap.* **26**, 199

Rich, R.M., Mould, J.R. 1991, A. J. **101**, 1286

Sandage, A. 1970, *Ap. J.*, **162**, 841

Smith, V.V., Lambert, D.L. 1989, *Ap. J. (Letters),* **345**, L75

Straniero, O., Chieffi, A. 1991, *Ap. J. Suppl.,* **76**, 525

Sweigart, A.V., Gross, P.G. 1978, *Ap. J. Suppl.*, **36**, 405

Sweigart, A.V., Greggio, L., Renzini, A. 1990, *Ap. J.*, **364**, 527

Turner, E.L. 1991, *A. J.*, **101**, 5

van Altena, W.F., Lee, J.T., Hanson, R.B., Lutz, T.E. 1988, in *Calibration of Stellar Ages*, ed. A.G.D. Philip (Schenectady: L. Davis), p. 175

VandenBerg, D.A. 1990, in *Astrophysical Ages and Dating Methods*, ed. E. Vangioni-Flan *et al.* (Gif sur Yvette: Ed. Frontières), p. 241

VandenBerg, D.A., Bell, R.A. 1985, *Ap. J. Suppl.*, **58**, 561

Weinberg, S. 1989, *Rev. Modern Phys.,* **61**, 1

Wyse, R.E.F. 1985, *Ap. J.*, **299**, 593

DISCUSSION

Y.-W. LEE: You mentioned the use of HST to observe M31 globular clusters to obtain the HB luminosity-[Fe/H] relationship directly. While I agree with you 100%, I like to remind you that there is another, much cheaper method already available. That is the use of RR Lyrae in ω Cen. Since there is a wide range in [Fe/H], and since all RR Lyrae in ω Cen are located at the same distance, we can use the apparent V mag vs. [Fe/H] relationship directly. The result (see Lee 1991, *Ap. J. (Letters)*, , June 1) supports the magnitude difference of 0.2 mag for 1 dex variation in [Fe/H], consistent with LDZ HB models.

RENZINI: You are right, ω Cen certaily helps, but its metallicity range (in total Z, not just Fe) may be rather modest and the precision with which one can derive the *slope* correspondingly reduced.

MATEO: I have two comments. (1) As you pointed out, some CGs have blue stragglers. As it turns out, some of these extend into the instability strip and are seen as pulsating dwarf Cepheids. These stars follow a P-L relation that can be trigonometrically calibrated by Hypparcos. So a 4th distance determination method for GCs may soon be available. (2) You may recall from my talk that I plotted the $(V-K)_o$-age relation using Persson *et al.* (1983) data and *new* age estimates for the clusters. I remind you that there was only one break in this relation at 10^9 yrs corresponding to the RGB phase transition; no feature is visible at the AGB transition which I assume will show up at about 10^8 yrs.

RENZINI: I think there are in LMC clusters in which the AGB is well developed, as testified by the presence of bright carbon stars, while the RGB is still lacking. One such case is NGC 2209, which unfortunately is not in the Persson *et al.* sample. Notice that only $\sim 2-3\times 10^8$ yr separate the development of the two branches (cf. Fig. 3), while SWB type 4 cluster are rather few and span a $\sim$ 3 to 4 times longer age interval. So we don't expect to find many clusters which are older than the AGB phase transition, but younger than the RGB one. Still, I agree with you that the AGB transition alone (at $t \simeq$ few 10^8 yr) cannot account for the whole jump in $V-K$, but it needs the assistence of the RGB transition.

MOULD: (1) You should be looking for the RGB tip discontinuity closer to $M_{\rm bol}=-3.6$, and at that luminosity the M31 bulge counts are very incomplete. (2) There are more than 10^5 stars brighter than $M_{\rm bol}=-4$ in a $10^{10}L_{V\odot}$ M31 bulge scaled to our field.

RENZINI: If the RGB tip is as faint as $M_{\rm bol}=-3.6$, then why at this luminosity the LF of Baade's window is so perfectly smooth? Frogel and Whitford 's counts are supposedly complete down to much fainter luminosities. The only clear drop is at $M_{\rm bol}=-4.2$, and thus I suspect that this is the RGB tip in Baade's window. Otherwise we have a serious problem with either the counts or the RGB theory. (2) OK, I'm looking forward to the detailed counts, and to their dependence on the assumed bottomline.

FROGEL: (1) The luminosity function for M31 that R. Davies, D. Terndrup and myself have determined has an even slower fall off above $M_{\rm bol}=-4.5$ than that of Rich and Mould. (2) The steep fall off in the bulge luminosity function undoubtely refers to the termination of AGB evolution, not of RGB evolution. I have no idea why the RGB tip is not seen in our Baades's window data. (3) It is really essential that models be made to predict terminations of AGB evolution as a function of mass and [Fe/H]. It is a mistake to keep using LMC clusters for galaxy models as the AGB stars in them have [Fe/H] significantly lower than the mean for E galaxies and spiral bulges.

RENZINI: (1) No comment. (2) I am not so certain. If the drop is due to the AGB then your Baade's window luminosity function should be dramatically incomplete already at the RGB tip, which has to be around $M_{\rm bol}=-4.0$. Again, it makes more sense to me to look at the $M_{\rm bol}=-4.2$ drop as due to the RGB tip. (3) I agree.

FREEDMAN: Jay Frogel has argued that in the Galactic Bulge the tip of the AGB occurs at $M_{\rm bol}=-4.2$ due to high metallicity. However the tip of the RGB has not been identified in the

Bulge luminosity function. As Alvio has pointed out, the tip of the RGB should be easy to detect, certainly in the case of the Galactic Bulge. The lack of a sharp cutoff in the Bulge LF at $M_{\rm bol} \gtrsim -3.7$ may be suggesting either that the slope of the RGB tip luminosity vs. metallicity relation is steeper at high metallicities (than measured for Globular clusters) or that the LF of the Bulge based on the Blanco M giant surveys is incomplete. In any case, it is still interesting that the M31 LF shows no sharp drop at $M_{\rm bol} = -4.2$ unlike the Galactic Bulge LF.

RENZINI: Yes, a range in metallicity will smooth out the RGB drop in the LF, but a sharp ridge line should still be recognizable in a well populated CMD.

PAGEL: (1) I do not claim $\Delta Y/\Delta Z = 6$ at high metallicity, only at low metallicity, while agreeing that the resulting trend is contrary to naïve theoretical expectations. (2) I believe that Th/Eu ratios in the few halo field stars for which they are available are an argument against an age as low as 10 Gyr; 15 Gyr seems more reasonable.

RENZINI: (1) Indeed, with a straight $\Delta Y/\Delta Z = 6$, in SMR stars with $Z = 5Z_\odot = 0.1$ hydrogen should be reduced to a trace element! I'm looking forward to see a non-naïve theory giving $\Delta Y/\Delta Z = 6$ at low metallicity, (2) 15 Gyr is also my preferred value.

SEARCHES FOR PRIMEVAL GALAXIES

S. Djorgovski and D. J. Thompson
Palomar Observatory
California Institute of Technology
Pasadena, CA 91125, USA

ABSTRACT. We review primeval galaxy searches based on the Lyα line emission. Simple arguments are given which suggest that primeval galaxies (interpreted here as ellipticals and bulges undergoing their first major bursts of star formation) should be detectable with present-day technology. Many active objects are now known at large redshifts, which may be plausibly interpreted as young galaxies, but there is so far no convincing detection of a field population of forming normal galaxies. This suggests that either primeval galaxies were obscured, and/or are to be found at higher redshifts, $z_{gf} > 5$.

Galaxy formation is one of the central problems of modern cosmology. It touches on many different subfields of astrophysics: formation and evolution of large-scale structure, chemical evolution and star formation history of galaxies and stellar populations, physics of star formation, etc. Discovery of normal galaxies forming at large redshifts will be an important milestone and would open many new exciting problems and fields of inquiry.

The purpose of this review is to describe some of the observational issues involved, to describe the results from modern searches for primeval galaxies (hereafter PGs), and the future prospects. Previous reviews include those by Davis 1980, Koo 1986, Spinrad 1987, 1989, Cowie 1988, 1989, Djorgovski 1988a,b, 1992, and numerous excellent papers in the proceedings edited by Frenk et al. 1989 and Bergeron et al. 1988; and also in Hewitt et al. 1987, Thuan et al. 1988, Kron & Renzini 1988, Kron 1990, and many others.

Many different things are meant by galaxy formation. At least two distinct kinds of phenomena are involved (Silk & Norman 1981; Silk 1987; Larson 1990; etc.): (1) assembling of the mass in galaxy-size units, e.g., via merging; and (2) conversion of gas into stars, and energy dissipation. They may, but need not coincide in time or in the location; neither need have had a well defined or even a unique peak epoch. The first is the purview of theories and numerical simulations. Unfortunatelly, since most of the mass is (and probably always was) dark, direct observations of (1) are difficult or impossible. The second is what may be observable, and is where most of the energy release probably occurs. It is perhaps better to think of galaxy formation not as a single, well defined event or phenomenon, but rather as an evolving sequence of processes, which continue to the present epoch. There is no sharp or natural boundary between galaxy formation and galaxy evolution.

Some galaxy formation clearly happens even today. First, star formation in the disks of most normal spiral galaxies has not changed very substantially over the Hubble time, and for the late types it may even be on the rise. This continuing star formation may be powered by a gradual infall of gas. A better term for it would be galaxy *growth*. There may be genuine young *dwarf* galaxies at $z \sim 0$, such as I Zw 18 (Kunth & Sargent 1986), or H I 1225+018 (Giovanelli & Haynes 1989). It is still not clear whether such systems are just begining their first major bursts of star formation, or whether they represent rejuvenated old dwarfs, and how many progenitors of such systems (i.e., large intergalactic gas clouds) there may be left at $z \sim 0$. In any case, such objects are rare, judging by the failure of many searches to find more of them. Whereas these may be true young galaxies, they

B. Barbuy and A. Renzini (eds.), The Stellar Populations of Galaxies, 337–347.

are probably distinct from the high-z progenitors of normal galaxies, and contain only a small fraction of the total baryonic mass. Finally, there are merger remnants and merger-stimulated starburst galaxies, including many ultraluminous IRAS objects, in which large star formation rates ($\sim 100\ M_{\odot}/yr$) are implied. Had any of them been discovered at large redshifts, they could have been readily called PGs. However, it is clear that at $z \sim 0$ these are mergers of preexisting galaxies, and as such, they are better called examples of galaxy *trans*formation, than a formation proper. For the purposes of this review, we define PGs as the progenitors of normal ellipticals and bulges, undergoing their first major bursts of star formation at large redshifts, a definition assumed by many observers.

The exceptional smoothness of the CMBR photosphere at $z \sim 1000$ ($\Delta T/T < 10^{-5}$ on galaxy and cluster scales, as of this writing) places an upper limit on the redshift of galaxy and large-scale structure formation. The absence of an obvious population of PGs at $z < 1$ places a lower limit to the characteristic redshift of galaxy formation, z_{gf}. We now know normal galaxies and rich clusters out to $z \sim 1$ (Gunn et al. 1986; Thompson & Djorgovski 1991), radio galaxies out to $z = 3.8$ (Chambers et al. 1990), and quasars out to $z = 4.9$ (Schneider et al. 1991). The redshift range $\sim 5 - 1000$ is still a *terra incognita*, in which the bulk of galaxy formation may have happened, or at least began.

Peebles 1989 lists several simple, general arguments which suggest that $z_{gf} \sim 10 \times 3^{\pm 1}$. The basic idea is that several lines of reasoning and evidence suggest that protogalaxies collapsed by about a factor of 10, which would make galactic halos adjacent at $z \sim 10$. On the other hand, most n-body models of hierarchical structure formation, and the CDM scenario in particular, suggest that galaxy formation happens late, i.e., $z_{gf} \simeq 2 \pm 1$ (e.g., Baron & White 1987; Silk & Szalay 1987; White 1989; etc.). What these models really predict is the peak merging epoch, which is then identified with the epoch of galaxy formation, which is reasonable, since merging of gas-rich protogalactic fragments may be the cause of the initial starbursts. A good argument for a higher z_{gf} may be the existence of copious quasars and radio galaxies at $z > 3$, and the absence of a sharp (if any) quasar number density cutoff out to $z \sim 5$. In order to form quasars at $z > 4$, it may be necessary to start formation of their host galaxies at $z > 10$ or beyond (Turner 1991). A reasonable upper bound on z_{gf} may be given by the requirement that protogalaxies are able to cool via the inverse Compton mechanism on the CMBR photons, which is ineffective at $z > 30$ or so. The plausible hunting ground for PGs is then at $2 < z < 30$. Most searches for PGs to date have concentrated to lower redshifts, $z < 5 - 7$, mainly because of technical limitations: beyond that, Lyα-bright objects become hard to observe in the visible light.

Fossil evidence about the star formation histories of normal galaxies is written in their color distributions at $z \sim 0$. For all age-sensitive colors, e.g., $(U - B)$ or $(B - V)$, the distributions are bimodal, with a broad blue peak corresponding to disks and irregulars, and a narrow red peak corresponding to ellipticals and bulges. For most stellar population synthesis models, these correspond roughly to the average stellar ages of $\sim 3 - 6$ Gyr and $\sim 8 - 20$ Gyr. For a reasonable range of cosmologies, these translate to the characteristic star formation redshifts of $\sim 1 - 3$ for disks, and $\sim 5 - 30$ for ellipticals and bulges. Copious faint blue galaxies seen in deep surveys, such as those by Tyson 1988, Cowie et al. 1988, or Lilly et al. 1991, are thought to be largely at $z < 3$, due to the absence of a Lyman limit break in their broad-band colors (Guhathakurta et al. 1990). Possibly, these optically faintest galaxies seen so far are disks and dwarfs undergoing some luminosity evolution at $z \sim 0.5 - 3$. The absence of any $z > 1$ galaxies in the spectroscopic surveys reaching down to $B \sim 22^m$ also suggests that we haven't yet seen a population of young ellipticals or bulges. Since the average stars in ellipticals and bulges are older than the average stars in disks, it follows that ellipticals and bulges probably form at $z > 3$.

Faint galaxy counts also provide a means of estimating the expected number density of PGs on the sky. There are about 10^6 galaxies/degree^2 down to $m \sim 27$, which can only be a lower limit on their total surface density: fainter objects probably exist, and many galaxies may also be obscured. Appeals to merging are unlikely to change these estimates by a large factor, since most galaxies today have disks which are dynamically fragile, and at least a half of the Hubble time old. If a galaxy spends at least a percent of its lifetime in a PG phase, then there should be at least $\sim 10^4$ PGs per degree^2, or about one every half arcmin. An alternative estimate may be made as follows: Integrating the comoving volume per solid angle for $z > 5$, say, gives $\sim 1 - 10$ Mpc^3/arcsec^2, depending on the cosmology. At $z \sim 0$, there are a few $\times 10^{-3}$ L_* galaxies/Mpc^3. Again assuming a negligible density evolution, we derive $\sim 10^{-2}$ PGs per arcsec^2, or about one every 10 arcsec, close to our previous estimate. There are thus enough PGs on the sky, with the r.m.s. well matched to the fields of our detectors. It is also possible that PGs are strongly clustered: in just about all reasonable scenarios for structure formation, the highest peaks of the density field collapse first, and such peaks should be highly correlated. With this in mind, we started a search for protoclusters around $z > 4$ quasars (Djorgovski et al. 1992).

There are three principal energy release mechanisms from forming galaxies. First, there is the binding energy, which must be released as a protogalaxy cools and collapses. If this process occurs at $z > 20$ or so, the dominant mechanism is the inverse Compton cooling on the CMBR, which in principle may be observable today as a distortion of the CMBR spectrum. Energy from dissipative shocks in collapsing protogalaxies may also be released in recombination lines, such as Lyα. Taking an average galaxy mass of $\sim 10^{45}$ g, and an average 1-D internal velocity of $\langle V^2 \rangle \sim (200 \text{ km/s})^2$, we get $E_{bind} \sim 10^{59}$ erg for a typical non-dwarf galaxy. A comparable amount of binding energy must also be released from collapsing protostars, but the form of its release depends very much on the physical state and chemical composition of the early ISM. The dominant energy source, at least for normal galaxies, is the nuclear burning in stars: $E_{nuc} \sim M_\star c^2 \Delta X \epsilon$, where $M_\star \sim 10^{43}$ g is the total baryonic mass consumed in the primordial starburst, $\Delta X = \Delta Z + \Delta Y \sim 0.05$ is the fraction of hydrogen converted into the heavier elements, and $\epsilon \sim 1$ Mev / 1 Gev = 0.001 is the net efficiency of the nuclear reactions. This gives $E_{nuc} \sim 10^{61}$ erg for a typical elliptical or a massive bulge. Finally, a young galaxy may develop an active nucleus, e.g., quasars at $z > 4$. For a characteristic AGN luminosity of $\sim 10^{45}$ erg/s $\sim 10^{12} L_\odot$, and a lifetime $\tau \sim 10^8$ yr, we get $E_{AGN} \sim 10^{60}$ erg, some fraction of which may be coupled to the gas within the host protogalaxy. Depending on the as yet poorly understood feedback processes in protogalaxies, these characteristic energies may be related to the characteristic masses or other global properties of galaxies (cf. Ikeuchi & Norman 1991).

The observationally relevant quantity is not the total released energy, but the luminosity, i.e., $L \sim E/(\text{time scale})$. There are roughly three relevant time scales: the free fall or a typical starburst duration time scale, $t_{ff} \sim t_\star \sim 10^7 - 10^8$ yr; a cluster crossing or a merging time scale, $t_{merg} \sim 10^9$ yr; and a time scale for a gradual infall over the Hubble time, $t_{inf} \sim 10^{10}$ yr. Dividing the characteristic energy of $\sim 10^{61}$ erg by these time scales, we derive the corresponding luminosities of $\sim 10^{12}$, 10^{11}, and 10^{10} $L_\odot$. The first one corresponds roughly to typical quasar luminosities, and such PGs, if they existed and were not confused with quasars (Meier 1976; Terlevich, this volume) presumably would have been detected by now. Formation of a large fraction of old stars over a time period as short as a single t_{ff} requires a remarkable synchronicity, and may be unlikely. The intermediate time scale (formation over a Gyr) looks physically more appealing, especially if vigorous merging of protogalactic fragments is involved, as is suggested by n-body simulations. The last time scale gives the luminosities comparable with those of normal galactic disks.

The absolute magnitudes corresponding to these luminosities are $M \sim -26, -23$, and -20. At the redshifts of interest, the distance moduli are $(m - M) \sim 47 \pm 3$, depending on the cosmology. Therefore, we may expect unobscured protoellipticals to have apparent magnitudes roughly $\sim 24 \pm 3$, which is not strongly dependent on the wavelength, since the UV continua of actively star-forming galaxies are fairly flat. This is well within our observational abilities in the optical, and, at the bright end, in the near-IR as well. We note also that at $z \sim 2 \pm 1$, the expected magnitudes of disk galaxies with $M \sim -20$ are roughly $m \sim 26 \pm 2$, comparable with those of the faint blue galaxies seen in deep field surveys.

The key issue of observability of PGs is whether they are dusty or not. If the primordial starbursts were completely shrouded, the energy would emerge at $\lambda_{rest} \sim 30 - 100\mu$m, and would be observable in the FIR/sub-mm region today. If they were unobscured, the bulk of the energy would be emitted in the restframe UV, and be observable today in the optical or NIR region. This choice determines the observing strategy.

There is a methodological choice of either searching for individual sources, or trying to detect collective effects of forming galaxies, such as diffuse backgrounds. Depending on the importance of dust, PGs could be sources of a detectable UV/optical/NIR background, or a FIR/sub-mm background (recall the vigorous activity which followed the claim of a sub-mm excess in the CMBR by the Nagoya-Berkeley collaboration), and could also contribute substantially to at least some parts of the x-ray background. Discussions of these issues can be found, e.g., in papers by Songaila et al. 1990, Djorgovski & Weir 1990, and numerous good reviews in the proceedings edited by Bowyer & Leinert 1990.

Whereas the energetics does not seem to pose a major problem in detecting PGs, there is still a problem of their recognition. One needs some sort of a spectral signature of young, actively star-forming galaxies at large redshifts. Except for the Lyman break at 912 Å, their continua should be relatively featureless and flat, thus containing little redshift information. So far, broad band color selection in deep surveys never produced any good PG candidates. (An equivalent problem exists if one wishes to search for obscured primeval galaxies at FIR or sub-mm wavelengths.) We are driven to search for emission line signatures, e.g., strong recombination lines such as Lyα. Lyα emission in PGs can be powered by photoionization by young, massive stars (they must be present, since there are metallic lines in high-z quasars, and most of the mass in old stellar populations at $z \sim 0$ is metal rich). Additional mechanisms include shock ionization from infalling and colliding protogalactic fragments, supernovæ(Shull & Silk 1979), cooling of the first stars (Silk 1977, 1985), and photoionization by early AGN, if any are present. Using the semiempirical conversion between the SFR and Hα luminosity by Kennicutt 1983, and the simple case-B photoionization models to convert to Lyα, we estimate the Lyα luminosity of $\sim 10^{42}$ erg/s for each $1 M_{\odot}$/yr of star formation with a normal IMF (an IMF biased towards the high-mass end would be of course more effective). The characteristic SFR in PGs is estimated at $\sim 10^{2\pm1} M_{\odot}$/yr, giving expected $L(\text{Ly}\alpha) \sim 10^{44\pm1}$ erg/s. At the characteristic luminosity distance of $\sim 10^{29}$ cm, corresponding to $(m - M) \sim 47$, the expected Lyα line fluxes are $\sim 10^{-16\pm1}$ erg/cm^2/s, which is observable with present-day technology.

There are indeed many Lyα galaxies at high redshifts now known. However, essentially all of them are associated in some way with active nuclei: powerful radio sources (e.g., 3C 326.1 at $z = 1.825$; McCarthy et al. 1987), companions of high-z quasars (e.g., PKS 1614+051 at $z = 3.215$; Djorgovski et al. 1985, 1987), etc. Discussions and further references have been given in the reviews by Spinrad 1987, 1989 and Djorgovski 1988a,b. Since then, more interesting objects have been discovered. Among the radio galaxies are 3C 294 at $z = 1.786$ (McCarthy et al. 1990), 3C 257 at $z = 2.474$ (Dickinson et al., in prep.), B2 0902+34 at $z = 3.395$ (Lilly 1988), and 4C 41.17 at $z = 3.800$ (Chambers et al. 1990), and several others. These objects are characterized by a strong, extended (~ 100 kpc) Lyα emission, sometimes with a very low ionization (suggesting that at least some fraction of the line emission is due to photoionization by young stars), and a clumpy, low surface brightness continuum morphology. These may be genuine forming galaxies, but the presence of powerful radio sources in them makes them suspect. Heckman et al. 1991 and Hu et al. 1991 discovered extended Lyα nebulosities or possible galaxy companions near several high-z radio loud quasars. Another interesting new quasar companion, C1548+0917, has been found by Steidel et al. 1991. There is at least one detection of Lyα emission from a damped Lyα absorber (towards H0836+113, at $z = 2.466$, by Wolfe et al. 1991). Lowenthal et al. 1991 discovered an active galaxy near a damped absorber at $z = 2.309$ towards the quasar PHL 957. (Another possible case at $z = 3.409$, found by Turnshek et al. 1991 still needs a confirmation.) Finally, a high-ionization emission line galaxy at $z = 2.286$, apparently associated with the IRAS FPS 10214+4724 was found by Rowan-Robinson et al. 1991. This object (or possibly a hidden companion thereof) has an estimated luminosity of $\sim 10^{14} L_{\odot}$; it may well be an obscured quasar, rather than an ultrapowerful starburst.

These discoveries are encouraging, and demonstrate that Lyα-luminous objects with properties which may be expected of PGs can be found with the existing technology today. Some or all of them may well *be* PGs, albeit containing active nuclei. However, it would be much better if a population of field objects, not associated in any way with active nuclei, was found at comparable or higher redshifts.

Emission-line based PG searches cover a parameter space whose axes are: (1) the redshift coverage, Δz; (2) the depth or limiting flux, F_{lim}; and (3) the solid angle coverage, $\Delta\omega$. The plane defined by (Δz,F_{lim}) describes the star formation history; the plane defined by (F_{lim},$\Delta\omega$) reflects the evolution of the luminosity function; and the plane defined by (Δz,$\Delta\omega$) is sensitive to the cosmology. Observing strategies can be designed on the basis of perceived coverage of the most relevant or the least explored portions of this parameter space. For example, slitless spectroscopy surveys, such as the grism surveys by Koo & Kron 1980 and Schmidt et al. 1986 cover a large $\Delta\omega$, and may have a large Δz, but are not very deep; spectroscopic long-slit surveys cover a small $\Delta\omega$, can be very deep, and with a good Δz; and narrow-band imaging surveys, be it with interference filters or with a Fabry-Perot, cover an intermediate $\Delta\omega$, go very deep, but have only a limited Δz. The surveys to date have concentrated on the Lyα line; in the future, equivalent surveys may be done in the IR, searching for the Balmer or oxygen lines (λ 3727 and 5007 Å), or even Lyα itself at $z > 10$.

Surveys prior to 1985 have been reviewed by Koo 1986. Several field surveys have been conducted since then. Particularly noteworthy are the narrow-band imaging surveys by Pritchet & Hartwick 1987, 1990, the narrow-band and long-slit surveys reported by Cowie 1988, and the long-slit survey by Lowenthal et al. 1990. Djorgovski et al. (in prep.) obtained deep, narrow-band images centered on the redshifted Lyα line of about 40 fields of quasars at $z \sim 1.9 - 3.8$. The fields were searched for objects showing Lyα emission in the bands with typical $\Delta z \sim 0.01$. No obvious candidates were found down to the typical $F_{lim} \sim 1.5 \times 10^{-17}$ erg/cm^2/s, although some very faint, possible line-excess objects were detected. We use the limits from these surveys, the earlier work referenced by Koo 1986, and the results to date from our own surveys, described below, in Figures 1 – 3.

Our first experiment (Thompson et al. 1990, 1992a,b) consists of deep imaging of selected fields in a series of adjacent narrow bands, with a spectroscopic follow-up of all objects which show a probable line emission excess in one or more bands. A special, low-resolution Fabry-Perot imaging interferometer was built for this purpose. Several high-latitude fields, which transit close to the zenith at Palomar, with minimal IRAS cirrus, away from bright stars, galaxies, known foreground clusters, and as "empty" as possible, were selected. The field of view is $\sim$ 5.5 arcmin square. Three-dimensional data cubes are built up by successive exposures stepped in wavelength an amount equal to the instrumental FWHM ($\simeq 10^3$ km/s in the restframe for Lyα, or $\sim$ 20 – 25 Å). We search in several redshift intervals chosen to avoid the night sky emission lines: 2.80 – 2.89, 3.27 – 3.45, 4.42 – 4.61, and 4.74 – 4.90. Data cubes corresponding to these redshift slices are obtained and searched for objects which show a possible emission line "excess". To date we have surveyed 3 fields (0.03 deg^2) in the Δz range 4.42 – 4.61, and 6 fields (0.06 deg^2) in the range 4.74 – 4.90, down to the limit of $AB_\nu \sim 23^m$. This corresponds to a surface density of $<$ 13 objects/deg^2 down to F_{lim}(Lyα) $\sim 10^{-16}$ erg/cm^2/s, declining to a surface density of $<$ 2600 objects/deg^2 down to F_{lim}(Lyα) $\sim 3.2 \times 10^{-17}$ erg/cm^2/s, for compact objects ($\leq$ 2 arcsec); for extended objects, the limits are less stringent (about a factor of two worse). The net surveyed comoving volume so far is $1.41 \times 10^5 h_{75}^{-3}$ Mpc^3 for $\Omega_0 = 0$, or $2.14 \times 10^4 h_{75}^{-3}$ Mpc^3 for $\Omega_0 = 1$. In a typical data cube, we find 3 or 4 excellent candidates, and up to 20 or more other faint emission line galaxies worth following up. So far we confirmed spectroscopically half a dozen starburst galaxies at intermediate redshifts, typically $z \sim 0.4 - 0.9$ (detected through their [O II] or [O III] emission), and a couple of low-z AGN. There are several intriguing faint objects where long-slit spectra have confirmed the reality of the faint line emission detected by the Fabry-Perot, but with a signal insufficient to determine the redshifts unambiguously. Some of them may be genuine young, star-forming Lyα galaxies, but more data are needed before we can be certain of their nature. If none of them are actually primeval galaxies, our preliminary limits may already be in conflict with the CDM model predictions by Baron & White 1987 (see Figure 1).

Our second experiment is a serendipitous long-slit spectroscopic search, using data obtained in the course of other projects. After two-dimensional sky subtraction, the spectroscopic CCD frames are examined carefully for any possible emission-line objects which may have been covered by the slit. We typically span the wavelengths 4000 – 8000 Å, corresponding to a Lyα redshift of 2.3 – 5.6. One-hour exposures reach F_{lim} comparable to the Fabry–Perot survey, depending in wavelength on the night sky spectrum. These exposures have a large, continuous redshift coverage, but they cover a much smaller area, typicaly $\sim$ 2 arcsec by 2 arcmin. To date, $\sim$ 50 long exposures have been examined, and we anticipate that several tens more will be covered in this search within a year or two. So far, at least a dozen interesting objects were found. They are mostly star forming or mildly active galaxies at moderate redshifts. Typical magnitudes are in the range $19^m - 23^m$, and redshifts in the range $z \sim 0.3 - 0.8$, with the median $z \simeq 0.6$. The most distant object found so far is an apparently normal $\sim 24^m$ field galaxy, G0333+3208, at $z = 1.018$ (Thompson & Djorgovski 1991). Optical luminosities and [O II] emission equivalent widths of these objects are consistent with those of $\sim L_*$ galaxies undergoing a relatively mild evolution, powered by star formation. Their optical continuum shapes are broadly similar to those of the numerous, faint blue or flat-spectrum galaxies found in deep field surveys.

Figures 1 – 3 illustrate the limits on the Lyα PGs from various modern surveys, assuming that none of the the faint candidates found so far are PGs, and not counting various active objects found at $z \sim 2 - 5$. We used a Friedman cosmology with $H_0 = 75$ km/s/Mpc and $\Omega_0 = 0.2$ in computing the distance-dependent quantities and models, but the results are not very sensitive on that choice of parameters, at least at the level of

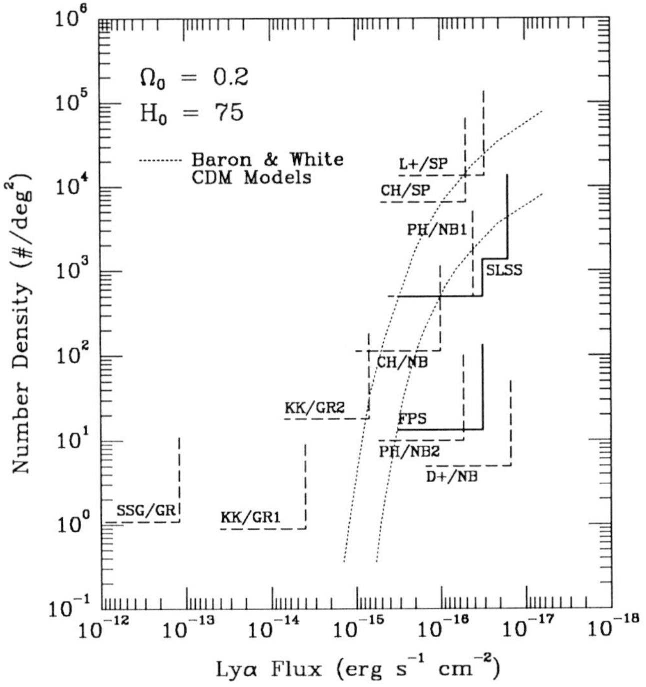

Figure 1. Limits on the surface number density of Lyα PGs, as a function of the limiting line flux. The limits are labeled as reference/method: KK = Koo & Kron 1980; SSG = Schmidt et al. 1986; PH = Pritchet & Hartwick 1987, 1990; CH = Cowie & Hu, rep. in Cowie 1988; L+ = Lowenthal et al. 1990; D+ = Djorgovski et al., in prep.; FPS = our Fabry-Perot survey; SLSS = our serendipitous long-slit survey; GR = grism (slitless spectroscopy); NB = narrow-band imaging; SP = spectroscopy. The limits shown are schematic, since the actual lines may be curved, reflecting a spread in sensitivity for a given survey. The dotted lines indicate the range of CDM-based models from Baron & White 1987, scaled for the $H_0 = 75$ km/s/Mpc, $\Omega_0 = 0.2$ cosmology. This diagram does not indicate the redshift coverage for each survey.

precision of the data. The limits were taken or computed on the basis of quoted papers, and uncertainties of the order of a factor of 2 may be typical. The limits reached in several independent modern surveys are such that normal forming galaxies, or at least E's and bulges, should have been found, if they existed at these redshifts, and were unobscured.

Possibly the most likely explanation for their absence is that PGs were dusty (see, e.g., van den Bergh 1990). If the restframe extinction was higher than $A_V \sim 1^m$, flux at 1216 Å would be depressed by more than a factor of 10. Several pieces of evidence support this hypothesis, but none of them is compelling, and for each one there are about equally good (or bad) counterarguments. First, it can be argued *a priori* that since the luminous starbursts seen at low redshifts are very dusty, so should their high-z counterparts. On the other hand, it is not obvious that the physical conditions in protogalaxies were sufficiently similar for this analogy to work; for example, the *first* stars must have formed in an environment without dust grains. Second, Lyα emission from low-z star-forming dwarfs is generally very weak, although some exceptions do exist (Meier & Terlevich 1981; Deharveng et al. 1985; Hartman et al. 1988). The observed Lyα/Hβ ratio in dwarfs appears to be a strong function of metallicity, suggesting that possibly the very initial phases of a protogalactic starburst might be clean, but it is not clear how long such a "window of observability" would last. It is also not obvious that star-forming dwarfs at low redshifts represent good analogs of the high redshift protogiants. Weedman 1991 estimated the

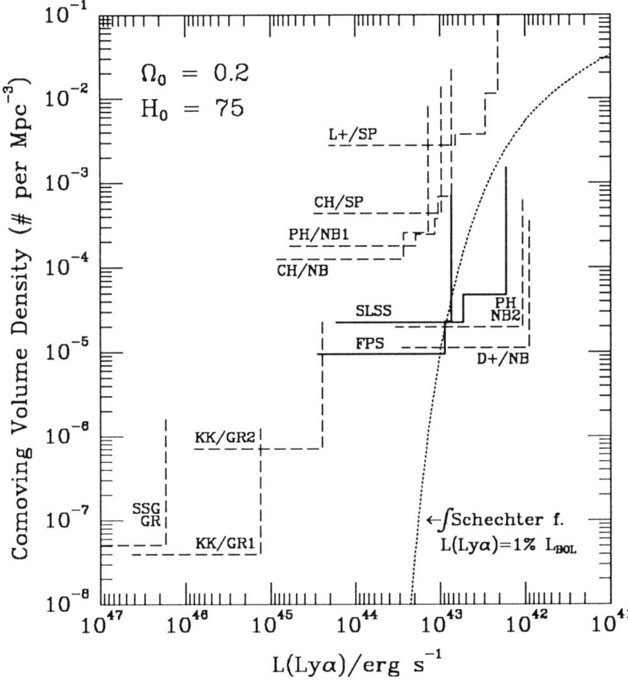

Figure 2. Limits from Figure 1, but with the observable coordinates translated into the comoving volume number density and restframe Lyα luminosity, using the same cosmology. The limits correspond to the central redshift in a given survey; accounting for the finite depth would tilt or curve the limit lines. The dotted line is an integral of the $z = 0$ Schechter luminosity function (i.e., no density evolution), assuming that there is no luminosity evolution and that 1% of the $z = 0$ bolometric luminosity is radiated in the Lyα line. This is probably a very pessimistic model, but even so, several surveys are already in conflict with its predictions.

fraction of escaping UV photons for a sample of Markarian starburst galaxies, and found that a large fraction of them must be relatively unobscured. There is some evidence that *disks* at $z \sim 2$ were at least moderately dusty, with the Lyα emission depressed (Pei et al. 1991; Elston et al. 1991; Charlot & Fall 1991). However, forming ellipticals and bulges could have been different, e.g., with the ISM fully reionized. Finally, we already know numerous Lyα-luminous and star-forming galaxies at large redshifts, with no signs of extinction in their extended Lyα and UV nebulosities.

Perhaps the best argument against completely obscured star formation in PGs is the absence of a detectable sub-mm background in the COBE data (Mather et al. 1990). A crude, but robust argument is as follows: energy density today, deposited in a hypothetical sub-mm background (FIR in the restframe) by a population of dusty PGs is $u_{smb} = 7 \times 10^{-14}(\Omega_\star/0.02)(H_0/75)^2(1+z_{gf})^{-1}$ erg/cm^3, where $\Omega_\star$ is the fraction of the critical density in the stars in ellipticals and bulges (Djorgovski & Weir 1990). This should be compared with the present limits from COBE, $u_{smb} < 0.003\ u_{CMBR} = 1.3 \times 10^{-15}$ erg/cm^3. Even stronger limits could be obtained by fitting predicted sub-mm backgound spectra with the

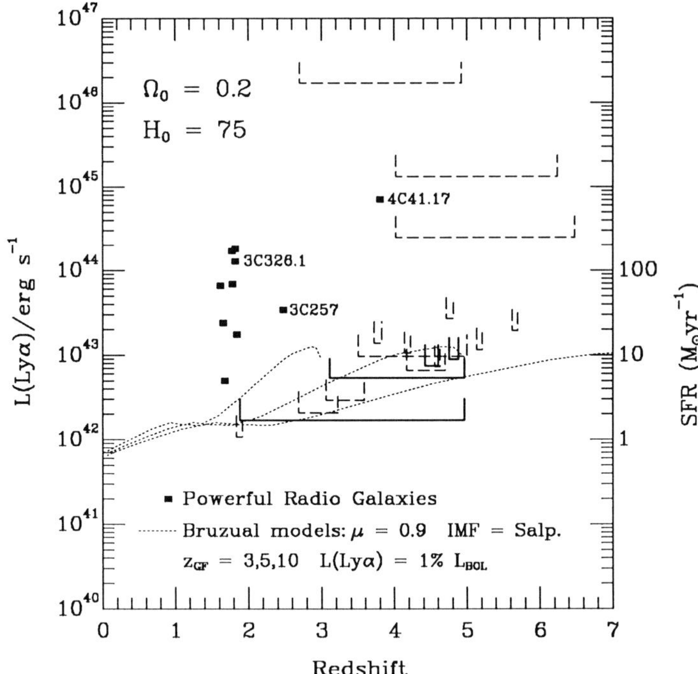

Figure 3. The sensitivity of PG surveys, expressed as the limiting Lyα line luminosity as a function of redshift, computed in the same way as in Figure 2. The solid lines are the limits from our long-slit (large z coverage) and Fabry-Perot (two small Δz windows), and the dashed lines are limits from other surveys. This figure contains no information on the area or volume coverage, and is complementary to Figure 2. Solid dots represent some of the powerful, Lyα luminous radio galaxies at high redshifts. Objects with line luminosities an order of magnitde or two lower could have been detected in the surveys to date. The dotted lines represent Bruzual $\mu = 0.9$ models for a galaxy with the initial $M_{gas} = 10^{11} M_\odot$ (roughly, a normal ellptical), assuming that only 1% of the total bolometric luminosity emerges in the Lyα line, and starting at z_{gf} = 3, 5, or 10. The corresponding Lyα line luminosities powered by star formation of 100, 10, or 1 $M_\odot$/yr are indicated on the right.

COBE limits as a function of wavelength, e.g., by reversing the procedure described by Djorgovski & Weir 1990. We conclude that at most a few percent of the observed stars in $z \sim 0$ ellipticals and bulges could have been formed in dusty PG starbursts similar to those in IRAS galaxies today, or cooler. If at least some dark matter is baryonic, these limits would be stronger. However, the limits can be relaxed somewhat if the starbursts happened at $z < 3$, and the dust in them was considerably hotter than in their present-day counterparts. If there were dusty PGs at high redshifts, they may be detectable from the ground using sub-mm telescopes with the next-generation detectors and bolometer arrays.

We are driven to other possible explanations for the absence of Lyα-luminous PGs in surveys to date. One possibility is that PGs were only slightly dusty – just enough to depress the Lyα below the present limits, but not enough to leave an imprint on the CMBR spectrum. Another possibility is that protoellipticals and protobulges are to be found at higher redshifts, $z_{gf} > 5$, which have been poorly explored so far. If either was the case, we may expect to see a new population of objects in deep near and mid-IR surveys, perhaps appearing at $K \sim 23^m$. This is just beyond the horizon (cf. Collins & Joseph 1988, or Cowie et al. 1990), and may be doable from the ground with the forthcoming 8 – 10 m telescopes, or with the HST/NIC. If $z_{gf} > 10$, deep surveys at 3 – 10μm with ISO and SIRTF may be able to detect them (cf. Franceschini et al. 1990).

We wish to thank John Trauger for the help in designing and building the Fabry-Perot imager, and in performing the observations. We also thank the staff of Palomar Observatory for their expert help during numerous observing runs. S.D. acknowledges support from the Alfred P. Sloan Foundation, the NSF PYI award AST-9157412, and a AAS travel grant.

References:

Baron, J., & White, S. 1987, ApJ, 322, 585
Bergeron, J., Kunth, D., Rocca-Volmerange, B., & Van, J.T.T. (eds.) 1987, High-Redshift and Primeval Galaxies, (Gif sur Yvette: Editions Frontières)
Bowyer, S., & Leinert, C. (eds.) 1990, The Galactic and Extragalactic Background Radiation, IAU Symp. 139 (Dordrecth: Kluwer)
Chambers, K., Miley, G., & van Breugel, W. 1990, ApJ, 363, 21
Charlot, S., & Fall, S.M. 1991, ApJ, 378, 471
Collins, C., & Joseph, R. 1988, MNRAS, 235, 209
Cowie, L. 1988, in Post-Recombination Universe, ed. N.Kaiser & A.Lasenby, (Kluwer), p.1
Cowie, L., Lilly, S., Gardner, J., & McLean, I. 1988, ApJ, 332, L29
Cowie, L. 1989, in Frenk et al. 1989, p. 31
Cowie, L., Gardner, J., Lilly, S., & McLean, I. 1990, ApJ, 360, L1
Davis, M. 1980, in IAU Symposium 92, ed. G.Abell & P.J.E.Peebles (Reidel),p.57
Deharveng, J.M., Joubert, M., & Kunth, D. in Star-Forming Dwarf Galaxies and Related Objects, ed. D.Kunth et al., (Gif sur Yvette: Ed.Frontières), p.431
Djorgovski, S., Spinrad, H., McCarthy, P., & Strauss, M. 1985, ApJ, 299, L1
Djorgovski, S., Strauss, M., Perley, R., Spinrad, H., & McCarthy, P. 1987, AJ, 93, 1318
Djorgovski, S. 1988, in Thuan et al. 1988, p. 401
Djorgovski, S. 1988, in Kron & Renzini 1988, p. 259
Djorgovski, S., & Weir, N. 1990, ApJ, 351, 343
Djorgovski, S., Smith, J.D., and Thompson, D.J. 1992, in The Space Distribution of Quasars, eds. D. Crampton and D. Durand, ASP Conf.Ser. in press
Elston, R., Bechtold, J., Lowenthal, J., & Rieke, M. 1991, ApJ, 373, L39
Franceschini, A., Toffolati, L., Danese, L., & De Zotti, G. 1990, MemSAIt, 61, 115
Frenk, C., Ellis, R., Shanks, T., Heavens, A., & Peacock, J. (eds.), The Epoch of Galaxy Formation, (Dordrecht: Kluwer)
Giovanelli, R., & Haynes, M. 1989, ApJ, 346, L5
Gunn, J., Hoessel, J. & Oke, J.B. 1986, ApJ, 306, 30
Guhathakurta, P., Tyson, J.A. & Majewski, S. 1990, ApJ, 357, L9
Hartmann, L., Huchra, J., Geller, M., O'Brien, P., & Wilson, R. 1988, ApJ, 326, 101

Heckman, T., Lehnert, M., van Breugel, W., & Miley, G. 1991, ApJ, 370, 78
Hewitt, A., Burbidge, G., & Fang, L.Z. (eds.) 1987, Observational Cosmology, IAU Symposium 124, (Dordrecht: Reidel)
Hu, E., Songaila, A., Cowie, L., & Stockton, A. 1991, ApJ, 368, 28
Ikeuchi, S., & Norman, C. 1991, ApJ, 375, 479
Kennicutt, R. 1983, ApJ, 272, 54
Koo, D., & Kron, R.G. 1980, PASP, 92, 537
Koo, D. 1986, in Spectral Evolution of Galaxies, ed. C.Chiosi & A. Renzini (Reidel),p.419
Kron, R.G. & Renzini, A. (eds.) 1988, in Towards Understanding Galaxies at Large Redshift, (Dordrecht: Kluwer)
Kron, R.G. (ed.) 1990, Evolution of the Universe of Galaxies, ASP Conf.Ser., vol. 10
Kunth, D., & Sargent, W. 1986, ApJ, 300, 496
Larson, R. 1990, PASP, 102, 709
Lilly, S. 1988, ApJ, 333, 161
Lilly, S., Cowie, L., & Gardner, J. 1991, ApJ, 369, 79
Lowenthal, J., Hogan, C., Leach, R., Schmidt, G., & Foltz, C. 1990, ApJ, 357, 3
Lowenthal, J., et al. 1991, ApJ, 377, L73
Mather, J., et al. (the COBE team) 1990, 354, L37
McCarthy, P., et al. 1987, ApJ, 319, L39
McCarthy, P., et al. 1990, ApJ, 365, 487
Meier, D. 1976, ApJ, 203, L103
Meier, D. & Terlevich, R. 1981, ApJ, 246, L109
Peebles, P.J.E. 1989, in Frenk et al. 1989, (Dordrecht: Kluwer), p. 1
Pei, Y., Fall, S.M., & Bechtold, J. 1991, ApJ, 378, 6
Pritchet, C., & Hartwick, F.D.A. 1987, ApJ, 320, 464
Pritchet, C., & Hartwick, F.D.A. 1990, ApJ, 355, L11
Rowan-Robinson, M., et al. 1991, Nature, 351, 719
Schmidt, M., Schneider, D., & Gunn, J. 1986, ApJ, 306, 411
Schneider, D., Schmidt, M., & Gunn, J. 1991, AJ, 102, 837
Shull, M., & Silk, J. 1979, ApJ, 234, 427
Silk, J. 1977, ApJ, 211, 638
Silk, J. & Norman, C. 1981, ApJ, 247, 59
Silk, J. 1985, ApJ, 297, 1
Silk, J. & Szalay, A. 1987, ApJ, 323, L107
Silk, J., 1987, in Hewitt et al. 1987, p. 391
Songaila, A., Cowie, L., & Lilly, S., 1990, ApJ, 348, 371
Spinrad, H. 1987, in Bergeron et al. 1987, p. 59
Spinrad, H. 1989, in Frenk et al. 1989, p.39
Steidel, C., Sargent, W., & Dickinson, M. 1991, AJ, 101, 1187
Thompson, D., Djorgovski, S., & Trauger, J. 1990, BAAS, 22, 1216
Thompson, D.J., & Djorgovski, S. 1991, ApJ, 371, L55
Thompson, D.J., Djorgovski, S., and Trauger, J. 1992a, in The Space Distribution of Quasars, eds. D. Crampton and D. Durand, ASP Conf.Ser. in press
Thompson, D.J., Djorgovski, S., and Trauger, J. 1992b, in Cosmology and Large-Scale Structure in the Universe, ed. R. de Carvalho, ASP Conf.Ser. in press
Thuan, T.X., Montmerle, T., & Van, J.T.T. (eds.) 1988, Starbursts and Galaxy Evolution, (Gif sur Yvette: Editions Frontières)
Turner, E. 1991, AJ, 101, 5
Turnshek, D., et al. 1991, preprint
Tyson, J.A. 1988, AJ, 96, 1
van den Bergh, S. 1990, PASP, 102, 503
Weedman, D. 1991, in Massive Stars in Starbursts, eds. C. Leitherer et al., (Cambridge: Cambridge Univ. Press), p. 317
White, S. 1989, in Frenk et al. 1989, p. 15
Wolfe, A., Turnshek, D., Lanzetta, K., & Oke, J.B. 1991, ApJ, in press

Discussion:

R. Terlevich: (1) Our work on nearby HII regions and HII galaxies indicate that the strongest signature in the UV spectrum of a young system is the stellar continuum and not the Lyα emission. Also, when performing magnitude-limited surveys, the first objects to be found are the brightest and *not* the L_* ones. So, your detection limits are too pessimistic by perhaps 3 mag or more. (2) Your claim during your presentation that no quasar shows an absorption line type of spectrum to the blue of Ly α (when you presented the 28,000 sec spectrum), is in contradiction with observations of deep Lyα forest depression in quasars at $z \sim 4$, about 20% of them showing the Ly limit cut-off.

Djorgovski: On (1): I basically agree with you that the most likely explanation for the absence of numerous Lyα galaxies at $z \sim 2-5$ is dust extinction, although that is not the only possibility. On (2): You misunderstood me, what I said is the cutoff at $\lambda_{rest} = 912$Å, which is characteristic of stellar populations, but is not strongly present in quasars.

Zinnecker: Has the redshifted Lyα line been detected in IRAS 10214+4724? In view of the large amounts of dust implied by being an IRAS source, this would be very surprising.

Lonsdale: There is a spectrum of it with a weak Lyα – it is significantly fainter than C IV. The formal reliability of the optical identification is probably about 80% based on the magnitude of the galaxy and the size and distance of the error ellipse. There is a firm radio detection of the galaxy which raises the reliability of the identification with the IRAS source.

Bershady: Is there an overwhelming evidence against the possibility that ellipticals and bulges formed their first generation of stars in more diffuse systems? And if so, it would be important to express the Fabry-Perot limits in terms of surface brightness.

Djorgovski: Certainly, the formation of ellipticals probably did involve lots of merging. However, the metallicity gradients and the high central luminosity and phase space densities argue that a substantial fraction of the initial star formation occured within relatively compact regions. If one wishes to consider amorphous, well-resolved objects akin to 3C326.1, then our flux limits are less stringent by about a factor of 2 or 3.

Spergel: The high central densities in the core of ellipticals and in spiral bulges argue for a high redshift of galaxy formation: $z \sim 10 - 100$. In non-gaussian models such as the textures, galaxy formation begins at these high z's. I would encourage observers to devise strategies for detecting galaxy formation at high z.

Djorgovski: Most searches to date were confined to the optical window (Lyα at $z \sim 2-6$) simply because of the technical limitations. I certainly agree that the next frontier is in the IR, corresponding to a high z_{gf}.

Rocca-Volmerange: Calculations of transfer through the Lyα emission line show that profiles could be very narrow at such distances, and so undetectable (Valls-Gabaud & Beujaffel 1991). What do you think about that?

Djorgovski: Lyα line widths are not very critical for most surveys; typical instrumental resolution is of the order of 1000 km/s. The most important selection effect is the line flux.

SOME INSIGHTS INTO THE PHOTOMETRIC EVOLUTION OF GALAXIES

PEETER TRAAT

European Southern Observatory, D-8046 Garching b. München, Germany
Tartu Astrophysical Observatory, 202444 Tõravere, Estonia

The photometric (spectral) evolution is but one aspect of a more universal central question in modern astrophysics — the complex problem of formation and evolution of galaxies. The method used to follow the photometric evolution is straightforward, consisting in computing on the basis of adopted prescriptions for star formation the stellar populations present at various moments and summing up the contributions of all the multitude of stars of different ages and evolutionary stages. The usual strategy is to neglect the dynamics and chemical changes of matter, although during the initial collapse of a galaxy and rapid enrichment of its matter both processes should be important contributors shaping the forming stellar populations. Other widely used simplifications are the rejection of interstellar absorption and the restriction of models to the volume element with constant star formation time scale in its borders (so-called "one-zone models", they stricly apply to the infinitely narrow ellipsoidal shells in real galaxies). An additional postulate that the region considered is closed and isolated, having not been subjected to interactions or mass exchange with the surroundings, is introduced in the simpliest closed "classical" or "canonical" models (Traat 1988), computed e.g. by Tinsley (1972), Searle *et al.* (1973), Traat and Einasto (1979), and others. Such assumption, however, is not always valid and some objects need to be modelled by burst models or more universal accretion models.

Star formation is the key process determining all structural (like the bulge-disk and stellar mass to gas ratios) and photometric properties of galaxies, but the factors governing the star formation on a global scale are still largely unknown. The widely used Schmidt law, $\mathcal{R}(t) = \gamma\rho_g^s/\rho_0$ (ρ_g — gas volume density, γ and n — constants, $t_0 = (\gamma\rho_0^{s-1})^{-1}$ — the characteristic time of star formation) what in fact states just that one needs matter to make stars and this is the easier, the more matter is available, is up to now the best parametrization for star formation rate (SFR). The observational values of n obtained from surface densities are between $1 \div 2$ with some tendency to cluster around ~ 1.5, but the actual value of index for volume density depends on density distribution. In a closed region a condition $\rho_g + \rho_s = \rho_0$ holds, by abandoning mass loss from evolved stars and integration of Schmidt law one gets for SFR the functional form

$$\mathcal{R}(t) = \begin{cases} e^{-\tau}/t_0 \quad , & s = 1, \\ \frac{1}{t_0}\left[1 + (s-1)\tau\right]^{\frac{-s}{s-1}} \quad , & s \neq 1, \end{cases}$$

with $\tau = t/t_0$ as dimensionless time. This means that the time evolution of a closed system is fixed by a single parameter — the star formation time-scale t_0. In the physically most plausible case $s > 0$ the resulting SFR is monotonically decreasing, with gas density and SFR being highest at the initial moment $t = 0$; in the absence of correlation ($s = 0$) the resulting SFR is constant.

The colors of canonical models, computed relative to V, start from a blue initial value and evolve monotonously redder. The luminosity curve of classical models is

B. Barbuy and A. Renzini (eds.), The Stellar Populations of Galaxies, 349–352.

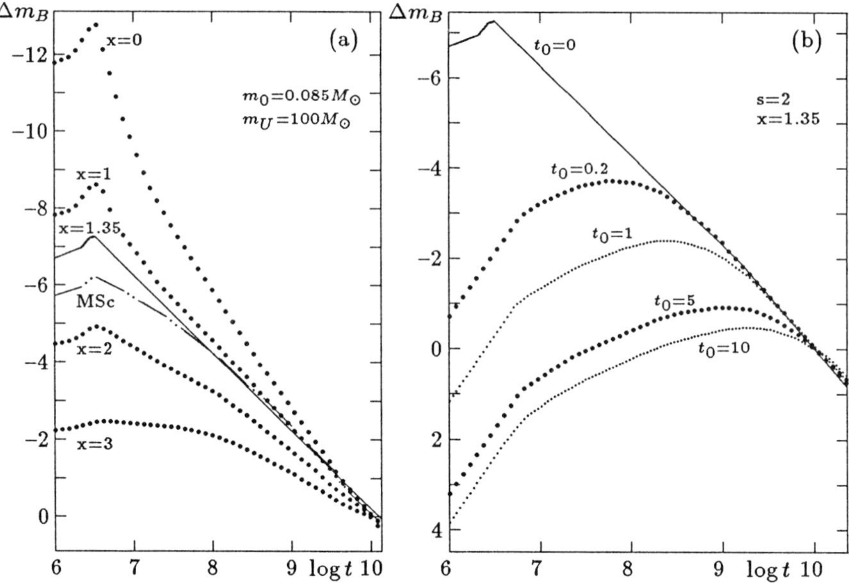

Figure 1: Past luminosity evolution: *a)* coeval stellar population with different IMF slopes; x=1.35 is Salpeter's original value, MSc denotes the lognormal solar neighbourhood IMF from Miller and Scalo (1979); *b)* stellar populations differing in characteristic formation time-scales, t_0=0 in the case of an coeval stellar population. In both cases *a)* and *b)* the curves are normalized at t = 10 Gy, and $Z = 0.03, X = 0.70$.

rising rapidly to an early maximum, caused by the arrival of massive luminous stars to the main sequence and followed by a continuous decrease. With the growth of t_0 this luminosity maximum shifts to the lower absolute values and progressively later ages (*cf* Fig. 1*b*). At the standard value, $x = 1.35$, of the slope of the power-law IMF $\psi(m) \sim m^{-(1+x)}$ and minimum stellar mass taken equal to $m_0 = 0.085 M_\odot$ (the smallest mass of a normal composition star igniting the hydrogen on the main sequence), our models with different t_0-s produce at the age 12 Gy M/L_B ratios between 4 and 10, which is just the right range observed for galaxies of different morphological types.

The left panel of Fig. 1 depicts the past luminosity behaviour of closed models depending on the value of slope x of the power-law IMF under the assumption of star formation in a single initial burst. The amplitude of luminosity is the larger the smaller is x, since with its decrease the fraction of high-luminosity massive stars in IMF is progressively growing. As evident from panel *b)*, some time dilution of star formation will substantially reduce these top amplitudes. The shortest time-scale for primeval ellipticals to make most of their stars equals to the free-fall time ($\sim$ 0.2 Gy), probably some additional time dilution will be caused also by the rising SFR during these initial dynamical stages. Therefore the model with t_0 = 0.2 Gy, what in the illustrated s = 2 case makes half of all its stars during these first 0.2 Gy, seems to be a reasonable upper limit to the top luminosity of the rapidly star-forming bright massive ellipticals. Since the absorption is ignored in our models, the primeval galaxies are probably dimmer yet, but even this adoption of a small duration of star formation reduces the past luminosity amplitude of galaxies about twice relative to the $t_0 = 0$

case ($\sim$ 30 times in absolute units), to less than 4 magnitudes in B. It is clear that since the initial-burst models enormously overestimate top luminosities of primeval galaxies, one should correct the widespread practice to use them in comparisons with young objects, using instead the models with non-zero t_0.

While the present properties of "normal" galaxies are in a crude general accordance with the results on classical, closed models, a bulk of galaxies is deviating in their properties, most notably by peculiar colors. Some galaxies reveal the ongoing star formation on the level which cannot be sustained by the available matter for no more than but a fraction of a galaxy's lifetime, others indicate just more modest deviations from the monotonic behaviour of the SF process. A tendency prevails that most of these active galaxies are members of systems and interacting with another near-by galaxy or a gas cloud, although the isolated objects are not rare.

Their evolution is described by composite models, first discussed by Larson and Tinsley (1978), in which an extra burst of star formation with small duration is added to an old canonical model to account for the intense young component. The path of these models on the two-color diagram is, as a rule, an almost closed loop, the length of which depends on the burst intensity and its duration. Any such galaxy with a constant SFR during the burst has the bluest colors at the end of the burst, if it is short; in the case of a long and low-intensity burst the bluest point is reached earlier and after that the colors will turn slowly redder. When the SFR differs from constant, the bluest colors occur after the maximum of the burst if it is short, but an increase in burst length makes them first to coincide with maximal SFR and then to precede it. When the burst is over, the original colors will be approximately restored after about 0.5÷1 Gy.

On Fig. 2*a* the evolution of a starburst galaxy is plotted. The underlying old red galaxy is having an age of 12 Gy, it develops star formation bursts with different strengths b, which equals to the mass fraction converted to the stars. Since the original galaxy is extremely void of young blue massive stars, only a little amount of matter has to be processed into stars to cause significant changes of integrated colors.

The burst models do not specify openly where the matter for extra star formation comes from. One of likely sources is its accumulation by infall from a near-by galaxy/gas cloud or intergalactic space. Short accretion will leave no lasting imprint on a galaxy, but in the case of open systems with a continuous exchange of matter with ambience it will become key factor driving their evolution.

The accretion models have been studied by Traat (1988), he used the accretion function $\alpha(t)$ to fix the mass balance in the area under consideration (the whole system or some part of it). This function was scaled per unit *initial* mass and per 1 Gy. If accretion is present, $\alpha(t) > 0$, in the case of mass loss $\alpha(t) < 0$, if there is no accretion or mass loss, $\alpha(t) = 0$.

In the case of the constant infall to the system ($\alpha > 0, s > 0$) after some time the equilibrium will develop between gas infall and star formation, with the total mass of gas in a galaxy remaining unchanged and star formation proceeding at the rate $\mathcal{R} = \alpha$ exactly balanced by infall. This state will be reached the earlier the greater α and the smaller t_0 are. Such equilibrium production of stars has long been suspected for a number of galaxies, notably for faint irregulars which have had constant SFR-s during the last $\sim$ 10 Gy (Gallagher *et al.* 1984). Models, exposed to heavy accretion, tend to retain quite high relative gas content. They can finish even with larger mass of gas than was the initial mass of the parent cloud, from which the galaxy formed. In the opposite case when a galaxy is losing mass, its gas supplies will be completely exhausted at some moment, after which $\mathcal{R} \equiv 0$.

The accretion models are illustrated in Fig. 2*b* where the evolution of B–luminosity

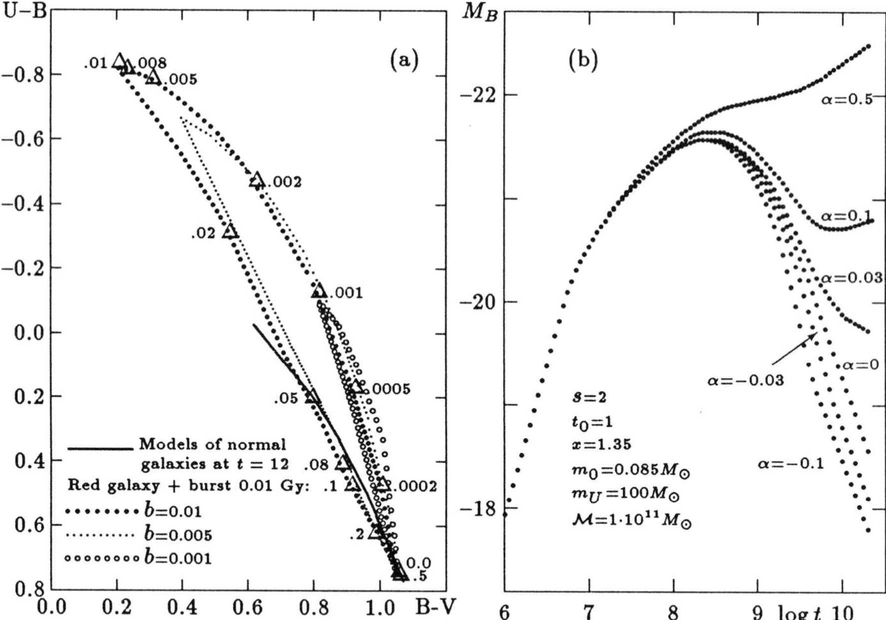

Figure 2: Models incorporating environmental effects: *a)* Color diagram of a very red model galaxy with age 12 Gy, developing a star formation burst with strength b. Triangles mark the burst age in Gy for the $b = 0.01$ case. *b)* Luminosity evolution of an stellar population (homogeneous galaxy) with SF time-scale $t_0 = 1$ and different levels of constant accretion of matter. Model $\alpha = 0$ represents the case without accretion, negative α values indicate mass loss.

of a model with $t_0 = 1$ and a number of accretion rates is plotted. The inclusion of mass exchange in the evolutionary models has the effect that every closed model with a fixed t_0 value will be replaced by a family of models with α as its parameter. Allowing a sufficient range for the accretion parameter, all the range of colors of normal galaxies can be explained with models differing only in accretion rate but having the same internal star formation time-scale. Since in young galaxies the accretion and mass loss are more probable, this can strongly influence any cosmological tests using very faraway and young galaxies. The main point to finish with is to stress that all the photometric evolutionary models of galaxies rely critically on their past SFR-s. Every revision or modification of assumptions involved in the specification of the SFR will typically cause radical changes in the output, what is also evident in the case of models incorporating environmental effects discussed above.

REFERENCES

Gallagher, T., Hunter, D., Tutukov A.V. 1984, *Astrophys. J.,* **284,** 544.

Larson, R.B., Tinsley, B.M. 1978, *Astrophys. J.,* **219,** 46.

Miller, G.E., Scalo, J. 1979, *Astrophys. J. Suppl.,* **41,** 513.

Searle, L., Sargent, W.L.W., Bagnuolo, W. 1973, *Astrophys. J.,* **179,** 427.

Tinsley, B.M. 1972, *Astron. Astrophys.,* **20,** 383.

Traat, P., Einasto, J. 1979, *Publ. Tartu Astrophys. Obs.,* **47,** 140.

Traat, P. 1988, *Tartu Astrofüüs. Obs. Teated* 91, 23.

TESTING THE WORLD MODEL THROUGH HIGH REDSHIFT GALAXIES IN CLUSTER

Alberto Buzzoni, Guido Chincarini, Emilio Molinari
Osservatorio Astronomico di Brera
Via Brera, 28 20121 Milano, Italy

ABSTRACT. We discuss the results of a cosmological test involving mean colors of the early-type galaxy population in distant clusters for tracing the World model. A global approach to the cosmological problem is attempted deriving the allowed combinations for the fundamental parameters (H_o, q_o, λ_o), and the redshift of galaxy formation (z_f).

1. Introduction

It is well known that a measure of the geometry of the Universe is the most direct and safe way to determine its curvature. However, when attempting a test resting on galaxies as standard candles, as the classical case of the Hubble diagram for example, the main set-back is that the evolution of the reference objects at high redshift could play a crucial role modulating our inferences about the cosmological model (Sandage 1988; Yoshii & Takahara 1988).

On the other hand, by learning how galaxies are and evolve we would get a powerful tool to estimate the geometry of the space. This can be done via stellar evolution applied to the study of the galactic stellar populations. So far, stellar evolution stands on solid grounds and we master in great detail the nuclear clock of stars. This will provide a safe link between the look-back time and the redshift allowing thus to determine which cosmological parameters fit better the observations.

Our approach to the cosmological problem is based on the observation and theoretical determination of the colors of cluster elliptical galaxies. Compared to magnitudes, colors are less dependent on aperture corrections and their direct determination does not depend on the distance. A relevant advantage comes then from considering elliptical galaxies, respect for instance to spirals, as to a large extent their evolutionary history is under control. Interstellar gas in ellipticals is scarce, and this suggests an extensive star formation which should have occurred early in the lifetime of the galaxies. It is a plausible working hypothesis, therefore, to consider their stellar populations as coeval with their age coinciding with that of the parent galaxy.

Finally, a last advantage using cluster galaxies as a set of objects located at the same distance is that due to a more favourable statistics, we can deduce the mean photometric properties of the whole sample with a larger confidence respect to single field objects.

2. The Observational Data Base

Seven clusters with redshift ranging from $z = 0.15$ up to 0.58 have been accounted for in our test. They are part of an ongoing observing programme undertaken by our group since 1986 at the ESO telescopes in La Silla (Chile). More extended details about the observations (taken in the Gunn g, r, i photometric system) and data reduction (using ESO MIDAS package) can be found in the contribution by Molinari *et al.* (this congress) where we present the results for three of the relevant clusters, and in Molinari, Buzzoni & Chincarini (1990) for a more general discussion of the procedures involved.

In the following calculations, mean colors for the population of elliptical galaxies in each cluster have been corrected for Galactic extinction according to Burstein & Heiles (1982) adopting $E(g-r) = 1.10E(B-V)$ and $E(g-i) = 1.73E(B-V)$.

B. Barbuy and A. Renzini (eds.), The Stellar Populations of Galaxies, 353–356.

3. Matching Theory and Observations

Although we cannot directly estimate the morphology of the galaxies in our distant clusters, it is reasonable however to identify ellipticals by means of their colors. As we showed in Molinari *et al.* (this congress) a majoritary population of galaxies systematically appears in the $(g-r)/(g-i)$ diagrams of the clusters clearly segregating as a clump of red objects. Moreover, the location of the red clump smoothly moves redward with increasing redshift and can be tracked quite confidently by evolving back in time present-day ellipticals. Therefore, this leads us to conclude that such red galaxies are indeed *bona fide* progenitors of present-day ellipticals. Based on Buzzoni's (1989) models for population synthesis, Molinari, Buzzoni & Chincarini (1990) derived the expected reference colors of redshifted early-type galaxies both for "passive" and "active" evolution (see quoted reference for further details).

Before proceeding on with our analysis, it is essential to clarify a preliminary question, possibly relevant for our aims, dealing with a safe determination of the metallicity of the elliptical galaxies. Since age and metallicity affect the galactic colors in the same way (Renzini & Buzzoni 1986) we need to disentangle the two effects in order to properly tune the stellar clock.

On the basis of a calibration of the Mg_2 spectral index, Buzzoni, Gariboldi & Mantegazza (this congress, 1991) studied the metallicity distribution for local early-type galaxies deriving typically [Fe/H] $= +0.2$, i.e. a value about 50% higher respect to the solar metallicity. Note that by using a solar [Fe/H] we would predict bluer colors at a given redshift, and consequently an older age for the Universe would be derived from the fit to the data. It is worth remarking that in no relevant way a spread in metallicity among cluster galaxies would affect our test through the well known Visvanathan & Sandage (1977) c-m effect. The net result would be in fact only a spread in the colors without affecting the mean values for the whole galaxy population on which rests in practice our fit.

4. The Cosmological Test

4.1 FRIEDMANN SOLUTIONS WITH $\lambda_o = 0$

In Fig. 1 we display a comparison between data and theoretical reference colors expected for two relevant choices of the deceleration parameter q_o pertinent to a (virtually) empty Universe ($q_o = 0$) and an Einstein–De Sitter model ($q_o = 1/2$). This procedure can be refined searching for the best fitting cosmological models through an iterative alogorithm. For a fixed value of q_o, in the range 0–1, we computed the expected colors for the clusters in our sample with varying H_o in the range 20–100 km/sec/Mpc. Minimization of the color residuals gave then the allowed combination in the (q_o, H_o) domain assuming different values for z_f. The best fits to the observations gave typically an rms of about 0.03 mag. The most direct result in the procedure is that a negative correlation exists between the Hubble constant and the deceleration parameter in the sense that smaller values for q_o require larger values for H_o. Assuming galaxies to be formed at $z_f = \infty$ we derive for the Universe a present age of $t_o = 16 \pm 2$ Gyr. A firm conclusion also deriving from Fig. 1, is that an Einstein–De Sitter model is not supported by the observations.

Accounting for all internal uncertaintes in the fit, an upper limit for the Hubble constant can be set at $H_o \leq 68$ (assuming $q_o = 0$). This is in agreement with the most recent determination of $H_o = 52 \pm 2$ given by Sandage & Tammann (1990), and with their lower limit at 45 ± 3. With $H_o = 52$ and assuming galaxies to have the same age of the Universe we derive $q_o = 0.06^{+0.14}_{-0.06}$ with a safe upper limit at 0.46.

Of course, the epoch of galaxy formation is still an open question widely discussed in the literature (Wyse 1985; Yoshii & Takahara 1988). A lower limit such as $z_f \geq 4$ can be derived from our test assuming for H_o the value by Sandage and Tammann.

4.2 INFLATIONARY SCENARIOS WITH $\lambda_o \neq 0$

A large amount of work went into solving some of the theoretical problems in the framework of the cosmological inflation theory with Ω_o =1. On the other hand, none of these attempts fully succeded in reconciling theoretical expectations with observations of the real galaxies (Yoshii & Takahara 1988; Fukugita *et al.* 1990; Guiderdoni & Rocca–Volmerange 1990).

A way out could be either the introduction of number counts evolution at high redshift (Rocca–Volmerange & Guiderdoni 1990) or a non-zero value for the cosmological constant λ_o (Fukugita *et al.* 1990). The latter possibility could be somewhat attractive because it preserves the present status of knowledge about galaxy evolution. In an inflationary scenario with zero–curvature space we have $\Omega_o + \lambda_o = 1$ providing to express the cosmological constant in normalized units such as $\lambda_o = 2/3\Lambda(c/H_o)^2$. A simple relationship links then the deceleration parameter: $q_o = 3/2\ \Omega_o - 1$.

When applying our previous analysis also to this case we have that even for the lower limit of H_o , the cosmological constant must be greater than 0.35 ($\Omega_o \leq 0.65$). For H_o =52 we derive $0.6 \leq \lambda_o \leq 0.85$ which leads to a preferred range of $0.15 \leq \Omega_o \leq 0.4$ for the density parameter. In this framework q_o is negative implying that the Universe scale factor is now increasing with positive acceleration. It is worth noting that our conclusions fully agree with the results of Fukugita *et al.* (1990) based on a completely independent analysis of deep galaxy counts.

As a final remark, we have that a small value for Ω_o would call for a value of H_o larger than 50 and possibly in the range 60-80 km/sec/Mpc. This might reconcile therefore a low-density Universe with a large value for the Hubble constant as it seems to derive for instance from the Tully-Fisher relation (Aaronson *et al.* 1980).

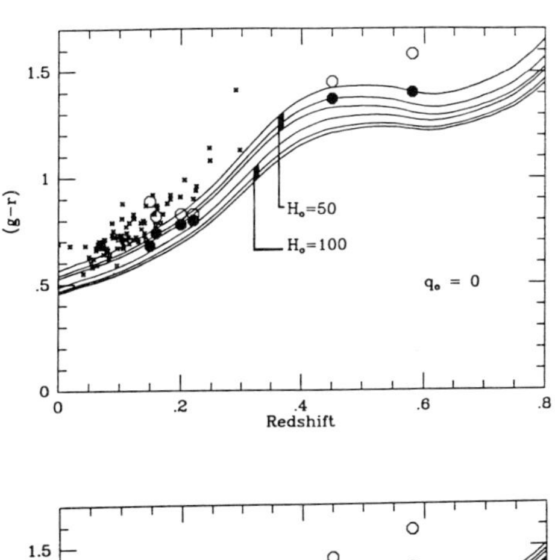

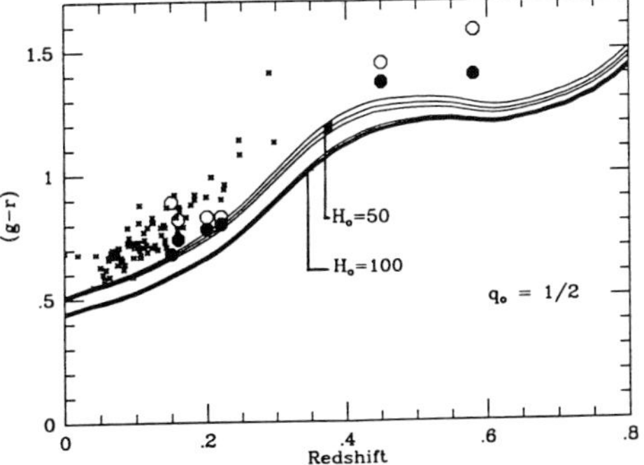

Fig. 1- Comparison between colors for cluster ellipticals in our observational sample and theoretical expectations for relevant cosmological models. In both panels filled dots represent the *mean* colors for ellipticals while open dots mark the colors for the *first-ranked* galaxies in each cluster. For comparison we reported also the observations by Schneider, Gunn & Hoessel (1983) (little crosses) referring only to first-ranked galaxies in their cluster sample. Theoretical color evolution is computed using Buzzoni's (1989) code for population synthesis assuming for galaxies [Fe/H] = +0.2. The two relevant values for the deceleration parameter q_o are displayed in each panel. Two families of curves have been computed according to H_o = 50 and 100 km/sec/Mpc, as labelled in the panels. For each value of H_o we assumed z_f = 3, 5 and ∞ (moving from the bluest to the reddest curve respectively).

REFERENCES

Aaronson, M., Mould, J., Huchra, J., Sullivan, W.T., Schommer, R., & Bothun, G. 1980, ApJ, 239, 12
Burstein, D., & Heiles, C. 1982, AJ, 87, 1165
Buzzoni, A. 1989, ApJS, 71, 817
Buzzoni, A., Gariboldi, G., & Mantegazza, L. 1991, AJ, submitted
Fukugita, M., Takahara, F., Yamashita, K., & Yoshii, Y. 1990, ApJL, 361, L4
Guiderdoni, B., & Rocca–Volmerange, B. 1990, A&A, 227, 362
Molinari, E., Buzzoni, A., & Chincarini, G. 1990, MNRAS, 246, 576
Renzini, A., & Buzzoni, A. 1986, in Spectral Evolution of Galaxies, eds. C.Chiosi and A.Renzini (Dordrecht: Reidel) p.195
Rocca–Volmerange, B., & Guiderdoni, B. 1990, MNRAS, 247, 166
Sandage, A. 1988, ARAA, 26, 561
Sandage, A., & Tammann, G.A. 1990, ApJ, 365, 1
Visvanathan, N., & Sandage, A. 1977, ApJ, 216, 214
Wyse, R.F.G. 1985, ApJ, 299, 593
Yoshii, Y, & Takahara, F. 1988, ApJ, 326, 1

Discussion

Djorgovski: What is, in more detail, the transformation you use to convert the Mg index into $[Fe/H]$?
Buzzoni: I used the calibration proposed in Buzzoni, Gariboldi & Mantegazza (see this congress). Briefly, the dependence of the Mg_2 index on [Fe/H] is $\partial Mg_2/\partial[Fe/H] = 0.135$ with a zero-point such as $Mg_2 = 0.28$ for stellar populations of solar metallicity.

Chokshi: Which colors do you predict for galaxies at $z = 0$, and are they consistent with those really observed?
Buzzoni: Synthetic colors for early-type galaxies at present time are found to be $(g-r) = 0.53 \pm 0.02$ and $(g-i) = 0.75 \pm 0.03$. To be consistent with our observations, these colors have been computed in the Gunn system as reproduced at ESO. While no difference occurs for the $(g-i)$ color respect to the standard system (in the range pertinent to galaxies), the $(g-r)$ is about 0.04 mag redder. Quite curiously, for what I know there are not extensive works reporting good photometry of local galaxies in the Gunn system. From observations of clusters at low redshift Schneider, Gunn & Hoessel (1983, ApJ, 264, 337) derive $(g-r) = 0.51$ and $(g-i) = 0.78$ (in the restframe and accounting for the little correction quoted above).

Bruzual: How confidently do your evolutionary models fit the first-ranked galaxies in the clusters? It seems to me that your predicted colors are systematically bluer respect to the observations. Is it a problem with the calibration?
Buzzoni: The synthetic models were *not* intended to fit first-ranked galaxies. To be consistent, we should enhance metallicity up to [Fe/H] = +0.5 predicting therefore redder colors. Our aim only was to reproduce the mean population of early-type galaxies on which relies the cosmological test.

Ferguson: To what extent are your constraints on H_o dependent on the match of your models to low-redshift galaxies? In other words: which constraints on H_o , q_o and λ_o do you still have by forcing your evolution models to fit present-day galaxies alone?
Buzzoni: The main conclusion accounting for the photometric properties of the present-day ellipticals converges toward an age about 15 Gyr and an IMF consistent with the Salpeter law. This is therefore fully consistent with the conclusions of the present work confirming that we are tracking confidently the history of the galaxies from $z = 0$ back in time.

UV TO IR MODELS OF GALAXY EVOLUTION AND COSMOLOGY

Brigitte ROCCA-VOLMERANGE
Institut d'Astrophysique
98 bis Bd Arago
F-75014 PARIS

Abstract

Recent improvements of spectrophotometric evolutionary models are described. New stellar libraries in the near-infrared (JHK) allow extension of the synthetic spectral Atlas of galaxies down to 10μm. From analyses in the far-UV and visible, observed colors and counts of faint galaxies are fitted by modelling a standard luminosity evolution and a low value of $\Omega_0(\simeq 0.1)$ while, in a Ω_0=1 Universe, models only fit data with a standard luminosity evolution and a number density evolution $\simeq (1+z)^{1.8}$: such a modelling is simulating a merging process. Another solution would be a tidally triggered star formation rate in a model in which galaxies form by hierarchical clustering of a dominant dark matter component. From evolution of M/L ratio, these models allow to link observed luminosity functions with mass distributions predicted from galaxy formation models and then to significantly connect evolution to formation models. Nevertheless these two models are not sufficient to fit some observational data such as the Hubble diagrams and faint galaxy counts in the near-infrared, the bright galaxy counts in visible and the Extragalactic Background Light. So new evolution scenarios are needed implying other constraints for cosmological parameters.

1. Introduction

Early evolution of galaxies makes up the keystone of cosmology : a link between the observable Universe and predictions from initial fluctuations. For the first time, a large variety of distant galaxies marks out with light or matter the way to the primeval phases. Individual sources or galaxy populations discovered through narrow pencil beams at distances up to now never reached, explore deep into the structures and likely test evolution. Galaxies turn up as stellar populations with typical spectral, dynamical and chemical signatures. However their appearance is depending on the geometry of the Universe and a model of galaxy evolution is needed to deconvolve the respective effects of evolution and cosmology. Such a model based on the physical processes driving birth and evolution

B. Barbuy and A. Renzini (eds.), The Stellar Populations of Galaxies, 357–366.

of stars or triggering emission of gaseous components, has to simultaneously interpret signatures of intervening, absorbing and diffusing matter. Significant results are essentially depending on three conditions : i) input data have to be mostly observational, ii) evolution scenarios fit various observational samples at z=0 on a large wavelength range (far-UV to IR) iii) when it is possible, details of data processing are taken into account in modelling. Several models have been built. Mostly they have similar basic principles and essentially differ from their input data. We shortly review some of them and their recent improvements. Then we show that at the present time the most convincing interpretation of the large redshift samples corresponds to luminosity and number density evolution which could be a phenomenological simulation of the merging process. However, as some observables are not well fitted by these models, our conclusion is a perspective of a better understanding of galaxy evolution by a simultaneous fit of observations from the far-UV to the infrared.

2. Present status of models

Historically these models were built in two phases. The first models of evolving stellar populations were simultaneously proposed by Tinsley, 1972 and Searle, Sargent et Bagnuolo, 1973. Respectively based on isomass tracks and isochrones following evolution of stars in the Hertzsprung–Russell diagram, they interpreted blue galaxies with bursts of massive stars. These photometric models were improved with nebular emission lines (Huchra, 1977) and with far–UV colors and metallicity effects (Rocca–Volmerange et al, 1981). Some recent models are still photometric, most of them take into account a simplified chemical evolution (Arimoto and Yoshii, 1986, Franceschini et al, 1991)

A second generation of models appeared when atlases of spectra replaced stellar colors as input data (Bruzual, 1983, Guiderdoni and Rocca–Volmerange, 1987 (GRV)), down to the far–UV from the IUE atlases (Wu et al, 1983, Heck et al, 1984). From then, several improvements were taken into account:

i) Recent improved stellar tracks have been published by Maeder and Meynet, 1988. Time duration phases may increase by a factor 3 compared to the Yale tracks, due to the overshooting of convective cores. Mass loss rates and opacities given by the most recent determinations (de Jager et al, 1988). Maeder's models essentially fit massive stars, typically $M \geq 1.5\ M_\odot$ (Maeder, private communication). Low mass stars are from Vandenbergh, 1985. These new tracks were introduced in our models by 1988 and the integration time step was refined down to 10^5 yr for calculating burst models. Most of recent results on ages of radiogalaxies (Rocca-Volmerange and Guiderdoni, 1990) and faint counts (Guiderdoni and Rocca-Volmerange, 1991) are based on these new input data. Maeder and Meynet, 1988 tracks were recently introduced in Bruzual's model (Charlot and Bruzual, 1991) and a more refined integration is carried out from a large ($\simeq$ 150) number of interpolated isochrones as used by Searle et al, 1973. Another model (Yoshii and Takahara, 1988) based its calculations in the far–UV from "so–called template" galaxies, observed with the ANS or IUE satellites. So their calculations of evolution in the far-UV are not explicit. This could explain why this model needs the help of a cosmological constant to fit Tyson's, 1988 observations in a low density universe (Fukugita et al, 1990). Detailed models with recent data of massive stars were proposed by Olofsson, 1989, Leitherer, 1990. Some models are extended in the near-infrared by

using stellar photometry or consistent black-body emission of stellar effective temperatures (Franceschini et al, 1991, Rocca-Volmerange and Gros, 1991).
ii) An extinction correction is proposed assuming stellar emitters mixed with gas. The optical depth through a spiral disk depends on inclination, gas content and metallicity (GRV).
iii) A nebular component (continuum plus lines) is calculated by estimating the number of Lyman continuum photons from stellar models (Clegg and Middlemass, 1987, Kudritski, 1990) and the current radiation field.
iv) The metallicity effect is taken into account in stellar evolutionaty tracks (Maeder and Meynet, 1991, in preparation). It corresponds to an increase of age with metallicity due to an increase of opacities. As a consequence, at a given age, the giant/dwarf number ratio is lower in metal-rich galaxies since giants take more time to reach the giant branch phase (Guiderdoni and Rocca-Volmerange, 1984). For other effects of increasing metallicity (lowering effective temperature and blanketing effect), Arimoto and Yoshii, 1986 gave rough estimates. Most uncertain are the evolutionary parameters of the Asymptotic Giant Branch (AGB) and post-AGB phases. If luminosities and effective temperatures can be derived from observations (Mould, this conference), phase duration can vary from 10^6 to 10^8 yrs according to models (see Renzini this conference).

3. Extension to the near-infrared

To extend spectrophotometric models to the near-infrared, two complementary libraries of stellar spectra have been observed and compiled:
i) Our far-UV and visible stellar library has been linked to the near-IR with observational infrared colors in the JHK bands. The caracteristics of this library are a wavelength range from 200Å to 10μm and a resolution $\Delta\lambda$ =10Å below 1μm. An extension of our Atlas of synthetic spectra of galaxies (Rocca-Volmerange and Guiderdoni, 1988, RVG), based on scenarios of evolution for 8 morphological types intends to reproduce colors of the Hubble Sequence: an example is shown on Figure 1. These spectra are used to calculate cosmological and evolutionary corrections (respectively k- and e- corrections) needed to predict apparent magnitudes and colors of distant galaxies according to:

$$m_\lambda(z) = M_\lambda(0) + (m - M)_{bol}(z) + k_\lambda(z) + e_\lambda(z)$$
$$C_\lambda(z) = C_\lambda(0) + k_{\lambda,c}(z) + e_{\lambda,c}(z)$$

with $M_\lambda(0)$ and $C_\lambda(0)$ respectively intrinsic magnitude and color at the present epoch. The important point is all the spectral UV to IR stellar data are observational, they cannot be the origin of the discrepancy between observations and model predictions in the near-infrared and so evolution scenarios for the Hubble sequence have to be revised.
ii) To study in more details the stellar populations of nearby and distant galaxies, Rocca -Volmerange and Maillard observed a new library of high resolution stellar spectra ($\simeq$40 dwarfs, giants and supergiants) with the Fourier Transform Spectrograph (FTS) at C.F.H.T. by 1989, 1990 through a large filter covering the H and K bands and with a high spectral resolution (2-3 Å) showing typical signatures of CO and H_2O bands, Brackett lines and many others. Data have been processed; an example of such spectra is given on figure 2 (Lancon et al, 1991).

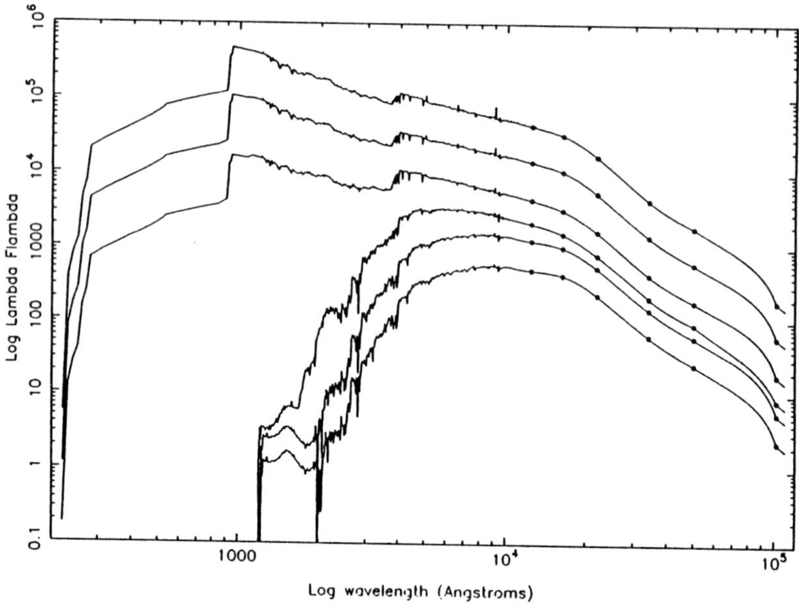

Fig. 1: Examples of synthetic spectra of galaxies extended from the far-UV (200Å) to the IR ($\simeq$10μm) calculated with an updated version of our model (Rocca-Volmerange and Gros, 1991).

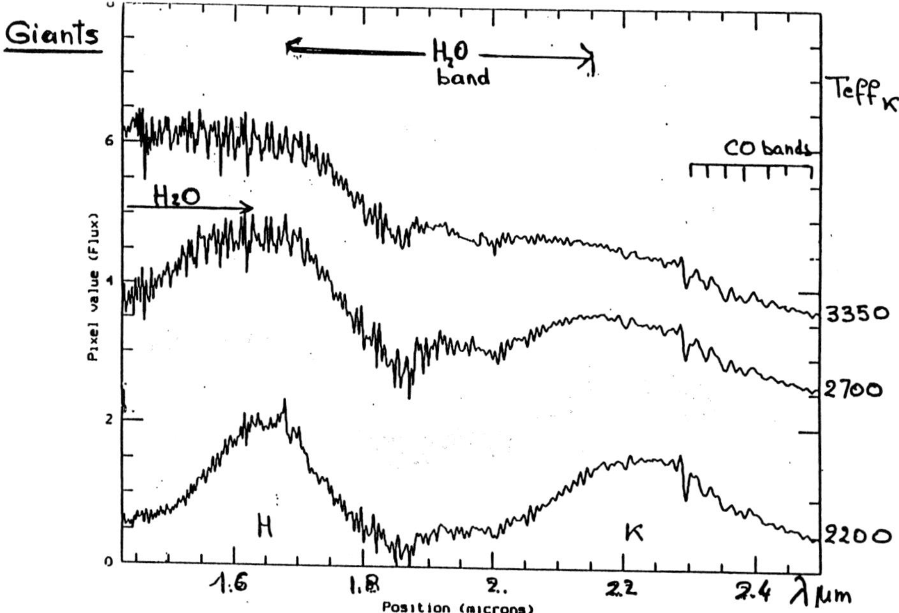

Fig. 2: High resolution spectra of the stellar library observed with the FTS instrument at C.F.H.T. through an H+K filter. Signatures of CO bands and H_2O bands are strongly varying with temperatures (Lançon et al, 1991).

4. Evidences of galaxy evolution

From the pionneer works of Butcher and Oemler, 1978, these last ten years confirmed a star formation evolution for distant galaxies providing vital clues to the process of luminosity evolution. Related to a starburst phenomenon, it has been firstly analysed in the optical from extremely blue colors (Searle, Sargent and Bagnuolo, 1973) and recently in the far-IR from the IRAS satellite (Sanders et al, 1988). In these last cases, an extreme evolution of the ultraluminous starbursts is associated to a dynamical interaction (Mirabel, 1991). At a lower level, the star formation activity in cluster galaxies has been related to dynamical processes to explain the morphological segregation (Dressler, 1980) and the discovery of A stars in elliptical galaxies (Dressler and Gunn, 1983).

The recent deepest surveys of faint counts of galaxies (Tyson, 1988, Cowie et al, 1990) strongly constrain scenarios of galaxy evolution and/or cosmological parameters. Evolution models fit number counts, redshift or color distributions on a large dynamic range down to B $\simeq$ 27 (Guiderdoni and Rocca-Volmerange, 1990, see also Guiderdoni, this conference). As number counts are strongly depending on the size of the covolume element, they give constraints on cosmological parameters: standard evolution models reproduce the data in a low density Universe while, by adding a number density evolution, data are compatible with models in a $\Omega_0 = 1$ Universe (Rocca-Volmerange and Guiderdoni, 1990, Guiderdoni and Rocca-Volmerange, 1991). However a large-scale structure survey of bright galaxies (the APM survey by Maddox et al, 1990) gives results in number counts, correlation functions, and other large scale properties in disagreement with the current models of galaxies. In particular, the evolution of galaxy number density between $m = 14$ and 19 is higher than predicted by models. Similar conclusions were given by Broadhurst et al, 1988 followed by Colless et al, 1990 who suggested a new population of blue galaxies vanished at the present time; we prefer another solution which would be new evolution scenarios, relaxing our initial hypotheses (Rocca-Volmerange and Gros, 1991).

Other signatures of galaxy evolution are given from the optical and infrared counterparts of powerful radiosources. The two most distant radiogalaxies: 0902+34 at z = 3.395 (Lilly, 1988) and 4C41.17 at z = 3.8 (Chambers et al, 1988) show signatures of stellar populations. Their typical features are: i) a large gap (up to a factor 10) of the observed flux from the far-UV plateau to the visible (rest frame) ii) the alignement of the radio and far-UV axes in most cases iii) an enormous Lyman-α (1215Å) line with an equivalent width of roughly 1000Å, implying a considerable star formation rate, about 1% of the total mass(Rocca-Volmerange and Guiderdoni, 1990) iv) different morphologies according to the wavelength range v) a narrow relation in the K-z Hubble diagram.

According to different authors, estimated ages spread from $\simeq$ 0.3Gyr to about 1.5 Gyr (Lilly, 1988, Rocca–Volmerange, 1988, Chambers and Charlot, 1990). In fact, arguments leading to a unique solution are not sound (Rocca–Volmerange and Guiderdoni, 1990) first because stellar evolutionary data of respectively supergiants, asymptotic giant branch stars and giants are not well established. Second, present status of observational data show that current star formation erases most of spectral features which could give a significant age (figure 3). Third, the infra-red emission associated to the radio jet (Joy et al, 1991) and possibly a dust contribution of these distant radiogalaxies could partly mask the stellar emission to be dated. Relative to the number density evolution, there is some evidence that radiogalaxies with $1 < z < 3.8$ are objects undergoing a rapid merging of the available clumps, as suggested by Djorgovski et al. 1988. The 16 3C radiogalaxies carefully imaged with sub-arcsecond seeing at the Canada-France-Hawaii Telescope all show a number of clumps, from 2 to 5 (Le Fèvre et al, 1988). Zepf & Koo,1989, used a deep survey of distant galaxies to estimate the relative frequency of close pairs at faint magnitudes ($B \leq 22$) relative to that for nearby bright galaxies. A statistically significant excess of faint pairs is consistent with an increase in the frequency of interactions $f = (1+z)^{4.0 \pm 2.5}$.

5. Star Formation in merging-driven evolution models

Among models of galaxy formation, more or less dissipative collapses (Larson 1975,1976) of clumpy turbulent protogalaxies with anisotropic stellar velocity (Carlberg, 1985) reproduce disks and spheroïds. On the other hand, dynamical friction acting on two or more building blocks could result in a merging process forming a condensed galaxy (Toomree and Toomree, 1972). Many observations favor, with a more or less high degree, a merging scenario. Fundamentally, a merging process is dynamical depending on the mass distribution of intervening blocks. For high masses of interacting components, a starburst strongly emits in the far-IR and/or in the far-UV. However such extreme cases are rare and the essential question is: what star formation rate (SFR) is induced by the bulk of interactions predicted by galaxy formation models? Likely it depends on intrisic and relative dynamical properties of components (mass, velocity,..) but we ignore on what way. Moreover the classical modelling $(1+z)^{\eta}$ increases the confusion because it is not clear if luminosity or number density are in fact multiplied by this factor.

According to the simultaneous approach of distant radiogalaxies and faint galaxy counts, Rocca-Volmerange and Guiderdoni, 1990, Guiderdoni and Rocca-Volmerange, 1991 proposed an unifying model in which all galaxies form from the merging of building blocks for which the spectral evolution is computed. The following scenario is suggested: the collapse and cooling of gas in the potential wells of dark-matter haloes lead to fragmentation into a number of clumps, say $N \simeq 10$ with a characteristic mass of an order of magnitude below the total baryonic mass. These clumps are the gas reservoir for star formation. The general rule for field galaxies would be that star formation begins at relatively high redshift in these clumps (z_{for}=10 to 5, derived from the colour distribution of faint galaxies as well as from the radiogalaxies). The clumps also “slowly” merge according to an ‘average’ $(1+z)^{\eta}$ law. Star formation continues in the merged objects from the residual gas. These objects would essentially be those observed in faint galaxy counts. Dynamical friction could be the process regulating merging and number evolution. In a small fraction of the massive, high-redshift objects, the merging of the available clumps could be more rapid, simply as a Poisson fluctuation. Since the clumps would then still be gas-rich, the star formation would be stronger, and the large amount of gas could also feed some central compact object and initiate the radiogalaxy phenomenon. The small number of high-redshift, active objects could be found from a search of radio sources, but would have no significant effect on faint galaxy counts. Finally, the subsequent growth of these objects is impeded by the large cooling time of very massive haloes (see e.g. Evrard, 1989). This scenario is compatible with most observations of radiogalaxies and faint galaxy counts and it has the essential property to save Ω_0=1 because the pure luminosity evolution models only fits the data in a low Ω_0 Universe incompatible with inflation. The number density evolution varies as $(1+z)^{1.8}$ and the luminosity evolution roughly follows our standard scenarios.

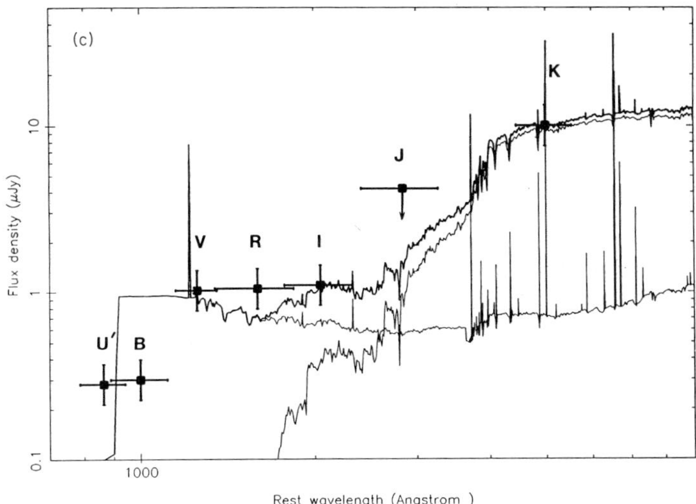

Fig. 3:Fit of the spectral energy distribution of 0902+34 in the U'BVRIJK bands at z=3.395. The spectra of two 0.1 Gyr and 2.5 Gyr are superimposed. Most of the evolution signs in the old burst are cancelled by the current ones.

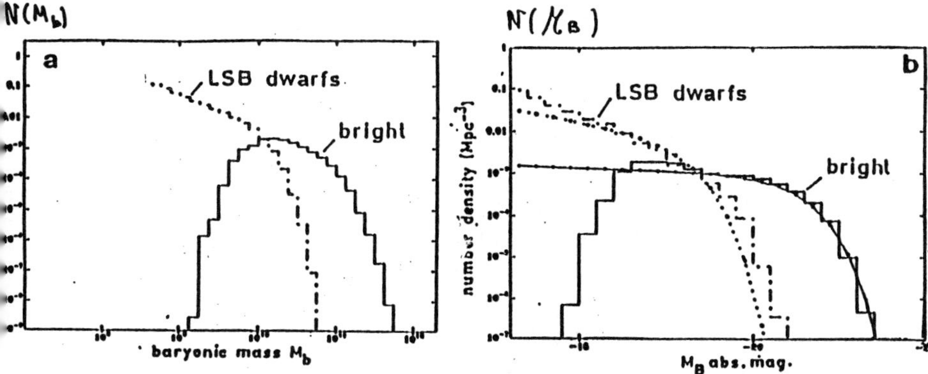

Fig. 4: Distribution of baryonic masses (a) compared to the corresponding luminosity functions (b) computed for bright galaxies and low-surface-brightness dwarf galaxies. The galaxy formation model and refinements of our evolution model are described in Lacey et al, 1991.

Another solution is a model in which galaxies form by hierarchical clustering of a dynamically dominant dark matter component and in which the rate of star formation is controlled by the frequency of tidal interactions with neighbouring galaxies. This has been proposed by Lacey and Silk, 1991. Owe to the mass-luminosity ratio M/L as a fruitful output of evolution models, luminosities, colours, surface brightnesses and circular velocities of nearby bright galaxies as well as distributions of redshifts, colours and numbers of distant faint galaxies are analysed with our spectrophotometric model (Lacey et al, 1991). Several constraints for the scenarios of galaxy evolution result from this complete study which will be extended to other galaxy formation models. Figure 5 shows the relation of mass-luminosity used in this model by adding a dwarf population. Early winds induced by supernovae explain their low surface brightnesses and the reason why this population is not detected at the present time (Silk, this conference).

These last solutions are compatible with the recent observations of bright galaxies (Maddox et al, 1990) only at a low level. Moreover, the blue (B) and red (K) number distributions are not fitted with the same models (Cowie et al, 1991). As a confirmation, we show in Rocca-Volmerange and Gros, 1991, that the JHK Hubble diagrams are not so well fitted that the corresponding blue diagrams. At evidence, new populations or new basic hypotheses of evolution scenarios have to be proposed. And only a significant confrontation of the blue and red will be avalaible to find the solution.

An excellent compilation of the present status of EBL measurements is given by Mattila,1991. These observational results can be compared to theoretical predictions for the background light due to galaxies. Evolution of galaxies has been calculated with a standard luminosity evolution in various cosmologies (Ω_0=0.1 and 1) and various redshifts of formation (z_{for}=2 or 30). All these curves are an order of magnitude lower than the observational values or upper limits in the UV, optical and IR. The important point is that merging models with number density evolution give roughly similar fits since we assume that the total comoving mass density is conserved. Similar conclusions were derived from Bruzual models and Yoshii and Takahara results. Figure 6 presents a strong near-IR emission due to a current starburst and two different extinction factors E_{B-V}=0.1 and 0.5, added to the global galaxy emission. the SED of the burst corresponds to a $(1+z)\lambda$ shift with 1+z=10. As an example, this population could be a solution for the EBL which has to be constrained by the new scenarios of galaxy evolution .

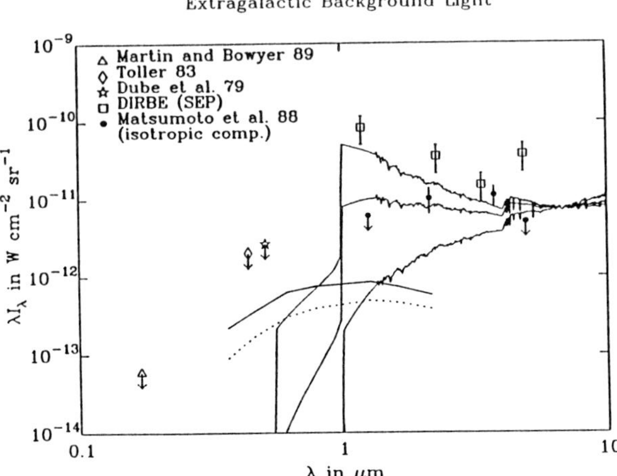

Fig. 5: Predictions of the global emission of galaxies with evolution (full line)and without evolution (dotted line) are far below observations. The contribution of a synthetic burst formed at z=9 with various extinction factors would fit the near- infrared recent data from DIRBE or Matsumoto et al, 1988.

References

Arimoto, N., Yoshii, Y., 1986, *Astron. Astrophys.*, **164**, 260
Bardeen, J.M., Bond, J.R., Kaiser, N., Szalay, A.S., 1986, *Astrophys. J.*, **304**, 15
Baron, E. & White, S.D.M., 1987. *Astrophys. J.* **322**, 585
Bithell, M. & Rees, M.J., 1990. *Mon. Not. R. astr. Soc.* **242**, 570
Broadhurst, T.J., Ellis, R.S. & Shanks, T., 1988. *Mon. Not. R. astr. Soc.* **235**, 827
Bruzual, G., 1983. *Astrophys. J.* **273**, 105
Butcher, H.R., Oemler, A., 1978, *Astrophys. J.*, **219**, 18
Carlberg, R.G., 1985, in *the Milky way Galaxy*, H. van Woerden, R.J. Allen, W.B. Burton (ed), Dordrecht, Reidel, p.165
Carlberg, R.G. & Couchman, H.M.P., 1989. *Astrophys. J.* **340**, 47
Chambers, K.C., Miley, G.K. & van Breugel, R.R., 1988. *Astrophys. J.* **329**, L75
Chambers, K.C. & Charlot, S, 1990. *Astrophys. J.* **348**, L1
Clegg, R.E.S., Middlemass, D., 1987, *Mon. N. Roy. astr. Soc.*, **244**, 408
Colless, M., Ellis, R.S., Taylor, K., Hook, R.N., 1990, *Mon. N. Roy. astr. Soc.*, **244**, 408
Cowie, L., Lilly, S., Gardner, J., 1991, preprint
de Jager, C., Nieuwenhuijzen, H., van der Hucht, K.A., 1988, *Astron. Astrophys.*, **164**, 260
Djorgovski, S., Spinrad, H., Marr, J., 1984, in New Aspects of Galaxy Photometry, J.L. Nieto (ed.), Springer–Verlag, p. 193
Djorgovski, S., Spinrad, H., McCarthy, P., Dickinson, M., Van Breugel, W. & Strom, R.G., 1988. *Astron. J.* **96**, 836
Dressler, A., 1980, *Astrophys. J.*, **236**, 351
Dressler, A., Gunn, J.E., 1983, *Astrophys.J.*, 270,7
Evrard, A.E, 1989. *Astrophys. J.* **341**, 26
Franceschini, A., Toffolati, L., Mazzei, P., Danese, L., de Zotti, G., 1991, preprint
Fukujita, F., Takahara, F., Yamashita, K., Yoshii, Y., 1990, *Astrophys. J.*, **361**, L1
Guiderdoni, B. & Rocca–Volmerange, B., 1984. *Astron. Astrophys.* **109**, 355
Guiderdoni, B. & Rocca–Volmerange, B., 1987. *Astron. Astrophys.* **186**, 1
Guiderdoni, B. & Rocca–Volmerange, B., 1990. *Astron. Astrophys.* **227**, 362
Guiderdoni, B., Rocca–Volmerange, B., 1991, in press
Heck, A., Egret, D., Jaschek, C., 1984, Battrick., B., ESA SP-1052.
Hintzen, P., Romanishin, W. & Valdes, F., 1991,*Astrophys. J.* **366**, 7
Huchra, J.P., 1977, Astrophys. J., **217**, 928
IUE Ultraviolet Spectral Atlas, 1983, Wu *et al.*, NASA N^{0}22
Joy, M., Harvey, P.M., Tollestrup, E.P., Sellgren, K., Mc Gregor, P.J., Hyland, A.R., 1991, *Astrophys. J.*, **366**, 82
Kudritski, R.P., in *Massive stars in starbursts*, C. Leitherer (ed), Baltimore, in press
Lacey, C., Silk, J., 1991, in press
Lacey, C., Guiderdoni, B., Rocca-Volmerange, B., Silk, J., submitted
Lançon, A., Rocca-Volmerange, B., Maillard, J.P., 1991, submitted
Larson, R., 1975, *Mon. Not. R. astr. Soc.* **173**, 671
Le Fèvre, O., Hammer, F. & Jones, J., 1988. *Astrophys. J.* **331**, L73
Lilly, S., 1988. *Astrophys. J.* **333**, 161
Maddox, S.J., Efstathiou, G. & Sutherland, W.J., 1990, *Mon. Not. R. astr. Soc.* **211**, 833
Lilly, S., 1989. *Astrophys. J.* **340**, 77

Le Fèvre, O., Hammer, F. & Jones, J., 1988. *Astrophys. J.* **331**, L73
Lilly, S., 1988. *Astrophys. J.* **333**, 161
Maddox, S.J., Efstathiou, G. & Sutherland, W.J., 1990, *Mon. Not. R. astr. Soc.* **211**, 833
Maeder, A. & Meynet, G., 1988. *Astron. Astrophys. Suppl. Ser.* **76**, 411
Mattila, K., 1991, in *The early universe from diffuse backgrounds*, Rocca-Volmerange, Deharveng, Tran Thanh Van (ed), Editions Frontières
Mirabel, I. F., & Sanders, D.B., 1988, *Astrophys. J.*, **340**, L53
Olofsson, K., 1989, *Astron. Astrophys. Supp. Ser.*, **76**, 317
Rocca–Volmerange, B., 1988. *The Messenger* **53**, 26
Rocca–Volmerange, B. & Guiderdoni, B., 1988. *Astron. Astrophys. Suppl. Ser.* **75**, 93
Rocca–Volmerange, B., Guiderdoni, B., 1990, *Month. Not. Roy. Astron. Soc.* , **247**, 166
Rocca–Volmerange, B. & Gros, L., 1991, in preparation
Sanders D.B., Soifer, B.T., Elias, J.H., Madore, B.F., Matthews, K., Neugebauer, G., Scoville, N.Z., 1988, *Astrophys. J.*, **325**, 74
Searle, L., Sargent, W.L.W., Bagnuolo, W., 1973, Astrophys. J., **179**, 427
Silk, J., 1978, *Astrophys. J.*, **220**, 378
Tinsley, B.M., 1972, *Astron. Astrophys.*, **20**, 383
Toomre, A., Toomre, J., 1972, *Astrophys. J.*, **178**, 623
Tyson, A., 1988. *Astron. J.* **96**, 1
VandenBergh, D., Bell, R.A., *Astrophys. Supp. Series*, **58**, 561
Yoshii, Y., Takahara, F., 1988, Astrophys. J., 326, 1
Zepf, S.E., Koo, D.C., 1989, Astrophys. J., 337, 34

Questions: J. Ostriker: You presented two quite different pictures. In the first, you had Ω_0=0.1 and $z_{for} \geq 10$. In the second, you had Ω_0=1 and $z_{for} \leq 3$ with extensive mergers. Am I free to believe either model? If so, I prefer the models without much recent merging since there are physical arguments against the latter. Recent merging would heat Sp disks too much. **B.R.V.**: Many observational arguments are in favor of old galaxies (high z_{for} and low Ω_0) even if merging is possible. Observations of large samples of galaxies at intermediate z will give nearly the answer.
G. Bruzual: I got the impression that you include chemical evolution in the evolutionary tracks but not in the stellar libraries. Could you comment on this point? **B.R.V.**: This is a current work.
A. Chokshi: Two comments and a question: i) 0902+34's K band data from Lilly is wrong. P. Eisenhardt's K band photometry drops the K-band data by about 1 magnitude. Given that I band data is an upper limit, the spectrum of 0902+34 is much flatter now. ii) The Hubble diagram is fairly insensitive to galaxy evolution models. iii) Can you list the key parameters, according to their importance for galaxy evolution models, if you try to reproduce the photometric properties of low z galaxies? For example, can one play off M_{high} in IMF versus amount of dust extinction or is it not allowed? **B.R.V.**: The question of key parameters is fundamental and not yet clearly solved. Your suggestion is difficult because of the predominant role of supernovae winds needed by most models. However the emission lines are correlated to the M_{high} and anticorrelated to the dust amount.
P. Whitelock: It has been known for a long time that M giants have strong H_2O absorption. However it is only the most luminous AGB stars-the Miras. Jay Frogel has told us that we don't expect these luminous AGB stars to make a significant contribution to the near-IR spectra of old metal-rich systems. If you find you need to include such objects in your spectral synthesis it would be very interesting as it would mean that the contribution from luminous AGB stars was much larger than predicted. **B.R.V.**: I agree that, for the first time, the H+K filter of the F.T.S. instrument allows a detailed estimate of the AGB stellar population in galaxies.

DARK POPULATIONS

Joseph Silk
Departments of Astronomy and Physics, and Center for Particle Astrophysics, University of California, Berkeley

ABSTRACT

After summarizing the evidence for dark baryonic matter in various environments, I review theoretical considerations relevant to the nature of such a component of the universe in regions where it is the dominant mass fraction. Phenomenological constraints are given on the primordial stellar initial mass function. The primordial rate of star formation is evaluated during the early stages of galaxy formation, utilizing simple theoretical considerations motivated by present day star formation rates. Applications are given to predominantly dark objects whose existence is inferred from galaxy formation theory, including low surface brightness giant and dwarf galaxies. Prospects for detection of dark populations that dominate our own galactic halo are described.

1. INTRODUCTION

Dark baryonic matter is an inescapable part of astronomical life. While negligible on the scale of the solar system, the local disk may contain a substantial dark component. Perhaps coincidentally, old stellar populations in galaxy spheroids and in ellipticals may contain a similar fractional component of dark matter. It is tempting to argue by continuity that regions where larger fractional amounts of dark matter are found, including galaxy halos and dE galaxies, which also consist almost exclusively of old stars, contain dark matter that is of essentially the same composition as that found elsewhere. Arguments of this type, treating dark matter as simply an extreme variation on the content of known, usually old, systems, will be developed here. It is of course possible that the dark matter content of dark halos consists of some qualitatively new population or even form that does not appear in luminous systems. Indeed I regard this possibility as likely for the uniform dark matter component postulated to give a critical closure density for the universe. However, nature would surely have to be perverse were the dark matter associated with luminous matter to drastically change its form as the dark matter content of a stellar system increases over a modest range.

This article is organized as follows. Section 2 summarizes the dark matter fraction in various astronomical systems, where I take dark matter to include any dark population whose presence is inferred from dynamical measurements. Sections 3 and 4 present various constraints from theory and from phenomenology on the initial mass-function of primordial stars that would have formed a baryonic dark halo. Heuristic expressions are given in Section 5 for the primordial star formation rate. Section 6 presents applications of the foregoing theory to dark systems expected in the context of models for the large-scale structure of the universe. The current status of searches for baryonic dark matter in our halo is reviewed in Section 7.

B. Barbuy and A. Renzini (eds.), The Stellar Populations of Galaxies, 367–378.

2. OBSERVED DARK MATTER FRACTION

A conservative estimate of the dark mass fraction in the solar system would be less than 0.1 percent. There is little support in our immediate vicinity for the contention that a substantial fraction of the dark mass might be in the form of planets, presumably Jupiter-like because of its hydrogen-rich composition. Moreover, the formation of Jupiter would be questionable in the absence of a rocky core.

In the old galactic disk, at least 10 percent of the mass is in the form of white dwarfs and neutron stars which accordingly may be considered as dark matter candidates. Only the very nearest cold white dwarfs (within a few parsecs of the sun) are detectable, and the old neutron stars are visible as radio pulsars, x-ray binaries, or perhaps gamma-ray bursters if they undergo substantial accretion. The white dwarf fraction comes from direct observation of the nearby white dwarf luminosity function, but would undercount white dwarfs with a large scale height. The neutron star (and black hole) fraction is estimated by assuming a past star formation rate in the disk, and is uncertain but likely to contribute considerably less than white dwarfs to the dark mass fraction.

An alternative approach to detecting the disk dark matter density near the sun relies on measuring the local gravitational acceleration perpendicular to the galactic plane. This has not led to any definitive result with regard to the need for a component of dark matter that is in excess of the known contributions from compact stellar remnants, gas and stars[1]. Up to fifty percent of the disk could be in dark matter without producing any excessive acceleration, and at least one analysis[2] has provided evidence for such a contribution. However uncertainties in the distribution and kinematics of the stellar populations used in the dynamic analyses[3] mean that the disk dark matter fraction satisfies $0.1 \leq f \leq 0.5$.

Globular clusters, as the oldest stellar population in the galaxy, would be a good laboratory for investigating the dark matter content of stellar populations were it not for the bias introduced by dynamical relaxation. Objects less massive than $1 M_{\odot}$ tend to evaporate, and more massive objects sink to the center. In the latter case, the observed distribution and dynamics of globular cluster stars provide a sensitive probe of the dark matter content. Modelling suggests a primordial neutron star mass fraction of between 0.1 and 1 percent, although white dwarfs could constitute up to 30 percent of the globular cluster mass in well-studied examples[4]. Binary formation, mass transfer, and stellar mergers are believed to be a continuing source of massive star formation in the cases of such globular clusters as 47 Tuc, where there are at least 12 millisecond pulsars as well as a centrally concentrated population of blue stragglers.

Modelling of the galactic spheroid results in about as much uncertainty as there is in the disk with regard to a possible dark matter component[5,6]. The situation should improve when the OH/IR stars are incorporated into the dynamical modelling. A similar uncertainty, allowing up to fifty percent of the dynamically measured mass to be dark, applies to elliptical galaxies within their characteristic half-light radii r_e. The mass-to-light ratio $M/L_B \approx 10h$, where $h = H_0/100$ km s^{-1}Mpc^{-1}, corresponds to that of an old stellar population which is normalized relative to the stellar population of the solar neighborhood. Our local mass-to-light ratio is between 2 and 4, and reflects the uncertainty in the contribution of local dark matter. One may conclude that, within the uncertainties, there is no dark matter problem in the luminous regions of galaxies, provided that one is prepared to tolerate the possibility of an initial mass function that results in up to fifty percent of the mass consisting either of compact stellar remnants or, less plausibly, brown dwarfs. Brown dwarfs have lower likelihood, in my view, because the required extrapolation from known dark matter contributions to well-studied regions is much larger than for white dwarfs. There is as yet no unambiguous evidence for the existence of even a single brown dwarf[7]. On larger scales, rotation curve measurements reveal that luminous spirals are embedded in dark halos that extend to more than 2–3 Holmberg radii. The dark mass fraction within this distance, which is only as far as the dark matter can be studied, exceeds 80 percent of the total enclosed mass. A similar result is believed to apply to ellipticals, although the evidence is far sparser, being based on rare cases of 21 cm rotation curves and a somewhat model-dependent interpretation of the x-ray luminosity as emission from diffuse hot gas that traces the halo potential well. Nearby dwarf ellipticals have also been the subject of intensive study, and these low surface brightness systems are dominated even in their central regions by dark matter. In these latter objects, the dark mass fraction amounts to about 90 percent for the extreme cases (where $M/L_B = 50 - 100$). A similar fraction is measured in galaxy clusters, where $M/L_B \approx 300h$, and about 10 percent of the mass is in diffuse hot gas.

An immediate reaction to the observed dark matter fraction in the regions (large-scale and dE's) where it is most extreme is that only a very modest extrapolation by about a factor of 3 (modest by astrophysical standards!) is required if the *entire* dark mass were to consist of white dwarfs rather than be characterized by the solar neighbourhood content of roughly 30 percent. A second reaction is that 90 percent is very different from the 99 percent dark fraction required for an inflationary universe, for which $\Omega = 1$, where we can write the ratio of mass to luminosity density in terms of Ω as $\Omega = (M/L_B)/(1500h)$. Thirdly, the success of the primordial nucleosynthesis model, combined with the LEP measurement of 3 neutrino species, in accounting for the abundances of ^{4}He, ^{2}H and ^{7}Li requires a substantial amount of baryonic matter in excess of that seen in luminous galaxies. Translating the limit $\Omega_b h^2 = 0.015$ into dark baryonic fraction f implies $0.3 \lesssim f \lesssim 0.8$. If this matter is in galaxy halos, then it cannot be diffuse gas: however, diffuse hot intergalactic gas can readily accomodate a density $\Omega \lesssim 0.1$ without producing an excessive x-ray background. Inhomogeneities generated at the quark- hadron phase transition can lead to deviations from the standard, homogeneous model predictions of light element abundances that allow a larger value of Ω_b. A factor of 2 increase in Ω_b cannot easily be excluded, so that $\Omega_b \lesssim 0.1$ and $f \lesssim 0.9$, with the upper limits possibly even being *required* by the nucleosynthesis modelling. Indeed, a recent analysis of lithium depletion via convection and diffusion in population II stars finds that $\Omega_b h^2 = 0.046$ is allowed, and possibly even favoured, by the observed dispersion in reddening–corrected ^{7}Li equivalent widths in metal–poor halo stars[8].

Thus I conclude that *all* of the dark matter that is directly *measured* may necessarily be baryonic. Since this is the dark matter that is associated with galaxies, either in halos or in clusters, such a conclusion is not very radical. Weakly interacting particle dark matter, a favored if controversial relic from the Big Bang because of the lack of any existence proof, is more likely to be associated with the relatively uniform dark matter distribution that is required in order to account for a critical density of matter $3H_0^2/8\pi G$ if the universe is bound or marginally bound.

3. BARYONIC DARK MATTER

Many candidates exist for baryonic dark matter (BDM) that range in mass from cometary scales to supermassive black holes. One can be confident that BDM formed from gas with zero metallicity at high redshift, and most likely in clouds of subgalactic mass. The successful bottom-up approach to large-scale structure via N-body simulation modelling of the galaxy correlation function and other characteristics of the galaxy distribution argues for initial clouds that were of scale comparable to those of dwarf galaxies. The earliest modern approach to galaxy formation, following the discovery of the CMB, utilized primordial isothermal fluctuations in density whose characteristic mass was identified with that of globular clusters and motivated by the post-decoupling Jeans mass[9]. Primordial adiabatic density fluctuations characterized by the radiative damping mass[1]0 led to a top-down galaxy formation model[11] that predicted excessive CMB anisotropies[12]. This model was revived with the advent of hot dark matter but has not resulted in a satisfactory model for large-scale structure. An early phase transition generally produces isocurvature rather than isothermal fluctuations, and the studied examples of such models result in fluctuation spectra with the largest fluctuations on the smallest scales, leading to a bottom-up evolutionary sequence. Cosmic strings and textures are examples of topological defects that generate non-gaussian isocurvature fluctuations.

It is unfortunate that given primordial clouds of baryons of mass, say, $10^6 - 10^8 M_\odot$, with no metals and, presumably, no magnetic fields, theory cannot distinguish between the following possibilities: fragmentation into substellar fragments, into stars of conventional mass, or into supermassive black holes. Hence, I will develop a phenomenological argument that is suggestive, if not overwhelming. It involves the Principle of Least Extrapolation. I require that BDM candidates should be known to exist and to constitute a significant fraction of the luminous regions of galaxies. The leading candidate, which is the greatest contributor in luminous regions, is the white dwarf. Other possibilities are neutron stars and black holes at the massive end of the primordial initial mass function (IMF), and degenerate dwarfs at the low mass end. One can immediately eliminate planets ($\lesssim 10^{-3}$ of luminous mass), supermassive black holes, and diffuse gas as BDM contenders.

The initial stellar mass function is only studied over a wide dynamical range in the solar vicinity. It represents all stars ever formed. It approximates a power-law with varying slope over the range 0.3 to 100 $M_\odot$, peaking at about $0.2M_\odot$ and declining towards the hydrogen-burning limit, $0.08M_\odot$. No brown dwarfs, defined to be objects below $0.08M_\odot$ but greater than the maximum mass of a planet of solar composition, about $0.002M_\odot$, have hitherto been definitively detected. The distinction between planet and brown dwarf is based on the partial lifting of degeneracy pressure in the core, resulting in a maximum planetary radius at about 2 Jupiter masses. Some energy release occurs for more massive degenerate dwarfs via slow gravitational contraction (hence *brown*[13]). An alternative measure of the minimum mass of a brown dwarf of primordial composition comes from considerations of opacity-limited gravitational fragmentation of a collapsing cloud. This leads to mass estimates of order $0.001M_\odot$, although the calculations assume spherical symmetry and are probably too idealized to be an adequate guide. Of course, above $\sim 1M_\odot$, one only measures the current epoch mass distribution, and a past star formation rate must be assumed in order to derive the IMF.

For BDM to consist of compact objects, either brown dwarfs or stellar remnants, the primordial IMF must differ drastically from the present IMF. One could not have formed many solar mass stars. Either the primordial IMF was predominantly brown dwarf-dominated, or else it was top-heavy, with strong suppression of star formation below $\sim 2-3M_\odot$. Theory has little to offer in the way of predicting the primordial IMF, but there are several suggestive phenomenological constraints that are described in the following section.

4. PRIMORDIAL IMF

4.1 Star formation theory

Theory is incapable of even estimating the characteristic stellar mass in primordial clouds. In the likely (but not inevitable) event that fragmentation occurred, the available analyses utilize linear perturbation theory to infer a minimum Jeans mass in an idealized collapse geometry. H_2 cooling in primordial clouds results in minimum fragment masses of $\sim 0.01M_\odot$. However, it has been argued that spherical collapse does not lead to formation of long–lived fragments, but that anisotropic collapse is the more generic initial condition that will lead to formation of a transient sheet. Such a sheet is unstable to fragmentation. Unfortunately, the non-linear evolution of the fragments is poorly understood. Numerical simulations suggest that a complex filamentary and clumpy structure develops[14]. The role of processes such as accretion and coalescence is unclear. The situation is murkier still if magnetic fields play a dynamical role as in conventional star formation where angular momentum transfer is regulated by coupling to magnetic fields. Angular momentum transfer in primordial clouds is likely to be controlled by disk instabilities. Moreover, it is entirely possible that bipolar outflows, prevalent in nearby star forming regions, develop around forming primordial protostars and provide a feedback mechanism that limits the growth of stellar masses.

4.2 Phenomenology: there were primordial stars

The existence of solar mass stars with metallicities $[Fe/H] < -4$ is a strong indication that ordinary stars formed in an essentially pristine environment. It takes very little in the way of element synthesis and ejection by massive stars, with their short evolutionary time-scale, to pollute to this level. Stars of primordial abundance produce a small amount of CNO by pp-cycle nucleosynthesis, so that stellar evolution with initial $[Fe/H] \lesssim -6$ is generally indistinguishable from zero initial [Fe/H]. However if $[Fe/H] < -4$, there is too little CNO for helium flashes, so that dredge-up is suppressed and yields are reduced for intermediate mass stars.

The study of heavy element abundance patterns in metal-poor stars, in statistical samples with $[Fe/H] < -3$, and in individual more extreme metal-poor examples, reveals the characteristic signatures of both r and s process abundance ratios. This indicates the presence of stellar precursors in the mass range $10 - 50M_{\odot}$ that are known to be the sites of these elements. Such stars must have existed very early in the proto-galaxy phase, *before* the most metal-poor observed stars had formed. These latter stars are of course the oldest stars in the galaxy and have masses of about $1M_{\odot}$.

Hence the protogalaxy must have contained stars that spanned essentially the same mass range as conventional stars. This does not of course allow us to infer anything about the mass fraction in such stars, the nucleosynthesis argument and direct observation only requiring a small fraction by mass, of order $Z/Z\odot$, in extreme metal-poor stars.

4.3 The primordial i.m.f. (probably) was different

Several phenomenological arguments attest to a primordial i.m.f. that differed from the solar neighborhood i.m.f. None of these are compelling, but there does seem to be a suggestive trend.

4.3.1 Alpha nuclei

The low dispersion in $\alpha-$nuclei abundances relative to Fe/H reported at this meeting for old disk stars, together with the radial gradient of declining α/Fe as a function of increasing mass and orbital radius, suggests that a change in the IMF has occurred between the epoch of disk formation and today. Of course, one may more directly infer a change in nucleosynthetic yield (heavy element fraction produced relative to net gas mass forming stars), and modification of the IMF is not a unique solution. The coupling of $\alpha-$nuclei, produced by massive stars, to iron abundance, produced in supernovae of type II, suggests that type II supernovae dominated Fe production in the inner galaxy and old disk. This could require a more top-heavy IMF than is presently observed. Alternatively, mass loss might have been less efficient at low metallicity, allowing intermediate mass stars to become supernovae. Such stars are sufficiently short-lived that their nucleosynthetic ejecta would be well coupled to the debris of massive stars. This provides a counter-example of how variable yield, rather than IMF, might be responsible for the observed effect. The enhancement of $\alpha-$nuclei, observed at $[Fe/H] \lesssim -1$ is usually attributed to the shorter evolutionary timescales of $\alpha-$producing stars relative to the SNI usually identified with Fe-producers. However the correlation of α with Fe, as demonstrated by the low dispersion, favors an interpretation via one of the mechanisms suggested above.

4.3.2 Gas content of spirals

The longevity of the gas reservoir in ordinary spirals requires either a modification of the IMF, enhancing the returned gas fraction via a low mass truncation, or else gas infall. The current star formation rate in a galaxy such as the Milky Way would have depleted the gas supply within a fraction of a Hubble time were the local IMF invariant with time and/or position in the galaxy. Tilting the IMF towards massive stars (a top-heavy IMF) or truncating it in regions of intense star formation below $\sim 2M_{\odot}$ at early times prolongs the gas supply, to give an effective e-folding time of several Gyr as is required. An alternative solution requires gas infall at a rate of $\sim 1M_{\odot}\mathrm{yr}^{-1}$ to balance the lock-up rate of gas into low mass stars. The current gas infall rate is unlikely to be in excess of $\sim 0.1M_{\odot}\mathrm{yr}^{-1}$, as inferred from observations of high velocity clouds and, particularly, diffuse galactic x-ray emission fluxes. At earlier epochs, however, infall may have played a greater role.

4.3.3 Starbursts, tidal interactions and galaxy formation

Modelling of galaxies undergoing intense formation, or starbursts, where the star formation rate per unit mass of gas is enhanced by up to two orders of magnitude relative to that in the Milky Way, suggests that the IMF must be top-heavy or deficient in low mass stars[15]. This enhances the gas return fraction from the local IMF value of $R \sim 0.3$ to ~ 0.9, and increases the star formation rate or efficiency, which is proportional to the net mass of gas consumed or $1 - R$, by an order of magnitude. Moreover, those galaxies undergoing the most extreme starburst activity, as measured by the ratio of $L_{60\mu}/L_B \sim 100$ relative to the Milky Way value of order unity, are almost invariably involved in a merger or close tidal interaction. Compelling theoretical modelling suggests that a merger of gas-rich systems strongly enhances the gas concentration as a consequence of tidally induced gas cloud inelastic interactions[16]. This is very likely to lead to the star formation bursts that are observed provided that the star formation efficiency per unit mass of gas is also enhanced. Gas flows to the central region of the merging system may suffice to enhance the efficiency and allow a star formation burst of long enough duration, or else the starburst may consist of transient episodes: the inference of an altered IMF offers a likely but not unavoidable resolution of the enhanced star formation efficiency.

The analogy with galaxy formation is intriguing for two reasons. Mergers and tidal interactions play an important role in hierarchical theories of structure formation: they are inevitable during the galaxy formation process. The inferred star formation rate in a protogalaxy can be inferred from population synthesis modelling of the observed light distribution and spectrum of a galaxy, and one finds that for an old, spheroid population, the protogalactic star formation rate per unit mass was enhanced by some two orders of magnitude relative to that observed today in galactic disks. The analogy with extreme starbursts is consequently rather suggestive. Indeed, the recent products of galaxy mergers are found to exhibit de Vaucouleurs-type radial light profiles, as do luminous ellipticals[17]. However, one cannot account for formation of most spheroids by mergers without involving a precursor population of gas-rich disks that are not seen today in sufficient number but could plausibly have been prevalent in the past.

4.3.4 Indicators of a bottom-heavy i.m.f.?

There are at least two pieces of evidence that give a weak indication of a primordial IMF enhanced in low mass stars. The absence of evidence for massive OB star formation in galaxy cluster cooling flows at no more than 10 percent of the rate concordant with the apparent rate of gas depletion, typically $\sim 100 M_\odot yr^{-1}$ in a rich cluster, under the assumption that the cooling gas forms stars with a solar neighbourhood IMF, has led various authors to deduce the presence of large numbers of very low mass stars ($< 0.1 M_\odot$) condensing from the cooling flows. The recent detection of as much as $\sim 10^{12} M_\odot$ of cold gas inferred to be in dense clouds in the cores of cooling flow clusters greatly weakens any conclusion about the inevitability of low mass star formation, since an alternative reservoir for the cooling flow gas is inferred[18]. Of course such clouds most likely undergo collisions and form stars, but one would now be reluctant to cite cooling flows as providing firm evidence for predominantly low mass star formation.

Another indication of a bottom-heavy IMF has emerged from studies of the faint end of the luminosity function in several globular clusters. A correlation is claimed between the IMF slope over the range $0.2 - 0.4 M_\odot$ and the dynamical age of the cluster[19]. Two clusters with long core relaxation time-scales ($\gtrsim 10^9$ yr) contain a steeper IMF ($dN/dlnm \propto m^{-x}$, $x \approx 3$) than is found for another two clusters with core relaxation times of $\sim 10^8$ yr, for which $x \approx 1$. Incompleteness corrections are large in this stellar mass range. However if future studies confirm this trend, one might be tempted to argue that those globular clusters which by virtue of low dynamical age come closest to sampling the initial IMF from which they formed, might be indicative of the primordial halo IMF. Of course, one would still require an enormous extrapolation towards lower stellar masses to obtain the required halo mass, while having formed the globular clusters themselves, which have low M/L, only from stars of mass $\gtrsim 0.1 M_\odot$.

5. PRIMORDIAL STAR FORMATION RATE

5.1 Protogalactic star formation rate: general considerations

Population synthesis modelling of galaxy spheroids, previously alluded to in the analogy with starbursts, implies that the characteristic star formation time is less than 1 Gyr. Globular clusters in the Milky Way halo differ in ages, however, by up to 2-4 Gyr. Moreover, the disk, as measured by the white dwarf luminosity function is only 9 Gyr old, and hence formed $\sim 5(\pm 2)$ Gyr after the oldest globular clusters. Evidently, our metal-poor spheroid, which only constitutes a small fraction of the *luminous* mass of the Galaxy, formed over a more extended time-scale, and hence with a lower specific star formation rate, than did the metal-rich spheroidal populations characteristic of the bulk of the light from luminous ellipticals. Metal-poor dwarf ellipticals also reveal evidence for a range in formation age, as inferred from the presence of carbon stars. The specific star formation rate directly observed in metal-poor spheroids is not dissimilar from that indirectly inferred by population synthesis modelling of disks in luminous spiral galaxies. Typical e-folding time-scales in disks are several Gyr.

There is an intriguing dynamical property of these systems. Luminous ellipticals tend to be slowly rotating, and supported by random stellar motions, whereas disks, low luminosity ellipticals, and spheroids are rotationally supported. Star formation rate and efficiency is evidently important in determining galaxy morphology. A high specific star formation rate means high star formation efficiency. In a merging hierarchy, this results in the formation of dense stellar subsystems that merge as a consequence of dynamical friction. There is considerable angular momentum transfer, and the merged core that forms is not rotationally supported. A low specific star formation rate means inefficient star formation. The resulting collapse is gas-rich and dissipative, with the result that a disk forms as angular momentum tends to be conserved by the contracting gas. Disks are likely to be generic units that form. Disk mergers provide a more violent trigger for a starburst, by virtue of the greatly enhanced tidal torques from the stellar components, that lead to a high specific star formation rate.

5.2 Star formation rates in disks and starbursts

To illustrate the physics that enters the star formation rate, I present a simple model for the rate of star formation in two distinct environments. Consider first a quiescent disk forming stars via non-axisymmetric gravitational instabilities that manifest themselves as spiral density waves and operate on the cold atomic and molecular gas components. The disk plausibly maintains itself in a state of marginal instability, with the Toomre parameter $Q \equiv \kappa\sigma/\pi G\mu$ of order unity. Here κ is epicyclic frequency, σ is gas velocity dispersion (or radial velocity dispersion of stars for a stellar disk), and μ is surface density. The disk forms stars if $Q < 1$, and massive stars evolve on a rapid time-scale ($< 10^7$yr) to heat the gas via winds and supernova explosions. This energy input tends to stabilize the disk, but if $Q > 1$, the gas dissipates and cools effectively in the absence of the OB stars to lower Q. At the edge of the disk, the decline in gas surface density suppresses star formation. These various effects are concisely represented in the following expression for star formation rate:

$$SFR(r) = \varepsilon\, \mu\, \Omega(1 - Q), \tag{1}$$

where ε is a parameter that can be identified with star formation efficiency: the fraction of the mass of a molecular cloud converted into stars. Here $\Omega(r)$ is the rotation rate of the disk: for a flat rotation curve, $\varepsilon \approx 2\Omega$. The Toomre parameter can be written as

$$Q = \mu_{crit}/\mu_{gas}; \quad \mu_{crit} \approx \sqrt{2}\Omega\sigma/(\pi G) \approx 8M_{\odot}pc^{-2},$$

the numerical value applying in the solar neighborhood where the observed cold gas surface density is about $13M_{\odot}\mathrm{pc}^{-2}$. Equation (1) provides an acceptable model for the radial dependence of the star formation rate observed in nearby disks, and yields the history of star formation and enrichment when appropriate assumptions about the initial mass function and infall are made.

Next, consider a situation in which the star formation rate is temporarily boosted above the steady rate value implied by (1), as might happen following a satellite merger that results in strong tidal torques being exerted on the interstellar gas clouds. The inelasticity of the ensuing gas cloud encounters generates a strong concentration of gas in the inner few kpc of a typical disk, and we may reasonably expect this to result in a starburst, an episode of enhanced star formation. The following argument suggests a simple estimate of the ensuing star formation rate.

Motivated by the three-component model for the interstellar medium[20], I assume that energy input from expanding supernova remnants self-regulates the interstellar gas via formation of a pervasive hot intercloud medium. The hot medium, when its volume filling factor is appreciable, pressurizes the interstellar clouds, which themselves collapse and form stars as they become destabilized. The condition that the porosity be of order unity suffices to constrain, and effectively determine, the star formation rate:

$$P = (SFR/M_{sn}) \times \nu_{sn},$$

where M_{sn} is the mean mass in stars per supernova and ν_{sn} is the 4-volume of a spherically symmetric supernova remnant. One finds that $M_{sn} \approx 200 M_{\odot}$ for a solar neighborhood IMF, and that $\nu \propto p^{-1.4} n^{-0.1}$, where p is the ambient pressure and n is the mean density of the interstellar medium. The self-regulation requirement $P \sim 1$ immediately leads to

$$SFR \propto p^{-1.4} n^{0.1} \ or \ \dot{M}_* \propto \sigma^{5.7}; \tag{2}$$

here SFR is the star formation rate per unit volume, $\dot{M}_*$ is the total star formation rate, and σ is the escape velocity. Now the rate of star formation cannot exceed the gas accretion rate in spherical free-fall, $\sim \sigma^3 G^{-1}$, from which one may infer from condition (2) that $\sigma \lesssim 50$ km s^{-1}. This suggests that star formation is inefficient in gas-rich dwarfs, by a factor of $\sim (\sigma/50 \text{ km s}^{-1})^{2.7}$ for $\sigma < 50$ km s^{-1}. A wind will be driven when the gas density has dropped low enough that cooling is slow over a dynamical time-scale. Massive galaxies undergo transient bursts of star formation at a rate that cannot exceed (2).

6. PROTOGALACTIC RELICS

6.1 Dim giants

The critical surface density below which star formation is suppressed in a disk suggests that there should be a population of giant disk galaxies, whose surface density is low. Candidates for such objects include galaxies such as Malin 1 and 2, and the damped Lyman alpha absorption clouds detected along the line of sight to high redshift quasars. Hierarchical formation models lead to an interesting prediction for such objects: they should have bright, essentially normal, bulges. At early epochs, the potentially dim giants are indistinguishable from normal galaxies. The late infall occurs at large radius in the former case, and this is what distinguishes the disk of a dim giant from that of a normal spiral. One can understand this distinction as a consequence of peak-background modulation: in a cluster or group environment, the late accretion is considerably enhanced onto a density peak of galactic scale relative to what one finds in the field[21].

6.2 Low surface brightness dwarfs

A huge dark dwarf galaxy population is generic to gaussian fluctuation models such as that of cold dark matter. The mass spectrum of density peaks that have just become non-linear contains approximately equal mass per logarithmic mass interval, so that

$$dN/dm \propto m^{-2},$$

whereas the observed dwarf luminosity function is $dN/dL \propto L^{-\alpha}$, with $1 \leq \alpha \leq 1.5$. These dark dwarfs are not seen in nearby galaxy populations, but may be luminous at

$z \gtrsim 1$. The deep B and K counts require either prolific merging of luminous galaxies[22] or a new population of blue dwarfs[23,24] at $z \gtrsim 1$. The multiple component absorption line systems seen along the lines of light to many quasars also have been interpreted in terms of a transient population of star-forming dwarfs, with some supporting evidence being provided by emission line detections[25].

A successful model which explains the proliferation and transient nature of dwarfs at modest redshift appeals to tidal interactions that occur as galaxy groups form[26]. This is a recent ($z \sim 1$) occurrence, at least in a universe with $\Omega = 1$ and relies on the hypothesis that tidal interactions induce starbursts. The dwarfs fade to invisibility by the present epoch: a typical faint galaxy with $B = 26$ at $z = 1$ fades to $M_B = -15, \mu_B = 26.5$ mag/arc-sec^2 at $z = 0$. Natural biasing is inevitable: uncollapsed regions which include most of the volume of the universe contain gas clouds that have formed few if any stars, and may plausibly be identified with the nearby unclustered population of Lyman alpha forest absorption clouds seen towards quasars. Starburst-induced winds heat the intergalactic medium: the dwarfs are sufficiently numerous that the volume filling fraction of the wind-heated IGM is large by a redshift of 5 in a CDM model. This suggests an alternative explanation of the Gunn-Peterson effect, involving collisional ionization and heating to $\gtrsim 3 \times 10^5$K of the IGM. The predicted luminosity function in groups and clusters is similar to that seen in Virgo and Fornax at low luminosities: $\alpha \approx -1.5$ for low surface brightness gas-poor dwarfs. The much larger number of gas-rich dwarfs in the field could plausibly have a sufficiently large effective cross-section to account for the Lyman alpha forest clouds.

7. PROSPECTS FOR DETECTION

7.1 Black holes and neutron stars

The least innocuous of BDM candidates, massive stars must have formed at sufficiently large redshift to have eluded detection via direct and, especially, via indirect searches. The diffuse background radiation signature is potentially observable, but adequate coverage in the infrared spectral region is only just now becoming available with the DIRBE detector on COBE. It is unlikely, however, that a BDM fraction $\Omega \sim 0.1$ would be detectable with current techniques. A more serious problem is pollution by heavy elements ejected as massive stars evolve and die. This may be avoided by appealing to sufficiently massive primordial stars ($\gtrsim 300 M_\odot$) whose yields are greatly suppressed, and/or by efficiently recycling the ejecta from neutron stars in dense star clusters at high redshift[27]. This latter possibility is feasible because Compton cooling helps quench any winds that might disperse the pollutants at $z \gtrsim 100$ from the dense, globular-cluster-like clouds that are the expected sites of primordial star formation. Some early enrichment is actually required, both in the intracluster medium and in the old disk, and BDM could plausibly provide the source of the pollutants.

7.2 White dwarfs

These are the most common of all BDM candidates. Fine-tuning of the primordial IMF is required if the BDM is almost exclusively white dwarfs, but this is hardly an objection. More seriously, one has to consider the overproduction of helium and astration of deuterium. Primordial white dwarfs are likely to be massive ($\gtrsim 1 M_\odot$), because of the larger characteristic mass and zero metallicity of their precursors. Mass loss in the late stages of stellar evolution, to the extent that it is driven by resonant absorption line scattering of radiation, should be less efficient than for conventional stars. The resulting helium yields are correspondingly uncertain. Of course, if most of the ejecta ($\sim$ 90 percent) end up in the intergalactic medium, the *disk* helium abundance is enhanced by $\triangle Y \lesssim 0.05$, an amount that can barely be tolerated within chemical evolution models and reconciled with primordial nucleosynthesis constraints.

White dwarfs as BDM candidates may still be visible if their age is less than about 15 Gyr, in the form of a low luminosity tail to the local white dwarf luminosity function.

These nearby old halo white dwarfs would also have high space velocities, and one expects to find[28] $\gtrsim 1$ per square degree with proper motion $\gtrsim 1$ arc-sec/yr at $m_I \lesssim 22$. A more speculative possibility is via the expected frequency of close white dwarf binaries, some of which should merge. If one identifies Type I supernovae with merging white dwarfs, one might anticipate a frequency of 1 merger/10 yr in the dark halo, were it to consist exclusively of white dwarfs with similar binary characteristics to disk white dwarfs[29]. While observations exclude such a high rate for conventional supernovae in dark halos, it is entirely possible that merging white dwarfs, especially of the massive, primordial variety, might result in "fizzlers", forming a neutron star but with greatly reduced optical signature because of the presence of a compact dense accretion disk. Disk fragmentation and ensuing fragment ejection provides a recoil mechanism that can produce neutron stars with high velocity. A speculative interpretation[30] of high velocity ($\lesssim$ 1000km/s) pulsars seen up to 1kpc above the plane and with predominantly, but not exclusively, downward motion would be in terms of a halo population of massive white dwarfs, some of which (about 0.1 percent) merge to form neutron stars.

7.3 Brown dwarfs

The cumulative infrared signature of massive (up to $0.1M\odot$ if primordial) brown dwarfs makes them potentially detectable in nearby dark halos as a diffuse glow. The nearest objects to the sun may also be detectable in a proper motion survey at near-infrared wavelengths. The most promising technique, however, for discovering brown dwarfs, as well as most other BDM candidates that span the mass range $10^{-8}M_\odot - 10M_\odot$, is via a gravitational microlensing survey of the Large Magellanic Cloud[31]. The probability of an event for an individual star is only 10^{-6}, but the starlight is amplified by 50 percent over a timescale $\sim 0.2M_x^{1/2}$yr, where M_x is the mass of the BDM candidate in units of a solar mass. Two experiments underway in 1991/92 will monitor up to 10^7 19th magnitude stars in the LMC every night to search for the achromatic, symmetric, non-repeating signal characteristic of a microlensing event.

8. CONCLUSIONS

Dark matter formation almost certainly preceded galaxy formation. Hence we had best understand dark matter before we can hope to understand how galaxies formed. Halo and galaxy cluster dark matter are inescapable as a dominant constituent of their environment. Dark matter in halos and in galaxy clusters is plausibly baryonic: it formed near baryonic matter concentrations, and is in an amount consistent with and, if H_0 is low, even required by primordial nucleosynthesis constraints. Baryonic dark matter most likely consists of a mixture of brown dwarfs, white dwarfs, neutron stars, and black holes, with a strong emphasis on either low mass objects (brown dwarfs $< 0.1M_\odot$) or compact remnants of massive ($\gtrsim 2M_\odot$) objects relative to the sun. Galaxy formation models, as well as observational evidence, suggests that there should be many dark, or at least low surface brightness, galaxies. These could be dim, gas-rich giants or diffuse gas clouds in the field, or gas-poor dwarfs in clusters.

References

1. K. Kuijken and G. Gilmore 1989, *M. N. R. A. S.*, **239**, 651.
2. J. Bahcall 1984, *Ap. J.*, **287**, 926.
3. A. Gould 1990, *Ap. J.*, **360**, 504.
4. G. Meylan 1989, *Astr. Ap.*, **214**, 106.
5. J. A. R. Caldwell and J. P. Ostriker 1981, *Ap. J.*, **251**, 61.
6. J. N. Bahcall, M. Schmidt and R. M. Soneira 1983, *Ap. J. Lett.*, **258**, L23.
7. G. Stringfellow 1991, *Ap. J. Lett.*, **375**, L21.
8. C. P. Deliyannis and M. H. Pinsonneault 1991, *Ap. J.*, submitted.

9. R. H. Dicke and P. J. E. Peebles 1968, *Ap. J.*,**154**, 891.
10. J. Silk 1968, *Ap. J.*, **151**, 459.
11. Ya. B. Zeldovich 1970, *Astr. Ap.*, **6**, 319.
12. M. L. Wilson and J. Silk 1981, *Ap. J.*, **243**, 14.
13. J. Tarter 1973, unpublished Ph. D. thesis, Univ. of Calif., Berkeley.
14. J. Monaghan 1991, in *Workshop on Star Formation in Different Environments*, Aust. J. Phys. (in press).
15. J. Scalo, in *Windows on Galaxies*, ed. A. Renzini *et al.* (Kluwer:Dordrecht) (in press).
16. J. Barnes and L. Hernquist 1991, *Ap. J. Lett.*,**370**, L65.
17. F. Schweizer *et al.* 1990, *Ap. J. Lett.*, **364**, L65.
18. D. A. White *et al.* 1991, *M. N. R. A. S.*, **252**, 72.
19. H. B. Richer and G. G. Fahlman 1991, in *The Formation and Evolution of Star Clusters*, ed. K. Janes (San Francisco: Astronomical Society of the Pacific), p. 120.
20. C. McKee and J. P. Ostriker 1977, *Ap. J.*, **218**, 148.
21. Y. Hoffman, J. Silk and R. F. G. Wyse 1991, in preparation.
22. B. Rocca–Volmerange and B. Guiderdoni 1990, *M. N. R. A. S.*, **247**, 166.
24. S. Cole, M.-A. Treyer and J. Silk 1991, *Ap. J.* (in press).
23. L. L. Cowie *et al.* 1991, Ap. J., (in press).
25. D. York *et al.* 1986, *Ap. J.*, **311**, 610.
26. C. Lacey and J. Silk 1991, *Ap. J.*, **381** (in press).
27. J. Silk, *Science*, **251**, 537.
28. F. Tamanaha *etal* 1990, *Ap. J.*, **358**, 164.
29. T. Smecker and R. F. G. Wyse 1991, *Ap. J.*, **372**, 448.
30. D. Eichler and J. Silk 1991, preprint.
31. B. Paczynski 1986, *Ap. J.*, **304**, 1.

DISCUSSION

FABER: Could I ask for a clarification? You gave us a justification for enhanced massive star formation during the early–merging phase of galaxies. However, I am confused about the physical motivation for brown dwarfs. Some of the observational evidence you cited comes from globular clusters, yet I associate them with the early–merging phase. Could you comment on circumstances that might favor an enhanced population of brown dwarfs?

SILK: I can only cite phenomenological evidence that is suggestive, but only weakly, of an enhanced low mass star population. Cluster cooling flows may reproduce physical conditions similar to those in an early phase of galaxy formation, and could be forming stars with a brown dwarf–dominated IMF – in fact, some authors argue this has to be the case. I do not know of any even half–way plausible physical argument that favours enhanced brown dwarf formation in either cooling flows or the early–merging phase of galaxy formation, but this is merely indicative of our lack of understanding of the star formation process itself.

PETERSON: In the lowest–metallicity stars found to date, with [Fe/H] about –4, the ratio of light even Z elements with respect to iron is not significantly different from what it is in stars of less extremely low metallicity. As far as these stars represent the earliest stage of (one solar mass) star formation, they constrain the nucleosynthesis mechanism – pointing towards explosive nucleosynthesis in massive stars whose IMF is not radically different from that responsible for subsequent halo enrichment.

SILK: That is exactly my point: the primordial IMF inferred for the first star contained massive stars with an essentially conventional mass function above, say, $10M\odot$. However, we can say little about the relative mass fraction in lower mass stars, other than that some existed so the primordial IMF could have been almost completely suppressed below a few solar masses.

RENZINI: You have mentioned the possibility of making CDM in clusters by cooling flows producing Jupiters. There is a constraint here that must be taken into account, and that is the amount of iron in the x-ray gas and locked into stars. With a Scalo IMF, type II SNe fall short by a factor of ~ 5 to produce the iron which is seen and by a factor ~ 50 if Jupiter CDM were to have the same iron content as the present day x-ray gas.

SILK: The iron abundance in intracluster gas may be an indicator of a top–heavy primordial IMF that produced at least some of the cluster dark matter in the form of stellar remnants. For example, if even 10 percent of the cluster dark matter were in this form, the inferred nucleosynthetic yield would account for the observed iron.

McNAMARA: You suggested that low mass stars forming in cluster cooling flows may be the form of the dark matter in clusters of galaxies. I wish to point out that the entire gas budget in x-ray emitting gas in clusters accounts for only $\sim 10 - 30\%$ of the virial mass of clusters. Of that gas, only $\sim 10\%$ at most will cool to form low–mass stars over the Hubble time, so there won't be enough Brown dwarfs to bind the cluster by about two orders of magnitude. However, I would agree that this is maybe a plausible mechanism for forming halos in many CD galaxies.

SILK: In order for cluster dark matter to entirely consist of low mass stars as inferred to form at the present epoch in cooling flows, one has to postulate that a much more rigorous cooling flow involving ~ 90 percent of the present cluster mass occurred when the cluster formed.

DJORGOVSKI: The microlensing search for hypothetical halo brown dwarfs strikes me as a very difficult experiment. First, there are many faint, variable objects on the sky. Second, there is a methodological problem: suppose you actually detect one, a star brightens, fades away, and then, by definition, **never does that again.** Now, repeatability is one of the basic requirements for a scientific experiments. How could you ever demonstrate that it was a microlensing event, and not something else?

SILK One would need a series of gold–plated events (with the expected achromatic and time–symmetric signatures) in different stars that were known from both prior and subsequent observation to be "normal" stars, i.e. stars that did not appear to be variables or have chromospheric activity.

THE NEUTRON STAR POPULATION IN THE GALAXY

G.S. BISNOVATYI-KOGAN
Space Research Institute, Moscow, USSR, and
Department of Astron. & Ap., University of Chicago, Chicago, IL, USA

ABSTRACT. Properties of neutron stars are discussed and their density in Galaxy is roughly estimated. Their input in average galactic density is not large and they cannot be a reasonable dark matter candidates.

1. Introduction

Neutron stars (NS) are observed as radiopulsars, strong X-ray sources in close binaries: X-ray pulsars, X-ray bursters, X-ray sources with quasi-periodical oscillations (QPO); most transient X-ray sources also contain NS.

As the lifetime of radiopulsars and most of the strong X-ray sourses is much less then the age of Galaxy, overwhelming majority of NS will be neither radiopulsars nor strong X-ray sources. Gamma-ray bursters (GRB) have been proposed as a candidates for old nearby NS (Bisnovatyi-Kogan et.al.,1975). Discovery of hard transient X-ray pulsar in the strongest GRB of 05 March 1979 (Mazetz et.al.,1979) and results of spatial distribution of faint GRB (Mazetz et.al.,1981) give evidences in favor of this interpretation. Here we estimate spatial density of NS on the base of theoretical analysis and observational data.

2. Stellar evolution and supernovae

Estimations of the number of NS in Galaxy from theory of stellar evolution with birth rate function (theoretically) and from the pulsar and supernovae statistics (observationally) were presented by Shapiro and Teukolsky (1983). Assume Sulpeter birth rate function

$$\psi_s dm = 2 \times 10^{-12} m^{-2.35} \quad dm \quad stars \quad pc^{-3} yr^{-1}, \qquad m = M/M_\odot . \tag{1}$$

This function is valid for mas interval $0.4 \leq m \leq 10$ and is steeper for larger masses. The birthrate of stars with masses in the interval $m_1 \leq m \leq m_2$ follows from (1) after integration

$$\Psi(m_1, m_2) = \int_{m_1}^{m_2} \psi dm = 1.48 \times 10^{-12} (m_1^{-1.35} - m_2^{-1.35}) \quad stars \quad pc^{-3} yr^{-1}. \tag{2}$$

The average mass of new born stars in mass interval $m_1 \leq m \leq m_2$ for Salpeter function is

$$\overline{m(m_1, m_2)} = 3.86 \frac{m_1^{-0.35} - m_2^{-0.35}}{m_1^{-1.35} - m_2^{-1.35}}. \tag{3}$$

B. Barbuy and A. Renzini (eds.), The Stellar Populations of Galaxies, 379–382.

Shapiro and Teukolsky (1983) obtained $\rho_{ns}/\rho_T= 0.02$ (ρ_{ns} is average density of the NS, $\rho_T = 0.14$ $M_\odot$ pc^{-3} is total barionic mass density in Galaxy) adopting that NS were born by stars with masses on main sequence (MMS) $4 \leq MMS \leq 10M_\odot$ and NS mass is equal to $1.4M_\odot$. Now the lower MMS for NS birth is taken equal to about $8M_\odot$ (see i.e. Bisnovatyi- Kogan,1989). The upper limit for MMS is not known. Taking into account progenitor mass of SN 1987A ($\sim 20M_\odot$) and lower metallicity of stars in LMC we adopt for estimations the value $M_{lim} = 35M_\odot$, for larger MMS collapse lead to formation black hole (BH). Masses of BH - stellar remnants, Cyg X-1 and A0620-00, are estimated between 4 and 15 $M_\odot$. For m_1=8 and m_2=35 we obtain with account of (2)

$$\rho_{ns}/\rho_T = 9.55 \times 10^{-3}. \tag{4}$$

For BH density these authors get ρ_{bh}/ρ_T=0.22, using m_{lim}=10 and M_{bh}=MMS with $\overline{m_{bh}}$=38.6, following from (3) at $m_2 = \infty$. Here we get $\rho_{bh}/\rho_T = 0.011$ for m_{lim}=35 and $\overline{m_{bh}}$ =10. Taking into account uncertanities, connected with stellar evolution in binaries, we obtain estimations

$$\rho_{ns}/\rho_T = 0.01 \div 0.03, \quad \rho_{bh}/\rho_T = 0.01 \div 0.03. \tag{5}$$

Corresponding birthrates in Galaxy will be

$$\dot{n}_{ns} = 0.01 \div 0.03 \quad yr^{-1}, \quad \dot{n}_{bh} = (1.5 \div 5) \times 10^{-3} \quad yr^{-1} \tag{6}$$

instead of values 0.021 and 0.0085 by Shapiro and Teukolsky (1983).

SN frequency in Galaxy is estimated as $R_{sn} = 1/28 \div 1/60$ yr^{-1} by different authors (Shapiro and Teukolsky, 1983) in accordance with theoretical estimations (6). In some SN explosions star is totally disrupted (presumably in SN I), so lower limit in (6) better coinside with SN statistics. It indicates that fate of massive stars in close binaries does not differ much from single stars.

3. Pulsar birthrate

Obtaining of pulsar birthrate from observations require a knowlege of luminosity evolution, form of the pulsar beam, magnetic field decay function. For exponential luminosity decay with time constant 4×10^6 yr, the minimum pulsar birthrate was found to be one every 230 years in Galaxy. Observational data show that beams are elongated in the latitude direction and beaming factor may be about 2. With account of uncertainties the birthrate of pulsars in Galaxy is estimated as one over $30 \div 120$ years (Taylor and Stinebring,1986) in exellent agreement with theoretical estimation (6).

4. X-ray sources and neutron star formation in binary systems

X-ray searches from the satellites lead to discovery of more then 100 bright sources, identified as NS in close binary systems. Total number of accreting neutron stars in Galaxy is of the same order, so their number is much less then total number of radiopulsars (about 10^5). Reminding that almost half of stars on the main sequence belong to binaries, we come to conclusion, that disruption of binary during formation of NS occures frequently and the number of NS remaining in pairs is about 10^{-3} of all NS in Galaxy.

Large percent of disrupted pairs and existance of pulsar velocities much larger then orbital in pairs show that Blaauw effect cannot explain this phenomena and indicate to their recoil origin. Bisnovatyi-Kogan and Moiseenko (1991) proposed mechanism of violation of mirror symmetry during formation of rotating NS with toroidal and poloidal components of magnetic field in progenitor. Consider dipole field in combination with symmetrical toroidal, having the same sign in both hemispheres. Nonuniform contraction during collapse lead to differential rotation and generation of additional toroidal field from poloidal one. Newly generated poloidal field have different signs in two hemispheres and the sum of original and generated toroidal fields have no mirror symmetry. Amplification of toroidal field by twisting lead to magnetorotational explosion (Bisnovatyi-Kogan,1989), whoose mirror asymmetry is determined by initial value of toroidal field. In order to get pulsar velocity ~ 300 km/s symmetrical part of toroidal field must be about 5×10^{14} Gs (Bisnovatyi-Kogan and Moiseenko, 1991). This value exeed the dipole pulsar magnetic fields ~ 500 times what is like the ratio of spot and large scale magnetic fields in Sun.

5. Recycled pulsars and LMXB

About 1/3 of bright X-ray sources belong to low-mass X-ray-binaries (LMXB), consisting of NS and low-mass star with $M \leq\sim M_{\odot}$. These objects are rather old (about 10^8 years) and have small magnetic fields. Accretion disk reaches a surface of NS and accellerates its rotation up to millisecond periods. After ceasing of mass transfer NS transformes into radiopulsar, which has a name of "recycled radiopulsar" (RR). The transformation of X-ray sources into radiopulsars was first considered by Bisnovatyi-Kogan and Komberg (1974).

Most numerous sample of RR originates from LMXB. About half of LMXB are situated in globular clusters leading to suggestion of the origin of LMXB in tidal capture of a red dwarf by NS (Fabian, Pringle and Rees, 1975). Among 42 known RR only 11 lay outside globular clusters. This is connected mainly with selection effect inherent to the search procedure. RR contain about 10 % of all known radiopulsars so their input in average galactic density is small. Radiopulsars are related to RR by one of three main properties:

1.belonging to globular cluster
2.belonging to close binary system
3.very small (less then 10 msec) period.

Usually RR obtain 2 or 3 of these properties. In addition all RR have magnetic fields 1 - 3 order of magnitude less then the smallest magnetic fields of ordinary radiopulsars. This confirms the suggestion of magnetic field decay during accretion made by Bisnovatyi-Kogan and Komberg (1974).

Recent discoveries of 5 RR in M 15 (Anderson et.al.,1990; Manchester,1990) and 11 RR in 47 Tuc (Manchester et.al.,1991) arises problems of evolution of LMXB and RR. The number of RR exeeds number of LMXB in one cluster about 10 times so the lifetime of LMXB must be 10 times less then of RR if their origin is in common. This explanation arises difficulties, connected with formation of large number of binaries in globular cluster by tidal capture. Calculations of Bisnovatyi-Kogan and Romanova (1982) give the number of possible binary much less then that observed in 47 Tuc. The possible explanation may be related with nonmonotonic evolution. In past cores of globular clusters could be more dense with larger binary birthrate. Formation of several binaries could reverse evolution and lead to expansion of the core. RR in 47 Tuc could be remnants of this dense core phase.

6. Gamma-ray burst sources

The number of registrated GRB is about 500 and is of the order of number of radiopulsars. If GRB originate on old single nearby NS there is a direct genetic connection between radiopulsars and GRB sources. In models of starquakes with or without nuclear explosion (Bisnovatyi-Kogan et.al.,1975; Tsygan,1975) all NS, including living and dead radiopulsars are the sources of GRB. If energy of explosion leading to GRB is limited by $10^{36} \div 10^{39}$ ergs then they could be observed by existing detectors vith sensitivity not better 3×10^{-7} $ergs/cm^2 = F_{lim}$ from distances less then $200 \div 5000$ pc. Taking into account that sensitivity of most detectors is less then F_{lim} it follows that average distance to observed GRB is about half thickness of the galactic disk in accordance with their isotropic distribution over the sky.

Assume that all NS in Galaxy $(2 \div 7) \times 10^8$ stars, are uniformly distributed in the disk with thickness 400 pc and radius 15 kpc and all observed GRB arrive from the part of the disk with radius 200 pc around Sun. Then the number of NS taking part in visible GRB production is equal to $(0.3 \div 1) \times 10^5$. Adopting the average frequency of visible GRB about one per day we obtain that old NS must give birth to GRB every $(0.3 \div 1) \times 10^5$ days or every $100 \div 350$ years.

7. Upper limits on the number of neutron stars in Galaxy

Estimate a density on NS not using extrapolation of modern birthrate to earliest stages. Consider two kinds of estimations:
1.Production of heavy elements.
It is established that elements beginning from carbon are ejected into interstellar medium in SN explosions and amount of heavy elements ejected in one explosion is not less then 0.1 $M_{\odot}$. For the density of heavy elements $\rho_Z \leq 0.03\ \rho_T$ and NS mass 1.4 $M_{\odot}$ we get

$$\rho_{ns}/\rho_T \leq 0.4 \qquad in \quad all \quad SN \quad explosions. \tag{7}$$

2. Neutrino background.
Formation of NS leaves a signature in form of neutrino background (Bisnovatyi-Kogan and Seidov, 1982). Binding energy of NS $\sim 0.1 M_{\odot}$ is transformed into middle-energy neutrino with $E_\nu = 5 \div 30$ Mev. It follows from estimations of Bisnovatyi-Kogan and Seidov (1982), that NS with average density $\rho_{ns} = 0.03\rho_T$ at constant production rate equal to the present one give the background in Cl - Ar Solar detector about 2×10^{-3} SNU. If most NS were produced in early epoch of violent starbirth, the energy of neutrino in background is less due to redshift and a possibility of detection falls. Using of Ga - Ge neutrino detectors may give better constrains on early formed neutrino background and lower absolute upper limit on density of NS. After estimation of spectra and amount of energy produced during formation of BH in relativistic collapse, one may get similar constrains on density of BH from stellar collapses.

8. References

Anderson S.V., Gorham P.M., Kulkarni S.R., Prince T.A., Wolszcan A. 1990, Nature, **346**,42.

Bisnovatyi-Kogan G.S. 1989, Physical Problems of the Theory of Stellar Evolution (in Russian), (Moscow: Nauka).

Bisnovatyi-Kogan G.S., Imshennik V.S., Nadyozhin D.K., Chechetkin V.M. 1975, Astrophys. Space Sci., **35**, 23.

Bisnovatyi-Kogan G.S., Komberg B.V. 1974, Sov. Astron. **18**, 217.

Bisnovatyi-Kogan G.S., Moiseenko S.G. 1991, Astron. Zh. (in press).

Bisnovatyi-Kogan G.S., Romanova M.M. 1982, Sov. Astron. **27**, 519.

Bisnovatyi-Kogan G.S., Seidov Z.F. 1982, Sov. Astron. **26**, 132.

Fabian A.C., Pringle J.E., Rees M.J. 1975, M.N.R.A.S. **172**, 15P.

Manchester R.N. 1990, private communication.

Manchester R.N., Lyne A.G., Robinson C., D'Amico H., Baies M., Lim J. 1991, Nature, **352**, 219.

Mazets E.P., Golenetskii S.V., Ilyinskii V.N., Aptekar' R.L., Guryan Yu.A. 1979, Nature, **282**, 587.

Mazets E.P., Golenetskii S.V., Ilyinskii V.N., Panov V.N., Aptekar' R.L. *et al.* 1981, Astrophys. Space Sci., **80**, 3.

Shapiro S.L., Teukolsky S.A. 1983, Black Holes, White Dwarfs and Neutron Stars (New York: John Wiley and Sons).

Taylor J., Stinebring D.R. 1986, Ann. Rev. Astron. Ap. **24**, 285.

DISCUSSION

HORVATH: In currently studied models of type II SN the final fate of the compact remnant turns out to be essentially determined by the mass of the iron collapsing core, which appears to be a non-monotonic function of the total progenitor mass. Would you predict a clear progenitor mass forming BH above and NS below from your magnetorotational driven explosions?

BISNOVATYI-KOGAN: The mentioned main sequence mass of 35 $M_{\odot}$ for the boundary between NS adn BH is very undefinite and may be considered as a personal opinion, based on experience in stellar evolution and supernova theory.

CHEMICAL ABUNDANCES OF GALACTIC PLANETARY NEBULAE

A. ACKER
Observatoire de Strasbourg, France
J. KÖPPEN
Institut für Theoretische Astrophysik, Heidelberg, Germany
B. STENHOLM
Lund Observatory, Sweden
and
G. JASNIEWICZ
Observatoire de Strasbourg, France

September 9, 1991

Abstract. The chemical composition of 86 planetary nebulae from the Strasbourg-ESO survey are analysed (Acker et al., 1989; Köppen et al., 1991). Strong correlations between O, S, Ar, as well as between N and N/O are found. Galactic radial gradients for the abundances of He, S, and Ar in the old disk nebulae are found in accord with the results of Faúndez-Albans and Maciel (1986). The S gradient is steeper than that of O, and the S/O ratio decreases with increasing distance from the galactic centre, which is quite different from the result deduced from HII regions.

Key words: Galaxy - Planetary nebulae - Chemical composition

Galactic abundance gradients

Only objects belonging to the galactic disk were analysed. They were selected to have a deviation from the local circular motion of less than $60 km.s^{-1}$, a helium abundance of less than 11.10, and a $lg(N/O)$ ratio of less than -0.3. In the following Table are given the constants a and b in the expression $lg(A(R)) = a + b(R - R_{\odot})$ (with the galactocentric distance R in kpc, $R_{\odot} = 7.8 kpc$), the correlation coefficient r, the errors s_a, s_b, and the number of nebulae used.

Elem.	*Sun*	a	b	r	s_a	s_b	*No.*
He	11.07	10.98	-0.011	0.81	0.02	0.003	11
N	7.99	8.05	-0.047	0.42	0.12	0.025	19
O	8.92	8.81	-0.014	0.20	0.08	0.016	21
S	7.23	6.89	-0.071	0.53	0.12	0.032	15
Ar	6.57	6.39	-0.014	0.46	0.05	0.010	10
N/O	-0.93	-0.70	-0.022	0.33	0.07	0.015	19
S/O	-1.69	-1.97	-0.053	0.61	0.08	0.020	15
Ar/O	-2.35	-2.49	-0.028	0.73	0.05	0.009	10

References

Acker A., Köppen J., Stenholm B., Jasniewicz G.: 1989, *Astron. Astrophys. Suppl.* **80**, 201
Köppen J., Acker A., Stenholm B.: 1991, *Astron. Astrophys.* , in press
Faúndez-Albans M., Maciel W.: 1986, *Astron. Astrophys.* **158**, 228

Line Asymmetry and Projection Factors in Cepheid Variables

Michael Albrow and P. L. Cottrell
Physics Department, University of Canterbury, Christchurch, New Zealand

ABSTRACT. New methods for displaying radial velocities of pulsating stars are presented. The profiles of weak spectral lines in Cepheids are observed to be asymmetric during the inward velocity part of the pulsation cycle but not the outward velocity part. A prescription is given for standardising the method of measuring radial velocity. New projection factors are calculated for Cepheid variables using this method and p=1.37 is recommended as a constant value. This should lead to a systematic error in radius determinations of $\sim 1\%$ from this source.

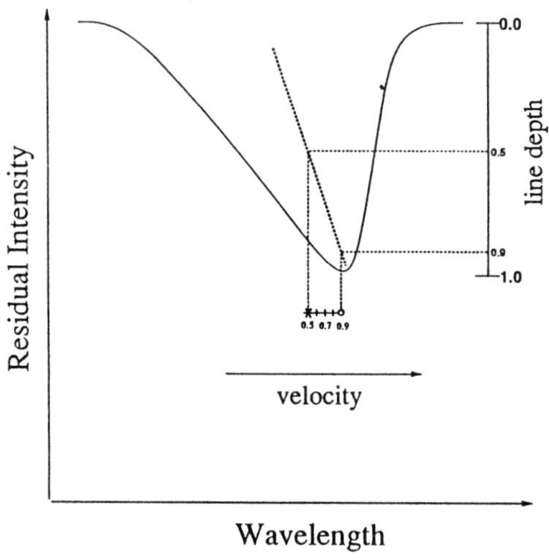

Fig 1. A schematic line profile and bisector showing the velocity as measured at different depths in the line profile. Individual metal line velocities are measured from the position of the bisector at depths 0.7, 0.8 and 0.9 in the line profile. Hα velocities are measured at a depth of 0.9.

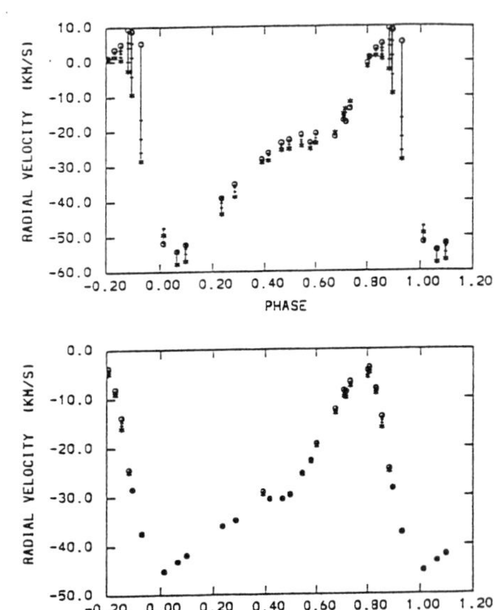

Fig 2. Radial velocity curves for W Sgr displaying the radial velocity at all points on the bisector between heights 0.5 and 0.9. a) Hα. b) Fe I $\lambda = 6546$ Å. This emphasises the degree of asymmetry in the spectral lines.

ORBITAL CHARACTERISTICS OF HIGH-VELOCITY STARS IN TWO GALACTIC MASS DISTRIBUTIONS.

Christine Allen[1,2], W. J. Schuster[1,3], and A. Poveda[1].

[1] Instituto de Astronomía, UNAM, Apartado Postal 70-264, 04510 México D.F. México.
[2] Dirección General de Servicios de Cómputo Académico, UNAM, 04510 México D.F., México.
[3] Instituto Nacional de Astrofísica, Optica y Electrónica, Apdo. Postal 51 y 216, 72000 Puebla, Pue., México.

ABSTRACT

Orbits for 615 halo and high-velocity disk stars have been numerically integrated in two different models for the galactic mass distribution, both satisfying recent observational constraints for the rotation curve and the perpendicular force (Allen and Martos 1986, Allen and Santillán 1991). In spite of major differences in the mathematical form of both models for the galactic potential, the orbital parameters of most of the computed orbits do not change appreciably. The greatest differences are found in the apogalactic distances reached by weakly bound stars, in the heights above galactic plane, and most importantly, in the total fraction of chaotic as opposed to semiperiodic orbits found for each galactic potential model.

CONCLUSIONS

1. A large number of orbits computed in two very different galactic mass models show very similar characteristics. This result is reassuring because it shows that the orbital parameters of the stars under study, and the conclusions drawn from these, do not depend sensitively on the details of the galactic model used.

2. A given orbit may or may not be of the same type in both models. Even so, the similarity of the orbital parameters persists, except in the case of some chaotic orbits.

3. Using the newer mass model a larger fraction of chaotic orbits is obtained (50% as opposed to 36%). This result is interpreted as being due to the fact that a larger volume of the central part of the galaxy is dominated by the spherical mass distribution.

REFERENCES

Allen, C. and Martos, M. A. 1986, Rev. Mexicana Astron Astrof., **13**, 137.
Allen, C. and Santillán, A. 1991, Rev. Mexicana Astron. Astrof., in press.

THREE-DIMENSIONAL PICTURE OF THE GALACTIC DISK AS DEFINED BY THE YOUNG STELLAR COMPONENT

EMILIO J. ALFARO, JESÚS CABRERA-CAÑO
ANTONIO J. DELGADO
Instituto de Astrofísica de Andalucía
Apdo. Correos 3004
E-18080 Granada
Spain

1. SUMMARY

Recently, Alfaro *et al.* (1991) [ApJ, 378, 106] have noted and stressed the clear spatial connection between the vertical structure of the galactic disk, defined by the young open cluster distribution (YOC), and the large scale star-forming activity in the Milky Way. In this context the estimate and study of the vertical pattern displayed by the Wolf-Rayet (WR) star population, as representative of the formation places of massive stars, could provide new insights on the connection between the morphological properties and the stellar-formation processes in our Galaxy.

With this aim in mind we have estimated a set of Z values for a grid of points on the X-Y plane from a sample of WR stars with measured Z, X, and Y coordinates (Vacca & Conti 1990 [AJ, 100, 431]) by using Automatic Cartography techniques (Kriging) (Cabrera-Caño *et al.* 1990 [in *Bias, Errors and Uncertainties in Astronomy*, Cambridge University Press, p. 101], Alfaro *et al.* 1991). The obtaining of this grid of points has allowed us to construct contour-maps and/or three-dimensional pictures of the galactic disk, as defined by this component, in the neighborhood of the Sun.

A comparative analysis between the two vertical patterns displayed by the YOC and WR distributions, has lead us to the following conclusions:

1.- At distances from the Sun lower than $\approx$ 1.5 kpc, both stellar populations show a very similar vertical behaviour.

2.- The WR distribution appears to be shifted down, $\approx$ 20 pc in average, with respect to the YOC Z-structure.

3.- The existence of a large and deep depression (*Big Dent*) in the galactic third quadrant (Alfaro *et al.* 1991), is corroborated by the distribution of the WR stars.

4.- At distance from the Sun larger than 1.5 kpc, both population show clear differences in their vertical patterns. Whether these discrepancies are mainly due to large errors in the distances of the remote WR stars or to actual differences in the Z-structure for both populations remains as an open question.

ASTRATION AND PRODUCTION IN CHEMICAL EVOLUTION

LILIA I. ARANY-PRADO*(1)* and RAMIRO DE LA REZA*(2)*
(1)Depto. de Astronomia - Observatorio do Valongo - UFRJ
Ladeira Pedro Antonio 43 - CEP 20080 - Rio de Janeiro - RJ
Brazil.
(2)Observatorio Nacional - CNPq
Rua Gal. Bruce 586 - CEP 20921 - Rio de Janeiro - RJ
Brazil.

Yokoi et al.[3] introduced a formalism for chemical evolution of the Galaxy which accounts for astration. Malaney et al.[1] used an analogous formalism. In both models however, the astration parameters, respectively χ^a and ξ, are defined only in the instantaneous recycling approximation. We intend to improve the theory by the development of more general equations in the sudden mass loss approximation. Our basic equations are that of Tinsley[2]. We suppose that: in the sudden mass loss approximation, the death of a star of inicial mass M (and lifetime τ_M) is defined only at the instant when all synthesis cease; the distribution of inicial i (any long life r-process radionuclide) is homogeneous at the birth of a star; this homogeneity is locally maintained throughout the life of the star. Our analysis of all depletion trajectories during astration is intended to give the definition of the *depletion reductor of i for the ejected region* (whith mass Q_{ej})

$$\Lambda_i(M) = 1 - \left[\frac{D^i(Q_{ej}(M)) + F^i(Q_{ej}(M))}{Q_{ej}\left[m_i^g(t-\tau_M)/m_g(t-\tau_M)\right]} \right] \quad \text{(that is analogous to } \chi^a \text{ or } 1\text{-}\xi\text{)}, \qquad (1)$$

where m_i^g/m_g is the i abundance on the ISM; D^i and F^i are, respectively the inicial i mass depleted and the result of the total "out minus in" inicial i mass in the region throughout the life of the star. Now, let $\mathcal{P}_k^s$ be the total production rate of a k nuclide due to all post main sequence stars with masses above a minimum M_k, and $\theta_M^k(t-t_M)$ the rate of convertion of stellar mass fraction on k, by an M star, $(t-t_M)$ before its birth at t_M. We can write

$$\mathcal{P}_k^s(t) = \int_{M_k}^{M_{max}} \int_{t-\tau_M}^{t-0.9\tau_M} M\,\theta_M^k(t-t_M)\,\psi(t_M)\,dt_M\,\phi(M)\,dM \quad , \qquad (2)$$

where ψ is the star formation rate and ϕ is the inicial mass function. It can be shown that, if: 1)$\psi(t_M) \approx \psi(t-\tau_M)$ in the short time interval $0.1\tau_M$ for each M ; 2)there is no depletion nor production of k in ISM; 3)there is no depletion of inicial k neither of new k during star evolution; 4)k abundance in the infall is negligible; 5)all locked new k is exausted in remnants; then, if k represents metals, from (2) and from a *general* equation of conservation of the nuclides, we can obtain the equation of Tinsley[2] for conservation of metals (however we should see this last one as an approximate equation).

References

[1]Malaney,R.A., Mathews,G.J. and Dearborn,D.S.P. (1989) Ap. J. **345**, 169-175.
[2]Tinsley,B.M. (1980) Fund. Cosmic Phys. **5**, 287-388.
[3]Yokoi,K., Takahashi,K. and Arnould,M. (1983) Astron. Astrophys. **117**, 65-82.

CAN THE FIRST STARS FORMED BE PRE-GALACTIC ?

J. C. N. de Araujo and R. Opher
Instituto Astronômico e Geofísico - U.S.P.
C.P. 9638, 01065 São Paulo, S.P. - Brasil

ABSTRACT. We show that in a scenario of isothermal density perturbations, stars $\gtrsim 50 M_\odot$ (Population III) at a redshift $z \gg z_{gal}$ (redshift of galaxy formation) can be formed. We take into account in the calculation the expansion of the Universe and a series of physical processes relevant to the primordial plasma: photon-drag, photon-cooling, recombination, collisional ionization, photoionization, Lyman-α cooling and molecular hydrogen cooling.

The formation of the structures of the Universe is yet an open question. The formation of structures requires the existence of a spectrum of perturbations.

The isothermal density perturbations can be nonlinear for sub-galactic scales (e.g. Hogan 1978, Lahav 1986, de Araujo & Opher 1988, 1989) and lead to the formation of Population III objects. It is also possible to produce nonlinear perturbations for scales $M < 10^8 M_\odot$ by isocurvature or adiabatic cold dark matter density perturbations (see de Araujo and Opher 1991).

The Population III objects can collapse directly or fragment, forming in this way the first stars (Population III).The galaxies, cluster of galaxies and voids, for example, can be generated in a scenario of successive explosions of Population III objects (or stars).

We study the formation of Population III stars that can be produced by the fragmentation of clouds of mass $M \sim M_j$ (Jeans mass at the beginning of the recombination era, $\sim 10^{5-6} M_\odot$). We use the fact that a perturbation for $M \ll M_j$, that can survive with some residual amplitude, can produce the fragmentation of clouds of mass $M \sim M_j$ in the first free fall time scale of M (see de Araujo & Opher 1989).

We obtained that the minimum mass that can fragment from the M_c cloud is $\sim 50 M_\odot$ at $z \sim 181$.

REFERENCES

de Araujo, J. C. N. & Opher, R. 1988 , MNRAS, 231, 923
de Araujo, J. C. N. & Opher, R. 1989 , MNRAS, 239, 271
de Araujo, J. C. N. & Opher, R. 1991, ApJ (in press)
Hogan, C. 1978, MNRAS, 185, 889
Lahav, O. 1986, MNRAS, 220, 259

HOW MANY BURSTS OF STAR FORMATION IN M82?

B.P. Artamanov[1], P. Traat[2]
1 Sternberg Astronomical Institute Universitetskij Prospekt 13, 119899 Moscow, U.S.S.R.
2Tartu Astronomical Observatory 202444 Tõravere, Estonia

1. Introduction

The galaxy M82 has in its central region ongoing a giant burst of star formation. Van van den Bergh (1971) first separated the bright semistellar objects in central region of M82, which in fact are young star clusters. Using subsecond seeing, Artamonov et al. (1990) managed to distinguish about 70 semistellar objects in M 82 in colour 'V'.

We have made estimates of the influence of different starbursts on the composite colour index of M 82 disk, using for the old disk component the "classical" models of photometric evolution with monotonically decreasing SFR from Traat (1988). Model data: $Z = 0.03$, power–law IMF with index $n = 2.3333$, stellar mass limits: $0.085M_{\odot} \div 100M_{\odot}$, young population formed in a single burst. The underlying old disk has been taken to have the age of 10^{10} yr. The brightness of young star clusters, formed in SF bursts, (L_1), was supposed to have different weights x relative to the brightness of the old disk, L_0, with the total brightness L being the sum of both: $L = L_0 + L_1$, $L_1 = x \cdot L_0$, $L = L_0(1 + x)$.

The table presents composite colours for star bursts of different ages t_{burst} and strengths x.

x	1.5	0.9	0.5	0.1	t_{burst}
$U-B/B-V$	$-0.66/+0.45$	$-0.47/+0.59$	$-0.24/+0.74$	$+0.33/+0.96$	10^6
$U-B/B-V$	$-0.34/+0.54$	$-0.22/+0.67$	$-0.02/+0.77$	$+0.48/+0.97$	$10^7 yrs$
$U-B/B-V$	$+0.15/+0.64$	$+0.26/+0.74$	$+0.36/+0.85$	$+0.58/+0.98$	$10^8 yrs$

Different areas of disk of M82 have the following mean U-B/B-V colors: A (nuclear region of M82) - 0.45/1.10; B (middle part of the disk) - 0.15/0.85 (Artamonov (1978), Bronkalla et al (1980)). Extinction in the centre of M 82 is approximately $A_v \sim 5^m$, in area B — $0^m.5 \div 1^m.0$.

From these colors, depending on the extinction, one gets for the burst population the limits for both areas A and B: $t_{burst} = 10^7 - 10^8$ yr, $x = 0.9 \div 1.5$. It is possible that the multiple SF bursts have been occurred in M82, but the rough estimates of burst ages doesn't allow to draw firm conclusions yet. We believe that detailed photometry of semistellar objects in M82 will help to solve this problem.

Artamonov B.P., 1978, Soviet Astr., 22, 7
Artamonov B.P., Novikov S.B., Shokin Ju.A., 1990, preprint 17 Sternberg Inst., Moscow
Bronkalla W., Notni P., Tiersch H.1980Astron. Nachr. 301, 217
Traat P., 1988,Tartu Astrof. Obs. Teated N$_o$., 91, 23
Van den Bergh S., 1971, A&A, 12, 474

PRE MAIN SEQUENCE CHROMOSPHERIC ACTIVITY

C.C. BATALHA
Departamento de Astrofísica
Observatório Nacional-CNPq
Rua José Cristino 77, CEP 20921
Rio de Janeiro-Brazil

ABSTRACT. We study the chromospheric evolution of low mass Pre Main Sequence Stars using the CaII line at 8542 Å as a probe of stellar activity. Our sample consists mostly of classical T Tauri Stars (CTTS - WHα > 10 mÅ) and a few weak TTS (WTTS - WHα < 10 mÅ) with simultaneous observations from 4500 Å up to 9000 Å . After the proper corrections for veiling, we measure the chromospheric flux of the Ca II emission core and compare it with Hα fluxes, veiling and the stellar age. Our main conclusions are the following:

1 - There is no trend between λ8542 and Hα fluxes for the whole set of stars, however a clear correlation is found for WTTS. This last finding is expected if both CaII and Hα are mainly formed in the atmospheric environment which might be the case for WTTS.
2 - Chromospheric flux does not correlate with stellar age for CTTS.
3 - We present a correlation between the accretion rate of the circumstellar disk (veiling) and chromospheric fluxes (see below). Accretion through magnetic loops crossing the circumstellar disk may provide additional emitting area to the CaII fluxes. This last finding demonstrates the strong linking between the disk and the stellar magnetic surface fluxes.

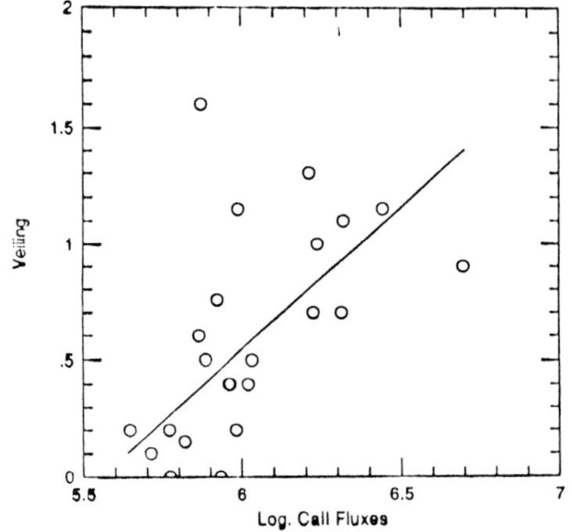

MORPHOLOGY AND COLORS OF DISTANT GALAXIES

M. A. Bershady [1,2]
Department of Astronomy & Astrophysics
University of Chicago
5640 S. Ellis Ave.
Chicago, IL 60637

ABSTRACT. We determine the correlation between spectral and morphological type for distant galaxies, defined by optical - near-infrared colors and image structure. As a point of departure, rest-frame colors of field galaxies observed to redshifts of 0.35 are shown to follow the trend in the UVK plane seen by Aaronson (1978) for local galaxies. Correlations are found between the deviations from this trend and absolute magnitude, compactness, and surface brightness. This works provides the foundation for investigations of galaxy structure at cosmologically interesting distances (Bershady, 1992).

1. Results

The distribution of measured ellipticity for red, or early spectral type galaxies is consistent with moderately elliptical to spherical systems. Late spectral type galaxies have an ellipticity distribution consistent with randomly oriented disks.

To first order, the gross dispersion in V-K for each spectral type, as well as the overlap between types can be parametrized by a linear color-magnitude (C-M) relation.

To second order, the C-M slope depends on galaxy spectral type, agreeing with results from studies of well resolved objects at lower redshift (Mobasher et al., 1986).

The sample contains a group of nearby, compact, dwarf galaxies, classified as late-type on the basis of their colors. Their distribution in the C-M diagrams implies, however, that they are early-type. Either the spectral classification is erroneous, or their is an additional parameter in the C-M relations, namely compactness.

The effects of dust reddening on the C-M relations are shown to be small, but measurable for galaxies of late spectral types.

References

Aaronson, M. 1978, *Ap. J. Lett.*, **221**, L103.
Bershady, M.A., 1992, Ph.D. Thesis, University of Chicago, in preparation.
Mobasher, B., Ellis, R.S., Sharples, R.M. 1986, *M.N.R.A.S.*, **223**, 11.

[1] Visiting Astronomer, Kitt Peak National Observatory, National Optical Astronomy Observatories, which is operated by the Association of Universities for Research in Astronomy, Inc., under contract with the National Science Foundation.
[2] NASA Graduate Fellow.

INTEGRATED UBV PHOTOMETRY OF 624 LMC CLUSTERS

E. Bica[1], J.J. Clariá[2], H. Dottori[1], J.F.C. Santos Jr.[1] and A. Piatti[2]
[1] Instituto de Física-UFRGS, Porto Alegre, RS, Brazil
[2] Observatorio Astronómico de Córdoba, Argentina

A compilation of integrated UBV photometry for 147 LMC clusters was published by van den Bergh (1981). We have increased by more than a factor 4 the sample of star clusters with integrated UBV photometry in the LMC, totalling now 624. The observations were carried out with the CTIO 0.6m and the 2.14m CASLEO telescopes. The sample is essentially complete up to V$\approx$ 13 but also contains fainter clusters up to V $\approx$ 14.5. The (U-B)vs(B-V) diagram has provided interesting results such as the detection of the helium flash gap in the color evolution (Bica et al. 1991). We have derived equivalent SWB types and study their spatial distribution. The spatial extent (Fig. 1) varies from diameter $\approx 6^o$ for the disk of HII region clusters (SWB 0) to $\approx 11^o$ at SWB VI, reaching 15° or more for SWB VII, which is probably a halo distribution. The coordinates in degrees are centered at NGC1928, a cluster close to the geometric center of the Bar. Groups SWB V or younger present inhomogeneities in their distributions. Group IVA (Fig. 1b) presents a strong concentration of clusters at the western extremity of the Bar (WEB). Group V is already at the 1-2 Gyr range and presents two clumps, one also coincident with the WEB and another with a structure known as the de Vaucouleurs' arm (Lynga and Westerlund, 1963). Fig. 2 shows a remarkable phenomenum in the Bar: SWB I clusters (dots) are shifted by an angle of 19° clockwise relative to those which trace the Bar, i. e. SWB II ones (triangles). This shift is centered at the WEB. These evidences point to an important mass concentration at the western extremity of the Bar. The age difference between SWB I and II and the spatial scale of the angular shift would indicate shock induced star formation for the SWB I group. The spatial distribution of age groups reveals phenomena which will certainly help to better understand the star formation history of the LMC and its dynamical evolution.

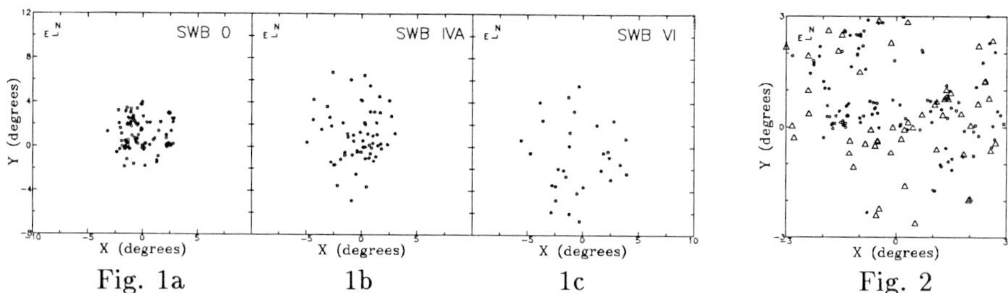

Fig. 1a 1b 1c Fig. 2

REFERENCES

Bica,E.,Clariá,J.J.,Dottori,H.D.,Santos Jr.,J.F.C.,Piatti,A. 1991, ApJ Letters, in press
Lynga,G., Westerlund,B. 1963, MNRAS, 127, 31
van den Bergh,S. 1981, AAS, 46, 79

Photometric Results for Stellar Fields Inside Supergiant Shell LMC4

D.J. BOMANS, A. VALLENARI, W. SEGGEWISS, T. RICHTLER, E.K. GREBEL, and K.S. DE BOER

Sternwarte der Universität Bonn, 5300 Bonn, F.R.Germany

September 24, 1991

Abstract. BVR photometry of stellar fields in the NW region of supergiant shell LMC4 has been performed. We present preliminary results on the age of the stellar populations in this part of the LMC.

Key words: Stellar Populations - Magellanic Clouds - Star Formation

Within the framework of the ESO Keyprogam on Coordinated Investigations of Selected Regions in the Magellanic Clouds (de Boer et al. 1989), we took Johnson-Cousins BVR CCD frames of 8 partly overlapping fields in Keyprogram Region E, which is located in the NW part of supershell LMC4 in the Large Magellanic Cloud (LMC). We obtained our data with the ESO-MPIA 2.2m telescope equipped with a Thomson 1024x1024 chip. The frames cover the region from the intermediate-age cluster NGC 1978 to the association NGC 1948.

To determine the age, we compared observational colour-magnitude diagrams (cmds) with theoretical models of stellar evolution. We used two different sets of tracks. One takes the effects of overshoot in a convective core into account (Bertelli et al. 1991) and the other one adopts a classical description of convective mixing in the core (Bencivenni et al. 1990). A nearly instantaneous star formation and a Salpeter IMF are assumed. As distance modulus to the LMC, we adopted 18.5 mag.

With the classical models, we derive 30 Myr as age of the field population, whereas it is 70 Myr when overshoot is considered. The ratio of main sequence to evolved stars (an indicator of the star formation history) indicates an enhancement of the global LMC star formation rate about 4 Gyr ago, which is compatible with the results of Mateo et al. (1990) and Bertelli et al. (1991) for other LMC fields.

The age of the association NGC 1948 is about 6 Myr from classical models or 20 Myr with overshoot. If this association is a result of self-propagating star formation, induced less than 15 Myr ago due to the shock wave from the supershell (Dopita et al. 1985), overshoot models predict a slightly too large age. On the other hand, the ages we derived are consistent with star formation progression in the LMC4 region derived by Reid et al. (1987).

For the globular cluster NGC 1978 we have only a preliminary age estimate of more than 2 Gyr, when compared with isochrones of Bertelli et al. (1991) and assuming $Z = 0.004$.

References

Bencivenni,D., Brocato,E., Buonanno, R., Castellani, C. 1990, ESO Preprint 729

Bertelli, G., Betto, R., Bressan, A., Chiosi, C., Nasi, E., Vallenari, A. 1990, A&AS, 85, 845

de Boer, K.S., Azzopardi, M., Baschek, B., et al. 1989, ESO Messenger, 57, 27

Dopita, M.A., Mathewson, D.S., Ford, V.L. 1985, ApJ, 297, 599

Reid, N., Mould, J., Thompson, I. 1987, ApJ, 323, 433

SYNTHETIC M/L_B RATIOS FOR E AND S GALAXIES

A. C. Borges, J. A. de Freitas Pacheco
IAG-USP, Depto. Astronomia
C.P. 9638, São Paulo 01065, Brazil

We report new calculations of M/L_B ratios for galaxies, which are considered to be closed homogeneous systems. A Salpeter's law was assumed for the initial mass function. The star formation rate was considered to be proportional to the amount of available gas. Spiral galaxies were considered to be constituted by two components: the bulge and the disk. Each component is characterized by its own timescale for the gas convertion into stars. We used the bulge-to-disk average light ratios by Koppen & Arimoto (1990) in order to derive the integral properties for the different morphological types. In our computatios we used the grid of models by Maeder & Meynet (1988) for stellar masses in the range $100 < m < 1$ and the evolutionary tracks compiled by Bruzual (1982) for lower masses.

The representative models for E-galaxies are characterized by an e-folding timescale for gas convertion of 1.36 Gyrs. The resulting properties are given in Table 1, and the main characteristics of our composite models for S-galaxies are given in Table 2.

Table 1 - E-galaxy properties

$(B-V)$	M/L_B	M_{HI}/M_{TOTAL}
0.85	15	1.6×10^{-5}

Table 2 - S-galaxy models

	Total		Disk			Bulge		
Type	$(B-V)$	M/L_B	$(B-V)$	M/L_B	τ(Gyrs)	$(B-V)$	M/L_B	τ(Gyrs)
Sa	0.71	6.5	0.67	5.0	4.1	0.80	10	2.1
Sb	0.63	5.1	0.55	3.0	7.3	0.80	10	2.1
Sc	0.53	2.4	0.50	1.6	9.6	0.81	11	2.0

Bruzual, G.: 1982, PhD thesis, University of California
Koppen, J., Arimoto, N.: 1990, A&A 240, 22
Maeder, A., Meynet, G.: 1988, A&AS 76, 411

NEW THEORETICAL ISOCHRONES

A. Bressan, G. Bertelli, C. Chiosi, F. Fagotto, E. Nasi
Department of Astronomy and Astronomical Observatory
Vicolo dell'Osservatorio 5
35122 Padova Italy

ABSTRACT Using evolutionary tracks computed by Alongi et al.(1991) for the two chemical compositions Z=0.020, Y=0.28 and Z=0.008, Y=0.25 we constructed theoretical isochrones from $20 \cdot 10^9$ yr to $3 \cdot 10^6$ yr.

1. The models

The models possess initial mass from 0.6 $M_\odot$ to 100 $M_\odot$, span the hydrogen burning phase, the helium burning phase (Horizontal Branch Phase for low mass stars) and end at the Thermally Pulsing AGB phase or at carbon ignition (in more massive stars). We follow explicitly H, ^{3}He, ^{4}He, ^{12}C, ^{13}C, ^{14}N, ^{15}N, ^{16}O, ^{17}O, ^{18}O, ^{20}Ne, ^{22}Ne, ^{25}Mg, ^{26}Mg, adopting updated reaction rates (Caughlan and Fowler 1988). Opacities are from the Los Alamos Opacity Library, implemented with the contribution by molecules. For each chemical composition, two sets of models have been computed, following the alternative treatments of the convective mixing, namely either the classical scheme with semiconvection during central helium burning phase or that accounting for mild overshoot during central hydrogen and helium burning phases.

2. The Isochrones

The isochrones are constructed with the procedure outlined by Bertelli et al. (1990), they account for mass-loss and include all phases up to either the stage of the planetary nebula formation, or the carbon ignition in a highly electron-degenerate core, or the quiet carbon ignition, depending upon the initial mass of the star. Conversions from luminosity and effective temperature to magnitude and colour are based on tables by Buser and Kurucz (1989). The tables provide the actual mass, luminosity, effective temperature, bolometric, U, B, V, R, I magnitudes and colours. Moreover they give the luminosity functions together with integrated magnitudes and colours along the isochrones, computed assuming a Salpeter IMF. A careful comparison with the observed Colour-Magnitude diagrams of some old open galactic clusters show that models which account for convective overshoot ought to be preferred.

Alongi M., Bertelli G., Bressan A., Chiosi C., Fagotto F., Greggio L., Nasi E., 1991a, A&A submitted
Bertelli G., Betto R., Bressan A., Chiosi C., Nasi E., Vallenari A. 1990, A&ASS
Buser R.& Kurucz R. L. 1989,private communication.
Caughlan G.R. & Fowler W.A. 1988 Atomic Data Nuc. Data Tables, 40, 283

THE HR DIAGRAM OF LMC SUPERGIANT STARS

A. Bressan, G. Bertelli, C. Chiosi
Department of Astronomy and Astronomical Observatory
Vicolo dell'Osservatorio 5
35122 Padova Italy

ABSTRACT. We analyze the HR diagram of LMC supergiant stars using a new set of evolutionary models for massive stars.

The synthetic HR diagram of LMC supergiant stars
New set of stellar models have been computed for massive stars (Alongi et al. 1991) with the following input physics: overshoot from the convective core and at the base of the convective envelope; Los Alamos Opacity Library modified according to Iglesias et al. 1990 around Log T $\simeq$ 5.3; density inversion inhibited by imposing a maximum temperature gradient ∇_T such that the density gradient $\nabla_\rho \geq 0$: $\nabla_{T_{Max}} = \frac{1-\chi_\mu \nabla_\mu}{\chi_T}$, where ∇_μ is the molecular weight gradient and χ_μ and χ_T have the usual meaning; H, ^{3}He, ^{4}He, ^{12}C, ^{13}C, ^{14}N, ^{15}N, ^{16}O, ^{17}O, ^{18}O, ^{20}Ne, ^{22}Ne, ^{25}Mg, ^{26}Mg, followed in detail (reaction rates from Caughlan and Fowler 1988); mass-loss by De Jager et al. (1988), or $10^{-3} M \odot yr^{-1}$ when LBV, or Langer (1989) when WR.
Due to envelope overshoot the evolutionary tracks for M $\leq$ 15 $M_\odot$ (Z=0.02) and M $\leq$ 20 $M_\odot$ (Z=0.008) exhibit a blue loop during the central He-burning phase. While in the loop, the models show CNO processed material at the surface, due to convective dredge-up. More massive stars either end helium burning as red supergiants or, because of mass-loss, enter the stage of WR star.
Using a Monte Carlo method we simulated the HR diagram of LMC supergiant stars (Fitzpatrick & Garmany 1990) adopting the composition Z=0.008 Y=0.25, a uniform age distribution and a Salpeter IMF. While the *Ledge* (Fitzpatrick & Garmany 1990) is well reproduced and corresponds to the *red side of the core He-burning band in the loop.* a major point of disagreement are the many stars falling in the so-called "Blue Hertzsprung Gap" predicted by the theory.
We found that opacity enhancements like those suggested by Iglesias et al. (1990) or binary evolution as proposed by Tuchman and Wheeler (1990) cannot remove the discrepancy. On the other hand we suggest that the gap could be masked by the uncertainties in the photometry, in the dereddening technique, and in the transformation from colours and magnitudes to effective temperatures and luminosities.

Alongi M., Bertelli G., Bressan A., Chiosi C., Fagotto F., Greggio L., Nasi E., 1991a, A.&A. submitted
Caughlan G.R., Fowler W.A., 1988, Atomic Data Nuc. Data Tables 40, 283
de Jager C., Nieuwenhuijzen H., van der Hucht K.A., 1988 AAS 72, 259
Fitzpatrick E.L., Garmany C.D., 1990, ApJ 363, 119
Iglesias C.A., Rogers FJ., Wilson B.G., 1990, ApJ 360, 221
Langer N., 1989, AA 210, 93
Tuchman J., Wheeler J.C., 1990, ApJ 363, 255

METALLICITY DISTRIBUTION OF ELLIPTICAL GALAXIES THROUGH A QUANTITATIVE CALIBRATION OF THE MAGNESIUM Mg_2 INDEX

Alberto Buzzoni, Giorgio Gariboldi
Osservatorio Astronomico di Brera
Via Brera, 28 20121 Milano, Italy

Luciano Mantegazza
Dip. di Fisica Nucleare e Teorica Università di Pavia
Via Bassi, 6 27100 Pavia, Italy

In this contribution we give a progress report for our work intending to approach in a more complete way the problem of a quantitative calibration of the Mg_2 index (Faber *et al.* 1977, *A.J.*, **82**, 941; Buzzoni, Gariboldi & Mantegazza 1991 submitted to *A.J.*). We have first investigated empirically the relationship between the index and the fundamental parameters for a wide set of Galactic standard stars deriving a detailed calibration for dwarfs and giants. This allowed to build up synthetic models for stellar populations exploring Mg_2 in the galaxies with varying overall distinctive parameters of the populations.

The global dependence of Mg_2 on [Fe/H] is found to be $\partial Mg_2/\partial[Fe/H] = 0.135$, in agreement with the empirical estimate derived by Brodie & Huchra (1990, *Ap.J.*, **362**, 503) considering both Galactic and M31 globular clusters. When applying our calibration to the exaustive sample of local ellipticals observed by Davies *et al.* (1987 *Ap.J. Suppl.*, **64**, 581) as shown in Fig. 1, we find that galaxies display a mean metallicity enhanced by 20–60% respect to the solar value spanning over one order of magnitude at the extreme edges of their assumed fiducial distribution.

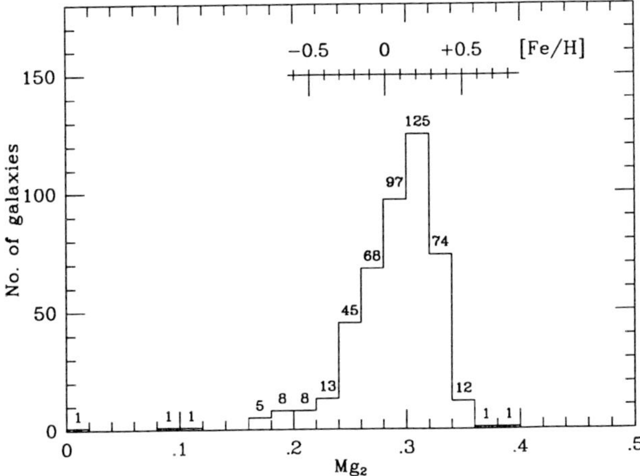

Fig. 1 - Metallicity calibration for elliptical galaxies in the Davies *et al.* (1987) sample. Galaxies are assumed 15 Gyr old with a canonical Salpeter IMF.

THE EVOLUTION OF THE GALAXY: THE ^{16}O GRADIENT AND THE SURFACE GAS DENSITY

LETICIA CARIGI
Centro de Inv. de Astronomía, CIDA
Apdo. Postal 264
Mérida 5101-A, Venezuela.

and

JOSE FRANCO
Instituto de Astronomía–UNAM
Apdo. Postal 70-264
04510 México D. F., México.

ABSTRACT: A series of simple analytical models for the evolution a galactic disk have been constructed. General solutions can be obtained under the assumption of pure radial gas flows, a star formation rate proportional to the nth power of the surface gas density, a constant IMF, and the instantaneous recycling approximation. Models with small radial flow velocities in the range 0.05 to 0.1 km s^{-1} and an initial exponential surface mass density can reproduce, for galactocentric radii larger than about 5 kpc, the Galactic ^{16}O abundance gradient and the present surface gas distribution.

1. Model Assumptions

The evolution of a two-component (gas and stars) galactic disk with a fixed total mass can be solved analytically. The radial distribution of mass in stars is coupled to the gas surface density distribution, $\sigma(r)$, via a power-law star formation rate, σ^n. The continuity equation, with a pure radial velocity term without infall, is then solved in a closed form. For the case of the Galaxy, the initial surface mass density is assumed exponential, with a scale length of 5 kpc, and with a hydrogen mass abundance $X = 0.75$ and null metallicity, $Z = 0$. The radial gas flow is assumed position-dependent, $v(r)$, but has no explicit time dependence. Also, the initial mass function is assumed constant during the whole evolution (either Scalo's IMF or Salpeter's IMF, with $x = 1.5$, and $m_l = 0.01$ M$_\odot$ and $m_u = 100$ M$_\odot$). The mass return and ^{16}O yields were taken from Köppen and Arimoto (1991), and the present age of the Galaxy is taken as 12 Gyr.

2. Results

The main results are as follows: *i*) The values of the exponent are bounded between $1 < n < 2.5$; for $n = 1$ there is no gradient and for $n = 2.5$ the SFR is outside the observed range (see Franco 1991). *ii*) For $n > 1$ and velocities proportional to either r^{-1} or to r, the gradient is produced by the initial surface density profile. For velocities that are constant or proportional to r, and bounded between 0.05 and 0.1 km s^{-1}, the main observational trends can be reproduced. *iii*) When the initial surface density is exponential, the star formation cutoff is important in the ^{16}O gradient but does not affect the final distribution of $\sigma(r)$.

References

Köppen, J. and Arimoto, N. 1991, *Astron. Ap. Suppl.*, **87**, 109.

Franco, J. 1991, in *Chemical and Dynamical Evolution of Galaxies* ed. F. Ferrini, J. Franco and F. Matteucci,(ETS Editrice; Pisa), p. 506.

The Outer Disk of the Galaxy

BRUCE W. CARNEY
University of North Carolina
PATRICK SEITZER
Space Telescope Science Institute
and
University of Michigan

Abstract.
The outermost gaseous disk of our Galaxy is warped (Burton and te Lintel Hekkert 1986; Wouterloot et al. 1990), and a stellar component may be present as well (Miyamoto et al. 1988; Djorgovski and Sosin 1989). If so, the outermost edges of the Galaxy's disk may be more readily studied using the warp than relying upon the terribly crowded and nearly to completely opaque mid-plane directions. The study of the outer disk, and of the warp, is crucial for several reasons, including estimating the core radius of the dark halo (Sparke and Casertano 1988), the disk metallicity gradient (which is a signature of the history of star formation efficiency), and of the ages of the oldest stars elsewhere in the Galactic disk (which can be used to determine if the disk grew slowly outward, as suggested by Gunn 1987).

We have begun a program to detect the old stellar population of the outer disk of the Galaxy by imaging fields at selected longitudes and latitudes where the H I maps of Burton and te Lintel Hekkert (1986) suggest the warp has reached maximal elevations above the plane as seen from the solar perspective. We selected control fields to estimate foreground (disk plus thick disk) contamination to have the same longitudes as the warp fields but opposite latitudes. All fields were found to have relatively clear lines of sight, based on the appearance of galaxies on the ESO/SRC or POSS charts. Only the southern hemisphere warp appears to have such clean lines of sight.

Four 8' x 8' warp fields and their four control fields were imaged through V and I_c filters with the Tek 1024 x 1024 CCD at the f/2.7 focus of the CTIO 4-meter reflector. Seeing was sub-arcsecond and 30 minutes (V) and 18 minutes (I_c) cumulative exposures were obtained, with limiting magnitudes $V_{lim} \geq 24$ in all cases. A total of 15,829 stars were measured in the four warp fields, and 11,722 stars in the four control fields. The excess is due partially to the Sun's location above mid-plane, but is mostly due to our detection of the outer disk's main sequence. The control fields were used to statistically subtract out the foreground components from the warp fields, leaving well-defined main sequences and in one field, a clear main sequence turn-off. We claim the old stellar warp population has been detected, and plan to continue its study using detailed reddening vs. distance maps, and searches for followed by spectroscopy of the outer disk red giants.

Key words: Milky Way Galaxy - Galactic Disk - Galactic Warp

References

Burton, W. B., and te Lintel Hekkert, P. 1986, A&AS, 65, 427.
Djorgovski, S., and Sosin, C. 1990, ApJ, 341, L13.
Gunn, J. E. 1987, in The Galaxy, ed. G. Gilmore and B. Carswell (Reidel: Dordrecht), p. 413.
Miyamoto, M., Yoshizawa, M., and Suzuki, S. 1988, A&A, 194, 107.
Sparke, L. S., and Casertano, S. 1988, MNRAS, 234, 873.
Wouterloot, J. G. A., Brand, J., Burton, W. B., and Kwee, K. K. 1989, A&A, 230, 21.

Systematic Differences Between the Field and Cluster Ellipticals

R.R. de Carvalho[1] and S. Djorgovski[2]
1. Observatorio Nacional, CNPq, Rio de Janeiro, Brasil
2. Palomar Observatory, Caltech, Pasadena, CA 91125, USA

We applied multivariate statistical techniques to study the possible systematic differences between the field and cluster ellipticals, in the framework of the fundamental plane (FP) of elliptical galaxies on the samples of elliptical galaxies from the general field and loose groups, and from rich clusters. The data set used here are the ellipticals from the "7 Samurai" (Burstein et al. 1987; Davies et al. 1987), limited to $cz < 6000$ km/s. There are 55 galaxies in the cluster sample (mostly from the Coma and Virgo clusters), and 57 galaxies in the field sample, which may include galaxies in loose and poor groups. We have also used the data set from Djorgovski & Davis (1987), with very similar results.

We find that the properties of ellipticals in rich clusters and in the field show systematic differences. The field ellipticals show a marginally higher statistical dimensionality than the cluster ellipticals; this difference is more prominent when the stellar population variables are included; i.e., the field ellipticals show more intrinsic scatter in their properties. This is also seen directly in the bivariate fits. In general, the field ellipticals seem to be a more heterogeneous family of objects. Pairwise (monovariate) correlations for the two samples are different; the correlations are systematically better for the cluster sample. This means that ellipticals in the two samples populate their fundamental planes in different ways. Bivariate correlations (equations of the FP) are also different for the two samples, implying that the two samples have different fundamental planes. This is especially true for the correlations which include the population variables Mg_2 and $(B - V)$, which are sensitive both to the enrichment history, and the star formation history. The field ellipticals are too blue, have too low Mg_2 and too high surface brightness at a fixed effective radius or luminosity, suggestive of the younger average ages.

These differences in scaling laws and correlations may be indicative of different formative histories. Specifically, they can be understood in the framework where cluster ellipticals form early, perhaps through numerous dissipative mergers of smaller fragments and the infall of gas, whereas at least some field ellipticals form through late, major mergers.

These systematic differences show again that one must be very careful in applying distance-indicator relations for galaxies defined in one sample to all galaxies of the given type: the field ellipticals do not follow the same distance-indicator relations as the ellipticals in rich clusters. We recall that the $D_n - \sigma$ relation of the "7 Samurai" (Dressler et al. 1987) was defined mainly using the rich clusters sample, and then applied indiscriminately to all ellipticals in their sample (a subset of which we have used here to demonstrate the systematic differences). It is important to understand the limits of accuracy of distance-indicator relations such as the various aspects of the FP (including the $D_n - \sigma$ relation) before credible claims of measurements of large-scale peculiar velocities can be made (cf. Djorgovski de Carvalho & Han 1988).

Related discussion and a complete list of references cited here can be found in Djorgovski & de Carvalho 1990, in Windows on Galaxies, ed. G. Fabbiano et al., (Dordrecht: Kluwer) p. 9. S.D. acknowledges support from the Alfred P. Sloan Foundation, the NSF PYI award AST-9157412, and a travel grant from the AAS.

METALLICITY OF THE STAR III-17 IN NGC 6553

S. Castro[1], B. Barbuy[1], S. Ortolani[2], E. Bica[3]
1 IAG-USP, CP 9638, São Paulo 01065, Brazil
2 Osservatorio Astronomico di Padova, Italy
3 UFRGS, D. Astronomia, CP 15051 Porto Alegre, Brazil

CCD échelle spectra in the wavelength range $\lambda\lambda$ 478 - 580 nm, were obtained at the 3.6m telescope at ESO for the star III-17 of the metal-rich globular cluster NGC 6553. This cluster was chosen because it is a relatively close bulge globular cluster, and it presented the possibility to be among the most metal-rich ones in the Galaxy.

In a previous work , BVRI CCD colour-magnitude diagrams (CMDs) were presented (Ortolani et al., 1990), where it was shown that, due to the high metallicity of NGC 6553, the red giant branch appeared to turn over for the cooler stars. The star III-17 was chosen for being among its brightest stars.

A major problem encountered for the derivation of stellar parameters, was the temperature determination due to a strong and uncertain reddening, and to the imprecisions in the relations colours-temperature for metal-rich giants. We used a method to disentangle the temperature from the metallicity effect: with the use of CMDs for 47 Tuc and NGC 6553, and the use of data for solar metallicity giants for which V, (B-V) and temperature are known, we could draw lines of constant temperatures in the V vs. (B-V) diagram for several clusters of different metallicities. The gravity was derived by a classical relation, correcting for the overionization effect. The metallicity was then determined by fitting synthetic spectra computed with different metallicities to the observed spectra.

The final parameters obtained for the star III-17 are: T_{eff} = 3850, log g = 0.4 and [M/H] = -0.2 $\pm$ 0.3.

CNO abundances were preliminarily also derived: [C/Fe] $\approx$ 0.0, [N/Fe] $\approx$ +0.6, [O/Fe] $\approx$ 0.0, indicating a possible CNO excess.

Ortolani, S., Barbuy, B., Bica, E.: 1990, A&A 236, 362

SYNTHETIC HORIZONTAL BRANCHES FOR GALACTIC GLOBULAR CLUSTERS

MÁRCIO CATELAN and MARIA LÚCIA QUARTA
Instituto Astronômico e Geofísico, Universidade de São Paulo
CP 9638, 01065 São Paulo, SP
Brasil

ABSTRACT. We present two wide sets of SHB models which have been computed for combinations of evolutionary parameters adequate to the GGC system. The effect of the enhancement of α-elements upon theoretical predictions related to the "Sandage effect" and globular cluster ages is also discussed.

Discussion

Using both the HB evolutionary tracks of Sweigart (1987, *Ap. J. Suppl.*, **65**, 95) and Lee & Demarque (1990, *Ap. J. Suppl.*, **73**, 709), we have computed wide sets of SHB models. For each combination of the evolutionary parameters Y and Z, we have varied the mean HB mass $\langle M_{HB} \rangle$ so as to yield from very blue to very red morphologies. We have considered three possibilities for the width σ_M of the (assumed) gaussian mass distribution: 0.01, 0.02, and 0.03. 50 runs (with 1200 stars each) of our new SHB code have been used to derive an extensive tabulation of morphological and pulsational parameters for each (Y, Z, $\langle M_{HB} \rangle$, σ_M) combination. Our models suggest that the lower envelope of the HB at the RR Lyrae level will differ from the ZAHB for both blue and red morphologies. Such a deviation makes the predicted HB "thickness" *at this color level* decrease for metallicities higher than [Fe/H] $\sim$ -1.3 (where several clusters with red HB morphologies are found), apparently at variance with the observations [see Catelan (1991, this volume)]. We have also studied the effect of the enhancement of α-elements upon the SHB models, through a simple re-scaling of the standard models in metallicity, as suggested by the calculations of Chieffi & Straniero (1990, private communication). Such α-enhanced models imply high ages ($\simeq$ 17 Gyr) at [Fe/H] $\simeq$ -2.2 (Y = 0.23, [α/Fe] = 0.48), practically identical to the case where [α/Fe] = 0, with important cosmological implications [cf. Catelan and de Freitas Pacheco (1991, *Astr. Ap.*, submitted)]. Moreover, a significant decrease of the predicted slope γ of the log$P$ - [Fe/H] relation results. Our models also show that reading the periods at logT_{eff} = 3.83 rather than 3.85 leads to appreciably higher γ values; accordingly, assuming [α/Fe] = 0 one obtains γ = -0.060 at logT_{eff} = 3.83 (-0.033 at 3.85), whereas assuming [α/Fe] = 0.48 one may derive γ = -0.023 at logT_{eff} = 3.85 (-0.043 at 3.83). All quoted values are for Y = 0.23.

Acknowledgment: Financial support by FAPESP (grant 89/3094-5) is acknowledged.

A POSSIBLE RECONCILLIATION AMONG DIFFERENT RR LYRAE ABSOLUTE MAGNITUDE CALIBRATIONS AND IMPLICATIONS FOR THE AGE-METALLICITY RELATION

MÁRCIO CATELAN
Instituto Astronômico e Geofísico, Universidade de São Paulo
CP 9638, 01065 São Paulo, SP
Brasil

ABSTRACT. A semi-empirical scenario is suggested which tentatively accounts for the (otherwise conflicting) slopes of the several available RR Lyrae absolute magnitude calibrations. This scenario implies a very small dependence of cluster ages upon metallicity.

Discussion

Sandage (1990, *Ap. J.*, **350**, 603) has shown that the HB "thickness" $\Delta M_V(HB)$ at the RR Lyrae color level increases monotonically with metallicity. After removing his inadequate (for our purposes) data points for 47 Tuc and ω Cen, one may obtain the regression equation $\Delta M_V(HB) = 0.751 + 0.235$ [Fe/H]. The lower envelope (LE) and "evolutionary mean level" (EML) $M_V(HB)$ calibrations are related through the expression $\eta = [M_V(LE)$ - $M_V(EML)]$ / $\Delta M_V(HB)$. I estimate a mean concentration parameter $\eta = 38$ % for 7 clusters. From the above considerations, the slopes α and zero-points β of the LE and EML expressions will differ by ~ 0.1 and ~ 0.3, respectively, with the LE values being higher. Thus, if the EML theoretical calibration of Lee (1990, *Ap. J.*, **363**, 159) is correct ($\alpha_{EML} = 0.19$, $\beta_{EML} = 0.97$), one will have ($\alpha_{LE} \simeq 0.28$, $\beta_{LE} \simeq 1.26$). The slopes (which are easier to treat) of the available calibrations seem to be consistent with these results, within the existing uncertainties, with only one *clear* exception: the "RGB bump" analysis [Fusi Pecci *et al.* (1990), *Astr. Ap.*, **238**, 95]. The results thereby obtained can, however, be questioned in terms of consistency arguments [see Catelan (1991, *Astr. Ap.*, submitted) for details and also an analysis of SHB predictions].

Based upon the "ΔV method" of age calibration, one can see that the above considerations lead to a unified scenario where globular cluster ages vary by only ~ 1 Gyr for a [Fe/H] variation as large as 1 dex, with important implications for our understanding of the formation of the galaxy.

Acknowledgments: I would like to thank Drs. M. L. Quarta, C. Cacciari, R. Buonanno, A. Renzini, and B. Barbuy for helpful comments. Financial support by FAPESP (grant 89/3094-5) is acknowledged.

STRUCTURE AND COLORS OF DWARF ELLIPTICAL GALAXIES IN THE FORNAX CLUSTER

S. A. CELLONE[1], J. C. FORTE[2], D. GEISLER[3]
1 Fac. de Cs. Astronómicas y Geofísicas
1900 La Plata, Argentina
2 C.C.67 S.28 Instituto de Astronomía y Física del Espacio
1428 Buenos Aires, Argentina
3 Cerro Tololo Interamerican Observatory
C. 603 La Serena, Chile

CCD photometry in the Washington system [3] for a sample of dwarf elliptical galaxies in the Fornax cluster is presented. The observations were done in October 1989 with the CTIO 0.9 and 1.5m telescopes; the galaxies being selected from the Fornax Cluster Catalogue [4].

As for other samples of dEs [1,2,6,7], the surface brightness profiles, although roughly exponential, show a variety of shapes. Analytical fits of the form

$$SB = SBo + 1.086 \cdot (R/a)^N,$$

were done, where SB is the surface brightness in elliptical annuli and R is the equivalent radius. In a few cases, a "shoulder" is evident in the outskirts of the nucleus.

The color profiles show no meaningful gradients, but in most cases the nuclei are marginally bluer, as it would be expected if they are the result of the last star forming event.

Integrated *(C-T1)* colors of these dwarfs fall within the range covered by the globular clusters around the Fornax cluster galaxy NGC 1399 [5].

FCC	T1(tot)	(C-T1)	So(T1)	b/a
82	16.12	1.54	21.88	0.83
118	17.10	1.43	22.77	0.85
135	15.73	1.46	20.73	0.51
188	15.33	1.56	21.46	0.93
195	16.44	1.50	21.69	0.59
203	15.23	1.59	20.83	0.54
314	15.62	1.49	22.63	0.92

Columns 1: Fornax Cluster Catalogue number (see [4]). 2: T1 integrated magnitude. 3: (C-T1) integrated color. 4: Extrapolated central surface brightness, in mag/sqarsec 5: semiaxial ratio.

References

1. Binggeli B., Sandage A., Tarenghi M. (1984). A.J. **89**, 64.
2. Caldwell N., Bothun G. (1987). A.J. **94**, 1126.
3. Canterna R. (1976). A.J. **81**, 228.
4. Ferguson H. (1989). A.J. **98**, 367.
5. Geisler D., Forte J. (1990) Ap. J. (Lett.) **350**, L5.
6. Ichikawa S., Wakamatsu K., Okamura S. (1986). Ap. J. Supp. **60**, 475.
7. Wirth A., Gallagher J. (1984). Ap. J. **282**, 85.

GALAXY FORMATION STUDY USING QSO ABSORBERS

Arati Chokshi
IPAC/Caltech
770 So. Wilson Ave.
Pasadena, CA 91125
U.S.A.

ABSTRACT. Observations of QSO absorbers at $Z \sim 2$ suggest that they represent a population of primeval, dynamically evolving systems which turn into the local galaxy population.

We analyze the published QSO absorption line data for Lyman α forest lines, metal line absorbers, and damped Lyman α systems in terms of their total hydrogen content (N_H) and the velocity dispersion (σ_ν). Lyman α forest absorbers form a contiguous sequence with metallic absorbers on this diagram. Thus, it is conceivable that the former represent low column density, low mass end of the larger metallic systems. For majority of the absorption systems, cooling appears to have played an important role on dynamical time scales at the epoch of observation, although most systems have not evolved sufficiently to resemble the local galaxy population. Thus they represent a population of primeval, dynamically evolving systems. We interpret the observations of large velocity dispersion in metal line systems as indicative of large scale star-formation process occurring in forming galaxies - possibly associated with the epoch when globulars made most of their stars. Their location on the cooling diagram indicates that, for most absorbers, large velocity dispersions cannot be attributed to galaxy-galaxy correlations. Of the two extreme cosmological scenarios, the data are more easily interpretable for a dark matter dominated universe than for a purely baryonic universe. For a dark to baryonic matter ratio of 10, observations suggest that the ultimate fate of all QSO absorbers is to evolve into the local galaxy population.

Warmers: Source of Ionization and N Enrichment in AGN

R. CID FERNANDES[1,3], H. A. DOTTORI[2], R. B. GRUENWALD[1] and S. M. VIEGAS[1]

[1] *IAG/USP, Caixa Postal 30.627 - 01051 São Paulo, Brasil*
[2] *IF/UFRGS, Caixa Postal 15051 - 90069 Porto Alegre, Brasil*
[3] *Institute of Astronomy, Madingley Road - CB3 0HA Cambridge, UK*

Abstract. We present a detailed study of the warmers phase in a starburst as a scenario for AGN (Terlevich & Melnick 1985). A numerical code ("ET") was developed to compute the time evolution of the total spectrum of an ionizing cluster. ET is based in Maeder & Meynet (1988) and Maeder (1990) tracks and in model atmospheres. The ionizing spectra were input into the photoionization code Aangaba (Gruenwald & Viegas 1991) to calculate the nebular emission for a grid of models covering different densities, ionization parameters, metallicities and cluster ages. The emission line spectrum go through an "active phase" beginning at $\approx$ 3 Myrs, which lasts for about 3 Myrs, when the population of warmers disapears. Metallicities between 1/2 and 2 $Z_\odot$ are necessary to generate AGN line ratios. The lower limmit is due to the fact that fewer warmers occur in metal poor regions (due to the smaller mass loss rates) and even if some do occur the line ratios would resemble an HII galaxy rather than a Seyfert 2 or a LINER. The upper limmit is set by strong line cooling of the emission nebula. The chemical pollution of the gas by the processed stellar winds from the very massive stars produce straightforwardly a N overabundance as that observed in many AGN. In Figure 1a and b, we show the process of heavy element enrichment as a function of time parametrized by γ—a quantity which measures the mass ejected by stars relative to the mass of gas surrounding the stellar association (see Cid Fernandes *et al.* 1991). The chemical yields were taken from Maeder (1990) models.

Key words: Active Galaxies - Photoionization Models - Chemical Evolution

References

Cid Fernandes, R., Dottori, H. A., Gruenwald, R. B. & Viegas, S. M.: 1991, *Mon. Not. R. astr. Soc.*, in press.

Gruenwald, R. B. & Viegas, S. M.: 1991, *Astrophys. J. Suppl.*, in press.

Maeder, A. & Meynet, G.: 1988, *Astr. Astrophys. Suppl.* **76**, 411.

Maeder, A.: 1990, *Astr. Astrophys. Suppl.* **84**, 139.

Terlevich, R. & Melnick, J.: 1985, *Mon. Not. R. astr. Soc.* **213**, 841.

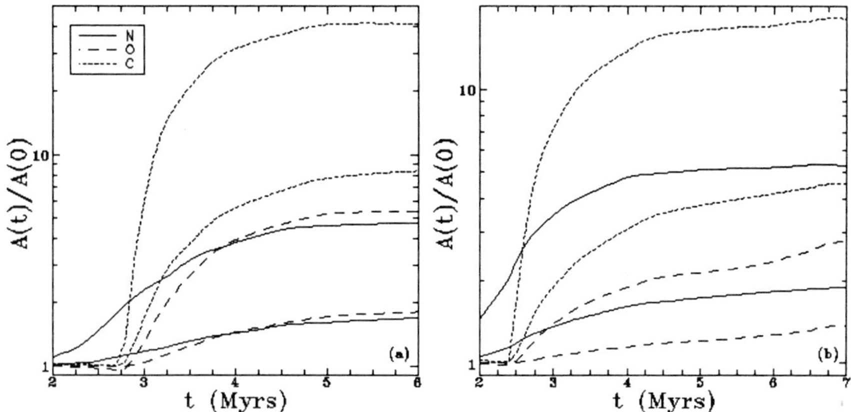

Fig. 1. Evolution of the C, N and O abundances with respect to their initial values for $Z(t=0) = Z_\odot$ (a) and $2Z_\odot$ (b). The two sets of curves in each figure correspond to $\gamma = 0.1$ (upper curves) and $\gamma = 1$ (lower curves).

THE M_V(RR)-[Fe/H] RELATION FROM THE BAADE-WESSELINK METHOD APPLIED TO FIELD RR LYRAE STARS

G. CLEMENTINI[1], C. CACCIARI[1], J.A. FERNLEY[2,3]

(1) Osservatorio Astronomico, Via Zamboni 33, I-40126 Bologna, Italy

(2) Dept. of Physics and Astronomy, University College London, London, England

(3) International Ultraviolet Explorer, VILSPA, Madrid, Spain

We have applied two slightly different formulations of the Baade-Wesselink (B-W) method : (i) the Infrared Flux (IF) method (Fernley *et al* 1990a), and (ii) the Surface Brightness (SB) method (Cacciari *et al* 1989a), to derive the absolute magnitude of the field RR Lyrae stars UU Cet, RV Phe and W Tuc. The photometric (BVRIJHK) and radial velocity data used in the analysis have been published in Cacciari *et al* (1987, 1989b, 1991) and Clementini *et al* (1990). The Walraven photometry used in the IF method is by Lub (1977). Typical accuracies for individual data points are ±0.01-0.02 mag for the magnitudes and colours, and ±1-2km/sec for the radial velocity data. Excluding RV Phe, which may be affected by Blazhko effects, the results obtained from the IF and SB methods are rather consistent especially when the V−K colour is used in the SB method. For details see Cacciari *et al.* (1991).

The B-W analysis, either SB or IF version, has been previously applied to a number of field RR Lyrae stars by many authors. The most recent compilations are given by Cacciari *et al.* (1991) and Jones *et al.* (1991), and include 24 stars that have been analyzed using infrared data. A least-squares fit weighted both in M_V(RR) and [Fe/H] of the results for these stars (where multiple M_V(RR) estimates have been averaged), leads to a relation between absolute magnitude and metallicity :

$$M_V(RR)=0.19(\pm0.03)[Fe/H]+1.01(\pm0.15)$$

The slope of this relation, however, can vary slightly depending on the criterium adopted to choose the stars and the magnitude determinations to use in the fitting procedure. In fact, since seven stars in the above list have been analysed with both IF and SB methods, and for 8 "problem" stars the absolute magnitude estimates are less accurate (Blazhko effect, high reddening, c-types), we have experimented with the weighting system according to the four different criteria listed below :

All stars (average M_V when multiple)	M_V(RR)=0.19[Fe/H]+1.01	16.1	19.5
All stars (IF results when multiple)	M_V(RR)=0.23[Fe/H]+1.04	16.1	18.6
All stars (SB results when multiple)	M_V(RR)=0.14[Fe/H]+0.98	16.2	20.9
Excluding the 8 "problem" stars	M_V(RR)=0.21[Fe/H]+1.01	15.9	18.8

In the last columns we have estimated the corresponding ages of Globular Clusters using the VandenBerg and Bell (1985) models for Y=0.23 and [Fe/H]=−0.80 (left) and [Fe/H]=−2.20 (right), on the assumption of ΔV(TO-HB)=3.50 and [CNO/Fe]=solar. If there is O-enhancement the ages decrease, [O/Fe]=+0.3 corresponding to a decrease of about 1 Gyr. If the O-enhancement is dependent on metallicity being larger for metal-poor clusters, then also the age spread decreases accordingly.

References

Cacciari, C., Clementini, G., Prévot, L., Lindgren, H., Lolli, M., and Oculi, L. 1987, *Astron. Astrophys. Suppl.*, **69**, 135

Cacciari, C., Clementini, G., Prévot, L., and Buser, R. 1989a, *Astron. Astrophys.*, **209**, 141

Cacciari, C., Clementini, and Buser, R. 1989b, *Astron. Astrophys.*, **209**, 154

Cacciari, C., Clementini, and Fernley, J.A. 1991, *Ap.J.*,submitted

Clementini G., Cacciari C. and Lindgren H. 1990, *Astron. Astrophys. Suppl.*, **85**, 865

Fernley, J.A., Skillen, I., Jameson, R.F., and Longmore, A.J. 1990, *M.N.R.A.S.*,**242**, 685

Jones, R.V., Carney, B.W., Storm, J., and Latham D.W. 1991, preprint

Lub, J. 1977, Ph. D. Thesis, University of Leiden

VandenBerg, D.A., and Bell, R.A. 1985, *Ap.J.Suppl.*,**58**, 561

SEPARATION OF HALO AND THICK DISC STARS IN TWO CATALOGS

J. COLIN, B. DAUPHOLE-FOUILLET, C. DUCOURANT
Observatoire de Bordeaux et URA 352, CNRS
BP 89
F-33270, Floirac
FRANCE

Recent studies (Gilmore and Reid 1983, Gilmore and Wyse 1985, 1986) concerning the halo stars have showed that a new structure, the thick disc, must be retained to describe the Galaxy. We suppose here that this structure exists and we deduce some results concerning the difference between the halo stars and the thick disc stars. The first assumption is that the inclination of the orbits of the halo stars must be randomly distributed. This must be true if the halo has been formed from a merger or from a non-rotating protogalactic cloud. Using the catalog of Carney et al.1990 (804 stars selected for their high velocities) and Norris' one 1986 (386 stars selected for their poor metallicity [Fe/H] < 0.6), we calculate the orbital inclinations of all the stars and we display them on an histogram (fig.1). The two samples (built on different criterions) show a flat distribution from 180° to 65° (fig. 1-a). From 65° to 0° there is clearly a mixing of halo stars and thick disc stars. If, as we assume, the halo stars have a flat distribution from 180° to 0°, we can separate the two population using another kinematical parameter : the eccentricity. We calculate it by using a logarithmic potential that gives a flat rotation curve. Now if we keep all the stars with an inclination between 180° and 65° and the stars with an inclination between 65° and 0° but with an eccentricity $e \geq 0.6$, we obtain a flat distribution of the inclination between 180° to 0° (fig. 1-b). In that case all these stars may be considered as halo stars. Plotting [Fe/H] against orbital inclination we obtain a gaussian symetrical distribution centered on -1.7. It seems that the apocenters of the selected halo stars reach a limit for 45 kpc.

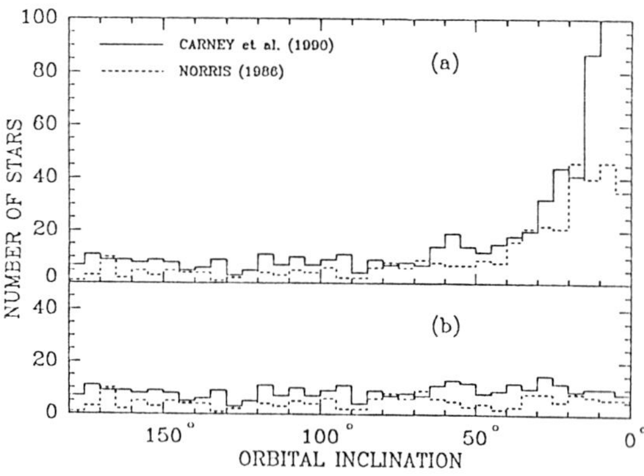

Fig. 1a-Histogram of the orbital inclinations for all stars. 1b-Same histogram for halo stars.

Carney, B.W., Aguilar, L., Latham, D.W., Laird, J.B. : 1990, Astron.J., **99**, 201-220
Gilmore, G., Reid, N. : 1983, Mon.Not.Roy.astr.Soc., **202**, 1025-1047
Gilmore, G., Wyse, R.F.G. : 1985, Astron.J., **90**, 2015-2026
Gilmore, G., Wyse, R.F.G. : 1986, Nature, **322**, 806-807
Norris, J. : 1986, Astrophys.J.Suppl., **61**, 667-698

SPECTRAL CHARACTERISTICS OF EARLY G-DWARF STARS TOWARDS THE GALACTIC POLES

C. J. CORBALLY
Vatican Observatory Group, Tucson, Az, USA
R. F. GARRISON
David Dunlap Observatory, Ont., Canada

We have obtained slit spectra of 299 candidate early G-dwarf stars from the objective prism and photometric surveys of Corbally and Garrison (1988a, 1988b) in the North and South Galactic Pole regions between 90 and 1700 pc from the sun. The slit spectra from reticon detectors on the S.A.A.O. 1.9 m and the Steward Observatory 2.3 m telescopes have a resolution of 2 Å and a signal-to-noise of around 35.

The spectra were classified against those of MK standards according to the methodology of Corbally (1987) which copes with weak-line spectra. That paper also describes the metallicity index, Δ_m, which is the numerical difference in spectral subtypes between the hydrogen line strength and the metallic line strength shown by a star.

Initial results show that 277 out of the 299 candidate stars are indeed dwarfs and so form a very homogeneous and unambiguous set of Galactic probes. One dwarf star is slightly strong-lined, while 28% and 7% have metallicities corresponding to intermediate ($\Delta_m > -10$) and extreme ($\Delta_m \leq -10$) Population II stars respectively. The remaining 65% have solar like abundances and are generally very normal looking. The most extreme weak-lined stars are CG-SGP-106 with a type of "K0 mF2" and CG-SGP-184 with "G5 mA6". These spectra are so weak-lined as to make luminosity determination uncertain.

No significant difference in the distribution of metallicities was detected between the north and south directions, nor did this distribution skew obviously with distance from the Galactic plane. If thin-disk stars outnumber halo/thick-disk stars up to a height of 2 kpc (Croswell *et al.* 1991), then these are expected results for this G-dwarf sample.

Radial velocities, obtained by the cross-correlation method with $\sigma \simeq 8$ km/s, show that, when the NGP and SGP data are combined, the distribution of velocities is similar for the two Galactic poles, but that more G-dwarf stars (67%) are streaming southward. These results may simply imply that the Sun is near the plane of the Galaxy and has a w velocity of about +15 km/s.

References

Corbally, C.J. 1987, AJ **94**, 161.
Corbally, C.J., and Garrison, R.F. 1988, AJ **95**, 739 (1988a).
Corbally, C.J., and Garrison, R.F. 1988, AJ **95**, 745 (1988b).
Croswell, K., *et al.* 1991, AJ **101**, 2078.

SEARCHING SIGNIFICANT SIGNATURES OF STELLAR POPULATION CHARACTERISTICS IN MULTIVARIATE STAR COUNT SAMPLES

M. CRÉZÉ, B. CHEN
Observatoire de Strasbourg,
CNRS URA 1280
11, rue de l'Université
67000 Strasbourg, France

A.C. ROBIN, O. BIENAYMÉ.
Observatoire de Besançon
41, Avenue de l'Observatoire
25000, Besançon, France

The limited observational performances which can be achieved on large collections of faint objects do not allow to derive intrisic stellar parameters such as distance, mass, age, space velocity, chemical composition of individual stars. However some information relevant to the distribution of these quantities is reflected in the n-dimensional distribution of observables : connecting observed distributions to the physical processes they come from is basically a multivariate problem for which some of us developped a synthetic approach of galaxy modelling (Robin, Crézé 1986, Bienaymé et al. 1987).
Most investigations of that kind, following Chiu (1980) use to project the data in a plot of reduced proper motions versus colour and eventually compare the observed distributions with the predictions of some stellar population model. In the present analysis, we try and use the multivariate analysis technics to search for meaningfull associations in the 5-dimensional space of observables between observed points and sets of simulated points generated from the above quoted galaxy model.
In practice, real stars are merged with model predicted ones in a single catalogue of 5-dimensional data (V, U-B, B-V, μ_l, μ_b) . Then a cluster analysis is performed providing a series of significant clusters. Different models can be used and the capability of models to represent the data is tested through the coincidence between real data and simulated ones in all clusters. Then the known intrisic properties of simulated stars can be used as tracers of the properties of associated real stars.
A preliminary investigation of the data obtained by Bienaymé et al 1991 in a field at intermediate galactic latitude, shows that a two component model of the Milky Way is not compatible with the data : an intermediate population between Halo and Disc is needed with clearly separate chemico-dynamical properties.

REFERENCES

Chiu L.T.G. 1980, ApJ Sup **44**,31.
Bienaymé O., Robin A.C., Crézé M. (1987) *Astron. Astrophys.* **180**, 94.
Bienaymé O., Mohan, V., Crézé M. 1991 to appear in *Astron.Astrophys*
Robin, A., Crézé M. (1986) *Astron. Astrophys.* **157**, 71.

ULTRAVIOLET STUDIES OF THE FACE-ON GALAXY NGC 2217

A. DANKS

ST Systems Corporation, Lanham, MD 20706, USA

M. PÉREZ

CSC/IUE Observatory, NASA-GSFC, Greenbelt, MD 20771, USA

and

B. ALTNER

CSC/GHRS, NASA-GSFC, Greenbelt, MD 20771, USA

Abstract. Recently a subgroup of S0 galaxies has been identified with external gas rotating in retrograde motion with respect to the stars (e.g. NGC 1216, NGC 4546 and NGC 7007). All of these galaxies are seen almost *edge-on* and present an excellent testing ground for studying the connection between the stellar formation and the counterrotating (non-primordial) gas. Recently, the almost *face-on* and gas-rich SB0 galaxy NGC 2217 was added to this list (Bettoni et al. 1990, AJ, 99, 1789), which presents a more favorable orientation for spectroscopic studies to determine its stellar population. We present three IUE long-wavelength region (LWP) observations of NGC 2217 taken at different orientations of the aperture centered on the galaxy nucleus. We have taken full advantage of the IUE spatial resolution capability to map the nuclear and bulge region. In addition, our multicolor CCD imagery obtained at CTIO allows us to identify some faint, but spatially resolved, dust extinction patterns which correlate with the weak IRAS fluxes.

1. Observations and Results

From CCD images of the disk galaxy (SB0) NGC 2217 we can identify a bright, nearly-circular nucleus of about 5 arcsec in diameter (for a distance of 14.9 Mpc, this corresponds to 400 pc), surrounded by an extended dusty region. Observations of this region were carried out using IUE in the long-wavelength camera (LWP), low-dispersion mode, through an oval aperture 10× 20.″ These images were analyzed by using the extended line-by-line (LBL) files which contain the spatial information. The width of the nuclear region corresponds to a Gaussian with the FWHM equal to the IUE point spread function; i.e., the nucleus appears as a point source. Our method separates the spectra of the central point-like source from the underlying extended component. The fluxes were corrected for extinction, however, we found that the central and extended components required different reddening corrections; the nucleus requiring A_V=0.29, and the extended region, A_V=0.53.

The data point to a marginally earlier stellar population around the central region. The UV light as a whole is dominated by a late-type stellar population of principally G and K stars. The almost *face-on* view of this galaxy appears optically thick to UV light. It is conceivable that with analogy to our own Galaxy, the stellar populations weakly detected in NGC 2217, are mostly halo and late-type stars in the center with and increasing contribution of dust and early stellar populations (so far undetected) as we move outward along the faint spiral arms. This result is contrary to our initial expectation, since the counterrotating gas does not appear to be enhancing star formation in this galaxy. The IRAS fluxes imply the presence of cool gas and whether any star formation is taking place is clearly insufficient to heat the observed dust.

NOVA LMC 1991: A SUPER-BRIGHT NOVA IN THE LARGE MAGELLANIC CLOUD

M. DELLA VALLE

European Southern Observatory, La Silla.

Abstract. We discuss the recent outburst of nova LMC 1991. By comparing its magnitude at maximum and rate of decline with those of 14 historical LMC novae, we show that nova LMC 1991 represents the first detection of a *super*-bright nova in the Large Magellanic Cloud.

Results

a) The analysis of the lightcurve of nova LMC 1991 has allowed us to derive its magnitude at maximum and rate of decline. The comparison between these parameters and the well established MMRD relation for LMC novae, shows that Nova LMC 1991 deviates toward the brighter magnitudes by more than one magnitude.

b) A critical review of the magnitude at maximum *vs.* rate of decline relationships for the most consistently investigated nova populations (Galaxy, LMC, M31, and Virgo; Della Valle 1991), shows that: 1) 90% of the novae of our sample is bounded inside a Δm strip of ± 0.5 mag. Moreover novae in this area seem to follow the same MMRD relationship. This sketch is entirely consistent with the prediction (Shara 1982), that a dispersion of MMRD relation should be observed as a result of the intrinsic scatter in the white dwarf luminosity before the outburst. The contribution to the scatter of the photometric errors is indeed less than 20%. 2) at least one object in each sample (9 objects in all) deviates strongly from the respective MMRD relationships. In particular, the M31 and LMC objects (uncertainty on the photometry $\simeq 0.1$ mag), are systematically 1 magnitude brighter than what is expected from their rate of decline. For the object in Virgo and the two galactic ones, the respective errorbars are still consistent with the 1 magnitude deviation. This also may suggests the presence of a parallel MMRD relationship for the *super*-bright novae, although in the framework of the present statistic we can not make definitive conclusions.

c) Though heavily limited by poor statistics, we can attempt an estimate of the frequency of *super*-bright outbursts in comparison with 'normal' events. It is apparent that the weight of this estimate is almost completely based on the 5 objects in M31, from which we obtain a frequency of $f \sim 0.06$. Nevertheless, it is interesting to note that consistent figures arise from the LMC population and more marginally from the galactic one. We can assume an indicative final value of $(N_{super})/(N_{normal}) \approx 0.07$.

Reference

Della Valle, M. 1991, *Astron. Astrophys.* in press

Shara, M. 1982, *Ap.J.*, **243**, 926

Stellar Population Changes in Post-Core-Collapse Globular Clusters

S. Djorgovski[1] and G. Piotto[2]
1. Palomar Observatory, Caltech, Pasadena, CA 91125, USA
2. Dipartimento di Astronomia, Università di Padova, Italy

ABSTRACT. Color gradients, in the sense of becoming bluer inwards, are found in post-core-collapse (PCC) clusters. No gradients are seen in clusters with King-model morphology. The gradients seem to be caused by the demise of red giants and/or subgiants, and possibly an increased number of blue stragglers or some other population of faint, blue objects. Practically all PCC clusters have blue horizontal branches, with faint blue tails. Bright red giants are clearly underabundant in the central regions of PCC clusters, whereas HB stars seem to be unaffected. At a fixed metallicity, PCC clusters also show blue FUV colors, as seen in the archival data from the IUE and ANS (Figure 1). This is consistent with our observations of their HB morphologies, but a presence of an additional hot population cannot be excluded. The gradients may be a consequence of stellar interactions during and after the core collapse, and the mechanism may be important for the formation of millisecond and binary pulsars and LMXB's. These effects represent a strong evidence that dynamical evolution of star clusters can physically modify their stellar populations. It is possible that similar effects may operate in at least some galactic nuclei. More details and further references are given by Djorgovski et al. 1991, ApJ, 372, L41. S.D. acknowledges support from the NASA grant NAG5-1173 and the NSF PYI award AST-9157412.

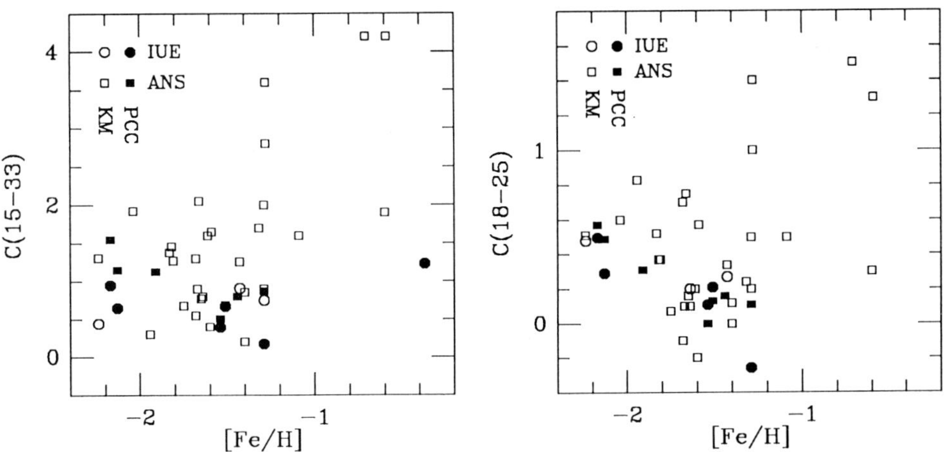

Figure 1: Extinction-corrected FUV colors of globular clusters measured with the IUE (circles) and ANS (squares), plotted vs. the metallicity. The solid symbols represent the PCC clusters, and the open symbols the clusters with the King model morphology. The cluster core structure appears to be at least as important as the metallicity in governing the UV colors, perhaps through the effect on the HB morphology.

A DEEP NEAR INFRARED SOUTHERN SKY SURVEY:

A new probe for hidden stellar populations in our Galaxy.

N. EPCHTEIN[1], F. GUGLIELMO[1] and W.B. BURTON[2].

[1] *Observatoire de Paris, France.*

[2] *Sterrewacht, Leiden, The Netherlands.*

September 27, 1991

We are currently implementing the instrumentation and the data analysis tools aimed at performing a deep, complete survey of the Southern Sky (project **DENIS**) in 3 near-IR photometric bands (I=0.9μm, J=1.25 μm, K=2.2μm). The anticipated 3σ limiting magnitudes for point sources in I, J and K are 18, 16 and 14, respectively. The spatial resolutions will be 3" in J and K and 1.5" in I. The survey will be carried out with a dedicated IR camera installed at the ESO 1 m telescope in Chile during 3 years and will start in mid '93. One expects the detection of several 10^7 sources. The final products that we plan to deliver to the community are, so far: a set of "cleaned" elementary images of $12' \times 12'$, a computer accessible data–bank ($\approx$ 3Tbytes) with appropriate software tools and specialized catalogs. One major impact of this survey will be to provide digitized maps, that will facilitate stellar counts and follow-up statistical investigations. Thanks to a better spatial resolution and a much lower sensitivity to cool dust emission, DENIS will be less limited by confusion in crowded regions (galactic plane, bulge, etc...) than was the IRAS [12μm] survey. The moderate interstellar extinction in the near-IR range compared to the optical range ($A_K \sim \frac{A_V}{10}$) will allow probing of stellar populations at low galactic latitudes and in optically obscured regions of our Galaxy. The near-IR range corresponds to the emission peak of frequent species of stars of low surface temperature. Knowledge of the faint end of the luminosity function is necessary for the determination of the contribution of low–mass stars to the missing mass in our Galaxy. The search for brown dwarfs is an especially important challenge. Modeling suggests that a few 10^4 of them could be detected, but the positive identification of even a few of them would be a major discovery. At the other end of the mass distribution, DENIS will detect all red giants out to distances halfway across the Galaxy, it will considerably increase the number of known AGB stars suffering high mass–loss effects and of Young Planetary Nebulae. Many major structural features of our Galaxy will benefit of this survey: stars in the molecular ring at 4 kpc will be detected, the radial and vertical scale lengths of the relevant spectral type stars will be improved. DENIS will provide new knowledge about the luminosity function of the populations of late and young stars and their distribution throughout the Galaxy. The bulge of our Galaxy will be more clearly revealed than in the IRAS survey. Of particular promise are studies of the distribution of late–type stars at large distances from the galactic center. It will be possible to trace aspects of the stellar distribution in the region characterized by the galactic warp and gravitationally dominated by dark matter. Finally the detection of all supergiants, secured by follow–up observations , will lead to new informations on the formation and evolution of massive stars and open clusters.

SYNTHETIC Mg_1, Mg_2 and Mgb INDICES

M. Erdelyi-Mendes, B. Barbuy, A. Milone
IAG-USP, Depto. Astronomia
C.P. 9638, São Paulo 01065, Brazil

A detailed study of strong spectral features is an important link between stellar spectroscopy and low-resolution spectroscopy or photometry of galaxies. The magnesium triplet at λ 517 nm is among the strongest features in the spectra of normal galaxies. A question arises regarding its use as a metallicity indicator for composite systems (e.g., Faber, 1973), given that it is also used as a gravity indicator for individual stars (Clark & McClure, 1979), indicating that the Mg feature shows a bi-parametric behaviour as a function of metallicity and gravity.

We have generated synthetic spectra in the wavelength region $\lambda\lambda$ 490-530 nm, for a grid of stellar parameters, in order to study the behaviour of the Mgb ($\lambda\lambda$ 516.2-519.3 nm), Mg_1 ($\lambda\lambda$ 507.1-513.4 nm), Mg_2 ($\lambda\lambda$ 515.6-519.7 nm), cf. Burstein et al. (1984) and the DDO "51" ($\lambda\lambda$ 490-530 nm) bandpasses, as a function of the stellar parameters effective temperature T_{eff}, gravity log g and metallicity [M/H].

Our calculations show that the Mgb and Mg_2 are practically insensitive to gravity, being very sensitive to metallicity, constituting therefore adequate metallicity indicators. Mg_1 and the DDO 51 filter on the other hand are sensitive to gravity, these indices showing a bi-parametric behaviour as a function of metallicity and gravity. The dependence of the indices as a function of temperature show otherwise a smooth increase for decreasing temperatures.

We conclude that the Mg_2 appears to be a trustful metallicity indicator, but the dependence of Mg_2 intensity as a function of metallicity for the metal-rich populations shows a non-linear behaviour.

Burstein, D., Faber, S.M., Gaskell, C.M., Krumm, N.: 1984, ApJ 287, 586
Clark, J.P.A., McClure, R.D.: 1979, PASP 91, 507
Faber, S.M.: 1973, ApJ 179, 731

INFRARED DETECTION OF SUPERNOVA REMNANTS IN THE NUCLEUS OF NGC 253 ?

DUNCAN A. FORBES
Institute of Astronomy, Cambridge, England

1. Results and Discussion

The high rate of gas consumption and evidence for massive stars in NGC 253, implies a prodigious star formation and supernovae rate. However there is no optical confirmation of these supernovae, presumably due to the enormous dust obscuration present. Radio observations (Antonucci and Ulvestad 1988) reveal compact radio sources i.e. young supernova remnants (SNRs).

Using a new PtSi 256 × 256 pixel near–IR array with $0.05''$/pix we obtained, at CTIO, an H image of the nuclear region which penetrates the dust. Our image (see Forbes, Ward and DePoy 1991 for details) reveals several regions of enhanced emission (hotspots) as well as the central nucleus. Assuming the H nucleus lies at the 6cm nucleus, we find very good spatial agreement between the hotspots and strong compact 6cm sources i.e. SNRs.

We spectulate that the hotspots are reradiation from dust associated with an expanding SNR shell. The IR colours of hotspot A are consistent with the observations of known type II SN some years after the explosion (Dwek 1983).

2. Conclusions and Future Work

With the available data we have not been able to positively identify the nature of the IR hotspots in NGC 253, although we consider SNRs to be the most plausible. The other possible sources – giant HII regions or clusters of RSGs – may be indirectly related to SNRs as the site of such events. One method of clarifying the nature of the hotspots is with high spatial resolution spectroscopic imaging. For example, strong CO absorption lines will indicate the presence of supergiants; whereas strong [FeII] ($1.26\mu m$ and $1.64\mu m$) and weak Brγ ($2.17\mu m$) will confirm that a SNR, and not an HII region, is the dominant source of the IR emission. Imaging in [FeII] may prove to be an excellent technique for revealing the remnants of massive stars, in the same way that Hα imaging reveals the birth of young stars. In the more distant future imaging in Co II at $10.52\mu m$ could be used to identify type II SN. The production of ^{56}Co is directly related to the initial stellar mass. Eventaully when a number of SN are observed in this line a mean IMF for starburst galaxies could be derived (van Buren and Norman 1989).

Acknowledgements

We thank the conference organisers for financial support.

References

Antonucci, R. R. J. and Ulvestad, J. S. (1988) *Ap. J. (Letters),* **330,** L97.
Dwek, E. (1983) *Ap. J.,* **274,** 175.
Forbes, D. A. Ward, M. J. and DePoy, D. L. (1991) *Ap. J. (Letters),* in press.
van Buren, D. and Norman, C. A. (1989) *Ap. J. (Letters),* **336,** L67.

RELATIVE ABUNDANCES OF THORIUM AND EUROPIUM IN HALO STARS

P. François
DASGAL,Observatoire de Paris, 61, Av. de l'Observatoire
F-75014 Paris

Recently, Butcher (1987) and Morell et al. (1991) used the abundance of the long lived isotope ^{232}Th (half life 14 Gyrs), compared to that of a stable element neodymium, to determine an upper limit for the age of the Galaxy. However, this method suffers from the fact that neodymium is partly s-process and partly r-process whereas thorium is pure r-process. Europium which is pure r-process may be used to avoid this problem. This poster presents new measurements of the thorium and europium abundances in halo stars.

Observations were carried out partly at ESO (La Silla) with the CAT+CESand partly at the Coudé focus of the 3.6m CFH telescope

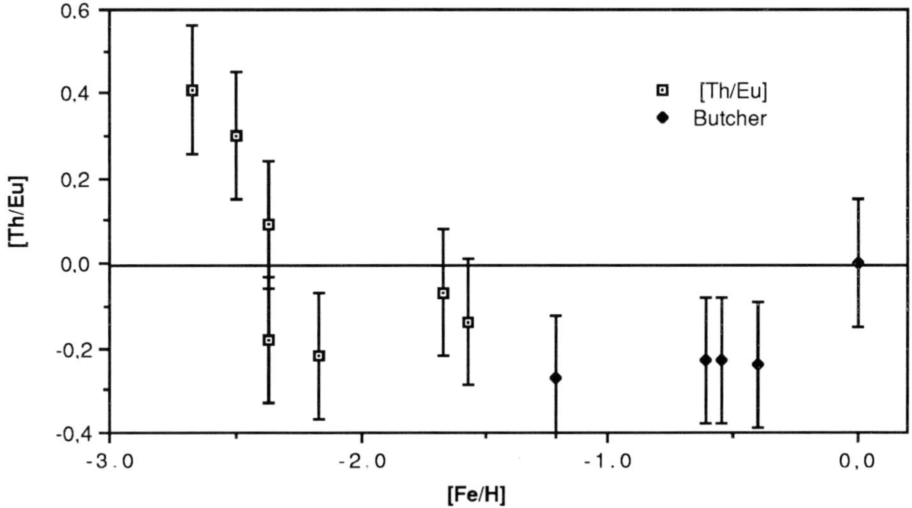

The Th/Eu radio-chronometer: Thorium and Europium are both r-process elements, so they alleviate one of the difficulties encountered with the Th/Nd chronometer. On figure 3, we plotted [Th/Eu] as a function of [Fe/H] used as a reference element. As Eu is a stable element whereas Th is not, we would expect a decreasing ratio Th/Eu as a function of decreasing metallicity (in the frame of a simple model) but we observe the contrary for the most metal poor objects. This could mean that effects of a different chemical history for these 2 elements are stronger than the effects of radioactive decay.

He ABUNDANCE IN RED GIANTS

J.A. DE FREITAS PACHECO and R.D.D. COSTA
Instituto Astronômico e Geofísico, Universidade de São Paulo, Brazil

Abstract. The He abundance in red giants is not well known. Moreover, some He enhancement in the stellar atmospheres is expected to occur as a consequence of distinct dredge-up episodes. In the present work is reported and discussed the He abundance for 5 symbiotic nebulae (Hen 2-38, RR Tel, RX Pup, HM Sge, KX TrA). Self-absorption effects in both hydrogen and HeI lines were taken into account.

Key words: Symbiotic stars - Chemical abundances

1. Introduction

The analysis of the helium emission lines in symbiotic stars, which are believed to be wide binary systems including a hot ionizing source, allow an estimate of that element in red giants.

The observational data were obtained at the National Laboratory for Astrophysics (Brazopolis - Brazil) and will be discussed elsewhere (de Freitas Pacheco and Costa 1991a).

2. Method of Analysis and Results

The determination of the helium abundance in symbiotics is quite problematic. First, because hydrogen Balmer line ratios often suggest self-absorption effects. They were taken into account following the scheme by Netzer (1975), which allows determination both the optical depth at $H\alpha$ and the colour excess $E(B-V)$.

The emissivity of each HeI line was estimated according to the results of Almog and Netzer (1989). The He^{+2} concentration was derived from the intensity ratio $HeII\lambda 4686$ with respect to $H\gamma$.

The present values for the helium abundance in red giants, which are components of symbiotic systems, are comparable to those observed in planetary nebulae.

RR Tel, RX Pup and V1016 Cyg have an average ratio $\frac{(C+N)}{O} = 0.64$, which compares quite well with the solar value. The average helium abundance for these stars is $\frac{He}{O} = 0.093$. These systems have probably undergone only the first dredge-up. On the other hand, HM Sge, KX TrA and HBV 475 have higher carbon abundance and, on the average, $\frac{(C+N)}{O} = 1.05$. The average helium abundance is also higher in these systems, namely $\frac{He}{H} = 0.11$. These characteristics suggest that these last three objects are more evolved, being probably AGB stars contaminated by He-burning products.

References

Almog,Y. and Netzer,H.(1989) Mon. Not. R. astr. Soc.**238**,57
De Freitas Pacheco,J.A. and Costa,R.D.D. (1991a) - submitted to publication.
Netzer,H. (1975) Mon. Not. R. astr. Soc. **171**, 395.

MASS LOSS FROM CENTRAL STARS OF PLANETARY NEBULAE WITH WC SPECTRUM

J.A. DE FREITAS PACHECO and R.D.D. COSTA
Instituto Astronômico e Geofísico, Universidade de São Paulo, Brazil

Abstract. We derived mass loss rates for 5 central stars (CS) having WC spectrum:NGC 5315, BD+30 3639, He2-99, He2-113, SwSt-1. The values are substantially higher than other previous estimates based on UV data. We consider that those stars have winds in which hydrogen is highly deficient (or absent) with helium and carbon being the dominant elements. The new mass loss rates reduce the timescale for the evolution along the horizontal track, but discrepancies with kinematic timescales still persist.

Key words: Planetary nebulae - Wolf-Rayet stars - Mass loss

1. Introduction

Post-AGB stars seem to have mass loss rates which are substantially higher than the predicted values using empirical formulae. UV spectra of CS's indicate that most of these stars are losing mass through a fast wind. The analysis by Cerruti-Sola and Perinotto (1985) led to the conclusion that all the CS's in their sample with a WR spectrum have a detected wind in the UV range.

In this work we estimate the mass loss rates for those 5 CS's considering that helium and carbon are the dominant elements in the chemical composition of the wind. Consequences for CS evolution are analysed later.

2. Mass Loss Rates and Results

If we assume that the helium lines are formed mainly by recombination and cascade, then the luminosity in a given transition can be expressed in function of the mass loss rate. Using such an equation and a velocity profile, we calculate the mass loss rate, finding 3.5×10^{-7}, 2.4×10^{-6}, 1.3×10^{-6}, 1.4×10^{-6}, 3.7×10^{-7} respectively for N5315, BD+30, He2-99, He2-113 and SwSt-1. These results show a clear trend in the sense that nebulae with a high $\frac{N}{O}$ ratio are associated with more massive CS's. We observed also a tendency of a decreasing mass loss rate as the gravity increases due to the contraction of the star, confirming the trend derived from UV data.

These rates of mass loss reduce the timescale along the horizontal track. A better agreement with the kinematic ages is found for some objects like N5315, but discrepancies still persist as Pottasch (1987) has emphasized.

An analysis of this problem with a larger sample including CS's with different spectral types is currently being done (Costa and de Freitas Pacheco 1991).

References

Cerruti-Sola,M., Perinotto,M. (1985), Ap.J. **291**, 237.
Costa,R.D.D. and de Freitas Pacheco,J.A. (1991) in preparation.
Pottasch,S.R. (1987) in *Planetary and Proto-Planetary Nebulae : From IRAS to ISO.* Reidel, Dordrecht pg. 79.

MODELS OF EMISSION LINE REGIONS AROUND CENTRAL CLUSTER GALAXIES

A. C. S. Friaça
Instituto Astronômico e Geofísico - U.S.P.
C.P. 9638, 01065 São Paulo, S.P. - Brasil

ABSTRACT. Hydrodynamical time-dependent models allows to follow the evolution of condensations in cooling flows until the phase of optical line emission.

The line emission flux ratio [NII]λ6583/Hα would allow to divide the optical filaments of cooling flows into two quite distinct classes, class I with $<$[NII]/H$\alpha >$ = 2.0 , and class II with $<$[NII]/H$\alpha >$ = 0.9 (Heckman *et al.* 1989). In this work we consider time-dependent models for the optical filaments including the effects of shocks and photoionization by soft X-rays produced in the cooling gas as well as by an OB star population formed in the cooling flow.

The evolution of the cooling condensations is obtained by time-dependent hydrodynamical calculations. The ionization state of the gas is obtained by solving the ionization equations during each time step using the code SUMA (Viegas-Aldrovandi and Contini 1989). When compared to the observations, the pure shock models give inconsistent ration lines and too weak luminosities. Therefore, two classes of models were built including the photoionization by the soft X-rays produced in the surrounding cooling flow: the class A models were applied to A1795 (host of class II filaments) at r = 20 kpc and the class B models to A496 (host of class I filaments) at r = 3 kpc. The values of the local X-ray flux and of the mass removal efficiency are derived from evolutive models for the cooling flows of A496 and A1795 (Friaça 1991). In class A models, we have also included the photoionization by a OB star population formed in the condensation. As a result, $L_{H\alpha}$ is greatly increased. The ratio for the [NII] emission is typical of class II filaments, and the other line ratios are also consistent with this class. Only a star formation efficiency much less than 10 % is required to fuel the low luminosity of the extended filaments, far from the center. In class B models, $L_{H\alpha}$ is not much increased, but it is adequate to class I filaments, and the derived [NII] line ratio has the correct value for this class. On the other hand, the higher X-ray flux in cooling flows with class II filaments could allow that also models with photoionization by only the ambient soft X-rays may be applicable to this class.

REFERENCES

Friaça, A.C.S. 1991, *A. & A.*, submitted.
Heckman, T.M., *et al.* 1989, *Ap. J.*, **338**, 48.
Viegas-Aldrovandi, S.M., and Contini, M. 1989, *A.& A.*, **215**, 253.

QSO ABSORPTION LINES AND THE GAS AND STAR CONTENT OF HIGH REDSHIFT GALAXIES

U. Fritze - v. Alvensleben, H. Krüger, K. J. Fricke
Universitätssternwarte
Geismarlandstr. 11
W–3400 Göttingen
Germany

ABSTRACT. We present a unified model for the detailed chemical evolution of individual elements from ^{12}C to ^{56}Fe **and** the photometric properties of galaxies via spectral evolutionary synthesis. Observations of narrow, heavy - element QSO absorption lines show an increase in the number of $MgII$ systems per redshift interval for redshifts $0 < z_r \lesssim 1.5$ and a decrease in the number of CIV systems for $1.3 \lesssim z_r \lesssim 4.1$. Both can be understood in terms of our galaxy evolution model accounting for SNI contributions which at the same time gives information about the structure of the Universe and about the IMF and star formation history in the intervening absorber galaxies. The spectrophotometric aspect of our unified model predicts spectral and photometric properties of these galaxies testable by optical identifications.

RESULTS. The abundances of both $^{24}Mg(z_r)$ and $^{12}C(z_r)$ trace the evolution of the global metallicity $Z(z_r)$ within the statistical error bars of the observational data, despite the partly different nucleosynthesis sites of these elements. Our model agrees well with the $MgII$ & CIV observations of Steidel et al. (1989) and Steidel (1990) for $(H_0,\ q_0,\ z_{form}) = (50,\ 0.5,\ 5)$. This shows that the metallicity maximum at some redshift $1.1 \lesssim z_r \lesssim 1.5$ should be real. The CIV data trace the chemical enrichment process while in $MgII$ a heavy element dilution effect is seen for $z_r \lesssim 1.4$ caused by the large number of low mass stars ($\sim 1\ M_\odot$) dying at late epochs giving back unenriched gas only. The constraints on the cosmological parameters, $50 \lesssim H_0 \lesssim 70$, $\Omega_0 \simeq 1$, $\Lambda_0 = 0$, and $5 \lesssim z_{form} < 10$, are consistent with those obtained from comparison of our spectrophotometric results with observations of high redshift galaxies ($z_r \lesssim 2$).
For the heavy element dilution effect to show up the gas content must be $<$ 10 %, the SFR must have been declining strongly enough with time : $\Psi \sim e^{-t/t_*}$ with $t_* \sim 1.0 \pm 0.2 Gyr$ and the IMF must be steep enough at high masses (of a Scalo rather than a Salpeter form) in the (halos of the) intervening galaxies. These results are independent of the cosmological model, the most distant galaxies seen in CIV absorption at redshifts $z_r \sim 4.1$ have evolutionary ages of only $\sim 1\ Gyr$.

Fritze - v. Alvensleben, U., Krüger, H., Fricke, K. J., Loose, H.-H. 1989, A&A **224**, L1
Fritze - v. Alvensleben, U., Krüger, H., Fricke, K. J. 1991, A&A **246**, L59
Steidel, C. C. 1990, Ap. J. Suppl. **72**, 1
Steidel, C. C., Sargent, W. L. W., Boksenberg, A. 1988, Ap. J. Letters **333**, L5

This work was supported in part by DFG grants Fr 325/28-1-2 and by the Verbundforschung Astronomie through BMFT grant WE-010 R 900 40.

METAL ABUNDANCES OF MAGELLANIC CLOUD CLUSTERS

D. GEISLER, N. SUNTZEFF
Cerro Tololo Inter-American Observatory
Casilla 603, La Serena, Chile

M. MATEO, J. GRAHAM
Observatories of The Carnegie Institution of Washington
813 Santa Barbara Street, Pasadena, CA 91101

ABSTRACT. High signal-to-noise ratio, medium resolution spectra have been obtained for ~8 giants in each of 18 LMC clusters with the CTIO 4-m multifiber ARGUS spectrograph. In addition, Washington CCD photometry has been obtained for ~50 SMC and LMC clusters with the CTIO 4-m and 1.5-m from which abundances can be obtained for ~25 giants per cluster. The derivation of metal abundances from these data will be discussed and some preliminary results presented.

1. Introduction

Magellanic Cloud clusters offer a unique opportunity to study chemical abundances in two galaxies whose chemical evolution and cluster formation histories are clearly different from that in our own Galaxy, as well as from each other. Unfortunately, good abundance determinations are still lacking for many MC clusters despite their crucial role in age-metallicity and distance calibrations. We have begun a large-scale program to determine abundances for many giants in each of a large sample of clusters using a variety of photometric and spectroscopic techniques.

2. Data

The CTIO 4-m ARGUS multi-object fiber-fed spectrograph is ideally suited to studying the composition of many MC cluster giants simultaneously. We have obtained data for 18 LMC clusters. The spectra range from ~4800 - 6400 Å, with a resolution of ~6 Å, and are of excellent signal/noise ratio. The strong absorption lines due to Mg, Fe and Na will be used to measure metal abundance, as outlined in Faber et al. (1985, ApJ Suppl., 57, 711). This method shows good abundance discrimination in most metallicity regimes. However, there does seem to be some confusion between intermediate-metallicity giants. Washington CCD photometry has been shown to be a useful technique for investigating metal-abundances of a large number of giants in MC clusters (Geisler 1987 AJ, 93, 1081). We have now obtained Washington CCD observations for a total of 16 SMC and 34 LMC clusters. Results indicate that abundances can be determined for a large number of giants (typically ~25) per cluster.

A COMPARISON BETWEEN A COLOR EVOLUTION MODEL AND A NEW SAMPLE OF LMC CLUSTERS: FORMATION RATE

L. Girardi and E. Bica
Instituto de Física-UFRGS, Porto Alegre, RS, Brazil

We present a preliminary comparison between a photometric cluster model and the enlarged sample of 624 LMC clusters recently observed in integrated UBV photometry by Bica et al. (1991). The model was computed with Maeder and Meynet (1991) isochrones in the mass range 120 – 0.8 $M_{\odot}$ with $Z_{\odot}$, and complemented down to 0.15 $M_{\odot}$ with VandenBerg et al. (1983) models. It includes all the essential phases of stellar evolution up to $\sim$ 1 Gyr, but lacks the HB and AGB phases for low mass stars. We considered IMF slopes $x = 2.5$ and 1.35, encompassing results from recent CCD data for LMC clusters (Richtler et al. 1991). The red supergiant (RSG) phase (t = 10 Myr) was reproduced on the observed diagram $(U - B)$ vs $(B - V)$, although shifted to the red by metallicity and/or mass loss effects. $x = 2.5$ describes better the RSG phase for the clump of massive clusters. On Fig. 1 the theoretical fading lines for $x = 1.35$ and 2.5 are superimposed on the data in the plane V vs. $\log t$ (SWB types). Mass strips are limited by fading lines corresponding to different cluster masses. We estimated the cluster formation rate (CFR) $\Delta N/\Delta t$ in two different ways (Fig. 2): 1) By counting the clusters in age bins in mass limited strips as shown in Fig. 1. 2) By counting the clusters above the observational cutoff at $V \simeq 13$, and correcting by the number of clusters below the cutoff which should be encompassed by a fading line starting at V = 13 for the youngest age bin. Fig 2. provides a comprehensive estimate of the CFR, although affected by dynamical effects such as evaporation of stars and tidal disruption (Wielen 1988). More work is necessary to disentangle the two effects.

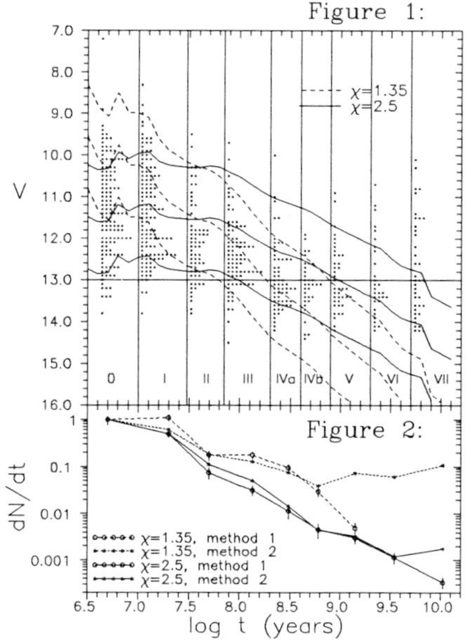

REFERENCES

Bica, E., Claria, J.J., Dottori, H., Santos Jr., J.F.C., Piatti, A. 1991, ApJ Letters, in press

Maeder, A., Meynet, G. 1991,A&AS, in press

Richtler, T., de Boer, K.S., Sagar, R. 1991, The Messenger, 64, 50

VandenBerg, D.A., Hartwick, F.D.A., Dawson, P., Alexander, D.R. 1983, ApJ, 266,747

Wielen, R. 1988, in Globular Clusters Systems in Galaxies, IAU Symp. 126, eds. J.E. Grindlay, A.G.D. Philip, (Reidel, Dordrecht), p. 393

Radial population synthesis and the ionization of gas in elliptical galaxies

Paul Goudfrooij and Bob van den Hoek
Astronomical Institute, University of Amsterdam
Kruislaan 403
NL-1098 SJ Amsterdam

Introduction

A remarkable discovery of recent years has been the detection of various kinds of interstellar matter (ISM) in elliptical galaxies. The presence of dust patches, ionized gas, and hot X-ray gas have proven to be quite common in ellipticals which were once thought to be simple structures, devoid of gas and dust. A review of ISM in ellipticals has recently been published by Forbes (1991). However, the origin and fate of ISM in ellipticals is not fully understood. Its origin may be stellar mass-loss, condensation of hot X-ray gas pervading the galaxy, or merging with gas-rich galaxies. Recently, the merger picture receives much attention, since (i) the ionized gas usually is dynamically decoupled from the stars, and (ii) the great variety in radial line-strength gradients found among ellipticals also favours merging as formation process (cf. Gorgas et al. 1990). To study the origin and fate of ISM in ellipticals we are currently undertaking an optical survey of a complete sample of ellipticals ($B_T^0 < 12$, cf. Goudfrooij et al. 1990). An important result of our extensive CCD imaging program is that a relevant fraction of the sample objects exhibits dust patches within Hα+[NII] line-emitting filaments. This common occurrence can be easily accounted for if the dust and gas have an external origin. In these cases, the extended line emission often has a peculiar distribution and is more sharply peaked at the nucleus than is the stellar continuum. Furthermore, all of these ellipticals exhibit a compact flat-spectrum radio source in their nucleus, suggesting that this nuclear activity also has an external origin. In this respect it would be interesting to know the excitation mechanism of the gas.

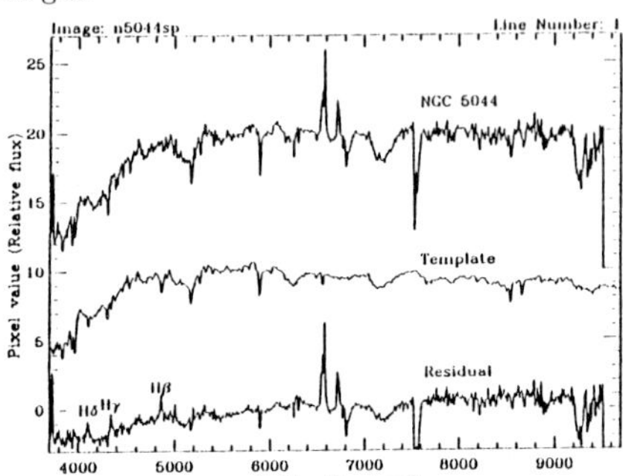

Fig. 1. *Nuclear spectrum of NGC 5044, its best-fitting template, and its residual pure emission spectrum. Notice the dramatic change in e.g. the Balmer line intensities. A suitable constant was subtracted from the template for visualisation*

Studying the Excitation Mechanism using Population Synthesis

Competing possibilities for the gas in ellipticals are shocks, photoionization by either an active nucleus, hot stars within the filaments, or coronal X-ray photons, and ionization by hot electrons in the coronal plasma. A major difficulty in deciding which mechanism is at work is the fact that the emission-line spectrum is superposed on a strong-lined stellar population, making the detection of e.g. faint Balmer line-emission a hard task. However, the use of a population synthesis program based on star cluster spectra[1] proves to be a powerful method to overcome this problem. In Fig. 1 a spectrum of the nucleus of the merger candidate NGC 5044 (cf. Goudfrooij 1991) is shown together with its best-fitting template, and the resulting pure emission spectrum. More information on the synthesis program can be found in van den Hoek & Goudfrooij (this conference). A major advantage of this method is the ability the trace radial changes in line intensity ratios while performing the radial population synthesis, probably leading to important clues concerning the excitation of gas in ellipticals. A more thorough account of this study will follow in due course.

References

Forbes, D.A. 1991, MNRAS, 249, 779
Gorgas, J., Efstathiou, G., Aragón, A. 1990, MNRAS, 245, 217
Goudfrooij, P. 1991, The ESO Messenger, 63, 42
Goudfrooij, P., Nørgaard-Nielsen, H.U., Hansen, L., Jørgensen, H.E., de Jong, T. 1990, A&A, 228, L9

[1] We gratefully acknowledge the ability to use the synthesis program written by E. Bica (e.g. 1988, A&A, 195, 76)

A CCD SEARCH FOR FAINT HIGH-LATITUDE CARBON STARS: DWARFS AMONG THE GIANTS

PAUL J. GREEN, BRUCE MARGON, SCOTT F. ANDERSON, and PETER M. GARNAVICH
University of Washington
KEM COOK
Lawrence Livermore National Laboratory
and
D. JACK MACCONNELL
Computer Sciences Corp., Astronomy Programs

ABSTRACT. We are acquiring a large-area sample of faint, high-latitude carbon star candidates for the study of halo dynamics by using an intermediate-band color system with CCDs in efficient survey modes. Except for one odd dwarf carbon (C) star, G77–61, it has long been assumed that these faint C stars are distant giants. However, we recently demonstrated that three more faint C stars are high proper motion objects, and therefore dwarfs. Now we are completing a proper motion survey of known faint high-latitude C stars to search for additional C dwarfs. The CCD and proper motion surveys together will place significant limits on the space density of C stars, be they dwarfs or giants.

Carbon stars are readily recognizable from their strong C_2 and CN absorption bands. A sample of faint high-latitude C giants would provide an excellent dynamical tracer of the outer halo, especially since other tracers (*e.g.*, K giants, RR Lyr stars, and globular clusters) yield poorly reconciled estimates of halo dynamical parameters. Several dozen faint high-latitude C stars have been detected from R=12 to 16 on low-dispersion objective prism plates of Sanduleak and Pesch (1988, ApJS, 66, 387). In a complementary survey at Kitt Peak, we have used the 0.9m telescope and CCDs to image 55 deg^2 of sky to V$\approx$ 18. This limit corresponds to a heliocentric distance about 40 kpc for the faintest objects thought to be typical of halo Pop I carbon stars. The photometric technique we use has been shown (Cook and Aaronson 1989, AJ, 97, 923) to efficiently distinguish C stars from other late-type stars by using intermediate-band filters, one ("77") centered on a region of TiO absorption at λ7750, and the other ("81") at λ8100 on a CN absorption band.

To date the only known flaw in the otherwise promising technique of employing faint C stars as outer halo tracers has been the puzzling existence of one dwarf C star: G77–61 (Dearborn *et al.* 1986, ApJ, 300, 314) has V= 13.9 along with a high proper motion, and thus is of main sequence luminosity. Recently, however, Green *et al.* (1991, ApJ Letters, 380, L31) recognized three more such dwarf C stars, at estimated distances of 170, 100, and 400 pc, respectively. Together with radial velocities (Bothun *et al.* 1991, AJ, 101, 2220), these distances imply space motions consistent with halo kinematics for all four dwarf C stars known to date. These distances also suggest space densities high enough that *C stars may be dwarfs more often than giants.*

Using the HST Guide Star Catalog and a digitization of the original Palomar Observatory Sky Survey, we are measuring proper motions for all faint high-latitude C stars of which we are aware. We will use spectroscopy of the resulting C dwarfs to estimate their metallicities, binarity, and possibly the source of their high carbon abundances. Carbon dwarfs of similar magnitude to G77-61 are detected in our CCD survey to about 300 pc. Although our spectroscopic followup is not complete at this writing, very low surface densities are already indicated. Considerable care will clearly be needed in future studies to determine which faint C stars are in fact distant objects. *JHK* colors for four presently known C dwarfs are consistent with the colors of other late-type field dwarfs, and so are likely to provide a convenient luminosity discriminant.

A Spectrophotometric Investigation of Dwarf Elliptical Galaxies

MICHAEL D. GREGG
Mt. Stromlo and Siding Spring Observatories
Private Bag, Weston Creek P.O.
A.C.T. 2611, Australia

Abstract.
Spectrophotometry of compact (M32-like) dwarf ellipticals (dE) and nucleated dwarf ellipticals (dE,N) in the Virgo and Fornax clusters has been used to investigate the stellar populations of these two types of dwarf galaxy and their relation to larger ellipticals.

The compact dwarf ellipticals have colors and Mg line strengths comparable to much brighter ellipticals yet have much lower central velocity dispersions. They have somehow managed to achieve high metallicities without the deep gravitational potentials of larger ellipticals. A striking example is the comparison of VCC 344 (Binggeli *et al.* 1985) with NGC 4472 which have identical Mg strength but their central velocity dispersions differ by a factor of 3-4. The structure of compact dE's has been attributed to tidal stripping by a large neighbor, as in the case of M32 and M31 (King 1962; Faber 1973). Arguments against this process have been presented by Nieto & Prugniel (1984). Tidal stripping does seem implausible in some circumstances where the Mg strength of the dwarf implies that it was as large or larger than its companion or when the dwarf exists in relative isolation as in the case of VCC 1627.

The nucleated dwarfs have relatively low Mg strength, similar to metal-rich globular clusters. Yet their 4000Å break is too small for their Mg strength compared to metal rich globular clusters. This can be interpreted as due to young or intermediate age populations, consistent with previous photometric studies and some limited spectroscopy (Caldwell & Bothun 1987, Zinnecker & Cannon, 1985, Mould & Bothun 1988). At least one dE,N in the present sample has a blue, strong Balmer line population and OII 3727 emission in its inner nucleus and a redder, weaker Balmer line population outside this region. This can be due only to a very young population, perhaps even ongoing star formation and suggests a rather recent origin for the nuclei. The similarity in appearance between the dE,N's and local group member NGC 205 (Zinnecker & Cannon 1985) may extend to their stellar content and star formation histories.

References

Binggeli, B., Sandage, A., & Tammann, G.A. 1985, AJ, 90, 1681
Caldwell, N., & Bothun, G.D. 1987, AJ, 94, 1126
Faber, S.M. 1973, ApJ, 179, 423
King, I.R. 1962, AJ, 67, 471
Mould, J.R. & Bothun, G.D. 1988, ApJ 324, 123
Nieto, J.L. & Prugniel, P. 1987, AA, 186, 30
Zinnecker, H. & Cannon, R.D. 1985, in *Star Forming Dwarf Galaxies and Related Objects*, ed. D. Kunth, T.X. Thuan, & J. Tran Thanh Van

QSO ABSORPTION-LINE SYSTEMS AND STAR FORMATION

Ruth B. Gruenwald, Sueli M. Viegas and Gustavo Detthow
Instituto Astronômico e Geofísico, Brazil

The properties of the QSO absorbing systems have been established by several large scale surveys. Observational results concerning the metal content, the distribution of the systems as a function of the redshift, and clustering, lead to two populations of intervening objects: (a) intergalactic clouds (Ly α systems), and (b) galaxies (heavy-element systems). More recently, observations have shown an analogy between QSO absorbers and gas-rich dwarf galaxies (York et al. 1990).

We have analyzed the origin for the QSO metal line systems by comparing observed blue lines to theoretical calculations of absorption in star forming HII regions. The HII region characteristics are obtained from the high excitation lines of C^{+3} and Si^{+3}; the low excitation line intensities (as CII, OI, NV, AlII, AlIII, SiII, SiIII and FeII) are used to check and constrain the model. Of the 20 systems analyzed (with $1.50 < z_a < 3.30$), 18 can be explained by this model. These 18 systems are associated with HII regions, ionized by an *O4* star, with a hydrogen density $n_H \leq 10$ cm^{-3} and chemical abundance $Z < Z_\odot/10$ (Viegas and Gruenwald 1991).

Now we present the results for 21 absorption systems at lower redshift ($1.20 < z_a < 2.30$) where the MgII and MgI lines are observable, in addition to CIV lines and (when available) CII lines. For 19 systems, the absorption line spectra can be fitted by HII region models with the same characteristics found in the previous sample.

The absorption spectra of gas-rich dwarf galaxies closely resemble those shown by QSOs (York et al. 1990). The absorption features of 11 galaxies have been analysed by the same method. For only 3 objects HII region models can reproduce the observed spectra if a higher density is assumed ($\sim 10^2$cm^{-3}). For most of the galaxies there must be a stellar contribution to the CIV absorption line.

Since the interstellar medium of the host galaxy seems to be mainly ionized, it is possible to make a rough estimate of a lower limit of $2M_\odot/yr$ to the star formation rate. This value is close to the average star formation rate for HII galaxies.

Although the model can be improved in its details, it is already clear that it offers a promising alternative origin for the QSO absorbing systems. They may provide a potential tool to probe the rate of star formation at high redshift as well as the chemical evolution of galaxies.

References

Viegas, S. M. and Gruenwald, R. B. 1991, ApJ, 377,39.
York, D. G., Caulet, A., Rybskii, P., Gallagher, J., Blades, J. C., Morton, D. C. & Wamstecker, W. 1990, ApJ, 351, 412

STELLAR ABUNDANCES IN THE OUTER GALACTIC DISK

Hugh C. Harris
U.S. Naval Observatory
P.O. Box 1149
Flagstaff, AZ 86002
USA

Current evolutionary models of the disk of the Milky Way incorporate radial inflows of (metal-poor) gas that affect the subsequent evolution of the inner disk. Therefore, knowledge of the metallicity of stars and/or gas as a function of radius and time is necessary for a complete understanding of disk evolution. Present data on stellar abundances (determined mostly from iron or iron-peak elements) for the outer disk are shown in Table 1. Results from different approaches have been converging and now give two important results: at $R_{GC} \sim 15$ kpc, the stellar disk is metal poor by about 0.6 dex relative to the solar neighborhood, and the data are consistent with a constant gradient of -0.07 kpc^{-1} from the solar neighborhood out to 15 kpc, where the data are becoming very sparse. Interestingly, the gradient of [O/Fe] measured with HII regions is the same. Because Cepheids have proven to be useful probes of abundances at large distances, efforts are underway to find additional distant Cepheids. The last line in the table indicates that stellar abundances in the Small Magellanic Cloud (with ages <5 Gyr) are remarkably similar to those in the outer disk, as had been tenatively suggested in many earlier studies.

TABLE 1. Mean Stellar Abundances

R_{GC}	Cepheids[1,2,3]			Open Clusters[4,5,6,7]			Supergiants[8,9,10,11,12]		
(kpc)	[Fe/H]	±	N	[Fe/H]	±	N	[Fe/H]	±	N
Solar Neighborhood:									
8.5	0.11			−0.04			0.13		
Outer Galactic Disk:									
12-13	−0.24	0.13	5	−0.50	0.10	5	...		
13-14	−0.37	0.12	8	−0.8	0.2	1	...		
14-16	−0.43	0.10	7	...			...		
>16	−0.88	0.23	2	−0.68	0.23	1	...		
Small Magellanic Cloud:									
	−0.54	0.15	45	−0.73	0.15	5	−0.7	0.1	11

[1] Harris (1981) AJ 86, 707.
[2] Harris and Pilachowski (1981) ApJ 282, 655.
[3] Harris (1981) AJ 86, 1192.
[4] Friel and Janes (1991) ASP Conf. Ser. 13, 569.
[5] Lennon et al. (1990) AA 240, 349.
[6] Geisler et al. (1991) preprint.
[7] DaCosta (1991) IAU Symp. 148, in press.
[8] Luck and Bond (1989) ApJS 71, 559.
[9] Russell and Bessell (1989) ApJS 70, 865.
[10] Spite et al. (1989) AA 222, 35.
[11] Dufton et al. (1990) ApJL 362, L59.
[12] Reitermann et al. (1990) AA 234, 109.

NGC 185 AND THE EXTENDED FABER-JACKSON RELATION

Enrico V. Held
Osservatorio Astronomico di Bologna

Tim de Zeeuw
Sterrewacht Leiden

Jeremy Mould & Alain Picard
Palomar Observatory, Caltech

It has recently been found that diffuse dwarf elliptical galaxies (dE), as well as low–luminosity normal ellipticals of relatively low surface–brightness, are supported by an anisotropic velocity distribution. New kinematic observations have been obtained of the dwarf elliptical galaxy NGC 185. The velocity dispersion is constant at 28 ± 8 km s^{-1} between $3''$ and $40''$, but may increase to about twice that value in the center. We find an upper limit of 10 km s^{-1} for the rotation along either axis, so that the velocity distribution is anisotropic. The derived M/L_B is ~ 3 in solar units.

Analysis of the kinematic and photometric data now available on dE's shows that giant ellipticals and dwarfs fall on a continuous sequence in the (L, σ)–plane. Figure 1 shows that most data points for dE's define a linear sequence extending the F–J relation for normal ellipticals (an exception is represented by the two faintest dwarfs, Draco and Ursa Minor). The slope corresponds to a relation $L \propto \sigma^{2.5}$, which is in close agreement with that predicted by supernova–driven galactic wind models without dark matter. More detailed results of this work will be published in The Astronomical Journal.

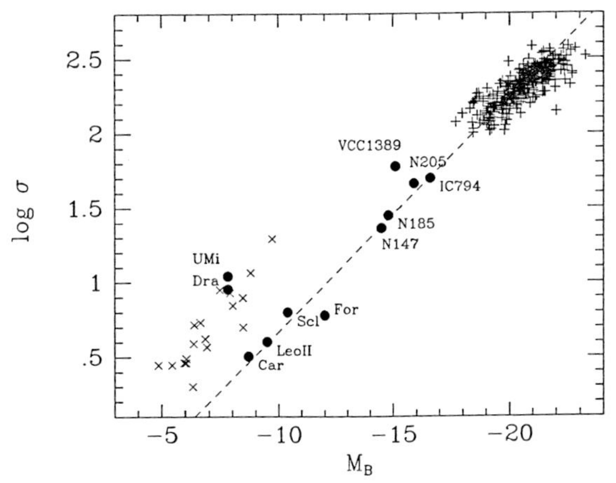

Figure 1: Velocity dispersion σ against absolute magnitude M_B, for dwarf elliptical and spheroidal galaxies (*filled dots*), normal ellipticals (*plus signs*), and globular clusters (*crosses*).
The *dashed line* represents a fit to Car, Leo II, Scl, For, NGC 185, NGC 205, NGC 147, and IC 794, and corresponds to the approximate relation $L \propto \sigma^{2.5}$.

STUDY OF STELLAR POPULATIONS USING NEURAL NETWORK TECHNIQUES

M. HERNANDEZ-PAJARES[1], E. MONTE[2]
1 Dp. Matemàtica Aplicada i Telemàtica
2 Dp. Teoría del Senyal i Comunicacions
Universitat Politècnica de Catalunya
Ap. 30002 Barcelona, Spain

1. Introduction

In this paper we consider a first application of the Learning Vectorial Quantification Neural method (LVQ) to the problem of studying and distinguishing between different populations within an stellar catalogue of the solar neighbourhood (a complete description can be found in Hernández-Pajares and Monte, 1991, *Artificial Neural Networks*, Ed. A.Prieto, *Lecture Notes in Computer Science* 540, Springer-Verlag, p.422). It consists, briefly, in the approximation of a set of vectors in a certain characteristic space that contains continuous elements. The representative points for every cluster are the centroids, calculated in such a way to minimize the distortion. Each of those can be labeled with integer numbers using a 2D representation that preserves the neighbouring property in the characteristic space: the Kohonen Map (Kohonen, 1988, IEEE Computer, **21** Nbr.3).

2. Calculations and results

The observational data considered is the Figueras (1986, Ph. Thesis, U. Barcelona) catalogue, which contains 12824 stars. The LVQ method has been applied working in the 14 dimensional characteristic space formed by 3-D position, velocity and residual velocity, jointly with spectral-photometric data. In the calculations we have taken a number of 8x8=64 14-dimensional centroids to be determined into the Kohonen map after 4×10^6 training iterations of the neural network.

A systematic, smooth and coincidental distribution of the distance and the residual velocity perpendicular to the galactic plane ($|W1|$) over the Kohonen map, appears as the main result. If we consider $|W1|$ in function of the distance to the galactic plane for the 64 centroids, we can distinguish between three groups of neighbouring centroids in the Kohonen map: (A) with $|W1| \leq 24$ Km/s, (B) with $24 < |W1| \leq 60$ Km/s and (C) with $|W1| > 60$ Km/s. These intervals deduced without any extra-statistical consideration, agree very well with the kinematic bins considered by Carney, Latham and Laird (1989, Astron. J. **97**, 423), related to the Galaxy populations: region (A) with a predominant Thin Disk component, (B) with the Thick Disk and (C) with the Halo population. In addition, jumps between the three intervals, especially the one between (B) and (C) could be concordant, without taking into account the bias of the sample, with a discrete nature of the Thick Disk; a matter of discussion nowadays (Gilmore, King and Van der Kruit, 1989, *The Milky Way as a Galaxy*, Geneva Obs., Switzerland).

THE SPHERICAL HARMONICS AS AN ALTERNATIVE TOOL FOR DETERMINING THE KINEMATICAL PARAMETERS OF THE LOCAL MILKY WAY

M. HERNANDEZ-PAJARES[1], J. NUÑEZ[2]
1 Dp. Matemàtica Aplicada i Telemàtica, U.P.C.
Ap. 30002 Barcelona, Spain
2 Departament d'Astronomia i Meteorologia, U.B.
Av. Diagonal 647,08028 Barcelona, Spain

1. Introduction

The main objective of this work has been the connection between the Ogorodnikov-Milne model (OM) and the Spherical Harmonic development (SH) for different kinematic observables. This new approach allow us to benefit from the high orthogonality, easy extension to high orders and no appearance of the distance in the SH expansions; and also to benefit from the intrinsic meaning of the OM parameters. This work (a complete description will appear in Hernández-Pajares & Núñez, 1991, Astron. J., submitted) completes the theoretical studies started in Hernández-Pajares & Núñez (1990a, Astrophys. Space Science **170**, 187 and 1990b, *Error, bias and uncertainties in Astronomy*, ed. Jaschek and Murtagh, Cambridge Univ. Press, 339).

2. Developments and results

When the velocity components are analyzed without taking into account the distance we are studying their projection on the celestial sphere. We have demonstrated that the angular part of those can be expressed as a SH finite series. Also to maintain that dependence after the projection is necessary and sufficient that the density of number of stars be a product of a radial and an angular function: i.e. the Separability hypothesis. Under this hypothesis we have established the analytical relationships between the SH coefficients and the OM parameters for six kinematic observables.

In order to know if a given stellar sample is separable we have designed a Test based on the Inertia Matrix. We have applied it to the Bright Star Catalogue (Hoffleit, 1982, 4th ed., New Haven, Yale University Observatory) with $m_V \leq 6$ and B-V ≥ 0, using the Bahcall and Soneira two-component model of the Galaxy (1980, Astrophys. J. Suppl. Ser. **44**, 73 and 1984, Astrophys. J. Suppl. Ser. **55**, 67). The result obtained clearly indicates that the complete sample fulfills the separability hypothesis. So it is feasible from an astronomical point of view.

Finally we have applied the new emergent strategy to a sample of more than 6000 stars with a residual velocity lower than 65 Km/s belonging the the kinematic catalogue compiled by Figueras (1986, Ph. Thesis, U. Barcelona). The results obtained are in agreement in several cases with others recent results of the literature and give additional estimations of galactic kinematic parameters. In the next future we will apply this procedure to small samples segregated in function of statistical and dynamic considerations (see for instance Hernández-Pajares and Monte, 1991, *Artificial Neural Networks*, ed. Prieto, *Lecture Notes in Computer Science*, Springer-Verlag, **540**, 422 and Cubarsí, 1990, Astron. J. 99, 1558.).

LUMINOSITY FUNCTIONS AND MASS FUNCTIONS FOR MASSIVE STARS: ASSOCIATIONS IN THE LARGE AND SMALL MAGELLANIC CLOUDS

Robert Hill
Department of Physics and Astronomy
Lousiana State University, Baton Rouge, LA, USA

Barry F. Madore
NASA/IPAC Extragalactic Database
IPAC/Caltech/JPL, Pasadena, CA , USA

Wendy L. Freedman
The Observatories
Carnegie Institution of Washington
Pasadena, CA , USA

UBV CCD photometry has been obtained for 14 OB associations in the Large and Small Magellanic Clouds. The data have been used to construct color-magnitude diagrams for the purpose of investigating the massive-star content of these extragalactic associations.

The color excesses derived for the various associations range from $E(B-V) = 0.01$ to 0.26 mag for the LMC, and from $E(B-V) = 0.06$ to 0.25 mag for the SMC associations. A detailed analysis and simulation of the effects of systematic and random errors in the photometry indicates that the observed scatter in the color excesses of individual stars can only be explained by the existence of differential reddening within many of the associations, in excess of a foreground Galactic component.

The main sequence luminosity functions of the Magellanic Cloud associations are remarkably similar over the luminosity range $-4 < M_V < -1$ mag. The slope of the luminosity function is flatter (0.3) than published luminosity functions slopes (0.7) for brighter ($M_V < -5$ mag) stars in the Milky Way and other Local Group galaxies.

The slope of the initial mass function (IMF) is $\Gamma = -2.0 \pm 0.5$ for masses $9 < M < 60 M_{\odot}$. There is no statistically significant evidence for a variation in the initial mass functions among the associations in any one galaxy. Nor is there any evidence for a difference between the slopes of the initial mass functions between the two galaxies; the implication of this last conclusion being that there is no dependence of the slope of the IMF on metallicity at this present epoch. There is however some suggestion that the slope of the IMF is leveling off for masses $M < 10 M_{\odot}$, but deeper and more complete studies are needed to confirm this trend.

This work was supported in part by the Jet Propulsion Laboratory, California Institute of Technology, under sponsorship of the National Aeronautics and Space Administration's Office of Space Science and Applications.

THE STELLAR POPULATIONS OF NEUTRON AND STRANGE STARS IN THE GALAXY

J.E.HORVATH[1] and G.A.FOGLIA[2]
[1]*Instituto Astronômico e Geofísico, Universidade de São Paulo. Av. M. Stéfano 4200 (04301) São Paulo SP BRASIL*
[2]*Instituto de Física, Universidade de São Paulo. Cx. Postal 20516, (01498) São Paulo SP BRASIL*

If strange matter [1] is the most stable state of cold, dense hadronic matter, the actual internal composition of compact stars would depend on the timescale τ_{ss} for the decay $n \rightarrow uds + energy$ to occur [2]. We have modelled the depletion of the neutron star population $N_{ns}(t)$ by the simple law

$$\frac{dN_{ns}}{dt} = SNII + AIC - BH - \frac{N_{ns}}{\tau_{ss}}$$

where $SNII$, AIC and $-BH$ represent birth in type II supernovae, accretion induced collapse of a white dwarf and collapse to black hole respectively. Assuming no decay of their magnetic fields (which would otherwise mask the features of a converted star), we have found two relevant solutions: $\tau_{ss} \gg 10^{10} yr$ (strongly suppressed conversion, all compact objects ns) and the other $\tau_{ss} \simeq N_{ns}(t_f)/K 10^{10} yr$, with $K = const.$ the net birthrate and $t_f \simeq$ age of the Galaxy. Under fairly different assumptions τ_{ss} turns out to be $\simeq 10^9 yr$ for the last case. Our conclusion is that the gross mismatch between τ_{ss} and the microscopic strangeness-changing reactions ($\simeq 10^{-8} s$) do not favour a mixed population, therefore suggesting a prompt birth of ss in $SNII$ explosions [3]. These objects should be then asked to provide a model for *any* pulsar observation, including glitches [4,5]. If, to avoid the above conclusion, we postulate accretion from a companion as the cause of the conversion we found that the critical density should lay within a factor ≤ 2 for any massive neutron star model. An extended version of this argument can be found in [6].

REFERENCES

1) Witten,E. (1984) *Phys.Rev.***D30**, 272.
2) Benvenuto,O.G. and Horvath,J.E. (1990) *M.N.R.A.S.* **247**, 584.
3) Benvenuto,O.G. and Horvath,J.E. (1989) *Phys.Rev.Lett.* **63**, 716.
4) Benvenuto,O.G.; Horvath,J.E. and Vucetich,H. (1990) *Phys.Rev.Lett.***64**, 713.
5) Alpar,M.A. (1987) *Phys.Rev.Lett.***58**, 2152.
6) Horvath,J.E. and Foglia,G.A. (1991) *Remarks on the relative populations of neutron and strange stars*, submitted for publication.

ABUNDANCES OF LONG-PERIOD VARIABLES: INITIAL RESULTS

SHAUN M.G. HUGHES AND PETER R. WOOD
Anglo-Australian Observatory, Coonabarabran, NSW 2357, Australia
Mt Stromlo and Siding Spring Obs., Australian National University, Canberra ACT, Australia

ABSTRACT. The kinematics of the long-period variables (LPVs), both in the LMC (Hughes et al. 1991 AJ 101, 1304) and in the Galaxy (eg. Feast 1963 MNRAS 125, 367) shows that there exists an Age-Period relation for these objects. Low-resolution (FORS) spectra have been obtained for a sample of SMC, LMC and Galactic LPVs using the Anglo-Australian Telescope, with the ultimate aim of using the bandstrengths of the TiO bands, in combination with a temperature indicator derived from the continuum, to estimate their metallicities in a method similar to that used by Mould and Bessell (1982 ApJ 262, 142), but using the TiO bandhead at 8480 Å. Metallicity estimates of these variables are essential, both in (1) verifying the finding of Wood et al. (1991 *The Magellanic Clouds*, eds. R.Haynes and D.Milne, (Kluwer: Dordrecht), p259) who theorise that the scatter in the mean Period-Luminosity and Period-Color relations of the LPVs in the LMC (Feast et al. 1989 MNRAS 241, 375) could be produced by a scatter in abundance, and (2) in investigating whether there exists a correlation between abundance, kinematics and period (age) for the LPVs in the LMC, as appears to be the case in the Galaxy.

Unfortunately, while the expected trend in bandstrength versus metallicity (Figure 1) is seen for the objects observed in 47 Tucanae, the LMC and the SMC, those objects observed in the solar neighbourhood run counter to this trend, indicating a problem with the method, most probably related to the continuum (temperature) estimate for very late M stars.

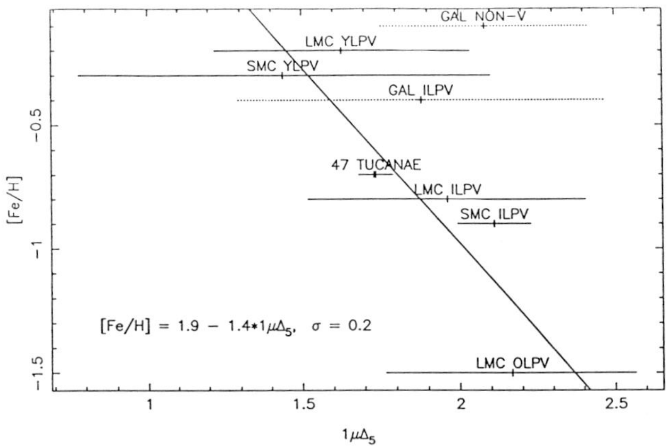

Figure 1. The distribution of observed continuum slopes (for a TiO bandstrength set at 0.5) against expected abundance [Fe/H], for various groups of LPVs and non-variable M giants (NON-V) in the solar neighbourhood (GAL), 47 Tucanae, the LMC and the SMC. The LPVs were grouped into classes of short period (~10Gy OLPVs), intermediate period (~3Gy ILPVs) and long period (~1Gy YLPVs). The error bars represent ± 1-σ (rms). A least squares fit to the LMC, SMC and 47 Tucanae data only (solid bars), is shown by the formula and the straight line.

THE TIME DEPENDENT RADIO SOURCES IN CEPHEUS A

V.A. Hughes
Queen's University
Kingston, Ontario K7L 3N6
Canada

The radio object Cepheus A is a known region of star formation, in a region of very heavy optical obscuration. It has been observed at the VLA since 1980, and is known to consist of 2 lines of about 13 objects. In particular, the central region has been monitored since 1988, at L, C, and U-band, with resolutions of 1."0, 0."3 and 0."1, corresponding to linear resolutions of 700, 200 and 70 au. The presence of a highly variable Source 8 had been known since 1980, and was known to appear and disappear in a period of less than 1 year. However, a new and much more intense Source 9 was seen 1988 November 2. Over the period 1990 March 13 – 1990 May 8, it showed increases in flux density at C-band by a factor of nearly six times over a period of 50 days. Spectra have been obtained, and show that the flux density peaks at about 5×10^9 Hz, and an increase in flux density is accompanied by an increase in the frequency of the peak. The variable sources are not resolved at 0."1 at U-Band. This strongly suggests gyrosynchrotron emission from a region of size $\sim$ 1 au, magnetic field $\sim$ 100 G, temperature $\sim 10^8$ K and density $\sim 10^6$ cm^{-3}. The variations in flux density are then due to increases in magnetic field, as might occur if the source were a protostar, or pre-main sequence star, which is rotating and shedding its magnetic field, by converting magnetic flux energy into particle energy. This would provide also a mechanism for molecular outflows. The observations and model will be described.

(See: Hughes, V.A., Astrophysical Journal, 1991 December 10).

Ages and metallicities of M31 star clusters

P. JABLONKA[1], D. ALLOIN[1] AND E. BICA[2]

1. Observatoire de Meudon, D.A.E.C, URA 173 CNRS, 92195 Meudon, France
2.Instituto de Fisica, UFRGS, Caixa Postal 15051, 9500, Porto Alegre, Brazil

October 16, 1991

Abstract. We present recent analysis of star star clusters in M31

Recently, Jablonka, Alloin and Bica (1991) (herafter JAB) aimed at studying the properties of M31's star clusters and at enlarging the data base of star cluster integrated properties (Bica and Alloin, 1987) at high metallicities, for population synthesis purpose.

We have obtained spectra for 9 clusters in M31, 7 globular clusters, G1, G78, G158, G170, G177, G219, G222 and 2 open clusters, C107 and C130. Those spectra cover a wide wavelength range (3500Å- 9800Å) which allows one to confidently disentangle age, metallicity and reddening effects.

Our main conclusions are the following:

1) We found evidence is found that the Galactic foreground reddening to M31 cannot be higher than $E(B-V) = 0.04$ ($A_B = 0.16$). The previous estimate of Burstein and Heiles (1984) ($A_B = 0.32$) have probably arisen from a residual contribution of the M31 disc due to an overlapping of the HI velocity profiles.

2) The luminous open clusters C107 and C130 are younger than 30 Myr.

3) G219 is found to to be either an intermediate age cluster or an old globular cluster with an anomalous, extremely strong blue horizontal branch, contrary to the previous hypothesis of a very old and metal poor cluster.

4) The cluster G170 is comparable in absorption-line strength to the bulge clusters NGC6553 and NGC6528 in our Galaxy. These, in turn, exhibit spectra which are comparable to those observed in most galaxy nuclei.

5) The inner bulge clusters G158 and G177 in M31 present metallic features as strong as those observed in the strongest-lined galaxy nuclei ever observed: i.e. the semi stellar nucleus of M31 and the nuclei of giant ellipticals. Their observation is of major importance for composite population synthesis, metallicity calibration and for the interpretation of metal rich stellar systems.

References

Bica ,E., Alloin ,D., 1986b, A&AS, **66**, 171
Burstein, D., Heiles, C., 1984, ApJS **54**, 33
Jablonka, P., Alloin, D., Bica, E., 1991, **in press**

Dark-to-luminous mass ratio in spiral galaxies

P. JABLONKA[1] AND N. ARIMOTO[2]

1. Observatoire de Meudon, D.A.E.C, URA 173 CNRS, 92195 Meudon, France
2. Physics Department, University of Durham, South Road, DH1 3LE, U.K.
Present address: Institut für Theoretische Astrophysik der Universität Heidelberg, Neuenheimer Feld 561, W-6900, Heidelberg 1, Germany

October 2, 1991

Abstract. We demonstrate that, contrary to the conventionnal view, late type spirals do not seem to contain more dark matter than early type ones

Tinsley (1981) has compared her predicted M/L_b vs $B-V$ variation for spiral galaxies with the observations. The slope of her relation was steeper than the observed one. She concluded that late type spirals contained more dark matter than late type ones . Vader (1984) reached the same conclusion comparing the M/L_h vs $B-H$ relation. Both of the authors used Larson and Tinsley's (LT) model, which may have strongly influenced their conclusions.

Arimoto and Jablonka (1991) (hereafter AJ) have developed a two-component bulge – disc population synthesis model for spiral galaxies. Three major improvements in the modelling have been included compared to LT model. First,the AJ model is constructed by taking into account halo pre-enriched gas infall onto the disc. Second, the AJ model has included the mass of the disc gas in the M/L_b ratio. Third, the effect of a change of stellar metallicity, due to chemical evolution, is explicitly taken into account in AJ model colours. We have compared theoretical M/L vs colour relations of the AJ model with those derived from 55 Sa-Sc galaxies (Jablonka and Arimoto, 1991). We found no clear evidence for a significant enhancement of dark matter towards later type spiral galaxies.

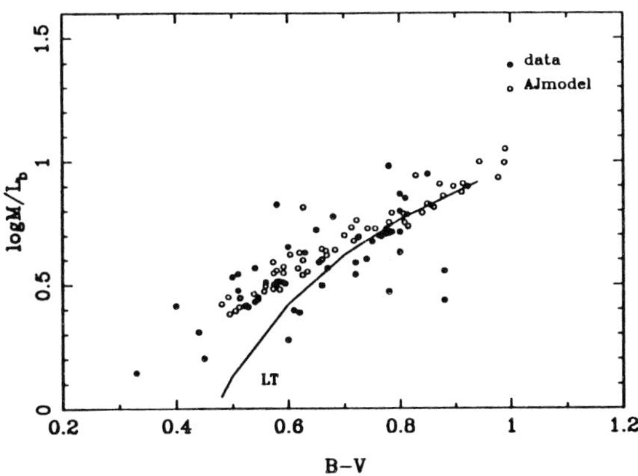

References

Arimoto,N., Jablonka,P., 1991, Astron. Astrophys. **in press**

Jablonka, P., Arimoto, N., 1991, Astron. Astrophys. **in press**

Tinsley, B.M.: 1981, Mon. Not. Roy. Astron. Soc. **194**, 63

Vader, J.P.: 1984, in Formation and evolution of galaxies and large scale structures in the Universe, eds. J.Audouze, J.T.T Van, D.Reidel Publishing Compagny, p.227

The Colour Gradient in M31 : Evidence for Disc Formation by Biased Infall ?

S.A.JOSEY

Astronomy Centre, Division of Physics and Astronomy, University of Sussex, Falmer, Brighton BN1 9QH, UK

N.ARIMOTO

Physics Department, University of Durham, South Road, Durham, DH1 3LE, UK

and

Institut für Theoretische Astrophysik der Universität Heidelberg, Im Neuenheimer Feld 561, D-6900 Heidelberg 1, F.R.G.

A photochemical evolution model has been used to investigate the chemical and photometric evolution of galactic discs which form by prolonged infall of halo material on a timescale that increases with radius, termed *biased infall*, (Josey & Arimoto, 1991). We find that the decline in the mean age and metallicity of the stellar population with radius generates significant colour gradients in the disc and suggest that biased infall may be responsible for the colour variations which have been observed in a number of spirals. Age variations are found to be the primary factor responsible for gradient production in the U, B and V bands while metallicity effects become increasingly important at longer wavelengths. Our model has been applied to M31, in which strong radial colour gradients have been observed, and we find that its chemical and photometric properties can be largely accounted for if its disc formed on a timescale that increased from 0.7 Gyr at the centre to 5 Gyr at a radius of 10 kpc (see Fig.1).

In our synthesis of the colour profile we have corrected for the bulge contribution and reddening by dust both in our own Galaxy and in M31. The bulge dominates the observed colours at small radii but its influence becomes negligible beyond ~ 5 kpc. The colour variation in the NE half of the disc of M31 is in good agreement with the model predictions, while that in the SW half is significantly redder which suggests that recent star formation in this part of the disc has been supressed.

References:

Josey, S.A. & Arimoto, N., 1991. *Astron. Astrophys.*, accepted.

Walterbos, R.A.M & Kennicutt, R.C., 1987. *Astron. Astrophys. Supp.*, **69**, 311.

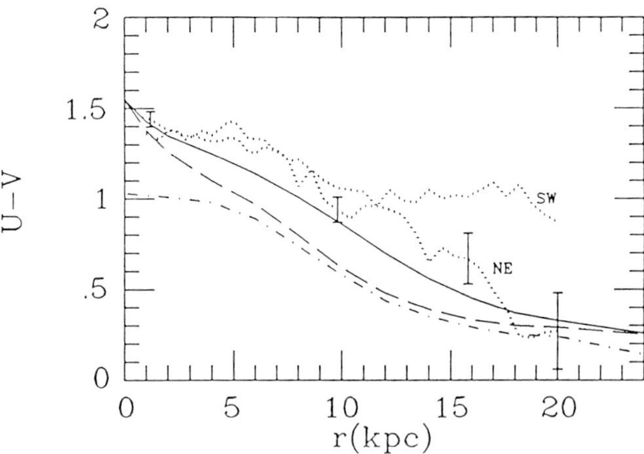

Fig. 1. Model colour profile of M31 disc (dash-dot line), with corrections for the bulge (dashed line) and internal extinction (solid line). Dotted lines are the observed U-V colour profiles of the NE and SW halves of the disc derived, from Walterbos & Kennicutt (1987) and corrected for Galactic extinction.

MOMENTS OF PECULIAR VELOCITIES IN STELLAR SYSTEMS WITH POINT AXIAL SYMMETRY

J.M. JUAN-ZORNOZA[1], J. SANZ I SUBIRANA[2]
1 Dp. Física Aplicada
2 Dp. Matemàtica Aplicada i Telemàtica
Universitat Politècnica de Catalunya
Ap. 30002 Barcelona, Spain

In this work we study a non-cylindrical point-axial symmetric stellar system model that verifies the Chandrasekhar postulates [1]. It explains the values for the centered moments of the peculiar velocities and the galactic rotation parameters, non possible with a cylindrical model [2].

In order to enable us to check the degree of validity of our model we have deduced relations between the centered moments and the kinematical parameters in the galactic plane, where the Sun is assumed to be situated. To prove the consistency of these relations we consider two stellar samples belonging to Population I [3]: *Sample A*, composed of 361 extreme population I stars (O0-B5) with distances less than 600 pc. *Sample B*, composed of 2.154 disk stars (A0-F5), with distances less than 500pc.

In both samples the obtained second order moment $\mu_{\tilde{\omega}\theta}$ is clearly non-zero (generally with values which are of one order of magnitude greater than the error) and the fourth order moments $\mu_{\tilde{\omega}\tilde{\omega}\tilde{\omega}\theta}$, $\mu_{\tilde{\omega}\theta\theta\theta}$, $\mu_{\tilde{\omega}\theta zz}$ likewise yield non-zero values. Furthemore, according to our model, in the galactic plane the moments $\mu_{\tilde{\omega}z}$ $\mu_{\theta z}$ $\mu_{\tilde{\omega}\tilde{\omega}\tilde{\omega}z}$ $\mu_{\tilde{\omega}\tilde{\omega}\theta z}$ $\mu_{\tilde{\omega}\theta\theta z}$ $\mu_{\tilde{\omega}\theta\theta z}$ $\mu_{\theta\theta\theta z}$ $\mu_{\tilde{\omega}zzz}$ $\mu_{\theta zzz}$ can be considerated zero, since the errors are of the same magnitude than the central value. As far as the third order moments are concerned, all of them except $\mu_{\tilde{\omega}\tilde{\omega}\theta}$ $\mu_{\theta zz}$and especially $\mu_{\theta\theta\theta}$, present values less than the double of the error. These three moments, which are not directly explained in our model, could predictably be obtained from overlapping stellar systems in the cylindrical and stationary case [4].

The relations between second and fourth order moments obtained from previous samples correspond to a Schwarzchild distribution function. Finally the results show that the C and K Oort's Constants have clearly non-zero values, according to our model.

REFERENCES:[1]Chandrasekhar S., 1942, "Principles of Stellar Dynamics", Univ. of Chicago/ [2] Sanz J. et al., 1987, Proceedings of the 10th ERAM of the IAU/ [3] Figueras F. et al., 1987, IAU Colloquium No. 100, edited by H.K.Eichhorn (Nac. Com. Astr.,Belgrade)/ [4] Cubarsí R., 1990, Astron. J. **99**, 1558.

CIRCUMNUCLEAR STAR FORMATION IN SEYFERT GALAXIES

W. KOLLATSCHNY[1,2], A. GOERDT[1], K.J. FRICKE[1]
[1]Universitäts-Sternwarte
Geismarlandstr. 11
D-3400 Göttingen, F.R.G.
[2]Dr. Remeis-Sternwarte
D-8600 Bamberg, F.R.G.

ABSTRACT: Using population synthesis methods the nuclear and circumnuclear spectra of Seyfert galaxies were modeled. The central nonthermal intensity and the surrounding star formation rate are correlated.

OBSERVATIONS AND POPULATION SYNTHESIS METHODS

We have obtained deep optical 2D-spectra of five nearby and face-on Seyfert/Starburst galaxies: NGC 1365, NGC 1566, NGC 1808, NGC 3081 and NGC 1097. The data have a spectral resolution of 3.5Å (covering the wavelength range from 3700-7200 Å) and a spatial resolution of a few hundred pc.

Population syntheses of the spectra were calculated using a modified Simplex algorithm developed by us. The presence of a possible nonthermal component was taken into account. The stellar library we used contains O5-M8 main sequence and giant stars. No initial solution is required and the nonnegativity condition for the various contributions is guaranteed.

RESULTS

The spatial extent of the nonthermal component peaks at the nuclei in the Seyfert galaxies as expected. No such component was found in NGC 1097 proving its starburst nature only. Generally, strong starburst components were found within 2 Kpc of the active galactic nuclei superimposed on an old stellar component. Furthermore, we found a correlation of the strength of the nonthermal flux coming from the central active nucleus with the strength of the circumnuclear starburst in the Seyfert galaxies. This is an indication of the induced nature of circumnuclear starbursts in Seyfert galaxies as expected by M. Begelman (1985, Ap. J. 297, 492).

This work has partly been supported by BMFT grant Verbundforschung Astronomie FKZ 50 OR 900 45.

STELLAR POPULATIONS IN SEYFERT 1 GALAXIES

J.K.KOTILAINEN
Institute of Astronomy, Madingley Road, Cambridge CB3 0HA, England

Abstract. We present a study of optical (BV) and infrared (JK) colours and colour gradients in a sample of Seyfert 1 galaxies. The galaxies belong to the hard X-ray selected complete sample of Piccinotti *et al.* (Ap.J., **253**, 485, 1982). We separate the luminosity profiles of the galaxies into an AGN, bulge and disk components (Kotilainen *et al.*, MNRAS, submitted), and determine the stellar BVJK colours in the nucleus and in an annulus around it. We use three colours, B–V, V–K and J–K, which are sensitive to different kinds of stars. We find no correlation between optical, optical/infrared and infrared colours in the nucleus or in the annulus. The recent star formation rate (SFR), which determines the B–V colour, must therefore vary considerably from galaxy to galaxy and be separated from the properties of the older stellar population responsible for the infrared colours. Most of the galaxies show a steep negative V–K and a flatter negative J–K gradient (i.e. redder colour towards the nucleus) and a slight positive B–V gradient. These colour gradients are steeper than in E/SO's and globular clusters. The optical and optical/infrared gradients are well correlated, whereas there is no correlation between B–V and J–K gradients. We compare a model (Arimoto and Yoshii, AA, **164**, 260, 1986) with varying initial mass function, star formation rate and age of the galaxy with the observed colours and gradients. While the models can account for normal galaxy colours, the colours of the Seyfert 1 hosts are generally much redder (in V–K and J–K). Explaining the colours and gradients of the sample galaxies requires a combination of internal differential reddening, thermal reradiation from hot dust grains, and metallicity, recent SFR and IMF that change with radius. Extremely red colours of some of the galaxies may need additional contribution from, for example, a very old red stellar population, extreme IMF or an extremely low lower mass cutoff (LMC).

STELLAR POPULATIONS OF BCD GALAXIES FROM SPECTRO-PHOTOMETRIC EVOLUTIONARY SYNTHESIS

H. Krüger, U. Fritze-v. Alvensleben, H.-H. Loose, K. J. Fricke
Universitätssternwarte
Geismarlandstr. 11
W–3400 Göttingen
Germany

ABSTRACT. Evolutionary synthesis models have been computed to construct the spectral energy distributions of BCD galaxies in the optical and NIR ranges (0.3 to $3.5\mu m$). Evolutionary tracks for stars having $Z = 1/10\, Z_{\odot}$ have been employed in order to match the observed low metal abundances of BCDs. Gaseous emission from H II regions has been included in the model. A starburst (of duration $5 \cdot 10^6\ yr$) is superimposed on an underlying component of red stars characterised by continuous star formation. Burst parameters, star formation rates and ionised hydrogen gas masses have been deduced by fitting the models to observed spectral energy distributions (SEDs) of BCDs.

Results

During strong bursts up to 50% of the total emission in the NIR may be produced by the nebular continuum. Gaseous emission lines do not contribute significantly to the total flux in the NIR. In the optical range line emission may produce up to 40% of the total flux (V and R bands), whereas the gaseous continuum provides only about 10%. Stellar emission in the NIR of both the underlying component and the starburst galaxy are mostly produced by red giants. The optical range (B band) is generally dominated by main-sequence stars during the starburst. Supergiants dominate for about $3 \cdot 10^7\ yr$ during strong bursts, especially in the NIR.

Observed SEDs of BCDs (Thuan 1983, Loose et al. 1991) can be well reproduced by models of varying burst strength and small amounts of internal extinction. If the IMF allows for the formation of stars between 0.04 and 120 $M_{\odot}$, typically $\sim 1 M_{\odot}\, yr^{-1}$ of gas is transformed into stars and burst parameters b lie in the range between 0.005 and 0.02. Supposing only stars more massive than 5 $M_{\odot}$ are formed, these numbers decrease by a factor of 8 (Krüger et al. 1991). Less than 1% of the hydrogen gas present in BCDs is ionised by hot massive stars.

References

Krüger, H., Fritze-von Alvensleben, U., Loose, H.-H., Fricke, K. J. 1991, A&A, **242**, 343
Loose, H.-H., Thuan, T. X., Freitag, V. 1991, in preparation
Thuan, T. X., 1983 ApJ, **268**, 667

THE HALO METALLICITY DISTRIBUTION

J. B. LAIRD
Bowling Green State University
Department of Physics and Astronomy
Bowling Green, OH 43403
USA

B. W. CARNEY
University of North Carolina
Department of Physics and Astronomy
CB 3255 Phillips Hall
Chapel Hill, NC 27599–3255
USA

D. W. LATHAM
Harvard-Smithsonian Center for Astrophysics
60 Garden Street
Cambridge, MA 02138
USA

The field dwarf sample is from the original Carney-Latham survey plus its recent extension to about 500 additional stars, a total of about 1450 stars. We have chosen a conservative criterion to isolate halo stars, namely retrograde orbits in the Galaxy, to minimize contamination by thin and thick disk stars, giving a halo sample of 144 stars.

The observed field dwarf sample is consistent with the simple model at the metal-poor end; there is no lack of very metal-poor stars. However, the observed sample has more metal-poor stars than predicted. A metal-rich excess persists even if the sample is limited to stars having $V < -250$ or -300 km/s, in which case contamination by disk stars is probably minimal.

The metallicity distribution of halo globular clusters shows a difference from the field dwarfs which is statistically significant. For a restricted sample of 49 halo clusters, a K-S test gives a 97% confidence level, with the largest difference occurring in the metal-poor tail. If the actual number of halo clusters is actually larger than 49, which seems likely, then the true difference is actually much worse since no additional extremely metal-poor clusters are known.

Monte-Carlo simulations of the cluster sample from a simple model distribution show that there is only a 0.1% chance that a sample of 49 objects would contain only 2 or fewer objects having log $z < -2.3$; the present cluster sample has only one. We conclude that the lack of very metal-poor clusters is statistically significant.

BLUE STARS IN THE DWARF ELLIPTICAL GALAXY NGC 205

Myung Gyoon Lee and Wendy L. Freedman
Carnegie Observatories
813 Santa Barbara Street, Pasadena, CA 91101, USA

Barry F. Madore
NASA/IPAC Extragalactic Database
IPAC/Caltech/JPL, Pasadena, CA 91125, USA

ABSTRACT. We present a study of stars in the central ($2.2' \times 3.5'$) area of NGC 205 using *BVRI* CCD photometry obtained at the prime focus of the CFHT 3.6m.

NGC 205 is a peculiar dwarf elliptical galaxy (SO/E5pec), located very close to the Andromeda galaxy. The presence of blue stars in this galaxy has been known since Baade (1951) counted about a dozen bright B stars on a deep U photograph of NGC 205.

The color-magnitude diagrams show a blue plume extending to $V \sim 20$ mag. The colors of the blue plume stars indicate that the reddening internal to the central area of NGC 205 is $E(B-V) = 0.1 - 0.3$. A significant fraction of the bluest stars and the brightest stars are concentrated inside 25″. Although the numbers of stars measured are small, to within the uncertainties, the slope of the V luminosity function of these stars for $20.0 < V < 21.5$ is not much different from that of other nearby galaxies.

In addition to the young stars, the color-magnitude diagrams show a first red giant branch (RGB) population, the tip of which is found at $I = 20.35 \pm 0.05$. This value is consistent with the result found by Mould et al. (1984): they detected a tip of the RGB in the field located at 9.5′ north from the center, giving $I = 20.4 \pm 0.1$. The bright stars located above the tip of the RGB are consistent with the presence of an intermediate-age population. The distance derived from the brightness of the tip of the RGB is $(m - M)_o = 24.37$, corresponding to 750 kpc. A value for the foreground reddening of $E(B-V) = 0.03$ from Burstein and Heiles (1984) was adopted in this study. The mean (V-I) color at $M_I = -3.5$ based on the photometry of Mould et al.(1984) is $(V - I)_{-3.5} = 1.76$, giving [Fe/H]$= -0.8$.

Surface photometry shows that the color gets bluer toward the center of the galaxy.

References

Baade, W. 1951, *Publ. Obs. Univ. of Michigan*, No.10, 7

Burstein, D., and Heiles, C. 1984, ApJS, 54, 33

Mould, J. R., Kristian, J., and Da Costa, G. S. 1984, ApJ, 278, 575

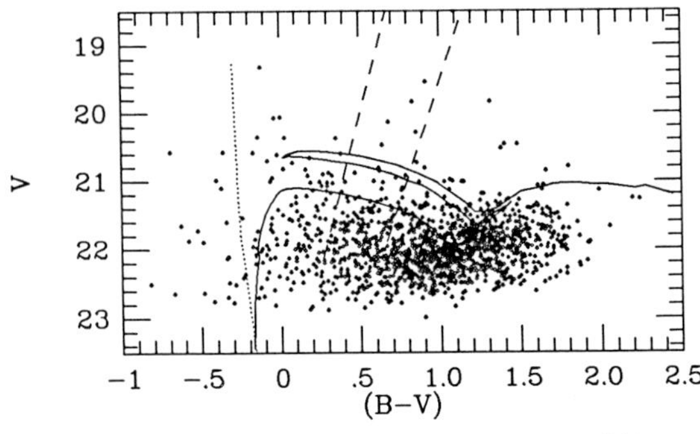

Fig. 1: $V - (B - V)$ diagram of NGC 205. The dotted line gives the ZAMS with solar metallicity. Two slanted parallel lines represent the position of the LMC Cepheid instability strip. The solid line shows the evolutionary track for Z=0.008 and $M = 5M_\odot$ incorporating convective over-shooting.

STELLAR POPULATIONS IN THE LOCAL GROUP GALAXY NGC 185

Myung Gyoon Lee and Wendy L. Freedman
Carnegie Observatories
813 Santa Barbara Street, Pasadena, CA 91101, USA

Barry F. Madore
NASA/IPAC Extragalactic Database
IPAC/Caltech/JPL, Pasadena, CA 91125, USA

ABSTRACT. We present a study of stars in the central ($2.2' \times 3.5'$) area of NGC 185 using *BVRI* CCD photometry obtained at the prime focus of the CFHT 3.6m.

NGC 185, a companion to the Andromeda galaxy in the Local Group, has been known to be a peculiar elliptical galaxy (dE3pec), because of the presence of blue stars, dust clouds, HI gas, and a possible supernova remnant.

The reddening of the bright foreground stars has been determined using the $(B-V)$–$(V-I)$ diagram to be $E(B-V) = 0.19 \pm 0.03$, which is the same as the value derived from the HI method.

The color-magnitude diagrams show three different stellar populations: (1) a small contribution of young stars having blue to yellow colors, (2) a first red giant branch (RGB) population, the tip of which is found to be at $I = 20.25 \pm 0.05$ and $(V-I) = 1.93$, and (3) a bright population of stars located above the tip of the RGB, possibly of intermediate age (consisting of extended asymptotic giant branch stars, carbon stars) in addition to long period variables, similar to those observed in old metal-rich globular clusters of our Galaxy.

The distance has been estimated using the brightness of the tip of the RGB following the recipe given in Da Costa and Armandroff (1990), giving $(m-M)_o = 23.90 \pm 0.15$, corresponding to 600 kpc. For comparison, the distance based on the photometry of RR Lyraes (Saha and Hoessel 1990) is $(m-M)_o = 23.96 \pm 0.25$, if one adopts $M_V(\mathrm{RR}) = 0.60$ mag.

The mean metallicity has been determined by using the median $(V-I)$ color of stars at $M_I = -3.5$, $(V-I)_{-3.5} = 1.71$, giving $[\mathrm{Fe/H}] = -1.3$. The observed dispersion in color $\Delta(V-I)_{-3.5} = 0.31$ (=0.18 after allowing for the photometric error (including crowding error) of 0.25), yields a metallicity range of $-1.9 < [\mathrm{Fe/H}] < -0.9$.

Surface photometry shows that the color becomes bluer slowly toward the center of the galaxy in the outer regions, getting rapidly blue inside $10''$.

References

Saha, A., and Hoessel, J. G. 1990, AJ, 99, 97

Da Costa, G. S., and Armandroff, T. E. 1990, AJ, 100, 162

EVIDENCE FOR AN OLD GALACTIC BULGE FROM RR LYRAE STARS IN BAADE'S WINDOW: THE INSIDE-OUT PICTURE OF GALAXY FORMATION

YOUNG-WOOK LEE
Yale Astronomy Department, 260 Whitney Avenue, New Haven, CT 06511

In their recent, high-quality spectroscopic observations of 59 RR Lyrae stars in Baade's window field of the Galactic nuclear bulge, Walker and Terndrup (1991) found that the metallicity distribution for their whole sample is sharply peaked at [Fe/H] = -1.0. Comparison of their data with the sample of halo RR Lyraes (Suntzeff et al. 1991) suggests that the metallicity at which the stellar population is most likely to form RR Lyraes increases with decreasing galactocentric distance.

The horizontal-branch (HB) model calculations indicate that this is what one would expect if the radial variation in HB morphology observed in the halo continues to the very center of the Galaxy. In particular, the observed peak of the bulge RR Lyrae abundance distribution at [Fe/H] = -1.0 can only be reproduced by the unique relationship between the HB type and [Fe/H] that is shifted to the right of the mean relationship for inner halo globular clusters (see Fig. 7a of Lee 1991).

If age is indeed the second parameter, as supported by recent work, this provides evidence, for the first time, that the oldest stellar population in the Galactic nuclear bulge is indeed older than that in the halo, perhaps by 1-1.5 Gyr. Other possibilities, such as high helium abundance (Y) or high core rotation rate for RR Lyrae stars in the bulge can be ruled out from the analyses of the periods of RR Lyraes. Also, the variations in CNO abundances affect the main-sequence turnoff and HB in the opposite way of what is needed to explain the observations of globular clusters in the halo (see Lee et al. 1988, 1991; Lee 1991).

This implies that the bulge may have been the first part of our Galaxy to form, and then have served as a nucleus around which the rest of the Galaxy was built up from the inside out (see Larson 1990).

REFERENCES

Larson, R. B. 1990, *Publ. A. S. P.*, **102**, 709.
Lee, Y.-W. 1991, in ASP Symp., *The Formation and Evolution of Star Clusters*, ed. K. Janes (San Francisco: ASP), p. 205.
Lee, Y.-W., Demarque, P., and Zinn, R. 1988, in *Calibration of Stellar Ages*, ed. A. G. D. Philip (Schenectady: Davis), p. 141.
Lee, Y.-W., Demarque, P., and Zinn, R. 1991, *Ap. J.*, submitted.
Suntzeff, N. B., Kinman, T. D., and Kraft, R. P. 1991, *Ap. J.*, **367**, 528.
Walker, A. R., and Terndrup, D. M. 1991, *Ap. J.*, **378**, 119.

INJECTION OF MASS AND ENERGY INTO THE ISM BY MASSIVE STARS

CLAUS LEITHERER, LAURENT DRISSEN, and CARMELLE ROBERT
Space Telescope Science Institute, 3700 San Martin Drive, Baltimore, MD 21218.

A large set of radiatively driven wind models for massive stars has been computed. We followed the stars from the ZAMS until they reach $T_{eff} = 15,000\ K$. The metallicity range is $0.1Z_\odot \leq Z \leq 3Z_\odot$. Power-law fits to the mass-loss rates and terminal velocities give:
$\log(\dot{M}/M_\odot yr^{-1}) =$
$2.20\log(L/L_\odot) - 0.68\log(M/M_\odot) + 1.38\log(T_{eff}/K) + 0.70\log(Z/Z_\odot) - 23.65 \quad (\sigma = 0.15);$
$\log(v_\infty/km sec^{-1}) =$
$-0.33\log(L/L_\odot) + 0.60\log(M/M_\odot) + 0.70\log(T_{eff}/K) + 0.15\log(Z/Z_\odot) + 1.00 \quad (\sigma = 0.12).$
We adopted Maeder's (A&AS **84**, 139 [1990]) tables for massive star evolution at different metallicities. These models were extended to lower-mass stars using the results of Maeder and Meynet (A&AS **76**, 411 [1988]). The kinetic energy flux, the momentum flux, and the total energy content due to stellar winds in *all* evolutionary phases, including supernova explosions, have been computed.

The two figures below show the kinetic energy flux of a population of stars forming with $SFR = 1\ M_\odot yr^{-1}$ for $Z = 2Z_\odot$ (left) and $Z = 0.1Z_\odot$ (right). A Salpeter IMF extending from 0.1 $M_\odot$ to 120 $M_\odot$ has been assumed. At ~4 Myr, OB and Wolf-Rayet (WR) stars are equally important for the energy flux. The energy flux scales nearly linearly with Z since $\dot{M} \propto Z^{0.70}$ and $v_\infty \propto Z^{0.15}$. The energy flux due to SN explosions is independent from Z. Therefore stellar winds are more important in a high-Z environment whereas SNe dominate in a low-Z environment. During the early phase of a starburst (< 3 Myr) stellar winds dominate the energetics. At later stages (depending on Z) SNe take over. For a typical starburst of age 10 Myr having solar Z both must be taken into account.

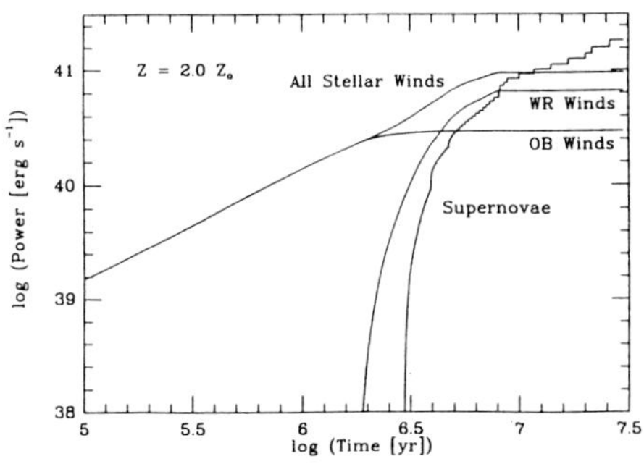

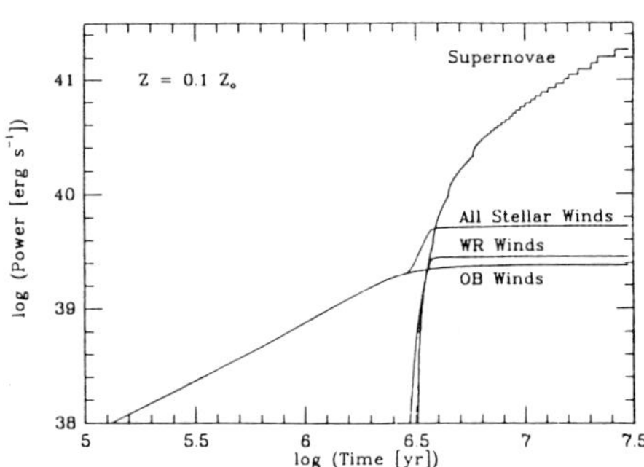

The distribution function for a new sample of OH/IR stars

PETER TE LINTEL HEKKERT — HERWIG DEJONGHE

Mt Stromlo and Siding Spring Observatories — Sterrenkundig Observatorium

Abstract. The new IRAS database of OH/IR stars (ages between 5 and 10 Gyr) show clearly different population characteristics from those used by Baud (1981) for his sample of OH/IR stars (ages under 1 Gyr) obtained by means of "blind" surveys. For the IRAS group the velocity dispersions are much higher and the z-scaleheight is larger than for Baud's sample.

Thus OH/IR stars span a range of ages and kinematical properties all the way from the youngest to the oldest stellar galactic population.

Using a quadratic programming method (Dejonghe 1989) we fit a database of Galactic orbits to the observed distribution of radial velocities, longitudes and latitudes of the IRAS sample of OH/IR stars. This sample is complete within a distance of 8 kpc. We distinguish two groups of OH/IR stars, one with high and the other with low expansion velocity of the circumstellar shell (te Lintel Hekkert 1991). The distribution functions for the two groups differ enough to suggest that they represent two different stellar populations.

The velocity dispersions in the solar neighbourhood can be used to estimate the averaged age for each group. We find 10 Gyr and 5 Gyr for each group. The ensuing ZAMS masses from these ages are $1 - 2\ M_{\odot}$ (Iben and Renzini, 1983). Corroborating evidence for such masses is found from the luminosities for AGB stars in the bulge and disc of around $5000\ L_{\odot}$. The velocity dispersions and z-scaleheights derived for the older group are fairly similar to those found for the so-called thick disc . The younger group fit well within the observed kinematical properties for the so-called old disc . Thus, the IRAS sample of OH/IR stars represents an old, low mass population.

We further find that outside $\varpi \approx 2$ to 3kpc the populations contain only stars rotating in a normal, co–rotating, way around the center, but not on circular orbits. Inside to $\varpi \approx 2$ to 3kpc many more radial and counter rotating orbits occur. Towards the galactic center the older group changes its dynamical properties from a disc to a spheriodal (see also te Lintel Hekkert, 1991, chapter 2). Since, the OH/IR stars in bulge and thick disc are very similar (*e.g.* similar metallicities and luminosities) and the kinematical transition from the thick disc to the Galactic bulge is a very smooth one we conclude that the thick disc and the spheriodal bulge are closely linked and presumably the spheriodal bulge can regarded as the result of the density concentration of the thick disc towards the Galactic centre and not as a seperate dynamical entity.

References

Baud, B., Habing, H.J., Matthews, H.E., Winnberg, A. 1981, *AA.* **95**, 171.

Dejonghe, H. 1989, *ApJ.* **343**, 271.

Iben Jr., I., Renzini, A. 1983, *Ann. Rev. A. A.* **21**, 271.

te Lintel Hekkert, thesis Leiden University, 1991.

UBVRI distances and metallicities for a sample of late-type HIPPARCOS stars

X. LURI, J. TORRA and F. FIGUERAS
Departament d'Astronomia i Meteorologia (Universitat de Barcelona)
Avinguda Diagonal 647 – 08028 Barcelona (Spain)

September 1991

Abstract. Using UBVRI photometry and proper motions we have developed a procedure to separate dwarfs and subdwarfs from stars of other luminosity classes. Three independent methods to take into account the effect of metallicity in photometric distance determinations have been applied, giving special attention to the use of infrared colours for stars with $(B-V) > 0.8^m$.

The application of these methods to a sample of 426 stars included in the HIPPARCOS Input Catalogue –for which we have obtained photometric data– allow us to analyze systematic differences.

Key words: Photometry – Metallicity – Luminosity Classes

As a previous step in the determination of photometric distances and metallicities of our sample of late-type stars, we have developed an algorithm which, using only UBVRI photometry and proper motions, classifies the stars into luminosity classes. The complete description of the adopted criteria –concerning photometry, kinematics, reddening and metallicity– as well as the methods used can be found in Luri et al. (1991). When applied to our sample, this program has given reliable luminosity classes for the most part of the stars, being the others classified as "peculiar" or unclassified.

Given a luminosity class, we have been able to assign a photometric parallax to the stars. For class III stars, interstellar absorption has been included after analyzing all the possible dereddening solutions. In the case of dwarfs and subdwarfs metallicity has been taken into account; up to three independent determinations have been obtained for each star, applying six different methods –depending on the color range–. A new method, using the R-I infrared index as a metal-free parameter, is proposed. We have detected systematic differences between some of the determinations, Luri (1991), which are due to the different hypothesis made when estimating the effects of metallicity on photometric colors as proposed by several authors. These differences reach, for very metal-poor stars, the 0.6^m level. There are no –nowadays– enough direct data on metal-poor stars to favour one determination or another; we must wait to have more reliable data –e.g. the trigonometric parallaxes coming from the HIPPARCOS mission– to decide which one is the best.

Acknowledgements

This work has been supported by the CICYT under contract ESP88-0731

References

Luri, X., Figueras, F. and Torra, J.: 1991, *Astron. J.* , (In preparation)
Luri, X.: 1991, *Tesis de licenciatura* , Universitat de Barcelona

COLOR DISTRIBUTION OF GALAXIES IN THE CORE OF S0400

D. MACCAGNI, B. GARILLI and D. BOTTINI
Istituto di Fisica Cosmica del CNR, Milano, Italy
and
G. VETTOLANI
Istituto di Radioastronomia del CNR, Bologna, Italy

September 16, 1991

S0400 is a cluster of galaxies of richness 1 and distance 6, morphologically classified as a regular Abell type. It is an X-ray source (Gioia et al. 1990). In October 1988, we observed S0400 with the ESO 3.6m telescope equipped with EFOSC. Spectroscopy of 4 galaxies yields a cluster redshift $z = 0.32$. *gri* photometry under 1 arcsec seeing conditions allowed us to classify all objects brighter than $m_r = 24$. Of the 224 galaxies detected in the $\sim$ 4x6 arcmin f.o.v., 193 have colors compatible with, or bluer than, the ones expected from synthesis models for E galaxies. These galaxies have been considered cluster members. The Figure shows their $g-r$ vs. $r-i$

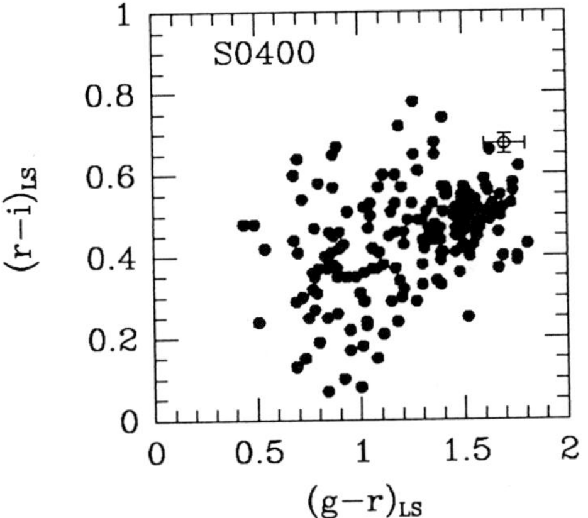

diagram (black dots). The cross represents the BCGs colors k-corrected to S0400 redshift (Schneider, Gunn & Hoessel 1983). Many objects (the cluster Es, including the BCG) clump around specific values of $g - r$ and $r - i$ which are bluer than expected provided the k-corrections are reliable enough. This could be interpreted as a sign of evolution in the E population of S0400.

References

Gioia, I.M. et al. 1990, ApJS, 72, 567

Schneider, D.P., Gunn, J.E., Hoessel, J.G. 1983, ApJ, 264, 337

PLANETARY NEBULAE AND STELLAR POPULATIONS

W. J. MACIEL, C. C. M. LEITE
Instituto Astronômico e Geofísico da USP
Av. Miguel Stefano 4200, 04301 São Paulo SP, Brasil

Recent work on galactic planetary nebulae (PN) has emphasized the complex nature of this galactic subsystem (W. J. Maciel, *31st Herstmonceux Conference*, ed. R. Terlevich, in press). A fairly large amount of data is now available, including distances, galactic distribution, kinematical properties, and chemical composition of the nebular gas, supplemented by data on the central stars for some objects. In this work, a large sample of PN is used to study the different stellar populations associated with the galactic nebulae. It is seen that the available data imply a physical description of the PN phenomenon that is consistent with an evolutionary model based on stellar populations.

Our total sample includes about 160 objects. The average height $< |z| >$ from the galactic plane and the height where half of the sample is contained $|z_{1/2}|$ vary according to the PN type, as follows: 150/110 pc (Type I); 280/210 pc (Type IIa); 350/250 pc (Type II); 420/300 pc (Type IIb); 660/400 pc (Type III); and 7200/6900 pc (Type IV). Although the absolute value of the scale given can be affected by the incompleteness of the sample, it is clear that the average distance from the plane increases according to the sequence I, IIa, IIb, III, and IV.

The average peculiar velocities of PN $< |\Delta V| >$, and the velocity where half of the sample is contained, $|\Delta V_{1/2}|$ also show a progression according to PN types: 19.9/18.2 (Type I); 21.3/18.0 (Type IIa); 20.7/19.3 (Type II); 20.1/20.5 (Type IIb); 79.7/66.1 (Type III); 172.8/167.2 (Type IV). Notice that for Types I and II the velocities are essentially the same, given the average uncertainty of about 10 km/s in the radial velocities.

The chemical composition analysis confirms the results given above, both for the elements synthesized by large mass stars (O, Ne, S, Ar, and Cl) and for the elements synthesized by intermediate mass stars (He, N, C). As a conclusion, a tentative classification for galactic PN would be, which updates the scheme given previously (W. J. Maciel, *IAU Symp. 131*, ed. S. Torres-Peimbert, Kluwer, 1989): Type I (Pop. I old); Type IIa (Pop. I old/I disk); Type IIb (Pop. I disk); Type III (Pop. II intermediate); Type IV (Pop. II extreme); Type V (Pop. I old/II extreme).

Work partly supported by CNPq and FAPESP.

EVOLUTIONARY SPECTRAL SYNTHESIS AND THE UV UPTURN IN ELLIPTICAL GALAXIES

GLADIS MAGRIS C., and GUSTAVO BRUZUAL A.
Centro de Investigaciones de Astronomía, CIDA
Apdo. Postal 264 - Mérida 5101A - Venezuela

In this work we analyze the UV upturn seen in elliptical galaxies using the evolutionary spectral library of Bruzual & Charlot (1991). We present models for 3 representative groups galaxies of the Burstein *et al.* (1988) sample with different levels of UV flux.

The spectrum of NGC 4649 is modeled with a 14 Gyr old stellar population (consistent with the age determination by Bertelli, Chiosi & Bertola (1989)). The star formation rate (SFR) corresponds to an initial burst of 1 Gyr (100 $M_\odot$ yr^{-1} in a 10^{11} $M_\odot$ galaxy) plus a 'residual' continuum star formation of 0.03 $M_\odot$ yr^{-1} This comparison indicates that if we consider only *normal* Post Asymptotic Giant Branch (PAGB) stars, included in our library, we cannot reproduce the observed spectrum for this galaxy. Classical PAGB's contribute to the total luminosity only for $\lambda \leq 2200$ Å, and NGC 4649 has an excess of flux in the range 2200 - 2600 Å (with respect to an old quiescent star system) which must be accounted for by a different stellar population. Other candidates have been explored by Greggio & Renzini (1990).

The spectrum of NGC 4472, with an intermediate value of the (1550-V) color, can be reproduced with an old population seen at 13.5 Gyr, which underwent a unique event of star formation, an initial burst of 1 Gyr of duration. The galaxies with this value of the (1550-V) color can be modeled by using only *normal* PAGB stars, resulting from the evolution of a quiescent stellar population.

For M32, the best model corresponds to a stellar system with a normal IMF that underwent two events of star formation: an initial 1 Gyr burst, and a second burst of the same duration at age 6 Gyr. This result, which agrees with O'Connell (1986), Bertelli, Chiosi & Bertola and others, has been questioned by Greggio & Renzini.

Our models are able to reproduce, in a consistent way, the observations of a large amount of quiescent E galaxies. The possibility of other evolutioned hot star candidates contributing to the UV flux is not excluded, but for many elliptical, the *normal* evolution of stars beyond the AGB is enough to account for their UV rising branch.

References

Bertelli, G., Chiosi, C., and Bertola, F. 1989, *Ap. J.*, **339**, 889.

Bruzual A., G., and Charlot, S. 1991, *Ap. J., Submitted.*

Burstein, D., Bertola, F., Buson, L. M., Faber, S. M., and Lauer, T. R. 1988, *Ap. J.* **328**, 440.

Greggio, L. y Renzini, A. 1990, *Ap. J.*, **364**, 35.

O'Connell, R. W. 1986, in *Stellar Populations*, ed. C. A. Norman, A. Renzini, y M. Tosi (Cambridge: Cambridge University Press), p.167.

Star Formation and Stellar Populations in Ring Galaxies

A. P. MARSTON
Department of Physics and Astronomy, Drake University, Des Moines, IA 50311, USA
and
P. N. APPLETON, M. LYSAGHT AND C. STRUCK-MARCELL
Department of Physics and Astronomy, Iowa State University, Ames, IA 50011, USA

Colliding ring galaxies provide a remarkable testbed for the study of star formation in perturbed galaxies. In the process of passing through a disk system, a small perturbing galaxy generates a density wave of stars and gas which expands into the host disk. This triggers a wave of star formation. As the star forming wave passes through the host galaxy, progressively older burst populations may be found interior to the ring. As part of a multiwavelength study of ring galaxies, we have performed optical and infrared imaging using the Kitt Peak 2.1m telescope. These images are used to explore the relation between stellar density wave amplitude and star formation rate. Color gradients are searched for which would indicate the presence of an aging burst population interior to the ring.

Results show that the star formation regions (indicated by Hα emission line morphology) are to be found exterior to the advancing stellar density wave (indicated by K band images). An example of this is shown in Fig. 1. These results suggest that cloud-cloud collisions are the dominant cause of the star forming bursts (Appleton and Struck-Marcell, 1987).

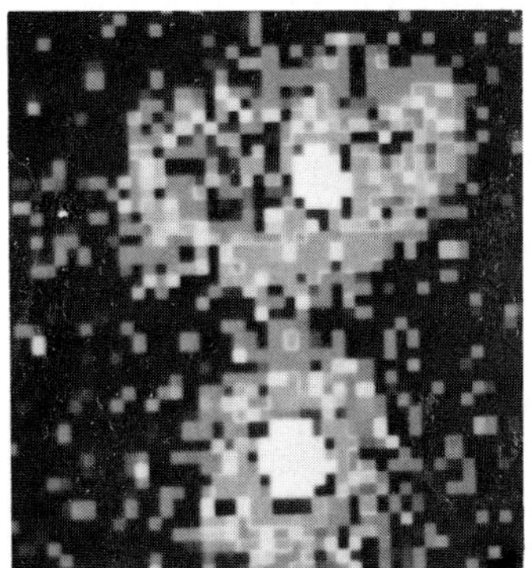

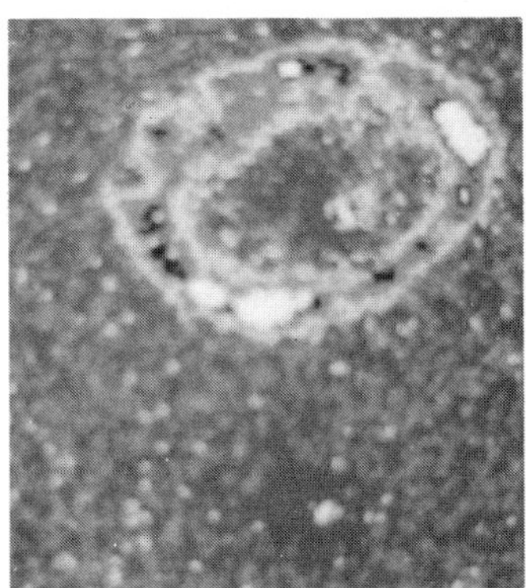

Fig. 1: K band, to the left, and Hα images of the ring galaxy LT41 showing the large ongoing star formation in the ring outside the stellar density wave peak.

Color gradients are also noted with progressively redder colors seen towards the centers of the ring. This is consistent with an evolutionary population sequence, with a progressively older burst population being found towards the centers of the ring systems.

References

Appleton, P. N. and Struck-Marcell C., 1987, *Ap.J.*, **318**, 103.

THE EVOLUTION OF THE RED GIANT CLUMP AND THE STRUCTURE OF THE SMALL MAGELLANIC CLOUD

Mario Mateo (OCIW), and Despina Hatzidimitriou (AAO)

The evolution of the red giant clump as a function of age has been investigated by analyzing the color-magnitude diagrams of a number of intermediate-age star clusters in the LMC. We find that the luminosity of the clump decreases systematically until about 1 Gyr. Thereafter, the mean apparent magnitude of the clump in LMC clusters remains constant at V $\sim$ 19.0 $\pm$0.1, although with significant scatter about this value in clusters with ages between 1 and 2 Gyr. The luminosity width of the clump also varies systematically with cluster age, being largest for clusters with ages near 1-1.5 Gyr. For six clusters older than 2 Gyr, the clump shows an intrinsic luminosity spread (defined as the FWHM of the best-fitting Gaussian) of 0.30 $\pm$ 0.07 mag. The systematic behavior of the color of the clump as a function of age was studied in detail by Hatzidimitriou (1991, *M. N. R. A. S.*, **251**, 545).

These results have been used to analyze the line-of-sight structure of the SMC based on the luminosity extension of the red giant clump. CCD photometry of two SMC fields – one in the West and the other in the Northeast – have been obtained with the Las Campanas 1m telescope for this purpose (see the figures below). The lack of bright main sequence stars in both fields indicate that neither contains a significant contribution from a stellar population younger than 1.5-2 Gyr. The red giant clump in the W field is only slightly more extended than expected from the observed vertical width of red giant clump in LMC cluster older than 2 Gyr; in contrast, the clump is significantly extended in the NE field. This implies a line-of-sight depth in the W SMC field of $\lesssim$ 5 kpc, while in the NE field the depth may be as large as 23 kpc, consistent with the recent study of Hatzidimitriou and Hawkins (1989, *M. N. R. A. S.*, **241**, 667; HH) based on photographic photometry. Alternatively, we can interpret the luminosity function of the red giant clump in the NE of the SMC as evidence for two distinct components separated by 12 kpc, and each about 7 kpc thick (HH). There is clear evidence that the NE region of SMC shows considerable depth along the line of sight, perhaps due to a recent tidal encounter with the LMC.

We acknowledge the help of L. T. Gardiner, and partial support from NASA grant # HF-1007.01-90A.

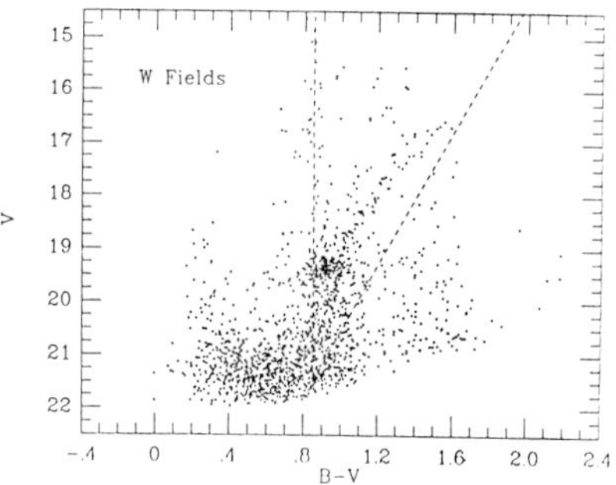

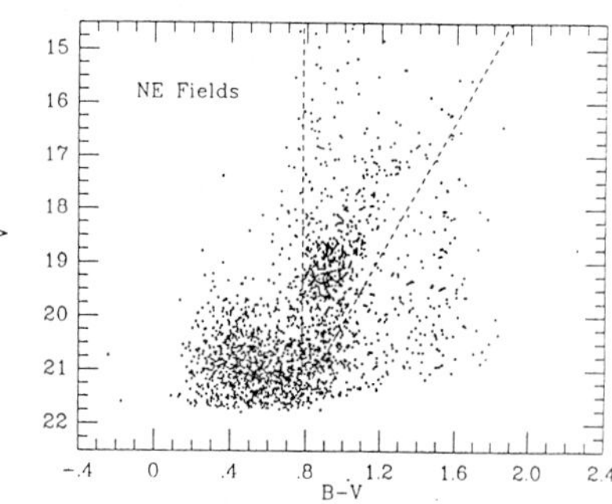

CMDs of the W (left) and NE SMC fields. The large luminosity extension of the clump in the NE field compared to the W field is evident.

THE STELLAR POPULATION AND INTERNAL KINEMATICS OF THE SEXTANS DWARF SPHEROIDAL GALAXY

Mario Mateo (OCIW), Nick Suntzeff (CTIO), James Nemec (WSU), Donald Terndrup (OSU), William Weller (CTIO), Edward Olszewski (Steward), Michael Irwin and Richard McMahon (Cambridge)

The Sextans dwarf Spheroidal (dSph) galaxy was discovered recently by Irwin, *et al.* (1990, *M. N. R. A. S.*, **244**, 16p). We report results concerning a number of the global properties of the stellar population of this system. Based on deep CCD photometry obtained with the prime-focus CCD camera on the CTIO 4m telescope, we find that the galaxy is dominated by an old stellar population, similar to that observed in the Draco, Sculptor, and Ursa Minor dwarfs. Some blue stragglers are also present; if these are associated with an intermediate age component, they indicate that only a tiny fraction of the stellar population of Sextans is younger than about 10 Gyr. Based on the apparent magnitude of the predominately red horizontal branch, we conclude that the true distance modulus of Sextans is 19.7 ± 0.3, and the total luminosity of the galaxy is $5.2 \times 10^5\ L_\odot$. A complete description of these results is given in Mateo, *et al.* (1991, *A. J.*, **101**, 892).

A series of CCD exposures obtained with the Las Campanas 1m telescope has been used to search for variable stars in Sextans down to the level of the horizontal branch. We have identified 41 RR Lyr variables and three anomalous Cepheids within about 0.6 core radii (10 arcmin) of the center of the galaxy. The identification of RR Lyr stars confirms the presence of an old population in Sextans. The ratio of the number of anomalous Cepheids and blue stragglers in Sextans is similar to that observed in the Ursa Minor dSph (Olszewski and Aaronson 1985, *A. J.*, **90**, 2221). The mean period of the AB RR Lyr stars is 0.61 days. From the mean color of the giant branch at the level of the horizontal branch, we derive a mean abundance for Sextans of [Fe/H] $= -1.9$; however, the large spread in the colors of the giants implies a significant heavy element abundance spread in the galaxy.

We also report spectroscopic results for 80 candidate red giants towards the center of Sextans. These data were obtained using the ARGUS fiber spectrograph at the CTIO 4m. Forty-four of these stars are likely members of the galaxy based on radial velocities measured using the near-IR Ca II triplet. From these data we derive a central velocity dispersion of 6.5 ± 1.7 km s^{-1}. Using the current best estimates of the structural parameters of Sextans yields a central M/L = 31, and a central mass density of 0.07 $M_\odot$pc^{-3} for an assumed isotropic distribution of stellar orbits. These results imply that a significant dark matter component is present in Sextans, although the central density of this component is considerably less than deduced for the Draco and Ursa Minor dwarfs. All three galaxies have similar total luminosities. The mean metallicity of Sextans is [Fe/H] $= -2.1$ based on the strength of the Ca II triplet for the seven giants with the best spectra. These data also support the existence of a significant abundance spread in the galaxy.

COLOR MAPPING CLUSTER COOLING FLOW GALAXIES

BRIAN R. MCNAMARA
Kapteyn Institute, Groningen

Abstract. Some results from a new CCD color imaging survey of centrally dominant cluster galaxies selected for their X-ray properties are discussed.

Key words: cD galaxies, color gradients, cooling flows

1. Discussion

We have obtained U-,b-,V-, and I-band CCD images of 19 centrally-dominant galaxies in cluster cooling flows and 4 without cooling flows. Using photometry through synthetic elliptical apertures, we find that many objects exhibit blue color-profile anomalies with respect to the control sample and to gE photometry from the recent literature. The anomalies are stronger and extend to larger galactic radius with increasing mass-accretion rate ($\dot{m}_{CF}$) estimated from X-ray observations (McNamara and O'Connell 1991, Preprint). Scale sizes for the anomalously blue regions are ~ 5 kpc for the modest accretors ($\dot{m}_{CF} \lesssim 100$ $M_\odot$ yr^{-1}) and $\gtrsim 20$ kpc for some of the largest accretors ($\dot{m}_{CF} \gtrsim 200$ $M_\odot$ yr^{-1}).

These correlations are likely to be due to recent star formation induced by the cooling flows, rather than tidally-induced starbursts. The prevalence of color anomalies suggests that star formation has been occurring continuously for a substantial fraction of the Hubble time. However, in no case are the blue color anomalies consistent with star formation with the Local Initial Mass Function at the rates estimated from X-ray observations. In addition, the color anomalies extend to only ~5–10% of the cooling radii estimated from X-ray observations. If the $\dot{m}_{CF}$'s are correct, then most of the accreting gas must reside in very low-mass stars and/or opaque gas.

Despite the harsh cluster environment, about half of our sample objects have dust patches or lanes. Most of these objects also have evidence for recent star formation.

The non-accreting centrally-dominant galaxies have red, metal-rich nuclei and mild blueward-rising color gradients to $r \sim 5$ kpc. These gradients are likely due to metallicity gradients resulting from dissipative processes during the early stages of galaxy assembly. The halo color gradients usually flatten to a roughly constant color for $r \sim 5$–15 kpc that is $\sim 0.1 - 0.2$ mag bluer than the nucleus. The flattening of the halo gradients may be due to stripping or mergers of cluster galaxies, or other dissipationless processes. Some objects appear to have red halos at large radii; however, additional observations which are optimized for halo photometry with large-format CCD's are needed for confirmation.

Additional observations along similar lines, in a uniform photometric system, will provide valuable constraints on the variations in age and metallicity and dust content of dominant cluster galaxies. In addition to their intrinsic interest, such data may be useful when using these objects to constrain cosmological models.

THE MORPHOLOGY AND STELLAR POPULATIONS OF THE DWARF AMORPHOUS GALAXIES NGC 216 AND NGC 2915

GERHARDT R. MEURER[1] *and* GLEN MACKIE[2]
[1] *Anglo-Australian Observatory*
P.O. Box 296, Epping NSW, 2121
Australia

[2] *Department of Astronomy*
University of Wisconsin
475 N. Charter St.
Madison, WI 53706
U.S.A.

ABSTRACT. The definitions of the amorphous and blue compact dwarf (BCD) classes of galaxies are very similar. One key difference is that BCDs are often selected for their *apparent* compactness (*i.e.* a small angular size), which selects against nearby objects, whereas amorphous galaxies must be extended. We present initial results of a project to determine the population distribution in dwarf ($M_B < -18$) amorphous galaxies (dAgs) and determine which dAgs can be classified as BCDs. We have used the 3.9m Anglo-Australian Telescope to obtain deep B and R CCD images of two dAgs: NGC 216 ($M_B = -17.3$) and NGC 2915 ($M_B = -14.1$). The morphology of NGC 216 is that of a dusty late-type edge-on disk galaxy, with a peculiar one-sided bar. It would not be classified as a BCD if seen face-on. However, NGC 2915 does have all the properties of a BCD, and can be classified so. It has numerous condensations near its center. Many of these are likely to be individual stars. We derive a distance to NGC 2915 of 5 Mpc if the brightest blue non-extended objects are blue supergiants, and if there is little internal extinction. Similar condensations are seen in the dAgs NGC 1705 (Meurer, *et al.*, 1989. *Astrophys. Space Sci.*, **156**: 141) and NGC 5253 (Caldwell and Phillips, 1989. *Astrophys. J.*, **338**: 789) which are also likely to be BCDs.

It is little wonder that NGC 216 and NGC 2915 have very different morphologies; they represent the extremes of the luminosity range spanned by our sample. It is likewise apparent that all dAgs are not BCDs. But, we have shown the effectiveness of using CCD imaging in identifying which dAgs are. The two galaxies are not totally dissimilar. They both have an exponential surface brightness profile with a central high surface brightness excess. Both also have strong color gradients with the hot blue stars more centrally concentrated than the cooler populations. This can be interpreted as due to either a recent increase in the central star formation rate relative to the that in the outer regions (*e.g.* a central star-burst) or a gradient in the initial mass function of stars formed.

CNO OVERABUNDANCES IN 6 STARS OF ω CENTAURI

A. Milone, B. Barbuy
IAG-USP, Depto. Astronomia
C.P. 9638, São Paulo 01065, Brazil
M. Spite, F. Spite
Observatoire de Paris-Meudon
92195 Meudon Pl. Cedex, France

1. Introduction

C_2 and CN bandheads in the wavelength region $\lambda\lambda$ 550-680 nm are synthesized for 6 stars of ω Centauri, in order to derive their carbon and nitrogen abundances. Oxygen abundances are also derived taking into account molecular associations with C and N compounded molecules, using the appropriate C and N abundances.

These determinations have been made for CO-rich and CO-normal stars according to the CO indices measured by Persson et al. (1980).

High resolution spectra were obtained at the 3.6m telescope of the *European Southern Observatory* (Chile), using a CCD and the Caspec spectrograph.

We derive CNO abundances for 2 CO-rich and 3 CO-normal stars. The star ROA 65 was not classified by Persson et al. The detailed analysis of these 6 stars were previously carried out by François et al. (1988).

The results for the CO-normal stars ROA (65), 74, 91, 256, reveal that they all show similar CNO excesses: [C/Fe] $\approx$ +0.25, [N/Fe] $\approx$ +0.5, [O/Fe] $\approx$ +0.15. These lower metal-abundance stars have a surprisingly high carbon abundance, suggesting that the initial C abundance was high.

As concerns the two "CO-rich" studied stars ROA 139 and ROA 270, they show very different CNO abundances: ROA 139 is C-poor and O-poor, and seems to show CNO abundances similar to O-poor stars in M13. ROA 270 shows the highest C abundance of our sample ([C/Fe] = +0.65) and it is also the richest in O ([O/Fe] = +0.75). These very high C and O abundances should reflect an initial overabundance in the gas from where it formed.

We conclude that the CO-normal stars have homogeneous CNO abundances, and the CO-rich stars show a pronounced variation, which might be interpreted as a self-enrichment or self-pollution, concerning at least the possibly second generation of stars (the more metal-rich) of ω Centauri.

François, P., Spite, M., Spite, F.: 1988, A&A 191, 267
Persson, S.E., Frogel, J.A., Cohen, J.G., Aaronson, M., Mathews, K.: 1980, ApJ 235, 452

KINEMATICS OF THE GALACTIC BULGE

D. Minniti, S.D.M. White**, E. Olszewski*, J. Hill* and M. Irwin***

**Steward Observatory, Univ. of Arizona, Tucson, U.S.A.*
***Institute for Astronomy, Cambridge, England*

We present spectroscopic observations of ~200 giant stars in two fields toward the Galactic bulge. The positions of fields 588 and 589 are at l,b = (8,7) and (12,3), respectively, which corresponds to a Galactocentric distance of ~1.6 kpc for the stars in the bulge. The K giants were selected from color-magnitude and color-color diagrams produced by scans of B, R and I plates from the APM machine. The spectra were obtained with the MX multi-object spectrograph at the Steward 90" telescope, and with the Red Channel spectrograph at the MMT. Radial velocities good to $\leq$ 10 km/s were obtained from cross-correlation techniques. A grid of ~100 standard stars was built to calibrate spectrophotometric indices that provide abundances and luminosities.

We also present infrared JHK photometry for ~5000 stars at field 588 reaching ~ 2 magnitudes fainter than the bulge HB at K. Metal abundances derived from JHK photometry and spectral indices show a wide distribution, similar to those found by other studies in fields along the minor axis of the Galaxy (*e.g.* Terndrup *et al.* 1991, Ap.J. **357**, 453).

The bulge of the Galaxy is found to be rotating, the mean velocities for fields 588 and 589 being 45 ± 10 and 78 ± 9 km/s, respectively. The bulge velocity dispersion decreases with increasing Galactocentric distance, we find $\sigma_{588} = 85 \pm 7$, $\sigma_{589} = 68 \pm 6$.

When dividing the samples into metal rich and metal poor stars, we find a correlation between [Fe/H] and σ, and possibly V_{rot}, in the sense that the metal rich population seems to have a lower velocity dispersion. Figure 1 shows the run of σ for K giants in the bulge from the present work and the observations of Rich (1990, Ap.J. 362, 604) at the Baade window field. This suggests that dissipational collapse could have played an important role in the formation of the bulge.

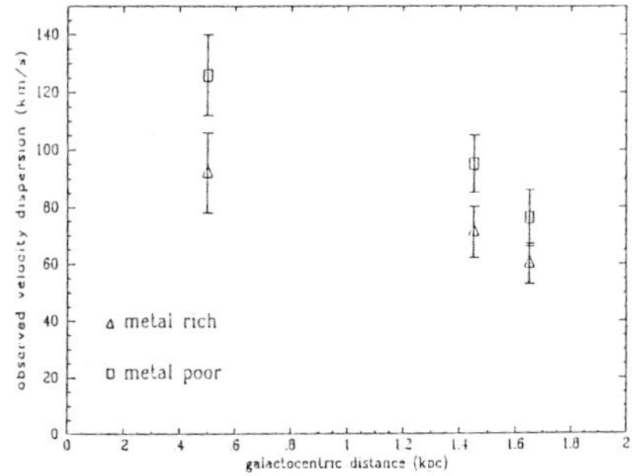

We study the shape of the velocity ellipsoid of the Galactic bulge on the basis of our radial velocities in the field at l,b = (8,7) and proper motions from Cudworth (1986, A.J. **92**, 348) in the field of the globular cluster M22 at l,b = (9.9, -7.6). The velocity ellipsoid at a Galactocentric distance of ~ 1.6 kpc is found not to be significantly flattened. This confirms our suggestion that the Galactic bulge is flattened by rotation rather than by velocity anisotropies. The proper motions also show the trend of higher σ for lower [Fe/H] seen on the radial velocity samples. If the bulge velocity ellipsoid is indeed nearly isotropic, we derive a rather small distance to the Galactic center ($R_\odot \approx 7$ kpc).

EVOLUTIONARY STATUS OF GALAXY POPULATION IN CLUSTERS AT INTERMEDIATE REDSHIFT ($z \sim 0.2$)

Emilio Molinari, Dolores Pedrana, Massimo Banzi, Alberto Buzzoni and Guido Chincarini

Osservatorio Astronomico di Brera
Via Brera, 28 20121 Milano, Italy

In this contribution we present observations for a set of three clusters of galaxies at intermediate redshift ($z \sim 0.2$) selected from the revised Abell catalog (Abell *et al.* 1989, *Ap.J. Suppl.*, **70**, 1) and observed in the Gunn photometric system at the ESO 3.6m telescope in La Silla (Chile) during various runs between 1986 and 1990. This is part of a more extended project addressing the study of the distant clusters, as outlined in more detail in Molinari *et al.* (1990, *M.N.R.A.S.*, **246**, 576).

Member galaxies and *bona fide* galaxy types have been determined considering isophotal radii and their location in the color-magnitude and two-color diagrams. A marked concentration of red galaxies around $(g-r) \sim 0.8$ clearly appear in the cluster population. They can be easily located redshifting the present-day colors of ellipticals with a little amount of blueing ($\Delta(g-r) \sim 0.03$ mag) due to quiescent evolution, as estimated on the basis of the models for stellar population synthesis (Buzzoni 1989, *Ap.J. Suppl.*, **71**, 817).

The evidence emerging from our work is that also distant clusters appear to be dominated by quite normal, i.e. quiescent, galaxy populations. If ongoing star formation in some ellipticals cannot be firmly excluded (as indicated for instance by emission–line or E+A galaxies) this could probably not fully account for the excess of blue objects in the Butcher & Oemler (1984, *Ap.J.*, **285**, 426) effect. In spite of the three clusters statistics, we note that f_B correlates with the Bautz-Morgan type that could be regarded as a measure of the dynamical relaxation of a cluster (White 1976, *M.N.R.A.S.*, **177**, 717). Thus gas cooling flows and/or galactic tidal stripping (Fabian *et al.* 1982, *M.N.R.A.S.*, **180**, 479) would probably help in leaving fewer blue objects in older (low-redshift) relaxed environments.

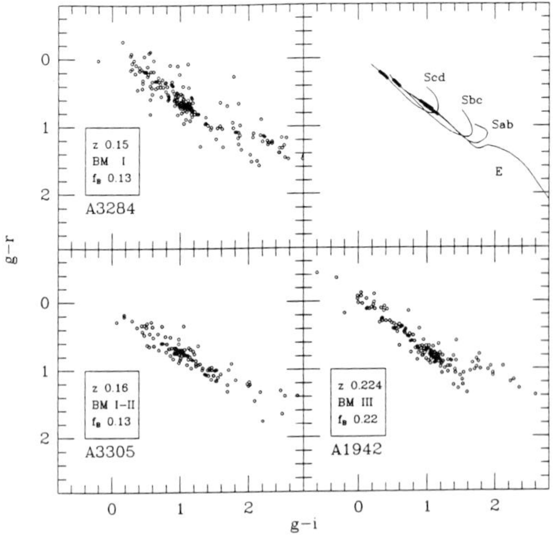

Fig. 1 – Two-color diagrams for cluster galaxy populations. Top right panel displays the locus expected for different galactic morphological types with varying redshift from $z = 0$ to 1 (in the sense of increasing $g - i$). Boldface marks the location expected for galaxies in our relevant range ($z = 0.15 - 0.22$). Loci for late-type galaxies have been calculated from Pence (1976, *Ap.J.*, **203**, 39) and Coleman *et al.* (1980, *Ap.J. Suppl.*, **43**, 393) while ellipticals are tracked by evolutionary models of Buzzoni (1989).

A STUDY OF THE PROPERTIES OF GALACTOCENTRIC ORBITS

A. A. Myullyari
University of Petrozavodsk, USSR

Yu. V. Nechitajlov
St Petersburg State University, 198904 Petrodvorets, Bibliotechnaya ploshchad' 2, USSR

S. Ninković
Astronomical Observatory, Volgina 7, 11050 Beograd, Yugoslavia

V. V. Orlov
St Petersburg State University, 198904 Petrodvorets, Bibliotechnaya ploshchad' 2, USSR

ABSTRACT. The galactocentric motion of 14 stars from the solar neighbourhood is studied. The stars are divided into nine groups according to their heliocentric-velocity components. In this way one can follow the motion of the stars belonging to the same group in the past.

We calculate the orbits for 14 stars from the solar neighbourhood; the Sun is among them. The data source is [1]. The stars are divided into nine groups according to the similarity in their heliocentric-velocity components. The galactic models applied in our calculations are [2], [3] and [4]. The aim of the study is to follow the motion of the sample stars to the past, to establish a possibility of belonging to a star stream and to examine the contours of their orbits. A preliminary result is that the selected stars belong to the thin disc, i. e. their galactocentric motion is possible to describe in Lindblad's epicyclic approximation.

REFERENCES

1. Gliese, W. (1969) 'Catalogue of Nearby Stars', Veroeffentlichungen des Astronomischen Rechen-Instituts, Heidelberg 22.
2. Sala, F. (1987) 'Galactic Kinematics: An Axi-Symmetric Time-Depending Model with Separable Potential', Rev. mex. astron. astrof. 14, 195-205.
3. Allen, C. and Martos, M. A. (1986) 'A Simple Realistic Model of the Galactic Mass Distribution for Orbit Computations', Revista mexicana de astronomia y astrofisica 13, 137-147.
4. Kutuzov, S. A. and Osipkov, L. P. (1989) 'Dvukhkomponentnaya model' gravitatsionnogo polya Galaktiki', Astronomicheskij zhurnal 66, 965-974.

A Color-Magnitude Diagram in Field #3 of the Palomar–Groningen Survey

Y.K. Ng[1], G. Bertelli[2], A. Bressan[2] & C. Chiosi[2]

[1]*Leiden Observatory, P.O. Box 9513, 2300 RA Leiden, The Netherlands*
[2]*Padova Observatory, Vicolo Osservatorio 5, 35122 Padova, Italy*

Field #3 of the Palomar–Groningen survey (PG3) is thought to be ideal for the study of stellar evolution in the Galactic Bulge (GB). In this field the work of Plaut (1973) and Wesselink (1987) is extended with the aim of constructing a Color Magnitude Diagram (CMD) for a great many stars. This is done by means of automated photographic photometry.

With 229 646 stars from a UKST blue (J7856) and red (R8491) plate a CMD is constructed for a $3\overset{\circ}{.}5 \times 3\overset{\circ}{.}5$ field, centered at $l = 1^o$ and $b = -11^o$. The plate limit for the blue and red plate is resp. $19\overset{m}{.}8$ and $18\overset{m}{.}7$. The estimated photometric accuracy is $\sim 0\overset{m}{.}10$.

With the theoretical isochrones including convective overshoot (Bertelli et al., 1990) synthetic CMDs (Chiosi et al., 1988) and Luminosity Functions (LFs) are composed, assuming (i) A Salpeter initial mass function (iia) A starburst from 14-10 $\times 10^9$ yrs ago for (Y,Z)=(0.28,0.02) (iib) A starburst from 14-12 $\times 10^9$ yrs ago for (Y,Z)=(0.25,0.0004).

We placed the stars in each of the synthetic CMDs at a distance of 8.0 kpc (Wesselink, 1987) and simulated a depth of 3 kpc for the GB along the line of sight. The Johnson B,R magnitudes are transformed to the UKST system (Wesselink, 1987). A gaussian spread in the photometric accuracy is assumed. For the reddening in the field we adopted E(B-R)=0.20±0.05 (Wesselink, 1987).

A comparison between the synthetic and photographic CMDs showed that the turnoff point in the photographic CMD is much more extended. This structure can be explained with a stellar population present between the GB and the Galactic Disk which has the same age and (almost) the same chemical composition as the GB stars !! We simulated the contribution from this stellar population with the same age and metallicity as the GB :

$$\int_0^1 dx = A_1 \int_{\log d_0}^{\log d_1} r dr + A_2 \int_{d_1}^{d_2} dr$$

With d_0 =1.0 kpc , d_1=6.5 kpc and d_2=9.5 kpc. A_1 and A_2 are normalisation constants. The ratio between the number of stars in- and outside the galactic bulge is a free parameter. For a ratio of approximately one the global shape of the synthetic CMDs and LFs is comparable with the photographic CMD and LF.

References

Bertelli, G. et al. (1990), Astron. Astrophys. Suppl. Ser. **85**, 845
Chiosi, C. et al. (1988), Astron. Astrophys. **196**, 84
Plaut, L. (1973), Astron. Astrophys. Suppl. Ser., **4**, 75
Wesselink, T. (1987), Ph.D. Thesis , University of Nijmegen, The Netherlands

ON THE LOCAL VELOCITY DISPERSIONS

S. NINKOVIĆ
Astronomical Observatory
Volgina 7
11050 Beograd
Yugoslavia

ABSTRACT. A new formula for the ratio of the planar velocity dispersions containing an asymmetric-drift term is proposed. It yields a good agreement with the flat rotation curve of the Galaxy.

One should bear in mind a few essential facts. Firstly, the stars of the galactic thin disc move around the galactic centre along nearly circular orbits and consequently one can derive a formula for the ratio of their planar velocity dispersions (e. g. [1]-p. 120). Secondly, the observations suggest that the ratio of the radial velocity dispersion to the transverse one is about 2.6 at the Sun (e. g. [2]). Thirdly, the galactic rotation curve seems fairly flat near the Sun (e. g. [3]).

What we measure is the deviation from the centroid velocity, not from the circular one. Therefore, the transverse residual velocity of a star must contain the asymmetric drift because on average the velocities of stars deviate systematically from the circular velocity by the amount of the asymmetric drift. If one accepts the value of about $14\,kms^{-1}$ for the local asymmetric drift concordant with the observations [2] and corrects the transverse velocity dispersion for its square, then one easily concludes that the rotation curve should be flat at the Sun.

REFERENCES

1. Binney, J. and Tremaine, S. (1987) Galactic Dynamics, Princeton University Press, Princeton, New Jersey.
2. Freeman, K. C. (1987) 'The Galactic Spheroid and Old Disk', Annual Review of Astronomy and Astrophysics 25, 603-632.
3. Fich, M., Blitz, L. and Stark, A. A. (1989) 'The Rotation Curve of the Milky Way to $2R_{\odot}$', Astrophysical Journal, 342, 272-284.

ON THE TOTAL KINETIC ENERGY OF OUR GALAXY WITH THE CONTRIBUTION OF THE POPULATIONS

S. NINKOVIĆ
Astronomical Observatory
Volgina 7
11050 Beograd
Yugoslavia

ABSTRACT. The total kinetic energy of our Galaxy is estimated to be $(0.55 - 3.3) \times 10^{16}$ $M_{\odot}\, km^2 s^{-2}$ and the specific one to about $4 \times 10^4\, km^2 s^{-2}$.

The kinetic energy of the Galaxy is calculated through the potential energy by assuming the virial theorem. In the calculation of the latter one it is assumed that there are three main contributors: the (central) bulge, the (thin) disc and the (dark) corona.

One also examines the contributions to the total kinetic energy of different galactic populations such as: thin disc, thick disc, bulge, halo, corona etc. According to what we know about the structure and kinematics of individual subsystems of the Galaxy (e. g. [1], [2], [3]) it seems that the kinematics of their internal parts (where the density is not negligible) suggests rms velocities of about 200 kms^{-1}, or slightly less. As a rough approximation one may use the statement that the fractions of the galactic populations in its total kinetic energy are equal to those of the mass. However, a more refined analysis discovers that the specific kinetic energy of the corona becomes the highest, especially in the case of a very high contribution of its to the total mass of the Galaxy (high local escape velocity). The case of the thin disc is, certainly, the most favourable since its kinematics can be approximated by a pure rotation and the structure by an exponential law. Since the effective boundary of the thin disc is much smaller than that of the corona, the extension of the latter one has no influence on the kinetic energy of the disc. There are no reasons to be different in the case of the other subsystems containing the "visible" matter. Thus one concludes that the specific kinetic energy of the dark corona is probably the highest among the galactic populations and therefore its fraction in the total kinetic energy of the Galaxy slightly exceeds that of its mass.

REFERENCES

1. Freeman, K. C. (1987) 'The Galactic Spheroid and Old Disk', Annual Review of Astronomy and Astrophysics 25, 603-632.
2. Blanco, V. and Terndrup, D. M. (1989) 'Longitude Distribution of Bulge M Giants: the Mass and Large-Scale Structure of the Spheroid', Astronomical Journal 98, 843-852.
3. Carney, B. W., Latham, D. W. and Laird, J. B. (1990) 'A Survey of Proper-Motion Stars.X.The Early Evolution of the Galaxy's Halo', Astronomical Journal 99, 572-589.

A MODEL OF THE GALAXY FOR STAR COUNTS IN THE INFRARED

ORTIZ, R.P., LÉPINE, J.R.D.
Instituto Astronômico e Geofísico
Universidade de São Paulo
P.O. Box 9638, CEP 01065
São Paulo, Brazil

We constructed a model for the Galaxy to reproduce near-infrared (J, H, K, bands) and mid-infrared 12 and 25 micron IRAS bands source counts for a given direction and sensitivity of detection. The Galaxy is conceived as being formed by two populations: a disk, with different scale heights for each spectral type, and a de Vaucouleurs spheroid. The interstellar extinction is considered to be proportional to the amount of gas along the line of sight; both neutral (HI) and molecular (H2) components were taken into account, using the HI density distribution of Burton & Gordon (1978) and the CO density distribution of Sanders et al.(1984) and Gordon & Burton (1976).

We investigated the presence of circumstellar dust shells around giant stars, using the sample of Hacking et al.(1985); the fraction of stars with infrared excess rises steeply beyond spectral type M4. To compute the absolute luminosity of these stars in the infrared we used a spherically symmetric model for the density and temperature of dust similar to that of Epchtein et al.(1990).

The model fits well the near-infrared and the 12 micron IRAS band source counts, but it fails at 25 micron due to the contamination of non-stellar galactic objects. The spiral arms of the Galaxy are detected on the isocounts maps.

References:

Burton, W.B., Gordon, M.A.(1978): A.&A., 63, 7

Epchtein, N., Le Bertre, T., Lépine, J.R.D.(1990): A.& A., 227, 82

Gordon, M.A., Burton, W.B.(1976): Ap.J., 208, 346

Hacking, P., Neugebauer, G., Emerson, J., Beichman, C., Chester, T., Gillett, F., Habing, H., Helou, H., Olnon, F., Rowan-Robinson, M., Soifer, B.T., Walker, D.(1985): P.A.S.P., 97, 616

Sanders, D.B., Solomon, P.M., Scoville, N.Z.(1984): Ap.J., 276, 182

METALLICITY GRADIENTS IN THE DISKS OF S0 GALAXIES[1]

A. Paquet, R. Bender, W. Seifert
Landessternwarte Königstuhl
D-6900 Heidelberg, Germany

The gaseous component in disks of spiral galaxies generally shows strong metallicity gradients of about 0.2 dex per scale length of the disk (e.g. Edmunds and Pagel 1984, MNRAS 211,507). Current models explain these gradients by a radially varying ratio of stellar density to gas density, possibly coupled with infall of gas from outside and radial gas flows in the disk (Lacey and Fall 1985, Ap.J. 290,154). It follows from these models that the final metallicity gradient (i.e. the gradient after infall has ceased and all gas has been locked into stars) should be quite shallow.
Observationally the question of the final metallicity gradient in disks can be accessed best via the investigation of disks in S0 galaxies. Although it is still a matter of debate whether S0's are just spirals in which the gas supply has run out several Gyrs ago or whether S0 galaxies are fundamentally different from spirals (e.g. with respect to D/B ratios), the metallicity gradients observed in the stellar disks of S0's may indicate the strength of the final metallicity gradients in the disks of spirals. For this purpose we derived radial Mg_2-profiles for a sample of S0 galaxies. Assuming that the mean age in the disks does not vary radially we can roughly relate Mg_2 to [Fe/H] following Terlevich et al. (1981, MNRAS 196,214) by $[Fe/H] = 3.9\ Mg_2 - 0.9$. From our data (see figure) we obtain a mean metallicity gradient in the disks of S0's of only 0.03 ± 0.03 dex in [Fe/H] per scale length. This average gradient is about a factor 6 weaker than the metallicity gradients measured in HII region analysis for the gaseous disks of spiral galaxies.

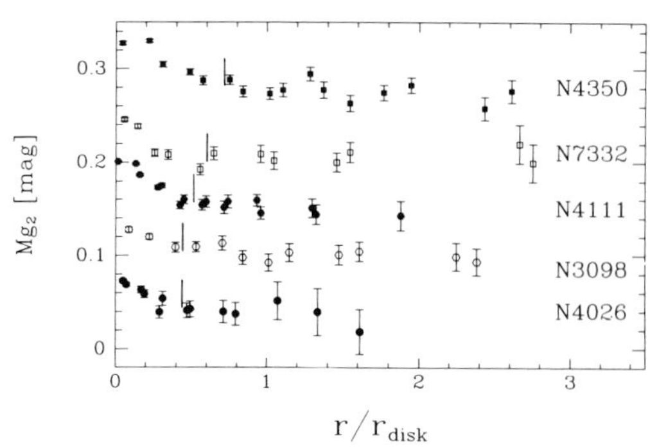

Figure:
Mg_2-index as a function of radius (normalized to the scale length of the disk r_D).The Mg_2-indices of the following galaxies have been shifted: NGC 4111 by -0.05, NGC 3098 by -0.11 and NGC 4026 by -0.2 [mag].The vertical bar indicates the radius inside which the bulge dominates the light.

[1] Based on observations obtained at the German-Spanish Astronomical Center, Calar Alto, operated by the Max-Planck-Insitut für Astronomie, Heidelberg, jointly with the Spanish National Commission for Astronomy.

30 DORADUS: THE STELLAR CONTENT AND IMF

JOEL Wm. PARKER
CASA, Campus Box 389
University of Colorado
Boulder, CO 80309-0389
USA

I have *UBV* photometry for roughly 1500 stars and *BV* photometry for an additional 900 stars in the 30 Doradus region as a part of my thesis work. Spectra and spectral classifications for over 200 stars have been obtained (by me and various other sources) for most of the early stars in the region. The reductions and classifications are nearly complete. It should be noted that the OB associations of 30 Doradus present unique problems in the analysis of the stellar contents due to the extreme crowding and highly variable nebulosity and reddening.

Using these preliminary data I obtain an initial mass function (IMF) for 30 Doradus having a slope of $\Gamma = -1.5$ (in this notation, the Salpeter slope is -1.35) for stars with masses $> 9\ M_{\odot}$. However, a single value for the IMF slope can be misleading, and should not be taken to imply that the slope is constant over the entire region. It is fairly clear that there have been several episodes of star formation in 30 Doradus during the last few million years, and one should not assume *a priori* that the slope of the IMF will be the same everywhere. More likely, the value quoted here is the average of different slopes in different regions.

This is supported by recent findings by Parker *et al.* (1991) for the OB associations Lucke-Hodge 9 and 10. Although these two associations are neighbors and supposedly share a similar environment, their slopes are $\Gamma = -1.6$ and $\Gamma = -1.0$ respectively. Parker & Walborn (1991) discuss that the morphologies of 30 Doradus and Lucke-Hodge 9 and 10 are very similar (two million years ago the region N 11 which contains Lucke-Hodge 9 and 10 could have looked very much like 30 Doradus does now). This leads to the possibility that in further analysis I will find that the inner region of 30 Doradus (corresponding to Lucke-Hodge 9 in N 11) will have a steeper IMF slope while the outer regions (corresponding to Lucke-Hodge 10 in N 11) will have flatter slopes.

REFERENCES

Parker, J.Wm., Garmany, C.D., Massey, P., and Walborn, N.R. 1991, AJ, submitted
Parker, J.Wm. and Walborn, N.R. 1991, ApJL, submitted

THE RSG AND WR CONTENT OF THE CIRCUMNUCLEAR HII REGIONS OF NGC 3310

M.G. PASTORIZA[1], H. DOTTORI[1], E. TERLEVICH[2], R. TERLEVICH[2] & A. DIAZ[3]
1- Instituto de Física-UFRGS, Porto Alegre - RS, Brazil. 2- Royal Greenwich Observatory, U.K. 3- Dpto de F. Teórica, Universidad Autónoma, Madrid, Spain.

We have analyzed the stellar content of the nucleus and circumnuclear HII regions of NGC 3310 (Figure1). The observations were made with the IDS and the blue coated GEC-CCD detector, at the INT telescope in La Palma, covering the spectral range 3600Å to 9700Å, with a spectral resolution of 4Å. We report the following results: *1-* The very nucleus is highly composed in ages, with similar contribution at 5500Å of an old-, metal-poor $Z \leq -0.5\ Z\odot$, an intermediate-, and a younger than 50 Myrs, population. In terms of Bica&Alloin library (1986) its stellar contents is represented by; $0.1HII + 0.18(Y1 + Y2) + 0.14(I1 + I2) + 0.38(G2 + G3 + G4)$ (Figure 2). *2-* The dominant stellar population of the circumnuclear HII regions is younger than 10 Myrs, with a strong presence of RSGs, specially in region L where we have demonstrated dinamically that the RSGs belong to the ionizing cluster. Figure 2-a and 2-b shows the observed and synthesized nuclear spectrum respectively, while 2-c and 2-d shows the same for region L. *3-* In region A (the JUMBO region) we have detected the WR feature at λ 4686Å. More than 200 WR are necessary to explain this emission. Two cycles of star formation constitute a good scenario to explain the whole spectral characteristics. A 15 Myrs old cycle furnishes the strong observed near-IR emission, and a 4.5 Myrs old one takes account for the WR feature. A one cycle scenario does not furnish the necessary ratio WR/RSG, within the framework of Maeder's evolutionary tracks.

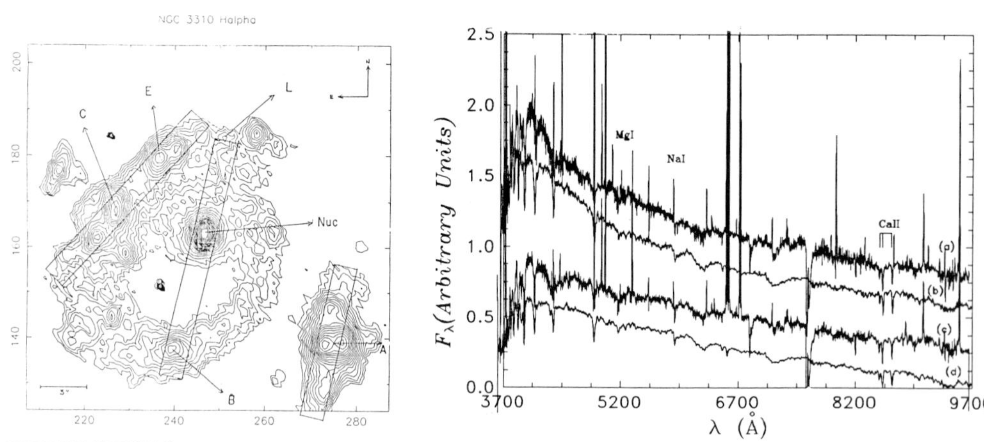

REFERENCES
Bica, E. and Alloin, D., 1986, A& Ap, 166, 83.
Terlevich, E., Diaz, A., Pastoriza, M., Terlevich, R. and Dottori, H., 1990, MNRAS 242,48p.

HOW TRANSPARENT ARE SPIRAL GALAXIES?
An analysis from a near-infrared perspective

R. F. Peletier and S. P. Willner
Harvard-Smithsonian Center for Astrophysics
60, Garden Street
Cambridge, MA 02138, USA

Until recently is was commonly thought that most spiral galaxies were transparent or contained only modest amounts of absorbing dust, making them optically thin in the B-band (Holmberg 1958). However Disney *et al.* (1989) showed that there was very little evidence for this. Using a very large optical database Valentijn (1990) showed that surface brightness in B is almost independent of inclination, implying that spiral galaxies in general are optically thick in B.

In an effort to find out what the distribution of the obscuring component is, we have investigated the inclination dependence of surface brightness of surface brightness in the H-band, where the effects of extinction are much less important than in B. We have taken a large sample of H-band aperture photometry from the literature (Aaronson *et al.* 1982), together with H-band images of a few edge-on galaxies. The selection criteria for this sample were independent of inclination. For each galaxy the surface brightness inside an isophote containing a fixed fraction of the galaxy light was determined by converting magnitudes in circular apertures to magnitudes in elliptical apertures, removing the surface brightness — mass dependence using the HI velocity width, and correcting for the fact that the fraction of the light of a galaxy inside an isophote of fixed surface brightness changes with inclination (see Peletier & Willner 1991 for details).

We have found a C-value (see Valentijn 1990) in H between 0.6 and 0.8, which impies that spiral galaxies in H behave in a semi-transparent way. Re-analyzing the optical data, after removing the most edge-on galaxies (5 %), C_B is found to be around 0.5. Using sandwich models (Disney *et al.* 1989) these numbers imply that the ratio of the scale height of the dust and stars has to be ≈ 0.6. For two nearly edge-on galaxies we find from near-infrared images that this is indeed the case, while in B this ratio is ≈ 0.3. This shows that dust with similar properties as in our galaxy is responsible for the obscuration in B. The sandwich models also show that for an average spiral galaxy within 0.3 D_{25} the optical depth in $B \approx 0.95$, and the face-on absorption ≈ 0.46 mag, almost $3 \times$ as high as previously assumed in the RC2.

REFERENCES:

Aaronson, M. *et al.* 1982, *Ap. J. Suppl.*, **50**, 241.
Disney, M. J., Davies, J. and Phillipps, S. 1989, *M.N.R.A.S.*, **239**, 939.
Holmberg, E. 1958, *Medn. Lunds Astr. Obs.* **2**, 136.
Peletier, R. F. and Willner, S. P., submitted to the *Astronomical Journal.*
Valentijn, E. A., 1990, *Nature*, **346**, 153.

THE HALO PLANETARY NEBULAE M2-29 AND BB-1

M. PEÑA, S. TORRES-PEIMBERT
I. Astronomía UNAM, Apdo. P. 70-264, México 04510 DF México

M.T. RUIZ
D. Astronomía, U. de Chile, Casilla 36-D, Santiago Chile

The physical conditions and chemical abundances were determined for M2-29 and BB-1 from optical spectrophotometric data and from available IUE material (Table 1). M2-29 has the lowest O/H ratio in PNe and it belongs to the Ar and S rich type IV PN group while BB-1 is very C, N, and Ne rich being Ar and S poor. Comparing with all known PNe of type IV it is found that C/N/O/Ne/Ar do not vary in lockstep, but C, N, O and Ne appear enriched in Ar poor objects (Table 2). C and N are expected to be enriched by the central star, however the anomalous O/Ne/Ar behavior is not understood unless O and Ne are also enriched by the progenitor star. The paper in full will appear in P.A.S.P. 1991.

Table 1. Derived Temperature, Density and Chemical Composition

	T_e	N_e	He	C	N	O	Ne	S	A
M2-29	24000	3000	10.97	—	6.98	7.31	6.72	5.91	5.26
BB-1	14500	3000	11.02	9.16	7.94	7.68	7.76	5.80	4.74

Table 2. Comparison with other objects (in 12 + log X/H)

object	He	O	C/O	N/O	Ne/O	S/O	Ar/O
M 2-29	10.97	7.3	—	-0.3	-0.6	-1.4	-2.0
BB-1	11.02	7.7	+1.5	+0.2	+0.1	-1.9	-3.0
K 648	11.02	7.7	+1.0	-1.2	-1.0	-2.5	-3.4
H 4-1	10.99	8.4	+0.9	+0.1	-1.7	-3.2	-3.7
NGC4361	11.02	7.8	+0.5	<-0.4	-0.2	—	-1.9
NGC2242	11.00	8.0	+0.4	-0.3	-0.2	—	-2.1
DDDM-1	11.00	8.1	<-1.0	-0.7	-0.7	-1.6	-2.3
PN06-41.1	10.96	8.1	<-0.8	—	-0.6	—	-2.3
PN242-37.1	11.03	8.4	<-0.8	<-0.4	-0.5	—	-2.0
Sun	—	8.9	-0.2	-0.9	-0.8	-1.6	-2.3

MOVING CLUSTERS AMONG GALACTIC HALO STARS

A. Poveda[(1)], C. Allen[(1,2)] and W. Schuster[(1,3)]

[1] Instituto de Astronomía, UNAM,
Apdo. 70-264, 04510 México, D. F.

[2] Dirección General de Servicios de Cómputo Académico, UNAM, 04510 México, D. F.

[3] Instituto Nacional de Astrofísica, Optica y Electrónica, Apdo. 51 y 216, 72000 Puebla, Pue., México.

ABSTRACT. It has been suggested that most of the primeval galactic globular clusters have been dissociated by encounters with molecular clouds, and by tidal effects in the vicinity of the galactic center; their stellar remnants would move preserving some memory of their original kinematical parameters and might resemble the well-known population I moving clusters. However it has been shown, by direct numerical integration, that a significant fraction of the halo stars have chaotic orbits and many others reach high z-values; hence the usual criterium of identifying group members by the similarity of their V-velocities breaks down. Instead, we use the constancy of the integrals of motion, i.e. the total energy and the z- component of the angular momentum, as well as the metallicity [Fe/H], as the identifying parameters for group membership of a given star. From the analysis of the metallicities and the integrals of motion of 206 halo stars, we have been able to identify several halo moving groups.

CONCLUSIONS

a) Because of the high incidence of chaotic orbits among halo stars, the similarity of the velocities, in particular the V-component, by itself does not seem to be a very good criterium to identifiy group members, b) The integrals of motion in an axis symmetric potential, the energy E and the z-component of the angular momentum h, as well as the iron abundance [Fe/H] are three independent parameters appropriate to identify moving groups in the halo population, c) We have identified 5 new possible moving groups among the sample of 206 halo stars, d) Halo moving groups must have originated in very rich and sparse parent clusters in the halo that have dissociated by tidal encounters with the galactic center. These clusters may be identified with globular clusters or dwarf satellite galaxies.

REFERENCES

Eggen, O.J., 1965, in: Galactic Structure, eds. A. Blaauw and Marten Schmidt. University of Chicago Press, p. 111.
Eggen, O.J., 1987, in: The Galaxy, eds. G. Gillmore and B. Carswell, Reidel Dordrecht, p. 211.
Schuster, W.J., Nissen, P.E. 1988, A&AS 73, 225.
Schuster, W.J., Nissen, P.E. 1989a, A&A 221, 65.
Schuster, W.J., Nissen, P.E. 1989b, A&A 222, 69.

POPULATION SYNTHESIS IN STARBURST GALAXIES

Mª Almudena Prieto
Space Telescope-European Coordinating Facility
Karl SchwarzschildStr 2
8046 Garching bei München
Germany
Affiliated to the Astrophysics Division, ESA

Abstract

Spectral energy distribution for a sample of starburst galaxies has been synthesized. The proposed sample contains galaxies whose optical light is mainly dominated by an early type galaxy population. A low excitation nebular spectrum is superimposed whose strength in the Halfa emssion line suggests the presence of very massive stars responsible of the observed ionization.
In order to reproduce their integrated spectral distribution a composite model is proposed. This esentially considers the contribution of an old population in which a new burst of star formation is undertaken. The burst is assumed to be a combination of ionizing clusters with ages between 3.5 and 10 Myr. The proposed model can adequately fit the observed both continuum distribution and Balmer absorption lines and accounts for the emitted Ha luminosity. The inferred mas of the ionizing cluster for the sample of galaxies analized is of the order of 10E7 Mo. The figure shows a representative spectrum of the sample. The derived model, superimposed in the figure, contains an EO population contributing 70% to the light at 5500 A, the remaining light been due to a combination of two young clusters, 4 and 10 Myr old.

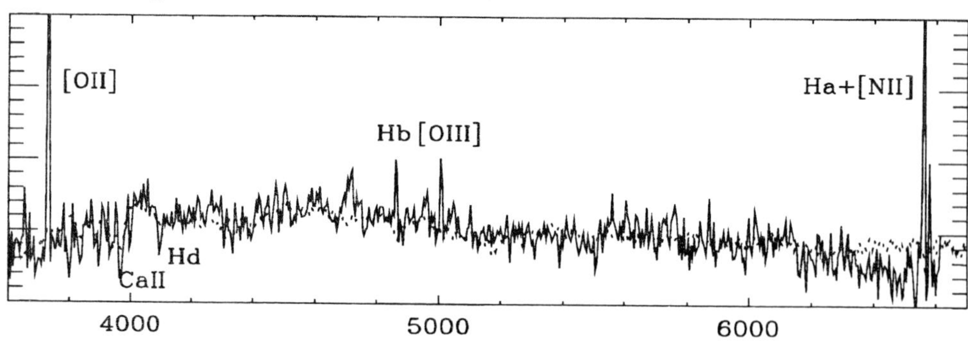

CARBON STARS IN THE SMALL MAGELLANIC CLOUD: POSITIONS, FINDING CHARTS AND SPECTROPHOTOMETRY

E. REBEIROT[1], M. AZZOPARDI[1], B.E. WESTERLUND[2]

[1]*Observatoire de Marseille, 2, Place Le Verrier*
13248 Marseille Cedex 4, France
[2]*Astronomical Observatory, Box 515*
S-751 20 Uppsala, Sweden

A survey for carbon stars in the Small Magellanic Cloud (SMC) has been carried out with the triplet camera at the prime focus of the ESO 3.6 m telescope using a Grism giving a dispersion of 2200 A/mm. 13 circular fields of 0.78 sq. degrees each, partially overlapping, cover the main body of the SMC (Rebeirot et al. 1987).

By combining a IIIa-J emulsion with a Schott GG435 filter the spectral range obtained on the plates is 4350-5300 A, thus keeping the number of overlaps as low as possible. Exposures of 5 min and 60 min allowed us to survey, even in the most crowded regions, the SMC carbon stars up to the limiting magnitude $m_{pg} \sim 20$; see for instance the field of the globular cluster NGC 419 (Azzopardi et al. 1986).

Visual inspection of the plates resulted in the identification of 1707 carbon star candidates. Comparison of our "green" spectral survey with the "near infrared" one by Blanco et al. (1980), for the regions that we have in common, shows that we have achieved a reasonable degree of completeness. This has been carefully discussed by Westerlund et al. (1986), McCarthy (1987), and Blanco and McCarthy (1990). The structure of the SMC derived from this carbon star survey has been shown by Azzopardi and Rebeirot (1991).

All the identified carbon stars have been scanned individually on the 60 min Grism plates with a PDS microdensitometer. The two dimensional displayed image of each Grism spectrum and of its surrounding region helped us in selecting some central scan lines (~ 10) to be averaged and normalized in order to obtain the related one-dimensional spectrogram. Density-to-intensity transformation, and data calibration using medium resolution slit spectroscopy of selected carbon stars, allowed us to get the following quantities for most of them (see Westerlund et al. 1986):

- a magnitude $m_1 = m(5220)$,
- a colour index, $m_3 - m_1 = m(4850) - m(5220)$,
- two measures of the strength of the C_2 band at 5165 A, namely the equivalent width and the depth of the band below the pseudo-continuum at $m_2 = 5030$ A.

The positions (Eq. 2000.0) for the 1707 carbon stars that we have discovered in the SMC are listed in a catalogue as well as the spectrophotometric measurements, remarks and cross-identifications, when available. Finding charts (2 arcmin sq. each) for all of them are provided (Rebeirot et al. 1992).

References

Azzopardi, M., Dumoulin, B., Quebatte, J. and Rebeirot, E. (1986), The Messenger 43, 12.
Azzopardi, M. and Rebeirot, E. (1991) IAU Symp. 148, in R. Haynes, D. Milne (eds.), The Magellanic Clouds, Kluwer Academic Publishers, Dordrecht, pp. 71-76.
Blanco, V.M. and McCarthy, M.F. (1990) Astron. J. 100, 674.
Blanco, V.M., McCarthy, M.F. and Blanco, B.M. (1980) Astrophys. J. 242, 938.
McCarthy, M.F. (1987) ESO conf. & Workshop Proc. No. 27, M. Azzopardi, F. Matteuci (eds.) p. 203.
Rebeirot, E., Azzopardi, M., Breysacher, J. and Westerlund, B.E. (1987), ESO conf. & Workshop Proc. No. 27, M. Azzopardi, F. Matteuci (eds.) p. 263.
Rebeirot, E., Azzopardi, M. and Westerlund, B.E. (1992) Astron. Astrophys. Suppl. (in preparation)
Westerlund, B.E., Azzopardi, M. and Breysacher, J. (1986) Astron. Astrophys. Suppl. 65, 79.

INFRARED PROPERTIES OF GALAXIES IN CLUSTERS: ABELL 194, ABELL 426 (PERSEUS) AND ABELL 2151 (HERCULES).

E. Recillas-Cruz [1,2] A. Serrano P.-G.[1,2] L. Carrasco[1,3] V. Ortega[1]

[1] Instituto de Astronomía, UNAM Apartado Postal 70-264, 0451 México, D.F.
[2] Programa Universitario de Investigación y Desarrollo Espacial, UNAM Apartado Postal 70-372, 04510 México, D.F.
[3] Observatorio Astronómico Nacional P.O. Box 439027, San Ysidro CA 92143-9027

ABSTRACT. We report here an observational study on the IR properties of members of the rich clusters of galaxies: Abell 194, Perseus and Hercules. Following the precepts described in previous papers concerning the manifold of the early-type galaxies from IR photometry for the Coma (Recillas-Cruz et al. 1990) and Virgo (Recillas-Cruz et al. 1991) cluster members; interstellar reddening and redshift corrections for Abell 194, Perseus and Hercules were estimated. Interstellar reddening corrections for Abell 194 and Hercules galaxy members were found to be small, except for Perseus cluster galaxies where extinction values are somewhat larger. IR redshift K-corrections were estimated from linear relations with z for (J-H), (H-K) and K (Persson et al. 1979). Corrected magnitudes and colors were then used to construct (J-H) vs. (H-K) diagrams for elliptical and S0 galaxies and color-magnitude diagrams (J-H), (H-K), (J-K), (B-K) and (V-K) vs. K.

Our main results are: (i) (J-H) vs. (H-K) color-color diagrams for early-type galaxies do indeed have colors typical of red giant bulge stars, (ii) In Abell 194, (J-H) vs. (H-K) diagram shows a larger dispersion in the (H-K) colors, as compared to similar diagrams for Perseus and Hercules clusters. (iii) Color-magnitude relations exists in the (B-K) and (V-K) colors vs. K for all morphological types, in the sense of luminous spheroidal components being redder in the IR. This result may reflect dynamically related differences amongst galaxies, or differences in the stellar luminosity function and/or environmental effects (Recillas-Cruz & Serrano 1991).

Based on observations obtained at Observatorio Astronómico Nacional (OAN), San Pedro Mártir, B.C., México.

REFERENCES

Persson, S.E., Frogel, J.A., Aaronson, M. 1979, ApJS, 39, 61
Recillas-Cruz, E., Serrano P.G., A. 1991, ApSS, in press.
Recillas-Cruz, E., Carrasco L., Serrano, A., Cruz-Gonzalez, I. 1990, A&A, 229, 64 (Paper I).
Recillas-Cruz, E., Carrasco L., Serrano, A., Cruz-Gonzalez, I. 1991, A&A, in press, (Paper II).

SYNTHETIC P-AGB POPULATIONS

ALVIO RENZINI[1] AND LETIZIA STANGHELLINI[2]
[1]Dipartimento di Astronomia, Università di Bologna
via Zamboni 33, I-40126 Bologna, Italy
[2]Osservatorio Astronomico di Bologna
via Zamboni 33, I-40126 Bologna, Italy

Extensive Montecarlo simulations of populations of post-Asymptotic Giant Branch (P-AGB) stars have been constructed. The time evolution of luminosity and effective temperature for any stellar mass has been approximated by suitable analytical approximations based on the P-AGB evolutionary tracks of Paczyński (1971, Acta Astron. 21, 417) and Schönberner (1983, A&A 79, 108). By constructing synthetic HR diagrams for P-AGB stars we explore the effects of various assumptions, such as the IMF, the initial mass-final mass relation, the AGB to PN transition time, etc. We have also investigated how the uncertainties in the various assumptions and observational errors in luminosity and temperature propagate into the inferred mass distribution of the P-AGB stars. As an example of the possible applications, we show in the Figure the mass distributions of two P-AGB populations, which differ only on the effective temperatures. We have assumed a systematic error in the *observed* temperature, such as $\log T_{\rm obs} = 1.01 \cdot \log T_{\rm true} + 0.01$. Such an error is very small in comparison to usual Zanstra temperature uncertainties for Planetary Nebula Nuclei, yet the two distributions are remarkably different. The broken line represents the *observed* distribution. The mean mass are respectively $< M/M_\odot >= 0.594$ for the *true* distribution (i. e. using $T_{\rm true}$), and $< M/M_\odot >= 0.628$ for the *observed* one (i. e. using $T_{\rm obs}$). This simulation emphasizes the need for accurate temperatures in order to obtain useful mass distributions for P-AGB stars, when using an *observed* distribution of P-AGB stars in the HR diagram. Further applications and details will be included in a forthcoming paper (Stanghellini & Renzini, *in preparation*).

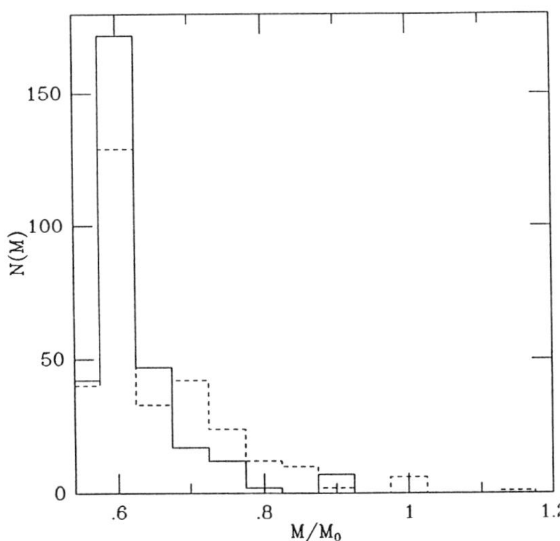

ANALYSIS OF VERY RICH K GIANT STARS

RAMIRO DE LA REZA and LICIO DA SILVA
Observatorio Nacional - CNPq
Rua Gal. Bruce 586 - CEP 20921 - Rio de Janeiro - RJ
Brazil.

The detection of a strong resonance line of Li I at 6708Å in the spectra of some apparently normal K giant stars, has roused what we can call the "K Giant Lithium Problem". In fact, according to the standard stellar evolution theory this element must be strongly depleted in the atmospheres of these stars. No satisfactory explanation of this peculiarity has been given to the present time. Anyway, to understand the involved mechanisms it is important to obtain the best possible values of the corresponding Li abundances. Our method use the secondary Li line at 6104Å which is very sensitive to abundances, specially when these are high. This is not the case of the strong and saturated resonance line. The weaker 6104 line is formed deeper in the stellar atmosphere than the resonance line, being less affected by non LTE effects. In this work we present partial results obtained with LTE theory (a general non LTE approach will be publish elsewhere) for the following stars: HD 787, HD 39853 and HD 19745. The first two are known bright K giant stars having respectively medium and high intensity 6104 lines. The last star is a member of a group of weak K giant stars recently discovered at the LNA (Brazil) in a search based on IRAS point sources. This star has the strongest 6104 line detected up to now in those kind of stars. The observations of the 6104 line were performed with the 1.4m Coude Auxiliary Telescope at ESO with a spectral resolution of 0.062Å. Each program star was observed at least twice and the S/N ratios of the combined spectra exceed 200. Other spectral ranges were observed to determine the Fe, O, C and N abundances. With the main parameters (Teff, log g, V turb.) for stars: HD 787 (3970, 1.7, 1.7); HD 39853 (3900, 1.2, 1.0); HD 19745 (4990, 2.1, 1.7); we obtain the following LTE abundances for Li by means of the 6104 line only (in the scale log H=12): HD 787, logLi=2.20; HD 39853, logLi=2.90; HD 19745, logLi=4.08. Concerning these three objects, HD 39853 is the only high velocity object and HD 19745 is the only IRAS source. The main conclusion is that the richest Li star HD 19745 seems to have a Li abundance larger than that of the mean interstellar medium (logLi=3.30). The explanation of this fact constitute a challenge to the classical theory of evolution of these objects. In fact, this apparently low mass star has not sufficient core mass to generate an extra quantity of ^{3}He which could explain a large ^{7}Li enrichment, by means of the ^{7}Be mechanism.

PHOTOMETRY OF GALACTIC GLOBULAR CLUSTERS OF THE DISK SYSTEM

T. RICHTLER, E.K. GREBEL, W. SEGGEWISS
Sternwarte der Universität Bonn, Germany

The ages of the galactic globular clusters belonging to the disk subsystem are still widely unknown. By considering the HB morphology Richtler et al. (1991) were led to the conjecture that some of the disk clusters may resemble the populous intermediate-age clusters in the Magellanic Clouds, but no main-sequence photometry was published at that time. We now present B,V photometry for the clusters NGC 6496, 6624, and 6637 (M69) which in all cases show the turn-off point and allow reliable age determinations.

The data were taken with the MPIA 2.2m telescope at ESO, La Silla in June 1990. Local standards in NGC 6624 and 6637 enabled a firm photometric scale, while the zero-point for NGC 6496 is still preliminary. DAOPHOT was used for the reduction of the CCD frames.

In all cases HB-TO, the magnitude difference between the horizontal branch and the turn-off, is 3.5±0.1 mag, thus placing our objects among the **old, "classical", globular clusters.**

Recently, Hatzidimitriou (1991) suggested that NGC 6342 could have a lower age limit of 5 ± 2 Gyrs, thus being the first representative of a class of intermediate-age clusters. We observed it recently at La Silla. The weather conditions, howewer, were very bad, but even with the limited quality of our V, (V-R) diagram we can reject an age of 5 Gyrs, while the minimum age for the clusters is about 8 Gyrs.

We also considered the morphological parameters $\Delta V_{1.4}$ and $(B-V)_{o,g}$ of the CMDs to estimate the metallicities of NGC 6624 and 6637. We got from $\Delta V_{1.4}$ -1.1 dex and -0.9 dex for NGC 6624 and 6637, resp., and from $(B-V)_{og}$ -0.9 dex and -0.8 dex. These values are considerably metal poorer than the values derived from integral photometry.

Other work also indicates that the metallicity scale of "metal rich" clusters is not yet firmly established: Friel and Geisler (1991) derived from Washington CCD-photometry -1.05 and -0.25 dex for NGC 6496 and 5927, resp. François (1991) got [Fe/H] = -1.08 dex for NGC 5927 (!) from high resolution spectroscopy. A new, homogeneous scale, based on individual stars rather than integrated properties is urgently needed. CCD photometry in intermediate/narrow bands is a promising approach.

REFERENCES
François, P. (1991) Astron. Astrophys. **247**, 56
Friel, E.D., Geisler, D. (1991) Astron. J. **101**, 1338
Hatzidimitriou, D. (1991) Monthly Notices Roy. Astron. Soc. **251**, 545
Richtler, T., de Boer, K.S., Vallenari, A., Seggewiss, W. (1991) in: IAU Symp. No. 148, The Magellanic Clouds, R. Haynes and D. Milne (eds.) Kluwer, Dordrecht, p. 198

THE WOLF-RAYET AND O STAR CONTENT OF VIOLENT STARBURSTS

CARMELLE ROBERT, LAURENT DRISSEN, and CLAUS LEITHERER
Space Telescope Science Institute, 3700 San Martin Drive, Baltimore, MD 21218.

New population synthesis models for starburst regions have been computed, with an emphasis on the W-R population and the output of ionizing photons.

Our models use Maeder's (1990, *A&AS* **84**, 139) latest evolutionary tracks which incorporate metallicity-dependent mass-loss rates. The most recent sets of model atmospheres, which consider the spherical extension of the envelope, NLTE and line-blanketing effects, are taken from Kudritzki *et al.* (1991, *Massive Stars in Starbursts*). Different values of the IMF slope and the lower and upper cut-off mass have been tested. Metallicities between $Z = 0.1Z_{\odot}$ and $Z = 2Z_{\odot}$ were considered. We also used small timesteps of the order of 10000 years.

Different durations of the burst were considered. This is particularly important in starburst galaxies where the hypothesis of an instantaneous burst is inconsistent with the size of the burst region. Adopting a burst duration of 5Myr (much smaller than the age of the parent galaxy but comparable to the lifetime of the most massive stars) we obtain, for example (figure 1), that the maximum WR/O ratio is smaller and that the W-R phase is stretched in time compared to the case of the instantaneous burst.

Some emission-line galaxies, especially at low metallicity, show the presence of narrow HeII 4686 emission (Campbell *et al.* 1986, *MNRAS* **223**, 811; Conti 1991, *Ap.J.*, **377**, 115) at a level of $F_{4686}/F_{H_{\beta}} \simeq$ 1-5%. Classical, hydrostatic model atmospheres clearly fail to reproduce such an emission. As shown in figure 2, the unified model atmospheres of Kudritzki *et al.* also fail to reproduce the observations (although there is an improvement with respect to the previous models). In the figure, different flux ratios are presented for different values of the IMF slope and the metallicity. The model predicts an increase of nebular 4686 emission during the W-R phase (starting around 3Myr), as well as an increase with metallicity and flatter IMF. On the contrary, the narrow 4686 line is observed in low-metallicity objects without W-R stars. The effect of dust has not been considered yet, but it is not expected to improve the situation. Other mechanisms (such as ionization by X-ray binaries, supernova explosions) may be invoked to produce the observed flux ratio.

FIGURE 1.

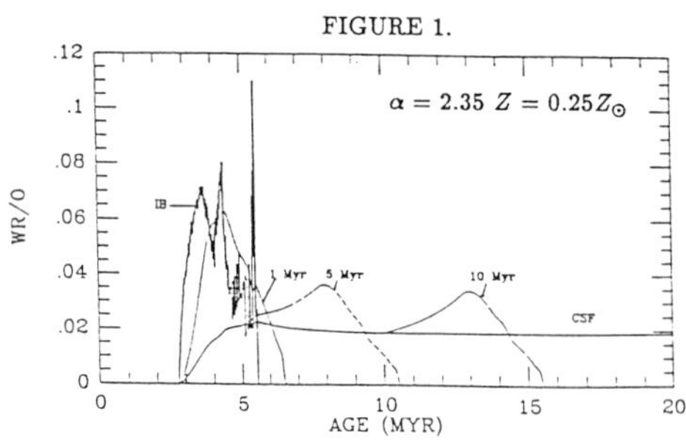

FIGURE 2.

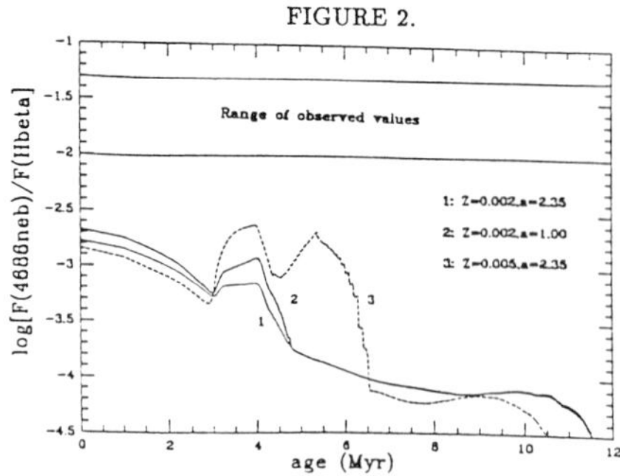

GRAIN COOLING IN COLLAPSING CLOUDS

S.C.F. Rossi, P. Benevides-Soares, B. Barbuy
IAG-USP, CP 9638, São Paulo 01065, Brazil
G. Pineau des Forêts
Observatoire de Meudon, 92195 Meudon Pl. Cedex, France

If the grains in pre-collapsing clouds are due to the contribution of winds from previous generations of stars and supernovae ejecta, they already contain a fraction if not all, of the heavy elements. The far infrared spectrum of Bok globules, which are at relatively early contraction stages, has been attributed to dusty material.

However, the actual grain content and distribution in contracting clouds can be very different from the initial one: during the contraction process itself, under low temperatures, there is a possibility of additional grain growth, owing to condensations and coagulation processes. Calculations of the grain growth in collapsing clouds are given in Rossi et al. (1991).

In the present work, grain cooling is computed and compared to molecular hydrogen cooling for clouds of different metallicities. We considered the isothermal phase of the diagrams (density, temperature) (n,T) for different metallicities, as simple as possible, adopting the relation by Silk (1977c) between gas temperature and metallicity: log T = 1 - 1/3(log z). A grain cooling function parametrized by Silk (1977b) was employed. H_2 cooling was computed assuming no UV external field, and that H_2 is formed on grain surfaces. The gas temperatures 10, 20, 50, 100, 215 and 415, for densities log n = 2, 3, 4 were considered.

The aim of these calculations was to find a temperature - density - metallicity condition where grain cooling and H_2 are of the same order. Silk (1977a) suggested a threshold value at $z = Z/Z_\odot \approx 4\,10^{-5}$, whereas Yoshii & Sabano (1980) suggested $z \approx 1.5\,10^{-5}$.

The present study shows that the threshold value depends critically on the dust temperature T_d: for a cloud of $z = 10^{-5}$, with gas temperature T = 465 K, for T_d = 20 K or 35 K, at log n = 2, grain cooling or H_2 cooling respectively will dominate. A detailed transfer work is necessary, where the external radiation will be a main parameter.

Rossi, S.C.F., Benevides-Soares, P., Barbuy, B.: 1991, A&A, in press
Silk, J.: 1977a, ApJ 211, 638
Silk, J.: 1977b, ApJ 214, 152
Silk, J.: 1977c, ApJ 214, 718
Yoshii, Y., Sabano, Y.: 1980, PASJ 32, 229

BAADE'S WINDOW PHOTOMETRY AND SPECTROSCOPY

A. Ruelas-Mayorga[1], P. Teague[2]
1 Universidad Autonoma de Mexico, Mexico
2 Laboratory for Experimental Astrophysics, Livermore, USA

An infrared scan of the Baade's Window (BW) is obtained. The cumulative K-counts function (hereafter CCF, which is the number of sources per square degree down to K = +13.5) is formed by combining 1.9m telescope scans with AAT scans. With the aid of a theoretical exponential disk model (e.g., Ruelas-Mayorga, A., Rev. Mex. Astr. Astrof. 22, 43, 1991) and observations at l = 20°, b = -5°, we decompose the observed CCF into disk-CCF and bulge-CCF components. The bulge CCF is steeper than the disk CCF in the range K = (+5.0,+11.0) showing a relative depletion of high mass stars with respect to the disk. The bulge CCF is compared with derived CCF's for the globular clusters 47 Tuc, M92, M3 and M13 and the open cluster M67. The similarity of the slope of the bulge-CCF to those of the globular clusters suggests that the stellar population of the bulge may be similar in age and perhaps also in metallicity characteristics of the stellar populations in globular clustes. Photometric studies of a bright-K subsample: 165 of the 578 sources found in BW down to K = +11.0 are made. Several sources with mild IR-excesses are found, and through spectroscopy, were confirmed to be Miras variables. The reddening E(J-K) = +0.27 agrees well with the value E(B-V) = +0.45 obtained by optical techniques. In a HR diagram most of the sources in our photometric subsample lie above the giant branch tips of ω Centauri, 47 Tuc and M92. If the giant branches (GB) of these clusters are extrapolated to higher K brightnesses, a sizeable fraction of our sample would lie between them. This also suggests that their metallicity may lie in the range between that of M92 and 47 Tuc. For those sources with redded J-K colours than the 47 Tuc GB, and with magnitudes brighter than K = +8.5, even higher metallicities are required.

Besides the photometric observations, spectroscopic CO and water observations of a sample of sources in the BW were obtained. We find it is convenient to divide the BW stars into groups according to the strengths of their CO bands. It is shown that those stars with normal and strong CO bands may be consistently interpreted as disk stars. We suggest that the CO weak stars may be true bulge members. The relative numbers of CO weak objects relative to the total number of stars is consistent with the bulge and disk CCFs data (see also Ruelas-Mayorga, A., Teague, P.: 1991, A&A, in press).

MASSIVE STAR FORMATION AND CHEMICAL EVOLUTION IN NGC 1313

S.D. RYDER
Mt Stromlo and Siding Spring Observatories
The Australian National University
Weston Creek P.O.
ACT 2611
Australia

Summary

The galaxy NGC 1313 is a late-type (almost Magellanic) barred spiral located midway between the Magellanic Clouds but at a distance of about 4.5 Mpc (de Vaucouleurs 1963). A comprehensive imaging and spectrophotometry program has been carried out in order to investigate the peculiar kinematics of NGC 1313 (Marcelin and Athanassoula 1982), as well as to study the relationships between the formation of massive stars and light element chemical abundances in spiral galaxies.

Flux-calibrated Hα CCD imaging has been used to study the distribution and rate of present-day massive star formation. Following the techniques of Kennicutt (1983), the total observed Hα luminosity of NGC 1313 implies a current formation rate for massive ($M > 10M_{\odot}$) stars of order 0.1 $M_{\odot}$ yr^{-1}, or a total star formation rate of around 0.6 $M_{\odot}$ yr^{-1}. This is comparable to the past average rate of 1–2 $M_{\odot}$ yr^{-1}. From surface photometry of V, I and Hα CCD images, we find that the scale lengths of the old and the younger stellar populations are both very similar to that of the current star formation, each being of order 1.4 kpc.

Pagel *et al.* (1980) concluded that NGC 1313 probably does not have an abundance gradient. We have obtained longslit spectrophotometry of 14 HII regions in the bar, spiral arms and isolated southern regions with the MSSSO 2.3m telescope, and determined their abundances using photoionisation models. Within the inner 100″ (2.2 kpc), there is only a weak or zero gradient, while beyond this the spiral structure ceases, and there are indications of a drop in metallicity. Despite its scruffy appearance, star formation in NGC 1313 has proceeded fairly steadily over the disk's lifetime, contributing to at most only a weak abundance gradient.

Acknowledgements

I wish to acknowledge generous financial support from both the Donovan Astronomical Trust and the Mt Stromlo and Siding Spring Observatories that made my attendance at this conference possible. I thank Mike Dopita for many helpful suggestions.

References

Kennicutt, R.C.: 1983, *Ap. J.* **272**, 54
Marcelin, M. and Athanassoula, E.: 1982, *Astr. Ap.* **105**, 76
Pagel, B.E.J., Edmunds, M.G. and Smith, G.: 1980, *M.N.R.A.S.* **193**, 219
de Vaucouleurs, G.: 1963, *Ap. J.* **137**, 720

STAR-FORMATION HISTORIES OF BLUE COMPACT DWARF GALAXIES

JOHN J. SALZER and RICHARD ELSTON
Kitt Peak National Observatory, P.O. Box 26732, Tucson, AZ 85726, U.S.A.

Abstract. We present the results of a multi-wavelength observational study of actively star-forming dwarf galaxies (also known as Blue Compact Dwarfs). We combine optical, infrared and H-alpha imaging with optical spectroscopy and HI data in an attempt to better understand the stellar population make-up and star-formation histories of this type of galaxy. In particular, we address the long-standing ambiguity concerning the source of the near-infrared flux in BCDs. We find in nearly all cases that there is clear evidence for an older population of stars. However, the currently active starburst often contributes a large fraction of the total near-infrared flux in the form of nebular continuum emission and/or light from hot main sequence stars.

Blue Compact Dwarf galaxies (BCDs) are low mass dwarf galaxies (typical luminosities are 10^7 - 10^9 $L_\odot$) containing one or more super star-formation region(s). The high surface brightness starburst region usually dominates the optical appearance of the galaxy, often to the extent of masking the presence of any underlying, older stellar population. Optical colors are very blue (mean B–V = 0.21 for our sample), and spectra are dominated by line emission. Nebular abundances are typically very low (1/5 - 1/40 solar).

An important question concerning BCDs is their star-formation histories. Their observed properties indicate a very young age. One way to check for an older underlying population is with near-infrared (NIR) photometry. Unfortunately, JHK colors alone don't give an unambiguous result, since they can't distinguish between the case of the IR flux being due mainly to a population of red giants (indicating previous star-formation in BCDs), or being due primarily to red supergiants that are slightly evolved members of the current star-formation episode. Other possible sources of NIR emission in BCDs that can further confuse the issue include hot dust, nebular continuum, and photospheric emission from the O and B main-sequence stars formed in the starburst.

Our approach to solving the ambiguity concerning the NIR emission is to look for differences in the spatial distribution of the blue (B band) and IR (H band) fluxes. This was done by creating calibrated B–H images and looking for color gradiants. A sample of roughly 20 BCDs were observed with both optical CCDs (UBVRI plus Hα narrow-band) and IR arrays (JHK). The galaxies chosen represent some of the more extremely active star-forming dwarfs known. In all cases (except I Zw 18), large changes in the B–H color between the star-forming region(s) and the outlying portions of the galaxy are visible, indicating that the extended NIR flux is coming from a population of stars that is distinct from the current starburst population. JHK colors of most BCDs are consistent with them being due to a mixture of light from a quiescent older population (colors like normal irregular galaxies) plus a sometimes very strong component due to the young hot stars. Nebular continuum emission is also a major contributor to the NIR fluxes in some objects. Red supergiants undoubtedly contribute to the total NIR flux as well, but in no cases does it appear to be the dominant source.

CNO EXCESS IN 47 *TUCANAE* FROM THE INTEGRATED SPECTRUM SYNTHESIS

J. F. C. Santos Jr., E. Bica, and H. Dottori
Instituto de Física-UFRGS, Porto Alegre, RS, Brazil

We have found evidence of a CNO/Fe excess in the integrated spectrum of 47 Tuc relative to a synthetic cluster built up of solar neighbourhood stellar spectra from Gunn & Stryker's library (GS, 1983). The 47 Tuc spectrum was synthesized between $3200 < \lambda < 9750$Å aided by the cluster color-magnitude diagram (Hesser et al. 1987) complemented with a low main sequence, which was simulated by a canonical initial mass function (IMF) of slope $x = 2.8$. If a flat IMF (Hesser et al. 1987) is attributed to stars earlier than M, then $x \approx 5$ is necessary for lower masses. A similar synthesis procedure was previously applied to the Galactic open cluster M11 (Santos Jr. et al. 1990). The residuals from the spectral synthesis were analyzed between $3200 < \lambda < 5400$Å, where a blanketing *stronger* in 47 Tuc than in the solar model remained. This is the integrated version of blue/violet excesses found in 47 Tuc individual giants (Hesser et al. 1977). It is well known that 47 Tuc has lower [Fe/H] than the solar model, suggesting non-solar [CNO/Fe] as responsible for the blanketing. As the localized bands CN, C_2 and CH are not enough to explain it, we have used 28 plausible diatomic molecular patterns (two examples are shown in Fig. 1) and a synthesis technique to suggest possible absorbers contributing to this blanketing. Fig. 2 presents the molecular model and the residuals shifted by a constant, where a NH localized band is clear. The CO molecule resulted the dominant distributed absorption ($\approx 50\%$ in flux of the total blanketing), followed by SiN (20%). Definite identifications of the absorbers need much higher resolution than that in GS's library (20-40Å). On the other hand, the methods proved to be very efficient for analysing the 47 Tuc population by means of stellar synthesis and molecular synthesis.

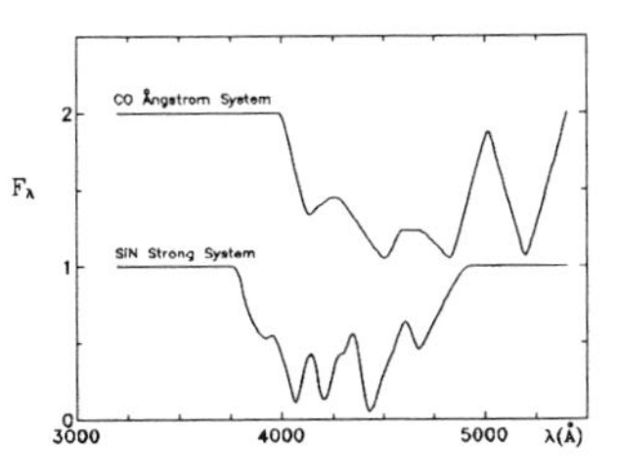

Fig. 1

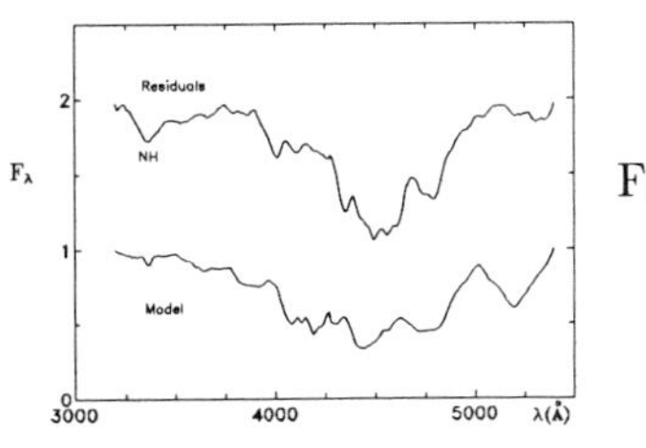

Fig. 2

REFERENCES
Gunn, J. E., and Stryker, L. L. 1983, ApJS, 52, 121
Hesser, J. E., Hartwick, F. D. A., and McClure, R. D. 1977, ApJS, 33, 471
Hesser, J. E., Harris, W. E., Vandenberg, D. A., Allwright, J. W. B., Shott, P., and Stetson, P. B. 1987, PASP, 99, 739
Santos Jr., J. F. C., Bica, E., and Dottori, H. 1990, PASP, 102, 454

Tests and discussion on the solution uniqueness of population synthesis methods

Alex A. Schmidt,[1,2] Marcus V.F. Copetti,[1,3] Danielle Alloin[4] and Pascale Jablonka[4]

[1] *Universidade Federal de Santa Maria, Depto. de Matemática, 97119 Santa Maria - RS, Brazil*
[2] *Astronomy Centre, University of Sussex, Falmer, Brighton, BN1 9QH, UK*
[3] *Royal Greenwich Observatory, Madingley Road, Cambridge CB3 0EZ, UK*
[4] *Observatory de Paris - URA 173 du CNRS, Département d'Astrophysique Extragalactique et de Cosmologie, F-92195 Meudon, France*

Population synthesis is a powerful tool to study stellar populations where the analysis of the stellar content of a composite system is based on the results of breaking down into components (of a given base) the spectrum of the observed system. Such process constitutes an inverse problem which can have a multitude of possible or "acceptable" solutions. This degenerative character of the synthesis rises mainly from observable errors and from the base of components itself with respect to its internal consistency and its (in)capacity to fully embrace all the free parameters involved.

In this work we present the results of a series of tests (26 pre-defined problems in terms of age and population content) for a constraint-free formulation of the synthesis problem based on a minimization technique that sweeps the entire space of solutions seeking for a representative solution instead of an optimal solution (for a complete discussion see Schmidt et al, 1991). A representative solution is obtained from a series of minimization cicles in which, through an analysis of the statistical distribution of the solutions, it is sought the reduction of the *natural domain* of each problem (obtained from random search) in a gradual convergent process. The star cluster data set from Bica (1988) (35 components, each representing a evolutionary stage in the age-metallicity plane), found to have a high degree of internal correlation allowing artificial combinations between components to mimic acceptable solutions, was reduced to 12 components after a multivariate analysis of the base (in order to minimize such effects) to compose our base.

The results of the tests can be summarized as follow: for old metal-rich dominant populations (elliptical galaxies, spiral galaxy bulges) the agreement between true and obtained proportions is very good; for young, intermediate-age and metal-poor dominat populations proportions less than 20% are not always recovered in a satisfactory way, however we note that the correct age is properly recovered while the metallicity determination is more uncertain. Nevertheless, strong contributions (larger than or equal to 30%) are easily discernible.

The degenerative effects due to measurement errors and internal dependence between base components constitute very serious difficulties in the solution of the population synthesis problem, and the multiplicity of solutions is an unavoidable characteristic that must be carefully taken into account. Nevertheless, the above results show that an accuracy better than 5% can be obtained for old metal-rich populations (composite systems most frequently observed), although less accurate results, up to a factor of 3, should be expected for cases when observable constraints are not restrictive enough, such as in blue composite populations. Yet a substancial improvement is to be expected from small observable errors and from the addition to the base of new features in the ultraviolet range to better represent the hot, blue contributions.

Bica, E., 1988, *Astr. Astrophys.*, **195**, 76.

Schmidt, A.A., Copetti, M.V.F., Alloin, D. and Jablonka, P., 1991, *MNRAS*, **249**, 766.

On Main Sequence Distances and the Local Distance Scale

R. A. Schommer
Cerro Tololo InterAmerican Observatory
Casilla 603
La Serena, Chile

E. W. Olszewski and M. A. Aaronson
Steward Observatory
University of Arizona
Tucson, AZ 85721

ABSTRACT. We analyze a deep CMD for the LMC cluster NGC1831, based on CTIO PFCCD frames. More than 4 magnitudes of main sequence (MS) are evident, as well as a prominent clump, and a few possible giant branch stars. We fit the MS to a series of models and Galactic clusters,and derive a distance of 18.3±0.1 and an age of 0.5±0.1 Gyr, for an abundance of about 2/3 solar ([Fe/H]∼−0.23), based on the VandenBerg models (Vandenberg and Poll 1989). A fit to the empirical Pleiades MS, assuming an intrisic modulus of 5.6, gives a LMC modulus of 18.25-18.35, after an adjustment for the relative abundance difference of −0.2.

We can also fit the cluster with convective overshoot models (Chiosi et al., 1988). Using solar abundances, we find an age of 0.8 Gyr with $(m\text{-}M)_o$ = 18.20–18.30. The main effect of the overshoot models is to increase the derived age, not to change the distance.

A more serious problem arises when we attempt to determine the distance using the Revised Yale Isochrones (Green *et. al*, 1987; hereafter RYI). We derive a distance of $(m\text{-}M)_o$ =18.4–18.6 for different abundances. A systematic difference between the VdB and RYI distance scale exists for Galactic clusters also, including the Hyades and Pleiades. We determine a distance modulus to the Pleiades of 5.75-5.8 based on RYI models, 0.15 to 0.20 longer than the VdB and Poll distance. High quality CMDs and abundance determinations can produce MS distances with internal accuracies of about 5%, when a specific set of models or zeropoints is assumed. The differences shown above, however, indicate that systematic errors at the 10% level still exist.

The true distance to the Pleiades is not known with high precision of course, and the RYI distance may be the proper one (we are not implying they are wrong). We do note however that the standard *Cepheid scale and the RYI scale differ by about 0.2mag*, with the Yale scale longer, and care should be used when comparing various local distances, because of often unstated assumptions on the true zeropoint. For example, some distances are still based on a Hyades modulus of 3.03 (Tammann 1987). Some recent Cepheid–based moduli imply an extremely bright absolute magnitude for RR Lyrae stars (*e.g.*, Sandage and Carlson (1983) use an apparent LMC modulus of 19.0, which would yield $M_v(RR) = 0.2$).

Despite recent claims in the literature that the local distance scale is known to ∼5% (Tonry 1991), we find that the zeropoints of all the local distances are uncertain by 10% or more (as discussed by Feast 1991).

E.O. acknowledges partial support from NSF grant AST86-11405.

KINEMATICS OF HALO AND HIGH–VELOCITY DISK STARS

W.J. Schuster[1,2], L. Parrao[1], and M.E. Contreras Martínez[1,3]
[1]Instituto de Astronomía, UNAM. Apdo.P. 70-264, C.P. 04510, México, D.F.
[2]I.N.A.O.E., Apdo.P. 51 y 216, C.P. 72000, Puebla, Pue., México.
[3]U.A.M., Unidad Azcapotzalco, México, D.F.

ABSTRACT. $uvby - \beta$ photometry for 1149 high–velocity and metal–poor stars has been obtained and is being used to derive metallicities, distances, and ages for these stars. The distances have been combined with proper motions and radial velocities from the literature to calculate the stars' galactic space velocities. Our ultimate aim is to reach a better understanding of the early dynamical and chemical evolution of our Galaxy by studying the possible existence of correlations between metallicity, age, and kinematics for these stars. In particular, the $V(rot)$, [Fe/H], Toomre energy, and W',[Fe/H] diagrams are being used to analyze the kinematic characteristics of this sample. Two distinct, separate populations are very clearly seen, and there is also some evidence for additional stellar components. Our kinematic and metallicity data argue quite strongly for mostly uncoupled evolutions for the disk and halo populations of the Galaxy.

CONCLUSIONS

a) Disk and Halo populations are very obvious in our sample. Even with the high-velocity, high-proper-motion, and low-metallicity selection criteria, the sample is still significantly contaminated by old thin disk stars, together with the thick disk and halo stars. Very low metallicity tails and very retrograde motions may indicate other components, or at least distinct physical processes during the Galactic evolution. b) The separation criterion of Paper V, a diagonal cut in the V(rot),[Fe/H] diagram, is a good one, but it is not perfect. The "halo" sample is nearly pure due to the small dispersions of the disk stars. The "high velocity disk" stars are more contaminated by the halo stars. c) The mean curve in the V(rot),[Fe/H] diagram for our total sample clearly shows a decoupling between the disk and halo evolutions. However, the mean curve for the $|W'| \geq 60$ km/s sample is different. The final interpretation of this diagram may not be so straightforward. d) $\sigma_{w'}$ for the thick disk is ~ 46 km/s corresponding to a scale height ≥ 1.4 kpc. This velocity dispersion has been obtained from a nearly pure sample of thick disk stars, and e) The W$'$,[Fe/H] diagram for halo stars, selected according to the diagonal cut in the V(rot),[Fe/H] diagram, indicates no metallicity gradient in the halo.

Reference

Nissen, P.E., Schuster, W.J. 1991, A&A, (in press). (Paper V).

HALO AND HIGH–VELOCITY DISK STARS

W.J. Schuster[1,2], C. Allen[1,3], and A. Poveda[1]
[1]Instituto de Astronomía, UNAM. Apdo.P. 70-264, C.P. 04510, México, D.F.
[2]I.N.A.O.E. A.P. 51 y 216, C.P. 72000, Puebla, Pue. México.
[3]D.G.S.C.A., UNAM. C.P. 04510, México, D.F.

ABSTRACT. Using the $uvby - \beta$ data of Schuster and Nissen (1988) to calculate photometric distances and using the galactic potentials of Allen and Martos (1986) and Allen and Santillán (1991), galactic orbits have been integrated for 615 halo and high-velocity disk stars. Correlations between orbital characteristics, metallicities, and ages are being studied. For the 206 halo stars of this sample, a particularly high percentage of the orbits (35-50%) show some chaotic behavior, due mainly to their low angular momenta which lead to small (< 1 Kpc) perigalactic distances. The newer potential of Allen and Santillán, which includes a spherical central bulge rather than a central mass point, gives the higher percentage of chaotic orbits. The effects of this chaos upon chemical gradients and upon the dynamical structure of the halo are being investigated.

CONCLUSIONS

a) Stars which pass near ($\leq$ 1 Kpc) the galactic center are likely to have chaotic orbits. The percentage of chaotic orbits in the halo may be as high as 50%. This chaos is a product of the central spherical mass distribution in the Galaxy, b) Chemical gradients are not found in the galactic halo, neither as a function of R_{Max} nor Z_{Max}. This holds even after those stars whose orbits show some evidence of chaotic behavior are removed from the sample, c) The chaotic "scattering" process near the galactic center produces vertical segregation of chaotic and non-chaotic orbits in the galactic halo; certain Z_{Max} 's are preferred by the chaotic orbits over others due to the conservation of the total orbital energy and due to the focusing of the "scattered" stars around families of tube orbits, and d) This vertical segregation may explain certain discrepant observations of the halo, such as, conflicting c/a values for the shape of the halo and unusual velocity dispersions or distributions near the galactic poles. Our results are in excellent agreement with Hartwick's two–component model for the halo.

References

Allen, C., Martos, M.A. 1986, Rev Mex Astr Astrofis **13**, 137-147.

Allen, C. Santillán, A. 1991, Rev Mex Astr Astrofis (in press).

Schuster, W.J., Nissen, P.E. 1988, A&AS **73**, 225-241.

CORRELATIONS BETWEEN UBV COLORS AND FINE STRUCTURE IN E + S0 GALAXIES

François Schweizer
Carnegie Inst. of Washington
5241 Broad Branch Rd. NW
Washington, DC 20015

Patrick Seitzer
Space Telescope Science Institute
3700 San Martin Drive
Baltimore, MD 21218

A study of 67 E and S0 galaxies located mostly in the field and in groups reveals that at any given luminosity the *UBV* colors become systematically bluer as the amount of fine structure (ripples, jets of luminous matter, X-structure, and boxy isophotes) increases. Figure 1 shows the resulting correlations between the color residuals $\Delta(U-B)_{e,0}$, $\Delta(B-V)_{e,0}$ (calculated as deviations from the mean color–luminosity relations) and the fine structure parameter Σ. These correlations closely resemble correlations found earlier between CN, Mg_2, and Hβ line strengths and the same parameter Σ in 36 ellipticals (Schweizer *et al.*, Ap.J. Letters **364**, L33, 1990). Both sets of correlations are most likely due to systematic variations in mean age, rather than mean metallicity, of the stellar populations in these early-type galaxies. We model the evolution of galaxies undergoing a major merger by convolving a single-burst model (Charlot & Bruzual, Ap.J. **367**, 126, 1991) with a star formation rate that declines exponentially with a long time constant (6–10 Gyr) before the merger and with a short time constant (0.1–0.5 Gyr) afterwards. Some of the model parameters are determined from observations of two 1–2 Gyr old merger remnants. Comparisons between the observed *UBV* colors and the models suggest that the bluest E + S0 galaxies in our sample formed through mergers only a few billion years ago, which also explains their high amount of fine structure.

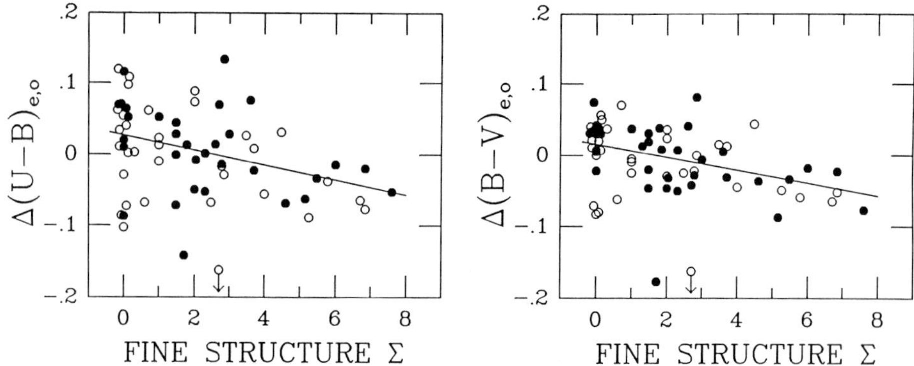

Figure 1. Color residuals $\Delta(U-B)_{e,0}$ (*left*) and $\Delta(B-V)_{e,0}$ (*right*) plotted versus the fine-structure parameter Σ. *Dots* mark E galaxies, *circles* S0 galaxies.

A METHOD TO DERIVE VELOCITY DISPERSIONS FROM COMPOSITE SPECTRA

R.E. de Souza, B. Barbuy, S. dos Anjos, M. Erdelyi-Mendes
IAG-USP, Depto. Astronomia
C.P. 9638, São Paulo 01065, Brazil

Different methods have been used to derive velocity dispersions of stellar populations in bulges of galaxies. In all cases, the spectrum of a K giant is used as template: the stellar spectrum convolved with the appropriate gaussian, the Fourier transform technique, or the cross-correlation technique (see Dalle Ore et al., 1991, for a review).

We propose to use as template a synthetic composite spectrum of a galactic metal-rich globular cluster. We base our method on the detailed studies of NGC 6553 by Ortolani et al. (1990). We have adopted a relative number of stars in each stage of evolution from a direct count in a V vs. (B-V) CMD. The main sequence population was derived using these counts, extented with a Salpeter IMF. NGC 6553 appear to have a solar metallicity (Barbuy et al., 1991).

The present sample of objects consists of 3 edge-on bright lenticular galaxies: NGC 0128, NGC 1381, NGC 1596, which are part of a program to study the properties of box-shaped galaxies (de Souza & dos Anjos, 1987).

The composite synthetic spectra of these galaxies was constructed using a library of synthetic spectra with stellar parameters cf. CMD of NGC 6553.

Our first preliminary results seem to show that the effects of velocity dispersion and metallicity can be disentangled. We estimate velocities of 170, 190 and 190 km s_{-1} for NGC 128, NGC 1381 and NGC 1596 respectively. The metallicities are all in the range $[M/H] = 0.0$ to $+0.5$; for NGC 1381, we estimate $[M/H] \approx +0.4$, in agreement with that of $[M/H] = +0.2$ derived by Pickles (1985).

Barbuy, B., Castro, S., Ortolani, S., Bica, E.: 1991, A&A, submitted
Dalle Ore, C., Faber, S.M., Jesus, J., Stoughton, R.: 1991, ApJ 366, 38
Ortolani, S., Barbuy, B., Bica, E.: 1990, A&A 236, 362
Pickles, A.: 1985, ApJ 296, 340
de Souza, R.E., dos Anjos, S.: 1987, A&AS 70, 465

LITHIUM IN POPULATION II STARS

F. SPITE, M. SPITE, R. CAYREL, S. HUILLE,
Observatoire de Paris
92195 Meudon Cedex
France

NEW data about dwarfs : a few more metal-poor dwarfs have been observed for lithium at the CAT + CES at La Silla (ESO). These new results, together with a few new measurements in the literature show again a constant lithium abundance, confirming the trend previously found (the plateau). Let us note however that two metal-poor dwarfs have recently been reported with a low lithium abundance. Obviously, more observations are needed.

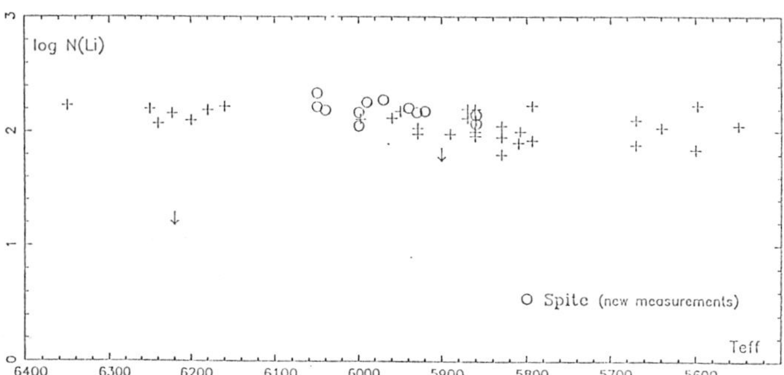

Abscissa : effective temperature (increasing to the left). Ordinate: log $(N_{Li}/N_H) + 12$. The lithium abundance is independent of the metallicity and of the temperature (and mass) of the metal-poor dwarfs. All measurements published before May 1991 are gathered in the figure.

INTERPRETATIONS : recent theories interpret the plateau by the combination of two depletion processes. Both processes deplete lithium with rates varying as a function of temperature (and mass), but these rates have opposite trends. The compensation is not perfect, and the theories predict a decrease of the lithium abundance at the hot end of the plateau. Since hotter (more massive) stars have evolved off, towards the giant branch, this decrease cannot be *directly* observed. However, the lithium abundance may be observed in the evolved stars, and the initial abundance in the dwarf progenitor can be computed if the dilution factor is known (when a dwarf evolves as a giant, the lithium of its atmosphere is diluted in lithium-free matter).

AN INTERESTING GIANT : we observed a very metal-poor giant with a lithium abundance lower than the value of the plateau by a factor of a little more than an order of magnitude. The dilution factor may be estimated also as a little more than an order of magnitude. This star, when it was unevolved was presumably hotter and more massive than the presently observed dwarfs. If this is true, this hotter star has only diluted (not destroyed) its "normal" lithium content. This is interesting, but NOT too much weight should be given to a single star : again, more observations are needed.

A Near-IR Imaging Survey of Interacting Galaxies

S.A. STANFORD
UC-Berkeley

and

H.A. BUSHOUSE
Computer Sciences Corporation

Abstract. We describe a near-infrared imaging survey of interacting galaxies.

We have obtained JHK broadband images of ~200 interacting galaxy systems. The goals of the project are to determine the relationship of the IR morphology of the old stellar component to interaction type and age, and to obtain near-IR colors of spatially-resolved components to determine if significant contributions to the observed IR light are made by thermal dust emission, obscuration or early-type stars, and investigate dependence of colors on interaction type and age.

The sample was selected from the Arp Atlas of Peculiar Galaxies (Arp 1966), because these systems are well-studied at many other wavelengths, and because they are mostly nearby, strong interactions which makes easier the study of possible morphological disturbances caused by interactions. At the mean survey redshift, the ~2 arcsec resolution of the imaging is ~600 pc. The images were obtained in the JHK bands using a SBRC InSb array on the 1.3 m telescope at KPNO. The pixel size was 1.35 arcsec giving a field of ~80 arcsec. Flatfielding is estimated to be good to 0.3%, and sky subtraction good to 0.1% of sky. We estimate the overall typical photometric error to be 0.1 mag.

Contour plots and nuclear and global photometry for the sample are presented in Bushouse & Stanford (1992), and a more detailed analysis of one subset, the disk-disk mergers, is presented in Stanford & Bushouse (1991).

We intend to continue with this project in three ways:

1. Further analysis of the IR imaging such as ellipse-fitting, comparison with other wavelength data, determination of possible group/cluster membership.

2. Optical images are being obtained to allow for the use of optical-IR colors in answering stellar population questions, and to provide morphological information on faint tidal tails, bridges and fans usually undetected by IR imaging.

3. Near-IR imaging of a normal galaxy sample with which to compare the peculiar galaxies of our survey is planned.

Acknowledgements

S.A. Stanford thanks the A.A.S. for a travel grant which made possible his attending this conference.

References

Arp, H.C. 1966, ApJS, 14, 1
Bushouse, H.A. & Stanford, S.A. 1992, ApJS, in press
Stanford, S.A. & Bushouse, H.A. 1991, ApJ, 371, 92

THE POPULATION OF PLANETARY NEBULAE IN THE GALACTIC BULGE

G. STASIŃSKA[1)]
A. ACKER, A. FRESNEAU, J.F. GAMEIRO, J. KÖPPEN, B. STENHOLM, R. TYLENDA
[1)] DAEC, Observatoire de Paris-Meudon
F-92195 Meudon Cedex
France

Planetary nebulae are one of the latest phases of evolution of low- and intermediate mass stars and can be used as probes of stellar populations and overall chemical evolution in galaxies. Bulge planetary nebulae offer the advantage of being a sample of objects at known distance. Therefore, their study is essential.

We present some results of an extensive study of planetary nebulae in the Galactic bulge, based on the Strasbourg-ESO spectroscopic survey of Galactic planetary nebulae, conducted by Acker and Stenholm.

Our sample of Galactic bulge planetary nebulae contains 270 objects, and is described in Acker et al. (1991).

From a comparison of a set of observational diagrams with simulations based on photoionisation models and taking into account observational errors and selection effects, we have found that the population of planetary nebulae in the Galactic bulge is well represented by a standard model in which the nebulae have a total mass of about 0.2 $M_\odot$, and are expanding around a central star which evolves according to the theoretical tracks of Schönberner (Stasińska et al., 1991a).

We have been able to derive the masses of the central stars for about 90 objects. The apparent mass distribution for these stars peaks around 0.59 $M_\odot$, and has a standard deviation of 0.025 $M_\odot$ (Tylenda et al. 1991).

The observed luminosity function of a complete subsample of Galactic bulge planetary nebulae has been compared to simulated luminosity functions obtained with different characteristics for the central star population and for the surrounding nebulae (Stasińska et al., 1991b). The results are consistent with the studies mentionned above. The total number of planetary nebulae in the galactic bulge is estimated to be about 700.

The next step will be to attempt a comparison of the main parameters of the bulge and disk PN population.

References

Acker, A., Köppen, J., Stenholm, B.: Raytchev, B. (1991) A&AS 89, 237
Stasińska, G., Tylenda, R., Acker, A., Stenholm, B. (1991a) A&A, 247, 173
Stasińska, G., Fresneau, A., da Silva Gameiro, G. F., Acker, A. (1991b) A&A in press
Tylenda, R., Stasińska, G., Acker, A., Stenholm, B. (1991) A&A 246, 221

CCD PHOTOMETRY OF HII GALAXIES

EDUARDO TELLES
Institute of Astronomy - Cambridge, UK
ROBERTO TERLEVICH
Royal Greenwich Observatory - Cambridge, UK
and
BERNARD E.J. PAGEL
NORDITA - Copenhagen, Denmark

We report the first results of a multicolour (broad V,R,I and narrow [OIII] bands) surface photometry study of a small sample of HII galaxies.

The data were obtained at the 2.5m Nordic Optical Telescope (NOT) at La Palma, Canary Islands under subarcsecond seeing conditions. Part of the data has 0.55 arc seconds resolution (FWHM). The data will allow us to investigate the bi-parametric behaviour of the luminosity vs line-witdh relation.

galaxies - starburst - extragalactic giant HII regions

HII galaxies are dwarf systems undergoing violent star formation. Their optical spectrum is indistinguishable from that of giant extragalactic HII regions.

The relations between Hβ luminosity, size, width of the emission lines and heavy element abundance of giant HII regions and HII galaxies (Terlevich & Melnick,1981, Melnick *et al.* 1987,1988) suggest that these systems are gravitationally bound in which the observed emission line widths represent the velocity dispersion of discrete gas clouds in the gravitational potential of the gas-star complex.

Dressler *et al.* (1987) found that the observed scatter in the luminosity line-width relation in elliptical galaxies was due to a bi-parametric behaviour, with the surface brightness being strongly correlated with the second parameter.

One very important clue for the origin of the bi-parametric behaviour of elliptical galaxies may lay in the above described correlations for the youngest galaxies, the HII galaxies. It is of fundamental importance to check if HII galaxies *also* have a bi-parametric behaviour with surface brightness. To answer this question we started a high resolution images survey of HII Galaxies with the NOT. We describe here the first results obtained in a successful run this year.

As we are interested in the structural properties of the burst, we concentrated on the high surface brightness component trying to make best use of the excellent seeing and pixel sampling. We have obtained effective and core radii for the best sampled objects. We found a systematic behaviour of the radius in the luminosity vs line width relation suggesting that indeed a bi-parametric behaviour may be already defined in these young systems.

References

Dressler et al.: 1987, '', *Astrophys. J.* , 42
Melnick,J.,Moles,M., Terlevich,R. & Garcia-Pelayo,J.M.: 1987, '', *Mon. Not. R. astr. Soc.* , 849
Melnick,J.,Terlevich,R. & Moles,M.: 1988, '', *Mon. Not. R. astr. Soc.* , 297
Terlevich,R. & Melnick,J.: 1981, '', *Mon. Not. R. astr. Soc.* , 839

Observations of Ly_α Emission in Young Galaxies

ELENA TERLEVICH and ROBERTO TERLEVICH
Royal Greenwich Observatory. Madingley Road, Cambridge, CB3 0EZ U.K.
and
ANGELES I. DíAZ and MARIA LUISA GARCíA VARGAS
Depto. de Física Teórica. U. Autónoma de Madrid. 28049-Madrid, Spain

Abstract. We report IUE observations of two extremely low metallicity HII galaxies. Lyman alpha emission was detected in both galaxies.

Key words: Primordial galaxies - Low metallicity galaxies - Ly_α emission

1. Results

Our result confirms previous findings that young or unevolved galaxies exhibit weak or absent Ly_α emission combined with a strong UV continuum, as well as the correlation between the Ly_α/H_β ratio and metal content. This correlation supports the hypothesis that in these systems Ly_α is destroyed by dust absorption. Given that very low metal content seems to be a necessary condition to detect Ly_α, the extreme rarity of galaxies with low metal content may explain the difficulty in detecting Ly_α emission associated with damped Ly_α systems in QSOs. Our results cast doubts on recent claims of detection of high redshift young galaxies associated with active galaxies, since they may represent very extended narrow line regions photoionized by a power law rather than by normal hot stars. Full account of this work is being published elsewhere.

Acknowledgements

Travel grants by the Royal Society and the IAU are thanked by ET and RT.

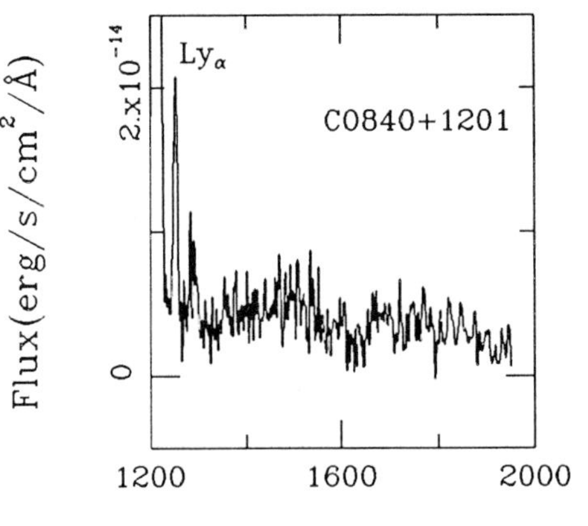

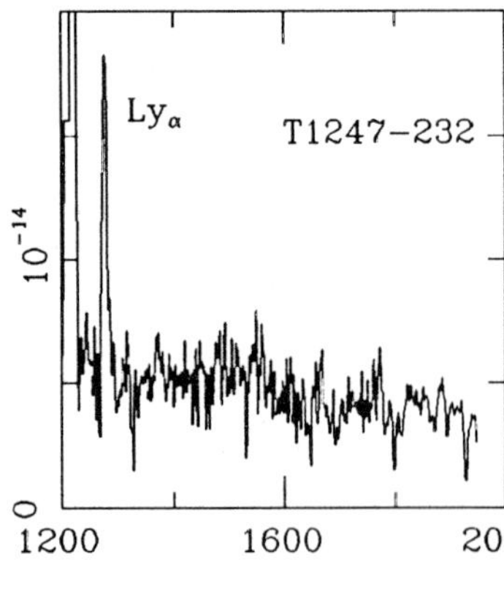

λ (Å)

Fig. 1. UV spectra of C0840+1201 and T1247-232; no emission lines other than Ly_α are detected and only a hint of absorption is seen in the spectrum of T1247-232 at the redshifted wavelenghts of CI and SiII.

TOOLS FOR A NEW APPROACH OF STELLAR POPULATIONS

F. THÉVENIN
Observatoire de la Côte d'Azur
G. JASNIEWICZ
Observatoire de Strasbourg
and
A. BIJAOUI
Observatoire de la Côte d'Azur

September 13, 1991

Over the past decade, a very considerable amount of effort in the understanding of stellar populations has gone. Because a large number of analysed data is the basis of this understanding, our efforts concerning this subject matter presented in this poster are mainly:

-Conception of spectrograph
Multiaperture spectrograph is now a reality on large telescopes. The spectrograph SFM has been achieved in the Observatoire de Marseille and is now at Calar Alto (Baranne et al., 1992). It is used on the German 3.5 m telescope. Such an instrument multiplicates by a large factor the number of observed stars.

-Numerical simulation of spectra of stars
This is an easy reality with fast computers today. Comparison with observed spectra gives the opportunity to use spectra at low resolution and low S/B ratio. An example concerning stars in the SMC and the LMC is given. We confirm in this way the previous published abundances, respectively -0.4 and -0.2 *dex* for the iron abundance of field stars in the SMC and LMC. These synthetic spectra can also be used to compute synthetic integrated light of globular clusters and spheroidal galaxies in order to interpret related observations.

-Spectroscopic data analysis
A new approach of spectra by synthetic one can be tried with the wavelet theory. This is developed to open a new way for the automatic classification of stellar spectra, and this could be an help for the realisation of a numerical spectrovelocimeter Coravel.

References

Baranne A., Blazit A., Foy R., Thévenin F.: 1992, to be published

Thévenin F., Jasniewicz G.: 1992, *Astron. Astrophys.*, to be published

THE BOLOMETRIC LUMINOSITY FUNCTION FOR THE LOWEST MASS STARS

CHRISTOPHER G. TINNEY
California Institute of Technology, 105-24
Pasadena, CA 91125

ABSTRACT. We present a luminosity function (LF) for the coolest stars in three POSSII fields - extending earlier LFs to much lower masses at higher precision than has been acheived previously.

1. The Survey

Although the subject of several studies in recent years, the form of the LF for extremely low mass stars ($M_{Bol} \gtrsim 12$ or $M \lesssim 0.1M_{\odot}$) remains poorly sampled. These stars are rare and emit most of their flux in the infrared making optical colour magnitude diagrams almost useless (cf. Monet *et al.* 1991, Fig.10). To remedy this, we are carrying out a survey covering 10 POSSII fields (over 300 square degrees) to a depth of I< 18. Lists of low mass candidates are compiled with R-I> 2.1 and all of these stars are being observed at K-band. The I-K colour so measured allows us to estimate M_{Bol} more precisely than can be done by purely optical studies. Our new measurement rules out the possibility of a LF steeply increasing towards the brown dwarf limit.

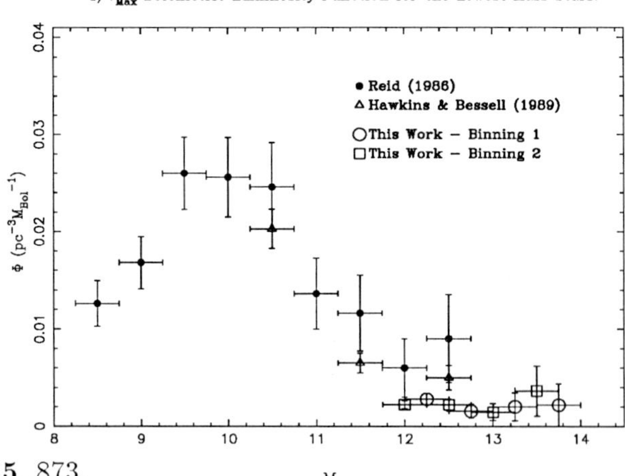

Figure 1. Previous functions by Reid (1986, solid dots) & Hawkins & Bessell (1988, open triangles) are shown. Two binnings of our function are shown.

4. References

Reid, I.N. 1986, *M.N.R.A.S.*, **225**, 873.
Monet, D. *et al.* 1991, *A.J.*, *submitted*, .
Hawkins, M. & Bessell, M. 1988, *M.N.R.A.S.*, **234**, 177.

ANALYSIS OF THE OPTICAL SPECTRA OF WOLF-RAYET GALAXIES

WILLIAM D. VACCA and PETER S. CONTI
Joint Institute for Laboratory Astrophysics
University of Colorado, Boulder, CO 80309-0440, USA

Wolf-Rayet (W-R) galaxies are a subset of emission-line galaxies in whose integrated spectra a broad (i.e., stellar in origin) He II λ 4686 emission feature has been detected. This line is a prominent emission feature in the spectra of WN stars. The presence of 10^2 to 10^5 W-R stars in these galaxies has been inferred from a comparison of the luminosity and equivalent width of this feature in the integrated galaxy spectra with those of the corresponding line in the spectra of Galactic and LMC WN stars. Most W-R galaxies exhibit other properties indicative of a very young starburst population, such as a relatively "blue" continuum and a strong nebular emission line spectrum due to photoionization by large numbers of hot, early-type stars. Their spectra are therefore very similar to to those of giant H II regions. There are currently about 40 W-R galaxies known.

As part of a systematic study of the properties of these galaxies, we have obtained moderate resolution (~ 3.5 Å) and S/N ($\sim 15 - 30$) optical spectra of ten W-R galaxies. The spectra were acquired during two observing runs on the 4-m telescope at CTIO and cover the wavelength ranges ~ 3120 to ~ 5400 Å and ~ 4500 to ~ 7000 Å.

The nebular emission lines have been used to determine reddenings, temperatures ($T_e \sim 12,000\ K$), densities ($N_e \sim 200$ cm^{-3}), and total ionizing fluxes in these galaxies. The observed emission line ratios are found to be similar to those of H II and starburst galaxies. In all ten of the W-R galaxies the He II λ4686 feature is resolved, with a typical FWHM ~ 15 Å. The spectra of several W-R galaxies exhibited a broad N III λ4638 emission feature as well. Both lines are characteristic of the spectra of Galactic and LMC WN stars. A few objects exhibit a broad emission feature of C IV λ5808, indicative of the presence of WC stars. Using the strength of the He II λ4686 feature, we have estimated the number of WN stars present in these galaxies to be $\sim 100 - 3000$. We have accounted for the contribution from the WN stars to the total ionizing flux and determined the WN/O star number ratios in these galaxies from the relative strengths of the stellar He II λ4686 and nebular $H\beta$ emission lines. We find ratios in the range $\sim 0.06 - 1.0$, much larger than those predicted by standard hot-star evolution models. This result suggests that the upper end of the initial mass functions in W-R galaxies may be very flat, the starburst episodes may be very short compared to massive star lifetimes, or the nebular regions may not be completely optically thick to Lyman continuum photons. Because evolutionary models indicate that the W-R phase lasts only ten percent of the O star main sequence lifetime, emission-line galaxies containing large numbers of W-R stars may represent the youngest phases of the starburst phenomenon, in which massive star formation is still occurring or has only recently ended ($\lesssim$ few $\times 10^6$ years ago).

Evolutionary Population Synthesis

Bob van den Hoek and Paul Goudfrooij
Astronomical Institute 'Anton Pannekoek', University of Amsterdam
Kruislaan 403, NL 1098 SJ Amsterdam, The Netherlands

Abstract Based on the population synthesis results[1] for the nuclear region of the elliptical galaxy NGC5044, possible theoretical Age Metallicity Relations (AMR) have been derived, reproducing the stellar luminosity distribution over age A and metallicity Z. A self-consistent method, used to constrain the star formation history and chemical evolution of a galaxy directly from its spectrum (i.e. from 3000 - 10000 Å) is presented. This method is predominantly based on the modelling of stellar luminosity contributions obtained by spectral decomposition methods with those theoretically calculated using a conventional galactic evolution model, incorporating a parametrised Star Formation Rate (SFR) and Initial Mass Function (IMF). Related model parameters were constrained by minimizing the χ^2- test defined by the corresponding luminosity distributions.

Results Preliminary model results are presented for the nucleus of NGC5044 and discussed with respect to the uniqueness of the derived SFR history (and AMR) and to the general properties of the applied method. In the left figure below we show the obtained stellar MS-luminosty distribution over age and metallicity based on the decomposition[1] of galactic absorption lines and continua using a base of star clusters covering a wide range in A and Z ($[Z/Z_{\odot}]$ indicated inside bins). The dominant part (i.e. $\approx$70%) of the stellar population present in the nucleus is found to be old, i.e. older than 10 Gyr, and has a metallicity $[Z/Z_{\odot}] = +0.6$. Furthermore, no significant young (i.e. $A<10^7$yr) stellar population is likely to be present, in agreement with specific emission line intensity ratios (see further Goudfrooij & Van den Hoek, this conference).

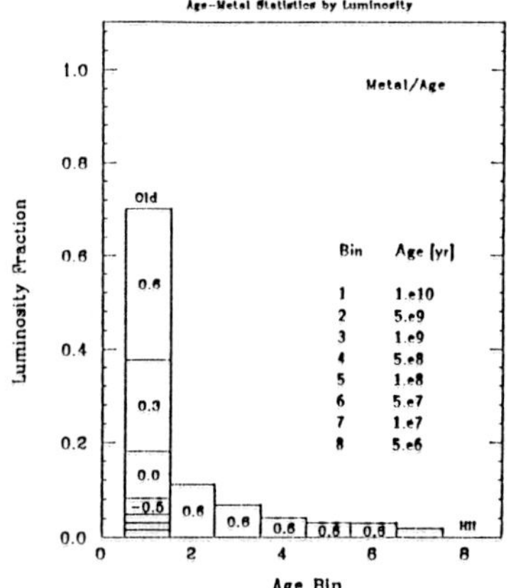

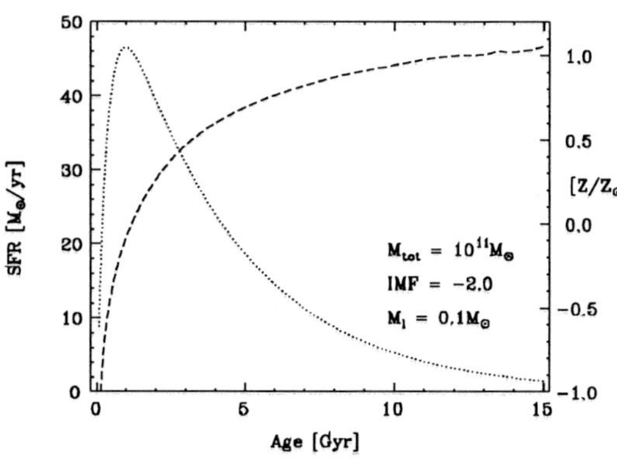

The global star formation history found for NGC5044 and the resulting AMR (dashed curve) as a function of galactic age are shown in the right figure above. The solution shown has been derived after $\approx$3000 minimalisation iterations[2]. Only a limited range of possible SFR histories, constraining the effective parameter-space of the double exponentional SFR, as well as the adopted power law IMF (and lower mass limit), has been found acceptable to model the stellar population distribution over A and Z, confirmed by the small number of possible solutions. Further promising results will be presented in due course.

[1]Results were obtained with an early version of the population synthesis model developed by Bica & Alloin (1986) see further e.g. Bica, E. & Alloin, D. 1986, A&A 162, p. 21, Bica, E. 1988, A&A 195, p. 76, and Bica, E., Alloin, D. and Schmidt, A.A. 1989, A&A 228, p. 23. [2]The effective parameter grid consists of more than $>10^7$ combinations!

The HI Surface Density in Low Surface Brightness Galaxies

J. M. VAN DER HULST
Kapteyn Astronomical Institute, Groningen, The Netherlands
E. D. SKILLMAN
Department of Astronomy, University of Minnesota, U.S.A.
G. D. BOTHUN
Department of Astronomy, University of Oregon, U.S.A.
and
T. R. SMITH
Department of Astronomy, University of Washington, U.S.A.

Abstract. The HI surface density of 8 low surface brightness galaxies falls below the critical density for star formation. This may explain why these galaxies appear so unevolved and are generally deficient in molecular gas.

Key words: low surface brightness galaxies, HI content of galaxies

Low Surface Brightnes (LSB) galaxies are galaxies whose light distributions are dominated by an exponential disk, with a central surface brightness at least 1.5 mag fainter than the canonical value of 21.6 mag arcsec $^{-2}$ exhibited by "normal" galaxies. One of the puzzling question regarding these objects is why they exhibit such a low current star formation activity despite their apparently large gas-to-star ratios. We observed 8 LSB galaxies in the 21-cm HI line at the VLA and briefly report the results here.

The HI distributions are fairly normal, but exhibit low peak column densities. Typical values are 4 - 6 $M_{\odot}$ pc^{-2}, a factor 2 lower than what is found in normal galaxies. The HI masses are, however, not abnormally low: the HI disks are simply extended. A comparison with optical diameters determined from radial surface brightness profiles shows that the HI to optical diameter ratios are systematically larger than found for galaxies with normal surface brightnesses.

We estimated rotation curves from velocity - position cuts along the major axis of the galaxies and calculated epicyclic frequencies as a function of radius. From these we evaluated the critical density for star formation as a function of radius following Kennicutt (1989). The observed HI surface densities fall below the critical density throughout the entire disk of the galaxies studied. This strongly suggests that the ISM does not fulfill the necessary conditions to sustain continuous, massive star formation, thus strengthening the idea that low surface brightness galaxies are objects in which active star formation ceased some few billion years ago. Moreover, it may be unlikely that molecular clouds can form under these low surface mass density conditions. This may explain the observations of Schombert et al. (1990) who noted significant CO deficiency in a sample of 12 LSB disk galaxies.

References

Kennicutt, R. C. 1989, Ap.J., **344**, 685.

Schombert, J. M., Bothun, G. D., Impey, C. D. and Mundy L. G. 1990, A.J., **100**, 1523.

ABUNDANCE GRADIENTS AND PHYSICAL PROPERTIES OF SPIRAL GALAXIES

M. B. VILA AND M. G. EDMUNDS
Department of Physics and Astronomy
University of Wales College of Cardiff
P.O. Box 913
Cardiff CF1 3TH, United Kingdom

The study of abundances in HII regions along the disc of spiral galaxies has shown the existence of negative gradients with higher abundance towards the centre (e.g. Pagel & Edmunds 1981; McCall, Rybski & Shields 1985; Garnett & Shields 1987; Edmunds 1989). The origin of these gradients is unclear but various chemical models with inflow, radial flows or star-formation cut-offs can produce them (e.g. Pagel & Edmunds 1981; Güsten & Mezger 1982; Díaz & Tosi 1984; Mayor & Vigroux 1981; Pitts & Tayler 1989; Clarke 1989). In the present study we have collected the chemical data available in the literature for some 30 spiral galaxies and we have calculated the abundances in a consistent way. The purpose is to carry out a thorough study of possible correlations or trends of abundance with many other galaxian properties, that can be followed up with further observations until a clearer picture for the origin of the gradients emerges. We summarize here some of the main results, which are fully included in Vila & Edmunds (1991). There is no prefered scale length (e.g. R25, disc effective radius, etc) which reduces the apparent variation of gradients between galaxies. The central abundances of spirals are correlated with their mass, barred spirals have shallower gradients and non-barred spirals show a correlation of gradient slope with morphological type. The correlation of abundance with mass surface density is confirmed and a weaker correlation is found with surface brightness. For those 9 galaxies where information on their HI, H_2, photometry and rotation curves is available, the gas fraction along the disc has been estimated. The initial results show an empirical yield uniformly decreasing with radius, possibly varying as a function of metallicity.

References

Clarke, C. J., 1989. *Mon. Not. R. astr. Soc.*, **238**, 283.
Díaz, A. I. & Tosi, M., 1984. *Mon. Not. R. astr. Soc.*, **208**, 365.
Edmunds, M. G., 1989. in *Evolutionaty Phenomena in Galaxies*, eds. J. E. Beckman and B. E. J. Pagel, Cambridge University Press, p356.
Garnett, D. R. & Shields, G. A., 1987. *Astrophys. J.*, **317**, 82.
Güsten, R. & Mezger, P. G., 1982. *Vistas in Astron.*, **26**, 159.
Mayor, M. & Vigroux, L., 1981. *Astr. Astrophys.*, **98**, 1.
McCall, M. L., Rybski, P. M. & Shields, G. A., 1985. *Astrophys. J. Suppl. Ser.*, **57**, 1.
Pagel, B. E. J. & Edmunds, M. G., 1981. *Ann. R. Astr. Astrophys.*, **19**, 77.
Pitts, E. & Tayler, R. J., 1989. *Mon. Not. R. astr. Soc.*, **240**, 373.
Vila, M. B. & Edmunds, M. G. E., 1991. *Mon. Not. R. astr. Soc.*, submitted.

COLOUR-DIFFERENCES AMONG GLOBULAR CLUSTER SYSTEMS

Stefan J. Wagner
Landessternwarte Heidelberg
Königstuhl
6900 Heidelberg
Germany

ABSTRACT. It is shown that the published colour-distributions of globular cluster systems surrounding nearby elliptical galaxies disagree with each other. While part of the discrepancies are introduced by zero-point errors, uncertainties in photometric transformations and other systematic differences between individual studies, independent measurements of the same cluster systems indicate that systematic errors are smaller than the observed off-sets. This implies that the latter differences are intrinsic.

Results

Globular Cluster Systems (GCS) are generally regarded as almost uncontaminated fossils of the earliest stellar populations in any galaxy. Their luminosity functions were found to be fairly universal, a result which might indicate that their formation processes were very similar in different galaxies.

By comparing the average colours of the globular clusters of the early-type galaxies which have been published in the literature, with the colour-distribution of the GCS of the Milky Way and M31, it is found that the former are significantly redder, in spite of being still bluer than the starlight of the corresponding host galaxies.

The different colours for early-type galaxies are to be expected since the disk and halo clusters of the Galaxy have different colour-distributions, and any other system is unlikely to contain the same mixture of GCS populations as found in the Milky Way.

One might expect that at least the colour-distributions of the globulars associated with the spheroidal component of early-type galaxies is similar in all objects. It is important to note however, that the colours and metallicities of the GCS of these large early-type galaxies are closer to those of the disk GCs of the Galaxy than to those galactic GCs associated with the halo.

Comparisons of different measurements of GC colour-distributions is hampered by technical problems (zero-point errors, insufficient transformations between different colour systems, etc.), which make an accurate absolute photometry of objects at 22 mag difficult. In addition different degrees of contamination by background galaxies (which have a broader colour-distribution), and radial colour-gradients in the GCSs may introduce off-sets between different studies. We have found, however that the latter complications introduce changes in the colour-distributions negligible in comparison with the empirically determined differences among observed GCSs. The colours derived by different groups using different material and even different photometric systems are often in better agreement with each other than the differences found in different early-type galaxies. Although the number of independent studies is still small, the present material indicates measurable differences among GCSs of different galaxies. The clusters would therefore be important tracers, but their properties are unlikely to be as universal as thought previously.

STELLAR EVOLUTION IN THE MAGELLANIC CLOUDS FROM STUDIES OF PLANETARY NEBULAE.

N.A. WALTON, M.J. BARLOW, & D.J. MONK
Department of Physics & Astronomy
University College London
Gower Street, London WC1E 6BT

R.E.S. CLEGG
Royal Greenwich Observatory
Madingley Road
Cambridge CB3 0EZ

We present the results of a spectroscopic study of planetary nebulae (PN) in the Magellanic Clouds. The optical survey of He, N, O, and Ne abundances by Monk *et al.* (1988) has been updated by higher S/N AAT optical data. In addition, carbon and other elemental abundances have been derived from the *IUE* spectra of 40 PN. Ionized nebular masses have been derived for 80 PN. The ionised mass versus nebular electron density plot shows that planetary nebulae become optically thin when their electron densities drop below 4500 cm^{-3}. Below this density, the mean nebular hydrogen mass found for non-Type I PN is 0.22±0.08 $M_\odot$. Using Zanstra and energy-balance methods, the mean central star mass found for 14 SMC and LMC PN is 0.59±0.02 $M_\odot$.

Th optical and UV Spectra of 40 planetary nebulae in the Magellanic Clouds have been analysed to derive abundances for He, C, N, O, Ne, S, Ar, and Si. The nitrogen abundances in the non-Type I nebulae are found to be consistent with the exposure of secondary nitrogen (produced by the CN cycle) by the first dredge-up, with 45% and 100% of the initial carbon having been converted to nitrogen in the LMC and SMC, respectively. All of the non-Type I PN have C/O ratios significantly larger than unity, consistent with the exposure of primary carbon by the third dredge-up. The carbon enhancements are largest in the SMC, the galaxy with the lower metallicity. The upper limit to the central star luminosities, $L \approx 8000\ L_\odot$, is similar to the observed upper limit to Carbon star luminosities in the LMC (Reid 1989).

Most of the Type I PN have C/O ratios below unity, which together with their high N abundances suggests that they mixed out freshly produced ^{12}C which was then converted to ^{14}N via 'deep envelope mixing'. There is some evidence for neon over-enhancements in the Type I PN, likely to be a result of the reactions $^{14}N(\alpha,\gamma)^{18}F(\alpha,\gamma)^{22}Ne$ in the He-burning zone during thermal pulses occurring in the late AGB evolutionary stage of the star.

We confirm the standard picture wherein a typical red giant of 1.3 $M_\odot$ produces a degenerate remnant with a core of 0.6 $M_\odot$, a surrounding nebula of 0.3 $M_\odot$ (now seen as a PN), and thus loses some 0.4 $M_\odot$ during earlier stages in its red giant evolution.

Full results of this work are to presented in Walton *et al.* (1991).

References

Monk, D. J., Barlow, M. J. & Clegg, R. E. S., 1988. *Mon. not. R. astr. Soc.*, **234**, 583.
Reid, N, 1989. *Astrophys. Sp. Sci.*, 156, 73.
Walton, N.A., Barlow, M.J., Monk, D.J., Clegg, R.E.S., 1991. *Mon. not. R. astr. Soc.*, submitted.

THE SHAPE OF THE BULGE FROM IRAS MIRAS

PATRICIA WHITELOCK
S A Astronomical Observatory
and
ROBIN CATCHPOLE
Royal Greenwich Observatory

ABSTRACT. Detailed observations of 103 IRAS Miras found in two strips across the Bulge with $-15^\circ < l < 15^\circ$ and $7^\circ < |b| < 8^\circ$ were discussed by Whitelock *et al.* (1991). Among other things they derived distance moduli for each object. In this work (which will be reported in more detail by Whitelock and Catchpole 1991) we examine the number of Miras per 0.3 mag distance modulus bin as a function of distance modulus and as a function of galactic longitude. The important aspect of this approach is that it provides us with a probe of the depth and structure of the bulge. If the number distribution as a function of modulus is examined separately for the two sides of the bulge, then it is very clear that the distribution of the Miras with $l > 0^\circ$ is narrower and peaks in front of the distribution of Miras with $l < 0^\circ$. This asymmetry can be understood if the Bulge is bar shaped, the bar is tilted at approximately 45° to the line of sight and the near end is in the first quadrant. This conclusion is consistent with those of Blitz & Spergel (1991) and Nakada *et al.* (1991). It is particularly interesting to note that Menzies' (1990) radial velocity data for a subset of the Miras discussed here show that the Bulge is rotating rapidly.

The observations were fitted with various models of the stellar density distribution. An X shaped Bulge model actually fits the data, particularly the longitude distribution, better than a simple triaxial ellipsoid. However given the limited data the details of the models are very speculative.

References

Blitz, L. & Spergel, D. L. (1991) *Ap. J.*, in press.

Menzies, J. W. (1990) 'Rotation of the Galactic Bulge from IRAS Miras', in B. J. Jarvis & D. M. Terndrup (eds.), *Bulges of Galaxies*, ESO Conf. & Workshop proc. 35, pp. 115–117.

Nakada, Y., Deguchi, S., Hashimoto, O., Izumiura, H., Onaka, T., Sekiguchi, K. & Yamamura, I. (1991) *Nature*, in press.

Whitelock, P. A. & Catchpole, R. M. (1991) In press.

Whitelock, P. A., Feast, M. W. & Catchpole, R. M. (1991) *M.N.R.A.S.*, **248**, 276–312.

THE PERIODS OF MIRAS IN THE BULGE AND THEIR LATITUDE DEPENDENCE

PATRICIA A. WHITELOCK
S A Astronomical Observatory
P O Box 9
Observatory
Cape 7935, SA

ABSTRACT. A comparison is made of the period distribution of Miras in the Bulge in fields centred at $b = -3^\circ.9$ and $b = -7^\circ.5$ The field centred at $l = 0^\circ.0$ $b = -7^\circ.5$, occupies 3.79 square degrees and is part of Baade/Plaut field 3. The Miras are those discussed by Wesselink (1987) and those discovered through follow up observations of IRAS objects. The sample is almost certainly incomplete as there are probably Miras of intermediate period which were not detected by either Wesselink or IRAS. The period distribution is shown in Fig 1a.

The other field is centered at $l = 0^\circ.9$ $b = -3^\circ.9$, it covers 0.33 square degrees of the Baade window around NGC 6522. The Miras were discovered by Lloyd Evans (1976) and include the 11 IRAS Miras within the area (Feast 1986). The IRAS survey may have been incomplete by as much as a factor of two this close to the galactic centre, but any missing Miras are unlikely to have long periods. The period distribution is shown in Fig 1b.

A comparison of the two distributions shows no clear differences. There is certainly no indication of an absence of long period objects in the high latitude field.

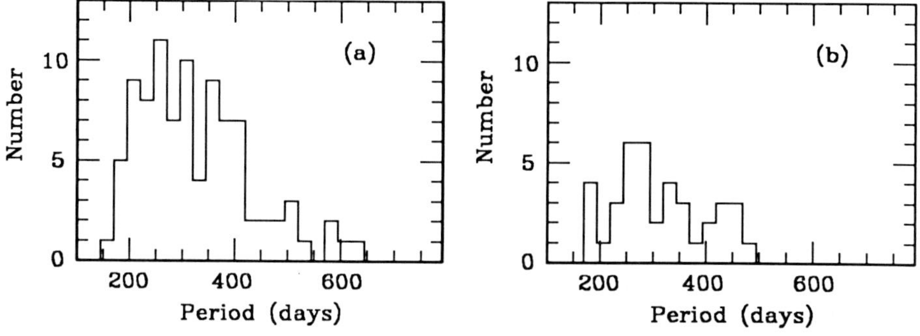

Figure 1: (a) All known Miras in the $b = -7^\circ.5$ field. (b) All known Miras in the NGC 6522 field.

References

Feast, M.W. (1986) In: *Light on Dark Matter*, ed F.P. Israel, Reidel, Dordrecht. pp 339–348.
Lloyd Evans, T. (1976) *M.N.R.A.S.*, **174**, 169–184.
Wesselink, Th. (1987) Ph D Thesis, Nijmegen.

NEW WHT+FOS OPTICAL SPECTRA OF WR STARS IN M33 & M31

A J WILLIS, H SCHILD & L J SMITH
Department of Physics & Astronomy
University College London
Gower Street
London WC1E 6BT, UK

We have secured new optical spectra of 11 WR stars in M33 and M31 using the WHT+Faint Object Spectrograph, covering $\lambda\lambda$4800–9800Å at 14Å and $\lambda\lambda$3500–5000Å at 7Å resolution. In our M33 sample five stars are classified as WC4–5 from their C IV λ5800/C III λ5696 line strengths (*viz* MC 6, MC 53, MC 79, MC 24 and MC 71), one as WC6 (MC 65), one is a WNE-CE star (MC 48) and one is possibly an intermediate WC4-WO star (MC 78). Of the two stars observed in M31, one is classified as WC4–5 (MS 11) and the other as WC7-8 (MS 4).

From our new data and earlier WHT+FOS spectra we find a wide spread in the FWHM of C IVλ5800 for WC4–5 stars in M33, from values $\sim$ 30Å (more typical of galactic WC8 stars) up to $\sim$ 110Å (typical for galactic WC4 stars). We confirm the corelation of the CIV FWHM with M33 galactocentric distance first reported by Schild, Smith & Willis (1990), suggesting a real physical relationship between WC4–5 stellar wind velocity and galactic location.

Smith & Maeder (1991) ascribed 5 stars in M33 as 'WO' (previously classified as WC) on the basis of the relative line strengths and line widths of the CIV λ5808 and C III-IVλ4650, but without access to the O VIλ3811, 3834 transitions, the great strength of which led to the identification of the WO sequence (Barlow & Hummer 1982). We have spectra of 3 of these stars covering the O VI region and do not confirm their conclusion. MC 48 is found to be a WNE-CE star and not a WO star. MC 6 shows only very weak OVI, and is clearly a WC4–5 star. MC 78 does show significant OVI emission but with a factor of $\geq$ 5 smaller intensity than the weakest-lined WO star. It appears to be intermediate between WC4 and WO. These results lower the known WO population in M33 to $\leq$ 2–3.

We have discovered that the 11th star in our new M33 sample, very close to MCA 1, is a likely Ofp/WN9 transition star, with its optical spectrum showing numerous, narrow emission lines in H I, He I, N II and N III, very similar to that of R 84 in the LMC. This is the first identification of such an object in M33.

A full report of our new WHT+FOS spectra will appear in *Astronomy & Astrophysics*

References

Barlow,M.J. & Hummer,D.G., 1982, *IAU Symp. No. 99*, p 387
Schild,H., Smith,L.J., & Willis,A.J., 1990, *Astron. Astrophys.,*, **237**, 169
Smith,L.F., & Maeder,A., 1991, *Astron. Astrophys.,*, **241**, 77

OB ASSOCIATIONS IN NGC 6822

CHRISTINE D. WILSON
Department of Astronomy
University of Maryland
College Park, MD, 20742 U.S.A.

ABSTRACT. An objective, automated group-finding algorithm has been used to re-identify the OB associations in the irregular galaxy NGC 6822. The properties of the OB associations, such as size, age, and mass, are compared with those of OB associations in M33 identified using similar data and techniques. These two data sets allow the first objective comparison to be made of the properties of OB associations in two quite different galaxies.

Determining the properties of OB associations in galaxies is important for understanding star formation. Unfortunately, comparing OB associations between different galaxies has been difficult due to the subjectivity involved in identifying OB associations by eye from photographic plates (Hodge 1986). With the advent of CCDs and automatic photometry programs, it is now possible to obtain accurate photometry for large samples of stars in Local Group galaxies; this photometry can then be fed into automated programs to search for groups of blue stars. The real power of such automated techniques is in comparing association properties between different galaxies, since essentially identical samples of stars can be selected and analyzed for each galaxy.

New CCD photometry of the Local Group irregular galaxy NGC 6822 has been obtained in the B and V bands at the Palomar 60" telescope. OB associations were identified using a "friends of friends" algorithm (Wilson 1991). All stars with $V < 21$ and $B - V < 0.5$ and a grouping radius of 26 pc were used to identify the associations. Thirteen associations were identified, of which nine correspond well to associations identified by Hodge (1977). The remaining four associations lie in the main body of the galaxy where there is a higher background density.

To match the selection criteria used for M33 (Wilson 1991), OB associations in NGC 6822 were re-identified using $V < 20.3$, $B - V < 0.5$, and a grouping radius of 22 pc. Four associations were identified using these criteria, compared to forty-one in the inner kiloparsec of M33. The number of stars, masses, and ages of the OB associations are very similar for both galaxies, while the median diameters are 30% smaller in NGC 6822 than in M33. Depending on the inclination of NGC 6822, the number of associations per unit area is 20-40% of that of the inner disk of M33, while the surface density of blue stars in NGC 6822 is 35-70% of that of M33.

References:

Hodge, P. W. 1977, *Ap. J. Suppl.* **33**, 69.
Hodge, P. 1986, in *Luminous Stars and Associations in Galaxies*, eds. C. W. H. de Loore, A. J. Willis, and P. Laskarides (Reidel: Boston), 369
Wilson, C. D. 1991, *A. J.* **101**, 1663.

CHEMICAL ABUNDANCES IN OLD POPULATIONS

GUY WORTHEY
Lick Observatory
Santa Cruz, CA 95064, USA

ABSTRACT. Preliminary single-burst population synthesis models are presented for weak and strong spectral features as a function of metallicity for old populations. Models agree with published globular cluster observations as well as theoretical calibrations. For small ellipticals, the galaxies and model predictions agree well in all indices. For the typical giant E, metallicity-insensitive features (G band, Hβ) continue to match the models, while the behavior of metallicity sensitive features (Fe, CN, Mg, Na D) diverges. In giant Es, with increasing metallicity, the light-element indices deepen relative to the iron-peak features far more rapidly than the models. This effect almost certainly indicates that, in the typical giant E, some light elements are enhanced with respect to the iron-peak elements compared to the solar ratios.

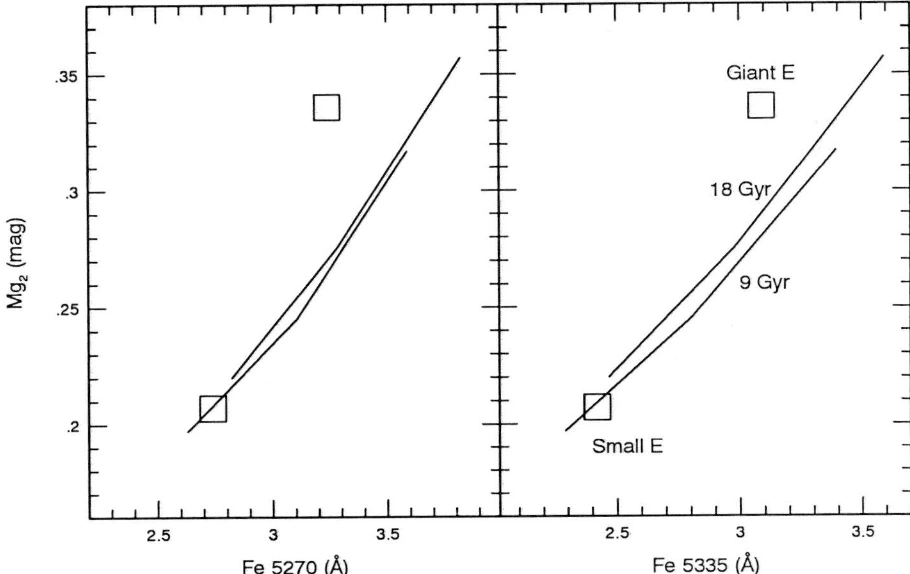

Figure 1. Mg_2 index shown as a function of two Fe indices. Median loci for small and giant ellipticals are shown as squares. Model predictions for burst ages of 9 and 18 Gyr are shown as lines with endpoints at ± .25 dex in [Fe/H]. No combination of age and metallicity can account for the enhancement of Mg_2 with respect to the iron indices in the giant E galaxies.

MORPHOLOGICAL AND SPATIAL DISTRIBUTIONS OF HIGH VELOCITY MOLECULAR OUTFLOWS

Yuefang Wu[1,2] **Maohai Huang**[3] **Jun He**[2]
1. CCAST (World Laboratory), P.O.Box 8730, Beijing 100080, P.R.China
2. Department of Geophysics , Peking University, Beijing 100871, P.R.China
3. Beijing Astronomical Observatory, Beijing 100080, P.R.China

ABSTRACT. Data of 136 high velocity outflows listed in a comprehensive catalogue and maps of three dimensional distribution are presented. The distribution of collimation factors and the formation rate is estimated.

After giving a brieve review, we present a catalogue of known high velocity outflows. The catalogue consists of 136 items compiled from a number of publications till Febrary,1991. Each item contains data of an object including the position in both the equatorial and the galactical coordinates, the outflow velocity (the bottom spectral linewidth at 0.1k or 0.2k above the zero temperature level), the collimation factor, the mass of the outflows and the luminosity (ref. Lada,1985). All references and materials providing data are also given.

Besides the collimation factors which are available directly from original articles, some are measured and calculated from their published contour circles. The scheme to calculate collimation factors is discussed.

We find that half of the collimation factors fall between 1.5 and 2.0. Analysed with a column modle, it is known that a large portion of the outflows have intrinsic collimation factors close to their observational ones. So the average intrinsic collimation factor of the sample outflows can be estimated at a bit less than 2, which means that every lobe of many bipolar sources roughly has a round feature, and sugests that there is a fairly large outflow velocity component which is perpendicular to the bipolar axis . It is also found that large mass (grater than $2M_\odot$) sources tend to have large collimattion factors.

We present maps of three dimenssional distribution of the outflows in two scales. Each chart shows spatial projection on different Galactic coordinate plane. The sun is the origin. Objects having different mass are plotted by different symbols.

105 outflows in the catalogue are not farther than one kiloparsec from the sun . The formation rate of the outflows is 7×10^{-4} $yr^{-1}kpc^{-2}$, if the avaerage life time of such object is assumed as 5×10^{4} years, an intermediate value between that of low and high mass outflows . That formation rate is close to the birth rate of the local stars.

THE STELLAR POPULATIONS AND DYNAMICS OF ELLIPTICAL GALAXIES IN COMPACT GROUPS

STEPHEN E. ZEPF

Dept. of Physics and Astronomy, Johns Hopkins University

and

Space Telescope Science Institute

Abstract. We present a study of the photometric and spectroscopic properties of elliptical galaxies in compact groups. We find that although some elliptical galaxies are affected by interactions and mergers, the current merger rate is small, even in compact groups of galaxies. We also find tentative evidence that the central velocity dispersions of elliptical galaxies in compact groups are lower than the velocity dispersions of similar galaxies in other environments.

1. Stellar Content of Elliptical Galaxies in Compact Groups

The compact groups of galaxies classified by Hickson (1982) provide a favorable environment for merging because they have high spatial densities and velocity dispersions comparable to galactic internal velocities. In our multicolor survey of 55 elliptical galaxies in these groups we find that four of them are unusually blue, indicative of recent interactions or mergers (Zepf et al. 1991). Although the discovery of blue ellipticals in compact groups is evidence that interactions and mergers are occurring, the relatively small number discovered indicates that the time scale for compact groups to evolve by mergers into a single elliptical galaxy is much longer than the observed crossing time. A slow evolution is also indicated by the small fraction of elliptical galaxies outside of compact groups which are unusually blue (Zepf and Whitmore 1991).

2. Dynamics of Elliptical Galaxies in Compact Groups

The elliptical galaxies in compact groups also offer an excellent opportunity to test the robustness of the "fundamental plane" to changes in environment. We have found tentative evidence that the elliptical galaxies in compact groups have a different zero point in the fundamental plane relations than elliptical galaxies in other environments. This offset is due to the velocity dispersions, which are lower in compact group ellipticals relative to similar ellipticals in other regions.

References

Hickson, P. 1982, *Ap. J.*, **255**, 382.
Zepf, S.E. and Whitmore, B.C. 1991, Ap.J., in press.
Zepf, S.E., Whitmore, B.C., and Levison, H.F. 1991, *Ap.J.*, in press.

THE PRE-MAIN SEQUENCE STELLAR POPULATION IN GALAXIES

H. ZINNECKER
Inst. Astronomie und Astrophysik
Universität Würzburg, Germany

A complete census of the stellar populations in galaxies must include not only the Main Sequence and various post-Main Sequence populations (SGB, RGB, AGB, HB, plus degenerate objects such as WD and NS) but also the pre-Main Sequence (PMS) stellar population.
These young stellar objects, predominantly of low mass ($M<2M_\odot$), derive their luminosity not from nuclear burning but from slow gravitational contraction and in part from the gravitational energy released during accretion of matter from circumstellar disks. However, some deuterium rather than hydrogen burning does occur for a brief period early on in PMS stellar evolution, either in the core for low-mass ($M<2M_\odot$) PMS stars (Mazzitelli & Moretti 1980) or in a shell for intermediate-mass (2-$8M_\odot$) PMS stars (Palla & Stahler 1990). The more massive young objects ($M>8M_\odot$) may not experience much of a Kelvin-Helmholtz phase and probably ignite core hydrogen burning before the end of their accretional growth.
Although the PMS population may be insignificant or of little importance in many galaxies (especially in old gas-poor systems, i.e. ellipticals), the PMS population may play a non-negligible role for population synthesis in gas-rich starburst systems and/or in very young, i.e. primeval galaxies (ages $<10^7$ yrs). Judging from 2 micron luminosity functions of young OB clusters (such as the Trapezium cluster in Orion) in which bright infrared sources (i.e. massive stars) are typically accompanied by hundreds of low-luminosity PMS members (see Zinnecker, McCaughrean, and Wilking 1991), it appears that about half the <u>integrated</u> 2 micron luminosity of such OB clusters actually comes from the low-mass PMS population. The same may be true for young galaxies. Moreover, the number of PMS stars in the star forming galaxies may be scaled from the estimated number of OB stars (about 1000 PMS stars for each massive ionizing star); for example, our Galaxy should contain a total of about 10^7 low-mass PMS stars younger than 10^7 yrs, assuming that most PMS stars form in OB clusters and that there are 10^4 O stars in the Galaxy.

References:

Mazzitelli and Moretti 1980, Ap.J. 235, 955
Palla and Stahler 1990, Ap.J.Lett. 360, L43
Zinnecker, McCaughrean, and Wilking 1991, Protostars & Planets III

STELLAR POPULATIONS OF A LOCAL SAMPLE FROM ITS VELOCITY DISTRIBUTION

R. Cubarsí
Department de Matematica Aplicada i Telematica
Universitat Politecnica de Catalunya, Barcelona, Spain

In the solar neighbourhood, we can assume that a stellar sample is composed of two stellar populations: thin disk and thick disk stars. If we assume as the only hypothesis that each stellar component is associated with a Schwarzchild velocity distribution function (e.g., Sanz, J., Juan, J.M., this symposium), then it is possible to determine the velocity distribution of both components (Cubarsí, R.: 1990, AJ 99, 1558). Thus, starting from the central velocity moments up to fourth order of the overall stellar sample we obtain the partial moments, the difference between centroid velocities and the percentage of mixture. Moreover, a set of 25 constraint equations between the total moments is determined. This method has been applied to some local stellar samples under the hypothesis of a two-component mixture (Hernandez-Pajares, M.: 1991, this symposium) and the obtained kinematical features of the stellar components are in agreement with kinematical properties of thin disk and thick disk stars described from other physical viewpoints (e.g., Sandage, A.: 1987, in *The Galaxy*, eds. G. Gilmore, R.F. Carswell (Reidel), p. 321).

THE GROUP OF Be/X-RAY SOURCES IN THE GALAXY

E. Janot-Pacheco
IAG-USP,Dept. Astronomia
C.P. 9638, São Paulo 01065, Brazil

The group of known X-ray sources associated with a Be star in a binary system consists of about 40 objects. They are the most abundant type of massive X-ray binary in the Galaxy. In this contribution we present an up-to-date picture of the main physical characteristics of these objects, including their evolutionary history, mass transfer mechanism, an interpretation of their behaviour in the orbital versus spin period diagram and the use of Be/X-ray sources as galactic tracers.

THE PROGENITORS OF PLANETARY NEBULAE

P. te Lintel Hekkert[1], A.A. Zijlstra[2]
[1] *Mt. Stromlo and Siding Spring Observatories*
[2] *Kapteyn Laboratorium, Groningen*

Based on a new sample of (IRAS based) OH/IR stars (te Lintel Hekkert et al., 1991, A&AS, in press), and a catalogue of planetary nebulae compiled by Acker (1983, A&AS, 54, 315), we show the relation between these two groups of objects, in terms of the kinematics and the Galactic distribution. In contrast with earlier analyses of samples of OH/IR stars, we find a close correlation between the kinematics of the planetary nebulae and the IRAS based sample of OH/IR stars. In particular, we find that the distribution of the planetary nebulae (PN) shows a good correlation with the OH/IR stars which have a low outflow velocity ($v_{exp} < 12.5$ km s^{-1}). Whether the high outflow velocity OH/IR stars also have a counterpart among PN is not clear.

The majority of the known PN thus appear to originate from low outflow velocity OH/IR stars. The ZAMS progenitor masses are probably in the range 1-1.5 $M_\odot$, and the stellar ages are $\approx$ 5-10 10^9 yrs. Only in the plane of the Galaxy may a significant fraction of the PN come from more massive progenitors. In the outer part of the Galaxy, a relatively larger fraction of PN appears to originate from carbon stars instead of OH/IR stars.

A BINARY EVOLUTIONARY MODEL FOR THE PROGENITOR OF SN 1987A

C. de Loore[1], D. Vanbeveren[2]
[1] *Astrophysical Institute, Vrije Universiteit, Brussels, Belgium*
[2] *Dept. Physics, Vrije Universiteit, Brussels, Belgium*

In this paper we explore the hypothesis that the blue progenitor of SN 1987A was a component of a close binary. It is shown that a blue spectral appearance at the end of core helium burning (and possibly also during the following evolutionary phases) is a natural phenomenon for a star with initial mass between 9 $M_\odot$ and 15 $M_\odot$ which was originally in a close binary with mass ratio $q \approx 1$ and who accreted at least 8 - 10 $M_\odot$ while already being a hydrogen shell burning star. This phenomenon does not depend on the exact treatment of the convective core overshooting, the stellar wind mass loss rate, and the metallicity. A blue SN progenitor in the Galaxy would therefore be a surprise.

STELLAR PARAMETERS IN THE BASEL FIELD SA 141

M.-N. Perrin[1], R. Cayrel[1], B. Barbuy[2], R. Buser[3]
[1] *Observatoire de Paris, France*
[2] *IAG-USP, C.P. 9638, São Paulo 01065, Brazil*
[3] *Universität Basel, Venusstrasse 7, Binningen, Switzerland*

A grid of 200 synthetic spectra was built in the wavelength region $\lambda\lambda$ 478-530 nm, and a method was developed for deriving the stellar parameters temperature, gravity and metallicity from low-resolution stellar spectra. The parameters of the observed spectra are derived by dividing the observed spectrum by that of a reference star. The resulting $\delta f(\lambda) = \log F_*/F_{ref}$ is then used, in conjunction with the grid of synthetic spectra, to derive the final parameters, through a perturbation method. The method was applied to a sample of 41 stars in the Basel field SA 141 (l = 245°, b = -85.8°). Since this direction points closely to the south galactic pole, the binning of stars into appropriate distance intervals provides a coarse view of the metallicity distributions associated with the major galactic populations: we find the thin disk, thick disk and halo components near the current best estimates of their scale heights above the galactic plane. The first results were presented in Cayrel et al. (1991a, A&A, 247, 108; 1991b, A&A, 247, 122). Further observations are in progress, with the aim of having a sample of about 1000 stars.

NEAR INFRARED IMAGING OF A GIANT RED ENVELOPE GALAXY

A. H. Prestwich
NASA-Marshall Space Flight Center, ES-65 Space Sciences
Huntsville, AL 35812, USA

Recently Maccagni et al. (ApJ, 334, L1, 1989) obtained g, r and i images of an X-ray luminous galaxy at the center of a poor cluster of galaxies (1E1111.9-3754). The images showed that the central galaxy was surrounded by a remarkable spatially extended red envelope, visible as a kink in the $r^{1/4}$ surface brightness profile at a radius of 100 kpc. Johnstone & Fabian (MNRAS, 237, 27p, 1989) suggested that the red envelope is composed of low mass stars formed in a cooling flow. This observation sparked considerable interest as the first direct detection of low mass star formation in cooling flows. We have obtained near infrared H-band (1.65 μ images of 1E1111.9-3754 from the Infrared Imaging Spectrometer on the 4-m Anglo-Australian Telescope. Even though low mass stars should be more prominent in the near infrared than at optical wavelengths, the 1.65 μ surface brightness profile shows no deviation from the $r^{1/4}$ profile out to a radius of 100 kpc. This observation argues against the hypothesis that the envelope is composed of low mass stars formed from cooled X-ray gas.

THE DISTRIBUTION OF LOW MASS STAR FORMING REGIONS

C.A. Torres[1], G.R. Quast[1], R. de la Reza[2]
[1] *Laboratório Nacional de Astrofísica, Minas Geraes, Brazil*
[2] *Observatório Nacional, Rio de Janeiro, Brazil*

Our knowledge of the distribution of low mass stars (T-associations) in the Galaxy, as an indicator of star forming regions (SFR), is badly known due to an observational bias. A correlation with high-mass stars (OB-associations) will not solve the problem, because some well known SFR as Taurus and Chamaleonis seem to preferentially form low mass stars. One way to tackle this question is by invoking the distribution of known T Tauri stars (TTS) and the distribution of selected IRAS sources that are potential candidates to be new TTS. We must be aware, however, that an important number of TTS are not IRAS sources. Considering these distributions, we have found a smooth distribution of SFR along the local galactic plane. The number of TTS in each SFR is very variable probably due to a variable efficiency of star formation. Also there are some isolated TTS from SFR which could indicate the dissipation of old SFR or that some TTS can be formed from isolated very small clouds.

Subject Index

Numbers in **boldface** indicate that the item is discussed throughout the entire article which starts on the **mentioned pagenumber.**

abundance, elemental: **123**
- distributions: **123**
- in globular clusters: 8, 130, 144
- gradients: 139, 164, 247, 383, 398, 428, 500

active galactic nuclei (AGNs): **271**, 340, 406
age:
- of bulge: 33
- of galactic disk: **75**
- of galaxies: 212, **255**, **325**, **337**, 361
- of globular clusters: 9, **15**, **325**
- stellar: 75

age-metallicity relation: 68, **103**, 120, 123, 136, 151, 403
age-velocity relation (AVR): 67, 77, 82
Astro-1 mission: **233**
asymptotic giant branch (AGB): 30, 37, 111, **181**, 183, 192, 218, 236, **291**, **321**, 329
α-elements: 9, 32, 36, 75, 136
associations: **93**

Baade window (BW): **29**, **41**, 48, 446, 480
baryonic mass: 361, **367**
binaries: 6
black hole: 85, 114, 374
blue compact dwarf galaxies (BCD): 442, 457, 482
blue stragglers: 37, 331
brown dwarfs: 375
bulge:
- of the Galaxy: **29**, **41**, **47**, **51**, **77**, 138, 145, 176, 202, 252, 330, 459, 462, 492, 503, 504
- globular clusters: **47**, 165
- structure: 35

bulges: 36, 162, 237, **245**, 291, 329
burst of star formation (see starburst)
BVRI Cousins: 47

carbon stars: 30, 192, **201**, 425, 473
CH stars: 153
Cepheids: 384
chemical evolution: 35, **75**, **123**, **133**, 150, 152, 278
- model: 113, **133**, 387

chemodynamical evolution models: **119**, 410
cloud fragmentation: 370, 479
cluster:
- stars (see globular clusters, young clusters)
- of galaxies: 299, 400, 404, 420, 450, 474

Cosmic Background Explorer (COBE): 34
colour-magnitude diagram (CMD): 49, 166, 170, 208, 219, 236, 454, 462
colour gradients: 8, 173, 239
convection: 225
cooling: 370, 479
cooling flows: 456
cosmic background radiation: 338
cosmological constant: 333, 354
cosmological density parameter (Ω_o): 341

dark matter: 194, 211, **367**
deceleration parameter (q_o): 354
disc:
  abundances: **75**, **123**, 145
  dynamics: **81**
  formation of: **119**
  Galactic: **65**, **75**
  galaxies: **119**, 212
  of M31, M33: 163
  old: 69, 123
  outer: 399
  scale-height: 61, **65**, 77, **81**
  thick: 4, 61, 66, 70, **77**, 82, **103**, **119**, 123, 408
  thin: 67, 82
dissipative galaxy formation: 119
30 Doradus: 98, 149, 216, 467
dust: 147, 339, 469, 479
dwarf compact galaxies: 170
dwarf ellipticals galaxies (dEs): **169**, 404, 426, 429
dwarf galaxies: 197, 427
dwarf irregulars (dIRRs): 197
dwarf spheroidal galaxies (dSph): **191**, **201**, 455
dynamics: **51**, **81**
dynamical evolution: **65**, 75

elliptical galaxies: **41**, **169**, 186, 213, 216, 237, 250, **255**, **267**, **271**, **281**, **291**, 327, **353**, 391, 394, 397, 400, **424**, 488, 507, 509
emission spectra: **337**, 419, 420
evolution of galaxies: **297**, **321**, **325**

faint galaxy counts: 338, 362

galactic bar: 33, 94
galactic bulge: (see bulge of the Galaxy)
Galactic chemical evolution (GCE): 3, 35, 69, **75**, **133**, 387, 398, 428
galactic collapse: 119
galactic poles: **61**, 409
galactic rotation: 62
galaxy formation: **255**, **337**, 354, 369, 446
γ-ray: 381
G-dwarfs: 138, 409
  problem: 3, 120, **133**
giant elliptical galaxies (gEs): 169, 258

giant molecular clouds (GMC): 84, **93**
globular clusters: 234, 368, 402, 403, 407, 413, 501
    abundances: 130, 143, 401
    ages: **15**, **325**
    bulge: 31, **47**, 143, 401
    disc: 4, 31, 477
    halo: **1**, **23**, 61, 202
    integrated spectra: 217
    in the Magellanic Clouds: (see Magellanic Clouds)
    in M33: 164
great attractor: 283
Gunn griz colours: **353**

Hβ: 44, 259, 282
HII region: 98, 212, 287, 427, 468, 493
halo:
    Galactic: **1**, 2, **51**, 62, 123, 143, 202, 385, 408, 443, 471, 486, 487
    of M31, M33: 164
halo-bulge connection: 38, 164
halo-disk connection: 4
heavy element abundances: 75, 129
    in the bulge: 36
helium abundance: (see also element index)
    solar: 110
    $\Delta Y/\Delta Z$: 34, 116, 133, 331
    primordial: 110, 133
hierarchical clustering: 256, 364
high redshift galaxies: 237
high-velocity stars: 486, 487
Hipparcos satellite: 15, 449
Hubble constant: 342, 354
Hubble diagram: 173, 353
Hubble sequence: 249, 359
Hubble Space Telescope (HST): 15, **233**, 326, 345
hydrodynamical models: 119

infall of gas: 337
Infrared galaxies: 38
infrared survey: 414, 465, 491
initial mass function: 97, 134, 208, 217, 256, 262, 316, 369
integrated spectra: 173, **215**, **255**
interacting galaxies: 491
intermediate age populations: 37, **181**, 192, 198, 329
Iras satellite: 34, 77, 262, 503
iron (see Fe in element index)
isochrones: **109**, 395

JHK colours: 175, 249, 298, 364, 491
Jupiters: 367

K giants: **29**, **41**
kinematics:
    of bulge stars: 33

stellar: **51**, **61**, 75, **81**, 430, 431, 461, 463, 464
thick disk: **61**
kinematics-age relation: **81**

lenticular galaxies: 241, 466, 488
library of stellar spectra: **215**, 257, **357**
long-period variables (LPVs): 33, 176, 183, 434
low mass stars: 496
low surface-brightness galaxies (LSBGs): **213**, 364, 499
luminosity:
function, of galaxies: 164, 211, 363
of globular clusters: 4, 181
Lyman α: **337**, 405, 494

M_{EC}: 110-111
M_{HeF}: 110-111
M_{up}: 110-111
M_{W}: 110-111
M giants: **29**, **77**
M31: 37, **161**, 219, 238, 252, 291, 304, 326, 330, 438, 505
globular clusters: 220, 436
M32: 35, **169**, 219, 238, 241, 318, 328, 426
M33: **161**, 505
M82: 389
Magellanic Clouds: 95, **147**, **181**, 432, 502
Large: **23**, 183, 202, 215, 218, 392, 396, 412
Small: **157**, 202, 454, 473
chemical evolution of: 147
clusters: **23**, 148, 150, 152, 392, 422, 423, 485
kinematics: 152, 159
magnesium index: 44, 257, **267**, **277**, 282, 354, 397, 415, 507
mass loss rate: 112, 181, 419
mass-to-light ratio (M/L): 150, 163, 194, 394, 437
mass-luminosity relation: 134
massive stars: **93**, 112, 447, 481
mergers: 81, 256, **267**, 281, 305, 362
metal-rich clusters: **47**, 145, 186, 221
metal-rich stars: **29**, **41**, **47**, 145
metallicity (Z): **109**, 196, 220, **277**, 322
of bulge stars: 31, **41**, **47**, 79, 145
distribution: 137
metallicity-scaleheight relation: 120
metallicity-luminosity relation: 196
Miras: 29, **77**, 503, 504
mixing, convective: 110
mixture models: **103**
model atmospheres: 216, **225**
molecular outflows: 508

NGC 147: 172
NGC 185: 171, 445
NGC 205: 171, 426, 444
neutron star: 367, 374, **379**, 433

novae: 379, 412

O stars: **93**, 478
OB associations: **93**, 506
    in the Magellanic Clouds: 97
odd-even effect: 128
OH/IR stars: 17, 30, 181, 183, 245, 253, 448
opacity: 109, **225**, 322, 395, 396
oxygen (see O in element index)
oxygen anomalies: 8, 15, 35, 116, 136, **143**, 213, 326
overshooting: 110, 207, 225, 396

planetary nebulae (PN): 383, 419, 451, 470, 492, 502
    in the bulge: 32
    in the Magellanic Clouds: 153
population III: 388
population synthesis: 173, **215**, 225, 246, **255**, 292, **321**, **325**, 424, 440, 472, 484, 495
    evolutionary: 251, **311**, **321**, **325**, 442, 452, 498
populations, stellar: 123
    models: **103**
post-AGB stars: **181**, **291**, 312, 419, 475
pre main-sequence stars: 390, 510
primeval galaxies: **337**
proto-globulars: 150
pulsars: **379**

quasars (QSO): **337**
    absorption spectra: 405, 421, 427

radiogalaxies: 332, 362
reddening: 43, 47, 48
redshift: **325**
    intermediate: **297**, 460
    high: 305, 324, **337**, **353**
RR Lyrae stars: 3, **15**, **23**, 33, **77**, 326, 403, 407, 446
    Oosterhoff type: 165
r-process: 35, 129

S0 galaxies (see lenticular galaxies)
S galaxies: (see spiral galaxies)
second parameter: 2, 3
Seyfert galaxies: 440, 441
spectral gradient: 169, 174, 267
spectral energy distribution (SED): 236, 321, 472
spectral evolution: 314
spectral synthesis: 144
spiral arms: 84, 88, **93**
spiral density waves: 84, 88..
spiral galaxies: **161**, 213, 216, 237, **245**, 252, 287, 291, 394, 437, 469, 500
s-process: 35, 76
starburst: 256, 371, 472, 478
star-gas interaction: 120
star formation:

bimodal: 256, 338
rate (SFR): 122, 134, 151, 159, 192, 209, 211, 251, 258, 298, 301, 317, 322, 372, 482
star forming regions: 236, 435, 481
stellar evolution: **109**, 379, 395, 396
strange stars: **433**
supergiants: **93**, 396
supernovae: 113, 136, 143, 379, 416
surface brightness: 87
surveys:
of galactic stellar populations: **61**
of SMC stellar populations: **157**
synthesis, populations (see population synthesis)

tidal interactions: 88, 159, 172, 371, 426
triaxiality: **51**
47 Tucanae: 23, 219, 221, 483

UBVRIJKLMN colours: 174, 218, 228, 249, 298, 312, **321**, 423, 432, 449, 488
uvby colours: 228
UV (ultraviolet): 218, **233**, 287, 411
excess: 162, 239, **291**, 316, **357**, 452
extreme (EUV): **233**

velocity:
dispersion: 33, 66, 83, 153, 195, **269**, 298, 429, 489
peculiar: **281**, 439

ω Centauri: 8, 235, 458
warmers: 406
warp: 399
white dwarf: 327, 367, 375
wind:
in galaxies: 209, 211
stellar: 114
Wolf-Rayet star: **93**, 110, 386, 478, 505
galaxies: 99, 497
scale-height: 94, 386

Y (see helium, He, abundance in element index)
yield: 115, 278
young clusters: 150, 386, 436
young disc stars: **65**, **75**, 81
young populations: 37, **93**

Element index

He: 34, 110, 116, 133, 331, 383, 418, 470, 502
Li: 124, 476, 490
Be: 124
C: 125, 458, 470, 483, 502
N: 383, 458, 470, 483, 502
O: 75, 125, 136, **143**, 383, 398, 428, 458, 470, 483, 502
Ne: 470
Na: 75, 127
Mg: 37, 75, 127,
Al: 37, 75, 127
Si: 37, 75, 502
S: 383, 470
Ar: 383, 470, 502
Ca-Ni: 37, 75, 128
Fe: 2, 3, 5, 15, 24, 33, 37, 75, **123**, **133**, **143**, 192, **225**, 326, 328
Sr: 129
Y: 75
Zr: 37, 75
Ba: 75, 129
La: 37
Eu: 37, 129, 417
Th: 417
even-Z: 127
odd-Z: 76, 127
iron-peak: 37, 75, 128
rare-earths: 37, 75, 129